Pathways into the Study of Ancient Sciences

Selected Essays by David Pingree

Pathways into the Study of Ancient Sciences

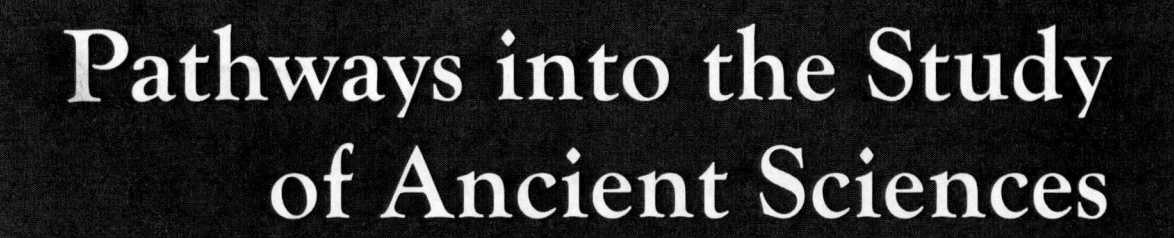

Selected Essays by David Pingree

Edited by
Isabelle Pingree
and
John M. Steele

The American Philosophical Society Press

Philadelphia • 2014

Transactions of the
American Philosophical Society
Held at Philadelphia
for Promoting Useful Knowledge
Volume 104, Part 3

ISBN: 978-1-60618-043-3
US ISSN: 0065-9746

Library of Congress Cataloging-in-Publication Data

Pathways into the study of ancient sciences : selected essays by David Pingree / edited by Isabelle Pingree and John M. Steele.
pages cm. – (Transactions of the American Philosophical Society held at Philadelphia for promoting useful knowledge ; volume 104, part 3)
Includes bibliographical references.
ISBN 978-1-60618-043-3
1. Science, Medieval. 2. Science–India–History. 3. Science–Islamic countries–History. 4. Astrology–History 5. Astronomy–History. I. Pingree, David Edwin, 1933-2005. II. Pingree, Isabelle, editor. III. Steele, John M., editor.
Q124.97.P38 2014
509–dc23

2014039654

Dedication

In loving memory of David Edwin Pingree,
from his wife Isabelle and his daughter Amanda.

The book is intended for use by undergraduates
and all persons who are interested in
the history of ancient science.

Dr. Alexander Jones and Dr. Pingree view Member publications at a meeting of the American Philosophical Society.

Contents

Preface *ix*

Acknowledgments *xi*

Information on Previous Publication *xiii*

Common Abbreviations *xvi*

Biographical Memoir by Charles Burnett *xvii*

General Studies

1. Hellenophilia versus the History of Science (1992) 3
2. Astrology (2001) 13
3. From Alexandria to Baghdād to Byzantium. The Transmission of Astrology (2001) 17
4. Zero and the Symbol for Zero in Early Sexagesimal and Decimal Place-Value Systems (2003) 53

Mesopotamia

5. Observational Texts Concerning the Planet Mercury (1975) [with E. Reiner] 61
6. A Neo-Babylonian Report on Seasonal Hours (1974–1977) [with E. Reiner] 67
7. Venus Phenomena in *Enūma Anu Enlil* (1993) 73

The Classical World

8. On the Greek Origin of the Indian Planetary Model Employing a Double Epicycle (1971) 91
9. The Recovery of Early Greek Astronomy from India (1976) 97
10. The Preceptum Canonis Ptolomei (1990) 113
11. The Teaching of the *Almagest* in Late Antiquity (1994) 135

India

12. Astronomy and Astrology in India and Iran (1963) 161
13. Precession and Trepidation in Indian Astronomy Before A.D. 1200 (1972) 179
14. The Mesopotamian Origin of Early Indian Mathematical Astronomy (1973) 189
15. Concentric with Equant (1974) 201
16. Indian Astronomy (1978) 205

17. The Purāṇas and Jyotiḥśāstra: Astronomy (1990) 209

18. Innovation and Stagnation in Medieval Indian Astronomy (1992) 217

19. Indian Reception of Muslim Versions of Ptolemaic Astronomy (1996) 225

20. Indian Astronomy in Medieval Spain (1996) 241

21. Nīlakaṇṭha's Planetary Models (2001) 251

22. The Logic of Non-Western Science: Mathematical Discoveries in Medieval India (2003) 261

Islam

23. Indian Influence on Sasanian and Early Islamic Astronomy and Astrology (1964–6) 273

24. The Fragments of the Works of Ya'qūb ibn Ṭāriq (1968) 283

25. The Fragments of the Works of al-Fazārī (1970) 313

26. The Greek Influence on Early Islamic Mathematical Astronomy (1973) 335

27. Al-Bīrūnī's Knowledge of Sanskrit Astronomical Texts (1975) 347

Byzantine, Medieval, and Renaissance Europe

28. Gregory Chioniades and Palaeologan Astronomy (1964) 365

29. The Indian and Pseudo-Indian Passages in Greek and Latin Astronomical and Astrological Texts (1976) 393

30. A New Look at *Melancolia I* (1980) 449

31. An Illustrated Greek Astronomical Manuscript (1982) 451

32. Plato's Hermetic *Book of the Cow* (1993) 463

33. Learned Magic in the Time of Frederick II (1994) 477

Bibliography of the Works of David Pingree *495*

Preface

In "Hellenophilia versus the History of Science," David Pingree defined science as follows: "science is a systematic explanation of perceived or imaginary phenomena, or else is based upon such an explanation." This broad view of science includes astronomy, mathematics and other sciences with which we are familiar today as well as those subjects deemed nonscientific by today's standards such as astrology and magic, but which as practiced in the ancient world were nevertheless based upon rational and mathematical foundations, and so are no less deserving of attention by a historian of science. Pingree's own research covered all of these fields—astronomy, mathematics, astrology, astral magic, and Hermetic knowledge—and he repeatedly demonstrated that not only were each of these subjects worthy of study in their own right, but that in the ancient and medieval period these fields were closely interconnected.

Pingree's expertise in the history of science encompassed not only the broad range of sciences practiced in the ancient and medieval period, but also the many cultures of what, for want of a better term, may be called the Western tradition, in which it was practiced: ancient Mesopotamia, the Classical World, India, the Islamic World, Byzantium, and Medieval and Renaissance Europe. His extensive knowledge of the languages and histories of these cultures enabled him to understand the flow of scientific knowledge from one culture to another: the careful study of the transmission of science is one of the most prominent features of Pingree's work.

The aim of this collection of essays is twofold. First, we hope that students and younger scholars entering the field of the history of science will be able to learn from the example of Pingree's work: how to study texts, how to treat ancient science with an open mind, and how to investigate the transmission of knowledge from one culture to another. Second, scholars who work on the history of science will find the collection of many of Pingree's most important papers a convenient reference work. Both audiences will come away with an overview of one person's perspective on the history of science: the approaches that have proved fruitful, the questions that are worth asking, and the results of one man's lifelong study of ancient and medieval scientific texts.

The selection of essays for this collection was based upon suggestions made by J. Lennart Berggren, Charles Burnett, Bernard R. Goldstein, Takao Hayashi,

Alexander Jones, and John Steele. In making the final selection we have attempted to obtain a balance between the different sciences, cultures, and languages on which Pingree worked, and between introductory or discursive essays that deal with broad topics and more focused textual studies or technical analyses of specific problems. We have sought this balance in order to serve readers who may come to this collection from a variety of backgrounds, whether as new students beginning to work on the study of early science who will benefit both from the broader overview papers and by learning from Pingree's methodological approach to studying ancient scientific texts, or established researchers who will have a convenient reference source for many of Pingree's most important papers.

The papers have been divided into six broad sections. The first section contains general studies on the history of science that discuss some of the central problems that underlie almost all of Pingree's work: the transmission of science and the study of ancient scientific texts. The subsequent sections contain essays on the sciences of five broad cultural groups: ancient Mesopotamia; the Classical World; India; Islam; and Byzantine, Medieval, and Renaissance Europe. Inspection of the titles of the essays in these sections, however, will reveal that many do not fit into one culture alone. Pingree's interest in and pioneering work on transmission led to many studies of the dependence of the science of one culture on that of another, and even the recovery of the lost science of one culture from preserved evidence in a later culture. Thus, readers who come to this collection interested in the science of a particular culture will be rewarded by reading the essays in the other sections. A bibliography of the words of David Pingree up to 2004 appeared in the volume *Studies in the History of the Exact Sciences in Honour of David Pingree* (Leiden: Brill, 2004), edited by C. Burnett, J. P. Hogendijk, K. Plofker, and M. Yano. We provide an updated bibliography of monographs, chapters in edited works, and journal articles at the end of this volume, based in large part upon the previously mentioned bibliography; for a list of encyclopedia entries and book reviews we refer the reader to that bibliography.

Isabelle Pingree
John Steele
Providence, RI
May 2014

Acknowledgments

It was such an exciting occasion when, at the American Philosophical Society's Benjamin Franklin Hall in Philadelphia, I met just the person I came to see, Mary McDonald, the society's editor and director of publications. I had an idea for a book and was trying for some weeks to determine how I would introduce the subject to her. For fear of becoming tongue-tied I wrote it out, and with it in my shaking hand I asked if she would be interested in publishing a book of my late husband Prof. David Pingree's essays that would be suitable for an undergraduate audience as well as the general reader. She said "Yes."

My first book. It has been a great and happy experience, and the generous aid of many friends and former colleagues of David's have helped to make it so. Later that day I told the news to Prof. Alexander Jones of New York University, himself a member of the Philosophical Society. He congratulated me and offered his help. And when I returned to Providence, Prof. John Steele of Brown University did the same. Both Alex and John, with their superior knowledge of the field of Ancient Science, have been the best of advisors and tutors.

My daughter Amanda Pingree has been a steady source of encouragement throughout this new experience. Her knowledge of the publishing world and her computer skills, being much stronger than mine, have made it possible, even though she lives in Switzerland, to call upon for advice or help in all matters. She had great facility with such important tasks during the planning phase of *Pathways* when she was able to find her father's colleagues' addresses and to set up a current mailing list for them.

Professor Takao Hayashi of Doshisha University, a former student of David's, made a major contribution to the essays in this book that deal with India. He carefully provided the appropriate titles, and then, when we needed "offprint copies" for publication, he was able to give us those as well from his own collection, some of which were gifts from his professor.

Senior Scholarly Resources Librarian William Monroe at Brown University, whose superior knowledge of the holdings of the John Hay Library, provided the "offprints" that were desperately needed when zerox copies were found to be inadequate.

Professor Charles Burnett of the Warburg Institute in London, a dear friend of many years has written the Introduction to this book that is both thorough and insightful. He provides us with a most accurate and detailed picture of David's scholarly legacy.

And to all of you who listened, and encouraged: many thanks!

Isabelle Pingree

* * *

I would like to thank J. Lennart Berggren, Charles Burnett, Bernard R. Goldstein, Takao Hayashi, and Alexander Jones for their help in selecting articles to include in this collection; Alexander Jones and Takao Hayashi for lending offprints of articles for scanning; W. S. Monroe for help identifying offprints in the Pingree collection of Brown University Library and arranging for their scanning; the editors and publishers who kindly provided permission to reprint these studies; and Mary McDonald for her help and patience in the preparation of this volume.

John Steele

Information on Previous Publication

Charles Burnett, "Biographical Memoir," *Proceedings of the American Philosophical Society* 152 (2008), 256–259; by permission of the American Philosophical Society.

1. "Hellenophilia versus the History of Science," *Isis* 83 (1992), 554–563; © 1992 The History of Science Society and by permission of the University of Chicago Press.

2. "Astrology," in P. Murdin, *Encylopedia of Astronomy & Astrophysics* (London: IOP Publishing, 2001); by permission of Taylor & Francis.

3. "From Alexandria to Baghdād to Byzantium: The Transmission of Astrology," *International Journal of the Classical Tradition* 8 (2001), 3–37; © 2001 Springer, with kind permission from Springer Science+Business Media B.V.

4. "Zero and the Symbol for Zero in Early Sexagesimal and Decimal Place Value Systems," in A. K. Bag and S. R. Sarma (ed.), *The Concept of* Śūnya (Delhi: Indira Gandhi National Centre for the Arts, Indian National Science Academy and Aryan Books International, 2003), 137–141; by permission of the Indira Gandhi National Centre for the Arts and the Indian National Science Academy.

5. "Observational Texts Concerning the Planet Mercury," *Revue d'Assyriologie* 69 (1975), 175-180 (with E. Reiner); © 1975 Presses Universitaires de France.

6. "A Neo-Babylonian Report on Seasonal Hours," *Archiv für Orientforschung* 25 (1974–1977), 50–55 (with E. Reiner); by permission of the editor.

7. "Venus Phenomena in *Enūma Anu Enlil*," in H. D. Galter (ed.), *Die Rolle der Astronomie in den Kulturen Mesopotamiens* (Graz: RM-Verlag, 1993), 259–273; by permission of RM-Verlag.

8. "On the Greek Origin of the Indian Planetary Model Employing a Double Epicycle," *Journal for the History of Astronomy* 2 (1971), 80–85; by permission of Science History Publications Ltd.

9. "The Recovery of Early Greek Astronomy from India," *Journal for the History of Astronomy* 7 (1976), 109–123; by permission of Science History Publications Ltd.
10. "The Preceptum Canonis Ptolomei," in J. Hamesse and M. Fattori (eds.), *Rencontres de Cultures dans la Philosophie Médiévale* (Louvain-la-Neuve: Cassino, 1990), 355–375; by permission of the Institut d'Études Médiévales de l'Université Catholique de Louvain.
11. "The Teaching of the *Almagest* in Late Antiquity," *Apeiron* 27 (1994), 75–98; by permission of De Gruyter.
12. "Astronomy and Astrology in India and Iran," *Isis* 54 (1963), 229–246; © 1963 The History of Science Society and by permission of the University of Chicago Press.
13. "Precession and Trepidation in Indian Astronomy Before A.D. 1200," *Journal for the History of Astronomy* 3 (1972), 27–35; by permission of Science History Publications Ltd.
14. "The Mesopotamian Origin of Early Indian Mathematical Astronomy," *Journal for the History of Astronomy* 4 (1973), 1–12; by permission of Science History Publications Ltd.
15. "Concentric with Equant," *Archives Internationale d'Histoire des Sciences* 24 (1974), 26–28; © 1974 International Academy of the History of Science.
16. "Indian Astronomy," *Proceedings of the American Philosophical Society* 122 (1978), 361–364; by permission of the American Philosophical Society.
17. "The Purāṇas and Jyotiḥśāstra: Astronomy," *Journal of the American Oriental Society* 116 (1990), 274–280; by permission of the American Oriental Society.
18. "Innovation and Stagnation in Medieval Indian Astronomy," in E. Benito Ruano and M. Espadas Burgos (eds.), *17 Congreso Internacional de Ciencias Historicas* (Madrid: Comité International des Sciences Historiques, 1992), 519–526; by permission of Comité International des Sciences Historiques.
19. "Indian Reception of Muslim Versions of Ptolemaic Astronomy," in F. J. Ragep and S. P. Ragep (eds.), *Tradition, Transmission, Transformation* (Leiden: Brill, 1996), 471–485; by permission of Brill.
20. "Indian Astronomy in Medieval Spain," in J. Casulleras and J. Samsó (eds.), *From Baghdad to Barcelona : Studies in the Islamic Exact Sciences in Honour of Juan Vernet* (Barcelona : Instituto "Millás Vallicrosa" de Historia de la Ciencia Arabe, 1996), 39–48; by permission of the Instituto "Millás Vallicrosa" de Historia de la Ciencia Arabe.

21. "Nīlakaṇṭha's Planetary Models," *Journal of Indian Philosophy* 29 (2001), 187–195; © 2001 Springer, with kind permission from Springer Science+Business Media B.V.

22. "The Logic of Non-Western Science: Mathematical Discoveries in Medieval India," *Daedalus* 132 (2003), 45–53; © 2003 American Academy of Arts and Sciences.

23. "Indian Influence on Sasanian and Early Islamic Astronomy and Astrology," *Journal of Oriental Research, Madras* 34-5 (1964–6), 118–126; by permission of the Kuppuswami Sastri Research Institute.

24. "The Fragments of the Works of Ya'qūb ibn Ṭāriq," *Journal of Near Eastern Studies* 27 (1968), 97–125; © 1968 The University of Chicago Press.

25. "The Fragments of the Works of al-Fazārī," *Journal of Near Eastern Studies* 29 (1970), 103–123; © 1970 The University of Chicago Press.

26. "The Greek Influence on Early Islamic Mathematical Astronomy," *Journal of the American Oriental Society* 93 (1973), 32–43; by permission of the American Oriental Society.

27. "Al-Bīrūnī's Knowledge of Sanskrit Astronomical Texts," in P. J. Chelkowski (ed.), *The Scholar and the Saint* (New York: New York University Press, 1975), 67–81; by permission of the editor.

28. "Gregory Chioniades and Palaeologan Astronomy," *Dumbarton Oaks Papers* 18 (1964), 133–160; © 1964 Dumbarton Oaks Research Library and Collections, Trustees for Harvard University.

29. "The Indian and Pseudo-Indian Passages in Greek and Latin Astronomical and Astrological Texts," *Viator* 7 (1976), 141–195; © 1976 The University of California Press.

30. "A New Look at Melancolia I," *Journal of the Warburg and Courtauld Institutes* 43 (1980), 257–258; by permission of the Warburg and Courtauld Institutes.

31. "An Illustrated Greek Astronomical Manuscript," *Journal of the Warburg and Courtauld Institutes* 45 (1982), 185–192; by permission of the Warburg and Courtauld Institutes.

32. "Plato's Hermetic *Book of the Cow*," in P. Prini (ed.), *Il Neoplatonismo nel Rinascimento* (Rome: Instituto della Enciclopedia Italiana Findata da G. Treccani, 1993), 133–145; by permission of Instituto della Enciclopedia Italiana.

33. "Learned Magic in the Time of Frederick II," *Micrologus* 2 (1994), 39–56; © SISMEL—Edizioni del Galluzzo.

Common Abbreviations

Planetary symbols

☽	Moon
☉	Sun
☿	Mercury
♀	Venus
♂	Mars
♃	Jupiter
♄	Saturn

Signs of the zodiac

♈	Aries	0° longitude
♉	Taurus	30°
♊	Gemini	60°
♋	Cancer	90°
♌	Leo	120°
♍	Virgo	150°
♎	Libra	180°
♏	Scorpio	210°
♐	Sagittarius	240°
♑	Capricorn	270°
♒	Aquarius	300°
♓	Pisces	330°

Planetary Phenomena

Outer planets

Γ	First visibility
Φ	First station
Θ	Acronychal rising
Ψ	Second station
Ω	Last visibility

Inner planets

Γ	First visibility in the east (morning)
Φ	Station in the east (morning)
Σ	Last visibility in the east (morning)
Ξ	First visibility in the west (evening)
Ψ	Station in the west (evening)
Ω	Last visibility in the west (evening)

Biographical Memoir by Charles Burnett

DAVID EDWIN PINGREE

2 JANUARY 1933 · 11 NOVEMBER 2005

PROCEEDINGS OF THE AMERICAN PHILOSOPHICAL SOCIETY

VOL. 152, NO. 2, JUNE 2008

DAVID EDWIN PINGREE, university professor, professor of the history of mathematics, and chair of the Department of the History of Mathematics at Brown University, died on 11 November 2005, from complications of the diabetes he had been suffering from throughout his life.[1] He died when he was at the height of his productivity, with many projects on the go, and is sorely missed by his colleagues and students and by the whole scientific community. A mark of the width of his interests and the international spread of his renown is the festschrift that was presented to him in January 2004, which includes, in its nine hundred–odd pages, sections on ancient Mesopotamia, India, the Arabic world, and the West, from antiquity until the Renaissance, and has contributions from scholars from the U.S., Canada, Great Britain, Austria, Germany, Italy, Belgium, Switzerland, Holland, India, and Japan.[2] It is impossible to classify him as a Sanskritist, a scholar of the culture of the Near East, or a Greek or Latin specialist. He was all of these, and much more. His merit was to see the whole canvas of the history of the science of the stars, spread out from the Western shores of Europe to the Eastern shores of India, and extending from the very beginnings of man's interest in the stars to the early modern era. The story consisted of many episodes and featured many characters, who spoke in several different languages, but in the end it was one story, and only David was able to tell it.

David was born in New Haven on 2 January 1933, the son of Daniel Pingree, who was at the time in the economics department at Yale University. He had two brothers, and one sister eighteen years younger than he was. He had no vision in one eye and only partial vision in the other, so he had problems seeing more than two feet in front of him. Consequently, instead of participating in sports at school, he read assiduously, supplementing the Latin and Greek he learnt at school with Sanskrit, which he taught himself. After graduating from Phillips Academy, Andover, he entered Harvard, where he successively took a B.A. *magna cum laude* in Classics and Sanskrit (1954), and completed a Ph.D. with the title "Materials for the Study of the Transmission of Greek Astrology to India" (1960). His fellow students remember him fondly under the nickname of "Marble Head," whose connotations embrace more than the village near Andover from which he came. A Fulbright scholarship enabled him to study Greek manuscripts at the Vatican Library

[1] I am grateful for information supplied by Isabelle Pingree, Masaaki Kubo, Wolfgang Hübner, Stephan Heilen, and Michio Yano.

[2] *Studies in the History of the Exact Sciences in Honour of David Pingree*, ed. C. Burnett, J. P. Hogendijk, K. Plofker, and M. Yano (Leiden: Brill, 2004). This book includes a list of publications up to July 2003.

(1954–55), and a Ford Fellowship allowed him to spend a year doing the same in India (1957–58); these were sandwiched between fellowships at Dumbarton Oaks (1956–57 and 1959–60). His manuscript studies laid the foundations for his subsequent research, based on his own editions of the material, much of which he had uncovered himself. In 1963 he married Isabelle Sanchirico, who became his lifelong companion and helpmate, as well as establishing a reputation as a specialist in early bookbindings; their only child, Amanda, was born in October 1965. In 1963 he was appointed assistant professor at the Oriental Institute in Chicago, where he established his firm links with Noel Swerdlow and Erica Reiner. He spent the academic year 1964–65 at the American University in Beirut, working with the indefatigable investigator of Arabic astronomy, Edward Kennedy. Then, in 1971, he joined the powerhouse of studies in the history of pre-modern astronomy: the Department of the History of Mathematics at Brown University in Providence. Here, brought together in the old and dusty building of Wilbour Hall under the shadow of the University Library, were Otto Neugebauer (the founder of the whole enterprise) and Aby Sachs, soon to be joined by Gerald Toomer, who knew each affectionately by names such as the Elephant, the Owl, and the Home-Ox (*homo oxoniensis*). Pingree, in turn, was known as "Abu Kayd" (the father of the comet), since his name resembled that of the pioneer in research on cometology, Alexandre Guy Pingré (1711–1796). In 1981, in recognition of his contribution to scholarship, he was awarded a prestigious MacArthur Fellowship, and from 1981 to 1986 he was the A. D. White Professor at Large at Cornell University. But Brown University remained his base for the rest of his life. He was attended by a succession of dogs, who accompanied him on his constitutional walk between his home and his office. This department acted as a beacon, attracting students and visiting scholars from throughout the world, who established links with London, Kyoto, Barcelona, Utrecht, and many other places. David built up his own library of books (some beautifully bound by Isabelle), microfilms, and copies of Sanskrit manuscripts written out by hand by Indian pundits. As the department gradually shrank due to the death or departure of its members, David was welcomed into the Classics department of the university, where he continued to teach until the year of his death, and where, fortunately, a post in the history of ancient sciences and one in Sanskrit will continue to be occupied.

David saw no sharp distinction between astronomy and astrology (or, for that matter, magic, in which astrology often played a significant part) and preferred to consider them together as part of the exact science of the ancients. He did not pursue a whiggish course of documenting "discoveries" and the earliest instances of doctrines that became part of modern universal science. Rather, he viewed each aspect of the science of the stars in respect to the community in which it arose and paid equal attention to the "important" and the run-of-the-mill doctrines in each context. He amassed shoeboxes full of horoscopes and transcribed texts at a rate that far outran the pace at which they could be published, just in order to get as full a picture as possible of the subject. His vast knowledge of detail was condensed—sometimes too drastically for readers used to a slower pace—in publications such as the biobibliographies of astrologers included in his edition of the *Yavanajātaka* of Sphujidhvaja (1978), his history of mathematical astronomy in India in the supplement volume of the *Dictionary of Scientific Biography* (1978), astrology and astronomy in Iran in the *Encyclopædia Iranica,* and his account of astrology in *The Dictionary of the History of Ideas* (1973–74). A masterpiece of succinctness and comprehensiveness is his history of astrology in *From Astral Omens to Astrology, from Babylon to Bīkānēr* (1997), originally delivered as a series of lectures in Rome, Bologna, and Venice. He wrote innumerable encyclopedia entries on astronomers of the past, many of which are little monographs in themselves.

His editions ranged from the Babylonian MUL.APIN (with Hermann Hunger, 1989), and *Babylonian Planetary Omens* (with Erica Reiner, 1975–2005), through the Greek astrological texts of Dorotheus (Greek fragments, and an Arabic version of the full text, 1976), Hephaestion (1973–74), Vettius Valens (1986), Gregory Chioniades (1985–86), and Arabic texts on astronomical tables (al-Hashimi, with Edward Kennedy, 1981), to Latin texts translated from Greek (Pseudo-Ptolemy, *Praeceptum Canonis*, 1997) and from Arabic (*Picatrix,* 1986, and *Liber Aristotilis* of Hugo of Santalla, with Charles Burnett, 1997). He conceded to Western scholars by providing English translations for Sanskrit and Arabic texts but assumed that Greek and Latin were still part of an educated person's accomplishments. In all cases his primary aim was to establish texts from all the manuscripts, and to provide complete word-indexes (the writer of this obituary remembers that he spent several weeks at the Warburg Institute, checking the index of his edition of Vettius Valens, which required altering every single page and line number to accommodate the Teubner page layout; no academic task was too menial for him).

His edition of *Picatrix* was his major contribution to the history of magic (sadly, his work on the related *k. al-nawāmis* of Plato, known as the *Liber vaccae* in Latin, never reached completion). Even here, the Indian element is present, for he recognized the Sanskrit words for the planets among the garbled lists of planetary names. This Latin text was compared word-for-word with the Arabic text, and included, in typical fashion, the extra or aberrant passages in all the extant manuscripts in a series of sixty-five appendixes. It completed the aim of Aby Warburg in 1914 to make available the Arabic and Latin versions of the text, with a translation in a modern European vernacular. Pingree's connection with the Warburg Institute was close: he held a Saxl Visiting Professorship there and contributed several articles to the Institute's journal, including one in which he ventured into the field of art history, making some acute observations about the interpretation of Dürer's *Melancolia I* (1981).

David was unstinting in the help that he provided to other scholars. Not only did he collaborate fruitfully with many of them (as the joint editions mentioned above attest), but he also carefully read typescripts before they were sent to the press, and made invaluable comments in his economic but minuscule handwriting. Above all, he wished to preserve for posterity the scientific products of the past, by making known the contents of Sanskrit manuscripts (*Census of the Exact Sciences in Sanskrit*, [1970–94], the Jyotisa MSS in the Library of the Wellcome Institute [London, 2004], and the establishment of the American Committee for South Asian Manuscripts, dedicated to cataloguing the manuscripts and texts in North American collections), and by ensuring that copies of these manuscripts were made.

David's papers (which include several editions of texts in many languages) and his reproductions of manuscripts will become available for future scholars by being deposited with the American Philosophical Society. His teaching and inspiration will live on in a legacy that encompasses several continents.

Elected 1975; Councillor 1980–83; Committee on Lewis Award 1980–88

CHARLES BURNETT
Professor of the History of Islamic Influences in Europe
The Warburg Institute

General Studies

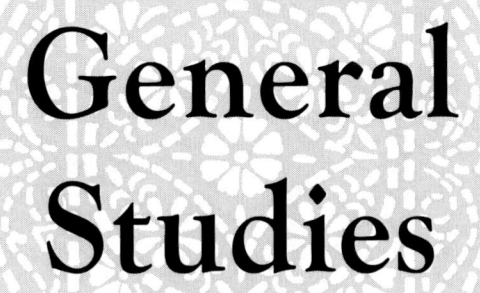

Hellenophilia versus the History of Science

By David Pingree†

THE GENESIS OF THIS PAPER lies in a conversation that I had with A. I. Sabra of Harvard on the perennial problem of the definition of science appropriate to a historian of science; its corruption (including the deliberately extreme mode of its expression) is entirely a result of my own labors. For the piece represents the attitudes toward the subject that I have developed over some three and a half decades of studying the history of the "exact" sciences (as I will persist in calling them despite the lack of exactitude in some of them), as practiced in ancient Mesopotamia, in ancient and medieval Greece, India, and the Latin-speaking West, and in medieval Islam. It is this experience, then, and the desire to reconstruct a complex history as accurately as possible, that motivates me—these two, and the wish to provide an apologia for my claim to be a historian of science rather than of quackery. For the sciences I study are those related to the stars, and they include not only various astronomies and the different mathematical theories they employ, but also astral omens, astrology, magic, medicine, and law (*dharmaśāstra*). All of these subjects, I would argue, were or are sciences within the contexts of the cultures in which they once flourished or now are practiced. As such they deserve to be studied by historians of science with as serious and thorough a purpose as are the topics that we usually find discussed in history of science classrooms or in the pages of *Isis*. This means that their intellectual content must be probed deeply, and not simply dismissed as rubbish or interpreted in the light of modern historical mythology; and that the intellectual content must be related to the culture that produced and nourished each, and to the social context within which each arose and developed.

In stating these opinions I may appear to have set myself up as a relativist, but I would deny the applicability of that epithet to my position since my interest lies not in judging the truth or falsehood of these or any other sciences, nor in discovering in them some part that might be useful or relevant to the present world, but simply in understanding how, why, where, and when they worked as functioning systems of thought and interacted with each other and with other systems of thought.

It is with these considerations in mind, then, that I have embraced the word employed in the title of this article, "Hellenophilia," as it is a most convenient description of a set of attitudes that I perceive to be of increasing prevalence

† Box 1900, Brown University, Providence, Rhode Island 02912.

This paper was originally delivered as a lecture at the Department of History of Science, Harvard University, 14 November 1990.

within the profession of the history of science, and which I believe to be thoroughly pernicious. I like "Hellenophilia" as a word because it brings to mind such other terms as "necrophilia," a barbaric excess that erupts as a disease from the passionate rather than from the rational soul; whereas the true love of the Greeks, Philhellenism, though also an attribute of barbarians such as are we—the epithet "Philhellene" was proudly borne by ancient Parthians, Semites, and Romans—arises preeminently from well-deserved admiration. A Philhellene is one who shares in what used to be, when children in the West still were taught the classics, a virtually universal awe of Greek literature, art, philosophy, and science; a Hellenophile suffers from a form of madness that blinds him or her to historical truth and creates in the imagination the idea that one of several false propositions is true. The first of these is that the Greeks invented science; the second is that they discovered a way to truth, the scientific method, that we are now successfully following; the third is that the only real sciences are those that began in Greece; and the fourth (and last?) is that the true definition of science is just that which scientists happen to be doing now, following a method or methods adumbrated by the Greeks, but never fully understood or utilized by them.

Hellenophiles, it might be observed, are overwhelmingly Westerners, displaying the cultural myopia common in all cultures of the world but, as well, the arrogance that characterized the medieval Christian's recognition of his own infallibility and that has now been inherited by our modern priests of science. Intellectually these Western Hellenophiles are still living in the miasma that permeated Europe until the nineteenth century, before the discovery of Sanskrit and the cracking of cuneiform destroyed such ethnocentric rubbish; such persons have simply not been exposed to the knowledge they would need to arrive at a more balanced judgment. But, sadly, I must report that many non-Westerners have caught a form of the disease Hellenophilia; they are deluded into believing that the greatest glory an Indian, a Chinese, an Arab, or an African scientist can have acquired is that gained by having anticipated either a Greek or a modern Westerner. So some Indians, for instance, busily reinterpret their divinely inspired *Ṛgveda* so that it teaches such modern hypothetical theories as that of relativity or the latest attempt to explain black holes, as if these transitory ideas were eternal complete truths. In doing this they are behaving as did those Christians who once believed it important to demonstrate that Genesis agrees with Greek science. These attempts do not enhance the brilliance of the authors or the reinterpreters of their sacred or scientific texts, but rather reveal a severe sense of cultural inferiority.

Parallel to this form of cultural denigration, practiced by the culture itself or by historians of science, is, say, the false claim that medieval Islam only preserved Greek science and transmitted it as Muslims had received it to the eager West. In fact Arab scientists, using Indian, Iranian, and Syrian sources as well as their own genius, revised the Greek sciences, transforming them into the Islamic sciences that, historically, served as the main basis for what little science there was in Western Europe in the twelfth and following centuries and for the amazing developments that happened three and four centuries later in Italy and Central Europe.

Another form that this Western arrogance takes is the naive assumption that other peoples in the world not only should be like us, but actually are or were—

"were" because this particular fallacy usually affects those who study Stone Age and other preliterate cultures that have been left defenseless in the face of modern reconstructions of their thoughts by their inability to record them in permanent form. In the history of the exact sciences the scholars who perpetrate wild theories of prehistoric science call themselves archaeoastronomers. The basic premise of some archaeoastronomers is that megalithic and other cultures in which writing was not known built stone monuments, some quite massive, in order to record their insights into the periodicity of celestial motions. This seems to me a trivial purpose to motivate such monumental communal efforts as the building of Stonehenge or the pyramids. There are many strong arguments to be raised against many of these interpretations. At this point, however, I wish only to point out that they go against the strong evidence from early literate societies that early man had little interest in the stars before the end of the third millennium B.C.; the cataloguing of stars and the recording of stellar and planetary phenomena are not a natural, but a learned activity that needs a motivation such as that which inspired the Babylonians, who believed that the gods send messages to mankind through the celestial bodies. The realization that some of these ominous phenomena are periodic can be dated securely in Mesopotamia to a time no earlier than the late second millennium B.C.; mathematical control of the relation between solar and lunar motion came only in about 500 B.C. The Egyptians also first began using selected stars as a sort of crude clock only in about 2000 B.C. and progressed no further in mathematical astronomy till they came under Babylonian influence. The earliest traces of a knowledge of astronomy in Greece and India seem also to be derived, in the early first millennium B.C., from Mesopotamia. I cannot speak of the astronomy of the early Chinese with authority because I am ignorant of their language, but I gather from what I have read that not even the beginnings of the system of the *hsiu* or lunar lodges can be dated before the late second millennium B.C. From the written evidence, then, it appears that an interest in the stars as omens arose in Mesopotamia after 2000 B.C. and started to develop toward mathematical astronomy in about 1200 B.C., but that the Babylonians began to invent mathematical models useful for the prediction of celestial phenomena with some degree of accuracy only in about 500 B.C. From Mesopotamia these astronomical ideas rapidly radiated to Egypt, later to Greece and India, and finally, perhaps, to China; in each of these cultures they were molded by the recipient scientists into something new, though still having recognizable Mesopotamian origins. The astral sciences spread from one civilization to another like a highly infectious disease. It is within the context of this documented history that I find implausible the suggestion that less advanced civilizations, without any known systems of writing or accurate record keeping, independently discovered complex lunar theories, or precession, or even an accurate intercalation cycle. The example of the Babylonians, with their need for a specific motive for observing stars and the fact that it took them a millennium and a half to arrive at a workable mathematical astronomy, and the examples of the Egyptians, Greeks, and Indians, if not the Chinese, who initially borrowed their astronomies from the Babylonians before each developed its science in its own way, seem to me to invalidate the theoretical basis for much of archaeoastronomy.

I return now to the four variants of Hellenophilia that I mentioned earlier.

Each, I would claim, distorts the history of science in two ways: passively, it limits the phenomena that the historian is willing or able to examine; actively, it perverts understanding both of Western sciences, from the Greeks till now, and of non-Western sciences. Thus those who still believe that the Greeks invented science either are altogether ignorant of, say, Babylonian mathematics and astronomy or else, though aware of them, fall into my second category and refuse to recognize them as sciences. The ignorance of the first group, of course, can and should be remedied through education; the obstinacy of the second in not acknowledging that Old Babylonian investigations of irrational numbers like $\sqrt{2}$, of arithmetical and geometrical series, or of Pythagorean triplets are science even though they are mathematically correct, or in asserting that the arithmetical schemes that they successfully used to control the many variables involved in the prediction of the time of the first visibility of the lunar crescent cannot bear the august name of *scientia* even though the predictions were essentially correct—this obstinacy is hard to deal with. It leaves the obstinate, however, in the awkward position of denying the status of science to one of the main contributors to the Greek astronomy that is the forebear of our positional astronomy. Such a person, of course, can name an arbitrary date at which positional astronomy comes to fit into his or her definition of science; but this cannot be accepted by a historian, as it is the historian's task to seek out the origins of the ideas that he or she is dealing with, and these manifestly lie, for astronomy, in the wedges impressed on clay tablets as well as in the observed motions of the celestial bodies. It is certainly possible to be a modern scientist without knowing history, or even with a firm belief in historical mythology; but can a historian of science function effectively under such disabilities?

While ignorance of Babylonian astronomy destroys the historian's ability to understand the origins and development of Greek and other astronomies together with their more modern descendants, it also tempts him or her to imagine that there is no other way to do astronomy than through the Greek and modern way of making observations and building geometric models. But Babylonian astronomy reveals how few observations are needed and how imprecise they may be if the astronomers are clever enough; and it also demonstrates that simple arithmetical models suffice for predicting the times and longitudes of periodic celestial phenomena. The Babylonian solutions are brilliant applications of mathematical structures to rather crude data, made purely to provide the possibility of prediction without any concern for theories of cosmological structure or celestial mechanics. The Greeks added the concern both for the geometrical structure of the universe and for the cinematics of the heavens, with a strong prejudice in favor of circles or spheres rotating with uniform motion; but they also, to a large extent, simply expressed the Babylonian period relations and arithmetical zigzag and step functions in a geometrical language, using observations to modify Babylonian parameters and to fine-tune their own geometrical models. A third variety of astronomy emerged from the synthesis of Babylonian arithmetical models, Hellenistic geometrical models, and local mathematical traditions that occurred in India in the fifth century A.D. In this astronomy questions of celestial cinematics receded into trivial mechanisms while computational finesse harnessed a broad range of mathematical techniques to the solution of astronomical problems, with the role of observations being limited to the confirmation, if possible, of accepted theory. There were historically many more astronomies, manifesting a variety of ways in which the same phenomena might be made predictable by mathematical means. These different astronomies reflect the different intellectual traditions of the various cultures as well as the specific problems that each society wished its astronomers to address—for example, the Babylonians were interested in certain horizon phenomena that they regarded as omens; the Greeks brought philosophical and physical problems into a science that had been purely mathematical, and as well introduced the more social aim of casting horoscopes; and the Indians devoted their efforts to the purely pragmatic goals of casting horoscopes, of predicting eclipses because of their significance as omens, and of evolving and regulating an extraordinarily complex calendar. I believe that those historians who limit themselves to the study of only one of these approaches to mathematical astronomy will blunder, as indeed many have, by not fully understanding the range of possibilities or the shaping force of purely cultural factors on the course that any science takes.

If it is evident that for a historian the proposition that the Greeks invented science must be rejected, it necessarily follows that they did not discover a unique scientific method. Indeed, they—and we—have not one, but many scientific methods; biologists, physicists, and astronomers went, and go, their separate methodological ways. I choose, therefore, to focus on the pride of Greek science: Euclidean geometry—which, of course, is purely logical and nonexperimental. This is often and justly praised for the rigor and power of its axiomatic system and for its ability to offer logical deductive proofs. Indeed, Babylonian and Indian mathematics are frequently criticized for relying not on proofs but on demonstrations. But without axioms and without proofs Indian mathematicians solved indeterminate equations of the second degree and discovered the infinite power series for trigonometrical functions centuries before European mathematicians independently reached similar results. These achievements amply demonstrate, I believe, that the Euclidean approach is not necessary for discovery in mathematics. Those who deny the validity of alternative scientific methods must somehow explain how equivalent scientific "truths" can be arrived at without Greek methods. And in their denial they clearly deprive themselves of an opportunity to understand science more deeply.

To depart briefly from the exact sciences, I would like to draw attention to the fascinating work of Francis Zimmermann on *āyurveda,* the Indian science of longevity, as it was and is currently practiced in Kerala. While he strongly supports the idea of the historical dependence of *āyurveda* on the Galenic theory of humors, Zimmermann is also keenly aware of the many ways in which Indian *vaidyas* have altered foreign notions while incorporating them into their own cultural traditions to create a theory of harmony and mutual influence between the humans, animals, and plants inhabiting any region and that region's terrain and climate. In Kerala these "ecological" ideas produced a local interpretation of *āyurvedic* doctrine that, it would appear, maintained a generally healthy human population, at least to that population's satisfaction, as long as the people were not subjected to wars, famines, or epidemics. The *āyurvedic* approach to medicine does not inspire its practitioners to make discoveries in molecular biology, but it is the correct medical science for the cultural context within which it operates. And it attempts to address psychological, social, and environmental aspects of health that our mechanistic medicine tends to ignore. Western doctors have something to learn about medical care from *āyurveda,* and so do Western historians of medicine.

The third fallacious opinion that I have associated with the Hellenophiles is that the only sciences are those that accredited Greeks recognized as such. This opinion generally takes the form of allowing Aristotle to define science for us, so that it excludes even the genuinely Greek sciences of astrology, divination, magic, and other so-called superstitions. This brings us squarely to the fundamental question of this paper: What is the proper definition of science for a historian of science? I would offer this as the simplest, broadest, and most useful: science is a systematic explanation of perceived or imaginary phenomena, or else is based on such an explanation. Mathematics finds a place in science only as one of the symbolical languages in which scientific explanations may be expressed. This definition deliberately fails to distinguish between true and false science, for explanations of phenomena are never complete and can never be *proved* to be "true." Obviously, this shortfall is as true of modern scientific hypotheses as of ancient ones. It is, therefore, inappropriate to apply a standard of truthfulness to the sciences, at least viewed as historical phenomena, for the best that modern scientists can claim—I cannot judge whether justly or not—is that they are closer to some truth than were their predecessors; nor, for the reasons I have already stated, can the methodologies of science be limited to just those employed by present-day scientists.

If my definition of science as it must be viewed by a historian is accepted, it is easy to show that astrology and certain "learned" forms of divination, magic, alchemy, and so on are "sciences." Some may regard this procedure for elevating superstition to the rank of scientific theory as arbitrary and unfair, but remember that modern science is the initial culprit in that it arbitrarily sets up its own criteria by which it judges itself and all others. If I am a relativist, then, it is precisely at this point where, as a historian, I refuse to allow modern scientists who know little of history to define for me the bounds of what in the past—or in the present—I am allowed to consider to be science. It pains me to hear some scientists, who have not seriously considered the subject, denounce astrology as "unscientific" when all that they mean is that it does not agree with their ideas about the way the universe functions and does not adhere to their concept of a correct methodology. It pains me not because I believe that astrology is true; on the contrary, I believe it to be totally false. But the anathemas hurled at it by some scientists remind me more of the anathemas leveled by the medieval Church against those who disagreed with its dogmas than of rational argument. In its persecution of heretics as in its missionary zeal and its tendency to sermonize and to pontificate, our scientific establishment displays marked similarities to the Church, whose place in our society it has largely usurped.

That Church, like modern science, condemned divination, astrology, and magic, though on the grounds that they limit God's power and human free will rather than that they fail to conform to our current "laws of nature." Both of these arguments, to my way of thinking, are arbitrary and irrelevant to a historian, who should remain free of either the Church's or modern science's theology.

To turn to history: Babylonian divination is a systematic explanation of phenomena based on the theory that certain of them are signs sent by the gods to warn those expert in their interpretation of future events; there is no causal connection, but only one of prediction so that appropriate countermeasures may be undertaken. In this system the stars and the planets, which are regarded as the manifestations of the individual gods in the sky, indicate by their changes in quality and in location the events that will befall mankind at large, or a country in general, or a specific king and his family. Omens that appear on earth are contrived by Šamaš, the god of the sun, to warn individual men of the coming of good or evil. In themselves, texts describing the rules of the interpretation of omens—*Enūma Anu Enlil* for celestial omens and *Šumma ālu* for terrestrial ones—are scientific; they provide systematic explanations of phenomena. The former is in addition closely allied to the origins of mathematical astronomy, for the tablets of *Enūma Anu Enlil* contain the first recorded realization of the periodicity of certain celestial phenomena and the first attempts to provide mathematical models for predicting the occurrences of such phenomena. This close linkage between divination and mathematical astronomy, as well as a linkage between divination and the observations necessary for constructing and refining mathematical astronomy, persisted in the first millennium B.C. in the cuneiform *Letters, Reports,* and *Diaries*. Moreover, as the mathematical models became more sophisticated and the descriptions of the observed phenomena became more precise, the rules for predicting terrestrial events from the celestial phenomena became more complex. Babylonia, then, provides us our first example of the fruitful interplay between the theoretical and the applied aspects of a science.

Astrology grew out of a union of aspects of advanced Babylonian celestial divination with Aristotelian physics and Hellenistic astronomy; this union—illicit, some may think—occurred in Egypt in the second century B.C. The product was the supreme attempt made in antiquity to create in a rigorous form a causal model of the *kosmos,* one in which the eternally repeating rotations of the celestial bodies, together with their varying but periodically recurring interrelationships, produce all changes in the sublunar world of the four elements that, whether primary, secondary, or tertiary effects, constitute the generation and decay of material bodies and the modifications of the parts or functions of the rational and irrational souls of men, animals, and plants. In other words, ancient Greek astrology in its strictest interpretation was the most comprehensive scientific theory of antiquity, providing through the application of the mathematical models appropriate to it predictions of all changes that take place in a world of cause and effect; it is not surprising, then, that it was called simply *mathēsis* or "science" by Firmicus Maternus and others.

But even within Greek astrology there was a movement toward a relaxation of the rigidity of the theory, both because of the frequent failure of the predictions and, more important, because of people's desire to circumvent unpleasant predictions—a practical rather than a theological demand for a modicum of free choice and self-determination. This trend toward an astrology that indicates predispositions instead of concrete inevitabilities was accentuated when the Greek form of this science was transmitted to India and transformed into a system that rapidly increased the complexity of the mathematical models in order to diffuse and mollify the inescapability of a simpler predictive scheme. The Indians attempted to match the bewildering variousness of real lives by an equally bewildering multiplicity of mathematically computable variants in astrology.

If Greek astrology is based on the idea that the motions and interrelations of the celestial spheres are ultimately the causes of all terrestrial phenomena, astral magic, which was concocted out of Babylonian and Indian liturgies and iconographies mingled with Greek astrology, Ptolemaic astronomy, and Hellenistic philosophy by the self-styled Sabaeans of Ḥarrān in the ninth century, assumes that the magus's soul is free of inhibiting stellar influences, so that, by manipulating terrestrial objects, he can reverse the processes of astrology and change the wills of the planetary spirits. In this way the magus can employ the astral influences defined by astrology to effect the changes he wishes in the sublunar world. This was a dream still dreamt by two founding members of the Royal Society, Kenelm Digby and Elias Ashmole.

These same Sabaeans invented also a second type of learned magic based on Plato's and Aristotle's theories of animal and human souls. In this magic, which I have dubbed psychic, the magus artificially creates new animals by uniting either within a womb or within a womblike chamber animal or human parts representing the material body and the particular part or function of the soul that he wishes his creation to be endowed with. The magus can then employ his artificial animal to accomplish wonders. Astral and psychic magic we may not wish to test in order to determine their validity, but as historians we must regard them as scientific, if for no other reason than because many Western scientists in the sixteenth and seventeenth centuries took them to be genuine sciences. And, of course, they do fall under the aegis of my definition of science.

The same status must be accorded, then, to alchemy—be it Greek, Arabic, Chinese, or Indian—and to other systematic theories that explain phenomena, whether the lapidaries and physiognomics that spread from Mesopotamia to Greece, to Iran, and to India, or the science of determining sites suitable for different types of buildings—a science found in different forms in China and in India—or the purely Indian analysis of the processes of converting thought into sound in order to produce intelligible speech. These and other sciences cannot be dismissed simply because they do not fall into the intellectual system favored by some Greek philosophers.

I have already, if I have been at all successful, persuaded you that the fourth variety of Hellenophilia, in which one defines science as that which modern Western scientists believe in and the methodologies with which they operate, is inappropriate to a historian, though it may be useful to a modern Western scientist. And I have already mentioned that among the advantages provided to the historian by looking outside of the confines of such a restricted definition are a realization of the potential diversity of interpretations of phenomena and of the actual diversity of the origins of the ideas that have developed into modern Western science among other sciences, and an objectivity born of an understanding of the cultural factors that impel sciences and scientists to follow one path rather than another. The loss of all these advantages is the price paid for suffering the passive effects of this form of Hellenophilia.

Its active form is more pervasive in and pernicious to history. This results in the attitude that it is the task of the historian not to study the whole of a science within its cultural context, but to attempt to discover within the science elements similar to elements of modern Western science. One example I can give you relates to the Indian Mādhava's demonstration, in about 1400 A.D., of the infinite power series of trigonometrical functions using geometrical and algebraic arguments. When this was first described in English by Charles Whish, in the 1830s, it was heralded as the Indians' discovery of the calculus. This claim and Mādhava's achievements were ignored by Western historians, presumably at first because they could not admit that an Indian discovered the calculus, but later because no one read anymore the *Transactions of the Royal Asiatic Society,* in which Whish's article was published. The matter resurfaced in the 1950s, and now we have the Sanskrit texts properly edited, and we understand the clever way that Mādhava derived the series *without* the calculus; but many historians still find it impossible to conceive of the problem and its solution in terms of anything other than the calculus and proclaim that the calculus is what Mādhava found. In this case the elegance and brilliance of Mādhava's mathematics are being distorted as they are buried under the current mathematical solution to a problem to which he discovered an alternate and powerful solution.

Other examples of this dangerous tendency abound. For instance, since the 1850s historians ignorant of Mādhava's work have argued about whether Indian astronomers had the concept of the infinitesimal calculus on the basis of their use of the equivalent of the cosine function in a formula for finding the instantaneous velocity of the moon, a formula that occurs already in a sixth-century Sanskrit text, the *Pañcasiddhāntikā*. I cannot tell you how that formula was derived, since its author, Varāhamihira, has not told me; but I find it totally implausible that some Indian discovered the calculus—a discovery for which previous developments in Indian mathematics would not at all have prepared him—applied his discovery only to the problem of the instantaneous velocity of the moon, and then threw it away. The idea that he might have discovered the calculus arises only from the Hellenophilic attitude that what is valuable in the past is what we have in the present; this attitude makes historians become treasure hunters seeking pearls in the dung heap without any concern for where the oysters live and how they manufacture gems.

One particularly dangerous form of this aspect of Hellenophilia is the positivist position that is confident that mathematical logic provides the correct answers to questions in the history of the exact sciences. I, of course, am not denying the power of mathematics to provide insights into the character and structure of scientific theories; obviously, Otto Neugebauer's brilliant analysis of the astronomical tables written in cuneiform during the Seleucid period gives us a profound understanding of how this astronomy worked mathematically, and it tells us something about some stages in the development of the science as recorded on the hundreds of tablets that he investigated. But it does not and cannot, as Neugebauer well knew, answer a whole range of historical questions. We do not know by whom, when, or where any Babylonian lunar or planetary theory was invented; we do not know what observations were used, or where and why they were recorded; we do not know much about the stages by which Babylonian astronomers went from the crude planetary periods, derived from omen texts, found in MUL.APIN to the full-scale ephemerides of the last few centuries B.C. Historians need to be very careful in assessing the nature of the questions the material at hand will allow them to answer with a reasonable expectation of probability; and they must hope for and search out new evidence. But the positivists jump in to claim that mathematical models (and they usually use quite simple ones) suffice to describe the idiosyncratic behavior of people and to account for the perverse quirks in their personalities. I will not here name names, but the number of historians of the exact sciences who suffer from this malady is appallingly large; they can be easily recognized by the characteristic trait that, in general, the remoter the time and the scantier the evidence, the more precise their computations. Millennia of history are made to depend on the measurement of an arc of a few minutes or degrees when it has not even been convincingly demonstrated that any arc was being measured at all.

So far I have been attempting to discredit Hellenophilia on the grounds that it renders those affected by it unable to imagine many significant questions that legitimately should be addressed by historians of science and that it perverts their judgment. Much of my argument has been based on the anthropological perception that science is not the apprehension of an external set of truths that mankind is progressively acquiring a greater knowledge of, but that rather the sciences are the products of human culture. But this viewpoint must be modified by a further consideration, to which I have from time to time alluded since it strengthens the arguments in favor of the definition of science that I proposed. This consideration is that, as a simple historical fact, scientific ideas have been transmitted for millennia from culture to culture, and transformed by each recipient culture into something new. This is particularly noticeable in the astral sciences that I study—astronomy, astral omens, astrology, and astral magic—but can be readily discerned in many others. The taproot and trunk of the tree of the astral sciences are buried in the Mesopotamian desert, with subsidiary roots in Egypt and China (I have lopped the Mayas off this arboreal image, as they are self-rooted). From Babylonia the tree branched out to Egypt, to Greece, to Syria, to Iran, to India, and to China; grafted onto different cultural stocks in each of these civilizations, it developed variant leaves, shoots, and flowers. The process of the intertwining of these diverse varieties of astronomies throughout Eurasia and North Africa was amazingly complex, as ideas, mathematical models, parameters, and instruments circulated rapidly over the vast expanse of divergent traditions. Out of this process modern Western astronomy sprang from a rather late branch that grew from and was fed by an incredibly complicated undergrowth. For very complex reasons this modern Western astronomy has choked off all of its rivals and destroyed the intellectual diversity that mankind enjoyed before it moved from simple communication to Western domination. We cannot know what the Islamic, Indian, or Chinese astral sciences might have become had this not happened, except that they would not have become what our culture has produced. But unraveling the intertwined webbing of these sciences is a fascinating and a rewarding task for a historian, and one in which much remains to be done. I strongly recommend to those of you who have the opportunity thus to broaden your perspectives to grasp it.

Astrology

Astrology is the theory that the planets, the Sun and the Moon, as well as the 12 'zodiacal signs', combine in various, ever-changing configurations with respect to each other and the local horizon to influence 'sublunar' events.

In astrology, as in classical astronomy before the acceptance of the Copernican hypothesis, it was assumed that the Earth was positioned in the center of a finite universe, and that the Earth's center was the center of the spheres of the celestial bodies—in upward order from the Earth, the Moon, Mercury, Venus, the Sun, Mars, Jupiter, Saturn, the 'fixed' stars, the ecliptic and, often, the 'prime mover'. The motion of these 10 spheres was, according to ARISTOTLE, circular and uniform in accordance with the nature of the element, ether, of which they consisted. The motions of the four elements that constitute the sublunar world were linear with respect to the center of the Earth, the elements earth and water moving 'down' towards the center, and the elements air and fire moving 'up' away from the center. The circular motions of the celestial spheres disturb the linear motions of the four sublunar elements, thereby causing change in and on the Earth and in the surrounding layers of water, air and fire. The resulting combinations of elements form the basic materials out of which terrestrial and atmospheric bodies are shaped, which are then continuously further affected by the motions of the celestial bodies and by subsidiary interactions of sublunar bodies. Thus the UNIVERSE is a vast but finite machine powered by the natural motions of its five constituent elements, among which the ETHER with its regular circular motion has the major effect, and astrology is the science that investigates the operations of this universal machine.

Celestial omens

In ancient Mesopotamia and in the many cultures influenced by it many celestial phenomena—eclipses, conjunctions of the Moon and the planets with each other and with the stars, a large number of phenomena due to the distortion of light as it passes through the atmosphere, and other appearances—were regarded as messages sent by the gods to warn the rulers of men or collective groups of men of impending disaster or good fortune. The interpretation of these celestial omens was also a science by which men of great learning strove to provide useful advice to kings and other government officials. The experience of these learned readers of omens led to the recognition of the periodic behavior of the Sun, the Moon and the planets, and ultimately to the development of mathematical methods of predicting the phenomena themselves (which, because of their predictability, should then have been, but were not, no longer regarded as messages sent by angry or well-pleased gods). It also led to the development of some techniques of interpretation of celestial omens that were carried over into astrology, including that of predicting the fate of a person from the positions of the planets at his or her birth or computed conception. In practice, these protohoroscopes of the last four centuries BC were not a part of astrology since the phenomena were ominous rather than effective and since the elaborate geometry used in astrology was not a factor in their interpretation.

The thema or horoscope

The basic tool used to interpret the influences of the celestial bodies at any given moment—say, that of a native's birth—was the diagram called the thema or horoscope. This diagram represents the ECLIPTIC, divided into 12 zodiacal signs of 30° each, as a circle or a quadrilateral with the four cardinal points marked on it—the ascendant on the eastern horizon, the descendant on the western horizon, the midheaven to the south at a point computed by using the local oblique ascensions, and the antimidheaven 180° opposite it to the north. Between and at these cardinal points are located the cusps of the 12 astrological places. The zodiacal signs are divided into various subdivisions: decans of 10° each, twelfths of 2°30′ each, and terms of varying length, usually five to a zodiacal sign. The longitudes of the planets are noted in the thema, which locations reveal their configurations (aspects) with each other—conjunction, sextile (to 60°), quartile (to 90°), trine (to 120°) and opposition (to 180°)—which may be computed by various mathematical formulae so as to be measured in oblique or right ascensions or so as to accommodate the planets' latitudes. Moreover, various lots (defined as the distance between two real or imaginary celestial bodies measured off from a third) and the prorogator (computed in various ways) are noted on the thema, as may also be such fixed points as the exaltations (points of maximum effectiveness) of the planets and the four triplicities (Aries, Leo, Sagittarius; Taurus, Virgo, Capricorn; Gemini, Libra, Aquarius; Cancer, Scorpion, Pisces).

Domains of influence

Each zodiacal sign is classified according to numerous categories: masculine–feminine; odd–even; animal–human; village–forest; etc. Moreover, there are zodiacal melothesias, assigning different parts of the human body to the zodiacal signs; topothesias, doing the same with types of terrain on the surface of the Earth; lists of the animals, the professions, the metals, the jewels, the colors, etc of the zodiacal signs. Each of the signs has especial influence over the categories assigned to it. The triplicities, being four in number, are each associated with a cardinal direction and an element.

The 12 astrological places each have dominion over a certain aspect of the native's life (bodily form; inherited wealth; siblings; parents; children; illnesses; marriage; manner of death; travels; profession; gain; loss) or of an enterprise undertaken. They are influential in their domains in accordance with the zodiacal signs they overlap and the presence within them or aspects to them of the planets and the lots.

The seven planets also (nine in India, where the ascending and descending nodes of the Moon are included) have each their own domains among physical types of humans and animals as well as psychological types, and each is related to types of terrain, of animals, of plants, of minerals, etc, and to an element and its humors (fire–hot and dry; water–cold and humid; air–hot and humid; earth–cold and dry). They are also classified as male or female, and as benefic, malefic or neuter. As the seven planets revolve through the zodiacal signs, and they in turn revolve about the stationary Earth, the influences they project into the sublunar sphere are constantly changing, creating ever-new events and transformations on Earth. These can, according to astrological theory, all be predicted on the assumption (alleged to possess some empirical foundations) that the accepted domains of the celestial bodies' influences are indeed correct. Historically, the system did not remain static; it was simply not complex enough to differentiate between the apparently limitless variations of human experience. In order to accommodate these variations, astrologers—especially those of India and Islam, who were duly followed in medieval Europe and in Byzantium—devised many more subdivisions of zodiacal signs in order to multiply the potential number of the effects of a planet, invented elaborate methods for assigning a numerical weight to the influence of each planet and expanded the number of effective bodies by introducing new lots.

Types of astrology

The paradigmatic type of astrology is genethlialogy, the science of interpreting the themata of nativities or of computed times of conception, probably invented in the 1st century BC in Ptolemaic Egypt. Such predictions can be and were also applied to animals and even plants. Although in theory the predictions are unique for each moment at a given locality, many astrologers wisely advised neophytes to learn something about the social and economic status of the native's family before attempting to describe his or her future life. A branch of genethlialogy is continuous horoscopy, which allows for a life-long usefulness of the astrologer. Since the planets and the ecliptic do not cease their revolutions, one can find their changing influence on the original life-pattern determined from the birth horoscope by casting a new horoscope for every anniversary of the birth, or for the beginning of every month or even day of the native's life, and one can also compute the time when each planet transits to another zodiacal sign or astrological place, thereby altering its effects, and one can watch the imaginary motion, in oblique ascensions, of the prorogator through different planets' terms and past different planets' bodies and aspects.

Catarchic astrology, invented simultaneously with the genethlialogical version, allows one to predict the course of some human undertaking from the horoscope of the moment of its inception. There are two major subdivisions of catarchic astrology, relating to marriage and to warfare. Some aspects of medical astrology are also catarchic, although other aspects depend on genethlialogy.

These two basic forms of astrology were transmitted from Greco-Roman Egypt to India in the 2nd century AD. In India interogational astrology was developed from catarchic astrology. In this form the astrologer answers specific questions posed by his client on the basis of the thema of the moment the question was asked. This clearly raises serious questions about the general theory of astrology; for either the client is exercising free will in determining when he or she will ask the question, so that there ceases to be a purely physical connection between the celestial bodies and the terrestrial event, or else his or her choice is determined by the stars and therefore could, in theory, be predicted by astrology and answered without recourse to a new thema.

These three types of astrology were transmitted from India to Sasanian Iran in the 3rd–5th centuries AD, as were the first two from the Roman Empire. The Sasanians added to them a fourth major type, historical astrology, which uses forms of continuous horoscopy (e.g. the horoscopes of the commencements of all or of selected years) as well as conjunctions of the two furthest of the then known planets, Saturn and Jupiter, or conjunctions of the two malefic planets, Saturn and Mars, in Cancer. Practitioners of historical astrology also cast the horoscopes of the coronations of kings and other historical beginnings, in these cases following the rules of catarchic astrology.

The question of astrology's validity

Clearly, whatever claims were made for an empirical basis, this whole theory is founded on arbitrary assumptions concerning the relationships between the individual zodiacal signs and planets and the physical parts and psychological aspects of sublunar objects and beings, as well as on a subsequently disproved theory of the COSMOS. In antiquity the arguments against astrology's validity were largely based on practical problems: how to explain the often great differences between the physical and mental attributes of twins and their separate lives; how to explain the simultaneous deaths of large numbers of people of different ages in a battle or natural catastrophe; how to be sure of the exact moment of birth and of the local ascendant at that moment. The Christian Church, and later Islam, frequently condemned astrology because it denies man or woman free will and so deprives him or her of the capacity to gain salvation through his or her own efforts and because it makes God irrelevant to the universe, which is an eternal and automatic machine. None of these arguments seriously damaged the credibility of astrology among many, nor has the collapse of the geocentric theory, although that has discouraged many scientifically minded persons from granting it any credence. Like several other current sciences that in fact continuously demonstrate their inability to predict the future accurately, astrology is still able to attract and to retain the patronage of innumerable clients.

Bibliography
Bouché-Leclercq A 1899 *L'astrologie Grecque* (Paris)
Koch-Westenholz U 1995 *Mesopotamian Astrology* (Copenhagen)
Pingree D 1978 *The Yavanajātaka of Sphujidhvaja* 2 vols (Cambridge)
Pingree D 1997 *From Astral Omens to Astrology, From Babylon to Bīkāner* (Rome)

David Pingree

From Alexandria to Baghdād to Byzantium. The Transmission of Astrology

DAVID PINGREE

It is argued in this article that a series of texts preserved in various Greek manuscripts are epitomes of an astrological compendium assembled by Rhetorius at Alexandria in about 620 AD. It is also demonstrated that this compendium was utilized and frequently refashioned by Theophilus of Edessa between 765 and 775 and was made available by Theophilus to his colleague at the ᶜAbbāsid court at Baghdād, Māshā'allāh. Māshā'allāh's works in turn strongly influenced the early development of Arabic astrology, and many of them were translated into Latin and Greek, thereby spreading Rhetorius' influence. A manuscript of Rhetorius' compendium was apparently brought to Byzantium by Theophilus' student, Stephanus, in about 790; from this archetype are descended the several Byzantine epitomes and reworkings of portions of this text; some of these—pseudo-Porphyry, Ep(itome) III, Ep. IIIb, and Ep. IV—passed through the hands of Demophilus in about 1000, while two of the remainder—Ep. IIb and *Ber.*—were the only ones to preserve the name of Rhetorius as their author.

The history of astrology offers its students two major challenges. The first is the complexity of its career in transmission from one cultural area to another and in transformation of its doctrines and methods to fit the interests and circumstances of its eager recipients.[1] The second is the complexity of the transmission of the individual texts within a single culture.[2] This second complexity is due to the fact that astrological texts tended to be copied by professionals interested more in gathering useful information than in preserving the *verba ipsa* of any author except the most authoritative. In this paper I shall attempt to solve problems related to both types of challenges

1. A number of examples of such transformations are discussed in D. Pingree, *From Astral Omens to Astrology, from Babylon to Bīkāner*, Serie Orientale Roma LXXVIII, Rome 1997.
2. One example consists of the disparate collections of omens found on the scattered tablets containing fragments of the 'Ishtar' section of *Enūma Anu Enlil*, some of which have been published in E. Reiner and D. Pingree, *Babylonian Planetary Omens*, part 1, Malibu 1975; part 2, Malibu 1981; and part 3, Groningen 1998. An example in Greek is Hephaestio's Ἀποτελεσματικά, which survives in an "original" form (*Hephaestionis Thebani Apotelesmaticorum libri tres*, ed. D Pingree, Leipzig 1973), and in four epitomes (*Hephaestionis Thebani Apotelesmaticorum epitomae quattuor*, ed. D. Pingree, Leipzig 1974).

David Pingree, Brown University, Department of the History of Mathematics, Box 1900. Providence, RI 02912, USA.

International Journal of the Classical Tradition, Vol. 8, No. 1, Summer 2001, pp. 3–37.

I. The Historical Background

Genethlialogy, the science of interpreting the horoscopic diagram representing the positions of the planets and the zodiacal signs at the moment of a native's conception or birth, was invented by Greeks, probably in Egypt, in about –100.[3] Very soon afterwards, in the last century B.C., there was derived from genethlialogy catarchic astrology or initiatives, which attempt to determine beforehand the most propitious moment for undertaking any enterprise.[4]

By the middle of the first century A.D., when Balbillus[5] wrote, a theory of continuous horoscopy[6] had been introduced into genethlialogy, and various particular applications of catarchic astrology had developed into quasi-independent sciences. Continuous horoscopy depends on an ἀφέτης or prorogator which travels through the native's horoscope, changing from time to time his fate as it passes by the planets, their aspects, and their terms; and it employs the horoscopic diagrams of nativity anniversaries to be compared with the nativity diagram itself to determine annual changes in the native's life. The subtypes of catarchic astrology included iatromathematics or medical astrology, offering prognoses from the horoscope of the time when the native became ill and specifying the best times in the future for employing the appropriate medical procedures;[7] marriage astrology, which determines the best moment for entering into a marriage and the features the native should look for in his or her future spouse;[8] and, by the early sixth century A.D. in the work of Julian of Laodicea, the beginnings of military astrology,[9] though that is largely an Indian and Iranian development. Greek genethlialogy and catarchic astrology were transmitted to India in the middle of the second century A.D. in the form of a Sanskrit prose translation of a Greek text apparently composed in Alexandria in the early second century. Already in the poetic version of this prose translation that we still possess, the *Yavanajātaka* composed by Sphujidhvaja in Western India in 269/270, catarchic astrology had been made the basis of the new science of interrogational astrology,[10] wherein the

3. The Babylonian nativity omens, published by F. Rochberg under the title *Babylonian Horoscopes* (Transactions of the American Philosophical Society LXXXVIII, 1, Philadelphia 1998), do not belong to the science of genethlialogy. For the relationship of the Babylonian omens to genethlialogy see Pingree, *Astral Omens* (as in n. 1), pp. 21–29.
4. The earliest extant full discussion of catarchic astrology is to be found in the fifth book of Dorotheus' astrological poem, written in about 75 A.D. (*Dorothei Sidonii Carmen astrologicum*, ed. D. Pingree, Leipzig 1976), but the method undoubtedly is at least a century older.
5. Manuscript *R* (for the manuscripts see section IV below, pp. 20–21) VI 60, edited by F. Cumont in *Catalogus Codicum Astrologorum Graecorum* (henceforth *CCAG*) VIII 3, Bruxelles 1912, pp. 103–104; and chapter 231 on fol. 80 of manuscript *B*, edited by F. Cumont in *CCAG* VIII 4, Bruxelles 1921, pp. 235–238.
6. The fully developed form is found in books III and IV of Dorotheus.
7. The oldest surviving iatromathematical text is probably the Ἰατρομαθηματικά attributed to Hermes, edited by J. L. Ideler in *Physici et medici graeci minores*, vol. I, Berlin 1841, pp. 387–396, repeated on pp. 430–440. There is a copy of this text on fols. 1–5 of manuscript *L*.
8. The oldest catarchic chapters on marriage are Dorotheus V 16 and 17.
9. There is a fragment of Dorotheus on a "Lot of Soldiering" (fr. II D on p. 432 [ed. Pingree, as in n. 4]) preserved in Hephaestio II 19, 22–26, but this is not catarchic. The chapters by Julian are preserved in VI 46–48 in manuscript *R*.
10. D. Pingree, *The Yavanajātaka of Sphujidhvaja*, Harvard Oriental Series XLVIII, 2 vols., Cambridge MA 1978, chapters 52–72.

results of any enterprise can be predicted by examining the horoscopic diagram of the moment the client puts his question to the astrologer; the same rules as are used in catarchic astrology are applied, not to find a future time appropriate for beginning the enterprise, but to judge the consequences of the enterprise, whether already undertaken or intended for the future, on the basis of the time at which the question about the enterprise is asked.[11] Also, military astrology was greatly developed in this early period of Indian astrology,[12] culminating in three books on the subject composed by Varāhamihira at Ujjayinī in the middle of the sixth century.[13]

During the Sasanian period in Iran astrological texts from both the Greek tradition (e.g., Dorotheus [ca. 75 A.D.] and Vettius Valens [ca. 175 A.D.]) and the Indian tradition (e.g., Varāhamihira) were translated into Pahlavī.[14] These traditions were mingled in Iran in such texts as those ascribed to Zarādusht[15] and to Buzurjmihr,[16] which are preserved in Arabic translations. In addition, Sasanian scholars developed from continuous horoscopy a science of historical astrology, based on annual horoscopic diagrams cast for the times of the Sun's entry into Aries, on *qismas* (points moving at the rate of 1° per year) and *intihā*'s (points moving at the rate of 30° per year) rotating through the ecliptic analogously to prorogators,[17] and on periodic conjunctions of Saturn and Jupiter in the four triplicities and of Saturn and Mars in Cancer.[18] As well they invented political horoscopy from catarchic astrology, taking the moment of enthronement, selected on astrological principles, or the preceding vernal equinox as the καταρχή of the regal or imperial initiative.[19] Beginning in about 750 some of the Pahlavī texts attributed to Zarādusht, as we have already mentioned, were translated into Arabic,[20] and later in the century original Greek works began also to be translated.[21]

11. Ibid., vol. 2, pp. 370–388.
12. Ibid, chapters 73–76 and vol. 2, pp. 388–402.
13. These are the *Bṛhadyātrā*, the *Yogayātrā*, and the *Ṭikanikayātrā*. Manuscripts and editions are listed in D. Pingree, *Census of the Exact Sciences in Sanskrit*, series A, vol. 5, Memoirs of the American Philosophical Society CCXIII, Philadelphia 1994, pp. 571a–572b.
14. Pingree, *Astral Omens* (as in n. 1), pp. 46–50.
15. Ibid., pp. 44–46.
16. C. A. Nalino, "Tracce di opere greche giunte agli arabi per trafila pehlevica," in *A Volume of Oriental Studies Presented to Edward G. Browne*, ed. T. W. Arnold and R.A. Nicholson, Cambridge 1922, repr. Amsterdam 1973, pp. 345–363 (pp. 352–356); and C. Burnett and D. Pingree, *The Liber Aristotilis of Hugo of Santala*, Warburg Institute Surveys and Texts XXVI, London 1997, pp. 8 and 140.
17. D. Pingree, *The Thousands of Abū Maᶜshar*, Studies of the Warburg Institute XXX, London 1968, pp. 59–60 and 65.
18. Ibid., pp. 70–121, and K. Yamamoto and C. Burnett, *Abū Maᶜšar On Historical Astrology*, Islamic Philosophy Theology and Science XXXIII–XXXIV, 2 vols., Leiden 2000.
19. D. Pingree, "Historical Horoscopes," *Journal of the American Oriental Society* LXXXII, 1962, pp. 487–502, and Pingree, *Astral Omens* (as in n. 1), pp. 57–62.
20. D. Pingree, "Classical and Byzantine Astrology in Sassanian Persia," *Dumbarton Oaks Papers* XLIII, 1989, pp. 217–239 (234–235).
21. Much Greek material is embedded in the Arabic treatises of Māshā'allāh and ᶜUmar ibn al-Farrukhān.

II. Rhetorius of Egypt

A. The Person

In our Greek sources Rhetorius is little more than an infrequent name. In a few manuscripts he is alleged to be the author of several chapters on astrology, and once of a collection of chapters selected from an epitome of an elementary treatise on astrology attributed to Antiochus of Athens (late second century).[22] Aside from this epitome, the material directly associated with Rhetorius' name found in manuscripts *B*, *Ber.*, and *R* (see section IV below, pp. 20–21) are on basic definitions (the zodiacal signs;[23] bright and shadowy degrees;[24] and masculine and feminine degrees[25]), and on genethlialogy (conception;[26] on the longevity of the parents[27]), and on initiatives.[28] The only statement in Greek relating anything of substance concerning Rhetorius is found in the poem entitled Εἰσαγωγὴ κατὰ μέρος ἀστρονομίας διὰ στίχου (*Partial Introduction to Astronomy in Verse*) which John Camaterus dedicated to the emperor Manuel I (1143–80):[29]

σοφός τις ἐκ τῶν παλαιῶν ῥήτωρ πεπυκνωμένος,
Ῥητόριος Αἰγύπτιος οὕτως ὠνομασμένος,
πρὸς ἐπιστήμην ἔμπειρος τῆς τῶν ἄστρων πορείας,
ἐνέγραψεν ἐπίσημα < ἐν τῷ > προχείρῳ λόγῳ
πρῶτον περὶ τὴν κίνησιν τῶν δώδεκα ζῳδίων.

> A certain wise man of the ancients, a terse orator, Rhetorius the Egyptian, so named, skilled in the science of the progress of the stars, wrote noteworthy things in the handy book, first about the motion of the twelve zodiacal signs.

If we accept Camaterus' information as trustworthy, Rhetorius was an Egyptian and the author of a work on the zodiacal signs. Franz Boll long ago[30] identified the work which Camaterus used with the chapter Ῥητορίου θησαυρὸς συνέχων τὸ πᾶν τῆς ἀστρονομίας ("Rhetorius's Treasure Containing All of Astronomy") which he also edited, though in a seriously contaminated fashion; genuinely Rhetorian is only the material found in manuscript *Ber.*[31] A number of sentences in this chapter are found also in manuscript *R*; the most important concern the longitudes of fixed stars found in

22. Epitome IIb of Rhetorius. The Epitomes are described in D. Pingree "Antiochus and Rhetorius," *Classical Philology* LXXII, 1977, pp. 203–223.
23. The excerpt in manuscript *Ber.*; see n. 31 below.
24. VI 17 in manuscript *R*.
25. VI 18 in manuscript *R*.
26. Epitome IV 23.
27. Chapter 221 in manuscript *B*, which corresponds to V 100, 6–9 in manuscript *R*.
28. VI 23, 18–29 in manuscript *R*.
29. Lines 92–96 in L. Weigl, *Johannes Kamateros* Εἰσαγωγὴ ἀστρονομίας, Würzburg 1907, p. 7.
30. F. Boll, *Sphaera*, Leipzig 1903, pp. 21–30.
31. In *CCAG* VII, Bruxelles 1908, pp. 194–213, where Boll mistakenly combines Rhetorius' chapter with expanded versions in two other manuscripts, one of which falsely attributes its conflated version to Teucer of Babylon.

book V 58 and 62 of manuscript *R*. The longitudes of the fixed stars or arcs of the zodiac harming vision given in V 62 and in manuscript *Ber*. are corrected from Ptolemy's positions to those they were assumed to have in about 480 by the addition of 3;26° of precession, while the longitudes of the thirty fixed stars in V 58 and in manuscript *Ber*. were corrected to the positions they were assumed to have in 505 by the addition of 3;40°.[32] Rhetorius, therefore, wrote in 505 or later. But exactly what he wrote is uncertain.

B. The Texts

In manuscript *R* book V bears the title: Ἐκ τῶν Ἀντιόχου Θησαυρῶν ἐπίλυσις καὶ διήγησις πάσης ἀστρονομικῆς τέχνης (*From the Treasures of Antiochus an Explanation and Description of the Whole Astronomical Art*). This title implies that what *R* contains is not the original text of the work of Antiochus of Athens, who apparently wrote in the late second century A.D.[33] This date is derived from the incomplete summary of his Εἰσαγωγικά (*Introductory Material*) preserved as VI 62 in *R* (Epitome I).[34] Antiochus refers to Nechepso, Petosiris, Hermes, and Timaeus, all of whom may be assigned to the first century A.D., and may be later than Ptolemy who wrote his Ἀποτελεσματικά (*Astrological Effects*) in the middle of the second.[35] The *terminus ante quem* is established by the citation of him by Porphyry (ca. 275 A.D.) in chapter 38 of his Εἰσαγωγή (*Introduction*).[36] The numerous verbal parallels between Antiochus and Porphyry, then, which include passages from chapters 3, 4, 7–11, 15, 20, 22–30, 35–39, 41, 44, and 45 of the latter,[37] demonstrate Porphyry's extreme dependence on the Athenian astrologer.

Closely related to both Antiochus' Εἰσαγωγικά and Porphyry's Εἰσαγωγή are chapters 1–53 of book V in *R*, which form Epitome II as preserved also in manuscript *L*.[38] For Epitome II discusses material contained in chapters 1–3, 5–14, and 16–18 of Epitome I, and was also, independently of that text, a source of chapters 31, 33, 34, and 40 of Porphyry.[39] Note that chapters 47–52 of the version of Porphyry's text available

32. D. Pingree, review of W. Hübner, *Grade und Gradbezirke der Tierkreiszeichen* (Leipzig 1995), in this journal (*IJCT*) VI, 1999/2000, 473–476 (476).
33. Pingree, "Antiochus and Rhetorius" (as in n. 22), pp. 203–205.
34. Ibid., pp. 205–206.
35. This guess depends on the statement by Hephaestio (II 10, 9 and 29) that Antiochus and Apollinarius agree with certain things stated by Ptolemy—certainly not definitive proof of their relative chronologies.
36. E. Boer and S. Weinstock, "Porphyrii Philosophi Introductio in Tetrabiblum Ptolemaei," in *CCAG* V 4, Bruxelles 1940, pp. 185–228 (210).
37. In what seem to be the genuine chapters of Porphyry, numbers 1 to 45, he refers to the ancients, the moderns, the Chaldaeans, Apollinarius, Petosiris, Ptolemy, and Thrasyllus as well as to Antiochus
38. Pingree, "Antiochus and Rhetorius" (as in n. 22) 206–208.
39. Compare Porphyry 31 and 33 with Epitome II 19, 34 with Epitome II 16, and 40 with Epitome II 1. Note also that Hephaestio I 13,1 is taken from Porphyry 7; I 14, 1 from Porphyry 11; I 14, 2 from Porphyry 13; I 15, 1–2 from Porphyry 14; I 15, 3 from Porphyry 15; I 16, 1–2 from Porphyry 24; I 16,3 from Porphyry 20; and I 17, 1–6 from Porphyry 29, but Hephaestio cites Porphyry from his commentary on Ptolemy's Ἀποτελεσματικά only in II 10, 23-27 and II 18, 15. Since these passages do not occur in Porphyry's Εἰσαγωγή, that work, as its contents also show, is not a commentary on Ptolemy despite its title and preface.

now in the recension prepared by one Demophilus, who was active at Constantinople in 990,[40] correspond verbatim to the epitome of Antiochus' work found in Epitome II, where they are chapters 10, 11, 12, 14, 15, and 46; they were inserted from this epitome into the archetype of the Porphyry manuscripts, presumably by Demophilus. That these six chapters could not have been borrowed from Antiochus by Porphyry is demonstrated by the fact that each of the first four of the relevant chapters in Epitome II refers to another section of *R*, or to *Ber.*, and that these references are retained in the Porphyry manuscripts where they lose their meaning. Thus in V 10,7 there is a reference concerning the παρανατέλλοντα ("simultaneously rising stars") of the decans to Teucer of Babylon; Teucer's system is summarized in the Rhetorius chapter in *Ber.* In V 11,8 it is said concerning the thirty bright stars: εὑρήσεις δὲ τὰ ἀποτελέσματα ἐν τοῖς ἑξῆς ("you will find their effects in what follows"); this points to V 58. In V 12,7 is written: τινὲς δὲ τῶν λαμπρῶν καὶ σκιαρῶν καὶ ἀμυδρῶν μοιρῶν οὐ μικρὰν ἔχουσι τὴν ἐνέργειαν εἴπερ οἱ ἀστέρες ἐν ταῖς λαμπραῖς μοίραις εὑρέθησαν τετυχηκότες, διὸ καὶ ταύτας ἐν τοῖς ἔπροσθεν ἠναγκάσθην καθυποτάξαι ("some of the bright, the shadowy, and the faint degrees have not a little effectiveness even though the planets were found to be in the bright degrees. Therefore I felt compelled to describe those [degrees] in what is yet to come"). This refers to the table in VI 17 which is entitled: περὶ λαμπρῶν καὶ σκιαρῶν μοιρῶν κατὰ Ῥητόριον ("on the bright and the shadowy degrees according to Rhetorius"). At the end of V 14 the author promises that he will explain clearly the matter of injuries and sufferings; he fulfils this promise in V 61. Incidentally, the chapters in the Porphyry manuscripts before and after the block inserted from Epitome II—chapters 46 and 53–55—are also not to be attributed to Porphyry.

We have seen, then, links between Epitome II and the latter half of book V (chapters 54–117) and to book VI of *R* as well as to *Ber.* These must be due to a redactor later than 505 A.D. and earlier than 990; the link to the text in *Ber.* suggests that this redactor may be Rhetorius. The linkage is strengthened by a few additional cases that do not involve Porphyry's manuscripts. In V 5, Περὶ ζῳδίων ἀσελγῶν ("On lustful zodiacal signs"), the reader is told: περὶ δὲ τούτων ἀκριβῶς ἐν τοῖς ἔμπροσθεν εὑρήσεις ("you will find concerning these precise information in what lies ahead"). This refers to V 67, which discusses a doctrine of lustful degrees. And the Rhetorius chapter in *Ber.* omits the terms of the zodiacal signs, but says of Aries: ἔχει ὅρια ε κατὰ Πτολεμαῖον καὶ Αἰγυπτίους καὶ μοίρας λαμπρὰς καὶ σκιεράς· καὶ τὰς μὲν λαμπρὰς ὑπετάξαμεν τοῖς ἔμπροσθεν σὺν ταῖς τῶν ἀναφορῶν τῶν κλιμάτων ("it has five terms according to Ptolemy and the Egyptians and bright and shadowy degrees; we subjoined the bright [degrees] to what lies ahead together with the [degrees] of the rising-times of the climes"). This is an unmistakable reference to V 12 and 13, whose titles are respectively: Περὶ τῶν ὁρίων κατ' Αἰγυπτίους καὶ Πτολεμαῖον καὶ λαμπρῶν καὶ σκιαρῶν μοιρῶν ("Concerning the terms according to the Egyptians and Ptolemy, and the bright and shadowy degrees"), and Περὶ τῶν ἀναφορῶν τῶν ἑπτὰ κλιμάτων ("On the rising times of the seven climes").

40. See his scholium or scholia on Porphyry 30 and D. Pingree, "The Horoscope of Constantinople," in Πρίσματα. *Naturwissenschaftsgeschichtliche Studien. Festschrift für W. Hartner*, ed. Y. Maeyama and W.G. Saltzer, Wiesbaden 1977, pp. 305–315 (306–308).

These examples provide one part of the evidence that Epitome II is not a pure text of Antiochus. Another part is the fact that the redactor incorporates into V 18 a reference to Paul of Alexandria, who wrote the second edition of his Εἰσαγωγικά (*Introductory Material*) in 378. It may be relevant to the argument that book VI is linked to book V, of which linkage one example has already been mentioned, that VI 24–40 is a paraphrase of Paul's book[41] supplemented by quotations from his commentator, Olympiodorus, who lectured on the Εἰσαγωγικά at Alexandria in the summer of 564.[42]

Epitome II, as has been indicated, consists of V 1–53 as presented in manuscripts *L* and *R*, and most of this must be a summary of a work by Antiochus. Parallel to the version in *L* is Epitome IIa,[43] which contains chapters 1–12, 15–48, and 52–53, and which was produced in the School of John Abramius in about 1380.[44] Another, more drastic revision is Epitome IIb,[45] which consists of chapters 1, 3, 4, 7–10, 16, 17, 21, 23–44, 46, and 47; the earliest of the nine manuscripts was copied in the fourteenth century. Its title is: Ῥητορίου Ἔκθεσις καὶ ἐπίλυσις περί τε τῶν προειρημένων δώδεκα ζῳδίων καὶ περὶ ἑτέρων διαφόρων ἐκ τῶν Ἀντιόχου Θησαυρῶν (*Rhetorius' Exposition and Explanation Concerning the Previously Mentioned Twelve Zodiacal Signs and Concerning Other Different Subjects from the Treasures of Antiochus*). There is no obvious reason for Rhetorius' name to appear here as the author of Epitome IIb unless he really was its author or the author of Epitome II. Unfortunately, the latter hypothesis, which I prefer, cannot be proved. If it were true, Rhetorius would appear to be the most plausible candidate for being the redactor of book V and at least part of book VI in *R*.

Chapters 54–117 of book V form a relatively consistent treatment of genethlialogical interpretation:[46] general indicators affecting the native (chapters 54–62); on his physical and moral attributes (chapters 63–81); on his profession (chapters 82–96); on his parents (chapters 97–102); on his siblings (chapters 103–108); on lunar configurations (chapters 109–112); and, as an example, the nativity of Pamprepius of Panopolis, who

41. E. Boer, *Pauli Alexandrini Elementa apotelesmatica*, Leipzig 1958, in which edition the unusual readings of *R* are cited under the siglum *Y*.
42. VI 40, 9–15 in *R*, though ascribed to Heliodorus, are found in Olympiodorus' commentary (ed. E. Boer, *Heliodori, ut dicitur, in Paulum Alexandrinum commentarium*, Leipzig 1962) on pp. 138–142. VI 5 is entitled: Σχόλια εἰς τὸν περὶ χρόνου διαιρέσεως ἐκ τῶν τοῦ Ἡλιοδώρου Συνουσιῶν, but sentences 1–3 and 5–7 are also found in Olympiodorus' commentary (pp. 127–128). But in the primary manuscript of the uncontaminated form of the commentary, our manuscript *W* (Boer's *A*), no author was named by the original scribe, though a later reader has written in the name Heliodorus. This attribution, however, is impossible since it was pointed out by Pingree (pp. 149–150a of Boer's edition) that the examples in the commentary can be dated between May and August of 564. The authorship of Olympiodorus was suggested by J. Warnon, "Le commentaire attribué à Héliodore sur les Εἰσαγωγικά de Paul d'Alexandrie," *Travaux de la Faculté de Philosophie et Lettres de l'Université Catholique de Louvain* II, 1967, pp. 197–217, and by L.G. Westerink, "Ein astrologisches Kolleg aus dem Jahre 564," *Byzantinische Zeitschrift* LXIV, 1971, pp. 6–21.
43. Pingree, "Antiochus and Rhetorius" (as in n. 22), pp. 208–209.
44. D. Pingree, "The Astrological School of John Abramius," *Dumbarton Oaks Papers* XXV, 1971, pp. 189–215.
45. Pingree, "Antiochus and Rhetorius" (as in n. 22), pp. 209–210.
46. Ibid., pp. 211–212.
47. D. Pingree, "Political Horoscopes from the Reign of Zeno," *Dumbarton Oaks Papers* XXX, 1976, pp. 135–150 (144–146). References to Pamprepius and the fragmentary remains of his poetry are assembled in H. Livrea, *Pamprepii Panopolitani Carmina*, Leipzig 1979.

was born on 29 September 440 (chapters 113–117).[47] Among the sources are mentioned Critodemus (early first century A.D.), Dorotheus, Ptolemy, and Vettius Valens. In V 110 is a horoscope that can be dated 24 February 601. This places the date of the redactor at about 620 at the earliest, shortly before 990 at the latest. The whole of book V was summarized in VI 61 under the title συγκεφαλαίωσις τῶν 'Αντιόχου Θησαυρῶν οἵτινες ἐπιλύσεις καὶ διηγήσεις τῆς ἀστρονομικῆς ἀπαγγέλλονται τέχνης (*Chapter-headings of Antiochus' Treasures which are called the Explanations and Descriptions of the Astronomical Art*). This is Epitome IIIa.[48] It could not have been the work of the redactor of books V and VI if the arguments set forth so far are correct; for that redactor would not have attributed the whole of book V to Antiochus. Presumably the other summaries of ancient astronomical works at the end of book VI—chapters 54–63—are also to be denied to his authorship.

There exists another epitome that unites book V with much of book VI. Epitome IIIb,[49] of which the oldest manuscript was copied in the late fourteenth century, contains material from V praefatio; 1, 2, 4–12, 15–17, 19, 22, 23, 29, 47, 48, 51, 53, 56, 59, 61–73, 75, 76, 79, 80, 85–88, 90–92, and 95–96; and from VI 1–5, 7, 9, 42, and 52. The incorporation of some verses composed by Theodore Prodromus in the twelfth century[50] into the version of VI 7 in Epitome IIIb provides a *terminus post quem*. The title of Epitome IIIb is identical to the title of book V; it is descended from the same manuscript as is *R*.

Three other manuscripts contain selections from both book V and book VI. It has already been noted that *L* preserves Epitome II; it also contains all or parts of VI 20, 25, 27, and 52. *V* was copied by several scribes in the fourteenth century, but its basic collection goes back to a twelfth-century original. It contains V 55 and VI 7, 22–23, and 46–50. On f. 156 it quotes several lines on the trine aspect of Saturn and Mars and attributes them to Rhetorius.[51] In the twelfth century, then, as is already clear from John Camaterus' poem, elements of Rhetorius' work that have not been incorporated into the epitomes we have so far discussed were in circulation in Byzantium.

Closely related to *V* through shared contents are two other fourteenth-century manuscripts. *B* was copied at the beginning of the century from a late eleventh-century compilation one of whose sources went back to the late tenth or early eleventh century.[52] At the beginning of *B* are the twenty-eight chapters which constitute Epitome IV.[53] The first eleven of these chapters correspond to V 57, 58, 59–60, 61–62, 64, 65, 66–76, 77, 78, 79–80, and 81, and the last five to V 97 and 99–101, 102, 104–105, 82–83, and 53. The eleven chapters between these two groups are mostly from an early sixth-century source on technical aspects of astrological computing; chapter 12 contains a horoscope dated 8 September 428, chapter 14 one dated 1 May 516, chapter 15 one dated December 400 or January 401 and another dated 2 April 488, and chapter 19 one

48. Pingree, "Antiochus and Rhetorius" (as in n. 22), pp. 212–213.
49. Ibid., pp. 213–215.
50. Στίχοι εἰς τοὺς δώδεκα μῆνας (*Lines On the Twelve Months*) edited by B. Keil, *Wiener Studien* XI, 1889, pp. 94–115.
51. This is derived from Dorotheus I 6, 2 and II 14, 4; see Dorotheus, ed. Pingree (as in n. 4), pp. 325–326. Quoted on the same folio are several verses of Dorotheus: from II 18, 2–3 (pp. 368–369) and from IV 1, 213 (p. 383).
52. D. Pingree, *Albumasaris De revolutionibus nativitatum*, Leipzig 1968, pp. VII–IX, and Idem, *Hephaestionis ... epitomae quattuor* (as in n. 2), pp. V–VIII.
53. Pingree, "Antiochus and Rhetorius" (as in n. 22), pp. 216–219.

dated 21 March 482. Chapter 23, on conception, is specifically attributed to Rhetorius; its title is Περὶ σπορᾶς ἐκ τῶν Ῥητορίου ("On Conception from the [Works] of Rhetorius"). Also the horoscope of 8 September 428 was known along with much from *R* to the Arab astrologer Māshā'allāh shortly before 800 A.D. I hypothesize, therefore, that these eleven chapters were part of the compendium that is found abbreviated in *R*.

It is clear, however, that they have been revised. The longitudes of the fixed stars in the reworking of V 58, Epitome IV 2, for instance, which deals with the astrological influences of the thirty classical παρανατέλλοντα, have been corrected from their longitudes in *R*, computed for 505 A.D., to longitudes computed for 884 A.D. (600 era Diocletian); this must be the approximate date of an early form of Epitome IV. A later redactor of the material in the first part (fols. 1–144) of *B* gives a horoscope dated 15 September 1006 and referring to 9 October 1019. This suggests that he may be the previously mentioned astrologer Demophilus.[54] In another passage, Epitome IV 26,1, this redactor reveals that he, like Demophilus, is from Constantinople by stating that he knows a lady in Byzantium who is the mother of twenty-four children. The frequent citations in Epitome IV from Dorotheus, Ptolemy, and Valens, with whose works we know Demophilus to have been familiar, and the references to Phnaēs the Egyptian in chapter 16, Antigonus in chapter 21, and Teucer the Babylonian in chapter 22, whose names occur in chapters of book V that Demophilus seems to have inserted into the manuscripts of Porphyry, strongly suggest that Epitome IV passed through the hands of that late tenth-century astrologer. In a later section of part 1 of *B* are presented reworkings of VI 6, 8, 9, 13, 15, and 45, and the substance of V 100 specifically attributed to Rhetorius.

In manuscript *E* the scribe, Eleutherius Zebelenus of Elis,[55] has copied his own compendium which he falsely ascribed to Palchus.[56] In this compilation he included or referred to VI 5, 6, 8, 14, 15, 16, 23, 42, 43, 44, 45, 46, 47, 48, and 50, as well as V 61, 50–51. Eleutherius has changed much in the chapters he copied.

Book V contains an epitome of an introductory work by Antiochus and part of a treatise on genethlialogy. Some chapters from this—V 48, 54, 58, 59, 97–98, 101, and 103—were translated into Latin in the thirteenth century as part of the *Liber Hermetis*.[57] Book VI contains for the most part summaries of or excerpts from the works of various authors, including Balbillus (ca. 50 A.D.), Dorotheus, Theon of Smyrna (early second century A.D.), Ptolemy's *Almagest* and *Phases*, Sarapion of Alexandria, Paul of Alexandria, the fifth-century Scholia on Ptolemy's *Handy Tables*, Heliodorus (ca. 500 A.D.), Eutocius (early sixth century), Julian of Laodicea (early sixth century), Olympiodorus (564 A.D.), and Rhetorius himself referring to his own epitomes of other astrologers' works. One of the chapters summarizing a doctrine of Julian of Laodicea contains an insert (VI 44, 21–23) made by Demophilus in which he quotes from Proclus. Since this

54. See n. 40 above.
55. Pingree, "Horoscope" (as in n. 40), pp. 306–314.
56. For the falsity of the attribution see Pingree, *Yavanajātaka* (as in n. 10), vol. 2, p. 437. Note that Eleutherius was a member of the school of John Abramius in which Epitome IIa of Rhetorius and the revised text of Olympiodorus (ascribed to Heliodorus) originated.
57. S. Feraboli, *Hermetis Trismegisti De triginta sex decanis*, Corpus Christianorum Continuatio Mediaeualis CXLIV, Turnhout 1994. These chapters constitute Epitome V; see Pingree, "Antiochus and Rhetorius" (as in n. 22), pp. 219–220.

insert is found in both *R* and *E*, the recension made by Demophilus in the late tenth century was the ancestor of both.

The long discussion in VI 52 of the horoscope for 28 October 497 excerpted from the astrological treatise of Eutocius nicely parallels the discussion of the horoscope of Pamprepius at the end of book V. I strongly suspect that Rhetorius, working in Alexandria in about 620, was the original compiler of books V and VI in *R*, and that he ended his compilation with VI 52. The next chapter is clearly an addition; it consists of rules for setting out an ephemeris, a subject already treated in two forms in VI 21, followed by a brief example of an ephemeris that covers the period from 1 to 6 and 8 to 13 January 796 (the days begin at midnight). The only person we know who might have prepared such an ephemeris in Byzantium in the 790's is Stephanus the Philosopher, whom I suspect to have brought the texts of Hephaestio, Rhetorius, and Theophilus from Baghdād to Constantinople in about 780.[58] For he states proudly in his Περὶ τῆς μαθηματικῆς τέχνης (*On the Mathematical Art*):[59]

> ὅ τε γὰρ Πτολεμαῖος τοῖς ἀπὸ τοῦ Ναβουχοδόνοσορ ἔτεσιν ἐχρήσατο καὶ μησὶν Αἰγυπτιακοῖς, ὁ δέ γε Θέων καὶ ῾Ηράκλειος καὶ ὁ ᾿Αμμώνιος τοῖς τοῦ Φιλίππου καὶ μησὶν Αἰγυπτιακοῖς, οἱ δὲ νεώτεροι τοῖς τῶν Περσικῶν ἡγεμόνων καὶ τοῖς Σαρακηνικοῖς ἔτεσιν. διὰ τοῦτο ἐξεθέμην κανόνιον κατὰ τὰ τοῦ κόσμου ἔτη καὶ τοὺς ἡμετέρους μῆνας καὶ τὰς μεθόδους ῥαδίας καὶ προχείρους.
>
> Ptolemy used the years from Nabuchodonosor [Nabonassar] and Egyptian months, Theon, Heraclius, and Ammonius the [years] of Philip and Egyptian months, and more recent [Muslim astronomers] the [years] of the Persian kings [e.g., Yazdijird III] and Arabic years. Therefore, we set out an astronomical table according to the years of the cosmos [the Byzantine era, whose epoch is –5508] and our months and [according to] easy and handy methods.

It is noteworthy that the only other references to Ammonius' tables are in VI 2, 13 and in the almanac composed by the tenth-century Andalusian scholar, Maslama al-Majriṭī.[60]

The valuable historical material preserved at the end of book VI seems to be independent of Rhetorius. Not one of the Epitomes and manuscripts that preserve Rhetorius refer to it. It could conceivably have been added by Stephanus; it begins with references to "Erimarabus, whom the Egyptians call a prophet and the discoverer of astronomy," to "Phorēdas (Bhūridāsa?) the Indian," and to "Hystaspēs Ōdapsus called the priest," three Oriental sages representing Egypt, India, and Persia of whom he might have learned during his stay in Muslim territory. However that may be, the author of the bibliographies and biographical sketches of Ptolemy, Paul, Demetrius,

58. Pingree, "Classical and Byzantine Astrology" (as in n. 20), pp. 238–239.
59. Edited by F. Cumont in *CCAG* II, Bruxelles 1900, pp. 181–186 (182). The same text is found in manuscript *V*.
60. J. M. Millas Vallicrosa, *Estudios sobre Azarquiel*, Madrid-Granada 1943–1950, pp. 72–237.

Thrasyllus, Critodemus, Callicrates, Balbillus, and Antiochus (VI 54–63) had access to an extremely rich library of astrological lore.

III. Theophilus of Edessa

A. The Person

Theophilus, the son of Thomas, was born in Edessa in about 695. For the first fifty years of his life we hear nothing of him, though we may conjecture, because of his interests in philosophy and science as well as pagan literature, that, though a devout Christian, he studied not only in Edessa, but as well in the nearby city Ḥarrān, where the Greek secular sciences—especially astronomy and astrology—were carefully cultivated.[61] His pious and philosophical interest in science is best expressed in the preface to the <Ἀποτελεσματικά> (*Astrological Effects*) that he addressed to his son, Deucalion[62] (Deucalion is presumably his Hellenic cover for Noah). In this preface he gives a Christian defense of astrology, as he says:[63] μαρτυρίαις ἱκαναῖς κεχρημένος ἔκ τε θείων γραφῶν καὶ τῶν σεμνῶν φιλοσόφων, τῶν τε θύραθεν καὶ τῶν τῷ ὀνόματι τοῦ χριστιανισμοῦ πεφωτισμένων ("having used sufficient witnesses from the divine scriptures and from the revered philosophers, from those outside the gate and those illuminated by the name of Christianity"). After explaining God's deeds on the seven days of creation as consonant with the astrological powers of the planets taken in the order of their lordships of the weekdays, he concludes[64] that there are two wisdoms between which the worthy may choose: the spiritual wisdom (σοφία πνευματική) of the perfect (τῶν τελείων) who do not use the mathematical sciences (τὰ μαθηματικά) and other sophistic complications, but who exercise the highest virtue with simplicity of manner together with holy chastity (μετὰ τῆς σεμνῆς παρθενίας); and the psychic wisdom (σοφία ψυχική) which searches for the cosmic philosophy through reason and action, holy marriage, the virtue of moderation, and blameless chastity (σεμνοῦ γάμου καὶ ἀρετῆς μέσης καὶ ἀμέμπτου παρθενίας). The sons of physicians, he continues, and of astronomers, Platonists, and Aristotelians pursue the latter wisdom without any blame or frenzy whatsoever. He concludes by referring obliquely to Paul (I Corinthians 15,41):[65] ὁ μὴ δυνάμενος γενέσθαι Ἥλιος γινέσθω Σελήνη, καὶ ὁ μὴ δυνάμενος γενέσθαι Σελήνη γινέσθω ἀστὴρ φωτεινός ("Let him who cannot be the Sun be the Moon, and let him who cannot be the Moon be a bright star"). The physicians, astronomers, Platonists, and Aristotelians who use τὰ μαθηματικά remind one strongly of the Neoplatonists of Ḥarrān, with whom Theophilus seems here to be associating himself while proclaiming his Christian faith.

61. D. Pingree, "The Ṣābians of Ḥarrān," forthcoming in this journal.
62. Edited by F. Cumont in *CCAG* V 1, Bruxelles 1904, pp. 234–238.
63. *CCAG* V 1, pp. 234–235.
64. Ibid., p. 238.
65. A. Souter, *Novum Testamentum Graece*, Oxford 1962: ἄλλη δόξα Ἡλίου καὶ ἄλλη δόξα Σελήνης καὶ ἄλλη δόξα ἀστέρων· ἀστὴρ γὰρ ἀστέρος διαφέρει ἐν δόξῃ ("The glory of the Sun is one thing, the glory of the Moon another, and the glory of the stars [yet] another; for one star differs from [another] star in glory").
66. *Ta'rīkh mukhtaṣar al-duwal*, Bayrūt 1958, p. 24, and E. A. Wallis Budge, *The Chronography of Gregory Abū'l Faraj*, 2 vols., Oxford 1932, vol. 1, pp. 116–117.

It was probably in the obscure period of his life, in the early eighth century, that he translated from Greek into Syriac, if we believe Gregory Bar Hebraeus,[66] the two books that Homer sang on the fall of Troy. He also translated into Syriac Aristotle's *Sophistici elenchi*[67] and Galen's *De methodo medendi*. Ḥunayn ibn Isḥāq (died 873 or 877 A.D.) calls the language of this translation of Galen "repulsive and bad,"[68] from which we may guess that Theophilus was more at home in Greek than in what one might have expected to be his native tongue.

Indeed, he had read in Greek and quotes in the astrological works that he wrote in that language after 760 the Σύνταξις μαθηματική (*Almagest*), *Handy Tables*, and Ἀποτελεσματικά (*Astrological Effects*) that Ptolemy had written in the middle of the second century A.D., as well as Dorotheus of Sidon's late first-century astrological poem, Vettius Valens' late second-century *Anthologies*, Hephaestio of Thebes' early fifth-century Ἀποτελεσματικά, and Rhetorius of Egypt's early seventh-century compendium.

He also seems to have learned Pahlavī. For he shows a knowledge of elements of Varāhamihira's *Bṛhadyātrā* in his work on military astrology,[69] and this text he probably gained access to through a Pahlavī translation of the Sanskrit original or an Arabic translation thereof. He also presents two theories that he ascribes to Zoroaster in two chapters of the so-called "second edition" of his Πόνοι περὶ πολεμικῶν καταρχῶν (*Labors Concerning Military Initiatives*), one on when an expected war will take place[70] and the other on determining the intent of someone who has sent you a letter.[71] The first is attributed to Zoroaster κατὰ πραξίδικον (which I take to mean "according to the avenger"). I suspect that by the "avenger" he means to refer obscurely to Sunbādh, the Ispahbadh or general under Abū Muslim who unsuccessfully attempted to avenge the latter's murder committed at the command of the Caliph, al-Manṣūr, in 755.[72] For Sunbādh was the patron of Saʿīd ibn Khurāsānkhurrah's translations of the five books of Zarādusht from Pahlavī into Arabic in about 750.[73] Theophilus may also have consulted the Pahlavī translations of Dorotheus and Valens that were available in Baghdād to Māshā'allāh,[74] though some of what Theophilus knew of the Sidonian seems not to have been included in the Pahlavī version.

Theophilus certainly wrote his surviving astrological treatises in Greek for their Greek versions cite his Greek sources verbatim. However, we find in Arabic writings

---

67. *Kitāb al-Fihrist li-'l-Nadīm*, ed. T. al-Ḥā'irī, Tehrān 1971, p. 310.
68. G. Bergträsser, *Ḥunain ibn Isḥāq. Über die syrischen und arabischen Galen-Übersetzungen*, Abhandlungen für die Kunde des Morgenlandes XVII 2, Leipzig 1925, p. 39 of the Arabic text. Concerning Ḥunayn see M. Ullmann, *Die Medizin im Islam*, Handbuch der Orientalistik 1. Abt., 6,1, Leiden 1970, pp. 115–119.
69. D. Pingree, "The Indian and Pseudo-Indian Passages in Greek and Latin Astronomical and Astrological Texts," *Viator* VII, 1976, 141–195 (148).
70. Πόνοι 31; edited by J. Bidez and F. Cumont, *Les mages hellénisés*, 2 vols., Paris 1938 (repr. New York 1975), vol. 2, pp. 225–226.
71. Πόνοι 39; ed. eidem, pp. 209–219.
72. D. Gutas, *Greek Thought, Arabic Culture*, London 1998, pp. 47–49 and the literature there cited.
73. Pingree, *Astral Omens* (as in n. 1), pp. 44–46.
74. D Pingree, "Māshā'allāh: Greek, Pahlavī, Arabic, and Latin Astrology," in *Perspectives arabes et médiévales sur la tradition scientifique et philosophique grecque*, ed. A. Hasnawi et al., Orientalia Lovaniensia Analecta LXXIX, Paris 1997, pp. 123–136.

many citations of those and other Theophilian works-especially one on genethlialogy that does not survive in its original Greek form. The Arabic authors who quote from Theophilus include Māshā'allāh, Sahl ibn Bishr, al-Qaṣrānī, and ᶜAlī ibn Abī al-Rijāl. Moreover, Abū Maᶜshar is cited by his gullible pupil, Shādhān, as having claimed that he had found in the *khazā'in al-mulūk* ("treasury of the kings"), by which he may mean the *bayt al-ḥikma* ("house of wisdom"), ancient books on astrology in Arabic, including a *Kitāb Thūfīl ibn Thūmā*—which must have been one of the earliest Arabic translations of a Greek scientific work.[75]

It is now time to try to establish the chronology of Theophilus's life from his first appearance in history until his death. He was present at the battle fought between Marwān II and the Khurāsānian army under ᶜAbdallāh ibn ᶜAlī on the left bank of the Greater Zāb between 15 and 24 January 750; for his eye-witness account is quoted by Agapius of Menbij.[76] He may have been serving Marwān II, whose headquarters were in Ḥarrān, as an astrological advisor; we do not know for sure. But, in any case, he soon was serving the victorious ᶜAbbāsids in that capacity. For, in the preface to the first edition of his Πόνοι, he speaks of his experience of warfare in a campaign waged ὑπὸ τῶν τηνικαῦτα κρατούντων ("by those then in power") against Margianē.[77] By the classical Margianē, of course, he means Khurāsān; and he is probably referring to the expedition led by al-Mahdī, the son of the Caliph al-Manṣūr, against the rebellious governor of Khurāsān, ᶜAbd al-Jabbār, in the winter of 758–759. Theophilus continued to serve the ᶜAbbāsids until the end of his life on 16 July 785.[78] The first edition of the Πόνοι, therefore, was probably written in the early 760's. Theophilus was able to use the chapters of Dorotheus on the effects of planetary aspects. He may have read these in the original Greek or in their Pahlavī translation; no Arabic version existed in the 760's, though Māshā'allāh made one a decade or two later.[79] In agreement with the hypothesis that he used the Pahlavī translation is the presence of echoes of Varāhamihira's Sanskrit *Bṛhadyātrā* in the Πόνοι, of which work as of other astrological texts by Varāhamihira a Pahlavī version seems to have existed.[80]

Theophilus wrote his work on historical astrology, which is also strongly influenced by Pahlavī sources, after 30 July 762, the date of the founding of Baghdād;[81] for he states in his Ἐπισυναγωγή (*Collection*):[82]

> ... κατὰ τὴν τῶν Σαρακηνῶν βασιλεύουσαν πόλιν, ὅπου ἀνείληφά μου τὴν ψηφοφορίαν, ἥτις ἐστὶν ἀνατολικωτέρα Βαβυλῶνος, τῆς δὲ Ἀλεξανδρείας

75. F. Rosenthal, "From Arabic Books and Manuscripts X," *Journal of the American Oriental Society* LXXXIII, 1963, pp. 454–456 (455).
76. L. Cheikho, *Agapius Episcopus Mabbugensis. Historia universalis*, Corpus Scriptorum Christianorum Orientalium LXV, Louvain 1954, p. 369.
77. Edited by F. Cumont in *CCAG* V 1, Bruxelles 1904, pp. 233–234 (234).
78. Pingree, *Yavanajātaka* (as in n. 10), vol. 2, pp. 443–444.
79. D. Pingree, "Māsha'āllāh's (?) Arabic Translation of Dorotheus," *Res Orientales* XII, 1999, pp. 191–209.
80. D. Pingree, "The Indian Iconography of the Decans and Horās," *Journal of the Warburg and Courtauld Institutes* XXVI, 1963, pp. 223–254 (252–254).
81. D. Pingree, "The Fragments of the Works of al-Fazārī," *Journal of Near Eastern Studies* XXIX, 1970, 103–123.
82. Edited by W. Kroll in *CCAG* I, Bruxelles 1898, pp. 129–131 (130).

> καθὼς ἐδοκίμασα ὥρας Γβ ς′ ιε′, ὅ ἐστιν μοῖραι ιγ λ′. ὀνομάζεται δ'αὕτη Εἰρηνόπολις, τῇ δὲ τῶν Σύρων διαλέκτῳ Βαγδαδᾶ.
>
> ... at the ruling city of the Saracens, where I made my calculations, which is further east than Babylon, but as I tested it east of Alexandria 2/3 1/6 1/15 of an hour, which is 13;30°. This city is called the City of Peace [Εἰρηνόπολις translates Dār al-salām], but in the dialect of the Syrians Baghdād.

Theophilus indicates his knowledge of Pahlavī sources in a passage that occurs a few lines before the one just quoted:

> οἱ δὲ κατὰ τὴν ἑῴαν ἅπασαν Περσῶν σοφίας ἐρασταὶ τὰς Ἑλληνικὰς βίβλους τῇ ἑαυτῶν μεταφράσαντες γλώττῃ τῷ ἐνιαυσιαίῳ καὶ μόνῳ κανόνι ἐχρήσαντο, ἤγουν τῷ ἐκ τῆς τοῦ Ἡλίου εἰς τὴν ἀρχὴν τοῦ Κριοῦ ἐποχῆς, καθάπερ Κριτόδημος καὶ Οὐάλης καὶ Δωρόθεος καὶ Τιμόχαρις καὶ οἱ περὶ αὐτούς.
>
> Those of the Persians in all the East who love wisdom, having translated the Greek books into their own tongue, used exclusively as their rule for the year that from the position of the Sun at the beginning of Aries, as do Critodemus, Valens, Dorotheus, Timocharis, and their associates.

It is well known that there were Pahlavī translations of Vettius Valens' *Anthologies*, Dorotheus of Sidon's astrological poem, and Ptolemy's *Almagest*. These Persians (or Theophilus) would have found Critodemus' name in Valens[83] and Timocharis' in the *Almagest*.[84]

Theophilus' next work must be the Περὶ καταρχῶν διαφόρων (*On Various Initiatives*) as it is referred to in a later work, which was written in about 770. The Περὶ καταρχῶν διαφόρων shows signs of the influence of Dorotheus and Hephaestio, but not as yet of Rhetorius. I take this to mean that in 765 or thereabouts, when Theophilus composed this work, he did not yet have access to Rhetorius.

In the preface to his <'Ἀποτελεσματικά>, Theophilus refers to the Περὶ καταρχῶν διαφόρων as the κοσμικαὶ ἀποτελέσεις καὶ αἱ κατὰ πεῦσιν καταρχαί ("cosmic effects and initiatives in accordance with interrogation"),[85] thereby signalling the relationship of catarchic to interrogational astrology. In this <'Ἀποτελεσματικά> we find Theophilus' epitomes of seven chapters from the fifth book of Rhetorius; in one of these (V 58), on the influences of thirty fixed stars, the longitudes of these παρανατέλλοντα have been increased to suit the year 770, which must be the date, within a year or two, of this

83. D. Pingree, *Vettii Valentis Antiocheni Anthologiarum libri novem*, Leipzig 1986, pp. 135 (III 5 and 6); 142 (III 9); 193 (IV 26); 221 (V 7); 223 (V 8); 288 (VIII 5); 316–317 (IX 1); and 334 and 336 (IX 9). For the Pahlavī version see Pingree, *Astral Omens* (as in n. 1), pp. 46–50.

84. J. L. Heiberg, *Claudii Ptolemaei Syntaxis mathematica*, 2 vols., Leizpig 1898–1903, vol. 2, pp. 3 (VII 1); 12 (VII 2); 17, 25, 28, and 29 (VIII 3); and 310–311 (X 4). For the Pahlavī translation see D. Pingree, "The Greek Influence on Early Islamic Mathematical Astronomy," *Journal of the American Oriental Society* 93, 1973, 32–43 (35–36).

85. Ed. Cumont (as in n. 61), p. 235.

book's composition. We can, therefore, pinpoint the date of Theophilus' acquisition of a copy of Rhetorius to the period between 765 and 770.

It was presumably in the 770's, then, that Theophilus completed the seventeen chapters of the second edition of his Πόνοι περὶ καταρχῶν πολεμικῶν, of which three are derived from the sixth book of Rhetorius. Indeed, one might even surmise that it was composed on the occasion of one of al-Mahdī's campaigns as Caliph, either that to suppress the rebellion of al-Muqannaʿ in Khurāsān between about 775 and 780 or those to defend the frontier with the Byzantines in 780 and 782.[86]

B. Greek Manuscript Tradition

The Greek manuscript tradition of Theophilus' works is both inadequate to preserving the text and instructive on how it was preserved. There are three main branches. The first consists of manuscript *L*, which was copied in about the year 1000, and manuscript *W*, which was copied a little before the year 1241. Both are descended from the same, now lost manuscript,[87] but the scribe of each omitted his own choice of undesirable chapters. The prototype contained all that we still have of Theophilus' Ἐπισυναγωγὴ περὶ κοσμικῶν καταρχῶν, which, despite its title, is on historical horoscopy of the Sasanian type; it contained also all of his untitled work concerning general astrology descended from Babylonian-type celestial omens combined with short chapters on catarchic, genethlialogical, and interrogational astrology, a collection which I call the <Ἀποτελεσματικά> ; a few chapters from the first edition of his Πόνοι περὶ καταρχῶν πολεμικῶν on military astrology; and a substantial chunk of his Περὶ καταρχῶν διαφόρων on catarchic astrology arranged by the twelve astrological places counted from the ascendent on the ecliptic. This last work breaks off in both manuscripts in the seventh place, but the Laurentianus has on another leaf the last chapter from the section on the ninth place and the beginning of the first from that on the tenth place.

The second class of manuscripts also has two members—manuscript *A*, a fourteenth century codex, and manuscript *Y*, copied from the same source in about 1400. They both contain an index covering the first and second editions of Theophilus' Πόνοι and most of his Περὶ καταρχῶν διαφόρων, breaking off at the sixth chapter of the tenth place, a little after the point where the Laurentianus broke off. Of Theophilus' actual text this class preserves only the first edition of the Πόνοι through its antepenultimate chapter and the entirety of the <Ἀποτελεσματικά>.

The third class is represented by one manuscript of the thirteenth century, manuscript *P*. This is a badly disturbed copy, but it still preserves both editions of the Πόνοι in their entirety except for the loss of 1 leaf; 2 single chapters from the <Ἀποτελεσματικά> ; and the four chapters concerning the sixth place in the Περὶ καταρχῶν διαφόρων.

86. On ʿAbbāsid history see H. Kennedy, *The Early Abbasid Caliphate*, London 1981.
87. This lost manuscript ended its main body of excerpts from Theophilus' Περὶ καταρχῶν διαφόρων at VII 1. Both manuscripts contain a statement: τοῦτο τὸ κεφάλαιον εὑρεῖς καὶ εἰς τὸ βιβλίον τὸ μαυρὸν εἰς φύλλον κ' ἀπὸ τῆς ἀρχῆς ("You will find this chapter in the black book on folio 20 from the beginning.") (f. 82v L, f. 117 W).

C. Theophilus' Sources

Classes one and three were instrumental also in preserving Hephaestio's text, and class one was also closely connected to the text of Rhetorius. These two, along with Dorotheus of Sidon whom Theophilus knew in a fuller version than that available in the Pahlavī translation, and therefore perhaps in the original hexameters or, possibly, in a fuller form of the prose paraphrase than is available to us in Hephaestio, were Theophilus' favorite authorities. He seldom names them, but in the first edition of his Πόνοι he transforms Dorotheus II 14–17, on aspects applied to genethlialogy, to a form in which they are applied to military astrology,[88] while in the second edition he bases three chapters on three from book six of Rhetorius which the latter had copied from the early sixth century author, Julian of Laodicea; they are on war and its dissolution.[89] In the Περὶ καταρχῶν διαφόρων Theophilus based one chapter, on medical astrology, on a chapter in Dorotheus's catarchic book five;[90] in the <'Αποτελεσματικά> one chapter, on whether an interrogation will concern a man or an animal, is a re-interpretation of a statement by Hephaestio about which astrological factors lead to the birth of a monster.[91] And a block of seven chapters consists of epitomes of material found in the fifth book of Rhetorius;[92] they are on the influences of the fixed stars, changed to refer to catarchic astrology from their earlier application to genethlialogy; on various undertakings; on the dodecatemoria; on people who are subject to demons or are epileptic; on thieves; on death; and on the lot of the killing place. But in the Περὶ καταρχῶν διαφόρων I find the influence of only Dorotheus and Hephaestio, and not of Rhetorius at all.

D. Theophilus and Māshā'allāh

Now that we have established at least the strong plausibility that Theophilus used Dorotheus—either in Greek or in Pahlavī—from the beginning of his career of writing on astrology in about 760, and that he began using Hephaestio in Greek in about 765, and Rhetorius a few years later, we can also remark that, so far as we as yet know, the chief or even sole users of Hephaestio, Rhetorius and Theophilus himself among astrologers writing in Arabic in the late eighth and early ninth centuries were Māshā'allāh, Theophilus' colleague at the Caliphal court, and Māshā'allāh's pupils, especially Sahl ibn Bishr and Abū ᶜAlī al-Khayyāṭ, though it is possible that they are to be joined by ᶜUmar ibn al-Farrukhān al-Ṭabarī, who does not name his sources in his lengthy *Kitāb al-masā'il* (*Book of Interrogations*), but who translated the Pahlavī version of Dorotheus into Arabic.

Māshā'allāh's use of Theophilus is proved by the compilations of his chapters on interrogational astrology preserved in Istanbul Laleli 2122bis, where Māshā'allāh also uses Hephaestio, and in Leiden Orientalis 891.[93] From the large number of horoscopes

88. Chapters 4–7.
89. Chapters 35–37.
90. VI 1–3. Numerous other chapters in the Περὶ καταρχῶν are influenced by Hephaestio.
91. Chapter 29.
92. Chapters 17–23.
93. Pingree, *Māshā'allāh* (as in n. 74).

preserved in the Leiden manuscript and in the numerous Byzantine Greek translations of Māshā'allāh's chapters on interrogations, we know that his examples were of queries addressed to him between 12 June 765 and 7 June 768. Thus Māshā'allāh's use in these books of Hephaestio and Theophilus fits in perfectly with our theory that Theophilus first relied on Hephaestio in the interrogational Περὶ καταρχῶν διαφόρων that he wrote in about 765.

The other part of that theory was that Theophilus did not have Rhetorius until shortly before 770. The main work that we know of in which Māshā'allāh used Rhetorius is the text which Hugo of Santalla translated into Latin in the 1140s as the *Liber Aristotilis de ducentis quinquaginta quinque Indorum voluminibus universalium questionum tam genetialium quam circularium summam continens* (*Book of Aristotle on the 255 Books of the Indians Containing the Totality of Universal Questions, both Genethlialogical and Revolutionary*), which may be the *Kitāb al-marḍā* (*Book of the Pleasant*) ascribed to Māshā'allāh by al-Nadīm.[94] In this one can identify fifty-three chapters from the fifth book and two from the sixth book of Rhetorius's compendium that Māshā'allāh drew upon.

This was indeed a massive influence. Moreover, the third book of the *Liber Aristotilis* follows the lead of Theophilus' Περὶ καταρχῶν διαφόρων in arranging its contents—here genethlialogical rather than interrogational—according to the twelve astrological places. Therefore, I hypothesize that the Arabic original of the *Liber Aristotilis* was composed after the period 765 to 770; I would further suggest, though I cannot prove it, that it was also composed after the interrogational books that Māshā'allāh wrote in about 770, and therefore also after Theophilus' use of Rhetorius in his <'Ἀποτελεσματικά>. The only other place where I am sure that Māshā'allāh used Rhetorius is in his *Kitāb al-mawālīd* (*Book of Nativities*) which survives in a Latin translation.[95] This contains twelve nativity horoscopes: three are drawn from Dorotheus, and the rest are from a sixth century source with nativity horoscopes datable to the period between 19 January 403 and 9 November 542. At least one of these nine, datable 8 September 428, is from Rhetorius; I suspect that the rest are also, though our corrupt Byzantine manuscript tradition of Rhetorius has not preserved them.

These twelve horoscopes were copied by Abū ᶜAlī al-Khayyāṭ in his *Kitāb al-mawālīd* (*Book of Nativities*).[96] The same author has quoted Eutocius in his *Kitāb al-masā'il* (*Book of Interrogations*);[97] and Eutocius is otherwise known as an astrologer only from the sixth book of Rhetorius[98] and from the Arabic translation of a commentary he wrote on the first book of Ptolemy's 'Ἀποτελεσματικά.[99]

Sahl ibn Bishr's *Kitāb al-masā'il* on interrogations, *Kitāb al-mawālīd* on nativities, and *Kitāb al-ikhtiyārāt ᶜalā al-buyūt al-ithnai ᶜashar* (*Book of Choices According to the Twelve Astrological Places*) on initiatives are all strongly influenced by Māshā'allāh and follow in general the corresponding works of Theophilus, including his lost work on nativities which is quoted in Sahl's *Kitāb al-mawālīd* and by later Arabic astrological anthologists.

94. Pingree, "Classical and Byzantine Astrology" (as in n. 20), pp. 227–239, and Burnett and Pingree, *Liber Aristotilis* (as in n. 16), pp. 3–9.
95. Edited by D. Pingree in E.S. Kennedy and D. Pingree, *The Astrological History of Māshā'allāh*, Cambridge, MA. 1971, pp. 145–165.
96. Ibid., pp. 166–174.
97. F. Sezgin, *Geschichte des Arabischen Schrifftums*, vol. 7, Leiden 1979, p. 121.
98. VI 40, 6; and VI 52.
99. Sezgin (as in n. 97), p. 48.

Through the works of Māshā'allāh, al-Khayyāṭ, and Sahl, in short, Theophilus exercised a profound influence on the rise of Arabic astrology and on its content. This influence is characterised by a synthesis of the Greek and Persian traditions, by continuing and expanding the Indian and Sasanian habit of transforming catarchic into interrogational astrology, and by the organisation of astrological treatises in conformity with the twelve astrological places. What he was not involved in was the invention of new astrological concepts and technical vocabulary with which to express them. Such new concepts and technical terms, the latter to some extent based on Pahlavī antecedents, appear first, I believe, in the first part of Sahl ibn Bishr's *Kitāb al-aḥkām ʿalā al-niṣbat al-falakīya (Book of Judgments According to the Relation of the Spheres)*, but were expressed in their classical form only by Abū Maʿshar in the middle of the ninth century. Theophilus' tendency to synthesize and Sahl's inventiveness turned Arabic astrology into a science recognisably different from its Greek, Sanskrit and Pahlavī predecessors; but for the first half century and more of its existence Arabic astrology, apart from the pure translations from Pahlavī or Greek, largely followed Theophilus' models.

IV. The Greek Manuscripts

A. Parisinus suppl. graecus 1241. A manuscript of 47 folia copied in the fourteenth century and purchased by Minois Mynas in the East. Described by F. Cumont in *CCAG* VIII 1, Bruxelles 1929, pp. 117–128. Contains an index of chapters, the first edition of Theophilus' Πόνοι περὶ καταρχῶν πολεμικῶν, and his <Ἀποτελεσματικά>.

B. Parisinus graecus 2506. A manuscript of 216 folia copied by two scribes in the fourteenth century. Described by F. Cumont in *CCAG* VIII 1, Bruxelles 1929, pp. 74–115. Contains Epitome IV of Rhetorius, 25 chapters of Epitome I of Hephaestio, eleven chapters related to book VI of Rhetorius, and some 45 chapters derived from Theophilus's various works.

Ber. Berolinensis graecus 173. A manuscript of 204 folia copied by several scribes in the fifteenth century. Described by F. Boll in *CCAG* VII, Bruxelles 1908, pp. 48–63. Ff. 48–49v contain an excerpt from Epitome IV 1 of Rhetorius, here ascribed to Hermes; ff. 139–144v contains the Ῥητορίου Θησαυρὸς συνέχων τὸ πᾶν τῆς ἀστρονομίας (*The Treasure of Rhetorius Containing All of Astronomy*), which is an epitome of something genuinely Rhetorius'; and ff. 145v–146v the preface of Epitome IIb of Rhetorius. In between these last two is a text derived from Valens I 1; it is published in *Vettii Valentis Antiocheni Anthologiarum libri novem*, ed. D. Pingree, Leipzig 1986, pp. 390–392.

E. Angelicus graecus 29. A manuscript of 346 folia mostly copied by Eleutherius Zebelenus of Elis on Mitylene in 1388. Described by F. Cumont and F. Boll in *CCAG* V 1, Bruxelles 1904, pp. 4–57. Eleutherius' compendium ascribed to Palchus is on ff. 92–151v; it contains chapters derived from both Rhetorius and Theophilus as well as from Māshā'allāh. A second scribe has copied 34 chapters from Theophilus on ff. 209–212, 314v, 335–338, and 343–346; these come from the Πόνοι περὶ καταρχῶν πολεμικῶν and the <Ἀποτελεσματικά>.

L. Laurentianus 28, 34. A manuscript of 170 folia copied in about 1000 A.D. It contains 12 chapters of Hephaestio, Epitome II of Rhetorius as well as some additional chapters, some excerpts from John Lydus' *De Ostentis* (*On Omens*), and numerous chapters from all of Theophilus's surviving texts.

P. Parisinus graecus 2417. A manuscript of 176 folia copied in the thirteenth century. Described by F. Cumont in *CCAG* VIII 1, Bruxelles 1929, pp. 9–20. It contains both editions of the Πόνοι and a few chapters from the <᾿Αποτελεσματικά> and the Περὶ καταρχῶν διαφόρων of Theophilus; it also is the primary manuscript preserving Hephaestio's <᾿Αποτελεσματικά> on ff. 39–175v.

R. Parisinus graecus 2425. A manuscript of 285 folia copied in the early fourteenth century. Described by P. Boudreaux in *CCAG* VIII 4, Bruxelles 1921, pp. 22–42. Ff. 4v–76 contain the four books of Ptolemy's <᾿Αποτελεσματικά>; this copy W. Hübner has placed at the head of his classis α in his edition (*Claudii Ptolemaei Opera quae exstant omnia*, vol. III, 1, *Apotelesmatika*) published at Stuttgart-Leipzig in 1998. Books V and VI on ff. 76–238 and ff. 1–1v contain the compendium of Rhetorius, the instructions and table of Stephanus, and important material for the history of astrology. At the end of the manuscript, on ff. 257–284, is an astronomical treatise composed between 1060 and 1072 by an unknown scholar using Arabic sources. Parts of this treatise are found in manuscript *V* and in the manuscripts of Epitome IIIb of Rhetorius. The treatise was edited by A. Jones, *An Eleventh-century Manual of Arabo-Byzantine Astronomy*, Corpus des astronomes byzantins 3, Amsterdam 1987.

V. Vaticanus graecus 1056. A manuscript of 244 folia copied in the fourteenth century, to a large extent from a twelfth-century source. Described by I. Heeg in *CCAG* V 3, Bruxelles 1910, pp. 7–64. It contains two chapters from the astronomical treatise found at the end of *R*; the Περὶ τῆς μαθηματικῆς τέχνης (*On the Mathematical Art*) of Stephanus the Philosopher; Epitome II of Hephaestio; and various chapters derived from Rhetorius, Theophilus, and Māshā'allāh. *V* shares a number of chapters with *E*.

W. Vindobonensis phil. gr. 115. A manuscript of 226 folia copied in the thirteenth century before 1241. Described by W. Kroll in *CCAG* VI, Bruxelles 1903, pp. 16–28. On ff. 25–117 are chapters almost in their entirety copied from Theophilus. This manuscript also contains five chapters from Epitome II of Hephaestio, and four from the original form of Hephaestio. It shares readings with both *L* and *V*. There is also found in it the original form of Olympiodorus' commentary on Paul of Alexandria to which the name Heliodorus has been attached by a later scribe.

Y. Vaticanus graecus 212. A manuscript of 152 folia, of which ff. 106–152 were copied at the end of the fourteenth or the beginning of the fifteenth century. Described by F. Cumont and F. Boll in *CCAG* V 1, Bruxelles 1904, pp. 64–71. This manuscript is a gemellus of manuscript *A*, but breaks off in chapter 12 of Theophilus' 'Αποτελεσματικά>. It was bought by Laudivius Zacchia in Crete, probably in 1475, and had entered the Vatican Library by 1481.

Appendices

Appendix I gives an example of the transference of chapters from book V of manuscript *R* into the archetype of the principal manuscripts of Porphyry. Note that pseudo-Porphyry is closer to *L* than to *R*, and that the promise in the first and last sentences, while fulfilled in manuscript *R*, is not fulfilled in pseudo-Porphyry. The chapter in Epitome IIIb, as is normally the case, is closer to *R* than to *L*, while Epitome IIa, which was produced in the School of John Abramius, has introduced many changes due to the epitomator's acquaintance with Ptolemy's and Hephaestio's astrological works.

Appendix II is intended to illustrate the general relationship between the text of manuscript *R* and that of Epitome IV, which greatly abbreviates its prototype, but adds the names of the ancient authorities when these are known to the epitomator.

Appendix III exemplifies the transmission of the chapter on the παρανατέλλοντα from its source in the chapter of the Anonymus anni 379, imperfectly preserved in manuscript *E*, through the various versions of the echoes of its lost Pahlavī translation, where it was attributed to Hermes, in Arabic and Latin and the adaptation of the

---

I. Rhetorius V 12

Ia. As preserved in manuscripts *L* and *R*.

Περὶ τῶν ὁρίων κατ᾽ Αἰγυπτίους καὶ Πτολεμαῖον[1] καὶ λαμπρῶν καὶ σκιαρῶν[2] μοιρῶν.

Πάλιν οἱ παλαιοὶ εἰς ἥμισυ χώρας ἢ τόπους τέμνοντες ἕκαστον ζῴδιον ὠνόμασαν ὅρια, οὐ κατὰ τὰς ἴσας μοίρας ὡς ἐπὶ τῶν δεκανῶν εἰρήκαμεν, ἀλλὰ διαφόρως κατὰ τὴν ἄλλην αἰτίαν ἥντινα ὑπέθεντο ἐν ταῖς τῶν ἀστέρων τελείαις περιόδοις (περιόδους δὲ λέγω ἅστινας ἐν τοῖς ἔμπροσθεν εὑρήσεις) [μετὰ δὲ τῶν ὁρίων]· αὐταὶ δὲ αἱ μοῖραι τῶν ὁρίων αἱ κατὰ ζῴδιον κείμεναι. τινὲς μὲν αὐτῶν εὑρίσκονται Διός, ἄλλαι δὲ ᾿Αφροδίτης, ἄλλαι Κρόνου. ὅταν οὖν τις[3] τῶν ἀστέρων εὑρεθῇ[4] οἴκῳ ἀγαθοποιοῦ καὶ ὁρίοις ἀγαθοποιοῦ[5] λόγον ἔχων πρὸς τὴν γένεσιν, ἀγαθοποιεῖ τὴν τύχην· εἰ δὲ εὑρεθῇ οἴκῳ μὲν ἀγαθοποιοῦ, ὁρίοις δὲ κακοποιοῦ, ἐλαττοῖ τὸ ἀγαθὸν τῆς τύχης· εἰ δὲ τύχῃ οἴκῳ[6] κακοποιοῦ καὶ ὁρίοις κακοποιοῦ, κακοῖ καὶ ἀμαυροῖ τὴν τύχην.[7] ἥ[8] οὖν τῶν ὁρίων ἐνέργεια ἐναλλοιοῖ τὴν τῶν ἀστέρων ἀποτελεσματογραφίαν καθὰ εἴρηται καὶ ἐπὶ τῶν προσώπων τῶν δεκανῶν. ὁ οὖν Πτολεμαῖος ἔν τισιν ὁρίοις οὐ συνῄνεσε τοῖς Αἰγυπτίοις, διὸ ἠναγκάσθη καὶ τούτων ὑπόμνησιν ποιήσασθαι. συμβάλλονται[9] δὲ τὰ ὅρια αὐτοῦ ἐν τοῖς αὐτῶν ἀποτελέσμασι μόνοις. τινὲς[10] δὲ τῶν λαμπρῶν καὶ σκιαρῶν[11] καὶ ἀμυδρῶν μοιρῶν οὐ μικρὰν ἔχουσι τὴν ἐνέργειαν εἴπερ οἱ ἀστέρες ἐν ταῖς λαμπραῖς μοίραις εὑρέθησαν τετυχηκότες, διὸ καὶ ταύτας[12] ἐν τοῖς ἔμπροσθεν ἠναγκάσθην[13] καθυποτάξαι.

1. πτολεμαίῳ *L* 2. σκιερῶν *R* 3. τι *L* 4. εὑρεθῇ om. *R* 5. ἀγαθοποεῖ *R* 6. οὐ *R* 7. τῇ τύχῃ *L*, τὴν ψυχήν *R* 8. εἰ *LR* 9. συμβαλλόντων *LR* 10. οἱ *LR* 11. σκιερῶν *R* 12. ταῦτα *L* 13. ἠναγκάσθη *R*.

Greek original of the Pahlavī translation by Rhetorius, to the rewriting in Epitome IV, and finally ending with Theophilus's adaptation of the same material, changing the emphasis from genethlialogical to catarchic astrology. The longitudes of the stars were corrected to their positions in 379 in the Anonymus, to 505 in the Hermes version, to 884 in Epitome IV, and to 770 by Theophilus (the manuscripts often give the wrong numbers). Note that Māshā'allāh has used the version found in manuscript *R*.

Appendix IV begins with Rhetorius' adaptation of one of Julian of Laodicea's chapters on war. Next is its virtually verbatim repetition by Theophilus (usually he changes much more). Then is an Arabic text by Māshā'allāh drawing on either Rhetorius or Theophilus, but mixing in other material, changing the emphasis from catarchic to interrogational astrology, and interpreting the words "raised up" (to exaltation) and "lowered" (to dejection) to refer to the superior and the inferior planets respectively. Finally there is presented the Byzantine translation of Māshā'allāh's text, made in about the year 1000.

I have translated the passages from Rhetorius and the Arabic passages from Māshā'allāh and Abū Ma^cshar.

On the terms according to the Egyptians and Ptolemy, and <on> the bright and shadowy degrees.

Again the ancients, dividing each zodiacal sign in half, called the regions or places "terms," <dividing them> not into equal <numbers of> degrees as we mentioned in the case of the decans, but differently in accordance with some other reason which they suggested on <the basis of> the perfect periods of the planets (I mean the periods which you will find in what lies ahead); these are the degrees of the terms lying in <each> zodiacal sign. Some of them are found belonging to Jupiter, others to Venus, others to Saturn. Therefore, when one of the planets should be found in the house of a benefic and the terms of a benefic, having a relationship to the nativity, it benefits <the native's> fortune; if it should be found in the house of a benefic, but the terms of a malefic, it diminishes the good of <his> fortune; but if it should be in the house of a malefic and the terms of a malefic, it injures and darkens <his> fortune. Therefore, the power of the terms changes the influence of the planets as was also said in the case of the "faces" of the decans. But Ptolemy did not agree with the Egyptians on some terms, wherefore he was compelled to mention them also. But his terms alone contribute in their influences. Some of the bright and shadowy and faint degrees have not a little power if the planets are found in the bright degrees, wherefore I was compelled to subjoin them in what lies ahead.

Ib. Psuedo-Porphyry 49 as preserved in manuscripts *S* (Monacensis gr. 419), *D* (Laurentianus 28, 20), and *M* (Marcianus gr. 314), without emendation.

Περὶ τῶν ὁρίων κατ' Αἰγυπτίους καὶ Πτολεμαῖον καὶ λαμπρῶν καὶ σκιαρῶν μοιρῶν.[1]

Πάλιν οἱ παλαιοὶ εἰς ἡμίσεις χώρας ἢ τόπους τέμνοντες ἕκαστον ζῴδιον ὠνόμασαν ὅρια, οὐ κατὰ τὰς ἴσας μοίρας ὡς ἐπὶ τῶν δεκανῶν εἰρήκαμεν, ἀλλὰ διαφόρως κατὰ ἄλλην αἰτίαν ἥντινα ὑπέθεντο ἐν ταῖς τῶν ἀστέρων τελείαις περιόδοις (περιόδους δὲ λέγω ἅστινας ἐν τοῖς ἔμπροσθεν εὑρήσεις)· αὐταὶ δὲ αἱ μοῖραι τῶν ὁρίων[2] αἱ κατὰ ζῴ διον κείμεναι. τινὲς μὲν αὐτῶν εὑρίσκονται Διός, ἄλλαι δὲ 'Αφροδίτης, ἄλλαι δὲ[3] Κρόνου, καὶ[4] ἄλλαι τῶν λοιπῶν πλανήτων. ὅταν οὖν τις[5] τῶν ἀστέρων εὑρεθῇ ἐν οἴκῳ ἀγαθοποιοῦ[6] καὶ ἐν ὁρίοις ἀγαθοποιοῦ λόγον ἔχων πρὸς τὴν γένεσιν, ἀγαθοποιεῖ τὴν τύχην· εἰ δὲ εὑρεθῇ ἐν οἴκῳ μὲν ἀγαθοποιοῦ[6], ὁρίοις δὲ κακοποιοῦ, κακοῖ καὶ ἀμαυροῖ τὴν τύχην. ἡ οὖν τῶν ὁρίων ἐνέργεια ἐναλλοιοῖ τὴν τῶν ἀστέρων ἀποτελεσματογραφίαν καθὰ εἴρηται καὶ ἐπὶ τῶν προσώπων τῶν δεκανῶν. ὁ οὖν Πτολεμαῖος ἔν τισιν ὁρίοις οὐ συνήνεσε τοῖς Αἰγυπτίοις, διὸ ἠναγκάσθη καὶ τούτων ὑπόμνησιν ποιήσασθαι, συμβαλλόντων[7] δὲ τὰ ὅρια αὐτῶν ἐν τοῖς αὐτῶν[8] ἀποτελέσμασι μόνοις. αἱ δὲ τῶν λαμπρῶν καὶ σκιαρῶν καὶ ἀμυδρῶν μοιρῶν οὐ μικρὰν ἔχουσι τὴν ἐνέργειαν εἴπερ οἱ ἀστέρες ἐν ταῖς λαμπραῖς μοίραις εὑρέθησαν τετυχηκότες, διὸ καὶ ταῦτα ἐν τοῖς ἔμπροσθεν ἠναγκάσθη καθυποτάξαι.

1. tit. om. *S.* 2. ζωρίων *S* 3. καὶ ἄλλαι *DM* 4. καὶ om. *S* 5. τινες *S* 6. καὶ–ἀγαθοποιοῦ om. *DM* 7. συμβάλλονται *S* 8. αὐτῶν om. *D.*

Ic. Ep. IIIb 10, without emendation.

Περὶ ὁρίων καὶ λαμπρῶν καὶ σκιερῶν μοιρῶν.

Πάλιν οἱ παλαιοὶ εἰς ἥμισυ χώρας ἢ τόπους τέμνοντες ἕκαστον ζῴδιον ὠνόμασαν ὅρια, οὐ κατὰ τὰς ἴσας μοίρας ὡς ἐπὶ τῶν δέκα εἰρήκαμεν, ἀλλὰ διαφόρας κατὰ τὴν ἄλλην αἰτίαν ἥντινα ὑπέθεντο ἐν ταῖς τῶν ἀστέρων τελείαις περιόδοις· αὐταὶ δὲ αἱ μοῖραι τῶν ὁρίων αἱ κατὰ ζῴδιον κείμεναι. τινὲς μὲν αὐτῶν εὑρίσκονται Διός, ἄλλαι δὲ 'Αφροδίτης, ἄλλαι δὲ Κρόνου. ὅταν οὖν τις τῶν ἀστέρων οἴκῳ ἀγαθοποιοῦ καὶ ὁρίοις ἀγαθοποιοῦ λόγον ἔχων πρὸς τὴν γένεσιν, ἀγαθοποιεῖ τὴν ψυχήν· εἰ δὲ εὑρεθῇ οἴκῳ μὲν ἀγαθοποιοῦ, ὁρίοις δὲ κακοποιοῦ, ἐλλατοῖ τὸ ἀγαθὸν τῆς τύχης. εἰ δὲ τύχῃ κακοποιοῦ καὶ ὁρίοις κακοποιοῦ, κἀκεῖ ἀμαυροῖ τὴν ψυχήν. ἡ οὖν τῶν ὁρίων ἐνέργεια ἐναλλοιοῖ τὴν τῶν ἀστέρων ἀποτελεσματουργίαν καθὰ εἴρηται καὶ ἐπὶ τῶν προσώπων τῶν δεκανῶν. ὁ οὖν Πτολεμαῖος ἔν τισιν ὁρίοις οὐ συνήναισε τοῖς Αἰγυπτίοις, διὸ ἠναγκάσθην καὶ τούτων τὴν ὑπόμνησιν ποιήσασθαι. οἱ οὖν τῶν λαμπρῶν καὶ σκιαρῶν καὶ ἀμυδρῶν μοιρῶν οὐ μικρὰν ἔχουσι τὴν ἐνέργειαν εἴπερ οἱ ἀστέρες ἐν ταῖς λαμπραῖς μοίραις εὑρέθησαν τετυχηκότες, διὸ καὶ ταύτας ἐν τοῖς ἔμπροσθεν ἠναγκάσθην καθυποτάξαι.

On the Terms according to the Egyptians and Ptolemy, and <on> the Bright and Shadowy Degrees

Again the ancients, dividing each zodiacal sign in half, called the regions or places "terms," <dividing them> not into equal <numbers of> degrees as we mentioned in the case of the decans, but differently in accordance with some other reason which they suggested on <the basis of> the perfect periods of the planets (I mean the periods which you will find in what lies ahead); these are the degrees of the terms lying in <each> zodiacal sign. Some of them are found belonging to Jupiter, others to Venus, others to Saturn, and others to the remaining planets. Therefore, when one of the planets should be found in the house of a benefic and the terms of a benefic, having a relationship to the nativity, it benefits <the native's> fortune; if it should be found in the house of a benefic, but the terms of a malefic, it injures and darkens <his> fortune. Therefore, the power of the terms changes the influence of the planets as was also said in the case of the "faces" of the decans. But Ptolemy did not agree with the Egyptians on some terms, wherefore he was compelled to mention them also, since they apply the terms in their influences only. Some of the bright and shadowy and faint degrees have not a little power if the planets are found in the bright degrees, wherefore I was compelled to subjoin them in what lies ahead.

On the Terms and the Bright and Shadowy Degrees

Again the ancients, dividing each zodiacal sign in half, called the regions or places "terms," <dividing them> not into equal <numbers of> degrees as we mentioned in the case of the ten, but different <numbers of degrees> in accordance with another reason which they suggested on <the basis of> the perfect periods of the planets; these are the degrees of the terms lying in <each> zodiacal sign. Some of them are found belonging to Jupiter, others to Venus, and others to Saturn. Therefore, when one of the planets should be found in the house of a benefic and the terms of a benefic, having a relationship to the nativity, it benefits <the native's> soul; if it should be found in the house of a benefic, but the terms of a malefic, it diminishes the good of <his> fortune; but if it should be <in the house> of a malefic and the terms of a malefic, in that case it blackens <his> soul. Therefore, the power of the terms changes the influence of the planets as was also said in the case of the "faces" of the decans. But Ptolemy did not agree with the Egyptians on some terms, wherefore I was compelled to mention them also. Therefore some <masculine!> of the bright and shadowy and faint degrees have not a little power if the planets are found in the bright degrees, wherefore I was compelled to subjoin them in what lies ahead.

Id. Epitome IIa 10, without emendation

Περὶ τῶν ὁρίων.

Πάλιν οἱ παλαιοὶ καὶ εἰς ἔτι λεπτότερα τμήματα ἕκαστον τῶν ζῳδίων διελόντες ε μὲν τῷ πλήθει, ἄνισα δὲ τῷ πηλίκῳ, τοῖς ε πλανωμένοις ἀπένειμαν καὶ ὅρια ἑκάστου τῶν ἀστέρων ὠνόμασαν, ἅτινα καὶ ἔκκεινται κανονικῶς ἐν τῇ τοῦ Πτολεμαίου Τετραβίβλῳ. ἔστι δὲ καὶ ἡ τούτων ἐνέργεια τοιαύτη. ὅταν τις τῶν ἀστέρων εὑρεθῇ ἐν οἴκῳ ἀγαθοποιοῦ καὶ ὁρίοις ἀγαθοποιοῦ λόγον ἔχων πρὸς τὴν γένεσιν, ἀγαθοποιεῖ τὴν τύχην· ἐὰν δὲ εὑρεθῇ ἐν οἴκῳ μὲν ἀγαθοποιοῦ, ὁρίοις δὲ κακοποιοῦ, ἐλαττοῖ τὸ ἀγαθὸν τῆς τύχης· εἰ δὲ καὶ ἐν οἴκῳ καὶ ὁρίοις κακοποιοῦ, κακοῖ καὶ ἀμαυροῖ τὴν τύχην. ἡ οὖν τῶν ὁρίων ἐνέργεια ἐναλλοιοῖ τὴν τῶν ἀστέρων ἀποτελεσματογραφίαν καθάπερ εἰρήκαμεν καὶ ἐπὶ τῶν δεκανῶν. αἱ δὲ τῶν λαμπρῶν καὶ σκιερῶν μοιρῶν φύσεις, περὶ ὧν ἐν τῇ ἀρχῇ τῆς τοῦ Ἡφαιστίωνος πραγματείας ἐν ᾗ περὶ τῶν δεκανῶν λεπτομερέστερον ὑποτετύπωται, οὐ μικρὰν ἔχουσι τὴν ἐνέργειαν ὁπόταν οἱ ἀστέρες ἐν αὐταῖς τυχόντες τοῦ γινομένου πράγματος κυριεύωσιν.

II. Rhetorius V 57, 123–125.

IIa. As preserved in manuscript *R*, without emendation.

Ἐμοὶ δὲ δοκεῖ καὶ ἐπὶ ἑκάστου οἴκου καὶ τόπου βλαπτομένου ὑπὸ κακοποιοῦ ἀποφήνασθαι τὴν βλάβην κατὰ τὴν φύσιν τοῦ οἰκοδεσπότου τοῦ ζῳδίου καθ' ὃ ὁ κακοποιός ἐστι καὶ τὰ σημαινόμενα ἀπὸ τοῦ τόπου.

Εἰ δὲ θέλεις λαβεῖν καὶ κλῆρον βίου, λάμβανε πάντοτε, νυκτὸς καὶ ἡμέρας, ἀπὸ οἰκοδεσπότου τοῦ δευτέρου τόπου ἐπὶ αὐτὸν τὸν τόπον τὸν δεύτερον, καὶ τὰ ἴσα ἀπόλυε ἀπὸ ὡροσκόπου· καὶ ὅπου ἂν ἐκπέσῃ ὁ κλῆρος, σκόπει τὸν κύριον τοῦ κλήρου καὶ αὐτὸ τὸ ζῴδιον ἐν ᾧ ὁ κλῆρος ἐνέπεσεν. καὶ εἰ καλῷ τόπῳ ἔτυχεν ὁ κλῆρος καὶ ὁ κύριος αὐτῷ δίχα τῆς τῶν κακοποιῶν ἀκτινοβολίας, καλὸν τὸν βίον λέγε· εἰ δὲ εὑρεθῇ ὁ κύριος τοῦ κλήρου ἑῷος δυτικὸς μετὰ ἑπτὰ ἡμέρας γενήσεται, ἑῷος ἀνατολικὸς κρυπτῶν καὶ λαθραίων πλοῦτον ποιήσει.

IIb. Ep. IV 1 110–112 as preserved in manuscript *B*, without emendation.

Ἐμοὶ δὲ δοκεῖ ὅτι καὶ ἐπὶ ἑκάστου οἴκου καὶ τόπου βλαπτομένου ὑπὸ Κρόνου, Ἄρεως, τὴν βλάβην γίνεσθαι κατὰ τὴν τοῦ οἰκοδεσπότου τοῦ ζῳδίου φύσιν ἔνθα ὁ κακοποιὸς ἕστηκεν, καὶ τὰ σημαινόμενα ἀπὸ τοῦ τόπου.

Κλῆρος βίου κατὰ Δωρόθεον καὶ ἡμέρας καὶ νυκτὸς ἀπὸ τοῦ κυρίου τοῦ β' τόπου ἐπ' αὐτὸν τὸν β', καὶ τὰ ἴσα ἀπὸ ὡροσκόπου. ἐὰν εὑρεθῇ ὁ κύριος τοῦ κλήρου μετὰ ζῳδίου ἡμέρας ποιῶν ἑῷαν ἀνατολήν, κρυπτὸν πλοῦτον ἕξει.

On the Terms

Again the ancients, dividing each zodiacal sign into even finer parts, 5 in quantity, but unequal in size, assigned <them> to the 5 planets and called <them> the terms of each of the planets, which also lie in tabular form in the *Tetrabiblos* of Ptolemy. Their power is such: whenever one of the planets should be found in the house of a benefic and the terms of a benefic, having a relationship to the nativity, it benefits the <native's> fortune; if it should be found in the house of a benefic and the terms of a malefic, it diminishes the good of <his> fortune; but if in the house and terms of a malefic, it injures and darkens <his> fortune. Therefore, the power of the terms changes the influence of the planets as we also said in the case of the decans. The natures of the bright and shadowy degrees, about which <it was written> at the beginning of Hephaestio's work in which it was outlined in detail about the decans, have not a little power whenever the planets, happening to be in them, control the matter that is current.

It seems best to me in the case of each house and <astrological> place that is aspected by a malefic to predict injury in accordance with the nature of the lord of the zodiacal sign in which the malefic is and <to predict> the things indicated from the <astrological> place.

If you wish to find the Lot of livelihood, take always, night and day, from the lord of the second <astrological> place to the second place itself, and count out an equal <number of degrees> from the ascendent; wherever the Lot falls, look at the lord of the Lot and the zodiacal sign itself into which the Lot fell. If the Lot and its lord happen to be in a good place without the aspect of the malefics, say that <his> livelihood is good; but if the lord of the Lot should be found eastern <and> setting, it will be after seven days, <but if> eastern <and> rising, it will produce a treasure of hidden and secret <things>.

It seems best to me that in the case of each house and <astrological> place that is harmed by Saturn <or> Mars injury <is predicted> to occur in accordance with the nature of the lord of the zodiacal sign where the malefic stands, and <to predict> the things indicated from the <astrological> place.

The Lot of livelihood according to Dorotheus: by both day and night from the lord of the second <astrological> place to the second <place> itself, and an equal <number of degrees> from the ascendent. If the lord of the Lot with the zodiacal sign is making <its> eastern rising, <the native> will have hidden treasure.

III. Rhetorius V 58, 38–44.

IIIa. Anonymus anni 379 as preserved in manuscript *E*, without correction (see *CCAG* V 1; 201–202).

ἐὰν δὲ ὡροσκοπῇ ἐπὶ γενέσεως ὁ ἐν τῷ ἡγουμένῳ ὤμῳ τοῦ Ὠρίωνος ὃς παρανατέλλη τῇ κζ′ μοίρᾳ τοῦ Ταύρου, ἢ ὁ Προκύων ὃς παρανατέλλει τῇ κζ′ μοίρᾳ τοῦ Καρκίνου, ἢ ὁ κοινὸς Ἵππου καὶ Ἀνδρομέδας ὃς παρανατέλλει τῇ κα′ μοίρᾳ τῶν Ἰχθύων, κράσεως ὄντες τῆς τοῦ Ἄρεως καὶ τῆς τοῦ Ἑρμοῦ, ποιοῦσι τοὺς οὕτως ἔχοντας μάλιστα ἐπὶ νυκτερινῶν γενέσεων στρατηγικούς, δεινούς, δράστας, πολυτρόπους, σοφιστικούς, πολυπράγμονας, λογίους, ἀπαραπεύστους, ὀξεῖς δὲ καὶ προσκορεῖς πρὸς τὰς ἐπιθυμίας, διαφορεῖς ὄντας παίδων καὶ παρθένων, ἐπὶ ὅρκους. ἐπὶ δὲ νυκτερινῆς γενέσεως ὡροσκοποῦντες ποιοῦσι τολμηρούς, ὠμούς, μεταμελητικούς, ψεύστας, κλέπτας, ἀθέους, ἀφίλους, ἐπιθέτας, θεατροκόπους, ἐφυβριστάς, μιαιφώνους, πλαστογράφους, γόητας, ἀνδροφόνους, οὐ καλῷ τέλει ἐνίοτε χρωμένους, μάλιστα, ὡς προείπομεν, ἐπὶ νυκτερινῆς γενέσεως.

IIIb. V 58, 38–44 as preserved in manuscript *R*, without emendation.

Πάλιν ἐν τῷ ἡγουμένῳ ἄμω τοῦ Ὠρίωνος, Ταύρου μοίρας κζ̄ μ′, νότιος, μεγέθους δευτέρου, κράσεως Ἄρεως καὶ Ἑρμοῦ.

Ὁ Προκύων, μοίρας τρεῖς δεῦ α′ Καρκίνου, νότιος, μεγέθους α′, κράσεως Ἄρεως καὶ Ἑρμοῦ.

Ὁ ἐν τῷ δεξιῷ ὤμῳ τοῦ Ὠρίωνος, Διδύμων μοίρας ε μ′, νότιος, μεγέθους β′, κράσεως Ἄρεως καὶ Ἑρμοῦ.

Ὁ κοινὸς Ἵππος καὶ Ἀνδρομέδα, Ἰχθύων μοίρας κᾱ λ′, βόρειος, μεγέθους β′, κράσεως Ἄρεως καὶ Ἑρμοῦ.

Ἵππου ὦμος, Ἰχθύων μοίρας ε ν′, βόρειος, μεγέθους β′, κράσεως Ἄρεως καὶ Ἑρμοῦ.

Οἱ τοιοῦτοι ἀστέρες ἰσομοίρως ὡροσκοποῦντες ἢ μεσουρανοῦντες, μάλιστα καὶ τῆς νυκτερινῆς γενέσεως, ποιοῦσι στρατηγούς, δεινούς, δράστας, πολυτρόπους, σοφιστικούς, πολυπράγμονας, λογίους, ἀπαρεγκλίτας, ὀξυφώνους, ἐξαπατητάς, ἐπιτευκτικούς, ὀξὺς δὲ ἅμα καὶ πρὸς κωρῆς πρὸς τὰς ἐπιθυμίας, διαφθήρωντας ἀθῴων καὶ παρθένων, ἐπιόρκους. ἐπὶ δὲ ἡμερινῆς γενέσεως ὡροσκοποῦντες ποιοῦσι τολμηρούς, ὠμούς, μεταμελητικούς, ψεύστας, κλέπτας, ἀθέους, ἀφύλλους, ἐπιθέτας, θεατροκόπους, ἐφυμβριστάς, μιαιφόνους, πλαστογράφους, ἀνδροφόνους, οὐ καλῷ τέλει ἐνίοτε χρωμένους, μάλιστα, ὡς προείπομεν, ἐπὶ ἡμερινῆς γενέσεως.

IIIc. Hermes, *De stellis beibeniis* VI (the Pahlavī translation of this treatise and the Arabic text of this portion of it are lost; therefore I substitute the Latin translation made by Salio of Padua in ca. 1220. I follow the edition by P. Kunitzsch, *Hermes Latinus* IV iv, Turnhout 2001, pp. 76–77).

Et est alia beibenia in 27 gradu 40 minutis Cancri, septentrionalis, exaltationis secunde, de natura Martis et Mercurii. et est alia beibenia in Cancro in 4 gradu et 4 minutis, septentrionalis, exaltationis prime, de natura Martis et Mercurii. et est alia beibenia in Pisce in

If the <star> in the advanced shoulder of Orion which rises simultaneously with the 27th degree of Taurus is in the ascendent in a nativity, or Procyon which rises simultaneously with the 27th degree of Cancer, or <the star> common to Pegasus and Andromeda which rises simultaneously with the 21st degree of Pisces, being of the mixture of Mars and Mercury, make those who are <born> thus, especially in the case of nocturnal nativities, fitted for command, dangerous, energetic, shifty, sophistical, busybodies, eloquent, not to be seduced (ἀπαραπείστους), sharp and sated in <their> desires, corruptors (διαφθήρωντας) of youths and maidens, perjurers (ἐπιόρκους). In the case of a nocturnal nativity they, rising, produce (those who are) bold, cruel, full of regrets, liars, thieves, atheists, friendless, imposters, courting applause, insolent, murderers, forgers, manslayers, sometimes not experiencing a good end, especially, as we have said, in the case of a nocturnal nativity.

Again <the star> on the advanced shoulder of Orion, at 27;40° of Taurus, southern, of second magnitude, of the mixture of Mars and Mercury.

Procyon, at 3;<10>° of Cancer, southern, of the first magnitude, of the mixture of Mars and Mercury.

The <star> on the right shoulder of Orion, at 5;40° of Gemini, southern, of the first magnitude, of the mixture of Mars and Mercury.

The <star> common <to> Pegasus and Andromeda, at 21;30° of Pisces, northern, of second magnitude, of the mixture of Mars and Mercury.

The shoulder of Pegasus, at 5;50° of Pisces, northern, of the second magnitude, of the mixture of Mars and Mercury.

These stars, when they rise or culminate precisely to the degree, especially <in the case of> a nocturnal nativity, produce generals, dangerous men, energetic, shifty, sophistical, busybodies, eloquent, inflexible, speaking sharply, deceivers, successful, but sharp at the same time and unsated (ἀπροσκορεῖς) in <their> desires, corrupting the harmless and virgins, perjurers. In the case of a diurnal nativity they, rising, produce <those who are> bold, cruel, full of regrets, liars, thieves, atheists, friendless (ἀφίλους), imposters, courting applause, insolent, murderers, forgers, manslayers, sometimes not experiencing a good end, especially, as we said before, in the case of a diurnal nativity.

There is another beibenia in 27 degree<s> 40 minutes of Cancer, northern, of the second rank, of the nature of Mars and Mercury. There is another beibenia in Cancer, in 4 degree<s> and 4 minutes, northern, of the first rank, of the nature of Mars and

21 gradu et 30 minutis, meridionalis, exaltationis secunde, de natura Martis et Mercurii. et est alia beibenia in Geminis in 5 gradu et 40 minutis, et est Manus Geminorum, septentrionalis, exaltationis prime, de natura Mercurii et Martis. et est alia beibenia in Virgine in 2 gradu et 21 minutis, meridionalis, exaltationis tertie, de natura Mercurii et Martis. et est alia beibenia in < ... > gradu Sagittarii, de natura Mercurii et Martis. et est alia beibenia in Pisce in 5 gradu et 50 minutis, septentrionalis, exaltationis tertie, de natura Martis et Mercurii.

Si ergo inveneris aliquam ex istis septem in ascendente vel in decimo in nativitate alicuius, et maxime si fuerit nativitas nocturna, natus erit magister militum, ponet se in negotiis, altus moribus, multi sensus, sapiens in omnibus et amat sapientiam, et famosus per multas terras, amator pecunie et habebit multam pecuniam, et diligit pueros contra naturam et puellas et est diversus nature, id est sodomita. sed cuius nativitas erit cum aliqua ex istis de die erit audax in omnibus negotiis, pauce pietatis, mendax, irosus, non diliget aliquem, deceptor, non est multum subdolus, loquitur de eo omnis male, occisor, falsus in suo testimonio, fictor nam aliud habet in corde et aliud loquitur ore. et deus est sapientior.

IIId. Māshā'allāh (lost in Arabic, but translated into Latin by Hugo of Santalla in the 1140's in his *Liber Aristotilis*, edited by C. Burnett and D. Pingree, London 1997, III ii 2, 31–33).

Fixarum item alia in xxvii gradu et xl punctis Tauri, in quarto item Cancri gradu et punctis iiii alia, in Geminis in v gradibus et punctis xl quedam, in Piscibus alia in xxi gradu et xxx punctis potestatis secunde; cetere sunt prime. omnes quidem australes, marcie et mercurialis sunt in complexionis statu. quarum quelibet in oriente vel medio celo sub nocturno potissimum natali exercitus propalabit ducem—magnanimus quidem erit, prudens, in omni discretus negocio, cum pueris et virginibus delectabitur et periurio gaudebit; diurno quidem existente natali, promptum, impium, mendacem, iracundum, amicorum expertem, subdolum, ociosum, derisorem, infamem, homicidam, incantatorem et omnia huiusmodi maleficia sectatur.

IIIe. Abū Maʿshar (787–886), *Kitāb aḥkām al-mawālīd* (*Book of Judgements of Nativities*) IX 1, VI, edited and translated by P. Kunitzsch, *Hermes Latinus* IV iv, Turnhout 2001, pp. 92–95.

wa unẓur ilā al-kawkab alladhī ʿalā mankib al-jawzā' fī NH daqīqa min al-jawzā' fī 'ṭ-ṭabaqat ath-thāniya min ash-sharaf mizāj al-mirrīkh wa ʿuṭārid, wa 'l-kawkab alladhī ʿalā yamīn al-minṭaqa fī 'ṭ-ṭabaqat al-ūlā min ash-sharaf mizāj al-mirrīkh wa ʿuṭārid, wa 'l-kawkab alladhī <fī> KD MH min al-ḥūt fī 'ṭ-ṭabaqat ath-thāniya min ash-sharaf mizāj al-mirrīkh wa ʿuṭārid, < ... >. fa-idhā wajadta aḥad hādhihī al-arbaʿ al-kawākib fī darajat aṭ-ṭāliʿ aw darajat wasat as-samā' wa lā siyyamā li-man yawladu bi-'l-layl fa-inna dhālika al-mawlūd yakūnu qā'idan al-juyūsh dakhālan fī 'l-umūr al-ʿiẓām ʿāqilan baṣīran bi jamīʿ al-ashyā' muḥibban li-'l-ʿilm mukhādiʿan muwāriban qalīl al-aṣdiqā' yuhza'u bi-'n-nās saffākan li-'d-dimā' ṣāḥib zūr jāmiʿan li-kull ḥubb.

Mercury. There is another beibenia in Pisces, in 21 degree<s> and 30 minutes, southern, of the second rank, of the nature of Mars and Mercury. There is another beibenia in Gemini, in 5 degree<s> and 40 minutes, and it is the hand of Gemini—northern, of the first rank, of the nature of Mercury and Mars. There is another beibenia in Virgo, in 2 degree<s> and 21 minutes, southern, of the third rank, of the nature of Mercury and Mars. There is another beibenia in the < ... > degree of Sagittarius, of the nature of Mercury and Mars. There is another beibenia in Pisces, in 5 degree<s> and 50 minutes, northern, of the third rank, of the nature of Mars and Mercury.

Therefore, if you find any of these seven in the ascendent or in the tenth <place> in the nativity of someone, and especially if the nativity is nocturnal, there will be born a leader of soldiers; he puts himself into enterprises; high in <his> manners, of much sense, wise in all <things>; and he loves wisdom; famous throughout many lands, a lover of money; and he will have a lot of money; and, contrary to nature, he loves boys as well as girls and is perverse in <his> nature, that is, a sodomite. But he whose nativity shall occur by day with any of these <stars> will be daring in all <his> enterprises, of little piety, a liar, angry; he doesn't love anyone; a deceiver, he is not very crafty; everyone speaks ill of him; a killer, false in his witness, a pretender (for he has one thing in his heart and speaks another with his mouth). And God knows better.

Also one of the fixed <stars> is in 27 degree<s> and 40 minutes of Taurus, also another in four degree<s> and 4 minutes of Cancer, a certain one in 5 degrees and 40 minutes in Gemini, another in Pisces in 21 degree<s> and 30 minutes, of the second power; the others are of the first. All are southern <and> in the situation of the mixture of Mars and Mercury. Anyone of these in the ascendent or midheaven, especially in a nocturnal nativity, will make manifest the leader of an army; he will be magnanimous, prudent, discrete in every enterprise; he will be delighted with boys and maidens and will rejoice in perjury. When the nativity is diurnal, <it will make him> ready, impious, a liar, irritable, lacking friends, crafty, lazy, a mocker, infamous, a murderer, an enchanter; and he pursues all evil deeds of this sort.

(this translation is based on Kunitzsch's)

Look at the star which is on the shoulder of Orion, in 58 minutes of Gemini, in the second class of distinction, a mixture of Mars and Mercury, and the star which is at the right of the belt, in the first class of distinction, a mixture of Mars and Mercury, and the star which is <at> 24;48° of Pisces, in the second class of distinction, a mixture of Mars and Mercury, < ... >. When you find one of these four stars in the degree of the ascendent or the degree of the midheaven, especially for one who is born at night, this native is a commander of troops, entering into grand projects, sensible, skillfull in all matters, a lover of science, a deceiver, a concealer, having few friends, mocking people, a shedder of blood, a master of falsehood, assembling all <sorts> of love.

IIIf. Excerpts in *Ber*.

Taurus
28. Πάλιν παρανατέλλει λαμπρὸς ἀστὴρ ὁ ἐν τῷ ἡγουμένῳ ὤμῳ τοῦ Ὠρίωνος, μοίρας κζ μ', νότιος, μεγέθους β', κράσεως Ἄρεως καὶ Ἑρμοῦ.
Gemini
42. Πάλιν παρανατέλλει λαμπρὸς ἀστὴρ ὁ ἐν τῷ δεξιῷ ὤμῳ τοῦ Ὠρίωνος, μοίρας ε, νότιος, μεγέθους α', κράσεως Ἄρεως καὶ Ἑρμοῦ.
Pisces
28. Παρανατέλλει δὲ τούτῳ λαμπρὸς ἀστὴρ ὁ ἐπὶ τοῦ Ἵππου, μοίρας ε ν', βόρειος, μεγέθους β', κράσεως Ἄρεως καὶ Ἑρμοῦ.
159. Πάλιν παρανατέλλει λαμπρὸς ἀστὴρ ὁ κοινὸς Ἵππου καὶ Ἀνδρομέδας, μοίρας κα, βόρειος, μεγέθους β', κράσεως Ἄρεως καὶ Ἑρμοῦ.

IIIg. Excerpts from Ep. V 25 (the *Liber Hermetis*)

Gemini
7. *In quinto gradu et minutis viginti sex oritur stella que est in dextro umero Orionis, nature Martis et Mercurii.*
Cancer
4. *In secundo gradu et minutis triginta sex oritur Anticanis cum radiis Canis erectus, nature Martis et Mercurii.*
Pisces
3. *In quinto gradu et minutis triginta sex oritur umerus Equi, nature Mercurii et Martis.*

IIIh. Ep. IV 2, 38–44 as preserved in manuscript *B*, without emendation.

κ'. Ὦμος ἡγούμενος Ὠρίων, Διδύμων α λ', νότιος ιζ λ', μεγέθους β', κράσεως Ἄρεως, Ἑρμοῦ, τῇ ς' μοίρᾳ Ταύρου.

κα'. Προκύων, Καρκίνου ζ ο' ιϛ λ', μεγέθους α', κράσεως Ἄρεως, Ἑρμοῦ, τῇ β' μοίρᾳ Καρκίνου.

κβ'. Ὦμος δεξιὸς Ὠρίων τῇ β' Διδύμων, Διδύμων θ λ', νότιος ιζ ο', μεγέθους α', κράσεως Ἄρεως, Ἑρμοῦ.

κγ'. ὁ καινὸς Ἵππου καὶ Ἀνδρομέδας τῇ κα' τῶν Ἰχθύων, Ἰχθύων κε κ', βόρειος κζ ο', μεγέθους β', κράσεως Ἄρεως, Ἑρμοῦ.

κδ'. Ὦμος Ἵππου, Ἰχθύων θ μ', βόρειος λα ο', μεγέθους β', κράσεως Ἄρεως, Ἑρμοῦ.

Καὶ οὗτοι πάλιν ἰσομοίρως ὡροσκοποῦντες ἢ μεσουρανοῦντες, μάλιστα νυκτὸς δέ, ποιοῦσι στρατηγούς, δεινούς, δράστας, πολυτρόπους, σοφιστικούς, πολυπράγμονας, λογίους, ὀξυφώνους, ἐξαπατητάς, ἐπιτευκτικούς, ὀξεῖς δ' ἅμα καὶ ἀπροσκορεῖς πρὸς τὰς ἐπιθυμίας, ἀσελγεῖς, ἐπιόρκους. εἰ δὲ ὡροσκοποῦσιν, τολμηρούς, ὠμούς, μεταμελητικούς, ψεύστας, κλέπτας, ἀθέους, ἀφίλους, ἐπιθέτας, θεατροκόπους, ἐφυβρίστους, μιαιφόνους, πλαστογράφους, γόητας, βιοθανάτους πολλάκις.

(Taurus). Again the bright star on the advanced shoulder of Orion rises simultaneously, at 27;40°, southern, of the second magnitude, of the mixture of Mars and Mercury.

(Gemini). Again the bright star on the right shoulder of Orion rises simultaneously, at 5°, southern, of the first magnitude, of the mixture of Mars and Mercury.

(Pisces). There rises simultaneously with this the bright star in Pegasus, at 5;50°, northern, of the second magnitude, of the mixture of Mars and Mercury.

Again the bright star common to Pegasus and Andromeda rises simultaneously, at 21°, northern, of the second magnitude, of the mixture of Mars and Mercury.

(Gemini). In the fifth degree and twenty-six minutes there rises the star which is on the right shoulder of Orion, of the nature of Mars and Mercury.

(Cancer). In the second degree and thirty-six minutes rises Anticanis erect in the rays of Canis, of the nature of Mars and Mercury.

(Pisces). In the fifth degree and thirty-six minutes rises the shoulder of Pegasus, of the nature of Mercury and Mars.

20. The advanced shoulder, Orion, Gemini 1;30, southern 17;30, of the second magnitude, of the mixture of Mars, Mercury, <rising simultaneously> with the 6th degree of Taurus.

21. Procyon, Cancer 7;0, <southern> 16;30, of the first magnitude, of the mixture of Mars, Mercury, <rising simultaneously> with the 2nd degree of Cancer.

22. Right shoulder, Orion, <rising simultaneously> with the 2nd degree of Gemini, Gemini 9;30, southern 17;0, of the first magnitude, of the mixture of Mars, Mercury.

23. The star common (κοινός) to Pegasus and Andromeda, <rising simultaneously> with the 21st <degree> of Pisces, Pisces 25;20, northern 26;0, of the second magnitude, of the mixture of Mars, Mercury.

24. Shoulder of Pegasus, Pisces 9;40, northern 31;0, of the second magnitude, of the mixture of Mars, Mercury.

These <stars> again which rise or culminate precisely to the degree, especially at night, produce generals, dangerous men, energetic, shifty, sophistical, busybodies, eloquent, speaking sharply, deceivers, successful, but sharp at the same time and unsated in <their> desires, lustful, perjurers. If they are rising, <they produce those who are> bold, cruel, full of regrets, liars, thieves, atheists, friendless, imposters, courting applause, insolent, murderers, forgers, magicians, frequently dying violently.

IIIi. Theophilus, <'Αποτελεσματικά> 17, edited by F. Cumont in *CCAG* V, 1, pp. 214–217; the passage cited is on p. 216.

Πάλιν ὁ ἐν τῷ ἡγουμένῳ ὤμῳ τοῦ Ὠρίωνος Διδύμων σ β', νότιος, μεγέθους β'· καὶ ὁ Προκύων Καρκίνου μοίρας ε μδ', μεγέθους β'· καὶ ὁ ἐπὶ τῷ δεξιῷ ὤμῳ τοῦ Ὠρίωνος Διδύμων μοίρας η κ', νότιος, μεγέθους α'· καὶ ὁ κοινὸς Ἵππου καὶ Ἀνδρομέδας Ἰχθύων μοίρας κδ, βόρειος, μεγέθους β'· καὶ Ἵππου ὦμος Ἰχθύων μοίρας η λ', βόρειος, μεγέθους β'. οὗτοι οἱ πέντε κράσεώς εἰσιν Ἄρεως καὶ Ἑρμοῦ.

Συγκεντρούμενοι οὖν καὶ συνανατέλλοντες τοῖς φωστῆρσι σημαίνουσιν ἐν ταῖς καταρχαῖς στρατηγέτας πολεμικούς, μηχανικούς, δολίους καὶ ὠμοὺς καὶ ψεύστας καὶ ἀναιρέτας καὶ ἀνδροφόνους καὶ οὐ καλῷ τέλει κεχρημένους.

IV. Julian of Laodicea on war.

IVa. VI 47, 1–2 as preserved in manuscripts *R, E,* and *V*.

Πάλιν οἱ μὲν ἐπιόντες ἀπὸ τοῦ ὡροσκόπου ληφθήσονται, οἷς δὲ ἐπίασιν ἀπὸ τοῦ δύνοντος,[1] τὸ αἴτιον ἀπὸ τοῦ μεσουρανήματος, ἡ ἔκβασις[2] ἀπὸ τοῦ ὑπογείου. τὰ φρούρια[3] δὲ ἑκάστου καὶ τὰς συμμαχίας οἱ[4] περιέχοντες τοὺς τόπους δηλώσουσιν· καὶ[5] ὑψούμενοι μὲν ἢ προστιθέντες πρόσθεσιν αὐτοῖς σημαίνουσιν, ταπεινούμενοι δὲ ἢ ἀφαιροῦντες καθαίρεσιν.

1. δυτικοῦ *E* 2. ἔκβασιν *R* 3. τοὺς φρουροὺς *E* 4. ε͟ἰ͟ *R* 5. καὶ om. *E*.

IVb. Theophilus, Πόνοι 36, 1–2 as preserved in manuscript *P*.

Οἱ μὲν ἐπιόντες ἀπὸ τοῦ ὡροσκόπου ληφθήσονται, οἷς δὲ ἐπίασιν ἀπὸ τοῦ δύνοντος, τὸ αἴτιον ἀπὸ τοῦ μεσουρανήματος, ἡ ἔκβασις ἀπὸ τοῦ ὑπογείου. τὰ ἀφρούρια ἑκάστου καὶ τὰς συμμαχίας οἱ παρέχοντες τόνδε τόπον δηλῶσιν· καὶ ὑψούμενοι μὲν ἢ προστιθέντες πρόσθεσιν αὐτῆς σημαίνει, ταπεινούμενοι δὲ ἢ ἀφαιροῦντες καθαίρεσιν.

IVc. Māshā'allāh 20 as preserved in Laleli 2122 bis.

Fī 'l-ḥurūb wa ghayrihā

qāla māshā'allāh: idhā su'ilta ᶜan ḥarb fa-kāna as-sā'il huwa al-mughnī bi-'l-amri yalzumuhu khayrihi wa sharrihi wa ṣalāḥihi wa fasādihi mithl al-malik wa 'l-qā'id. fa-unẓur li-hi min aṭ-ṭāliᶜ wa ṣāḥibihi wa li-ᶜadūwihi min an-naẓīr wa ṣāḥibihi. thumma ujᶜul al-munṣarif ᶜanhu al-qamar ᶜawnan li-'l-naẓīr wa ṣāḥibihi. thumma unẓur ilā al-kawākib ayyahumā ajwad mawḍiᶜan wa aqwā dalālatan wa akthar shihādatan fa-huwa al-ghālib. wa iᶜlam anna an-nujūm al-ᶜulyā aqwā min an-nujūm as-suflā.

Again <the star> on the advanced shoulder of Orion at Gemini 0;2°, southern, of the second magnitude; Procyon at Cancer 5;44°, of the second magnitude; and <the star> on the right shoulder of Orion at Gemini 8;20°, southern, of the first magnitude; and <the star> common to Pegasus and Andromeda at Pisces 24°, northern, of the second magnitude; and the shoulder of Pegasus at Pisces 8;30°, northern, of the second magnitude. These five are of the mixture of Mars and Mercury.

Therefore, being in a cardine and rising with the luminaries they indicate in initiatives warlike generals, resourceful, treacherous and cruel and liars and murderers and manslayers and not experiencing a good end.

Again those attacking will be found from the ascendent, but those whom they attack from the descendent, the cause from the midheaven, the outcome from the <cardine> beneath the earth. Those <planets> occupying <these astrological> places will indicate the forts and the allies of each; and <the planets> raised up <to their exaltations> or increasing <in their equations> indicate an increase for them, but those lowered <to their dejections> or diminishing <in their equations> a loss.

Those attacking will be found from the ascendent, but those whom they attack from the descendent, the cause from midheaven, the outcome from <the cardine> under the earth. Those <planets> occupying this <astrological> place indicate the ungarrisoned <places> and the allies of each; and <the planets> raised up <to their exaltations> or increasing <in their equations> indicate an increase of it (feminine), but those lowered <to their dejections> and diminishing <in their equations> a loss.

On Wars and the Like

Māshā'allāh says. If you are queried about a war, the querist who is capable in the matter—his good, his evil, his health, and his corruption cling to him—is like a king and a leader. Look for him from the ascendent and its lord, and for his enemy from the opposite <sign> and its lord. Then make <the planet> from which the Moon is receding a helper for the opposite <sign> and its lord. Then look to the two planets—whichever of them is better in its position and more powerful in its indication and more abundant in its aspects, it is the victor. Know that the superior planets are more powerful than the inferior planets.

IVd. Māshā'allāh 20 in a Byzantine translation as preserved in manuscript *V*.

Εἰ ἐρωτηθεὶς παρὰ στρατηγοῦ ἢ παρά τινος ἔχοντος φροντίδα τοῦ πολέμου, λάβε τὸν μὲν καταρχόμενον ἀπὸ τοῦ ὡροσκόπου (ἤγουν τὸν ἐρωτῶντα), ἀπὸ δὲ τοῦ ἑβδόμου τὸν ἐχθρόν. καὶ εἰ μὲν ἀπορέει τινὸς ἀστέρος ἡ Σελήνη, συνάπτει δὲ ἑτέρῳ ἀστέρι, λάβε τὸν μὲν ἀφ' οὗ ἀπέρρευσεν ἡ Σελήνη ἀστέρα εἰς τὰ τοῦ ὡροσκόπου, ᾧ δὲ συνάπτει εἰς τὰ τοῦ δύνοντος. Γνῶθι δὲ καὶ τοῦτο, ὡς οἱ ὑψηλοὶ ἀστέρες δυνατώτεροί εἰσι τῶν περιγείων.

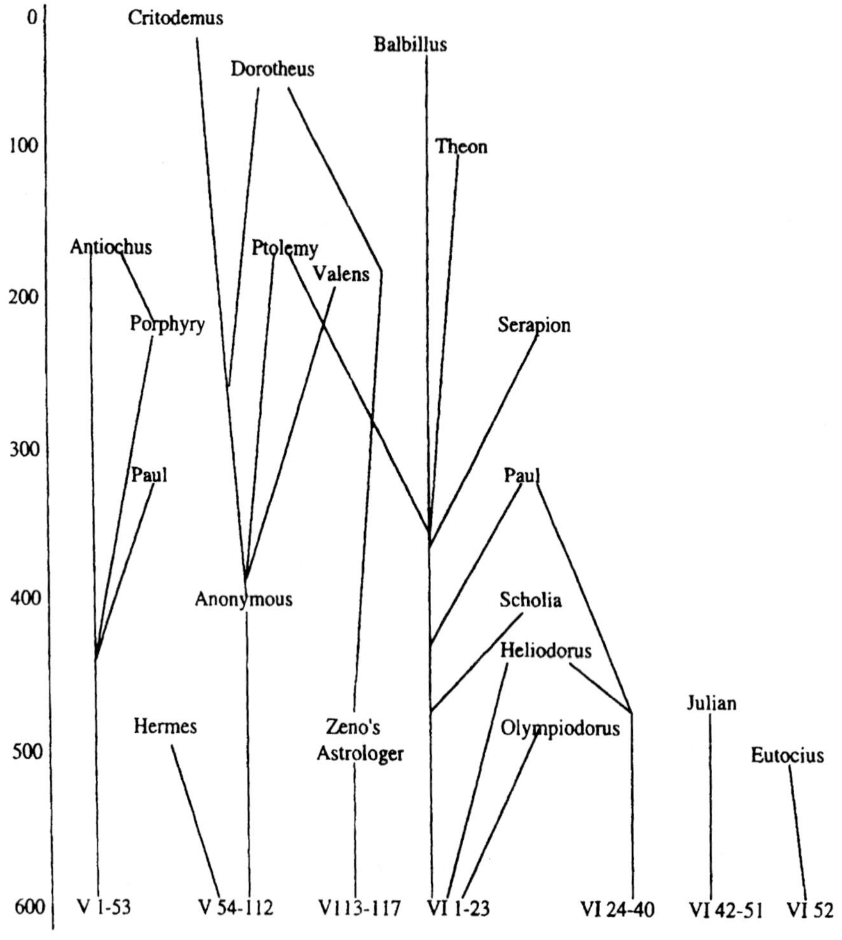

Schematic overview of Rhetorius' sources

If you are asked by a general or someone who is anxious about war, take the initiater (that is, the querist) from the ascendent, the enemy from the seventh <astrological place>. And if the Moon is receding from some planet and is conjoining with another planet, take the planet from which the Moon was receding for the things <indicated by> the descendent. And know this also, that the superior planets are more powerful than those near the earth.

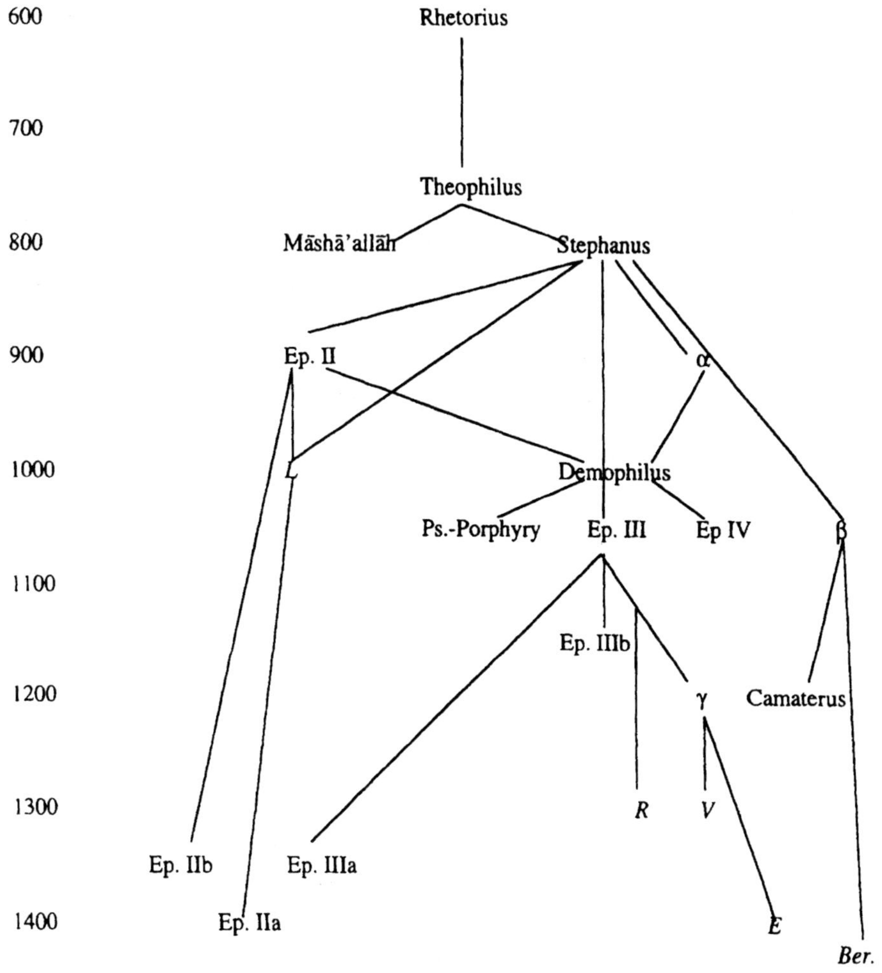

Schematic overview of the reception of Rhetorius

Zero and the Symbol for Zero in Early Sexagesimal and Decimal Place-Value Systems

DAVID PINGREE

In the early second millennium B.C. Mesopotamian mathematicians used a place-value system of notation with base sixty. Using two wedges impressed on clay — a vertial 𒁹 for one and a sideways 𒌋 for ten, they could express by the addition of symbols every number from one to nine and from ten (plus one to nine) to fifty (plus one to nine); after fifty-nine — 𒌋𒌋𒌋𒌋𒌋𒁹𒁹𒁹𒁹𒁹𒁹𒁹𒁹𒁹 — they returned again to 𒁹, meaning now 1 × 60. Since these numerical forms could be extended infinitely both as integers and as fractions, the symbol 𒁹 could mean, depending on context, 1, 1 × 60, 1 × 60^2, 1 × 60^3, . . . 1 × 60^n, or 1/60, 1/60^2, 1/60^3, . . . 1/60^n. In the earliest usage of this system, there was no expressed symbol corresponding to an empty sexagesimal place; instead, the Babylonians often avoided confusion by simply leaving an emtpy space between two symbols: e.g., 𒌋𒁹𒁹 would be read 12, but 𒌋 𒁹𒁹 10, 2 = 602. There was no way to write the equivalent of 10, 0, 2 = 36,002; such a number could only be recognized from context.

In the period between the late eighth and the late sixth centuries B.C., some scribes began to use a punctuation mark, 𒌋𒌋 or, 𒌋𒌋𒌋, that indicated a stop or separation as the separator of ten or any multiple of ten and of one or any other number up to nine when they were in two sexagesimal places; thus, 10, 2 would be written 𒌋𒌋𒌋 𒁹𒁹. But they also began to use this same symbol to change, for example, 𒁹𒁹𒌋 which was either 2, 10 = 130 or 2, 0, 10 = 7210, etc. into the unambiguous 𒁹𒁹 𒌋𒌋𒌋, 2, 0, 10. Here was first developed a concrete symbol indicating that there was no number occupying a place in a place-value system (see Neugebauer, 1934-1935, vol. 1, pp. 72-73 [no. 27], and Neugebauer-Sachs, 1945, pp. 34-35 [no. 33]).

This symbol was widely though not universally employed in Babylonian astronomical tables of the last three centuries B.C., in both meanings (see Neugebauer, 1941 and 1955), but the scribes often assumed that the arithmetical structure of the table would guide the reader to the correct interpretation of an ambiguous number. When it was necessary to name an empty sexagesimal place, the Akkadian words used were *nu tuk*, meaning "nothing", the exact equivalent of the Greek οὐδέν which translates it, and of the Latin *nihil*, but not of the Sanskrit *śūnya* and its Arabic derivative, *ṣifr*, both of which signify "empty".

Beginning in the second century B.C. Babylonian astronomical tables and parameters written in this fashion began to be translated into Greek. The Greeks used the letters of their alphabet to indicate numbers, with one letter corresponding to each number from one to nine, one letter for each number from ten in multiples of ten to ninety, and one letter for each number from one hundred in multiplies of a hundred to nine hundred, so that twenty-seven letters expressed all numbers up to 999, after which the sequence repeated itself. Thus, the three letters τ ξ ε (300 + 60 + 5) meant 365, for which the Babylonians wrote 𒐋 𒐊 (6, 5). The Greeks continued to use their traditional system to write integers in astronomical tables, but for the sexagesimal fractions they wrote the number in each sexagesimal place in their alphabetical notation; thus for $365 + 14/60 + 48/60^2$, which modern historians write in the form 365; 14, 48. Greek papyri of the period immediately preceding and following the beginning of the common era demonstrate that, by using vertical dividing lines such as are also used in Sanskrit manuscripts to separate the sexagesimal places, they removed all ambiguities: τξε|ιδ|μη But they did not leave an empty sexagesimal place blank; rather, they filled it with an adaptation of the Akkadian symbol for zero, 𒑱 ; this adaptation looks like a circle with a bar over it: ō or ō (see Jones, forthcoming). This, then, is the origin of the form of the symbol for zero as a circle. It is necessary to note that this form of zero was known, through a translation of Greek astronomical tables into Latin, in Western Europe before the introduction of the Arabic adaptation of the Indian numerals (see Pingree, forthcoming). The Romans, following the Greek practice, wrote the integers as decimal numbers, the fractions as sexagesimals; thus, ρνβ ō ι = CLIIO X = 152; 0, 10.

In India there is evidence in Buddhist and Jaina texts of uncertain date, but near the beginning of the common era, that a decimal place-value system was in use (see Gupta, 1995), but there is no certain evidence that a symbol for zero was in place before the fifth century A.D. However, it is likely that the notational system used in the tables versified in the *Pañcasiddhāntikā* of Varāhamihira — e.g., in the Sine table in IV 6-15, in which the Sines are expressed as units and (sexagesimal) minutes and seconds with zeros in empty places, or the tables of solar and lunar equations in VIII 3 and 6, from the *Romakasiddhānta*, using the same system — reflects the translation of these tables from Greek into Sanskrit in the third or fourth century A.D. (see Neugebauer-Pingree, 1970-1971). This hypothesis of a Greek origin is highly plausible not only because the contents of the tables and their contexts are Greek, but also because the system of notation used by Varāhamihira precisely corresponds to that found in the Greek papyri and medieval manuscripts. All later Sanskrit manuscripts of astronomical texts and tables also universally employ the same system. Since these tables in the *Pañcasiddhāntikā* were based on translations from the Greek, they in all likelihood used the Greek form of the symbol for zero in their sexagesimal fractions.

Exactly when this symbol began to be used in a decimal place-value system is difficult to know because of the lack of documents from the crucial period and the uncertainties regarding the dates and the interpretation of the texts that may be referring to such a symbol. The first indisputable reference, I would suggest, is in the *Paitāmahasiddhānta* of the *Viṣṇudharmottarapurāṇa* (see Dvivedin, 1912), where zero in decimal numbers is indicated not only by two words meaning "empty" — *śūnya* and *kha* (the empty sky) — but also by two words that imply rotundity — *pūrṇa*, "the full (moon)," and *puṣkara*, "a lotus". I have argued elsewhere that the date of the *Paitāmahasiddhānta* is about A.D. 425 (see Pingree-Morrissey, 1989, and Pingree, 1993).

The following facts concerning the steps in the development of a decimal place-value system in India can, I believe, be accepted by all.

1. The numbers in Sanskrit, as in other Indo-European languages, are on base ten.

2. A decimal place-value system in which counting high numbers was achieved by placing counters in a "units pit" indicating units, in a "tens pit" in which they indicated ten each, in a "hundreds pit" in which each indicated a hundred, etc., was being employed in India, especially among Jainas and Buddhists, towards the beginning of the common era (see Gupta, 1992 and 1995).

I would tentatively suggest some further steps.

3. The number of counters in each "pit" would naturally be indicated by a number between one and nine even though each counter in, say, the "hundreds pit" stood for a hundred. This would encourage the writing of those numbers with a sequence of numerals such as 365. However, some "pits" would be empty (*śūnya*). If the numbers were written down as words or as numerals, these empty "pits" would be called by a word meaning "empty", such as *śūnya*, which is itself the symbol for zero.

4. At the beginning of the common era Greeks appeared in India in large numbers, bringing with them, among other things, astronomical tables in which, in the sexagesimal fractions, the "empty" places were occupied by a circular symbol for zero such as are found in the Greek papyri of that period.

5. In translating these tables into Sanskrit, the symbol was taken over as a circle (*pūrṇa*) or a dot (*bindu*).

6. At some time in the period before A.D. 400, someone writing numbers in the decimal place-value system with numerals, used the sexagesimal symbol for zero instead of writing *śūnya*.

At this point India had two place-value systems using the same numerals and the same symbol for zero, while in the West there was only one place-value system, the sexagesimal and it was employed only to represent fractions in astronomical texts and tables. Moreover, the numbers other than zero were written by the Greeks with twenty-seven alphabetical symbols and by the Romans with seven symbols that they came to regard as letters (I V X L C D M). The Arabs in the ninth century learned the Indian numerals with zero and used them in mathematics, but they continued to use their abjad (alphabetical) symbols and sexagesimal fractions with zero in astronomical texts and tables. Similarly when Western Europeans learned the Indo-Arabic numerals with zero in the twelfth century, they replaced the Roman numerals with them in all contexts, but they continued to use sexagesimal fractions in astronomical texts and tables. What they forgot was that those astronomical fractions had been for centuries written with zeros.

REFERENCES

Dvivedin, 1912. V. Dvivedin, *Jyautiṣasidhāntasaṅgraha*, 2 fascs., Benares Sanskrit Series, **39**, Benares 1912, fasc. 2, part 1.

Gupta, 1992, R.C. Gupta, "The First Unenumerable Number in Jaina Mathematics", *Gaṇita Bhāratī*, **14**, 1992, 11-24.

Gupta, 1995. R.C. Gupta, "Who Invented the Zero", *Ganita Bhāratī*, **17**, 1995, 45-61.

Jones, forthcoming. Alexander Jones, *Astronomical Papyri from Oxyrhynchus.* Neugebauer, 1934-1935.

O. Neugebauer, *Mathematische Keilschrift-Texte*, 2 vols., Berlin: Springer, 1934-35.

Neugebauer, 1941. O. Neugebauer, "On a Special Use of the Sign 'Zero' in Cuneiform Astronomical Texts", *Journal of the American Oriental Society*, **61**, 1941, 213-215.

Neugebauer, 1955. O. Neugebauer, *Astronomical Cuneiform Texts*, 3 vols., London: Lund Humphries, 1955.

Neugebauer-Pingree, 1970-71. O. Neugebauer and David Pingree, *The Pañcasiddhāntikā of Varāhamihira*, 2 Vols., Kobenhavn: Det Kongelige Danske Videnskabernes Selskab, 1970-71.

Neugebauer-Sachs, 1945. O. Neugebauer and A. Sachs, *Mathematical Cuneiform Texts*, New Haven: American Oriental Society, 1945.

Pingree, 1993. David Pingree, "Āryabhaṭa, the Paitāmahasiddhānta and Greek Astronomy", *Studies in History of Medicine and Science*, NS **12**, 1993, 69-79.

Pingree-Morrissey, 1989. D. Pingree and P. Morrissey, "On the Identification of the *Yogatārās* of the Indian *Nakṣatras*", *Journal for the History of Astronomy*, **20**, 1989, 99-119.

Pingree, D., *The Preceptum Canonis Ptolomei*, forthcoming.

Mesopotamia

OBSERVATIONAL TEXTS CONCERNING THE PLANET MERCURY

by David Pingree and Erica Reiner

Our forthcoming edition of the planetary omens of the series Enūma Anu Enlil will include Tablet 63, the so-called Venus Tablet. This tablet derives predictions from the dates of the disappearances, durations of invisibility, and dates of the first visibilities of Venus. Among the fragments preliminarily assigned to this tablet on the basis of the terminology used, there was one which contained a period of invisibility that Asgar Aaboe, to whom it was first shown, recognized to be too short for the planet Venus, but possible for Mercury. Further search among similar material identified two more fragments, and A. Sachs drew our attention to BM 37467. It became clear not only that these tablets do not pertain to Venus, but also that they contain no omens. Therefore, and also for their intrinsic interest for the history of astronomy, we deemed it more appropriate to publish them separately.

The four fragments are BM 37467 and K.6153, in Babylonian script, and Rm.2,303 and Rm.2,361, both in Assyrian script, and parts of different tablets. The last three belong to the Kuyunjik collection of the British Museum; all four are here published with the permission of its Trustees.

The texts record in each section (in months and days) a date of the first visibility, the duration of visibility, and a date of disappearance of a planet. In some cases the subsequent duration of invisibility is also recorded. Since the first visibilities occur alternately in the East and West, the planet must be either Venus or Mercury. And, since the attested periods of visibility and invisibility range from 22 days to 45 days, the planet must be Mercury.

In Tablet 56 of EAE there is a section parallel to these texts (TCL 6 16:35-43 and dupls., see provisionally ZA 52 252:96-104). Therein the protasis of the omen (lines 96-98) involves the first visibility of Saturn, its period of visibility (one year), and its disappearance, and predicts a period of invisibility of twenty days, if the planet is red. The next section (lines 99-104) gives a similar omen involving Mercury. Our texts are probably to be connected with this type of astral omen.

In addition to the dates of first visibility, the texts edited here give the time-interval between first visibility and sunrise in the East, and between sunset and first visibility in the West. This is expressed in the hitherto unattested phrases—time being expressed in UŠ, *i. e.*, thirtieths of a double hour *(bēru)*—x UŠ GÙB ^{d}UTU GAR-*ma*, "the sun being x UŠ to the left", for the first case, and KI.TA(-*nu*) ^{d}UTU x UŠ GAR-*ma*, "the sun being x UŠ below", for the second. Presumably these directions are given with respect to Mercury, though it is not impossible that they are given with respect to the horizon. Although the exact Akkadian reading of these phrases is uncertain—note that "left" occurs between the number of UŠ's and "the sun", while "below" precedes both "the sun" and the number of UŠ's—this interpretation imposes itself, and must be posited also for the slightly divergent variant x UŠ *ina* GÙB ^{d}UTU [GAR-*ma*] of Rm.2,361 rev. 2.

Since three of these tablets come from Kuyunjik, even though no dated colophon is preserved, these observational texts are older than the seventh century B.C. Though they show that astronomers of this early period were interested in the phenomena of Mercury that became the structural basis of the mathematical theory of that planet in the Seleucid period, the data in the texts could not be used to construct such a mathematical model for Mercury's motion because, at least in the preserved portions, they are not precisely dated with respect to any epoch-date. Unfortunately the fragmentary state in which they are preserved makes it impossible for us to date them astronomically. The observational texts concerning Mercury in *LBAT* (nos. 1368-1385) differ from those here edited in giving the precise date which is needed for mathematical astronomy.

TEXTS

1. Rm.2,361. Assyrian script. Possibly a two-column tablet; only the right-hand columns are preserved. Obverse and reverse have been assigned according to content, though the side here taken as reverse is also flat.

§ I obv. ii? 1′ [x ūmī ina AN-e uḫ-ḫ]a-r[am-ma]

§ II 2′ [DIŠ ina ITI MN ^{d}GUD.UD UD.X.KAM ina ^{d}UTU.ŠÚ.A
KI.T]A-nu ^{d}UTU 10 UŠ [GAR-ma]

3′ [IGI-ir (. . .)]
22 u_4-mi ina ^{d}[UTU.ŠÚ.A DU-ma]

4′ [UD.X.KAM ša ITI MN] ina ^{d}UTU.ŠÚ.A TÙM-ma

rev. i? 1 [x ūmī ina AN]-[e] uḫ-ḫa-[ram-ma]

§ III 2 [DIŠ ina ITI MN ^{d}GUD.UD UD.X.KAM ina ^{d}UTU].È 5
UŠ ina GÙB ^{d}UTU [GAR-ma]

3 [IGI-ir (...) IT]I UD.8.KAM ina ^{d}UTU.È [DU-ma]
4 [UD.X.KAM ša ITI MN] ina ^{d}UTU.È [TÙM-ma]
5 [x ūmī ina AN-e] uḫ-ḫa-r[am-ma]
§ IV 6 [DIŠ ina ITI MN ^{d}GUD.UD UD.X.KAM ina ^{d}UTU.ŠÚ.A
KI.TA]-nu ^{d}UTU $\frac{1}{2}$ KAS.GÍD [GAR-ma]
7 [IGI-ir (...) x ūmī ina ^{d}UTU.ŠÚ].[A] DU-ma [TÙM$^{?}$]
8 [8 ... ^{d}GU]D$^{?}$.UD šá MU.AN.NA
9 traces
break

Translation

§ I Visibility in the East; broken, but traces suggest similar structure as in the other paragraphs.

§ II [In MN on the nth day Mercury becomes visible in the West], the sun [being] 10 UŠ [bel]ow, [and remains] in the [West] for 22 days, and then disappears in the West [in MN on the nth day]; it remains invisible [for n days].

§ III [In MN on the nth day Mercury becomes visible] in the East, the sun [being] 5 UŠ to the left, [and remains] in the East for one month and 8 days, [and then disappears] in the East [in MN on the nth day]; it remains invisible [for n days].

§ IV [In MN on the nth day Mercury becomes visible in the West], the sun [being] one-half *bēru* (*i. e.*, 15 UŠ) [bel]ow, and remains in the West [for n days].

Subscript: [8 sections... of] Mercury for the year.

Notes

Note that the phrase indicating the duration of invisibility, *i. e.*, the phrase "it remains invisible for n days", is set off from the observation record by a dividing-line on the tablet. That this phrase belongs with the preceding is shown by the connective particle *-ma* at the end of the preceding line which is preserved in § II (obv. 4′) and has accordingly been restored in the corresponding line rev. 4 (§ III).

The phenomena of section IV were not completed at the end of the year, and therefore that sentence differs from the preceding. As each pair of complete sections constitutes a synodic period of Mercury, which would average about 116 days, there must be 6 complete sections in a year plus a fraction. Our § IV represents part of that fraction; unless Mercury was first visible in the West at the beginning of the year, the first paragraph for the year would be the end of a section on a visibility in the East followed by a period of invisibility.

2. K.6153. Babylonian script. Part of obverse and traces of two lines on the reverse preserved.

§ I 1′ [DIŠ ina ITI MN ^{d}GUD.UD UD.X].KAM ina $^{⌈d⌉}$[UTU.È X UŠ GÙB
^{d}UTU GAR-ma IGI-ir]
2′ [x ūmī ina ^{d}UTU.È DU-ma UD.X.KAM ša] ITI ŠU ina ^{d}UTU.⌈È⌉
[itbal x ūmī ina šamê uḫḫaramma]
§ II 3′ [DIŠ ina ITI MN ^{d}GUD.UD UD.X].KAM ina ^{d}UTU.ŠÚ.A KI.TA
[^{d}UTU X UŠ GAR-ma IGI-ir]

Revue d'Assyriologie, LXIXe 23

4′ [x ūmī ina] ^{d}UTU.ŠÚ.A DU-ma UD.15.[KAM ša ITI MN ina ^{d}UTU.ŠÚ.A itbal x ūmī ina šamê uḫ̮ḫaramma]

§ III 5′ [DIŠ ina ITI MN ^{d}GUD.UD UD].26.KAM ina ^{d}UTU.È 5 u[š GÙB ^{d}UTU GAR-ma IGI-ir]

6′ [IGI.6.GÁL KAS.G]ÍD ana KUR ^{d}UTU KUR-ḫ̮a 1 UŠ (read ITI) ina ^{d}[UTU.È DU-ma UD.X.KAM]

7′ [ša ITI MN ina ^{d}]UTU.È it-bal 1 ITI UD [ina šamê uḫ̮ḫaramma]

§ IV 8′ [DIŠ ina ITI MN ^{d}GUD.UD] UD.13.KAM ina ^{d}UTU.ŠÚ.A K[I.TA ^{d}UTU x UŠ GAR-ma IGI-ir]

9′ [. . .] EN.NUN ^{d}U[SAN . . .]

10′ [ ] ⌈it?⌉-[bal . . .]

break

Translation

§ I [In MN on the n]th [day Mercury becomes visible] in [the East, the sun being *n* UŠ to the left, and remains in the East for n days, and then disappears] in the East [on the nth day] of month IV; [it remains invisible for n days].

§ II [In MN on the n]th [day Mercury becomes visible] in the West, [the sun being *n* UŠ] below, and remains in the West [for n days, and then disappears in the West] on the 15th [of MN; it remains invisible for n days].

§ III [In MN on the] 26th day [Mercury becomes visible] in the East, [the sun being] 5 UŠ [to the left], and it rises [one-sixth of] a *bēru* (*i. e.*, 5 UŠ) before (lit. toward) sunrise; 1 UŠ (read month) in the [East it remains], and then disappears in the East [in MN on the nth day; it remains invisible] for one month [(and n days?)].

§ IV [In MN] on the 13th day [Mercury becomes visible] in the West, [the sun being *n* UŠ belo]w, [. . .] the evening watch [. . .] disappears [. . .].

Notes

§§ III and IV contain supplementary records of the time of first visibility not preserved elsewhere in this tablet or in the other texts of the group, with the possible exception of Text 3 § III. There is, therefore, some uncertainty in our restorations of these sections.

3. Rm.2,303. Neo-Assyrian script. Two-column tablet, only one side preserved.

§ I col. i 1′ traces

2′ [x ūmī ina ^{d}UTU.ŠÚ.A DU-ma UD.X.KAM ša ITI MN ina] ^{d}UTU.ŠÚ.A it-ba[l]

3′ [x ūmī ina šamê u]ḫ̮-ḫ̮a-ra-am-ma

§ II 4′ [DIŠ ina ITI MN ^{d}GUD.UD UD.X.KAM ina ^{d}UTU].⌈È⌉ 10 UŠ GÙB ^{d}UTU GAR-ma

5′ [IGI-ir x ūmī ina ^{d}UTU.È DU]-ma UD.26.KAM ša ITI.ŠU

6′ [ina ^{d}UTU.È itbal] 1 ITI 15 u_4-mi ina AN-e ZAL-ma

§ III 7′ [DIŠ ina ITI MN ^{d}GUD.UD UD.X.KAM ina ^{d}UTU.ŠÚ].A KI.TA ^{d}UTU 10 UŠ GAR-ma IGI-ir

	8′	[...] x ŠÚ 23
		u_4-mi ina ^{d}UTU.ŠÚ.A DU-ma
	9′	[UD.X.KAM ša ITI MN ina ^{d}UTU.ŠÚ].⌈A⌉ it-bal 22 u_4-mi
		ina AN-e ZAL-ma
§ IV	10′	[DIŠ ina ITI MN ^{d}GUD.UD UD.X.KAM ina ^{d}UTU.È][1] 7 UŠ
	11′	[GÙB ^{d}UTU GAR-ma IGI]-⌈ir?⌉
	break	
§ V col. ii	1′	x [
	2′	DIŠ [
	3′	B[E-ma
§ VI	4′	DIŠ ina ITI [
	5′	^{d}U[TU
	6′	e-[
	7′	BE-ma [
§ VII	8′	DIŠ ina I[TI
	9′	BE-⌈ma⌉ [
	break	

Translation

§ I [In MN on the nth day Mercury becomes visible in the West, the sun being *n* UŠ below, and remains in the West for n days, and then] disappears [in] the West [in MN on the nth day]; it remains invisible [for n days].

§ II [In MN on the nth day Mercury becomes visible in the E]ast, the sun being 10 UŠ to the left, [and remains in the East for n days], and then [disappears in the East] on the 26th of month IV; it remains invisible for one month and 15 days.

§ III [In MN on the nth day Mercury] becomes visible [in the W]est, the sun being 10 UŠ below, [...] ..., [and] remains in the West for 23 days, and then disappears [in the W]est [in MN on the nth day]; it remains invisible for 22 days.

§ IV [In MN on the nth day Mercury becomes visible in the East, the sun being] 7 UŠ [to the left]--

Notes

Although the two preserved sections comprise three lines each, the division of the content among the three lines differs from § II to § III; § IV may have comprised four lines. Therefore it is impossible to restore the missing half line in § III. The fully preserved sign is ŠÚ, and the preceding partly preserved sign (the top half of a vertical wedge) is possibly also ŠÚ or the remains of the sign UTU. A further problem is posed by the line beginnings of column ii; again, the sections seem to be of unequal length (the fully preserved middle section has 4 lines, the others are only partly preserved). The last line of §§ V and VI, and the last preserved, possibly last, line of § VII begin with BE-*ma, i. e., šumma* "if". It is of course uncertain whether column ii still deals with Mercury.

1. Traces are compatible with È.

4. BM 37467. Neo-Babylonian, transliterated by A. Sachs, collated.

§ I	1′	[ ]	⌈DU⌉-ma [ ]
§ II	2′	[ ]	37 u_4-mu ina AN-⌈e⌉ [ ]
	3′	[ ]	12 u_4-mu ina AN-e ZAL-[ma]
§ III	4′	[ ]	IGI ana IGI BAR SAL+[ ]
	5′	[ ]	x x [ ]
	break		

Notes

§ II. The figure of 37 days refers to the duration of visibility of Mercury, followed by a statement concerning a period of invisibility of 12 (possibly 22) days. The preceding § I may have contained the record of visibility in the West; the last preserved lines do not fit readily into any of the observed patterns of this text group.

A Neo-Babylonian Report on Seasonal Hours

By David Pingree (Providence) and Erica Reiner (Chicago)

K. 2077 + 3771 is described in Bezold's *Catalogue* as «part of an astrological text» and was consequently studied for possible inclusion in Erica Reiner's forthcoming edition of *Enūma Anu Enlil*. Bezold had already drawn attention to the «squares» with figures, eight on the reverse, but only two preserved on the obverse, and these «squares with figures» pointed to a text more astronomical than astrological in character.

David Pingree, when Erica Reiner showed him her transliteration of the text, recognized that the text deals with the lengths of seasonal hours, and that the «squares» represent a table of the lengths of seasonal hours for every fifteenth day of an «ideal» solar year of 360 days. The missing portions of the table could therefore be reconstructed. When subsequently the fragment K. 11044 was joined to K. 2077+, the newly gained portions confirmed Pingree's reconstruction not only of the table, but also of its description in the first four lines of the reverse. The difficulties presented by the remaining ten lines of the reverse, however, remain substantial.

The text is here presented in transliteration and translation. The astronomical commentary is by David Pingree. Moreover, the presentation and the philological commentary have greatly benefited from discussions with David Pingree.

This edition is dedicated to the memory of Professor Ernst Weidner, whose pioneer work on the series *Enūma Anu Enlil* had aided and inspired us in our undertaking.

K 2077+3771+11044
Obv., Rev., Lower Edge

K. 2077 + 3771 + 11044

obv. 1. [ ] MUL.AL.LUL DU-*az* *ù* ^{d}*Sin*
2. [ ] x MUL.SUḪUR.MÁŠ.ḪA DU-*ma* 8 KAS.BU UD-*mu* 4 KAS.BU MI

---

3. [ ] ⸢2/3⸣ KAS.BU *qaq-qa-ru* 10 UŠ UD-*mu ik-te-ri*
4. [DU]-*ak šá-lul-tú* UD.15.KAM 18 UŠ 20 GAR DU-*ak*
5. [*a-šar*]-*ri šá* 3 UD.15.KAM.MEŠ DIŠ 2/3 KAS.BU ⸢*šá* x x⸣ ⸢*a*⸣-*lak šá* ^{d}UTU
6. [ ] x^{a}-*ú* 11-*ú* 12-*ú* 10+[UŠ] LÁ-*ṭi*

---

7. [ ] GUR-*ár*
8. [ ] x-*da-ma*
9. [ ] ITI.APIN
10. [ ] x x

break of 5 lines

16. [	12 UŠ 30 GAR]	[10 UŠ]	[12 UŠ 30 GAR]	[15 UŠ]	[17 UŠ 30 GAR]
17. [	11 UŠ 40 GAR]	⸢10 UŠ 50 GAR⸣	⸢13 UŠ 20⸣ [GAR]	⸢15 UŠ⸣ [50 GAR]	18 UŠ [20 GAR]
18. [	10 UŠ] ⸢50 GAR⸣	11 UŠ 40 GAR	14 UŠ 10 GAR	16 UŠ 40 GAR	19 UŠ [10 GAR]
19. [	3 UD.15].KAM.ME	3 UD.15.KAM.ME	3 UD.15.KAM.[M]E	3 UD.15.KAM.ME	3 U[D.15.KAM.ME]
20. [	*šá* 5 KAS U]D-*mu*	*šá* 4 KAS UD-*mu*	*šá* 5 KAS UD-*m*[*u*]	*šá* 6 KAS UD-*mu*	*šá* 7 KAS UD-*mu*

lower edge (in Assyrian script): 1. [ ] *tal-lak-tú šá* ^{d}UTU [x x] *di-ib-bu*^{b}
2. [ ]-⸢^{d}⸣*Gu-la* LÚ.A.⸢ZU⸣
3. [*li*]*m-mu* ^{I} ^{d}EN-KUR-*u-a*

4*

rev. i only GAR preserved at ends of lines (for lower edge, see p. 51).

rev. ii 1. *an-nu-ú tal-lak-tú šá* ^{d}UTU TA KASKAL^{II} *šu-ut* ^{d}*En-lí*[*l*]
2. EN KASKAL^{II} *šu-ut* ^{d}*É-a* TA KASKAL^{II} *šu-ut* ^{d}*É-a*
3. EN KASKAL^{II} *šu-ut* ^{d}*En-lil* TA ^{d}UTU.È EN ^{d}UTU.ŠÚ.A
4. TA ^{d}UTU.ŠÚ.A EN ^{d}UTU.È 12 KAS.BU *qaq-qar mi-šiḫ-ti a-šar-ri*
5. *ki-ṣip-ta-šú šá-lim-ti áš-ṭur qaq-qar ul ma-al-la a-ḫa-meš šú-ú*
6. *ut-ru ù muṭ-ṭi-e li-ik-ṣi-pu-ma liq-bu-nim-ma*
7. *ina pi-i lu-še-eš-mi* LUGAL *i-di ki-i dib-bi an-nu-tim*
8. *ina ṭup-pi la šaṭ-ru ù ina pi-i* UN.ME *la ma-šu-*⌜*ú*⌝
9. *ina ṭup-pi* LÚ.ŠAMAN.LÁ *ul i-šem-mi-i ú*-x-[
10. LÚ.SAG.LUGAL *Ḫat-tu-ú lu-kal-li-mu* x [
11. *mi-šiḫ-ti* KI.MEŠ *ù bi-rit* x [
12. *an-na-a-ti ina pi-i lu-šá-*[
13. *tal-lak-ti* ^{d}*Sin* ^{d}UTU ^{d}UDU.IDIM.MEŠ [x x]
14. *di-ri u na-dan* GISKIM *ina* ŠÀ-*bi in-nam-*[(*ma*)]-*ru*

15. 2/3 KAS.BU	15 UŠ	10 UŠ	15 UŠ
16. 19 UŠ 10 GAR	14 UŠ 10 GAR	10 UŠ 50 GAR	15 UŠ 50 GAR
17. 18 UŠ 20 GAR	13 UŠ 20 GAR	11 UŠ 40 GAR	16 UŠ 40 GAR
18. 17 UŠ 30 GAR	12 UŠ 30 GAR	[12] UŠ 30 GAR	17 UŠ 30 GAR
19. 16 UŠ 40 GAR	11 UŠ 40 GAR	[13 U]Š 20 GAR	18 UŠ 20 GAR
20. 15 UŠ +50[c] GAR	10 UŠ 50 [GAR]	[14 U]Š 10 GAR	19 UŠ 10 GAR

[a] Traces are not those of the numeral 10.

[b] If the word is *dibbū* (and not, for instance, *dibdibbu*, «clepsydra»), it may appear here in the technical sense used in reports, see Oppenheim, *Centaurus* 14 128 note 9.

[c] Text has 20, not 50.

Translation

obv. 1. [... the Sun] stands in Cancer and the Moon
2. [...] stands in Capricorn, then the day is 8 *bēru* (long), the night 4 *bēru*.

3. [... from?] 2/3 *bēr qaqqar* the day shortened to 10 uš,
4. [...] goes, the third fifteen-day (period) goes 18 uš 20 gar
5. [... *asar*]*ri*? per? three fifteen-day (periods) to? 2/3 *bēru* of .... the travel of the Sun
6. [...] ...th, eleventh, twelfth, [...] decreases [by x uš]

7. [...] returns
8. [...] ....
9. [...] Arahsamna
10–16 broken
16–20: see Astronomical Commentary

rev. ii 1. This is the course of the Sun from the Path of Enlil
2. to the Path of Ea, from the Path of Ea
3. to the Path of Enlil. From sunrise to sunset,
4. from sunset to sunrise 12 *bēr qaqqar* is the measurement of the *asarru*.
5. I wrote down its complete computation. The *qaqqar* is not equal.
6. Let them compute the excess and the deficiencies and let them tell me,
7. then I will announce? it. The king should know that these matters
8. are not written in (any) tablet, but? they are not forgotten in the mouth of people;
9. the apprentice scribe cannot hear it from a tablet (or: it is not read from the tablet of the apprentice scribe), ...
10. I? will show it to? the Hittite *ša rēš šarri* official [...]
11. the measurements of the *qaqqaru*-s and the interval of [...]

12. these I will [...]
13. The course of the Sun, the Moon, the planets [...]
14. intercalations and giving of signs will be found in it.

15—20: see Astronomical Commentary.
Subscript and colophon: 1. [...] the course of the Sun [...] statements?
2. [...]-Gula, the scribe
3. [...] eponymy of Bēl-šadūa

While the astronomical interpretation of the tables of the text, and their usefulness in attaining one of the stated purposes, the determination of intercalations, is clear (see Astronomical Commentary), the text contains a number of difficulties, and its format is without parallel.

The tablet is written in Neo-Babylonian script and dialect, except for the subscript and colophon on the lower edge, which are written in Assyrian script [1]. According to the colophon, it was written in the eponymy of Bēl-šadūa; if this eponym is identical to Bēl-Ḫarrān-šadūa, the date is 649 B. C. [2].

The tablet has the oblong shape associated with the reports of scholars to Esarhaddon and Assurbanipal [3]; almost the entire left half is missing, as can be calculated from the disposition of the table on the bottom of the obverse. The reverse is divided by a vertical ruling into two columns. The right half is almost complete, but of the left, only a few ends of lines, seemingly all ending in g a r, are visible; it seems likely that these lines contained tables or computations.

On the obverse, the lines run across the entire width of the tablet, and can be only partially restored. The first two lines state the length of daylight on the longest day of the year, and the next four lines seem to give the explanation of the table (see Astronomical Commentary). From lines 7 to 17 the surface of the tablet is worn off, and therefore the introduction to the table in obv. [16]—20 is not preserved.

The almost completely preserved right column of the reverse contains, as do the reports, a communication to the king. It states, first, that the text deals with the (yearly) course of the Sun from the Path of Enlil to the Path of Ea and back, and that from sunrise to sunset and from sunset to sunrise the day (a nychthemeron) contains 12 *bēru* (lines 1—4). The writer goes on to say that he has made certain computations, and gives instructions for further computations to be made, presumably in view of finding intercalations (lines 5—6). There follow claims about the novelty of the subject matter or procedure: the writer informs the king that these matters had not been written on a tablet but had been preserved in the oral tradition (lines 7—8). The remainder (lines 8—12) is unclear in its details. The writer ends with the statement that in the text may be found not only intercalations, but also «giving of signs», and refers not only to the course of the Sun but also to those of the Moon and the planets, which do not otherwise appear in the preserved part of the tablet.

The interpretation of this «report» is hampered by the occurrence of unique terms. A suggestion for the meaning of *bēr qaqqar* (line 4, beside *qaqqar* lines 5 and 10) is made in the Astronomical Commentary, p. 55. A hit herto unattested word, *asarri* or *ašarri*, appears in rev. 4, and probably in obv. 5; it may designate the «seasonal hour» or the nychthemeron. The term *kiṣiptu*, here translated «computation», in rev. 5 is likewise unattested so far, but the verb *keṣēpu* (*kaṣāpu*) from which it is derived seems to mean «to count, to compute» (see the dictionaries).

Another obscure reference is that to the LÚ. SAG.LUGAL *ḫattû*, the «Hittite» *ša rēš šarri*-official. The term «Hittite» in this period should refer to Syria, but we know of no other mention of a Syrian *ša rēš šarri* [4]. If this reference is to the oral transmission of the new computation, it may indicate, as suggested in the Astronomical Commentary, a derivation from Egypt via Syria.

---

1) The only other Babylonian tablet known to me that has a subscript added in Assyrian script is K. 159, published by Klauber, *Politisch-Religiöse Texte* ... no. 105.

2) The writer is unlikely to be Urad-Gula, a LÚ.MAŠ.MAŠ, who is the author of the letters LAS 223—25 (S. Parpola, *Letters from Assyrian Scholars* ...), because this scholar writes in Assyrian, not in Babylonian. Possible restorations are Arad-Gula (mentioned in the Neo-Babylonian report Thompson, *The Reports* ... 90 r. 6 (= Harper, *Assyrian and Babylonian Letters* ... 1109) in broken context, or Rīmūt-Gula (see Hunger, *Babyl. und assyr. Kolophone* no. 140).

3) On these reports see A. L. Oppenheim, *Centaurus* 14 (1969) 97—135, especially pp. 97f.

4) For *ša rēši* and *ša rēš šarri* in Neo-Babylonian, see Oppenheim, *The Gaster Festschrift* (*The Journal of the Anc. Near Eastern Soc. of Columbia Univ.*, vol. 5, New York, 1973), pp. 329ff.

Astronomical Commentary
By David Pingree

The table that is given twice on this tablet records the length of seasonal hours for every fifteenth day of an «ideal» solar year, where a seasonal hour is the twelfth of the length of daylight. These lengths of seasonal hours are measured in uš and gar for which it holds that

60 gar = 1 uš = 4 minutes
30 uš = 1 *bēru* = 120 minutes.

The entries form a linear zigzag function — that is, a continuous function $y(x)$ which increases and decreases periodically with constant slope between two limits m and M and which can be represented by a «zigzag» graph[1]. The period of this function is a year of 360 days in a schematic calendar, and the values $y(x)$ are given for 24 equidistant points — every fifteenth day — on the axis x. The maximum, M, is equal to 2/3 *bēru* or 20 uš at entry (1) (the Summer Solstice); the minimum, m, is equal to 10 uš at entry (13) (the Winter Solstice); and d, the constant difference, is equal to 50 gar or 0;50 uš. If these entries are multiplied by 12 they result in the parameters listed in obverse 1–3: M = 2/3 *bēru* · 12 = 8 *bēru* (the length of daylight on the longest day); m = 10 uš · 12 = 120 uš = 4 *bēru* (the length of daylight on the shortest day); and d = 0; 50 uš · 12 = 10 uš (the amount by which the daylight is said to decrease in obverse 3, *scilicet* in every 15 days).

The mean value, μ, of this zigzag function ($\mu = 1/2\ (M+m)$), equal to 15 uš, occurs at entries (7) (Fall Equinox) and (19) (Vernal Equinox); again, 15 uš · 12 = 180 uš = 6 *bēru*, so that both the day and the night contain 6 *bēru* each.

The table is divided in both its occurrences (on the obverse and the reverse) into eight sections (Bezold's «squares») of which each contains three 15-day periods or 1+1/2 30-day «months». The layout of these two occurrences differs, however.

On the reverse the complete table is preserved; its eight sections form four vertical columns and are to be read, as any cuneiform tablet divided into columns, first, down, and then continuing with the column adjacent to the right. In the middle, between columns two and three, i. e., between sections 4 and 5, there is a vertical double dividing line.

On the obverse, each section by itself occupies a column; there is again a double dividing line running down the middle, between columns 4 and 5. The eight sections or columns filled the entire width of the tablet, and in its present fragmentary state only the four sections (5–8) to the right of the double dividing line are completely preserved; to the left of the double dividing line, part of the fourth section remains. This table on the obverse also contains, in addition to the eight times three entries that are found in the table on the reverse, two more lines for each section, which give the length of daylight at the beginning of each 45-day period, in the formulation «three 15-days periods of n *bēru* daylight». In the preserved right half of the table n increases from the minimum, 4 *bēru* (at entry 13, the Winter Solstice) to 7 *bēru* (at entry 22, 45 days before Summer Solstice, or 45 days after the Vernal Equinox); in the now lost left half of the table n decreases from the maximum, 8 *bēru* (at entry 1, the Summer Solstice), to 5 *bēru* (at entry 10, 45 days before the Winter Solstice, or 45 days after the Fall Equinox). The summary entries in obverse 20 thus form a linear zigzag function wherein M = 8 *bēru*, m = 4 *bēru*, and d = 1 *bēru*.

This simple structure, then, is based on the well-known Babylonian parameter for the ratio of the longest to the shortest day in the year, 2 : 1, attested in Tablet XIV of *Enūma Anu Enlil* and elsewhere[2]. The division of the length of daylight into twelve equal parts or seasonal hours is much less well attested.

This division is first attested in Egypt, in a text from the cenotaph of Seti I (1303–1290 B. C.)[3]; in Mesopotamia, the text here published provides the first undisputable evidence. The only other occurrence hitherto recognized — but not uniformly interpreted — is found on faces C and D of an ivory prism in the British Museum, BM 123340[4]. These faces, when

[1]) See O. Neugebauer, *Astronomical Cuneiform Texts* (London, 1955), vol. 1 pp. 28ff.

[2]) O. Neugebauer, *Isis* 37 (1947), 37–43.

[3]) See O. Neugebauer and R. A. Parker, *Egyptian Astronomical Texts*, vol. 1 (Providence 1960), pp. 116–121.

[4]) S. Langdon, *Babylonian Menologies* ... (London, 1935), pp. 55–64 (copy on p. 55); S. Smith, *Iraq* 31 (1969), 74–81.

complete, gave the lapsed time since sunrise for the ends of the twelve seasonal hours of a day or a night in six pairs of months. The new text confirms this interpretation of the ivory prism, and attests to the fact that a division of the day and the night each into twelve hours was known in Mesopotamia in the seventh century B. C. The rather obscure reference to the «Hittite official» (rev. 10) raises the possibility that this division was derived from Egypt via Syria [5].

The purpose of the table is stated in rev. 14 to include the finding of intercalations [6]. This permits one interpretation of the procedure described in rev. 5—7. Here *qaqqar* can refer either to the separate entries in the table or to the length of daylight given in *bēru*-s in obv. 20. The inequality of the *qaqqar* (note the use of the singular in rev. 5) is connected by the text with solar motion in obv. 1—2 and rev. 1—3. If the length of daylight is observed on a particular day, that can be translated into an approximate solar longitude — or rather, a particular date in the Babylonian calendar; for obv. 1—2 seems to inform us that the first entry, with the maximum length of the seasonal hour, occurs on the fifteenth day of Tammuz — literally, at the opposition of the Sun and the Moon (full moon), the Sun being in Cancer. The computation of the excess or deficiency of the observed length of daylight with respect to the tabular length of daylight for one's calendar-date will immediately indicate when intercalations are necessary; a difference of 20 uš or 2/3 *bēru* in the length of daylight denotes a lag of 30 days of the civil calendar behind a luni-solar calendar.

One further innovation in this text is its use of the term *bēr qaqqar* to refer, not to a linear measurement, but to a period of time; this meaning is clear in obv. 3, when compared with rev. 4 and 5. *Bēru* is always used in its normal chronological sense — a twelfth of a nychthemeron (see obv. 2, 5, and 20), so that rev. 1—4 cannot be giving a linear measurement (a 12 *bēr qaqqar*) for the distance travelled by the Sun each day and night along the ecliptic from the Path of Enlil to the Path of Ea and back, or above and below the horizon, but rather simply states the fact that, wherever the Sun is on the ecliptic, each nychthemeron contains 12 *bēru*.

[5]) For the seasonal hours of the Egyptians and Greeks see also R. Borger, JEOL VI/18 (1964) 329f.

[6]) Beside intercalations (*diri*) the text mentions «giving of signs»; this may refer to expository material, see CAD 7 (I) 305b sub *ittu* mng. 1c—1'.

Venus Phenomena In *Enūma Anu Enlil*

David Pingree – Providence

Out of the vast number of cuneiform tablets identified during the last century as containing planetary omens, and therefore as connected with the series *Enūma Anu Enlil*, and specifically with its fourth and last section, commonly called Ištar, the largest group is that devoted to Venus. This fact immediately differentiates the importance of this planet in the omen tradition from the place that it holds in the mathematical astronomy of the Seleucid period, where Venus is rather poorly represented. However, two tablets – an atypical text from the second half of the fifth century B.C.[1] and an ACT procedure text[2] – both from Babylon, indicate that rather complex theories of Venus involving subdivisions of the arcs and times between its characteristic phenomena were already current in the early Achaemenid period and continued in use in the Seleucid period. These are quite sufficient to show that Venus was not neglected by those who were constructing the mathematical theories of the planets, even though its orbit as seen from the earth did not fit the models that were developed for the superior planets.

But, while Venus' phenomena were not amenable to description by step-functions or zig-zag functions, these same phenomena provided the earliest mathematical theory of any planet that we have; this is found in the δ section of Tablet 63 of *Enūma Anu Enlil*, the so-called Venus Tablet of Ammiṣaduqa, a crude theory by which approximate dates for the successive occurrences of four phenomena – first visibilities and invisibilities in the East and in the West – could be anticipated.[3] In general, however, the scribes of *Enūma Anu Enlil* were not interested in

---

1 BM 36301 published by O. Neugebauer and A. Sachs, „Some Atypical Astronomical Cuneiform Texts. I", *JCS* 21 (1967), 183-218, esp. 194-198.

2 BM 34221+ in O. Neugebauer, *ACT*, London 1955, vol. 2, pp. 396-403.

3 E. Reiner and D. Pingree, *The Venus Tablet of Ammiṣaduqa*, BPO 1, Malibu 1975, p. 24.

these Greek-letter phenomena except as parts of more complex omens – and even then, at least in my understanding of what they wrote, never directly mentioned the other two phenomena that are included in the later mathematical theory, first and second station, nor even the arc of retrogression that separates them. But in the planetary omens it is really only Mars' rather extreme and Jupiter's retrograde arcs that drew attention, as each entered a constellation and then backed out of it before plunging into it again. The exceptions in *Enūma Anu Enlil* outside of Tablet 63 to this general neglect of the Greek-letter phenomena are the rather banal protases: „Venus in the winter rises in the East, in the summer in the West“ and its opposite;[4] then, „Venus in the winter rises in the East and does not set“ and its permutations;[5] and „Venus stands in the West and sets – on the 7th day it rises in the East“, and the impossible transposition of this correct period of invisibility at inferior conjunction to the period of Venus' invisibility at superior conjunction.[6]

Given, then, that the Venus phenomena considered ominous in *Enūma Anu Enlil* – and in the Reports and Letters sent by the diviners to Esarhaddon and Assurbanipal[7] – were *not* in general the Greek-letter phenomena, what were they? This is a question Erica Reiner and I have been wrestling with for some time as we have sifted through, ordered, and attempted to impose some meaning on the numerous fragmentary texts that will be included in the next fascicle of *Babylonian Planetary Omens*.[8] This paper is an attempt to summarize our tentative conclusions.

As far as can be determined, either five or six Tablets of some canonical *Enūma Anu Enlil* were devoted to the enumeration of omens in which the main element of the protasis was the planet Venus; these were numbered either 58 or 59 to 63. The last, the Venus tablet of Ammiṣaduqa, states in its colophon that it is the 63rd Tablet, and then gives the catchline of a Tablet devoted to Jupiter. On the other side of the Venus block Tablet 56

[4] VAT 10218, omens 94-95 with parallels.

[5] VAT 10218, omens 96-99 with parallels.

[6] VAT 10218, omen 104 with parallels.

[7] The Letters have been published by S. Parpola, *Letters from Assyrian Scholars to the Kings Esarhaddon and Assurbanipal*, AOAT 5, 2 vols., Neukirchen-Vluyn, 1970-1983. A new edition of the Reports by H. Hunger has appeared as SAA 8, Helsinki 1992.

[8] This fascicule – *BPO* 3 – should appear shortly.

concerns various planets, and Tablet 57 seems to have obscure contents. We know nothing of the omens contained in Tablet 58. But Tablets 59 and 60 record Venus omens arranged by months, the first six on Tablet 59 and the last six on Tablet 60. The beginning of the sequence of omens was missing already in antiquity since various derivative texts knew nothing of the omens from the beginning of month 1; needless to say the extant copies of Tablet 59 are also broken at the beginning, but this is an accidental coincidence. The frame-omens of each month are very simple affairs drawn from the Venus omens included in Labat's edition of *Iqqur īpuš*,[9] while many of the interior omens are compositions of several omens found elsewhere in *Enūma Anu Enlil*. To these circumstances we can add the facts that the omens used to construct the composite omens in Tablets 59 and 60 are drawn primarily from what we regard as the older stratum of Venus omens, and that copies of Tablets 59 and 60 were deposited in Assurbanipal's library, so that these two Tablets antedate the seventh century. From these data it is plausible to conclude that an original collection of Venus omens arranged by months and closely connected to Venus omens included in Labat's edition of *Iqqur īpuš* existed towards the end of the second millennium B.C., and that the present Tablets 59 and 60 of *Enūma Anu Enlil* were compiled towards the beginning of the first millennium B.C.

While Tablet 58 may have contained Venus omens, our remaining material, after we have assigned the appropriate fragments to Tablets 59, 60, and 63, falls naturally into two piles; the first, which I will call Group A, contains the most archaic omens, while in the second, which I will call Group F, are found several innovations indicative of a date not too long before -700. It is not impossible, then, that the „original" forms – of course we only have broken and contaminated copies – of Groups A and F were Tablets 61 and 62; indeed, the colophon of one manuscript of Group F, K. 148, states that it contains a commentary on Tablet 61.

I must now say a few words about the six Groups into which we have divided the Venus omens other than those found in Tablet 63. There are eleven representatives of material belonging to Group A, headed by two commented texts (commentaries were useful because the surface meaning of many Group A protases are difficult to relate to reality), VAT 10218

[9] R. Labat, *Un calendrier babylonien des travaux des signes et des mois*, Paris 1965, par. 82-86 on pp. 164-171.

10218 and K. 148, which together provide most of the basis for our understanding of the Venus phenomena. The protases of Group A have the following characteristics, which I believe point clearly to their antiquity:

1. They are generally simpler phenomena than many in Tablets 59 and 60 (our Group C); and, in fact, as I hope to demonstrate elsewhere, several omens in Group C appear to have been composed by combining omens found in part in Group A.

2. Their wording is frequently very obscure, causing difficulties not only to us, but to the ancient commentators as well. The difficulty in understanding what phenomena were being referred to experienced by the commentators of the Neo-Assyrian period points to a break in the tradition of using Venus omens, which are attested for the Old Babylonian period by Tablet 63. Perhaps this break occurred during the Kassite period. If this line of reasoning is correct, at least a number of Group A omens may be presumed to have been first formulated in the Old Babylonian period.

3. The protases of the Group A omens do not refer to the factors introduced into the protases of the Group F omens – primarily Venus' *ašar niṣirti* in Pisces, which the Greeks later called its ὕψωμα and which was certainly known in Mesopotamia at the beginning of Esarhaddon's reign;[10] and the paths of Enlil, Anu, and Ea, which were already referred to in texts from the second millennium.[11]

4. None of the monthly Venus omens that are included in Labat's edition of *Iqqur īpuš* and that characterize Groups C, D, and E are found in the texts belonging to Group A.

Another observation important to be made with respect to the Group A omens is that the majority of Venus omens in the Reports and Letters to Esarhaddon and Assurbanipal come from Group A, as do those that can be reasonably identified with Akkadian omens among the Venus omens

[10] See H. Hunger and D. Pingree, *MUL.APIN*, *AfO Beiheft* 24, Horn 1989, pp. 146-147.

[11] See E. Reiner and D. Pingree, *Enūma Anu Enlil, Tablets 50-51 BPO* 2, Malibu 1981, pp. 3-8.

in the Sanskrit *Gargasaṃhitā*, which were transmitted to India in the fifth century B.C.[12]

Group B consists of the remains of thirteen excerpt tablets compiled from Group A, but frequently in rearrangements and with the insertion of omens from extraneous sources. These insertions, however, never include references either to Venus' *ašar niṣirti* or to the paths of Enlil , Anu, and Ea, both of which characterize the tablets assigned to Group F.

Group C consists of Tablets 59 and 60, concerning which I have already said enough.

Group D is formed by four tablets which contain monthly Venus omens in paragraphs 83, 86, 84, and 85 of Labat's edition of *Iqqur īpuš*. Group D is linked to Group C and *Enūma Anu Enlil* only by the fact that the appropriate omens from its first two sections, Labat's paragraphs 83 and 86, are the frame-omens for each of Group C's monthly sections. However, the significance of this apparent common ordering of these omens in Groups C and D is clouded by the fact that one cannot tell from Labat's edition either what the order of the paragraphs in the section of *Iqqur īpuš* devoted to Venus ought to be, or even whether that section is legitimately a part of *Iqqur īpuš* or not.

Group E consists of ten tablets which contain conflations of omens drawn from Groups A and C, among which have been interspersed omens from other sources. Obviously, its usefulness lies mainly in supplying lost omens and restoring broken ones in its two principle sources.

Finally, Group F, represented by twenty-one texts, combines some older omens derived from Group A with new ones such as those involving Venus' *ašar niṣirti* and the paths of Enlil, Anu, and Ea.

Now that the material and our organization of it have been described, we can turn to the phenomena themselves. Those that I believe that I understand fall into three categories: Venus' motion with respect to other planets and to the constellations; Venus' motion with respect to the

12 D. Pingree, „Venus Omens in India and Babylon", in *Language, Literature, and History*, ed. F. Rochberg-Halton, *AOS* 67, New Haven 1987, pp. 293-315.

horizon; and various light phenomena. A fourth category embraces those phenomena which I certainly do not understand.

A number of protases from Group A omens involve Venus' appearing with the Moon.[13] Clearly, these phenomena can only occur a few nights before the Moon's conjunction with the Sun, when the Moon and Venus may be together above the eastern horizon before Sunrise, and a few nights after conjunction, just after Sunset in the West. In both of these situations the Moon is crescent-shaped, so that it is not strange that many of the conjunctions of Venus with the Moon involve the latter's horns. The phenomena described include Venus' being above, below, before, or behind the Moon, or entering into either its right or its left horn, or standing between them. As in many other cases, the same phenomenon may be described in several different ways in different protases; often these different protases are all followed by the same or similar apodoses. This situation presumably reflects the fact that the sources used by the compiler of Group A included representatives of different local traditions, that also may have evolved over different time-spans. Without further documentation it will be difficult if not utterly impossible to work out a detailed history of the spread and transformation of these omens.

At least one Venus-Moon omen in Group A appears to be astronomically impossible. In its protasis Venus is said to enter into the Moon and to stay there beyond one watch.[14] It should be possible in winter for Venus at its greatest elongation from the Sun to remain above the western horizon for more than a third of a night; but the Moon in that time will certainly have passed beyond the planet long before it sets. Perhaps the ever useful clouds need to be invoked to provide cover both for Venus and for the Babylonian observors. Another omen in a Group F text is more obviously impossible; it requires Venus to enter into the Moon on the 15th day of the month.[15]

Among Group A protases three refer to Venus in conjunction with the Sun.[16] They are: „Venus reaches the Sun and enters the Sun"; „Venus enters the Sun and comes out"; and „Venus in the morning stands in front

[13] VAT 10218, omens 3 and 25-47 with parallels.

[14] K. 3111 + 10672, omen 13.

[15] K. 3601 rev. 34.

[16] VAT 10218, omens 48-50 with parallels.

of the Sun". For one planet to be „in front of" another or a star means that it rises first of the two; therefore, „in front of" near the eastern horizon signifies „above", near the western horizon „below". This third omen, then, with Venus being seen above the Sun in the morning, would make sense if the Sun's light were sufficiently obscured that both it and the planet were visible. The first two Venus-Sun omens, however, refer to phenomena that occur at every conjunction; however, unless the text is referring to Venus transits of the Sun, they are not directly observed, but must be inferred from observations over several nights or, in the case of superior conjunctions, several months. Indeed, it is characteristic of many of the planetary omens in *Enūma Anu Enlil* that they refer not to single events, but to processes that occur over substantial periods of time.

The planet with which Venus is most commonly linked in *Enūma Anu Enlil* is Jupiter.[17] Normally, the phenomena are simply that they appear close to each other or that they pass one another; more rarely Venus „enters into" or occults Jupiter. A more complex phenomenon involves Venus and Jupiter being on opposite sides of the Moon. The only other planet named in a protasis is Mars, whose occultation by Venus is referred to in a Group A text.[18] It is not at all clear to me why so much attention was paid in this material to Jupiter to the utter neglect of all the other planets with the exception of Mars mentioned in just one omen.

However, if one of the ancient commentators was correct, they are present, though disguised under a metaphor. This is the AGA or „tiara" which Venus is said to wear – a configuration that the commentators generally take to mean that something is „in front of" Venus, that is, rises before it and, in the morning, is above it. One set of three omens in Group A reads: „Venus wears the tiara of the Sun"; „Venus wears the tiara of the Moon"; and „Venus wears two tiaras". The commentator takes them to mean that, respectively, Saturn, Mercury, and two planets stand in front of Venus.[19] Whether or not this interpretation is correct with regard to the original meanings of these omens, it is of interest to note that in the Indian tradition also Saturn and Mercury are associated with the Sun and the Moon respectively; for they are regarded as the sons of the two luminaries. In another place also the same commentator interprets

[17] VAT 10218, omens 24 and 51-59 with parallels.

[18] VAT 10218, omen 63 with parallels; and Sm. 1354, omen 4.

[19] K. 148, omens 15-17.

interprets protases in which the tiara of Venus is black, white, green, or red to mean that in front of Venus stand Sirius (or Saturn?), Jupiter, Mars, or Mercury.[20] From more traditional color symbolism one would expect black to represent Saturn; and green and red to represent Mercury and Mars rather than the other way around. But at least Venus' wearing of a tiara was often understood to mean that another planet stands in front of her.

In the texts of Group A only six constellations are associated with Venus – MUL.MUL (the Pleiades), SI.PA.ZI.AN.NA (Orion), ŠUDUN (Boötes, but understood to mean Jupiter), MAR.GÍD DA (Ursa Maior), AŠ.GAN (Pegasus), and the Star of Eridu (Puppis).[21] It is, of course, remarkable that most of these constellations are too distant from the ecliptic for Venus ever to enter into them; rather, the omen is usually that it is in front of them – that is, rises before they do or, in the case of MAR.GÍD.DA, is higher in altitude in the morning than the constellation is. In the texts of Group F, on the other hand, the names of eight more constellations occur, and most of these are close enough to the ecliptic that the protases can correctly indicate that Venus enters into them. These eight constellations are: AB.SIN (Virgo), APIN (Triangulum), KU_6 (Piscis Austrinus), MAŠ.TAB.BA (Gemini), UR.GU.LA (Leo), *Enmešarra* (the northern part of Taurus), GU_4.AN.NA (Taurus itself), and GÍR.TAB (Scorpio).[22] Clearly there has been a shift in interest from more or less simultaneous risings – this, the earlier interest, is also reflected, I suspect, in the TE-omens, perhaps originating in the Old Babylonian period, in which two constellations or a planet and a constellation TE or approach each other – a shift, in any case, from an older interest to a later interest in Venus' passing by stars in the path of the Moon – an interest ending eventually in some of the entries in the *Astronomical Diaries*.

Another innovation relative to Venus' motion with respect to planets and constellations that was introduced in Group F involves its entry into its *ašar niṣirti* – perhaps meaning „secret place" – near the end of Pisces.[23]

---

[20] K. 148, omens 10-13.

[21] VAT 10218, omens 6-7; 60-62; and 126-129; and Sm. 1354, omen 3, all with parallels.

[22] K. 7936 : 11, rev. 8, rev. 13-17 and rev. 26-29 with parallels.

[23] D.T. 47 : 28-32 with parallels.

The omens in which Venus' *ašar niṣirti* occurs include: „Venus becomes visible in the North and reaches its *ašar niṣirti* and disappears". The commentator interprets this to mean: „it becomes visible in the path of Enlil or in the path of Anu and completes nine months and goes lower".[24] I take the text and its commentary to mean that, with the path of Enlil being that part of the eastern horizon that is North of about 20° North of the East-point,[25] Venus becomes visible in the morning when the Sun is in Gemini or Cancer, that is, in June or July in the Gregorian calendar roughly; and nine months later, in February or March, it first reaches its *ašar niṣirti*, and then becomes invisible. This interpretation is perhaps supported by another commentary in which the protasis' statement: „Venus reaches its *ašar niṣirti*" is glossed by the statement: „it reaches the Lion";[26] I presume that this means that Venus' first visibility in the evening occurs in Leo, in Gregorian August, its last visibility in the evening at the end of Pisces or the beginning of Aries, in late March or April.

The other type of motion associated with Venus is with respect to the horizon, and this is in two modes: a motion to the North and the South along the eastern and western horizons of the points at which it appears and disappears (this motion must be observed over a number of mornings or evenings), and its motion in altitude, that is, how high it reaches above the eastern horizon before Sunrise and how high it is above the western horizon at Sunset (this motion can be observed either in a single morning or evening, or over a sequence in order to determine a trend). Clearly

[24] K. 2346 : 23 with parallels.

[25] See J. Koch, *Neue Untersuchungen zur Topographie des babylonischen Fixsternhimmels*, Wiesbaden 1989, p. 125. For his azimuth I would substitute the equivalent idea of the rising amplitude, η, known along with polar coordinates and other elements from early Indian astronomy. These Indian concepts are modifications of Hellenistic ideas that in turn originated in Babylonian astronomy. The rising-amplitude is found with the formula:

$$\mathrm{Sin}\,\eta = \mathrm{R}x\frac{\mathrm{Sin}\,\delta}{\mathrm{Sin}\,\bar{\phi}}$$

The number connected with the boundary between the paths of Anu and those of Enlil and of Ea in Group F texts is 2/3 *bēru* or 20° of rising-amplitude; see, e.g., K. 7936 : rev. 8 and D.T. 47 : 21-27. Then, with R = 1 and $\bar{\phi}$ = 90°-36° = 54° , δ is about 16°.

[26] D.T. 47 : 27.

or evening, or over a sequence in order to determine a trend). Clearly both of these motions are influenced by the angle at which the ecliptic crosses the local horizon and the latitude of the planet, while the primary cause of the first motion is the apparent motion of the Sun in the ecliptic, and the primary cause of the second is the motion of Venus in its proper orbit. The combination of these four factors will lead to a great variety of observed phenomena, which are reflected in our omens.

Erica Reiner and I had already concluded in fascicle 2 of *Babylonian Planetary Omens* that the word KI.GUB, „place" or „position", is the technical term for the position of a star or a planet when it is first visible on a particular night; at Sunset it might be anywhere in the visible sky, but during the night the stars and planets that appear have their KI.GUB's on the eastern horizon.[27] In these Venus omens the KI.GUB also means, in addition to the two meanings given above, the points along the western horizon at which it sets. Let me quote some Group A protases that support this interpretation:[28]

„Venus changes its KI.GUB for nine months in the East, for nine months in the West". Here, of course, the nine months constitute Venus' period of visibility according to *Enūma Anu Enlil*, a period during which, if a morning star, its KI.GUB or rising point moves North and South along the eastern horizon, but, if an evening star, its KI.GUB or setting point moves similarly along the western horizon. Another protasis is: „Venus stands in the KI.GUB of the Moon".[29] I take this to mean that Venus rises in the morning or sets in the evening over the same point on the horizon that the waning or crescent Moon had previously crossed.

The protases of four successive omens in Group A have Venus stand in its KI.GUB to the North, to the South, to the East, and to the West. To stand precisely to the East and to the West should mean to rise above the East-point or to set below the West-point on the horizon – in other words, as other omens put it, to split heaven in half.[30] But for Venus' KI.GUB literally to be to the North or to the South would be astronomically impossible; rather, we must interpret these omens in the same way as we

---

27 *BPO* 2, pp. 17-18.

28 The KI.GUB omens are found in VAT 10218, omens 109 -117.

29 E.g., VAT 10218, omen 112 with parallels.

30 E.g., VAT 10218, omen 12 with parallels.

interpret such passages as that in MUL.APIN (II i 9-24): „On the 15th of Du'uzu ... the Sun which rose towards the North with the head of the Lion turns and keeps moving down (*ultanappal*) towards the South ... On the 15th of Tešrītu the Sun rises in the Scales in the East ... On the 15th of Ṭebetu .... the Sun which rose towards the South with the head of Aquarius (reading GU.LA for the text's erroneous UR.GU.LA), turns and keeps coming up (*undanaḫar*) towards the North ... On the 15th of Nisannu ... in the evening ... the Sun stands in the West in front of the Stars (i.e., of the Pleiades)". This passage describes the changing of the KI.GUB of the Sun, and thereby also of Venus, and informs us of the sense in which it appears to the North, to the South, to the East, and to the West. One ancient commentator , indeed, interprets the phrase „changes its KI.GUB" to mean: „it goes higher, variant: it goes lower".[31] I take „goes higher" (NUM) and „goes lower" (*uttaḫḫas*) in this context to be parallel in meaning to MUL.APIN's „keeps coming up to the North" and „keeps moving down towards the South"; I shall return to this later.

One final pair of omens that place the KI.GUB on the horizon state that: „Venus' KI.GUB is red" and „Venus KI.GUB is green".[32] Though the ancient commentator, presumably on analogy with the way in which he had understood different colored tiaras of Venus to be planets, interprets these phrases to mean that Mars or Saturn stands with Venus as its KI.GUB is respectively red and green, it seems to me that the limitation of the possible color of a KI.GUB to just these two means that the phenomena being described are the red flash and the green flash that are only produced when Venus' light is refracted by the atmosphere as it rises or sets – i.e., is on the horizon.

Another phenomenon for which there are several different expressions is that Venus does not change its KI.GUB, that is, that it rises above or sets below the same point on the horizon that it had crossed the previous night. This is the interpretation that the ancient commentator places on the protasis of the first omen in Group A: „Venus stands still (*ikūn*) in the morning".[33] Some modern commentators have taken this to refer to Venus' having reached its eastern – that is, second – stationary point. I, however, am convinced that the ancient commentator is correct.

31 K. 6021 : rev. 11 with parallels.

32 K. 35 : 16-17. K. 35 is a Group B manuscript.

33 VAT 10218, omen 1 with parallels.

Stationary points have no place among the planetary phenomena viewed as ominous by the authors of *Enūma Anu Enlil*; only retrogressions out of one constellation into its predecessor appear in the omen texts, and then only in the case of the superior planets, never of Venus.

The validity of our interpretation of the phrases „it goes higher“ and „it goes lower“ to mean that Venus' rising or setting point moves to the North or to the South along the eastern or western horizon is supported, I believe, by a set of three successive protases in Group A:[34] „Venus at its appearance moves towards its front“, „Venus at its appearance halves heaven and stands“, and „Venus at its appearance goes higher“. The second of these phenomena, as we have already seen, refers to those occasions at which Venus rises or sets over the East-point or the West-point on the horizon; the other two apparently mean that its KI.GUB is moving to the North. The first is of a type that I shall return to later; but the third, „Venus goes higher“, is repeated in Group F, where it is interpreted by the ancient commentator to mean: „it appears in the East in the path of Ea and goes higher calmly towards the path of Enlil“[35] – that is, its KI.GUB moves from the South to the North.

The three paths themselves do not appear in the protases of the Venus omens belonging to Group A, but they are a prominent feature of those of Group F.[36] As was pointed out in fascicle two of *Babylonian Planetary Omens*, the paths were originally arcs along the eastern horizon above which the constellations and the planets rise. There are three basic omens of which each has three variants as the phenomenon occurs in each of the paths. They are: „Venus appears in the path“, „Venus follows the path ... for six months and stands“, and „Venus in the East and in the West stands in the path“. The second of these, in which Venus is said to remain for six months in each of the three paths, seems to me to be astronomically impossible, if its presence in a path is intended to be continuous. The Sun will remain in the path of Anu, which stretches along the horizon about 20° North and 20° South of the East-point, for about 90 days or three months; it will remain in either the path of Ea or that of Enlil for about the same time. I cannot imagine a configuration of Venus, the Sun, and the Earth such that Venus will remain visible within the same path for six

[34] VAT 10218, omens 11-13 with parallels.

[35] K. 2346 : 12 with parallels.

[36] K. 7936 : 1-9 with parallels, etc.

months. The third omen, in which Venus remains in one of the paths in the East and in the West, would seem to mean that its last visibility on one horizon and its subsequent first visibility on the opposite horizon both occur in the same path. This, of course, is more likely to occur at inferior than at superior conjunctions.

Whereas in Group A the phenomena of Venus' KI.GUB moving towards the North or towards the South was described in terms of its going higher or lower, in Group F the expressions are:[37] „Venus at its appearance has a head", and „Venus at its appearance has a rear". The ancient commentators interpret the first of these, „Venus ... has a head", to mean that it appears first in the path of Anu and that, during its period of visibility, its rising-points become more northerly until it appears in the path of Enlil; and they interpret the second, „Venus has a rear", to mean contrarily that its rising points become more southerly until it appears in the path of Ea.[38] Other commentators add as a variable that Venus does or does not complete 2/3 *bēru* or 2 *bēru*.[39] In this case I believe one must consider that they are referring to arcs along the horizon measured in *bēru* and UŠ. Then 2/3 *bēru* or 20 UŠ = 20° is the approximate measure of the arc between the East-point and the northernmost or southernmost boundary of Anu; and 2 *bēru* = 60 UŠ or 60° approximates the arc over which the KI.GUB of the Sun ranges.[40]

The second motion that Venus has with respect to the horizon is along circles of altitude. This can be thought of in a trivial sense as occurring every night that it is visible, in a more significant but still periodic sense as its being seen on successive nights having an increasing or a decreasing altitude – i.e., rising higher or lower before Sunrise or being higher or lower at Sunset. I take it that the latter phenomena, being slightly less obvious, are the ones referred to in Group A by the verb *ullāt* in the phrase „Venus rapidly goes high" and by the verb *uštaktit* in the phrase „Venus goes down and sets".[41] More problematical is a protasis

[37] K. 7936 : 15-19 with parallels.

[38] K. 2346 : 8-11.

[39] D. T. 47 : 20-26.

[40] Defined by Koch as between azimuths of 240° and 300°, corresponding to rising-amplitudes of ± 30°. With our previous formula and with $\delta = \varepsilon \approx 24°$, Φ is about 36°.

[41] VAT 10218, omen 4 with parallels; and K. 229 : 29 with parallels.

found in Group F: „Venus ascends to the *ziqpu*".[42] Schaumberger offered the explanation that Venus could be seen at the zenith if the light of the Sun were sufficiently dimmed by clouds.[43] But, in light of our determination that „to go upwards" can mean that Venus' KI.GUB becomes more northerly, I would suggest that the omen may be that Venus rises (or sets) over the point on the horizon that had previously been crossed by a *ziqpu*-star. This would be possible in the cases of several of the *ziqpu*-stars listed in AO 6478;[44] these are, with their declinations for -700, α Herculis with a declination of +18;30°, α Leonis with one of 22°, and ϑ Cancri with one of 22;40°.

There remain those omens in which the phenomenon is the distortion of either Venus' own light or that of some other shining object. According to the texts of Group A Venus may have a *mešḫu* – also called a *šabiḫu* – at its right, its left and crosswise in its middle; and these *mešḫū/šabiḫū* may be red, white, or green.[45] In the same Group A Venus is said to produce – *imšuḫ* – a *ṣirḫu*.[46] These three, the *mešḫu*, the *šabiḫu* and the *ṣirḫu*, seem to be mirages,[47] but we can say little more about them.

Another set of omens in Group A represents Venus as being red, black, white, green, or green and red.[48] The same colors are referred to in the Indian adaptations of Mesopotamian Venus omens. If we can take „white" and „black" to be the equivalents of „bright" and „obscure", then the red, the green, and their combination would refer to the refraction of the light of Venus into a red band (the red flash) and a green band (the green flash) when it is on the horizon. Perhaps the same sort of explanation can be offered for the green, white, and red *mešḫū*, except that it does not seem, as far as I know, that the flash could be sent to the right or the left as the text indicates.

---

42 K. 3601 : 23 with parallels.

43 J. Schaumberger, *Ergänzungsheft* III, Münster 1935, pp. 291-292.

44 J. Schaumberger, „Die *Ziqpu*-Gestirne nach neuen Keilschrifttexten", *ZA* 50 (1952), 214-229.

45 VAT 10218, omens 80-88 with parallels.

46 VAT 10218, omens 118-123 with parallels.

47 *BPO* 2, p. 19.

48 VAT 10218, omens 15-18 with parallels.

Brightness, dimness, and scintillation are other phenomena associated with Venus in *Enūma Anu Enlil.* These require little comment, except that brightness and dimness should not be assumed necessarily to be concommittant with inferior and superior conjunction respectively.

But more remarkable are a group of protases that as yet make no sense at all to us. These include: „Venus at its appearance flickers like a bull";[49] the verb here hesitatingly translated „flickers" is *ištappu.* The simile perhaps reflects some esoteric myth which I am ignorant of. Other undeciphered omens are: „Venus appears in the East and is female"; „Venus appears in the West and is male"; „Venus' brilliance falls to the ground"; and „Venus alone is perfect" – *ediššišú gítmalat.*[50] Would that our understanding of these texts were as perfect as is she!

49 VAT 10218, omen 5 with parallels.

50 K. 3601 : rev.31-32; K. 800 : 14; and K. 3601 : rev.48 respectively.

The Classical World

ON THE GREEK ORIGIN OF THE INDIAN PLANETARY MODEL EMPLOYING A DOUBLE EPICYCLE

DAVID PINGREE, University of Chicago

The development of geometric models to explain the inequalities of planetary motion in terms of combinations of circular motions is characteristic of Greek astronomy, and derives its initial bias from statements of the Pythagoreans, Plato, and Aristotle regarding the structure of the universe and, in the case of Aristotle, the difference between the mechanics of the sublunar centre of the cosmos and that of the celestial spheres. It is hardly necessary to point out here that among the devices invented to account for these inequalities are the eccentric circle and the epicycle, whose properties were investigated by Apollonius of Perge in about 200 B.C.[1] The occurrence of these devices in Sanskrit astronomical texts of the fifth century A.D. immediately suggests some Greek influence. And the supposition of such an influence is greatly strengthened by the fact that Greek adaptations of Babylonian linear astronomy[2] and Greek treatises on genethlialogy[3] were translated into Sanskrit between the second and fourth centuries A.D.; it is virtually confirmed by the fact that the earliest Sanskrit *siddhântas* to employ epicyclic models, the older *Romaka* (for the luminaries only[4]) and *Pauliśa* (for the Sun only[5]), are largely of Greek or Greco-Babylonian origin. It is my intention here to investigate the Greek background of the common Indian model for the star-planets which involves two concentric epicycles.

In the *Paitâmahasiddhânta* of the *Viṣṇudharmottarapurâṇa*,[6] which is our earliest extant exponent of the Indian double-epicycle planetary model (it was probably composed in the first half of the fifth century A.D.), the pattern was set for all later texts except for those belonging to or aware of the *audayaka* System of the Âryapakṣa. The orbits of the planets are concentric with the centre of the Earth. The single inequalities recognized in the cases of the two luminaries are explained by *manda*-epicycles (corresponding functionally to the Ptolemaic eccentricity of the Sun and lunar epicycle respectively), the two inequalities recognized in the case of each of the five star-planets by a *manda*-epicycle (corresponding to the Ptolemaic eccentricity) and a *śîghra*-epicycle (corresponding to the Ptolemaic epicycle). The further refinements of the Ptolemaic models are unknown to the Indian astronomers.

If one tries to imagine the geometric model utilized by the *Paitâmahasiddhânta*, one immediately realizes that it cannot be cinematic; the planet cannot ride simultaneously on the circumferences of two epicycles. These two epicycles must be regarded simply as devices for calculating the amounts of the equations by which the mean planet on its concentric orbit is displaced to its true position. This interpretation is confirmed by the explanation offered in early texts of the mechanics of the unequal motions of the planets: demons stationed at the *manda* and *śîghra* points on their respective epicycles pull at the planets with chords of wind.[7] The computation of the total effect of these two independent forces upon the mean planet varies somewhat from one school (*pakṣa*) of astronomers to another, or even from astronomer to astronomer within a *pakṣa*. But the fundamental concept remains clear: the planet is always situated on the circumference of a deferent circle concentric with the centre of the Earth, while two epicycles (one each for the Sun and Moon) revolve about it. As the planet progresses with its mean velocity about this deferent circle, at each instant it is pulled by the two epicycles away from its mean to its true longitude. These instantaneous true longitudes are subject to computation, but a true course of the planet over a period of time can only be conceived of as a series of such instantaneous true longitudes.

This is not to deny that the equivalence of an eccentric to an epicyclic model had also been transmitted from Greece to India, though it would seem that the transmission was effected by a different text from that from which the double-epicycle theory is derived. Though it is true that in one passage of the *Âryabhaṭîya*[9] Âryabhaṭa, who was born in 476, clearly implies a double-epicycle model by giving the circumferences of the *manda* and *śîghra* epicycles, in another[10] he correctly describes an eccentric-epicyclic model and indicates the different directions a planet must travel on an epicycle to produce the differing effects of the equation of the anomaly and the equation of the centre. Though a number of later Indian astronomers acquainted with the *Âryabhaṭîya* or derivative texts of the Âryapakṣa refer to the eccentric model,[11] it seems seldom to have been used in computation.

The problem, then, is to explain the development of the double-epicycle theory. Initially, as noted above, we have strong reasons for believing it to be of Greek origin, and we have observed that its most striking effect is the maintenance of a constant distance of each of the five star-planets from the Earth (we shall discuss later the varying distances of the luminaries which the computation of the eclipses necessitates). Let us now examine the Greek tradition to locate where in it such a planetary model might have originated.

In Greek intellectual circles there was a clear conflict between the physical theory of motion posited by Aristotle and the cinematic models of planetary motion developed by later astronomers. In the *De caelo* Aristotle claims that the motions of the celestial spheres, which consist of the fifth element, must be uniform and circular about the centre of the universe, that is, the centre of the Earth.[12] The principle of concentricity is thus added to the principles of uniformity and circularity enunciated by Plato.[13] In the *Metaphysics* Aristotle succeeds in constructing, on the basis of Callippus's modifications of Eudoxus's hypothesis of homocentric spheres, a mechanical model of the celestial spheres satisfying the principles laid down in the *De caelo*, providing for a transfer of energy from the outermost sphere to that of the Moon, and qualitatively accounting for the planetary phenomena of retrogression and latitude;[14] he could not, however, explain the varying distances of the planets from the Earth made evident by the varying diameters of the Moon and the Sun.[15]

Greek astronomers in the Hellenistic period generally ignored Aristotelian physics, as did also the Stoic philosopher Posidonius who evidently followed Hipparchus in accepting the eccentricity of the planetary spheres.[16] Pliny's source also utilized eccentricity to account for the equation of the centre, but employed as a mechanism to produce the second inequality rays of fire emanating from the Sun and pushing the superior planets further from the Earth near opposition.[17]

But by the late first century A.D. Aristotle's physics was again taken seriously into consideration, and the problem of deciding between concentricity and eccentricity became a basic one. Ptolemy in the *Syntaxis* refers respectfully to Aristotle's division of theoretical science into three branches: physical (sublunar), mathematical (celestial), and theological.[18] When he comes to discuss the inequality of planetary motion, however, he states that the only two principles that must be preserved are uniformity and circularity.[19] Here he follows Plato rather than Aristotle, though he agrees with Aristotle against Plato that the celestial bodies are eternal. Of course, Ptolemy later, in hypothesizing the equant, abandons even uniformity of motion. In the *Hypotheses*, where he suggests a physical actualization of the mathematical models of the *Syntaxis*, he still insists that uniformity and circularity of motion are the only prerequisites for the eternity of the celestial bodies, though he has clearly read Book Λ of the *Metaphysics*.[20]

The Peripatetic scholars of the second century were much less convinced of the priority of mathematics over physics in determining the true nature of the planetary orbits, though they evidently were interested only in qualitative results. Alexander of Aphrodisias rejected both epicycles and eccenters as being opposed to the doctrine of concentricity, and asserted the validity of the Eudoxan system as described by Aristotle.[21] Alexander's teacher, Sosigenes, saw the difficulties in rejecting eccentricity much more clearly, but was too good a Peripatetic to abandon Aristotle. He attempts to interpret the system of eccenters and epicycles as not being entirely opposed to Aristotelian physics.[22] But earlier in the century, or perhaps at the end of the first century A.D. (at any rate, before Ptolemy), another Peripatetic philosopher proposed a theory that bears some similarity to the second book of Ptolemy's *Hypotheses*, and allows a qualitative explanation of the phenomena including apparent varying distances without abandoning concentricity.

Adrastus suggested for each planet, aside from a concentric hollow sphere producing the daily rotation from east to west, two spheres: a large concentric and hollow double-sphere rotating from west to east at the velocity of the planet's mean motion, and a small solid sphere rolling about between the inner surfaces of the large double-sphere. If the planet is simply the perceptible portion of the surface of the small solid sphere, one can account qualitatively for the latitude of the planet, its variation in distance and velocity, its stationary points, and its retrogressions. For, as Adrastus and his summarizer, Theon of Smyrna, point out, the motion of the planet as a point on the surface of the small solid sphere as it spins about its own axis and is carried about by the large, hollow, concentric double-sphere is equivalent either to motion on an epicycle or on an eccentric.[23] Quantitatively, the trouble with Adrastus's model is that it can generate only one inequality at a time while the planets are known to have two.

We have already seen that Adrastus's successors in the second century gradually abandoned all hope of saving the appearances and Aristotle too, with Alexander finally preferring the philosopher to the phenomena. The debate was not engaged again, so far as we can tell, till the fifth and sixth centuries, and by then Neoplatonism and Christianity had entered the fray. Proclus, a pagan Neoplatonist, tried to take a middle view, sustaining Plato while accepting a modified Aristotelianism and relegating the astronomers to the role of mere calculators. In Proclus's opinion only the sphere of the fixed stars must naturally move in a simple circular motion about the centre of the Earth; the planetary spheres, lying in between a body moving in simple circular motion about the centre and the four elements in the sublunar sphere moving in simple straight motion toward and away from that centre, naturally move in a mixed fashion which combines the two types of motion. As the planets are ensouled, this mixed motion is voluntary, but it is also periodic. The astronomers' eccentres and epicycles are human constructs enabling man to predict planetary positions, but they bear no resemblence to the real construction of the heavens.[24] Philoponus, a Christian Neoplatonist, assumed the validity of the astronomers' hypotheses in order to be able to employ them in his arsenal of weapons against Aristotle's doctrine of the eternity of the celestial spheres. Simplicius rejects Philoponus's arguments without a very serious discussion, and seems to incline toward Sosigenes's view which allows him essentially to ignore the problem after an extended examination of the dilemma.[25]

From this brief review it is clear that the Indian model which employs a double epicycle fits most closely into the attempts of Peripatetics in the late first and second centuries to preserve concentricity while explaining some of the phenomena. Within the context of the history of Greek astronomy proper also it naturally falls between Hipparchus and Ptolemy.[26] One would expect, then, that the model originated between about A.D. 50 and 150 among astronomers trying to accommodate Aristotelian physics or among Peripatetics trying to accommodate Apollonian and Hipparchan eccentres and epicycles. The actual Greek text translated into Sanskrit may well, of course, have been written as much as two centuries later.

The main problem with this hypothetical reconstruction of events is that the Indian model does not seem to explain variation in distance as Adrastus had tried to do, and as Sosigeses thought it important to do; even Aristotle himself was worried by this problem.[27] But we know that Alexander of Aphrodisias ignored this matter, and that Polemarchus of Cyzicus, Eudoxus's pupil, thought it unimportant.[28] Other Aristotelians might well have done likewise, and, in any case, any attempt at conciliating Aristotle with the astronomers had to make some sacrifices.

But Indian astronomical texts do, of course, preserve a method of calculating planetary distances, and it is based on the good Platonic and Aristotelian principle that every planet travels the same absolute distance in the same interval of time,[29] rather than on the Ptolemaic principle that the structure of eccentres and epicycles fills all the space between the Moon and the sphere of the fixed stars.[30] For the Indians, then, planetary distance varies inversely with planetary velocity.[31] The inferior planets, in this theory, present a difficult problem: their mean motions are equal to the Sun's while the anomalistic motion of Venus is less than the Sun's mean motion. If, then, one uses the mean motions of the inferior planets, they are at the same distance from the Earth as is the Sun, whereas if one uses their anomalistic motions Venus's orbit must lie between that of the Sun and that of Mars. But in the Indian double-epicycle model, in which the angles at the centre of the epicycles (i.e. the mean planet) are measured with respect to a line parallel to the direction of the sidereally fixed first point of Aries from the centre of the Earth, the parameters for the mean motions of the inferior planets are given in terms of the velocities of those planets' *śîghras*—that is, the sums of their respective anomalies and the mean motion of the Sun. By this device, whether of Greek or Indian origin, the orbit of Venus is properly positioned between those of Mercury and the Sun. In any case, the distances of the planets from the Earth calculated by this Indian method are appreciably different from those computed by Ptolemy, and there is no contact between their spheres.

The Indians had still to take into account the problem of the varying distances of the Sun and the Moon whose computation is essential for the prediction of eclipse magnitudes. These distances they made to vary inversely with the true instantaneous velocities of the luminaries.[32] Thereby, of course, as was inevitable, strict concentricity was lost. This fact, however, does not militate against the theory of the Peripatetic origin of the Indian double-epicycle model.

The effect of the rule for determining the varying distances is that the ratio of the diameter of the disc of the luminary to its velocity remains constant. That law is apparently first given in the *Romakasiddhânta*[33] and is thereby associated with the introduction of a Greek epicyclic model for the Sun and Moon into India. This text did not deal, so far as we know, with the problem of planetary motion, and therefore nowhere mentions the double-epicyclic or any other sort of planetary model. The Indians, moreover, were not concerned with the problem of preserving Aristotelian physics and would have seen no inconsistency in combining a double-epicycle model for the five star-planets derived from one Greek source with a method for computing the varying distances of the Sun and Moon derived from another Greek source. The transmission of Greek astronomical theories to India, then, was a complex process. Many texts were translated, and they represented divergent trends in Hellenistic science. The Indian astonomers could not be expected to understand the philosophical or historical bases of what they received; they simply did their best to use this new information in conjunction with their own traditions to construct functional systems allowing them to make reasonably accurate predictions.

REFERENCES

1. See O. Neugebauer, "The equivalence of eccentric and epicyclic motion according to Apollonius", *Scripta mathematica*, xxiv (1959), 5–21.
2. See D. Pingree, "Astronomy and astrology in India and Iran", *Isis*, liv (1963), 229–246, and O. Neugebauer and D. Pingree, *The Pañcasiddhântikâ of Varâhamihira* (Copenhagen, 1971), *passim*.
3. See D. Pingree, *The Yavanajâtaka of Sphujidhvaja*, to appear shortly in the *Harvard oriental series*.
4. *Pañcasiddhântikâ* VIII, 1–6.
5. *Pañcasiddhântikâ* III, 1–3. The arguments against a Greek origin presented by B. Chatterjee, most recently in her edition of Brahmagupta's *Khaṇḍakhâdyaka* (2 vols., Calcutta 1970: vol. i, pp. 264–295), are unconvincing if for no other reason because she assumes that Ptolemy is representative of all Greek astronomers.
6. D. Pingree, "The *Paitâmahasiddhânta* of the *Viṣṇudharmottarapurâṇa*", *Brahmavidyâ*, xxxi–xxxii (1967–68), 472–510 (see III, 10–11 and IV, 7–12).

7. *Sûryasiddhânta* II, 2; Pingree, *op. cit.* (ref. 2), 242; and *Bundahishn* as translated by D. N. Mackenzie, "Zoroastrian astrology in the Bundahišn", *Bulletin of the School of Oriental and African Studies*, xxvii (1964), 511–529 (see especially 516–517).
8. For a graphical representation of this model see *Pañcasiddhântikâ* II, 101, fig. 59.
9. Daśagîtikâ, 8–9. The Daśagîtikâ was not always included in the *Âryabhaṭîya*; see, e.g., the recension of Nîlakaṇṭha.
10. Kâlakriyâ, 17–20.
11. E.g., Brahmagupta (628), *Brâhmasphuṭasiddhânta* XIV, 10–11 and XXI, 24; Bhâskara (*ca* 629), *Mahâbhâskarîya* IV, 19–23 and 48–54; and Lalla (*ca* 750), *Śiṣyadhîvṛddhida*, Golâdhyâya, Chedyakâdhikâra 8–12 and Grahagolabandha 3.
12. *De caelo* I, 2–3 (268b–269b) and II, 5–8 (287b–290b).
13. Eudemus *via* Sosigenes quoted by Simplicius, *In De caelo* II, 12 (p. 488, Heiberg) = fr. 148 in F. Wehrli, *Die Schule des Aristoteles*, viii: *Eudemos von Rhodos* (Basel, 1955), 68. Cf. also *Phaedrus* 246E–247C; *Republic* 616B–617D; *Timaeus* 34B; 36B–D; 38C–39D; and 40A–D; and *Laws* 821B–822C.
14. *Metaphysics* Λ, 8 (1073a–1074b).
15. Cf. ref. 27 below.
16. The eccentric model of the Sun is described by Geminus, Posidonius's pupil, in *Introductio* I, 31–41; see also Cleomedes, I, 6 (the Sun) and II, 5 (all the planets).
17. *Historia naturalis* II, 59–73.
18. *Syntaxis* I, 1.
19. *Syntaxis* III, 3.
20. *Hypotheses*, p. 114, Heiberg.
21. J. Freudenthal, *Die durch Averroes erhaltenen Fragmente zur Metaphysik des Aristoteles* (Berlin, 1885), 111; Simplicius, *In De caelo* I, 2 (pp. 31–32, Heiberg).
22. Simplicius, *In De caelo* II, 12 (pp. 509-510, Heiberg).
23. Theon of Smyrna, *τὰ κατὰ τὸ μαθηματικὸν χρήσιμα εἰς τὴν Πλάτωνος ἀνάγνωσιν*, ed. E. Hiller (Leipzig, 1878), 181–6. A contemporary qualitative epicyclic model is found in P. Mich. 149; see A. Aaboe in *Centaurus*, ix (1963), 1–10.
24. Proclus, *In Platonis Timaeum commentaria*, iii, 56-57; 96; and 146-149, Diehl.
25. Simplicius, *In De caelo* I, 2 (pp. 31–33, Heiberg) and II, 12 (pp. 504–7, Heiberg).
26. Theon of Smyrna asserts (p. 188, Hiller) that Hipparchus (and Plato) preferred the epicycle over the eccentric because it is more natural. The authority of this assertion is questionable.
27. Simplicius, *In De caelo* II, 12 (p. 505, Heiberg) = fr. 211 in V. Rose, *Aristotelis fragmenta* (Leipzig, 1886).
28. Simplicius, *In De caelo* II, 12 (p. 505, Heiberg).
29. Plato, *Timaeus* 38E–39B; Aristotle, *De caelo* II, 10 (291a–291b). The Indian ascending order before Greek influence was felt was apparently Sun, Moon, *nakṣatras*, and planets; see W. Kirfel, *Das Purâṇa vom Weltgebäude* (Bonn, 1954), 48–49 (where, however, the planets are listed in the order Mercury, Venus, Mars, Jupiter, and Saturn, indicating a conflation of the Greek tradition with the older Indian).
30. B. R. Goldstein, *The Arabic version of Ptolemy's "Planetary Hypotheses"* (Philadelphia, 1967), 6–7; see also N. Swerdlow, *Ptolemy's theory of the distances and sizes of the planets*, unpublished dissertation submitted to Yale University in 1968.
31. See, e.g., *Paitâmahasiddhânta* III, 6–7 and D. Pingree, "The Later Pauliśasiddhânta", *Centaurus*, xiv (1969), 172–241 (fr. P58–P61).
32. See e.g., *Paitâmahasiddhânta* V, 2–4.
33. *Pañcasiddhântikâ* VIII, 15.

THE RECOVERY OF EARLY GREEK ASTRONOMY FROM INDIA

DAVID PINGREE, Brown University

The recent publication by O. Neugebauer of his monumental *A history of ancient mathematical astronomy*[1] provides us with penetrating analyses of the remnants of Greek astronomy. Those remains are dominated by "the Greatest", Ptolemy's *Σύνταξις μαθηματική*,[2] which appears to have been so enormously successful that his principal successors—Pappus,[3] Theon,[4] and Stephanus[5]—limited their activities to the writing of commentaries on it and on the *Handy tables*[6] that are largely based on it. And of his predecessors' works virtually all that survive intact are the elementary treatises on spherics by Euclid, Autolycus, and Theodosius,[7] that were incorporated in the early Byzantine period into a collection used for instruction in the schools. To supplement Ptolemy's accounts of the work of Apollonius,[8] Hipparchus,[9] and other early Greek astronomers, historians have had to rely on disparate and often desperate sources: on the handbooks written by Geminus[10] in about 50 AD and by Cleomedes[11] in about 370 AD; on the encyclopedias of Pliny[12] and of Martianus Capella;[13] on the philosophical treatise of Adrastus pillaged by Theon of Smyrna and Chalcidius; on the summaries and commentaries of Proclus and of Simplicius; on the astrological compendia of Vettius Valens, pseudo-Rhetorius, and pseudo-Heliodorus; and on fragmentary inscriptions and papyri.

However, one of those civilizations that were profoundly influenced by Greek culture has preserved a number of texts (composed in the second through seventh centuries AD) that represent non-Ptolemaic Greek astronomy. This civilization is that of India, and the texts are in Sanskrit.[14] It is certain that Greek astronomical texts were translated into Syriac and into Pahlavī, as well as into Sanskrit, but of the former we still have but little, and of the latter almost nothing; and in both cases we must rely for much of our knowledge on late accounts in Arabic.[15] The Sanskrit texts, however, though often either incorrectly or not at all understood by those who have transmitted them to us, formed the basis of a scientific tradition that only in this century has been destroyed under the impact of Western astronomy. The object of this paper is to characterize the Greek astronomy transmitted to India and to determine the times and the places in which this transmission was effected, in so far as that is possible.

The transmission was certainly very complex. It involved many levels and periods of Greek astronomy: adaptations of Babylonian lunar and planetary theories; the year-length of Hipparchus, an adaptation of his coordinate-system for the fixed stars, and his theories of precession and trepidation; tables of chords transformed into tables of sines; Peripatetic planetary models employing double epicycles and concentres with equants; non-Ptolemaic planetary models combining an eccentre with an epicycle; the solution of problems in spherical astronomy by means of gnomons and analemmata; the computation and, probably, the projection of eclipses; the essential data for computing planetary parameters; models for determining planetary latitudes; and the basic theory used in determining planetary distances. And this transmission extended over several centuries; it apparently began in the second century of our era, and continued till the late fourth or early fifth century. The locations of the recipient Indians indicate Western India as the point of entry of these various Greek theories; there exists literary, epigraphic, archaeological, and numismatic evidence for a massive Greek influence on this area in precisely the period of this transmission.[16]

Since none of the original translations of Greek texts into Sanskrit survives, we must try to disentangle the transmitted material from the adaptations of later authors. Often we have not sufficient evidence to be very confident about particular details in the historical process that led to the creation of the astronomy of the siddhāntas, though the general trend of events is now quite clear. The Sanskrit works on which we must especially rely are the following:

The *Yavanajātaka* (or *Greek Genethlialogy*) is a poem on horoscopy written by Sphujidhvaja in or near Ujjayinī in Western India in 269/270 AD.[17] This poem is based on a translation of a Greek astrological text written in Egypt—and probably in Alexandria—in the early second century AD. The translation was made by Yavaneśvara, "the Lord of the Greeks", in the Kṣatrapa kingdom in Western India in 149/150. The seventyninth chapter of the *Yavanajātaka* is devoted to astronomy.

The *Paitāmahasiddhānta*, which is the fundamental text of the Brāhmapakṣa in Indian astronomy, was probably composed in the early fifth century AD.[18] It has survived only because of its incorporation in the sixth or seventh century into the enormous *Viṣṇudharmottarapurāṇa*.

Āryabhaṭa of Kusumapura or Pāṭalipura, modern Patna on the Ganges East of Benares, was born in 476.[19] The epoch that he uses in his *Āryabhaṭīya* is 3,600 (= 1,0,0) of the Kaliyuga, or 499 AD. The *Āryabhaṭīya* is the foundation of the Āryapakṣa. He also wrote a work now lost that is the basis of the Ārdharātrikapakṣa.

Varāhamihira, an astrologer of Persian ancestry who lived at or near Ujjayinī in the middle of the sixth century, wrote a *Pañcasiddhāntikā*[20] in which he summarized five astronomical texts: a *Paitāmahasiddhānta* whose epoch is 80 AD and which is based on the elements of Lagadha's *Jyotiṣavedāṅga*, which was written in the late fifth century BC under Mesopotamian influence; a *Vasiṣṭhasiddhānta*, of which an earlier version was referred to by Sphujidhvaja but which was known to Varāhamihira in a recension of 499; a *Romakasiddhānta* (*Roman astronomical system*) and a *Pauliśasiddhānta* (*Paulus's astronomical system*), both in the recension of Lāṭadeva, a pupil of Āryabhaṭa; and finally Lāṭadeva's adaptation of a *Sūryasiddhānta* to conform to the Ārdharātrikapakṣa founded by Āryabhaṭa.

Bhāskara wrote a *Bhāṣya* or commentary on the *Āryabhaṭīya* at Valabhī in Saurāṣṭra in 629, and a *Mahābhāskarīya* before 628; he also wrote a shorter version of the latter entitled *Laghubhāskarīya*.[21] These works all belong to the Āryapakṣa.

Brahmagupta, a resident of Bhillamāla, modern Bhinmal near Mt Abu in southern Rājasthān, composed the *Brāhmasphuṭasiddhānta* in accordance with the tenets of the Brāhmapakṣa in 628 AD, and the *Khaṇḍakhādyaka* on the basis of the Ārdharātrikapakṣa in 665 AD.[22]

We have already mentioned that Lagadha's *Jyotiṣavedāṅga* was influenced by Mesopotamian astronomy. But it only dealt with calendaric problems, involving solar years, synodic months, the variation in the length of daylight, and the mean motions of the Sun and the Moon through the nakṣatras.[23] Babylonian astronomy, however, had also evolved methods of predicting planetary positions, particularly the longitudes and times of the occurrences of first and last visibilities and of first and last stations. And it could determine the true longitude of the Moon. The basic devices employed by the Babylonians were period relations, which represent mean motions, and so-called zig-zag and step functions, which reflect the deviations from the mean.[24] In this matter, as in all others that have been described elsewhere, I will avoid most details in this paper. What is significant for us is only that the Greeks were familiar with many of the Babylonian period relations,[25] and had developed a method (which is preserved now only in some papyri and late astrological texts) of arriving at approximate planetary longitudes from the mean elongations of the planets from the Sun at the times of the occurrences of the Greek-letter phenomena.

The Indian adaptations of Greek texts had adopted a technique of computing planetary longitudes based on two steps. The days since epoch—the ahargaṇa in Sanskrit—when divided by the days in successively smaller period relations yield a remainder that indicates how much has elapsed of the current smallest period, which is an anomalistic month in the case of the Moon, its synodic period in the case of each of the five star-planets. Such rules for the Moon are preserved in Greek papyri,[26] and, at a primitive level, for the planets in the *Anthologies* of Vettius Valens.[27] Then the longitudinal increment to the Moon since the beginning of the current anomalistic month is found by means of the summing of a zig-zag function of lunar velocity. Such a zig-zag function of daily lunar motion is found in the *Introduction* of Geminus.[28] The longitude of a planet is found from the days since the beginning of its synodic period by dividing that period into sections between phenomena, and assigning a velocity to the planet in each section as it falls in specified arcs of the eliptic. Such methods again are attested in Demotic and Greek papyri.[29]

These techniques as preserved in the Sanskrit texts were certainly not invented in India, which lacked the astronomical tradition necessary for their development. Nor were they introduced directly from Mesopotamia since they first appear, in a crude form, in the *Yavanajātaka*, which is based on the translation of a Greek text made three-quarters of a century after the last dated cuneiform ephemerides was inscribed. The full forms of these techniques are found in India only in texts, the *Vasiṣṭha* and the *Pauliśa*, of the third or fourth century which represent translations of Greek texts other than that translated by Yavaneśvara. The papyri, Geminus, Vettius Valens, and the three Sanskrit texts all attest to the popularity of this Greco-Babylonian astronomy in the Roman Empire in the first three or four centuries of our era.

Another Greek adaptation of Babylonian material that was transmitted to India is System A of the rising-times of the zodiacal signs, which we know in

Greek from a number of texts,[30] but which appears first in India in the *Yavanajātaka* and is often repeated thereafter. This also, as it employs the zodiacal signs introduced into India by Greek sources (as is indicated by their Greek names and iconography) in the second century AD, could not have come directly from Mesopotamia even though it is based on the same Babylonian ratio of the longest to the shortest day in the year, 3:2, that was used in the *Jyotiṣavedāṅga* and its derivatives.

Other material that reached India from Greece between the second and fourth centuries seem related to the Hipparchan tradition. The length of a year in the *Jyotiṣavedāṅga*, for instance, had been 365 days—an Egyptian parameter also used, apparently, in Achaemenid Iran. Sphujidhvaja describes a yuga of 165 years, which he calls Greek, in which the year is tropical and contains 6,5;14,32 days. Elsewhere he gives a year-length of 6,5;14,47 days; a trivial emendation of the unique manuscript produces 6,5;14,48 days, which is the value established by Hipparchus and adopted by Ptolemy. This Hipparchan year-length was also used in the *Romakasiddhānta*, where it was combined with the Metonic 19-year cycle to produce a yuga of 2,850 (150·19) years, in which there is an integer number of days (1,040,953). This yuga or annus magnus, though not recorded by Censorinus, must be of Greek origin since the Metonic cycle is not otherwise known in India.

Another Greek year-length that is found in the Sanskrit texts of this period is 6,5;15 days—a Julian year—which is attested in the *Vasiṣṭhasiddhānta*. The occurrence in this text of the parameter that was adopted in Rome in −44 and in Egypt between −25 and −22 supports the theory that its other basic contents—the Greco-Babylonian methods of computing lunar and planetary longitudes—also came from a Greek source.

But since the fifth century AD the normal Indian year has been sidereal rather than tropical. And the earliest sidereal year that is attested in the Sanskrit texts also belongs to our material. It is 6,5;15,30 days, found in the *Pauliśa-siddhānta*; this parameter is derived by equating 120 years with an integer number of days—43,831. It is found, as far as I am aware, in one other source: al-Battānī attributes it to "the Egyptians and Babylonians". Al-Battānī could only have derived this information, whether it is in any sense correct or not, from a Syriac or Arabic source that had in turn derived it from a Greek original. Therefore, I believe that we can be confident that the Pauliśa's year-length occurred in a Greek text. Again, the *Pauliśa*, like the *Vasiṣṭha*, preserves Greco-Babylonian methods of computing lunar and planetary longitudes; its Greek year-length supports the theory that they also were developed in Greece.

The difference between a sidereal and a tropical year, of course, is precession, which Hipparchus established as 1° in between 77 and 100 years. The *Romaka*, we are informed by Bhāskara in his *Āryabhaṭīyabhāṣya*, argued against the fixed colures of the *Jyotiṣavedāṅga* in favour of a motion of the solstices; it proposed a precession of approximately 1° in 60 years.[31] The parameter does not seem to be Greek, but the idea certainly is. The same can be said of the theory of trepidation. It was known among the Greeks to Theon in the fourth century and to Proclus in the fifth, though it seems in fact to be another Hipparchan hypothesis;[32] it occurs first in India in association with one Maṇindha, a name which is the Sanskrit transliteration of Manetho. In

Maṇindha's theory the amplitude of the trepidation is 27° on either side of sidereal Aries 0° and the rate is 3° in 200 years, which is precisely the value of precession that Yaḥyā ibn Abī Manṣūr claimed to have established on the basis of an autumnal equinox that he observed on 19 September 830. The sidereal and tropical zodiacs were considered by Maṇindha to have coincided at the beginning of the Kaliyuga in −3101; in his theory they will coincide again in precisely 3,600 years, or in 499 AD. As we shall see later, this choice of epochs dates Maṇindha to the beginning of the Greek period of Indian astronomy—the fifth century AD—and is a most important indicator of the way in which the Indians derived their parameters for the mean motions of the planets. The earlier Greco-Babylonian period, then, was probably familiar only with precession.

As we have seen, the *Vasiṣṭha* and *Pauliśa* compute the lunar longitudinal increase within an anomalistic month by means of the summing of a linear zig-zag function based on lunar velocity. But the *Romaka*, which has no planetary theory, tabulates, for the Sun and the Moon, their equations of the centre for intervals in mean motion of 15°—that is, of the βαθμός or half-sign familiar from Hellenistic Greek astronomy.[33] These tables as preserved by Varāhamihira are corrupt, but it is clear that the solar apogee was placed at Gemini 15° (a choice determined by the use of the βαθμοί); that the maximum solar equation is 2;23,23° (Hipparchus's and Ptolemy's is 2;23°); and that the maximum lunar equation is 4;46°. This last number is one of Hipparchus's estimates of the radius of the Moon's epicycle expressed in sixtieths of the radius of the deferent. Varāhamihira, therefore, may have misunderstood a table of sines of the equation, with R equalling 60, as a table of the equations themselves. It is not clear what model was used to compute these equations, but the structure of the tables using βαθμοί and the Hipparchan values of the solar and probably lunar equations suffice to prove their Greek origin.

Similar to these tables in the *Romaka* is the table of the equation of the centre of the Sun found in the *Pauliśa*. In this table the equations are given for the beginnings of the zodiacal signs, and the solar apogee is located at Gemini 20°; the maximum equation is 1;12°, which the text should have instructed one to double in order to produce 2;24°, a close approximation to Hipparchus's value. Again we seem to be dealing with a Greek table.

The rules for computing lunar latitude and parallax in Varāhamihira's summaries of the *Romaka* and *Pauliśa* assume the use of a sine-table in which $R = 120$. We do not know whether or not this value was also used in the two original siddhāntas, but it strongly suggests a derivation from a Greek chord-table in which $R = 60$ since $\sin_{120}\theta = \text{chrd}_{60}2\theta$. We have such a Greek chord-table with $R = 60$ very accurately computed by Ptolemy. But Varāhamihira's sine-table is not derived directly from Ptolemy since it is less accurately computed; in six out of twentyfour cases it deviates from the true sine by a sixtieth of a part, whereas a comparable sine-table with $R = 120$ derived from Ptolemy's table of chords never deviates by more than a sixtieth of a sixtieth of a part. This leads one to suspect that the origin of the Indian sine-table with $R = 120$ is a pre-Ptolemaic, or rather pre-Menelaus, chord-table with $R = 60$. I believe a Greek origin is also indicated by the choice of the interval in the Indian table: 3;45°. For this is a fourth of a βαθμός. The pre-Menelaus Greek chord-table, then, would have been computed for intervals of 7;30° or half a βαθμός, as Hipparchus's apparently was. Since Ptolemy did not use a sine-table, nor did any of his Greek successors that we know of, it is more likely that an Indian thought of using that function than that it had already been tabulated in Greece.

Another early Indian sine-table uses 3438 as the value of R; it first appears in the *Paitāmahasiddhānta* in the early fifth century, and is the closest integer corresponding to R in the formula $\pi = \frac{C}{2R}$ if $C = 21{,}600$, the minutes in a circle. This Indian sine-table is closely related to Hipparchus's chord-table as reconstructed by Toomer,[34] in which R also is 3438 and the interval is 7;30°.

The actual methods of computing longitudinal and latitudinal parallax in the Sanskrit texts differ from those of Ptolemy and from what we know of Hipparchus, but the determination of the cause and magnitude of lunar and solar parallax and its analysis into two vectors strongly suggest a Greek origin for the Indian computations that is post-Hipparchan and non-Ptolemaic. The computation of the durations of eclipses and eclipse-magnitudes in the *Romaka* and *Pauliśa* are based on a model in which the lunar orbit is assumed to be parallel to the ecliptic. Such approximative models could also be Greek in origin, though there seems to be no extant evidence of their existence in Greece. Ptolemy, of course, computes the elements of an eclipse on the basis of the correct model with inclined lunar orbit.

Furthermore, the *Pauliśa* presents a method of projection for eclipses that is used for determining the directions of impact and release relative to an "east-west" line. This information, like that of the colours of eclipses also discussed in the *Pauliśa*, was used in making predictions of terrestrial events as it had also been so used in the Babylonian astral omen-series *Enūma Anu Enlil* and in Greek omen texts ultimately derived therefrom.

The computation of the deflection of the "east-west" line from the ecliptic in these projections as described in later texts and that of lunar visibility in chap. 5 of the *Pañcasiddhāntikā* involve a modified Hipparchan coordinate-system—polar longitudes and polar latitudes.[35] These determine the distance of a celestial body from the ecliptic along the meridian that passes through it, and the longitude of the point of intersection of that meridian and the ecliptic. This system of coordinates is commonly used in later Indian texts in solving problems of the visibilities and conjunctions of planets and stars. It is certainly Greek; perhaps also ultimately Greek is the Indian method of converting ecliptic into polar coordinates.

Finally, the *Pauliśa* is the earliest Sanskrit text to utilize an analemma and the characteristic day-circle, Earth-sine, and ascensional difference to solve problems relating to local time. This kind of analysis was greatly developed by later Indian astronomers, but its origins also go back at least to Hipparchus.

This summary has covered virtually everything that can be recovered of Indian astronomy from the period between the second and fourth centuries AD. There seems to be very little in it that one can positively assert to have originated in India, and an overwhelming accumulation of evidence pointing to at least four Greek sources, transformed in Western India into the *Yavanajātaka* and the siddhāntas of Vasiṣṭha, Romaka, and Pauliśa. They represent an astronomy that owed much to the Babylonians and to Hipparchus, but nothing to Ptolemy. It would be as foolish to believe that the Indians have not modified their borrowings, as it would be difficult to determine precisely what those modifications were. But there can remain little doubt that much of the character of this non-Ptolemaic Greek astronomy has survived the transmission. But, while Indian astronomers at the end of this Greco-Babylonian period could compute many celestial events, they had not the theoretical methodology or observational material necessary to devise the cinematic planetary models that appear in the next, "Greek" period.

This Greek period begins in the fifth century AD, and is characterized as far as planetary theory is concerned by the use of the yuga system (inspired by the Greek anni magni and realized in the Keskinto Inscription[36]) for determining mean longitudes, and by the use of geometrical models to explain the planets' deviations from those mean longitudes. These planetary models have been discussed elsewhere: the eccentric with epicycle is pre-Ptolemaic, and is found in an early Arabic text of Māshā'allāh, to whom it was transmitted through a Syriac source;[37] the double epicycle that solves a Peripatetic problem by accounting for two inequalities while preserving the concentricity of the planets;[38] and the concentric with equant which also generates the proper unequal motion of a planet on a concentric.[39] All of these models are clearly Greek; the existence of three suggests the translation of several Greek astronomical texts into Sanskrit under the Guptas in the late fourth and early fifth centuries. The problem is that the Indian parameters of mean motions and of epicycle dimensions appear to be unique.

A solution to the mean motion problem has recently been offered by Roger Billard.[40] He has demonstrated that of the three original Indian astronomical systems, in the two originated by Āryabhaṭa, the Āryapakṣa and the Ārdharātrikapakṣa, the planetary mean longitudes are generally in error but converge on correct values in about 510 AD, when also approximately the Indian sidereal zodiac coincided with the tropical, whereas the mean longitudes in the Brāhmapakṣa never converge as satisfactorily. From these data, but with the neglect of much other relevant evidence, Billard concludes that Āryabhaṭa's work is based on his own careful observations, and that he was the true inventor of Indian planetary theory.

In fact, we know from several sources that the Brāhmapakṣa is older than Āryabhaṭa, who in fact refers to it; a Brāhmapakṣa text, the *Paitāmahasiddhānta*, existed in the early fifth century, while Āryabhaṭa was born in 476 AD. This fact alone disproves Billard's theory; further doubt is cast upon it by the impossibility of it happening that an observer who used Āryabhaṭa's method

TABLE 1.

Planet	λ	Distance from ζ Piscium
Saturn	290;48°	−29;45°
Jupiter	325;04°	+4;27°
Mars	301;55°	−18;42°
Sun	314;38°	−5;59°
Moon	323;02°	+2;25°
Ascending node	153;39°	−166;58°

TABLE 2.

Planet	Ptolemy	Āryapakṣa		Ārdharātrikapakṣa	
	λ	*R*	λ	*R*	λ
Saturn	48;40°	146,564	49;12°	146,564	49;12°
Jupiter	188;06°	364,224	187;12°	364,220	186°
Mars	7;08°	2,296,824	7;12°	2,296,824	7;12°
Venus's śighra	356;45°	7,022,388	356;24°	7,022,388	356;24°
Mercury's śighra	184°	17,937,020	186°	17,937,000	180°
Moon	283;30°	57,753,336	280;48°	57,753,336	280;48°
Ascending node	−7;11°	−232,226	−7;48°	−232,226	−7;48°

of computing true longitudes could derive correctly the planets' mean longitudes from observations of their true longitudes. Moreover, what we know of Indian astronomical observations does not lead us to expect any success of the sort that Billard attributes to Āryabhaṭa. There were no adequate instruments for measuring celestial angles, and those who discuss observations—e.g., Bhāskara I and Brahmagupta—are interested in demonstration rather than discovery, and so are content with very crude and makeshift procedures. All Indian astronomers have been content to operate with a tradition of longitudes and latitudes, polar or ecliptic, of so-called 'junction stars'. These do not in fact correspond to the actual distribution of any set of stars in the heavens. Since the Indians employ a sidereal reference-system for all longitudes, it is difficult to see how they could both have been in the habit of observing and have been content with their traditional star-catalogues.

But if Āryabhaṭa did not observe, how did he arrive at mean motions that produce results more correct for his own time than for any other? The answer is extremely simple: he used Greek tables of mean motions to compute the mean longitudes for a specific time, and thence derived the rotations in a Mahāyuga.

The procedure followed by Āryabhaṭa can be reconstructed as follows. The Brāhmapakṣa assumes an approximate mean conjunction of the planets at sidereal Aries 0° (where Aries 0° is approximately the longitude of ζ Piscium) at sunrise on 18 February −3101. Indeed, if we compute with the *Almagest* this estimate is not horrendously bad.

The interval from the beginning of the Kaliyuga till Ptolemy's epoch of noon 26 February −746 is 860,172¼ days, or 2356 Egyptian years, 7 months, 22 days and 6 hours, while at Ptolemy's rate of precession the longitude of ζ Piscium was 320;37°. The mean longitudes of the planets and their distances from ζ Piscium according to Ptolemy at the beginning of the Kaliyuga are given in Table 1.

Āryabhaṭa wishes to be able to avoid the use of the Kalpa of 4,320,000,000 years that is the basis of the Brāhmapakṣa's computation of the mean longitudes of the planets; therefore he chooses the shorter period of a Mahāyuga of 4,320,000 years, in which he seeks to find an integer number of rotations, *R*, for each planet. He further restricts his choice by dividing the Mahāyuga into four equal parts, of which the last—the Kaliyuga—begins with the mean conjunction of −3101. Therefore all his *R*'s (except those for the lunar mandocca and node, which were not at Aries 0° at the beginning of the Kaliyuga) must be divisible by four.

TABLE 3.

Brāhmapakṣa		Rotations in 3600 years	R'	R
Saturn	146,567,298	122+ 48;40°	146,562+	146,564
Jupiter	364,226,455	303+188;06°	364,227	364,228[1]
Mars	2,296,828,522	1,914+ 7;08°	2,296,823+	2,296,824
Venus's śīghra	7,022,389,492	5,851+356;45°	7,022,389+	7,022,388
Mercury's śīghra	17,936,998,984	14,947+184°	17,937,013+	17,937,012[2]
Moon	57,753,300,000	48,127+283;30°	57,753,345	57,753,344[3]
Ascending node	−232,311,168	−(193+ 87;11°)	−232,223+	−232,226[4]

Notes

1. Āryapakṣa: 364,224; Ārdharātrikapakṣa: 364,220.
2. Āryapakṣa: 17,937,020; Ārdharātrikapakṣa: 17,937,000.
3. Āryapakṣa and Ārdharātrikapakṣa: 57,753,336.
4. This number must be divisible by two but not by four so that the longitude of the node is 180° at the beginning of the Kaliyuga.

He determined the values of R, I would suggest, by computing from his Greek table the mean longitudes of the planets at noon of 21 March 499, which was precisely 3600 or 1,0,0 of Āryabhaṭa's years of 6,5;15,31,15 days since the beginning of the Kaliyuga. As 1,0,0 years is $\frac{1}{20,0}$ of a Mahāyuga, the rotations of each planet in 1,0,0 years multiplied by 20,0 will give their R's. An initial estimate of the rotations in 3600 years would have been based on the Brāhmapakṣa parameters; these would then have been corrected on the basis of the computed mean longitudes for 499.

In Table 2 I have computed the mean longitudes of the planets, λ, at noon of 21 March 499 according to Ptolemy and, from the Āryapakṣa's and the Ārdharātrikapakṣa's R's, the corresponding λ's for the end of 3600 years of Kaliyuga. Note that Āryabhaṭa is the author of both the Indian pakṣas, and that the śīghra is the sum of the mean motions of the Sun and the planet's anomaly. The śīghra is also used in the Keskinto Inscription. In Table 3 I compute from the Brāhma's R's in a Kalpa and Ptolemy's λ's for 21 March 499 the nearest R's in a Mahāyuga that are divisible by four.

These tables do not prove that Āryabhaṭa used Ptolemy, but rather that he used some other Greek source close to Ptolemy. They also demonstrate how trivial Āryabhaṭa's task was once he had a Greek set of tables of planetary mean motions, and explain his apparent accuracy in the early sixth century and increasing inaccuracy as one moves away from his own lifetime; the first reflects the accuracy of the Greek tables, the second the inaccuracy of his assumption of a mean conjunction in −3101. Finally, it shows that his parameters are simple modifications of those of the Brāhmapakṣa, so that his statement at the end of the *Āryabhaṭīya* that his system is derived from Brahma's is, in fact, correct.

The author of the Brāhmapakṣa, who wrote some 75 years before Āryabhaṭa, did not use a Greek table to establish his values of R; as I reconstruct it, he only used period relations such as those familiar from Babylonian times and transmitted to India by the Greeks. Note that such sidereal relations are presented, though inadequately, in the Keskinto Inscription and by Achilles (third

TABLE 4.

	Ptolemy		Paitāmaha		Āryapakṣa		Ārdharātrikapakṣa	
	c_μ	$2e$[1]	c_μ	e	c_μ	e	c_μ	e
Saturn	42;48	7;08	30	5	58;30 – 40;30	9;45 – 6;45	60	10
Jupiter	32;18	5;23	33	5;30	36 – 31;30	6 – 5;15	32	5;20
Mars	78;42	13;07	70	11;40	81 – 63	13;30 – 10;30	70	11;40
Sun	15	2;30	13;40	2;16,40	13;30	2;15	14	2;20
Venus	15	2;30	11	1;50	18 – 9	3 – 1;30	14	2;20
Mercury	36	6	38	6;20	31;30 – 22;30	5;15 – 3;45	28	4;40
Moon	31;30	5;15	31;36	5;16	31;30	*5;15*	31	5;10

	c_σ	r	c_σ	r	c_σ	r	c_σ	r
Saturn	39	6;30	40	6;40	40;30 – 36	6;45 – 6	40	6;40
Jupiter	69	11;30	68	11;20	72 – 67;30	12 – 11;15	72	12
Mars	237	39;30	243	40;30	238;30 – 229;30	39;45 – 38;15	234	39
Venus	259	43;10	258	43	265;30 – 256;30	44;15 – 42;45	260	43;20
Mercury	135	22;30	132	22	139;30 – 130;30	*23;15 – 21;45*	132	22

1. This is the first approximate distance between the equant and the centre of the Earth for Saturn, Jupiter, Mars, and Venus, and between the centre of the small circle bearing the centre of the deferent and the centre of the Earth for Mercury. The values for the Sun and the Moon are e, the distances between the centres of their deferents and the centre of the Earth. If one uses Ptolemy's final values for the $2e$'s of the superior planets, one has

	$2e$	c_μ
Saturn	6;50	41
Jupiter	5;30	33
Mars	12	72

The c_μ of Jupiter is that of the Paitāmaha, the c_μ of Mars is the mean of those of the Āryapakṣa. This must be fortuitous.

century?). The author of the Brāhmapakṣa computed from the transmitted Greek relations the rotations, R', of the planets over a Kalpa of 4,320,000,000 years so that the R' of each is in the hundreds of millions or in the billions; if a planet makes a sidereal rotations in b years, then $R' = \frac{a}{b} \,.\, 4{,}320{,}000{,}000$.
Adjustments then can be made on the order of several thousand without observably altering the yearly mean motions. In fact, a difference of 3600 rotations in a Kalpa means a difference of about one minute in the yearly mean motion.

Somewhere within the expanse of the Kalpa lies the beginning of the Kaliyuga, at which time there occurred approximately a mean conjunction of the planets. It cannot be in the middle because the lunar mandocca had a longitude of 90° on 18 February −3101, but it must be located fairly far within the Kalpa so that small adjustments to R' will produce sufficiently large corrections. Moreover, it was one of his goals to have a mean conjunction of the Sun and the Moon at the beginning of the Kaliyuga, so that the Moon's mean longitude must be 0°.

If we call the interval between the beginnings of the Kalpa and the Kaliyuga, which theoretically must be a multiple of 432,000, d, and the years in the Kalpa—4,320,000,000—Y, the mean longitude of the planet in rotations at the beginning of the Kaliyuga is given by the formula

$$R \,.\, \frac{d}{Y} = r \,.$$

For the Moon, $R = 57.753{,}300{,}000$, and since $d = a \,.\, 432{,}000$,

$$\frac{d}{Y} = \frac{a.\, 432{,}000}{4{,}320{,}000{,}000} = \frac{a}{10000}.$$

From this it follows that

$$R \,.\, \frac{d}{Y} = a \,.\, 5{,}775{,}330 \,,$$

i.e., that r is an integer and the mean longitude of the Moon is indeed 0°.

The problem remains to select a. Our astronomer chose 4,567, so that $d = 1{,}972{,}944{,}000$ years. Now for all the planets except the Sun and the Moon, $R' \,.\, \frac{d}{Y}$ will produce an integer number of revolutions plus a residue, c'. This residue must be reduced nearly to 0 (the Brāhmapakṣa seeks only an approximate mean conjunction at Aries 0° at the beginning of the Kaliyuga) by changing R' to R. If the difference between R' and R be denoted x, the problem reduces to an indeterminate equation:

$$4567x - 10000y = c' \,.$$

The solution of such an equation by the kuṭṭaka or pulverizer (a development from the so-called Euclidean algorithm and therefore another adaptation of material transmitted from Greece[42]) is familiar to us from early Indian astronomical works. In the *Mahābhāskarīya* and the *Brāhmasphuṭasiddhānta* it is applied exclusively to problems of planetary mean motion; Āryabhaṭa, who

APOGEES

	Ptolemy	Brāhmapakṣa (628)	Āryapakṣa	Ārdharātrikapakṣa
Saturn	233°	260;55°	236°	240°
Jupiter	161°	172;30°	180°	160°
Mars	115;30°	128;24°	118°	110°
Sun	65;30°	77;55°	78°	80°
Venus	55°	81;15°	90°	80°
Mercury	190°	224;54°	210°	220°

NODES

	Ptolemy	Brāhmapakṣa (628)	Āryapakṣa	Ārdharātrikapakṣa
Saturn	97;3°	103;12°	100°	100°
Jupiter	81;25°	82;01°	80°	80°
Mars	34;54°	31;54°	40°	40°
Venus	59;43°	59;47°	60°	60°
Mercury	25;52°	21;11°	20°	20°

TABLE 5.

does not need it for deriving his *R*'s, gives it a more generalized expression. The solution of

$$4567x_0 - 10000y_0 = 1$$

yields $x_0 = 8903$, $y_0 = 4066$. With the multiplier 8903 and c''s adjusted to produce convenient results, one can compute the new *R*'s that produce an approximate mean conjunction in −3101 and reasonably good results in the fifth and sixth centuries AD. But there will indeed be no pattern of the different mean motions converging in accuracy at some specific time, as that is not a goal built into the computation. Thereby the difference observed by Billard to pertain between Āryabhaṭa's two pakṣas and the Brāhmapakṣa can be explained by the difference between their methods of deriving the *R*'s of the planets.

Once the *R*'s were obtained it remained a problem to determine the dimensions of the epicycles. The Indians give these as the circumferences measured in units of which there are 6,0 in the deferent. Therefore, since Ptolemy gives the eccentricities and radii of the epicycles measured in parts of the radius of the deferent, which is 1,0, we can compare the Indian values with Ptolemy's by multiplying the latter or dividing the former by six. The circumferences of the epicycles are given in Table 4 with the corresponding values of the eccentricity and radius in Ptolemaic units. Precise identities are in italics; the mean values of the Āryapakṣa's parameters are used for this purpose.

Again, there is no evidence whatsoever that the Indians had an appropriate methodology for deriving these parameters, all of which are identical with or very close to Ptolemy's. If the models are of Greek origin, so must be their parameters. The most serious differences between the Ptolemaic and the Indian sets of parameters are located in the longitudes of the apogees (Table 5), where the Indian values are all higher than Ptolemy's. Clearly, as Ptolemy's apogees (except for the Sun's) are fixed with respect to the fixed stars, the addition to his longitudes of the difference between his longitude of ζ Piscium and Aries 0° in 138 AD, which is 7°, will cause the longitudes of his apogees to be measured from the same sidereal point as are the Indians'. However, this correction obviously is insufficient. But the same observations ought to be used to determine the directions of the apsidal lines and eccentricities. Therefore, the most plausible hypothesis seems to be that the parameters of the Brāhma and Ārya pakṣas are the results of two independent Greek derivations of these elements, and that the astronomers who made them did not depend on Ptolemy. The parameters of the Ārdharātrikapakṣa are not independent; they differ from the others because of their mode of expression.

Another part of the Indian planetary theory that is derived from a Greek source is that relating to latitude. The Indians assume, as does Ptolemy in the *Handy tables*, that the planes of the epicycles of the superior planets are parallel to the plane of the ecliptic, while the planes of their deferents and those of the epicycles of the inferior planets are inclined with respect to it. The Indians, however, employ a far simpler method of computation than does Ptolemy.

Finally, the problem of the geocentric distances of the planetary orbits is solved by the Indians by applying the Greek idea that each of the planets travels the same linear distance as do all the others in a given period of time. Since a Kalpa is one such given period, the distance travelled by each planet within it is the product of its R and the circumference of its orbit in some units; the Indians use yojanas. There will be one product, O, for all of the planets, so that the orbit of each in yojanas is $\frac{O}{R}$.

To translate this theory into numbers the Indians chose to assign a value to the length in yojanas of a minute in the orbit of the Moon; this is clearly related to the estimate of the circumference of the Earth since lunar horizontal parallax, estimated by the Indians to be about 0;53°, is simply the radius of the Earth seen at the distance of the Moon. The Brāhmapakṣa assumes that a minute in the orbit of the Moon equals 15 yojanas, which would make the Earth's radius about 795 yojanas and its circumference 4993; the texts of the Brāhmapakṣa give the Earth's circumference as 5000. Āryabhaṭa, on the other hand, assumes that each minute in the Moon's orbit contains 10 yojanas so that the Earth's radius is about 530 and its circumference about 3328; in fact, he gives the diameter as 1050 and the circumference as 3200, which implies a horizontal parallax of 0;52,30°. Simple multiplication and division will produce the radii and circumferences of the remaining planetary orbits. By using the R's of the śīghras of the two inferior planets rather than those of their anomalies the Indians succeed in locating the orbit of Venus below that of the Sun. They had no reason for desiring this result, but the question had been much discussed among the Greeks, most of whom had agreed on the order adopted by the Indians. It is tempting, therefore, to see the Indian method of computing planetary distances and their use of the śīghras of the inferior planets as parts of a unified Greek cosmological scheme.

In summary, then, we can point to evidence indicating that at least four Greek texts expounding Greco-Babylonian astronomy were transmitted to Western India in the second, third, and fourth centuries when the area was ruled from Ujjayinī by the Western Kṣatrapas, whose territories maintained close commercial ties with the Roman Empire and were hosts to large numbers of Greek expatriates. Furthermore, it is apparent that some Greek texts (probably two) on planetary theory that were strongly influenced by Peripatetic notions of cosmology and a set of astronomical tables were transmitted to the same area of India shortly before or after 400 AD, when it had come under the domination of the Guptas. If one exercises some care in separating from this material the Indians' modifications of it, there remains a great deal of genuine Greek astronomy that helps to remind us that the *Almagest* did *not* immediately cause the loss of non-Ptolemaic astronomy.

REFERENCES

1. Otto Neugebauer, *A history of ancient mathematical astronomy* (3 vols, Berlin–Heidelberg–New York, 1975); hereafter cited as *HAMA*.
2. *HAMA*, 21–273 and 834–8.
3. *HAMA*, 968–9.
4. *HAMA*, 965–8.
5. *HAMA*, 1045–51.
6. *HAMA*, 969–1028.
7. *HAMA*, 750–69.
8. *HAMA*, 262–73.
9. *HAMA*, 274–343.
10. *HAMA*, 578–87.
11. *HAMA*, 959–69.
12. *HAMA*, 802–5.
13. *HAMA*, 1030.
14. See D. Pingree, "An Essay on the History of Mathematical Astronomy in India", to appear in the supplement to the *Dictionary of scientific biography*.
15. See D. Pingree, "The Greek Influence on Early Islamic Mathematical Astronomy", *Journal of the American Oriental Society*, xciii (1973), 32–43.
16. See D. Pingree, "Astronomy and Astrology in India and Iran", *Isis*, liv (1963), 229–46. Much new archaeological material has come to light since that article was written.
17. D. Pingree, *The Yavanajātaka of Sphujidhvaja*, to appear in the *Harvard oriental series*.
18. D. Pingree, "The *Paitāmahasiddhānta* of the *Viṣṇudharmottarapurāṇa*", *Brahmavidyā*, xxxi–xxxii (1967–68), 472–510.
19. *Āryabhaṭīya*, ed. H. Kern (Leiden, 1875). For further information see D. Pingree, *Census of the exact sciences in Sanskrit*, Series A, i (Philadelphia, 1970), 50b–53b, and ii (Philadelphia, 1971), 15b.
20. O. Neugebauer and D. Pingree, *The Pañcasiddhāntikā of Varāhamihira* (2 vols, Copenhagen, 1970–71).
21. See D. Pingree in *Dictionary of scientific biography*, ii (New York, 1970), 114–15.
22. See D. Pingree, *ibid.*, 416–18.
23. D. Pingree, "The Mesopotamian Origin of Early Indian Mathematical Astronomy", *Journal for the history of astronomy*, iv (1973), 1–12.
24. *HAMA*, 347–555.
25. *HAMA*, 309–12 and 601–7.
26. *HAMA*, 808–15.
27. *HAMA*, 793–801.
28. *HAMA*, 602–3.
29. *HAMA*, 790–1 and 946–8.
30. *HAMA*, 715–21.
31. See D. Pingree, "Precession and Trepidation in Indian Astronomy before AD 1200", *Journal for the history of astronomy*, iii (1972), 27–35.
32. *HAMA*, 631–4.
33. *HAMA*, 669–74.
34. G. J. Toomer, "The Chord Table of Hipparchus and the Early History of Greek Trigonometry", *Centaurus*, xviii (1973), 6–28.

35. Hipparchus used polar longitude as one coordinate, $90° - \delta$ (usually) as the other. The polar latitude is the difference between the declinations of the celestial body itself and the point on the ecliptic measured by its polar longitude.
36. *HAMA*, 698–705.
37. D. Pingree, "Māshā'allāh: Some Sasanian and Syriac Sources", *Essays on Islamic philosophy and science* (Albany, 1975), 5–14.
38. D. Pingree, "On the Greek Origin of the Indian Planetary Model Employing a Double Epicycle", *Journal for the history of astronomy*, ii (1971), 80–85.
39. D. Pingree, "Concentric with Equant", *Archives internationales d'histoire des sciences*, xxiv (1974), 26–29.
40. Roger Billard, *L'astronomie indienne* (Paris, 1971).
41. *HAMA*, 950–2.
42. *HAMA*, 1116–25.

DAVID PINGREE

THE PRECEPTUM CANONIS PTOLOMEI

Toward the middle of the sixth century the former secretary of the Ostrogothic king, Theoderic, the Senator Cassiodorus composed for the monks of his monastic foundation, Vivarium, a guide to sacred and profane literature[1]. The last substantive chapter, before the conclusio, of the second book of the *Institutiones* is on the last of the seven liberal arts, *De astronomia*[2]. Cassiodorus begins this chapter with a brief compilation of appropriate passages from Scripture, and then proceeds to define the sixteen topics that he believes astronomy to consist of : «spherica positio, sphericus motus», and so on[3]. Certain technical terms that occur among these sixteen topics he not only defines, but for each provides the Greek equivalent. So for the forward motion of the planets, which he calls in Latin «praecedentia vel antegradatio», he gives the Greek «propodismos» ; for the planets' retrograde motion, «remotio vel retrogradatio» in Latin, he gives the Greek «ypopodismos aut anapodismos» ; and for their stations, «status» in Latin, he provides the Greek «stirigmos». The Latin terms were copied by Isidore of Seville in his *Etymologiae*[4], and thence were borrowed by numerous medieval

[1] *Cassiodori Senatoris Institutiones*, ed. R.A.B. MYNORS, Oxford, 1937, reprinted 1961, henceforth cited as *Inst.*

[2] *Inst.*, p. 153-157.

[3] This list, on p. 154 of *Inst.* with definitions on p. 154-155, is a very inadequate summary of the topics normal to astronomy. It includes spherics ; mean motions of the planets ; direct and retrograde motions and stations of the planets ; the computus, which presumably includes solar and lunar theory ; the sizes of the Sun, the Moon, and the earth ; and solar and lunar eclipses. The section on spherics does *not* include right or oblique ascensions ; the section on the planets seems *not* to include a direct discussion of the two inequalities of planetary motion ; it is unclear how much of solar and lunar theory is included under computus ; and there is no reference to the fixed stars. So far as I am aware Cassiodorus' list corresponds to the contents of no known astronomical text in Greek or Latin.

[4] *Etymologiae* III 68-70 in *Isidori Hispalensis Episcopi Etymologiarum sive Originum Libri XX*, ed. W.M. LINDSAY, 2 vol., Oxford, 1911, henceforth cited as *Etym.* Cf. HRABANUS MAURUS, *De computo* 37, in *Rabani Mogontiacensis Epis-*

authors ; but in the Latin literature before Cassiodorus they do not occur — except for retrogradatio or retrogradus[5] — in Pliny, Firmicus Maternus, Macrobius, Calcidius, or Martianus Capella. The Latin terms, then, seem to be Cassiodorus' own translations of the Greek terms that he cites.

These Greek terms — προποδισμός, ὑποποδισμός, ἀναποδισμός, and στηριγμός — are common in Greek astronomical and astrological texts. The question, then, is apparent : did Cassiodorus know enough Greek and enough astronomy to be able to peruse, say, Ptolemy's Σύνταξις or Proclus' Ὑποτύπωσις, to pick out these terms, and to understand their meaning ? Or did he find them in a Latin text other than those we have mentioned ? Though Cassiodorus was certainly capable of translating Greek[6], he does not give any evidence of having an advanced knowledge of astronomy sufficient to allow him to read Ptolemy or Proclus. But there is evidence that he consulted the Latin version of Ptolemy's Κανόνες πρόχειροι or *Handy Tables*, that is, the *Preceptum canonis Ptolomei*, in whose instructions Greek technical terms are presented in transliteration[7]. Though the parts of the *Preceptum* that dealt with the forward and retrograde motion and the stations of the planets are no longer present in the manuscripts that have survived of that curious work, we shall soon see that their absence can be explained.

But now we must return to the chapter *De astronomia* in book II of the *Institutiones*. Cassiodorus proceeds after his enumeration and defi-

copi De computo, ed. W.M. STEVENS, Turnhout, 1979, p. 163-331, esp. p. 249 (Corpus Christianorum, Continuatio Mediaevalis, XLIV) ; this edition is henceforth cited as *De comp.*

[5] Retrogradus occurs eight times in Firmicus Maternus' *Mathesis* ; see *Iulii Firmici Materni Metheseos Libri VIII*, ed. W. KROLL, F. SKUTSCH, and K. ZIEGLER, 2 vol., Leipzig, 1897-1913, 2, p. 368 and 379. And Martianus Capella uses the word retrogradatio once, in VIII, 881 ; see *Martianus Capella*, ed. J. WILLIS, Leipzig, 1983, p. 334.

[6] See P. COURCELLE, *Late Latin Writers and their Greek Sources*, translated by H.E. WEDECK, Cambridge, Mass., 1969, p. 356.

[7] It was assumed by E. HONIGMANN, *Die sieben Klimata und die Πόλεις Ἐπίσημοι*, Heidelberg, 1929, p. 103, and by A. VAN DE VYVER, *Les plus anciennes traductions latines médiévales (X^e^-XIe siècles) de traités d'astronomie et d'astrologie*, in *Osiris*, 1 (1936), p. 658-691, esp. p. 687, that Cassiodorus refers to the Greek Κανόνες πρόχειροι rather than to the *Preceptum* ; it will be argued below that this assumption is unfounded.

nition of the sixteen topics of astronomy to describe the literature pertinent to that science[8] : «De astronomia vero disciplina in utraque lingua diversorum quidem sunt scripta volumina ; inter quos tamen Ptolomeus apud Graecos praecipuus habetur, qui de hac re duos codices edidit, quorum unum minorem, alterum maiorem vocavit astronomum». Ptolemy's «minor astronomus» may be the μικρὸς ἀστρονομούμενος to which the scholiast to the Vatican manuscript of Pappus, Vat. gr. 218, refers at the beginning of book VI of the Συναγωγή[9], and the μικρὸς ἀστρονόμος that Theon is said in an anonymous text on isoperimetric figures to have commented on[10]. This latter text is part of the Εἰσαγωγή to Ptolemy's Σύνταξις that Mogenet has shown to have probably been composed by Eutocius[11], who wrote in about the year 500. The *Little Astronomy* was probably the collection of mainly Hellenistic texts on astronomy and geometry that are preserved as a corpus in Byzantine, Arabic, and medieval Latin manuscripts[12]. None of these texts, however, is by Ptolemy, so that it remains problematical why Cassiodorus should have claimed that his «minor astronomus» was by the great Alexandrian[13]. The «maior astronomus» of Ptolemy, however, must be his Σύνταξις. Though it was known, probably already in the third century, as mgstyk ī hrōmāy in Pahlavī[14] — it is from this Pahlavī corruption of the Greek μεγιστή that the Arabic al-majisṭī is derived, whence our *Almagest* — the earliest reference to it in Greek with a title corresponding to Cassiodorus' «maior» is found, as far as I am aware, in the commentary on Archimedes' Κύκλου μέτρησις written by the same Eutocius who mentioned the μικρὸς ἀστρονόμος ; there is mentioned the Μεγάλη σύνταξις of Claudius Ptolemaeus[15]. All of this hints at some connection between the source that Cassiodorus used for the sentence

[8] *Inst.*, p. 155-156.

[9] *Pappi Alexandrini Collectionis quae supersunt*, ed. F. HULTSCH, 3 vol., Berlin, 1875-1878, reprinted Amsterdam, 1965, 2, p. 474 ; henceforth cited as *Coll.*

[10] *Coll.*, 3, p. 1142.

[11] J. MOGENET, *L'Introduction à l'Almageste*, Bruxelles, 1956 (Mémoires in -8° de l'Académie Royale de Belgique, 2e s., 51, 2).

[12] See D. PINGREE, in *Gnomon*, 40 (1968), p. 13-17.

[13] It is at least possible that already in late antiquity, as in medieval Islam, the «Little Astronomy» was regarded as a prelude to the *Almagest.*

[14] So the *Dēnkart* as quoted by H.W. BAILEY, *Zoroastrian Problems in the Ninth-Century Books*, Oxford, 1943, reprinted, 1971, p. 86.

[15] In the commentary on the third theorem in *Archimède. IV. Commentaires d'Eutocius et fragments*, ed. C. MUGLER, Paris, 1972, p. 144.

about Ptolemy's two works that we have quoted and Eutocius ; perhaps the link was Eutocius' editor, Isidore of Miletus, or the latter's pupil[16], either of whom Cassiodorus might have met during his stay in Constantinople in the late 40's and early 50's of the sixth century[17].

Be that as it may, Cassiodorus' next couple of sentences are crucial to our suggestion that he consulted the *Preceptum canonis Ptolomei*[18]. «Is etiam [*i.e.*, Ptolomeus] canones, quibus cursus astrorum inveniantur instituit». These *canones* are surely the Κανόνες πρόχειροι[19]. Cassiodorus continues : «ex quibus, ut mihi videtur [this implies, I believe, that Cassiodorus has seen them], climata forsitan nosse, horarum spatia comprehendere, Lunae cursum pro inquisitione paschali, Solis eclipsin, ne simplices aliqua confusione turbentur, qua ratione fiant advertere non videtur absurdum». These topics, among others, are all dealt with in the *Preceptum*, though that text does not indicate that the «Lunae cursus» might be used for investigating the date of Easter. The next sentence in Cassiodorus, however, directly reflects the first sentences of the *Preceptum.* The Senator writes[20] : «sunt enim, ut dictum est, climata quasi septem lineae ab oriente in occidentem directae, in quibus et mores hominum dispares et quaedam animalia specialiter diversa nascuntur ; quae vocitata sunt a locis quibusdam famosis [*i.e.* the Greek πόλεις ἐπίσημοι], quorum primum est Merohis, secundum Sohinis, tertium Catochoras, id est Africa, quartum Rodus, quintum Hellespontus, sextum Mesopontum, septimum Borysthenus». The opening words of the *Preceptum* are : «Intellectus climatum. Polis episeme»[21]. The following

16 Isidore of Miletus was the editor of both the two books of Eutocius' Ὑπόμνημα on Archimedes' Περὶ σφαίρας καὶ κυλίνδρου and his Ὑπόμνημα on Archimedes' Κύκλου μέτρησις according to notes added by Isidore's pupil at the end of each of these three books.

17 On Cassiodorus' visit to Constantinople see, *e.g.*, COURCELLE, *Late Latin Writers...*, p. 335.

18 *Inst.*, p. 156.

19 Ptolemy's introduction to the tables, Προχείρων κανόνων διάταξις καὶ ψηφοφορία, was edited by J.L. HEIBERG in *Claudii Ptolemaei ... Opera astronomica minora*, Leipzig, 1907, p. 157-185 ; the tables themselves, with Theon's *Little Commentary* and various other texts, were published by l'Abbé HALMA, 3 vol., Paris, 1822-25.

20 *Inst.*, p. 156.

21 The second phrase, a simple transliteration of πόλεις ἐπίσημοι such as are characteristic of the *Preceptum*, has often been misread «poli sepissime».

chapter is a commentary on the first seven tables of the *Handy Tables* which give the oblique ascensions for the latitudes of the seven climata[22]. The Latin text agrees with the latitudes and with the names of the cities given in these tables[23]. The cities are the same as those named by Cassiodorus : Meroe, Syene, dia tes catho coras (*i.e.*, a simple transliteration of the Greek), Rodos, Hellesponto, meso Ponto, and Boristene. The list is not found elsewhere in Latin, but only in Cassiodorus and the *Preceptum* ; Martianus Capella, for instance, substitutes Alexandria for κάτωχώρα, injects Rome after Rhodes, omits the middle of the Pontos, and adds an eighth clime between Lake Maeotis and the Riphaean mountains[24]. Nor does the *Preceptum*'s list appear in the Greek instructions for using the *Handy Tables*, neither those by Ptolemy himself[25] nor the two sets by Theon, the Ὑπόμνημα in five books addressed to Eulalius and Origen[26] and that in one book addressed to

[22] The seven climata of the *Preceptum* are those of the Ptolemaic system, which ga back to a Hellenistic source ; see O. NEUGEBAUER, *History of Ancient Mathematical Astronomy*, 3 vol., New York, 1975, p. 725-726. The headings of the seven tables of the *Handy Tables* in Halma's edition (2, Paris, 1823, p. 2-57) are :

I. ἀναφοραὶ τῶν διὰ Μερόης ὡρῶν ιγ, μοιρῶν ις, ἑξηκοστῶν κζ´.
II. ἀναφοραὶ τῶν διὰ Συήνης κλίματος ὡρῶν ιγ L´, μοιρῶν κγ να´.
III. τρίτον κλίμα διὰ τῆς κάτω χώρας ὡρῶν ιδ, μοιρῶν λ κβ´.
IV. κλίμα τέταρτον διὰ Ῥόδου ὡρῶν ιδ L´, μοιρῶν λς.
V. κλίμα πέμπτον τὸ δι᾽ Ἑλλησπόντου ὡρῶν ιε, μοιρῶν μ νς´.
VI. ἕκτον κλίμα διὰ μέσου Πόντου ὡρῶν ιε L´, μοιρῶν μς α´ (error for με α´).
VII. κλίμα ἕβδομον διὰ Βορυσθένους ὡρῶν ις, μοιρῶν μη <λβ´>.

[23] The seven climata and their latitudinal boundaries in the *Preceptum* are :

I. Meroe. 15;15° to 23;51°
II. Syene. 23;51° to 30;22°
III. in Egypto dia tes catho coras. 30;22° to 36°
IV. Rodos. 36° to 40;56°
V. Hellesponto. 40;56° to 45;1°
VI. meso Ponto. 45;1° to 48;32°
VII. Boristene. 48;32° «usque ad id quod excesserit in septimo climate».

[24] *De nuptiis Philologiae et Mercurii*, VIII, 876, p. 332, ed. WILLIS.

[25] He mentions the seven climata on p. 160, ed. HEIBERG, but names none of them. In Σύνταξις μαθηματική II, 13, however, Ptolemy heads the tables of ecliptic angles with the names of the seven places and their latitudes.

[26] Book I was edited by J. MOGENET and A. TIHON, *Le «Grand Commentaire» de Théon d'Alexandrie aux Tables Faciles de Ptolémée. Livre I*, Città del Vaticano, 1985 (Studi e testi, 315). In chapter 4, p. 101, Meroe is mentionned as lying on the first parallel, but no other of the seven climata is named.

Epiphanius[27]. The list occurs in Greek only in the headings to the tables themselves in the *Handy Tables*. This is not a likely place for Cassiodorus to have looked for information — the only details he records — concerning Ptolemy's work. I believe there can be little doubt that he rather took these details from the only other source that would have been available, the *Preceptum canonis Ptolomei*, wherein the names of these seven localities are given at the very beginning of the text.

If the *Preceptum* was used by Cassiodorus when he wrote the *Institutiones*, it must be older than 550. Internal evidence allows us to date it much more precisely. Most of the *Preceptum* parallels Theon's monobiblos on the *Handy Tables*, the so-called *Little Commentary*, composed after his *Large Commentary* in or shortly before 377. Of the 25 chapters of the *Little Commentary* I have found in the *Preceptum* parallels to chapter I concerning the 5 κεφάλαια (called gnomones in Latin) into which the time since the epoch of the *Tables*, 1 Thoth of the year 1 of Philip (12 November -323), is divided, and the computation of the time that has elapsed since then ; to chapter 2 on computing the longitude of the Sun ; to chapter 3 on computing local time from the tables of oblique ascensions ; to chapter 4 on computing the time difference between two localities ; to chapter 6 on computing the ascendent ; to chapter 7 on computing the midheaven ; to chapter 8 on computing the longitude of the Moon ; to chapter 10 on computing the longitudes of the five planets ; to chapter 13 on the declination of the Sun ; to chapter 20 on the syzygies of the Sun and the Moon (in one of his several discussions of this topic the author of the *Preceptum* cites Theon by name) ; to chapter 21 on lunar eclipses ; and to chapter 23 on solar eclipses. Thus only about half of the chapters needed to explain the uses of the *Handy Tables* are available in the surviving manuscripts of the *Preceptum*, and, though some verbal echoes prove that the author of the *Preceptum* used Theon's *Little Commentary*, the *Preceptum* presents the material in a different and confused order, with many abbreviations, expansions, and repetitions. The *Preceptum* is a translation from the Greek as is shown by the fact that many phrases in it — sometimes whole clauses — are simply transliterations of the Greek original, but

[27] Edited by A. TIHON, *Le «Petit Commentaire» de Théon d'Alexandrie aux Tables Faciles de Ptolémée*, Città del Vaticano, 1978 (Studi e testi, 282). In chapter 3, p. 212, Theon mentions that the latitude of the fourth clime is 36°, of the fifth 40;56°, and the sixth 45;1°, but he names none of them.

the author of the Greek original used more than Theon *and* the translator himself made significant additions.

In one manuscript the first half of the text is divided into 25 chapters numbered 1 to 9 and 11 to 26, and the second into 20 chapters, of which the first are numbered 1 to 11 and the remaining 9 are unnumbered[28]. There is no reason to believe that these divisions, especially in the second part of the text, correspond to the translator's intentions, but I shall retain them for their convenience in the rest of this paper. More precise references can be given when the edition is published.

The last four chapters of the first section, numbered 23 to 26, are only partially paralleled in Theon, but give alternative instructions for finding the day of the lunar month, the zodiacal sign in which the Moon is, the longitude of the ascendent, and the longitude of midheaven. The first two chapters are related to a method falsely called Hipparchan by Vettius Valens[29]. It attempts to employ the «Metonic» cycle of nineteen years since Augustus -9 to determine the zodiacal sign in which the Moon lies on the false assumption that the annual epact is ten days exactly. The *Preceptum* in chapter 23 uses as an example 8 days before the Kalends of September in Augustus 383, which is 25 August 354. This, of course, falls within the lifetime of Theon, though it is certainly a method of which he would not approve. He gives in his *Little Commentary*[30] the more correct approximation according to which the years from -9 of Augustus are divided by 19 ; the accumulated epact, then, is 11/30 of the remaining number of years — that is, Theon's epact is 11 tithis instead of the crude 10 days of the *Preceptum* and of Valens. These four chapters were certainly translated from Greek because they contain not only transliterations of the Greek forms of the names of the Egyptian months and of the Greek technical terms «epagomenas», «panselenos», and «anaforas», but as well of the Greek equivalent, «deuteru», of the normal Latin «secundi».

28 The manuscript in which at least some of the chapters are numbered in the way described is the relatively late Chartres 498, the second volume of Thierry of Chartres' *Heptateuchon*.

29 *Vettii Valentis Antiocheni Anthologiarum libri novem*, ed. D. PINGREE, Leipzig, 1986, I, 17 : Ἱππάρχειον περὶ ψήφου Σελήνης ἐν ποίῳ ζῳδίῳ ; see also Appendix VIII on p. 397-398, and O. NEUGEBAUER, *A History of Ancient Mathematical Astronomy*, p. 824-826.

30 *Little Commentary*, p. 257-258.

This excerpt from an «anonymus anni 354», however, is not the only place in which the *Preceptum* introduces the era of Augustus. In chapter 4 of the first part, when the translator is discussing the 25-year tables of mean motion in the *Handy Tables*, called in Greek εἰκοσαπενταετηρίς (which the *Preceptum* transliterates as icosapenteeteris), the translator says that you take the years since Augustus and add to them the years of Cleopatra, which are 294. Indeed, Ptolemy's κανὼν βασιλέων in the *Handy Tables* correctly gives the years from the era of Philip till the death of Cleopatra as 294[31] ; and Theon's κανὼν ὑπάτων Ῥωμαίων lists the years ἀπὸ Ἀλεξάνδρου (from -323) and the years ἀπὸ Αὐγούστου[32]. Presumably the *Preceptum* had a similar table of Roman consuls.

The years between Philip and Augustus, and those between Augustus and Diocletian, are used in an example found in the chapter on the panselenos, number 15 in the second part of the *Preceptum*. This chapter is independent of Theon, and contains no traces of a Greek original ; it is most likely the work of the translator. The example he gives is for pridie of the Ides of September in the year 244 of Diocletian, which was 14 September 528. The interval from Philip to Augustus is given as 294 years, and that from Augustus to Diocletian is given as 313 years ; this gives a total from Philip to Diocletian of 607 years, a number also found in Theon's *Little Commentary*[33]. The total, then, from Philip to the date for which the computation was made is 751. The difficulty that the translator had in providing this example *suo Marte* is indicated by his final two sentences : «Merito ne aliquando contigerit ut sciat qui rationem novit qualiter casus istos observare debeat ut non laboret sicut et mihi evenit. Et cum magno labore hoc potui investigare quia ratio ista multas habet difficultates».

But the first chapter of the second part of the *Preceptum* is decisive in establishing its date. This is part of a recapitulation of the rule for finding the longitude of the Sun already given in chapters 4-7 and 12 of

31 HALMA, I, p. 139 ; see also J. BAINBRIDGE, *Procli Sphaera*, London, 1620, p. 49.

32 *Fasti Theonis Alexandrini*, ed. H. USENER in *Chronica Minora saec. IV.V.VI.VII*, ed. T. MOMMSEN, 3, *MGH AA*, 13, Berlin, 1898, p. 359-381 esp. p. 375-381, headings of columns 1 and 2.

33 *Little Commentary*, p. 204.

the first part. But here an example is given for finding the icosapenteeteris and the ete apla (a transliteration of the Greek ἔτη ἁπλᾶ). We are first told that from Philip to Diocletian there are 607 years ; «et a Diocletiano usque in consulatum quartum domini Iustiniani imperatoris, hoc est in diem V Kalendarum Septembris, anni sunt ccl. Fiunt omnes anni dccclvii». The day in question was 28 August 534, the last day of year 857 of Philip ; and the problem will be to find the longitude of the Sun during the next year, from 29 August 534 till 28 August 535. One may note in passing that the fourth consulate of Justinian did indeed cover the year 534. In the next chapter the translator, following the rule given by Theon in his *Little Commentary*, finds the intercalary days that constitute the difference between the Julian calendar used for ordinary purposes and the Egyptian calendar used in the *Handy Tables*. To compute this he first takes the years from the fifth year of Augustus to Diocletian, which are 308. To these he must add the years that have passed since Diocletian ; as he expresses it : «et a Diocletiano usque nunc anni sunt ccli»[34]. So our translator clearly indicates that for him «nunc» means the Alexandrian year that began on 29 August 534 ; and the *Preceptum* was written just two decades before Cassiodorus composed the *Institutiones*.

This fact brings to mind the fact that in the praefatio to book I of the *Institutiones*[35] Cassiodorus recalls : «nisus sum cum beatissimo Agapito papa urbis Romae ut, sicut apud Alexandriam multo tempore fuisse traditur institutum, nunc etiam in Nisibi civitate Syrorum Hebreis sedulo fertur exponi, collatis expensis in urbe Romana professos doctores scholae potius acciperent Christianae, unde et anima susciperet aeternam salutem et casto atque purissimo eloquio fidelium lingua comeretur». This pious plan, wrecked, as Cassiodorus observes, «per bella ferventia et turbulenta», must have been plotted during the brief pontificate of Agapetus I, between 13 May 535 and 22 April 536 ; the date can be narrowed even further, because Agapetus left Rome in November or December of 535 on a mission to Constantinople from which he never returned[36]. Was the *Preceptum* commissioned by the Senator and Pope

[34] *Little Commentary*, p. 204.

[35] *Inst.*, p. 3.

[36] See J.B. BURY, *History of the Later Roman Empire*, 2 vol., London, 1923, repr. New York, 1958, 2, p. 172, fn. 1.

as the text from which astronomy was to be taught in their proposed Christian Academy in Rome[37] ?

In the *Preceptum* three places are mentionned in addition to the seven πόλεις ἐπίσημοι that give their names to the κλίματα, and Alexandria for which the *Handy Tables* were computed. These three are Apamea, Carthage, and Rome. In all three cases the city is named in the context of computing the time difference between it and Alexandria. Apamea is mentioned in chapters 19 and 21 of the first part as being 9 1/2° or about 1/2 and 1/7 of an hour East of Alexandria. Apamea is known to have been a place where the *Handy Tables* were studied ; this is attested to by a scholion dated in the year 179 of Diocletian (that is, in 462/3) copied onto f. 298v of the principle manuscript of Theon's *Large Commentary*, Vaticanus graecus 190[38]. Carthage is mentioned in chapter 6 of the second part of the *Preceptum* as being 26° or 1 1/2 and 1/4 hours West of Alexandria. I know of no other evidence for the presence of the *Handy Tables* in Carthage, though it is quite reasonable to suppose that the local astronomers and astrologers possessed copies of such a useful work. Finally, Rome is referred to in chapters 2, 19, and 21 of the first part as being 24;10° or about 1 1/2 and 1/10 hours West of Alexandria. Theon in his *Little Commentary*[39] gives the same difference in degrees, but converts it into time more accurately as 1 1/2 and 1/9 hours. Since the example using Rome is already found in Theon, it will not suffice to demonstrate that the translator was working there.

However, a person translating from Greek into Latin in the sixth century would scarcely be working in Apamea, though both Carthage and Rome would be possible ; in both cities, due to Belisarius' successful campaigns, Justinian was recognized as Imperator in 534-535. But the text accords us one final clue that allows us to discriminate between the two rivals. At the end of the second section is a group of chapters discussing various problems in astrology. In the seventeenth chapter, which deals with the computation of the ascendent when one is informed that a nativity occurred on a particular hour of the day or the night, the following example is given «Verbi causa : invenimus Solem

[37] The idea that Cassiodorus sponsored the translation of the *Preceptum* was dismissed by VAN DE VYVER, *Les plus anciennes traductions*..., p. 687, fn. 145, on the grounds that he does not claim that responsability.

[38] *Large Commentary*, p. 73-78.

[39] *Little Commentary*, p. 215.

in Ariete partes iiii minutas xxiiii. et quia nocturna fuit genesis, induximus in diametro Solis, id est in Libra, in climate quinto. ergo induximus in Libra partes iiii minutas xxiiii, et invenimus adiacere in horarum minutis partes xiiii minutas xlv». In the tables of oblique ascension for the fifth clime in the *Handy Tables* 14;45 hours is indeed what one would find by interpolation as the time correponding to a longitude of Libra 4;24°[40]. So the translator-astrologer was active in the fifth clime, which in the *Handy Tables* and the *Preceptum* extends between 40;56° and 45;1° of northern latitude. Since the latitude of Rome, according to the *Handy Tables*, is 41;20° North[41], the *Preceptum* was written in that city.

However, it is precisely this passage that, while it allows us to pinpoint Rome as the place of composition of the *Preceptum*, with others like it forces us to conclude that, though Cassiodorus referred to the *Preceptum* when he was writing his *Institutiones* at Vivarium, he probably had nothing to do with commissioning this work. For astrology was clearly the profession of the translator, and he assumes it is the profession of his reader. Thus, in chapter 8 of the first part he refers to the «clima in quo quis natus est» ; and in the second part, chapter 10 instructs the astrologer on how to find the Lot of Fortune, and chapter 20 contains a section on the influences of aspects, which may be an interpolation, or a misplacement. In light of Cassiodorus' vigorous attacks on astrology in both the *Institutiones*[42] and in his commentary on Psalm 148[43], it seems very unlikely that he would have wished the genethlialogical applications of astronomy described in the *Preceptum* to be taught to the students of his Christian Academy.

The *Preceptum*, then, was composed at Rome in 535, primarily in order to provide Western astrologers with the means to cast horoscopes. It did so by presenting a Latin version of the *Handy Tables* and instructions for their use ; these instructions were derived from parts of Theon's *Little Commentary* on the *Handy Tables*, used in a version that had perhaps been taught at Apamea and at Carthage ; from an astro-

40 HALMA, 2, p. 38.

41 HONIGMANN, *Die sieben Klimata...*, p. 197 (Vat. gr. 1291) ; it is mistakenly given by Leid. 78 as 41;50° (p. 213).

42 *Inst.*, p. 156-157.

43 *Magni Aurelii Cassiodori Expositio Psalmorum*, ed. M. ADRIAEN, Turnhout, 1958, p. 1321 (Corpus Christianorum, Series Latina, 97-98).

nomical treatise of the year 354 ; and from the author's own experience as a practicing astrologer. The usefulness of the *Preceptum* as a guide to astrologers ignorant of Greek is, however, very limited ; because instead of being a translation of the Greek it often is merely a transliteration of it. It is not so bad that the headings of the tables themselves are usually simple transliterations, which are used in the instructions to refer to those specific tables. Thus, in the chapters on computing the position of the Moon, we frequently are told to look up the centron autes Selenes or the boriu peras, by which are meant columns 4 and 5 of the lunar mean motion tables, or even the canonion megistus eliu ce selenes for computing the magnitudes of eclipses. More puzzling to the poor Latin readers must have been sentences like the following wherein not only technical terms and phrases, but even common words are simply transliterated Greek words : «queres similiter in horon cronis apo mesembrias in epiciclo Lune partes x quot horas faciant, et invenies contra eas in prima selide horas x engista»[44] ; or, «et horum utrorumque canonum, id est megistu apostematos et elacistu apostematos, hiperocam sumes vel ex dactilis vel ex emptosi vel ex mone, si habuerit. dehinc numerum illum centron autes Selenes purgatum inferes in procanonion loxoseos»[45] ; or, finally, «mesos cronos enim dicitur hora illa vel momentum hore quo hisomiri Sol et Luna facti sunt. et si, verbi causa, de sexta hora noctis quando equinoctium est deduxeris emptoseos et mones horas duas, incipiet hora noctis quarta eclipsis fieri, id est pro duo horon hisemerinon tu mesonictios. item ipsas emptoseos et mones horas si quas addas sex horis to meson cronon, erit purgatio eclipsis Lune in hora noctis viii, hoc est in meta duo horas hisemerinas tu mesonictios. ut autem cognoscas quota hora incipiat mones cronos»[46]. The text goes on and on with this weird and motiveless alternation of Latin and Greek. To a large extent, the *Preceptum* must have seemed pure nonsense — though very learned nonsense — to those not *in utraque lingua periti*. One must repeat the complaint made by Pope Gregory the Great to Eulogius, the Patriarch of Alexandria, in August of 600 about translators who resort to transliteration[47] : «indicamus praeterea quia gravem hic

[44] I 12.
[45] I 20.
[46] I 20.
[47] X 21 in *Gregorii I Papae Registrum epistolarum*, 2, ed. L.M. HARTMANN, in *MGH, Epistolae*, 2, Berlin, 1899, p. 256-258, esp. p. 258.

interpretum difficultatem patimur. dum enim non sunt qui sensum de sensu exprimant, sed transferre verborum semper proprietatem volunt, omnem dictorum sensum confundunt. unde agitur ut ea quae translata fuerint nisi cum gravi labore intellegere nullo modo valeamus».

Its esoteric learning for an age bereft of any other meaningful form of astronomical knowledge was surely the feature of the *Preceptum* that preserved it, though imperfectly, for us to read. There presently exist six manuscripts of the *Preceptum* ; two others that were at Chartres were destroyed during the Second World War (of one of which microfilm copies survive) ; and a ninth manuscript that once contained the *Preceptum* is now missing the relevant section. All nine manuscripts were copied between the years 1000 and 1250. Briefly, I would reconstruct the history of the manuscripts thus.

The oldest manuscript, Harley 2506 in the British Library[48], was copied in about the year 1000, possibly at Winchester[49], though a better claim can be made for Fleury[50]. It was copied from a defective manuscript since : 1) after chapter 10 of the second part is written : «desunt folia iii», and the following three colums are left blank ; 2) at the end of the second part, before the tables, is written «desunt folia iiii», and the following three colums are left blank ; and 3) of the tables themselves all that are left are the mean motion tables of the Sun, the table of the equation of the Sun, the mean motion tables of the Moon, the table of the equation of the Moon, the table of the solar declination and lunar latitude, and some of the lunar eclipse tables, though the instructions refer to tables of oblique and right ascension, the full set of eclipse tables, and

[48] The *Preceptum* is on f. 55v-69 with the tables on f. 70-73v.

[49] See, *e.g.*, F. SAXL and H. MEIER, *Verzeichnis astrologischer und mythologischer illustrierter Handschriften des lateinischen Mittelalters*, 3, London, 1953, p. 157-160.

[50] E. TEMPLE, *Anglo-Saxon Manuscripts. 900-1066*, which is vol. 2 of J.J.G. ALEXANDER, ed., *A Survey of Manuscripts Illuminated in the British Isles*, London, 1976, no. 42. The connection with Fleury, of course, is through Abbo and Berno ; but Abbo had been to England before Harley 2506 was copied. It appears that the same illuminator decorated several manuscripts copied at Ramsey Abbey and Fleury as well as Harley 2506, so that the last might have been copied at Ramsey as well as Winchester or Fleury ; see D. GREMONT and L. DONNAT, *Fleury, le Mont Saint-Michel et l'Angleterre à la fin du Xe siècle et au début du XIe siècle à propos du manuscrit d'Orléans n° 127 (105)*, in *Millénaire monastique du Mont Saint-Michel*, 1, Paris, 1966, p. 751-793, esp. p. 775-777.

all of the planetary tables, and they imply the existence of the tables of Roman consuls and of the πόλεις ἐπίσημοι. None of the extant manuscripts contains more than does Harley 2506, so that the defective copy from which it was derived is the ancestor of all of them ; but the Harley manuscript contains several genuine phrases found in no other manuscript, so that the ancestor of all the other copies was probably a defective copy of its archetype.

That archetype, which was probably in England in circa 1000, is associated with Abbo of Fleury[51], the expert on computus who studied astronomy at Paris and Rheims in the late 960's and early 970's[52] and was sent to Ramsey Abbey to teach between 986 and 988[53] ; one of his pupils, indeed, was the monk Byrhtferth[54], whose *Manual* of 1011 is evidence both of the fact that Abbo taught astronomy and of the general ineptitude of student or teacher or both. Part of Harley 2506 was copied from Harley 647[55], a manuscript transcribed in France in the middle of the ninth century which was probably sent to England by Abbo ; and Harley 2506 contains a copy of an astronomical work written by Abbo in which he mentioned the year 978[56].

From the ancestor of Harley 2506, which I will dub α, was made a second copy, now lost, β. This second copy of α was apparently in the hands of a follower of Adelard of Bath[57] ; all of its early descendants are associated with one or another of Adelard's works. One of these descendants is now Avranches, Bibliothèque municipale 235, a manus-

51 See P. COUSIN, *Abbon de Fleury-sur-Loire*, Paris, 1954, and, most importantly, A. VAN DE VYVER, *Les oeuvres inédites d'Abbon de Fleury*, in *Revue Bénédictine*, 47 (1935), p. 125-169.

52 AIMOINUS FLORIACENSIS, *De vita et martyrio sancti Abbonis*, in *PL*, 139, c. 390. For a date before 973 see VAN DE VYVER, *Les oeuvres...*, p. 160.

53 *PL*, 139, c. 391-392.

54 *Byrhtferth's Manual*, ed. S.J. CRAWFORD, 1, in *EETS*, 127, London, 1929, p. 232, where Abbo's oral teaching is recorded. See also C. HART, *The Ramsey Computus*, in *English Historical Review*, 85 (1970), p. 29-44.

55 SAXL and MEIER, p. 149-151.

56 A. VAN DE VYVER, *Les oeuvres...*, p. 146, and *Les plus anciennes traductions...*, p. 677-678.

57 The association of his translations with this lost manuscript must be dated after his return to England in ca. 1120, and cannot have occurred much later than 1140 in light of the several manuscripts containing his works and the *Preceptum* copied in the middle of the twelfth century.

cript copied in the twelfth century[58]. It was formerly in Mont-St.-Michel, and *may* have been brought there from Winchester[59]. Avranches 235 once contained Adelard of Bath's translation of Thābit ibn Qurra's *Maqāla fi-'l-ṭalismāt* as the *Liber prestigiorum*[60], and still preserves a version of Adelard's translation of Abū Ma'shar's *Kitāb al-mudkhal al-ṣaghīr* as the *Isagoga minor*[61].

A gemellus of Avranches 235 is the section of Oxford, Corpus Christi College 283 that contains the text of the *Preceptum* with none of the tables[62]. This manuscript was copied in the late eleventh century, but was at some time — presumably in the twelfth century — brought to Chartres, and there joined with various texts that were popular at that great center in the twelfth century, including the revision of Adelard's version of al-Khwārizmī's *al-Zīj al-Sindhind* that is associated with Petrus Alfonsus.

From a third copy of β, which I shall call γ, are descended the Chartrian manuscripts, all of the twelfth century. The first is Chartres 214, now destroyed, which may, of course, be γ itself ; it contained Adelard's original version of al-Khwārizmī[63]. Next is Chartres 498, the

[58] The *Preceptum* is on f. 1-16 and 17-26.

[59] For the close connections between Winchester and Mont St-Michel in the twelfth century see G. NORTIER, *Les bibliothèques médiévales des abbayes bénédictines de Normandie*, Caen, 1966, p. 140.

[60] For this treatise on talismans by Thābit see D. PINGREE, *The Diffusion of Arabic Magical Texts in Western Europe*, in *La diffusione delle scienze islamiche nel medio evo europeo*, Roma, 1987, p. 59-102, esp. p. 74-75. An edition is being prepared by C. Burnett.

[61] See F.J. CARMODY, *Arabic Astronomical and Astrological Sciences in Latin Translation*, Berkeley-Los Angeles, 1956, p. 98. An edition of this also is being prepared by C. Burnett.

[62] The *Preceptum* is on f. 65-81v, the al-Khwārizmī in a later hand on f. 114-145, for which see O. NEUGEBAUER, *The Astronomical Tables of al-Khwārizmī*, København, 1962, p. 133-234. For the manuscript's connection with Chartres, and particularly with Chartres 214, see C. BURNETT, *The Contents and Affiliation of the Scientific Manuscripts Written at, or Brought to, Chartres in the Time of John of Salisbury*, in *The World of John of Salisbury*, Oxford, 1984, p. 127-160, esp. p. 130-132, 137, 141, and 146.

[63] OMONT, MOLINIER, COUDERC, and COYECQUE, *Catalogue général des manuscrits des bibliothèques publiques de France : Départements*, XI : Chartres, Paris, 1890, p. 109-110. The *Preceptum* was on f. 1-13, al-Khwārizmī on f. 41-102. Chartres 214 is manuscript C in A. BJØRNBO, R. BESTHORN, and H. SUTER, *Die astro-*

second volume of Thierry of Chartres' *Heptateuchon*[64], also destroyed ; in this copy the tables of the *Preceptum* are intermingled with those of Adelard's version of *al-Zīj al-Sindhind*[65]. And finally comes the second part of Paris BN latin 14754[66]. Closely related to the text in Chartres 498 is that in Firenze, Conventi soppressi J.IX.39[67] ; and closely related to the text in Paris BN latin 14754 is that in Hannover IV 394[68]. Both the Firenze and the Hannover manuscripts were copied in the late twelfth or early thirteenth century.

From this it is clear that six of the nine known manuscripts of the *Preceptum* are connected with Chartres, though their source most likely came from England ; and that that source, which was associated with the translations of Adelard of Bath, was descended from the manuscript from which were copied the Harley and Avranches manuscripts, a source which was probably brought (or sent) to England in the late tenth century by Abbo of Fleury. Abbo presumably found his manuscript in northern France, perhaps at Paris or Rheims.

The ninth manuscript, Palatinus latinus 1417 in the Vatican, was copied in southern Germany in the middle of the eleventh century[69]. It now contains only the *Liber Nimrod*, of which it is the oldest surviving copy ; but on folio 1 a fifteenth century scribe has written a table of contents which includes : «libellus seu tractatus Ptolomei regis ad scien-

nomischen Tafeln des Muḥammad ibn Mūsā al-Khwārizmī, København, 1914 (*Mem. Acad. Roy. Sci. et Lett. de Danemark*, series 7, Letters 3, 1).

64 See E. JEAUNEAU, *Le «Prologus in Eptatheucon» de Thierry de Chartres*, in *Medieval Studies*, 16 (1954), p. 171-175, repr. in his *Lectio philosophorum*, Amsterdam, 1973, p. 87-91. Thierry taught at Chartres between about 1120 and 1153, and died in 1157.

65 The *Preceptum* is on f. 174-189v. I have used a copy of the microfilm belonging to the Pontifical Institute of Medieval Studies at Toronto, to whose generosity I am greatly indebted.

66 On f. 233-255. Concerning the history of this manuscript see S.J. LIVESEY and R.H. ROUSE, *Nimrod the Astronomer*, in *Traditio*, 37 (1981), p. 203-266, esp. p. 224.

67 On f. 13-32. This manuscript once belonged to Coluccio Salutati (1331-1406) ; see P.L. ROSE, *The Italian Renaissance of Mathematics*, Genève, 1975, p. 27 and 57. It was later willed to San Marco by Niccolò Niccoli (*ca.* 1364-1437) ; see ROSE, p. 33 and 62.

68 On f. 39-63v.

69 LIVESLEY and ROUSE, *Nimrod*..., p. 217.

dum horas diei et noctis», that is, the *Liber horologii Ptolomei regis*, and : «tractatus de distinctione climatum mundi et de terminis septem climatum», which precisely describes the first chapter of the *Preceptum*. If this identification of the last item in the Palatinus with the *Preceptum* is correct[70], it would be exceedingly useful to find either it or a descendant since it is perhaps independent of the manuscripts from England and Chartres.

That the *Preceptum* was known early in southern Germany is made likely by a chapter in a computistical treatise preserved of f. 41v of manuscript 248 at St. Gall[71] which was copied in the first third of the ninth century. The title of this chapter is : «Si vis scire a Septembre usque ad Decembrem, hoc est ab initio anni Aegyptiorum». This is a clear reference to the Alexandrian calendar as employed in the *Preceptum* with the substitution of Roman month names for Egyptian — *e.g.*, of September for Thoth. I know of no other Latin source than the *Preceptum* from which the St. Gall computist could have acquired his knowledge of this Romanized form of the Alexandrian calendar.

The second piece of evidence is also from the first third of the ninth century, from Fulda where Hrabanus Maurus wrote his *De computo*. Hrabanus states[72] : «modo autem, id est anno dominicae incarnationis DCCCXX mense Iulio nona die mensis, est Sol in XXIII parte Cancri, Luna in nona parte Tauri, stella Saturni in signo Arietis, Iovis in Librae, Martis in Piscium, Veneris quoque stella et Mercurii, quia iuxta Solem in luce diurna modo sunt, non apparet in quo signo morentur». In fact, on the afternoon of 9 July 820, the Sun was at Cancer 21°, the Moon at Taurus 30° (it was at Taurus 9° at about sunset of 7 July), Saturn at Aries 19°, Jupiter at Libra 29°, Mars at Pisces 16°, Venus at Cancer 26°, just 5°

70 VAN DE VYVER, *Les plus anciennes traductions...*, p. 686, fn. 140, suggests that the *De distinctione climatum mundi* refers to chapters 18-19 of the *De utilitatibus astrolabii* edited by Bubnov (see fn. 79).

71 A. CORDOLIANI, *Les manuscrits de comput ecclésiastique de l'Abbaye de Saint Gall du VIIIe au XIIe siècle*, in *Zeitschrift für Schweizerische Kirchengeschichte — Revue d'histoire ecclésiastique Suisse*, 49 (1955), p. 161-200, esp. p. 175. The names of the Egyptian months are to be found with the dates in the Roman calendar with which they begin in BEDE, *De temporum ratione*, 11, in C.W. JONES, *Bedae Opera de temporibus*, Cambridge, Mass., 1943, p. 205 ; cf. HRABANUS, *De comp.*, 30, p. 234-235.

72 *De comp.*, 48, p. 259.

from the Sun and so invisible, and Mercury at Leo 16°, some 25° from the Sun and so, presumably, visible briefly after sunset, though Hrabanus apparently did not see it. In any case, his method of reporting the longitude of the Sun and the Moon to the degree while only giving those of the superior planets in their zodiacal signs, taken together with his statement about the invisibility of the inferior planets, indicates that Hrabanus computed the longitudes of the two luminaries, and observed those of the planets. The only text for computing longitudes available in the West in 820 was the *Preceptum* ; and, though it originally had planetary tables, since at least the tenth century it has contained only the tables for the two luminaries. Using the *Handy Tables* I compute the longitude of the Sun at noon in Alexandria of 9 July 820 to be Cancer 23;19°, that of the Moon Taurus 26;16°. Since Rome is, according to the *Preceptum*, 1 1/2 and 1/10 hours West of Alexandria, at noon of 9 July 820 in Rome the longitude of the Moon would be computed to be about Taurus 28°. Fulda is a little bit more than 12°, or about 48 minutes of time, West of Rome, which would make little difference in the Moon's longitude. I don't know whether Hrabanus could account for the longitudinal difference between Fulda and Rome. It is tempting to correct Hrabanus' 9° of Taurus to 29° ; but the accepted reading is supported by a manuscript copied by Walahfrid Strabo at Reichenau within about five years of the composition of the *De computo*[73], and another that was collated by Walahfrid at Fulda in the period between 827 and 829[74]. There remains the strong possibility, given the opaque nature of the instructions in the *Preceptum*, that Hrabanus made a mistake. This problem cannot at present be definitively solved. However, given that he must have had a set of astronomical tables to arrive at *any* longitude expressed in such precise terms, the probability is that Hrabanus had a copy of the *Preceptum* at Fulda ; and the further probability is that it had already lost the planetary tables — *i.e.*, that it was closely related to the manuscript in Northern France that was the ancestor of all our extant copies.

There is one other early horoscope that is pertinent to the history of the *Preceptum*. This is preserved in an illustration on f. 93v of Vossianus Q.79 in Leiden, a manuscript copied in northern France in the first half of the ninth century. The positions of the planets were probably

[73] Saint Gall 878 ; see *De comp*, p. 191-192.
[74] Oxford Can. misc. 353 ; see *De comp.*, p. 190-191.

cast with the help of the *Preceptum*, the only set of astronomical tables known to have existed in Latin before the twelfth century. Eastwood dated[75] the horoscope to 28 March 579, with Saturn in Aquarius as the diagram shows ; Jupiter just leaving the indicated Gemini ; Mars having retrograded 8° back from Scorpius, where it is shown to be, into Libra ; the Sun properly in Aries and the Moon properly opposite it in Libra ; but Venus in Aries when it is shown on the border of Taurus and Gemini, and Mercury in Pisces when it is shown on the border of Pisces and Aries. If Eastwood is correct, the *Preceptum* still had its planetary tables in the late sixth century, which is not surprising. But another date for the horoscope is possible — 14 April 816[76], near the beginning of the reign of Louis the Pious (814-840) who was noted for his interest in astronomy[77]. At this time Saturn was again in Aquarius and Jupiter at the end of Gemini ; Mars had just retrogressed out of Scorpius into Libra ; the Sun was at the end of Aries, and the Moon was in Libra ; Venus was in Gemini, and Mercury in Aries. This is not a great improvment, if any at all, because of the peculiar and ambiguous manner the artist has chosen for depicting the positions of Venus and Mercury in relation to the Sun, the earth, and the zodiacal signs ; this method, incidentally, reflects not a theory of the heliocentric orbits of the two inferior planets, but the simple fact that in Ptolemaic astronomy the centers of the epicycles of Venus and Mercury have a longitude equal to that of the mean Sun. The horoscope represents the mean longitudes of the su-

75 B.S. EASTWOOD, *Origins and Contents of the Leiden Planetary Configuration (MS. Voss. Q. 79, f. 93v), an Artistic Astronomical Schema of the Early Middle Ages*, in *Viator*, 14 (1983), p. 1-40, esp. p. 2-4.

76 The apparent date in the illustration on f. 93v of Vossianus Q. 79 and the date as computed by means of B. TUCKERMAN, *Planetary, Lunar, and Solar Positions. A.D. 2 to A.D. 1649 at Five-day and Ten-day Intervals*, Philadelphia, 1964, are as follows :

Planet	*Manuscript*	*28 March 579*	*14 April 816*
Saturn	Aquarius	Aquarius 7°	Aquarius 25°
Jupiter	Gemini	Cancer 1°	Gemini 30°
Mars	Scorpius	Libra 22° retr.	Libra 29° retr.
Sun	Aries	Aries 10°	Aries 29°
Venus	Taurus/Gemini	Aries 15°	Gemini 21°
Mercury	Pisces/Aries	Pisces 23°	Aries 16°
Moon	Libra	Libra 10°	Libra 12°

77 See, *e.g.*, *Vita Hludowici imperatoris*, ed. G.H. PERTZ, *MGH, Scriptores*, 2, Hannover, 1829, p. 604-648, esp. chapter 58 on p. 642-643.

perior planets and the luminaries, the anomalies of the inferior planets. But, if the horoscope was cast in the Spring of 816, shortly before the Leiden manuscript was produced, that fact would suggest that a complete copy of the *Preceptum* was still available in northern France in the early ninth century — at Louis' court, if that is where the codex was produced — but an already truncated copy was all that was available to Hrabanus at Fulda. I suspect, therefore, that Eastwood's date in the late sixth century is correct, though the early ninth century date remains a possibility.

Few indeed are the medieval scholars aside from those whom I mentioned who can be shown to have read the *Preceptum*. It is clear, for instance, that the statement made by Hugh of St. Victor, who died in about 1141, that : « Ptolomaeus rex Aegypti» write «canones ... quibus cursus astrorum invenitur» is based on Isidore of Saville's expansion of Cassiodorus rather than directly on the *Preceptum*[78]. But it is probably the *Preceptum* that is referred to in the *De utilitatibus astrolabii* written by Abbo's contemporary, Gerbert (972-1003)[79]. And Rudolph of Bruges seems to have used it to compute the longitude of the Sun at Béziers on 24 April 1144[80]. With the introduction of serious astronomical tables translated from or based on Arabic works available in Spain in the early twelfth century, the *Preceptum* became generally useless except as an antiquarian or bibliographical curiosity. As such it appears for the last time in medieval literature, as far as I am aware, in the *Speculum astronomiae* of Albertus Magnus[81]. Earlier still, in the middle of the twelfth

78 LIVESLEY and ROUSE, *Nimrod...*, p. 235.

79 Edited as *Gerberti liber de astrolabio* by N. BUBNOV, *Gerberti postea Silvestri II papae Opera Mathematica*, Berlin, 1899, p. 109-147 ; for Ptolemy see I, 2 (p. 116), XIII, 2 (p. 135) and XVIII, 1 (p. 139). On the authenticity of the attribution to Gerbert see P. KUNITZSCH, *Glossar der arabischen Fachausdrücke in der mittelalterlichen europäischen Astrolabliteratur*, Göttingen, 1983, p. 455-571, esp. 479-480 (Nachrichten Akad. Wiss. Göttingen, Phil.-hist. Kl., 1982, 11).

80 R. MERCIER, *Astronomical Tables in the Twelfth Century*, in *Adelard of Bath*, ed. C. BURNETT, London, 1987, p. 87-115, esp. p. 114-115, shows that Rudolph most likely used Ptolemy's tables to compute the solar longitude, but he wrongly states that it would have been impossible to use these tables. Men of intelligence, such as Hrabanus and Rudolph, clearly were able to overcome the difficulties the *Preceptum* presented to its readers.

81 *Alberto Magno. Speculum astronomiae*, ed. S. CAROTI, M. PEREIRA, and S. ZAMPANI under the dir. of P. ZAMBELLI, Pisa, 1977, p. 10-11.

century, Thierry of Chartres could do nothing better when he copied the *Preceptum* into his *Heptateuchon* than playfully to retransliterate its Latin transliterations back into the Greek alphabet. Unfortunately, this attempt was a total disaster, rendering the already virtually impenetrable text even more obscure. For Thierry did not know Greek well enough to be able to select the correct Greek letters for his retransliteration, especially confounding aspirate with nonaspirate consonants and long with short vowels.

The *Preceptum* was almost a complete failure as a translation because far too much of it was simply a transliteration. For a half millennium, then, before the eagerly sought after Arabic astronomical tables were translated, the Latin West had possessed copies of a set of tables superior to al-Khwārizmī's *al-Zīj al-Sindhind* translated by Adelard, but in a form whose opaqueness most medieval scholars were unable to penetrate.

Brown University, Providence

The Teaching of the Almagest *in Late Antiquity*

David Pingree

My objectives in this paper are two: to reveal the richness of the tradition of the study of the *Almagest*[1] between the date of its composition in the middle of the second century AD[2] and its appearance simultaneously in Byzantium[3] and Baghdād[4] in the late eighth century, and to hypothesise about the origins of one of several commentaries on the *Almagest* that were composed during this period.

The earliest documented attempt to discuss critically any portion of the *Almagest* was written by an otherwise totally obscure Artemidorus in the latter half of the second century or at the beginning of the third. A passage regarding the lunar theory of Books IV and V of the *Almagest* is quoted from this Artemidorus in the anonymous commentary on the *Handy Tables* edited by Jones and shown by him to contain an example

1 Heiberg 1898-1903; translated into English by Toomer 1984.

2 For the date, see Toomer 1984: 1.

3 It is referred to by Stephanus the Philosopher c. 790: see *Catalogus Codicum Astrologorum Graecorum* 2, 1900: 182. For the date of Stephanus, see Pingree 1989.

4 See Pingree 1973. For the extant versions in Arabic, see the introductions in Kunitzsch 1986; 1991.

that can be dated 24 April 213.[5] Artemidorus established a tradition — fortunately not followed universally — of failing to understand or of misrepresenting Ptolemy's statements in the *Almagest*.[6]

It is clear that Artemidorus, however deficient his understanding, approached the *Almagest* as an astronomer interested in the way in which Ptolemy's solutions to problems work mathematically. The same can be said of the two Alexandrian scholars who commented on the *Almagest* during the course of the fourth century. Pappus composed his Σχόλιον[7] after 18 October 320, the date for which, in his commentary on *Almagest* VI 4, he computed the time of a mean conjunction of the sun and the moon.[8] Of Pappus's *Scholion* there survive in their entirety only Books V and VI; as we shall see, some fragments of other books — it is not yet clear how many — can be recovered from other sources. Rome conjectured that Pappus followed a tradition established by some earlier commentator of dividing each of the thirteen books of the *Almagest* (excluding, presumably, the Star Catalogue in VII and VIII) into sections called θεωρήματα; echoes of this practice survive in the commentary of Theon[9] and in the later scholia.[10] These divisions may have been more useful in the teaching of the *Almagest*, which was strongly directed toward the students' acquiring computational skills, than were Ptolemy's chapter divisions, which reflect his concern with the logical development of astronomical theory. Despite his emphasis on computations, Pappus makes in Books V and VI about a dozen errors in calculation, and in eight cases seems to have deliberately falsified his results in order to make them agree with the *Almagest*.[11] His students could not have been very alert since they seem to have let him get away with his slipshod habits.

5 Jones 1990.

6 Jones 1990: 11-12.

7 The commentary on Books V and VI is edited by Rome 1931.

8 Rome 1931: 180-183, cf. x-xiii.

9 Rome 1931: xviii-xx.

10 Traces occur, for example, in the scholia of Vaticanus Graecus 1594 discussed below.

11 In Rome 1936: lxxxv.

Theon composed his Ὑπόμνημα[12] on the *Almagest* most probably in the 360s and 370s. Now that Professor Tihon has discovered most of Theon's commentary on Book V in the margins of Vaticanus Graecus 198,[13] we have all of the Ὑπόμνημα except for Book XI. Of the eighteen manuscripts of the commentary known to Rome the oldest is Laurentianus 28.18, copied in the ninth century and at present preserving only Books I to IV and VI. In fact, ten of the manuscripts contain nothing beyond Book VI, and they, without the Laurentianus, form Rome's Class II. So there existed an edition of Theon that broke the text at the end of Ptolemy's discussion of spherical trigonometry, and the theories of the sun, the moon, and eclipses. This is also the dividing point between the two volumes of Heiberg's critical edition. One wonders if the survival of Books V and VI of Pappus's Σχόλιον may not be due to their being the end of a first volume of an edition of which the second volume in its entirety, the first in its beginning have been lost.

Theon assumes a fairly low level of mathematical ability and accomplishment in his students. They need, for instance, to be instructed at great length on the multiplication and division of sexagesimal fractions and on a variety of geometrical theories, while in general Theon teaches them nothing beyond the replication of Ptolemy's calculations. If the students had gone on to study the five books of Theon's so-called *Great Commentary* on the *Handy Tables*,[14] they would have learned how Ptolemy transformed the parameters and models of the *Almagest* into astronomical tables, but from neither Pappus nor Theon would they have learned how to conduct research that would lead to improvements in either Ptolemy's parameters or his models.

If Cameron's interpretation is correct, Theon not only commented on the *Almagest*; he edited the text or at least Books I and II, while his daughter, Hypatia, edited at least that of Book III.[15] It remains to be determined

---

12 Books I and II were edited by Rome 1936, and Books III and IV in Rome 1943. Books I-II, IV-X, and XII-XIII together with Book V of Pappus were published with the *Almagest* as *Theonis Alexandrini In Claudii Ptolemaei Magnam Constructionem Commentariorum lib.* XI (Basle, 1538); but this edition contains only a fragment of Theon on Book V.

13 Tihon 1987.

14 Books I-III have been edited by Tihon 1985-1991 (the first volume with J. Mogenet).

15 Cameron 1990.

precisely in what ways they modified Ptolemy's text, or in what ways Hypatia changed her father's Ὑπόμνημα if, indeed, she did; some possibilities are given by Cameron; others have been pointed out by Tihon.[16]

Questions concerning the situation in which Pappus and Theon taught the *Almagest* and what sort of students they might have attracted present problems whose solutions we can only guess at. The *Suda*, to be sure, states that Theon belonged to the Museum at Alexandria,[17] but both the accuracy of this report and even the existence of the Museum in the late fourth century are in doubt. Pappus, Theon, and Hypatia all bore the title of philosopher, but this need not indicate an official teaching post. Theon remarks in the preface to his Ὑπόμνημα:[18] 'He who is called to astronomize should not bring forward the μακρολογίαι from philosophy'; this statement presumably reflects the attitude of all three to the relation of philosophy to astronomy. Moreover, the known works of all three are devoted almost exclusively to the mathematical sciences. They seem, then, to have been private teachers of quadrivial subjects to students who had received a basic education. Those students who studied the *Almagest* under their tutelage were interested primarily in learning the details of Ptolemy's mathematical models, and had no practical motive such as the practice of astrology. Astrologers operated with far less sophisticated material; at best they learned how to manipulate the *Handy Tables*, a skill also taught by Theon.[19]

So astronomy in the fourth century was merely the culmination of the quadrivium, a course of study whose pursuit would produce an educated man; and it was considered to have been perfected in the *Almagest*, which an advanced student could hope to understand, but certainly not to improve. A different attitude was manifested by Neoplatonist philosophers in the fifth century; for Proclus took very seriously indeed the position of the planetary spheres in the universe as intermediaries between the supercelestial, intellectual world, and this sublunar world of sensation. In his exposition of the *Almagest*, therefore, the Ὑποτύπωσις τῶν ἀστρονομικῶν ὑποθέσεων, while displaying a full command of its

16 Tihon 1992: esp. 131-135.

17 Θ 205 (2.702 Adler).

18 Rome 1936: 319.

19 Tihon 1978.

technical details, Proclus's opening words are: 'Great Plato, O comrade, deems it fit that he who is truly a philosopher, casting aside the senses and the whole wandering substance, say farewell to them, astronomize beyond heaven, and consider abstract slowness and swiftness there in their true number'.[20] It is, then, only with reluctance that Proclus undertakes his explanation of the *Almagest*; and he ends the Ὑποτύπωσις by casting doubt on the reality of epicycles and eccenters, by pointing out that the astronomers have not ventured to assign causes for their motions, and finally by criticizing them because: 'Proceeding backwards, they do not decide upon what follows from hypotheses as the other sciences do, but they try to fashion the hypotheses from the conclusions'.[21] Despite such criticisms, Proclus and all succeeding Neoplatonists believed astronomy to be necessary to theology, and the *Almagest* to be the supreme achievement of astronomical studies.

A student of Proclus named Hilarius of Antioch is stated in the *Hypotyposis* to have proved (or to possess a proof of) the equivalence of the epicyclic and eccentric models.[22] Damascius in his *Life of Isidore* informs us that Hilarius was dismissed from the Academy because Proclus could not stand his gluttony.[23] And Marinus, Proclus's successor as the head of the Academy, is known to have commented on the *Handy Tables*, citing Pappus on Book V of the *Almagest*.[24] At the end of Photius's summary of Damascius's *Life of Isidore* we are told that Damascius studied geometry and arithmetic and the other mathematical sciences under Marinus at Athens, but that Ammonius in Alexandria was his instructor in Plato's works and in Ptolemy's *Syntaxis*.[25] Indeed, preceding the *Almagest* in several manuscripts is a report from the hand of Heliodorus of several observations made by himself and his brother Ammonius in Alexandria between 498 and 510.[26] These notes are accom-

20 *Hypotyposis* 1.1 (Manitius 1909: 2).

21 *Hypotyposis* 7.55-57 (Manitius 1909: 238).

22 *Hypotyposis* 3.73-88 (Manitius 1909: 76-84).

23 Photius, *Bibliotheca*, Codex 242.266 (6.50-51 Henry).

24 Tihon 1976.

25 Photius, *Bibliotheca*, Codex 181 (2.192 Henry), cf. Codex 242.145 (6.36 Henry).

26 Heiberg 1907: xxxiv-xxxvii, cf. Neugebauer 1975: 2.1038-1041.

panied by an anonymous *Prolegomena to the Almagest*, which Mogenet attributed to Eutocius,[27] though that attribution has been questioned by Knorr.[28] Thus a large number of Proclus's students and immediate successors studied the *Almagest*.

There remains one substantial, but little studied witness to late antique teaching of the *Almagest*. This was an apparently full-scale commentary out of which was fashioned a set of scholia.[29] The oldest copy of these scholia is one of the oldest copies of the *Almagest* itself. Vaticanus Graecus 1594[30] is a magnificent folio volume of 284 leaves, copied in an early minuscule in the ninth century, normally with two columns per page. A note on fol. 284v written in the thirteenth century indicates that it then consisted of 315 folia; but, despite its loss of 31 folia since that time, it still contains, in addition to the *Almagest* (fols. 9-263v), Ptolemy's Φάσεις (fols. 264-272), Περὶ κριτηρίου καὶ ἡγεμονικοῦ (fols. 272v-277), and Ὑποθέσεις (fols. 278-283), for each of which it is one of the most important witnesses. This collection is preceded, on fols. 1-8v, by the Προλεγόμενα, though incomplete at the end; the lost folia clearly contained the record of the observations made by Ammonius and Heliodorus at Alexandria in the early sixth century. This connects the text, in some undefined way, with the *Almagest* tradition of Alexandria. The likelihood is that its archetype emanated from that city in the sixth century.

But we are interested not in the text itself, but in the copious scholia written in its margins and between its columns. Most of these scholia were written by the original scribe in the ninth century; they appear to be extracted from a single commentary or course of lectures. Others were copied in perhaps the twelfth century; they contain, among many other things, important references to Islamic astronomy and to Theon's *Great Commentary* on the *Handy Tables*.[31] Another scribe, of the thirteenth century, records at the end of the *Almagest* (fol. 263v) that the codex belonged to an unknown Leo the astronomer. And on fol. 1 is a note, written, if

27 Mogenet 1956.

28 Knorr 1989: 155-211.

29 Mogenet 1975.

30 Giannelli 1950: 223-225.

31 Mogenet 1962; 1975: 307-311; Tihon 1989.

Ševčenko is correct, by Nicephorus Gregoras, who must have had access to this beautiful manuscript at Constantinople in the early fourteenth century. Three centuries later, in 1604, it was in the possession of Lelius Ruinus, from whose library it was purchased for the Vatican in 1622.

Fortunately, because the margins whereon the scholia were written are frequently in such a damaged state and the ink so faded that help is desperately required in deciphering them, most of both the ninth century and the twelfth century scholia were copied in the thirteenth century onto fols. 25-80 of Vaticanus Graecus 184,[32] a manuscript of 220 folia copied by several contemporary scribes. A good idea of its date is furnished by the presence on fols. 2-8 of the oldest surviving copy of the Κατ' Ἰνδοὺς ψηφηφορία ἡ λεγομένη μεγάλη that was composed in 1252,[33] following which is a copy of a ψηφηφορία of 906 copied in 1270/1.[34] Between this text and the scholia are the Προλεγόμενα, the *Canobic Inscription*, and the report of the observations of Ammonius and Heliodorus, while following the scholia is the text of the *Almagest* (fols. 82-220),[35] corrected in part directly from Vaticanus Graecus 1594.[36] It should not surprise us, then, that the set of scholia copied onto fols. 25-80 of Vaticanus Graecus 184 were copied directly from Vaticanus Graecus 1594, with no distinction being made, of course, between the ninth and the twelfth century scholia.

The title of this collection of scholia is: Ξέωνος Ἀλεξανδρέως σχόλια πάνυ χρήσιμα εἰς τὴν Μεγάλην Σύνταξιν. It is true that Theon has contributed heavily if not heedfully to the composition of these scholia, but his words are often changed, and often supplemented by a later commentator's sometimes silly remarks. Still, the scholia are at times useful for reconstructing Theon's text, especially since it provides evidence for the state of that text more than two hundred years before the copying of the Laurentian manuscript, to which it is closer than to the other manuscripts of Theon. The distance in time and culture of our commentator from Theon is perhaps indicated at one of the few points at which the scholiast dares

32 Mercati and Franchi de' Cavalieri 1923: 210-212.

33 Allard 1977.

34 Pingree 1986: 398-406.

35 Heiberg 1907: xcv-cxxvi.

36 Heiberg 1907: cxvii-cxx, cf. xxxii-xxxiii.

to contradict his predecessor; for, when he criticizes some no longer extant tables of compound ratios produced by Theon,[37] he calls him sarcastically: ὁ μέγας φιλόσοφος, and remarks: 'Behold, the tables of Theon have been preserved even though they are not correct'.

Our commentator also frequently quotes from Pappus's commentary, naming him directly;[38] these quotations are scattered throughout the *Almagest*, and so inform us of at least some of Pappus's comments on books other than the fifth and the sixth of the *Almagest*. Undoubtedly many other quotations from Pappus exist among the scholia without direct attribution; they will be difficult indeed to identify. Our commentator also quotes from, among others, Proclus and Marinus; this means that he must have written in the sixth century at the earliest.

There seem to me to be basically two different ways in which these scholia could have come into being. The first requires a single intelligence interpreting the text in a more or less consistent manner — consistent both with itself and with the author's mathematical, astronomical, and philosophical points of view. But, if this were the case, it must be admitted that the author, in lecturing on the *Almagest*, has been content to copy from his predecessors many passages that seemed to him to be useful, though he occasionally expanded upon or added to them. The alternative hypothesis is to assume that some teacher sat down with several *Almagest* commentaries and compiled his own commentary out of them, again occasionally interjecting his own comments, but not striving for consistency. This hypothesis would help to explain the triviality and/or disconnectedness of many of the mathematical scholia as well as the contradictions or incompatibilities of some of the astronomical scholia with each other. But this hypothesis also has the necessary consequence that there existed in late antiquity several more commentaries on the *Almagest* than those of Theon and Pappus. I wish now briefly to examine the writings of those sixth and early seventh century scholars whom we know of and who can be shown to have studied the *Almagest* in order to assess the possibility of their having commented on it and to compare their known views with those found in our scholia.

Seven scholars — Heliodorus, Eutocius, John Philoponus, Simplicius, Olympiodorus, Rhetorius of Egypt, and Stephanus of Alexandria — come

37 Fol. 23v of Vat. gr. 1594 (V), fol. 30 of Vat. gr. 184 (v).

38 E.g., fols. 31v and 189 of V, fols. 35 and 67v of v.

immediately to mind as students of Ptolemaic astronomy and as potential candidates for the authorship either of the commentary itself that lies behind our scholia or of commentaries that contributed to it. Let us briefly consider each one, as this exercise will bear ample witness to the remarkable extent to which the *Almagest* was taught in late antiquity, as well as providing the necessary background to the solution to the problem of the authorship of the scholia that I shall propose at the end of this paper.

The first candidate is Heliodorus, whose report of a series of seven astronomical observations was prefaced to the archetype of some of the more important manuscripts of the *Almagest*, including Vaticani Graeci 184 and, once, 1594. Heliodorus made an observation in May 498 by himself, one in February 503 with his brother, Ammonius, and one in March 509 by himself. Heliodorus was probably also the observor in June 509 who compared his observation of a conjunction of Mars and Jupiter with computations from the Κανών (i.e., the *Handy Tables*) and the Σύνταξις; clearly Heliodorus was an interested and reasonably competent astronomer. But, though Tannery attributed to him the anonymous Προλεγόμενα[39] and Boll some simple planetary schemes[40] borrowed by Rhetorius from the equally anonymous Σχόλια to the *Handy Tables* together with some other Rhetorian pieces,[41] his only known publication was a commentary on Paul of Alexandria's astrological Εἰσαγωγή, which commentary Heliodorus apparently entitled Διδασκαλία ἀστρονομική; aside from a few fragments in scattered manuscripts[42] and some citations by Rhetorius,[43] it is imperfectly known to us through its having been used by Olympiodorus as the basis for his commentary on Paul.[44] In any

39 Tannery 1894.

40 *Catalogus Codicum Astrologorum Graecorum* 7, 1908: 119-122, cf. Neugebauer 1958. This is *Epitome* III b 28 of Rhetorius, to be published in the second volume of the forthcoming edition by Pingree.

41 Rhetorius 6.2.

42 Warnon 1967: esp. 212-217. The question of Heliodorus's commentary needs further investigation.

43 Rhetorius 6.5.1-11.

44 Boer 1962, cf. Warnon 1967; Westerink 1971.

case, there is nothing in this material to indicate that Heliodorus might have been the author of a commentary on the *Almagest* as well.

The second scholar is Eutocius of Ascalon, who dedicated his Ὑπόμνημα on Archimedes's Περὶ σφαίρας καὶ κυλίνδρου[45] to Ammonius, who was probably his teacher. Eutocius wrote primarily on mathematics, commentaries on works of Archimedes and Apollonius, but he also composed a treatise on genethlialogy, the Ἀστρολογούμενα; from this work Rhetorius has preserved a lengthy discussion of a horoscope that can be dated 28 October 497.[46] As is usual in astrological texts, the positions of the planets were computed by means of the *Handy Tables*.

But Eutocius is quite familiar with the *Almagest* as well. At the end of his commentary on Archimedes's Ἡ τοῦ κύκλου μέτρησις he refers to Ptolemy's method of computing chords as expounded in the Μαθηματικὴ Σύνταξις,[47] and earlier in the same commentary he mentions that methods of finding the square-root of a given number are described by Heron in his Μετρικά and by Pappus, Theon, and many other commentators on the Μεγάλη Σύνταξις of Ptolemy.[48] We do not know who these other commentators might be; but it is of interest to note that the methods of both Heron and Theon are summarized in the anonymous Προλεγόμενα which has been characterized by Mogenet as a commentary on the first book of the *Almagest*.[49] Mogenet further notes that Eutocius, in his commentary on the second book of Apollonius's *Conics*,[50] states that he has discussed compound ratios 'in the commentary published by us on the fourth theorem of the second book of Archimedes's Περὶ σφαίρας καὶ κυλίνδρου and in the σχόλια of the first book of Ptolemy's Σύνταξις.' Mogenet takes this commentary by Eutocius on the first book to be identical with the Προλεγόμενα and not to have been extended to the rest of the *Almagest* (an unprovable hypothesis); he buttresses this

45 Heiberg 1915: 2.

46 Rhetorius 6.52. Toomer 1976: 18 n. 2 surmises that this may be Eutocius's own horoscope.

47 Heiberg 1915: 260.

48 Heiberg 1915: 232.

49 Mogenet 1956: 22-26.

50 Heiberg 1893: 218.

theory of Eutocius's authorship by comparing the treatment of compound ratios in the Προλεγόμενα with Eutocius's treatments of this subject in his commentaries on the Περὶ σφαίρας καὶ κυλίνδρου and on the *Conics*. It is on the basis of a re-examination of this comparison that Knorr denies that Eutocius could have written the Προλεγόμενα, suggesting instead the obscure Arcadius whom also Eutocius mentions as the author of συντάγματα along with Theon and Pappus.[51] For us it suffices to remark that what little concerning compound ratios our scholia contain is derived from or influenced by Theon, not by Eutocius or the Προλεγόμενα; and that the explanation of the multiplication and division of sexagesimal fractions given in our scholia as well as other material is different from those presented in the Προλεγόμενα. I conclude that neither Eutocius nor the author of the Προλεγόμενα wrote our commentary nor was followed in his mathematics by our commentator. One can only hope that parts of Eutocius's lost commentary, whether it was just on Book I or on all of the *Almagest*, may be preserved in the scholia that adorn the margins of other *Almagest* manuscripts.

John Philoponus, who demonstrates his interest and competence in mathematical astronomy in his Περὶ τῆς τοῦ ἀστρολάβου χρήσεως καὶ κατασκευῆς,[52] clearly studied the *Almagest* (presumably under his teacher, Ammonius) since in his commentary on Aristotle's Μετεωρολογικά, in his *De opificio mundi*, and in the *De aeternitate mundi contra Proclum* he refers to Hipparchus's and Ptolemy's hypothesis of a ninth, starless sphere, which can be deduced from *Almagest* VII 3, and to their value for the rate of precession.[53] Philoponus's statement that Ptolemy attributes to Aristotle the view that the planets have no proper motion of their own comes from the second book of the *Planetary Hypotheses* rather than from the *Almagest*, but the refutation of Aristotle's counteracting spheres does not.[54] There is no reason at all to believe that Philoponus wrote a

---

51 Heiberg 1915: 120, cf. Knorr 1989: 166-168. Nothing, of course, makes us certain that Arcadius's σύνταγμα was a commentary on the *Almagest*.

52 Segonds 1981.

53 John Philoponus, *In Meteora (CAG* 14.1) p. 110 Hayduck; *De opificio mundi* pp. 15-16, 113, 115-117 Reichardt; *De aeternitate mundi contra Proclum* p. 537 Rabe.

54 John Philoponus, *In Meteora (CAG* 14.1) p. 109 Hayduck; *De opificio mundi* pp. 114-115 Reichardt.

commentary on the *Almagest*; and in any case his Monophysite tendencies conflict with the theology of our scholiast.

At this point I shall break the chronological sequence and, skipping temporarily over Simplicius, turn to Olympiodorus. He, as we have seen, utilized Heliodorus's commentary on Paul of Alexandria's astrological Εἰσαγωγή as the basis for his own lectures on that text delivered in Alexandria during the summer of 564.[55] The original version of this commentary refers only to the *Handy Tables*; and Olympiodorus's explanations of astronomical matters does not rise above the rather low level achieved by Paul. However, in his commentary on Aristotle's Μετεωρολογικά Olympiodorus does demonstrate some familiarity with the *Almagest*. What he repeats from that text are quite specific things: the ratios of the distances of the sun, the moon, and the tip of the earth's shadow from the center of the earth;[56] the fact that the size of the sun is 170 times that of the earth;[57] that the greater a celestial body's parallax, the closer it is to the earth;[58] and that the fixed stars move with precession 1° in 100 years.[59] In a commentary on the Μετεωρολογικά, of course, there is little room for the discussion of mathematical astronomy; still it must be conceded that Olympiodorus nowhere in his extant writings displays the knowledge of or interest in mathematics that a commentator on the *Almagest* ought to possess. Therefore I conclude that he is unlikely to have been the author of any part of our scholia.

We turn now to our fifth candidate, Rhetorius of Egypt, who wrote an enormous astrological compendium, entitled Θησαυροί, in the early seventh century, most likely at Alexandria.[60] In the first four chapters of the sixth book he assembles astronomical information that he believes will be useful to astrologers. His sources include the anonymous fifth or sixth century scholia on the *Handy Tables*,[61] Theon of Smyrna's Τὰ κατὰ τὸν μαθηματικὸν χρήσιμα εἰς τὴν Πλάτωνος ἀνάγνωσιν, and Ptolemy's Φάσεις

---

55 See the works cited in note 43.

56 Olympiodorus, *In Meteora (CAG* 12.2) p. 68 Stüve, cf. *Almagest* 5.15.

57 Olympiodorus, *In Meteora (CAG* 12.2) p. 19 Stüve, cf. *Almagest* 5.16.

58 Olympiodorus, *In Meteora (CAG* 12.2) p. 75 Stüve, cf. *Almagest* 5.17.

59 Olympiodorus, *In Meteora (CAG* 12.2) p. 76 Stüve, cf. *Almagest* 7.2.

60 Pingree 1977.

61 Tihon 1973.

ἀπλανῶν ἀστέρων. He also at several points cites the Σύνταξις — at one point, for instance, to the effect that the solar parallax is 2 minutes (Ptolemy's maximum is 0;2,51°)[62] and, most surprisingly, that both Venus and Mercury manifest a parallax. This seems a deduction from the computed distances from the earth of the two inferior planets which we will discuss later as it occurs in our commentary, but there without any reference to the parallax of the inferior planets. Note that Ptolemy explicitly denies a parallax of any (observable) size for them.[63] This suggests that Rhetorius is relying on some elaboration and correction of the *Almagest*. However, the general crudeness of Rhetorius's astronomical knowledge confirms our conclusion that he cannot be an authority for our commentary, but he does bear witness to the fact that the *Almagest* was still being studied at Alexandria a decade or two before the Muslim conquest in 642.

Our next to last candidate is Stephanus of Alexandria. He has been identified, ever since Usener's pamphlet of 1880,[64] with the author of the commentary on the *Handy Tables* of which all of the evidence within the body of the commentary proclaims the author to have been the emperor Heraclius himself. The earliest reference that we possess to Stephanus's activities as a commentator on the works of Ptolemy is in the extravagant and fraudulent introduction to the Ἀποτελεσματικὴ πραγματεία composed by a certain Stephanus the Philosopher in 775.[65] This Stephanus, claiming rightly or wrongly to be an Alexandrian and asserting that he is writing shortly after Muḥammad's Hijra in 622, states that he has explained, among many other texts, τὰς πτολεμαϊκὰς (lacuna) καὶ Συντάξεις καὶ ὀργανικὰς αὐτοῦ μαγγανείας. Into the lacuna Usener suggests inserting the word κανονογραφίας, thereby making the later eighth century Stephanus a witness to the early seventh century Stephanus's authorship of the 'Heraclian' commentary on the *Handy Tables*. While the later Stephanus may not be an altogether credible witness to this[66] (though it is

62 Rhetorius 6.3.87, cf. *Almagest* 5.18.

63 *Almagest* 9.1.

64 Usener 1914.

65 See Usener 1914: 266-289, esp. 267; for Stephanus the Philosopher, see note 3 above.

66 In his later astronomical work he attributed the *Handy Tables* to Heraclius (*Catalogus Codicum Astrologorum Graecorum* 2, 1900: 182).

certainly likely that Stephanus did write the commentary for — even perhaps with — the emperor), he is a witness to Stephanus's fame as an astronomer in Byzantium in the late eighth century, and later Byzantine sources confirm the reality of this reputation. Stephanus of Alexandria cannot, therefore, be excluded from our list of potential commentators on the *Almagest*, especially since Stephanus the Philosopher expressly states him to have written such a commentary. However, I have found nothing yet in our scholia to link them with Stephanus of Alexandria.

But I believe that Stephanus the Philosopher will lead us back to the milieu in which the commentary was written. I have shown elsewhere that he was a student of Theophilus of Edessa in Baghdād in the 760s, from whom he would have learned Ptolemaic astronomy as well as the new Indo-Irano-Greek astronomy of the early 'Abbāsids, based in part on an earlier Syrian tradition.[67] Theophilus had studied Greek science in Greek — including, clearly the *Almagest*[68] — in the early decades of the eighth century at Edessa, and perhaps also at nearby Ḥarrān. We can by no means provide a continuous chain of scholars of the *Almagest* linking Theophilus to the Neoplatonist philosophers such as Damascius, Priscianus Lydus,[69] and Simplicius who, according to almost entirely convincing evidence assembled by Tardieu,[70] had settled at Ḥarrān after their return from Persia (i.e., Ctesiphon) in 533. Severus Sebokht, who was educated at nearby Nisibis in Syriac, Greek, and Pahlavī in the early seventh century (he was an old and ailing man in 661/662) certainly knew his *Almagest* well;[71] but the name of no other astronomer from north-eastern Syria comes to mind till we get back to Sergius of Reshaina, an older contemporary of Simplicius.

It is only in our imaginations, then, that we can reconstruct a continuous school of Ptolemaic astronomy flourishing in Syria between the sixth and the ninth century, in which century that region produced

---

67 Pingree 1975; see also Pingree 1973.

68 Pingree 1989: 237.

69 Priscianus used the *Almagest* in composing his *Solutiones ad Chosroem*, perhaps at Ḥarrān in the mid-530s (*CAG: Supplementum Aristotelicum* 1.2, p. 42 Bywater).

70 Tardieu 1986; 1987; 1990.

71 See esp. Nau 1929-1931.

such highly competent astronomers as Thābit ibn Qurra of Ḥarrān, Qusṭā ibn Lūqā of Ba'albak, and al-Battānī of Raqqa. Their existence in combination with that of Severus and Theophilus is the best evidence so far available for the continuity of the teaching tradition of the *Almagest* in Syria. Thābit and al-Battānī were members of the religious community that called itself Ṣābian, while Qusṭā, like our proposed seventh and eighth century links, Severus and Theophilus, was a Christian, though the three were of different persuasions. We must not suppose that such doctrinal differences raised an insuperable barrier to the transmission of scientific knowledge; Severus is known to have had contacts, though not necessarily friendly, with anti-Monophysite Greeks on Cyprus through his student Basil, and had argued against the Maronites before the Caliph Mu'awīya in 659; and Theophilus, a Maronite, was active as an astrological advisor to the 'Abbāsid court from the 750s till his death in 785, and at that court shared his knowledge of Greek astrology and astronomy with Māshā'allāh, a Persian Jew.[72]

Let us turn now to the first link in this chain that we have so hypothetically forged, Simplicius of Cilicia. We are informed by the philosopher himself[73] that he was present in Alexandria when his teacher, Ammonius, observed the longitude of Arcturus with an armillary sphere, and thereby confirmed the *Almagest*'s constant of precession; such a conclusion requires Ammonius to have made an error of about 1;30° in his determination of Arcturus's longitude.[74] Still, the report suffices to show us that Simplicius, like his colleague Damascius, studied the *Almagest* under Ammonius; and his familiarity with that text is amply demonstrated in his commentary on the *De caelo*.[75] We also know that Simplicius was a careful student of Euclid, as was the author of our scholia, since his commentary on the beginning of the *Elements* is exten-

72 See Pingree 1973; forthcoming.

73 Simplicius, *In Cael.* (*CAG* 7, 1894) p. 462 Heiberg.

74 This represents Ptolemy's initial error in determining Arcturus's tropical longitude.

75 Simplicius, *In Cael.* (*CAG* 7, 1894) pp. 32, 411, 462, 474 (where Simplicius attributes the theory behind the distances of the moon, Mercury, Venus, and the sun found in Proclus's *Hypotyposis* and our scholium to the Σύνταξις itself: see notes 85, 86), 506, 539, 542 Heiberg.

sively quoted in its Arabic translation by al-Nayrīzī.[76] It is possible, then, that Simplicius either wrote the commentary which is the source of our scholia or else a commentary that was utilized by the compiler of the commentary on which those scholia are based.

And there are indeed statements in the scholia that may emanate from our philosopher. We begin with the observation of Arcturus undertaken by Ammonius and witnessed by Simplicius. This is based on *Almagest* VIII 3, where Ptolemy describes the construction of a solid sphere or celestial globe on which are marked the stars, with, according to Ptolemy, Sirius being the reference star;[77] its longitude in 1 Antoninus or 137 AD was, according to Ptolemy, Gemini 17 2/3°,[78] or 12 1/3° before the summer solstice. This distance, according to Ptolemy's theory of precession, will decrease by 1° in every 100 years. Ammonius chose to test precession — which had been vigorously denied by Proclus[79] — not by observing Sirius's distance from the summer solstice, but Arcturus's from the autumnal equinox. In the *Almagest* the longitude of Arcturus is given as Virgo 27°,[80] *i.e.*, 3° from the equinox. Our commentator,[81] however, reverts to Ptolemy's use of Sirius as the reference star (in this, as in some of his wording, he follows Theon), and remarks: 'Having found the Dog-star at the beginning of the reign of Antoninus to be distant from the summer solstice 12 1/3 degrees, after 500 years we shall have it distant 7 1/3 (degrees)'. This indicates not only that our commentator sustains Ammonius against Proclus, but that he wrote after 537 and before 637. The earlier years of this century would nicely fit into the active life of Simplicius, the latter years into that of Stephanus. Of course,

76 Besthorn and Heiberg 1893-1932: 8-40. For the Latin translation by Gerard of Cremona, which fills a lengthy lacuna in the Arabic, see the inadequate edition by Curtze 1899: 1-42. The source of al-Nayrīzī's notion that the purpose of studying Euclid is to prepare oneself to read the *Almagest* (Besthorn and Heiberg 1893-1932: 6) might be from Simplicius or from al-Hajjāj.

77 See Neugebauer 1975: 2.890-892.

78 *Almagest* 8.1.

79 *Hypotyposis* 7.45-47 (Manitius 1909: 234).

80 *Almagest* 7.5.

81 Fol. 169 of V, fol. 62v of v.

considering the amount of interest in the *Almagest* during this period we cannot limit our choice of author to only these two.

Another significant astronomical theory to which our commentator refers is that of the order and distances of the planets from the earth. Ptolemy had shown in the *Almagest* that the maximum distance of the moon from the center of the earth is 64;10 earth radii,[82] and that the mean distance of the sun is 1210 earth radii.[83] In the second half of the first book of the *Planetary Hypotheses*[84] Ptolemy calculated, by the approximate ratios to each other of rounded values derived from the parameters of the *Almagest*, the minimum and maximum distances of the planets; this computation is summarized more or less accurately for the moon, Mercury, Venus, and the sun by Proclus in his commentary on the *Timaeus*;[85] but in the *Hypotyposis*[86] he presents a computation based on other parameters equal or very close to those that appear in the *Almagest*. Precisely the same numbers, though embedded in a far more elaborate exposition than is presented by Proclus in the Ὑποτύπωσις, are found in our commentary, in a scholium on *Almagest* IX 1.[87] This seems to represent the source of Proclus's report in the *Hypotyposis*,[88] which was written apparently before he composed his commentary on the *Timaeus*, by which time he knew the last half of the first book of the *Planetary Hypotheses*, which we know otherwise only from its ninth century translation into Arabic.

The *Hypotyposis* was apparently also written before Proclus's commentary on the *Republic* in which Ptolemy's criticism of the Aristotelian adaptation of Eudoxus's theory of homocentric spheres is reported.[89]

---

82 *Almagest* 5.13.

83 *Almagest* 5.15.

84 Goldstein 1967: 28-30 (Arabic); 6-7 (English); Morelon 1993: esp. 62-71; see Neugebauer 1975: 2.919-922.

85 Proclus, *In Timaeum* 3.62-63 Diehl. Proclus ascribes this computation to Ptolemy in the *Hypotheses*.

86 *Hypotyposis* 7.19-23 (Manitius 1909: 220-224).

87 Fol. 174 in V, fol. 63v in v.

88 This must go back to a commentary written before Proclus, conceivably that by Pappus.

89 Proclus, *In Rempublicam* 2.230 Kroll.

This criticism is found in the second book of the *Planetary Hypotheses*, also lost in Greek, but preserved in Arabic.[90]

Hitherto only Proclus and Simplicius among Greek scholars have been known to have read the second book of the *Planetary Hypotheses*; to them we have joined Philoponus[91]. Now we can add to their number the author of our commentary on the *Almagest*.[92] The relevant scholium, relating to *Almagest* XIII 2, describes the mechanism employed by Ptolemy to account for planetary latitude ἐν τῇ δευτέρᾳ τῶν Ὑποθέσεων αὐτοῦ δύο βίβλίων. This scholium leads me to conjecture that the scholiast has derived at least this comment and probably others from a commentary by Simplicius or one of his followers in Syria. As we shall see, there are other passages in the scholia whose authorship it would be impossible to ascribe to Simplicius; but Syria seems to be the right place since it is there, apparently, that the Greek manuscript of the entire *Planetary Hypotheses* survived to be translated into Arabic. That translation is said in one of its two copies to have been corrected by Thābit ibn Qurra;[93] though the poor quality of the Arabic makes this claim doubtful, it is clear that Thābit knew the second part of Book I.[94] Moreover, one of the two classes of the Greek manuscripts that contain the first half of Book I consists of Vaticanus Graecus 1594 and its descendents,[95] so that the history of this text and of our scholia are closely bound together. I would guess that its ancestry goes back through Syria to Alexandria.

Among the scholia are several that seem too mathematically primitive to be worthy of Simplicius; but, of course, we do not know what level of competence may have been attained by his pupils. More seriously against his authorship of the commentary as a whole is a series of three

---

90 Goldstein 1967: 36 ff.

91 Simplicius, *In Cael.* (*CAG* 7, 1894) p. 456, cf. Heiberg 1907: 110 (Ptolemy); for Philoponus, see the passages cited in note 53.

92 Fol. 247v in V, fol. 75 in v.

93 Leiden arab. 1155: see Goldstein 1967: 13.

94 *Tashīl al-majasṭī*, in Morelon 1987: 13-15. This was translated into Latin under the title *De hiis que indigent expositione antequam legatur Almagesti*: the passage in question is in Carmody 1960: 137.

95 Heiberg 1907: clxxi-clxxiii.

theological scholia[96] on *Almagest* I 1, where Ptolemy remarks: 'The first cause of the first motion of all things, if one were to consider it simply, one might think to be an invisible (ἀόρατον) and unmoving (ἀκίνητον) deity, and the type (of philosophy) concerned with investigating this to be θεολογικόν, since such an activity is above somewhere in the loftiest parts of the universe (ἄνω που περὶ τὰ μετεωρότατα τοῦ κόσμου). In this statement Ptolemy follows Aristotle's description of the prime mover in, e.g., *Metaphysics* Λ 7.

The first theological scholium is as follows: 'He spoke well (when he said): "Unmoving divine (θεῖον instead of θεόν)". Everything moving is imperfect (ἀτελές); it wishes to arrive at perfection (τέλος). God is not imperfect, but is perfect (τέλειος)'. This could have been written by an Aristotelian expanding on the Stagirite's theory that the stars have appetetive souls.[97] There is no comment here on God's invisibility.

The second scholium is: 'Ptolemy says that God is above somewhere in the loftiest parts of the universe. Some are puzzled at this, saying that the divine is everywhere (ἁπανταχοῦ). We say that "above" indicates "superiority" (ὑπεροχήν), not a location there'. The 'some' who are puzzled are presumably the commentator's pupils, Christians voicing a common Christian dogma. The commentator's response, while feeble (for 'the loftiest parts of the universe' surely indicate a location), reveals his own adherence to Christianity. But of what variety?

The answer to that question, I believe, is to be found in the scholium[98] on a later passage in *Almagest* I 1, where Ptolemy states that one should speak of theology (τὸ ... θεολογικόν) as 'more conjecture than knowledge' because of its being completely unmanifest (ἀφανές) and ungraspable (ἀνεπίληπτον). The scholiast misunderstands this to refer to God rather than to theology, and reacts sharply: 'It is not (the case) that divine things (τὰ θεῖα) are unmanifest; indeed, this is blasphemy. But that they appear to be unmanifest because of our weakness as they do not fall under our sense-perceptions, just as the sun appears as darkness to bats because they are unfit (to see it)'.

96 The first two are found on fol. 9v in V, fol. 25 in v.

97 See Wolfson 1962.

98 Fol. 10 in V, fol. 25v in v.

This is an argument that would appeal, I believe, only to a Nestorian, for whom Jesus Christ, through whom the divine was revealed to mankind, has two substances (οὐσίαι) and two concomitant natures (φύσεις), one pair divine (the λόγος) and one pair human (the ἄνθρωπος), but they are united by a single πρόσωπον.[99] The human nature is what was perceived by Christ's contemporaries, the divine was known only by words and by deeds. Nestorius himself in his *Bazaar of Heracleides* quotes with approval the words of John I 18: 'No man has ever seen God'.[100] Later in the same work Nestorius says: 'Man indeed is known by the human πρόσωπον (here meaning simply "character" or "person"), that is, by the σχῆμα of the body and by the likeness, but God (is known) by the name which is more excellent than all names and by the adoration of all creation and by the confession of him as God'.[101]

If, then this teacher of the *Almagest*, whose comments are preserved in our scholia, was a Nestorian, and since his instruction was in Greek, where and when did he teach? It must have been outside of the Byzantine Empire, in a region where Greek survived as a language of learning, in the late sixth (assuming that he did use Simplicius) or early seventh century, before 637, the five hundredth anniversary of Ptolemy's star catalogue. The only possibility seems to be Nisibis, just across the Euphrates from Edessa and Ḥarrān, in Persian territory, and the seat of the principle school for training the Nestorian clergy — the school, in fact, which served as a model, along with Alexandria, for the Christian Academy at Rome planned by Cassiodorus and Pope Agapetus I in 535; in this Roman Academy, it appears, Ptolemy's *Handy Tables* and *Almagest* were to be taught.[102] However, the *Almagest* was most likely taught outside of the Nestorian theological seminary; the presence of such instruction, in Greek and in Nisibis, in the decades around 600 is indicated by the career of Severus Sebokht, a Monophysite in contrast to our Nestorian. The history of the school of Nisibis nicely conforms to this chronology; its director from about 570 till about 610 was Ḥenānā of

99 Anastos 1962.

100 Nestorius, *The Bazaar of Heracleides*, trans. G.R. Driver and L. Hodgson (Oxford, 1925), 51.

101 Nestorius: *Bazaar* 61.

102 Pingree 1990.

Ḥadiab, a man of strong Monophysite leanings.[103] His heretical views split the Nestorians of Nisibis, especially in the years after 596. The controversy culminated in a document issued by the orthodox Nestorians in 612 anathematising Ḥenānā's writings along with those who read them. It appears that in this heated controversy our commentator remained faithful to Nestorius, while Severus favored the position of Ḥenānā, abandoned Persia either before or in 612, and successfully pursued a career in the Syrian Monophysite church within the Byzantine Empire. He was the last teacher of the *Almagest* that we know of before Theophilus of Edessa, whose student, Stephanus the Philosopher, might possibly have carried the forebearer of Vaticanus Graecus 1594 to Constantinople shortly before 775.

Bibliography

Mathematical, astronomical and astrological texts are normally cited by the name of the modern editor whose text has been used with the date of the edition and the relevant page. References to Rhetorius are keyed to the forthcoming edition by D. Pingree, *Rhetorii Aegyptii Thesauri* (Leipzig). Philosophical and other ancient texts are cited by author's name and conventional title (with the editor's name when cited by page). Ancient commentaries on Aristotle are quoted from the Berlin corpus of *Commentaria in Aristotelem Graeca* (1882-1907) (abbreviated *CAG*).

A. Allard 1977. 'Le premier traité byzantin de calcul indien', *Revue d'histoire des textes* 7, 57-107.

M.V. Anastos 1962. 'Nestorius was Orthodox', *Dumbarton Oaks Papers* 16, 121-140.

R.O. Besthorn & J.L. Heiberg 1893-1932. *Codex Leidensis 399 1: Euclidis Elementa*. Copenhagen.

E. Boer 1962. *Heliodori, ut dicitur, In Paulum Alexandrinum Commentarium*. Leipzig.

Alan Cameron 1990. 'Isidore of Miletus and Hypatia: On the Editing of Mathematical Texts', *Greek, Roman, and Byzantine Studies* 31, 103-127.

F.J. Carmody 1960. *The Astronomical Works of Thabit b. Qurra*. Berkeley & Los Angeles.

M. Curtze 1899. *Anaritii In decem libros priores Elementorum Euclidis commentarii*. Leipzig.

103 Vööbus 1965: 234-317.

C. Giannelli 1950. *Codices Vaticani Graeci. Codices 1485-1683.* Vatican City.

B.R. Goldstein 1967. *The Arabic Version of Ptolemy's Planetary Hypothesis.* Philadelphia.

J.L. Heiberg 1893. *Apollonii Pergaei quae graece exstant* 2. Leipzig.

Heiberg 1898-1903, 1907. *Claudii Ptolemaei Opera* 1.1-2, 2. Leipzig.

Heiberg 1915. *Archimedis Opera omnia* 3. Leipzig.

A. Jones 1990. *Ptolemy's First Commentator.* Philadelphia.

W. Knorr 1989. *Textual Studies in Ancient and Medieval Geometry.* Boston.

P. Kunitzsch 1986, 1991. *Claudius Ptolemäus. Der Sternkatalog des Almagest. Die arabisch-mitteraltliche Tradition* 1, 3. Wiesbaden.

C. Manitius 1909. *Procli Diadochi Hypotyposis astronomicarum positionum.* Leipzig.

G. Mercati & P. Franchi de' Cavalieri 1923. *Codices Vaticani Graeci* 1. Rome.

J. Mogenet 1956. *Introduction à l'Almageste.* Brussels.

Mogenet 1962. 'Une scolie inédite du Vat. gr. 1594 sur les rapports entre l'astronomie arabe et Byzance', *Osiris* 14, 198-221.

Mogenet 1975. 'Sur quelques scolies de l'Almageste', *Le monde grec. Hommages à Claire Préaux.* Brussels, 302-311.

R. Morelon 1987. *Thābit ibn Qurra. Oeuvres d'astronomie.* Paris.

Morelon 1993. 'La version arabe du Livre des Hypothèses de Ptolémée', *Mélanges de l'Institut Dominicain d'Études Orientales du Caire* 21, 7-85.

F. Nau 1929-1931. 'Traité sur les "constellations" écrit en 661 par Sévère Sebokt, évêque de Qennesrin', *Revue de l'Orient chrétien* 27, 327-410; 28, 85-100.

O. Neugebauer 1958. 'On a Fragment of Heliodorus (?) on Planetary Motion', *Sudhoffs Archiv* 42, 237-244.

Neugebauer 1975. *A History of Ancient Mathematical Astronomy.* 3 vols. New York.

D. Pingree 1973. 'The Greek Influence on Early Islamic Mathematical Astronomy', *Journal of the American Oriental Society* 93, 32-43.

Pingree 1975. 'Māshā'allāh: Some Sasanian and Syriac Sources', *Essays on Islamic Philosophy and Science.* Albany, 5-14.

Pingree 1977. 'Antiochus and Rhetorius', *Classical Philology* 72, 203-223.

Pingree 1986. *Vettii Valentis Antiocheni Anthologiarum libri novem.* Leipzig.

Pingree 1989. 'Classical and Byzantine Astrology in Sassanian Persia', *Dumbarton Oaks Papers* 43, 227-239.

Pingree 1990. 'The Preceptum Canonis Ptolomaei', *Rencontres de cultures dans la philosophie médiévale.* Louvain-la-neuve & Cassino, 355-375.

Pingree (forthcoming). 'Māshā'allāh: Greek, Pahlavī, Arabic, and Latin Astrology', *Arabic Philosophy and Science.*

A. Rome 1931, 1936, 1943. *Commentaires de Pappus et de Théon d'Alexandrie sur l'Almageste* 1, 2, 3. Studi e Testi 54, 72, 106. Rome, Vatican City.

A.P. Segonds 1981. *Jean Philopon. Traité de l'astrolabe.* Paris.

P. Tannery 1894. 'Sur un fragment inédit des Métriques de Héron d'Alexandrie', *Bulletin des sciences mathématiques* 2, 18, 18-22. Reprinted in his *Mémoires scientifiques* 2, Toulouse-Paris, 1912, 451-454.

M. Tardieu 1986. 'Sābiens coraniques et "sābiens" de Ḥarrān', *Journal Asiatique* 274, 1-44.

Tardieu 1987. 'Les calendriers en usage à Ḥarrān d'après les sources arabes et le commentaire de Simplicius à la Physique d'Aristote', *Simplicius: sa vie, son oeuvre, sa survie.* Berlin & New York, 40-57.

Tardieu 1990. *Les paysages reliques: routes et haltes syriennes d'Isidore à Simplicius.* Paris.

A. Tihon 1973. 'Les scolies des Tables Faciles de Ptolémée', *Bulletin de l'Institut Historique Belge de Rome* 43, 49-110.

Tihon 1976. 'Notes sur l'astronomie grecque au Ve siècle de notre ère', *Janus* 63, 167-184.

Tihon 1978. *Le 'Petit Commentaire' de Théon d'Alexandrie aux Tables Faciles de Ptolémée.* Studi e Testi 282. Vatican City.

Tihon 1985-1991. *Le 'Grand Commentaire' de Théon d'Alexandrie aux Tables Faciles de Ptolémée.* Studi e Testi 315, 340. Vatican City (Vol. 1 was jointly edited with J. Mogenet.)

Tihon 1987. 'Le livre V retrouvé du Commentaire à l'Almageste de Théon d'Alexandrie', *Antiquité Classique* 56, 201-218.

Tihon 1989. 'Sur l'identité de l'astronome Alim', *Archives internationales d'histoire des sciences* 39, 3-21.

Tihon 1992. 'Propos sur l'édition de textes astronomiques grecs des IVe et Ve siècles de notre ère', *Les problèmes posés par l'édition critique des textes anciens et médiévaux.* Louvain-la-neuve, 113-137.

G.J. Toomer 1976. *Diocles On Burning Mirrors.* New York.

Toomer 1984. *Ptolemy's Almagest.* London & New York.

H. Usener 1914. 'De Stephano Alexandrino', *Kleine Schriften* 3. Leipzig, 247-321. Originally published as *De Stephano Alexandrino commentatio* (Bonn, 1880).

A. Vööbus 1965. *History of the School of Nisibis. Corpus Scriptorum Christianorum Orientalium* 266. Louvain.

J. Warnon 1967. 'Le commentaire attribué à Héliodore sur les Εἰσαγωγικά de Paul d'Alexandrie', *Travaux de la Faculté de Philosophie et Lettres de l'Université Catholique de Louvain* 2, 197-217.

L.G. Westerink 1971. 'Ein astrologisches Kolleg aus dem Jahre 564', *Byzantinische Zeitschrift* 64, 6-21.

H.A. Wolfson 1962. 'The Problem of the Souls of the Spheres from the Byzantine Commentaries on Aristotle through the Arabs and St. Thomas to Kepler', *Dumbarton Oaks Papers* 16, 67-93.

India

Astronomy and Astrology in India and Iran

By *David Pingree* *

ONLY in recent years have the interrelationships of Babylonian, Greek, and Indian astronomy and astrology become a subject which can be studied meaningfully. This development is due to several factors: our greatly increased understanding of cuneiform material made possible by the scholarship of Professor O. Neugebauer;[1] the discovery of Babylonian parameters and techniques not only in the standard Greek astronomical texts,[2] but in papyri and astrological treatises as well; and the finding of Mesopotamian material in Sanskrit works and in the traditions of South India. Unfortunately, a lack of familiarity with the Sanskrit sources and a failure to consider the transmission of scientific ideas in the context of a broad historical perspective have recently led one scholar to the erroneous conclusion that Sasanian Iran played a crucial role in the introduction of Greek and Babylonian astronomy and astrology to India and in the development of Indian planetary theory.[3] It is my purpose in this paper to survey briefly the influence of foreign ideas on Indian gaṇakas so as to make clear the creative use they made of their borrowings in devising the yuga-system of astronomy; and then to examine the character of Sasanian astronomy and astrology, pointing out their almost complete lack of originality.

The earliest Indian texts which are known – the Vedas, the Brâhmaṇas, and the Upaniṣads – are seldom concerned with any but the most obvious of astronomical phenomena; and when they are so concerned, they speak with an obscurity of language and thought that renders impossible an adequate exposition of the notions regarding celestial matters to which their authors subscribed. One may point to the statement that the year consists of 360 days as a possible trace of Babylonian influence in the Ṛgveda,[4] but there is little else which lends itself to a similar interpretation. It has often

* Harvard University.

1 See especially his *Astronomical Cuneiform Texts*, 3 vols., London, 1955 (hereafter *ACT*), and *The Exact Sciences in Antiquity*, 2nd ed., Providence, R. I., 1957, chap. 5 (hereafter *Exact Sciences*).

2 See A. Aaboe, " On the Babylonian Origin of Some Hipparchan Parameters," *Centaurus*, 1955-1956, *4*: 122-125, and Neugebauer, *Exact Sciences*, pp. 157, 183. The basic survey of the transmission of astronomy is O. Neugebauer, " The Transmission of Planetary Theories in Ancient and Medieval Astronomy," *Scr. Math., N. Y.*, 1956, *22*: 165-192.

3 B. L. van der Waerden, *Vjschr. Naturf. Ges. Zürich*, 1960, *105*: 140-143.

4 On the Vedic year, see G. Thibaut, *Astronomie, Astrologie und Mathematik*, Grundriss der Indo-Arischen Philologie und Altertumskunde 3, 9, Strassburg, 1899, pp. 7-9.

ISIS, 1963, VOL. 54, PART 2, NO. 176.

been proposed, of course, that the list of the twenty-eight nakṣatras which is given for the first time at the beginning of the last millennium before Christ in the Atharvaveda and in various Brâhmaṇas is borrowed from Mesopotamia.[5] But no cuneiform tablet yet deciphered presents a parallel; the hypothesis cannot be accepted in the total absence of corroborative evidence.

However, the nakṣatras are useful in the tracing of Indian influence on other cultures. The oldest lists [6] associate each constellation with a presiding deity who is to be suitably propitiated at the appointed times. It became important to perform certain sacrifices only under the benign influence of particularly auspicious nakṣatras.[7] The roster of activities for which each was considered auspicious or not was rapidly expanded,[8] and, in particular, the nakṣatras came to be closely connected with the twelve or sixteen saṃskâras or purificatory rites. Thereby they gave rise to the most substantial part of muhûrtaśâstra, or Indian catarchic astrology,[9] traces of which are to be found in Arabic, Byzantine, and medieval Latin texts.[10] The Indians also combined the twenty-eight nakṣatras with the Babylonian arts of brontology and seismology [11] in a form which, for some unknown reason,

5 See the most recent, J. Needham, *Science and Civilization in China*, vol. 3, Cambridge, 1959, pp. 252-259. S. Weinstock, "Lunar Mansions and Early Calendars," *J. Hellenic Studies*, 1949, *69*: 48-69 is based on a series of misinterpretations.

6 See, e. g., Taittirîyasaṃhitâ 4, 4, 10.

7 P. V. Kane, *History of Dharmaśâstra*, vol. 5, pt. 1, Poona, 1958, pp. 506-507.

8 *Ibid.*, pp. 523-525.

9 On muhûrtaśâstra, the oldest works seem to be: one of at least four versions of the Gargasaṃhitâ, that preserved in MS 210 of 1883/1884 of the Bhandarkar Oriental Research Institute, Poona, and MS 9277 of the Oriental Institute, Baroda; the Ratnakośa of Lalla (seventh century) in MS 27 of 1880/1881 of the Bhandarkar Oriental Research Institute and MS 1203 of the Viśveśvarânanda Vedic Research Institute, Hoshiarpur; and the Ratnamâlâ of Śrîpati (eleventh century), ed. by K. M. Chattopâdhyâya, Calcutta, 1915. See also P. Poucha, "La Jyotiṣaratnamâlâ ou guirlande des joyaux d'astrologie de Śrîpatibhaṭṭa," *Arch. Orientální*, 1946, *16*: 277-309, and M. G. Panse, *Jyotiṣaratnamâlâ of Śrîpatibhaṭṭa*, *Bull. of the Deccan College Res. Inst.*, 1956, *17*: 237-502, reprinted in the Deccan College Monograph Series, Poona, 1957. Besides these three works, I know of more than 100 other Sanskrit texts, not including their commentaries, on the same subject.

10 E. g., the text on the twenty-eight lunar stations comparing the theories of the Indians, the Persians (Sasanians using Indian sources), and Dorotheus (compiled from the fifth book of his Pentateuch, where the material is arranged under zodiacal signs, not lunar mansions, as is also the case in the poem of Maximus, which is largely derived from Dorotheus; of the Pentateuch there survives a late eighth-century Arabic translation of a third-century Pahlavi version in MS Yeni Jami 784 and MS Or. oct. 2663 of Berlin, now in Marburg). The Arabic original of this text is to be found in MS Add. 23,400 of the British Museum; the Greek version has been published by S. Weinstock in *Catalogus Codicum Astrologorum Graecorum*, ed. F. Cumont *et al.*, 12 vols. in 20 parts, Bruxelles, 1898-1953, vol. 9, pt. 1, pp. 138-156; the first five books of the Old Catalan version of 'Alî ibn abî 'r-Rijâl, who includes this text in his treatise, have been edited by G. Hilty, *El libro conplido en los iudizios de las estrellas*, Madrid, 1954; for editions and manuscripts of the Latin, which is a translation of the Old Catalan, see F. Carmody, *Arabic Astronomical and Astrological Sciences in Latin Translation*, Berkeley–Los Angeles, 1956, pp. 150-152. Cf. also John of Seville's Epitome astrologiae 4, 18 cited by J. M. Millás Vallicrosa, *Las traducciones orientales en los manuscritos de la Biblioteca Catedral de Toledo*, Madrid, 1942, pp. 157-158, and Carmody, *op. cit.*, p. 70.

11 For the Babylonian origin of these two methods of divination, see C. Bezold and F. Boll, *Reflexe astrologischer Keilinschriften bei griechischen Schriftstellern*, Sitz. Heidelberger Akad. Wiss. Phil-hist. Kl. 1911, Abh. 7, Heidelberg, 1911, pp. 45-52, and P. Hilaire de Wynghene, *Les présages astrologiques*, Übersicht über die Keilschrift-literatur, Heft 3, Roma, 1932, p. 56.

became immensely popular among the followers of Buddha.[12] Their works spread these superstitions throughout Central Asia and the Far East.[13]

The relative seclusion from the West which the Aryans had enjoyed in northern India for centuries after their invasions was broken shortly before 513 B. C., when Darius the Great conquered the Indus Valley. In the ensuing six centuries, save for a century and a half of security under the Mauryan emperors, North India was subjected to the successive incursions of the Greeks, the Śakas, the Pahlavas, and the Kuṣâṇas. An important aspect of this turbulent period was the opportunity it afforded of contact between the intellectuals of the West and India. This opportunity was not missed.

In the period from 500 to about 230 B. C. — under the Achemenid occupation and during the reigns of Candragupta Maurya, Bindusâra, and Aśoka — Indian astronomy was introduced for the first time to some reasonable Babylonian methods, and astrologers were led to show an interest in more significant phenomena than the nakṣatras.[14] A luni-solar calendar was propounded in the Jyotiṣavedâñga of Lagadha,[15] who probably wrote in the fifth century B. C. This calendar is described also in the Arthaśâstra of Kauṭilya,[16] which seems to be a Mauryan document; in the Jaina Sûryaprajñapti,[17] which probably preserves a Mauryan system; in the oldest version of the Gargasaṃhitâ,[18] which may have been written in the first century A. D.; and in the earliest version of the Paitâmahasiddhânta,[19] which uses as epoch 80 A. D. The period relation employed in this calendar — sixty-two synodic months in 1830 days — is extremely crude and, so far as I know, not Babylonian; but the attempt is analogous to the more accurate eight-year cycle introduced into Greece by Cleostratus of Tenedos towards the end of the sixth century B. C.

An important feature of the Jyotiṣavedâñga is its use of the tithi, or thirtieth of a synodic month, as a standard unit of time. Tithis, of course, play a similar role in the Babylonian linear astronomy of the Seleucid period.[20] It seems likely that the Indians borrowed the concept from Mesopotamia, though the exact origin of the tithi still remains obscure.

12 See, *inter alia*, the Śârdûlakarṇâvadâna of the Divyâvadâna, ed. S. Mukhopadhyaya, Santiniketan, 1954, reprinted in the Divyâvadâna, ed. P. L. Vaidya, Buddhist Sanskrit Texts 20, Darbhanga, 1959.

13 For the Chinese material, see M. Zemba, "On the Astronomy and Calendar of the Buddhist Books," *J. Indian and Buddhist Studies*, 1956, *4*: 18-27, kindly translated from the Japanese for me by Professor Shoren Ihara of Kyushu University, Fukuoka, Japan.

14 For the Achemenid influence on art in Mauryan India, see, for example, R. E. M. Wheeler in *Ancient India*, 1948, *4*: 92-101 and the appendix by Stuart Piggott, *ibid.*, pp. 101-103.

15 The Jyotiṣavedâñga of the Yajurveda and of the Ṛgveda with the commentary of Somâkara on the former were published by S. Dvivedî, Benares, 1908; see also the edition and translation of the text belonging to the Ṛgveda by R. Shamasastry, Mysore, 1936.

16 Kauṭalîyârthaśâstra 2, 20, ed. N. S. Venkatanathacharya, Oriental Res. Inst. Sanskrit Series 103, Mysore, 1960.

17 Ed. J. F. Kohl, Bonner Orientalistische Studien 20, Stuttgart, 1937; 10, 22 *et passim*.

18 Quoted by Somâkara on Jyotiṣavedâñga 10.

19 Summarized by Varâhamihira in chap. 12 of his Pañcasiddhântikâ, ed. G. Thibaut and S. Dvivedî, Benares, 1889; reprinted Lahore, 1930. See also M. P. Kharegat, *J. Bombay Branch of the Roy. Asiatic Soc.*, 1896, *19*: 109-141.

20 See Neugebauer, *ACT*, vol. 1, p. 40. Despite Kane, *op. cit.*, pp. 62 ff., it cannot be said that the tithi was conceived of as a thirtieth of a synodic month before Lagadha.

In their methods of measuring the time of day, the Indians of this early period also showed a knowledge of what the Babylonians had devised.[21] One method depends on the length of the shadow cast by a śaṅku or gnomon. This shadow, of course, varies during any half-day with the changing altitude of the sun; it also varies from day to day throughout a half-year as the sun travels along the ecliptic. In tabulating the increase and decrease of the noon-shadow throughout the year, the Indians employed a linear zigzag system which is clearly of Babylonian origin. But, more than this, they used 3:2 as the ratio of the longest to the shortest day of the year, a well-known Babylonian parameter [22] which is not applicable to any part of India except the extreme Northwest.[23] The other Indian method of telling time, by means of a ghaṭa, or pot with a small hole in the bottom through which water flows at a fixed rate, is also known to have been employed in Babylon.[24] In connection with the śaṅku it may be added that if one can believe their claims as recorded by Eratosthenes and Hipparchus via Strabo,[25] both Megasthenes and Daimachus, the Seleucid ambassadors at the Mauryan capital, Pâtaliputra or Palibothra, made gnomon-observations in India.

Babylonian influence in astrology was equally great; in fact, the planets first appear in Indian literature because of it. Venus is mentioned as the "Star of Plants" (Osadhitârakâ) in an early Buddhist text, the Majjhimanikâya; [26] and Kauṭilya asserts that the sun, and Jupiter and Venus in their risings, settings, and stations, cooperate in furthering the growth of plants.[27] The heliacal risings and settings of the planets and their stationary points are the so-called Greek-letter phenomena upon which the structure of Babylonian linear planetary theory is based.

In the Indian epics, the Râmâyaṇa and the Mahâbhârata, the planets also appear in an astrological context, their influence depending on their conjunctions with the constellations, on their retrogressions, and on their transits.[28] This type of astrology is termed gocâra; it is mentioned in a

[21] For the early Indian techniques of telling time, see H. Jacobi, "Einteilung des Tages und Zeitmessung im alten Indien," *Zeitschr. der Deutschen Morgenländischen Gesellschaft*, 1920, *74*: 247-263.

[22] See O. Neugebauer, *Osiris*, 1936, *2*: 517, and *Exact Sciences*, p. 183.

[23] B. R. Kulkarni, "Some Astronomical References from the Arthashastra and their Significance," *J. Univ. Bombay* (History, Economics and Sociology 33), 1948, *17, 1*: 1-3, tries to use this fact to prove that Kauṭilya wrote in Kashmir. But he is wrong; Kauṭilya is simply copying blindly. In fact, other evidence in the Arthaśâstra indicates that it was written in Bihar; see G. D. Tamaskar, "The Country of Kauṭilya's Arthaśâstra," *Siddha-Bhâratî or The Rosary of Indology*, vol. 2, Hoshiarpur, 1950, pp. 226-229.

[24] O. Neugebauer, "Studies in Ancient Astronomy, VIII. The Water Clock in Babylonian Astronomy," *Isis*, 1947, *37*: 37-43.

[25] Strabo 2 C 76-77, ed. A. Meineke, Leipzig, 1903. This passage is not included in D. R. Dicks, *The Geographical Fragments of Hipparchus*, London, 1960, p. 68. Though the observations were not of a very high order of accuracy, still some interest was shown in the subject by the Greeks in India; cf. Diodorus Siculus 2, 35, ed. F. Vogel, Leipzig, 1888; Baeton cited in Pliny, Naturalis Historia 6, 69 and 2, 184 (with Onesicritus), ed. C. Mayhoff, Leipzig, 1906, with which compare Martianus Capella 6, 694, ed. A. Dick, Leipzig, 1925, Pomponius Mela 3, 61, ed. C. Frick, Leipzig, 1880, and Solinus 52, 13, ed. T. Mommsen, 2nd ed., 1895, reprinted Berlin, 1958.

[26] Majjhimanikâya 2, 3, 7, ed. R. Chalmers, *Pali Text Society*, vol. 2, London, 1898, pp. 14, 34; cf. Buddhaghoṣa's Papañcasûdanî, ed. I. B. Horner, *Pali Text Society*, pt. 3, London, 1933, p. 274.

[27] Arthaśâstra 2, 24.

[28] Kane, *op. cit.*, pp. 531-532.

Buddhist anti-caste tract, the Śârdûlakarṇâvadâna,[29] which was probably written in the first century A. D. and is described in detail in the Gargasaṃhitâ [30] and in the sixth-century Bṛhatsaṃhitâ of Varâhamihira.[31] It represents an earlier stage of planetary astrology than does Hellenistic horoscopy; in fact, it is a method familiar from the reports of the astrologers of Babylon and Ninevah.[32] From a Babylonian source also comes the order in which the celestial bodies are named in the early second-century Nasik cave-inscription set up by his mother in honor of Gautamîputra Śâtakarṇi [33] and in a common Paurâṇika passage; [34] according to these sources the sun and moon precede the five star-planets.[35] There is no hint, however, that the Indians had learned a method of computing planetary positions in this period.

I have mentioned previously the Śaka or Scythian invasions of North India; it is necessary now to return to them. A family of Śakas, the Kṣaharâtas, established a kingdom in western India at the beginning of the first century A. D.[36] Their capital was Mînanagara,[37] but their source of wealth, Bhṛgukaccha, the modern Broach, was one of the main emporia for the brisk trade between India and the Mediterranean; it was known to the Greeks by the name Barygaza.[38] The Periplus maris Erythraei, a document written between 60 and 80 A. D., mentions the Kṣaharâta king Nahapâna,[39] whose riches are extolled in Jaina traditions and proved by the vast hoard

29 P. 31 Mukhopadhyaya.

30 Quoted by Bhaṭṭotpala on Bṛhatsaṃhitâ 4 and 6-10, ed. S. Dvivedî, Vizianagram Sanskrit Series 10, 2 vols., Benares, 1895-1897. This subject is also treated in the version of the Gargasaṃhitâ cited above in fn. 9 and in another found in MS 542 of 1895/1902 of the Bhandarkar Oriental Research Institute, MS G 8199 of the Asiatic Society of Bengal, Calcutta, MS 122 of the Jyotiṣa collection of the Sanskrit College, Benares, and MS fonds sanscrit 245 (1) of the Bibliothèque Nationale, Paris.

31 Chaps. 4 and 6-10.

32 See R. C. Thompson, *The Reports of the Magicians and Astrologers of Ninevah and Babylon in the British Museum*, London, 1900, *passim*.

33 Ed. E. Senart, *Epigraphia Indica*, 1905-1906, *8*: 60-65 (hereafter *EI*).

34 See J. F. Fleet, "A Note on the Purâṇas," *J. Roy. Asiatic Soc.*, 1912, 1046-1053 (hereafter *JRAS*), and "The Purâṇic Order of the Planets," *JRAS*, 1913, 384-385; and W. Kirfel, *Das Purâṇa vom Weltgebäude*, Bonner Orientalistische Studien, NS 1, Bonn, 1954, p. 278.

35 For the Babylonian order, see F. Boll in *Realencyclopädie der classischen Altertumswissenschaft*, 1912, *14*: cc. 2561-2570.

36 On the Śakas, see especially S. Chattopadhyaya, *The Śakas in India*, Viśva-Bharati Studies 1, Santiniketan, 1955, and J. N. Banerjea in *A Comprehensive History of India*, vol. 2, ed. K. A. Nilakanta Sastri, Bombay-Calcutta-Madras, 1957, chap. 9.

37 Periplus maris Erythraei 41 (Μινναγάρα), ed. H. Frisk, *Göteborgs Högskolas Årsskrift* 33, Göteborg, 1927, and Ptolemy, Geography 7, 1, 63 (Μινάγαρα), ed. L. Renou, *La Géographie de Ptolémée: L'Inde (VII, 1-4)*, Paris, 1925. There are also two other Scythian cities called "City of the Mînas"; one, in Seistan, is mentioned by Isidore of Charax 18 (Μὶν πόλις), ed. W. H. Schoff, Philadelphia, 1914, and the other, near the mouth of the Indus, in Periplus 38 (Μινναγάρ) and in Ptolemy 7, 1, 61 (Βιναγάρα).

38 On this trade, see now M. P. Charlesworth, "Roman Trade with India: A Resurvey," *Studies in Roman Economic and Social History in Honor of Allan Chester Johnson*, Princeton, 1951, pp. 131-143, and U. N. Ghoshal in *A Comprehensive History of India*, vol. 2, pp. 439-446.

39 The date of the Periplus and the reference to Nahapâna have recently been questioned by J. Pirenne, *Le royaume sud-arabe de Qatabân et sa datation*, Bibliothèque du Muséon, 48, Louvain, 1961, chap. 5. Her arguments are not entirely convincing; but even were they to prove correct, the conclusions reached in this paper would not be affected. Whether the Periplus refers to Nahapâna or not, I would not connect him with the Śaka era of 78 A. D.

of 14,000 silver coins found at Jogalthembi. This wealth depended largely on the Roman trade. The Śakas exported, besides many other articles, Chinese silk carried across Central Asia, through Kuṣâṇa territory down the Indus, across to Ujjain, and down the Narmadâ valley to Broach. In return they received, along with other useful products of Roman industry, dancing-girls and jugs of wine. In innumerable sites in Gujarât, Saurâṣṭra, and northern Mahârâṣṭra there have been unearthed fragments of Roman pottery and its imitations, copies of Roman bullae, Roman beads, and Roman statuettes, all of which date from the first to fourth centuries A. D.[40] A large number of inscriptions of the first and second centuries found in the Buddhist caves along the trade-routes of the western Ghâṭs record the donations of the Yavanas, or Greeks of Dhenukâkaṭa.[41] Indeed, one can date the Greek settlement in the area back to Mauryan times if one is willing to accept as sufficient evidence the Junâgaḍh inscription which mentions the Yavanarâja Tuṣâspa, who was Aśoka's governor of Kâṭhiâwâḍ.[42] It does not seem likely, however, despite Tarn's strenuous efforts,[43] that the Greek kingdom of Demetrius and Menander ever extended this far south, though Greeks from Gandhâra may well have made commercial trips to Bhṛgukaccha and Ujjain and augmented the Yavana community in Gujarât. Ptolemy's source was perhaps taking note of the Greek settlements when he placed towns with such un-Indian names as Byzantion on the coast below Barygaza.[44]

By Ptolemy's time, however, the Kṣaharâta dynasty had been overthrown by the Sâtavâhana Gautamîputra Śâtakarṇi,[45] whom we have had occasion to mention before, and he in turn had succumbed to a new Śaka dynasty, the western Kṣatrapas, in Gujarât and Saurâṣṭra. The greatest of the Kṣatrapas was Rudradâman I, who ruled from about 130 to about 160 A. D. His empire at one time extended over most of Central India, stretching as far as Kauśâmbî in the North and Kaliṅga in the East.[46] His capital was Ujjain, which for this reason became the Greenwich of Indian astronomers and the Arin of the Arabic and Latin astronomical treatises; for it was he and his successors who encouraged the introduction of Greek horoscopy and astronomy into India.

In 150 A. D. Yavaneśvara, the Lord of the Greeks, translated into Sanskrit prose a Greek astrological text which had been written in Alexandria in the preceding half-century. This translation is now lost, but there is pre-

[40] See Appendix 1 for the beginning of a bibliography of this material (sites in South India such as Arikamedu and Chandravalli are excluded).

[41] See D. D. Kosambi, "Dhenukâkaṭa," *J. Asiatic Soc. of Bombay*, 1955, *30*, pt. 2: 50-71.

[42] Ed. F. Kielhorn, *EI*, 1905-1906, *8*: 39-49.

[43] W. W. Tarn, *The Greeks in Bactria and India*, 2nd ed., Cambridge, 1951, pp. 147-150, 230; contrast the more sensible account in A. K. Narain, *The Indo-Greeks*, Oxford, 1957, pp. 91-95.

[44] Ptolemy 7, 1, 7.

[45] See fn. 33. The inscription associates the Greeks and the Pahlavas in Nahapâna's defeat.

[46] D. Pingree, "The Empires of Rudradaman and Yaśodharman: Evidence from Two Astrological Geographies," *J. Amer. Oriental Soc.*, 1959, *79*: 267-270 (hereafter *JAOS*). An inscription of one Rudradâmaśrî, dated paleographically to the third or fourth century, has recently been found in Mirzapur District, UP; see *Indian Archaeology 1959-60 — A Review*, New Delhi, 1960, p. 61; now published by D. C. Sircar in *EI*, 1961-1962, *34*: 244-245.

served in an early thirteenth-century palm-leaf manuscript in Kathmandu [47] a versification of it made in 270 by the Yavanarâja Sphujidhvaja. In the second century another Greek text on the same subject was translated into Sanskrit; this text and Yavaneśvara's were both used by a third-century author named Satya. Unfortunately, the second translation from the Greek is lost, and Satya's work is known only from the citations of later astrologers and in what appears to be a fairly recent forgery. However, there has survived a work based on both Sphujidhvaja and Satya; this is the Vṛddhayavanajâtaka of Mînarâja.[48]

The name Mînarâja connects its owner with the Mînas whom we have already come across in Mînanagara, the Kṣaharâta capital. Two other Mînanagaras are known, and they also are Scythian cities. Mînarâja, then, must have been a Śaka. But he also calls himself the Yavanarâja, or King of the Greeks, a title used by Tuṣâspa, Yavaneśvara, and Sphujidhvaja. One can now also cite an early fourth-century inscription discovered at Nâgârjunakoṇḍa [49] which mentions the Śaka of Ujjain, Rudradâman II (*c.* 335 – *c.* 345), and the Yavanarâja of Sañjayapuri (Sañjayapuri is probably the same as Sañjayantî, the modern Sañjân near Bombay, which Ptolemy, who included it within the kingdom of the western Kṣatrapas, calls Sazantion).[50] Yavanarâja, then, was an official title in the Śaka administration. As the Śakas were overthrown by Candragupta II shortly after 389,[51] and as the Vṛddhayavanajâtaka copies many ślokas from Sphujidhvaja, it is safe to date Mînarâja in the early fourth century; and we have recovered two astrological poems presenting almost purely Greek horoscopy in Sanskrit.

Indian genethlialogy is largely dependent on the teachings of Yavaneśvara and Satya, though elaborations have been indulged in from time to time; and it, in turn, has influenced Sasanian, Arabic, Byzantine, and western European astrology. But it is more important for our present purpose to examine the planetary theory given at the end of Sphujidhvaja's Yavanajâtaka than the spread of the science of astrology. The system and the parameters in this planetary theory are precisely identical with those found on cuneiform tablets of the Seleucid period. It is clear, then, that Babylonian linear astronomy was transmitted to India by the Greeks.[52] Normally, of course, Greek astronomical texts are devoid of these methods; but van der Waerden

47 Durbar Library 1180A.

48 On the interrelationships of these texts, see my unpublished thesis, *Materials for the Study of the Transmission of Greek Astrology to India*, submitted at Harvard in 1960, and forthcoming editions of Sphujidhvaja and Mînarâja.

49 *Indian Archaeology 1958-59 — A Review*, New Delhi, 1959, p. 8; *Indian Archaeology 1959-60 — A Review*, New Delhi, 1960, p. 54. The inscription cannot be dated in the year 30 of the Kalacuri Era (278/279 A. D.) as suggested by D. C. Sircar in *Indian Historical Quart.*, 1960, *36*: 24 f. (hereafter *IHQ*). This inscription has now been published by D. C. Sircar and K. G. Krishnan in *EI*, 1961-1962, *34*: 20-22.

50 See H. Raychaudhuri in *The Early History of the Deccan*, ed. G. Yazdani, vol. 1, London-Bombay-New York, 1960, p. 55; Mahâbhârata, Sabhâparvan 28, 47, ed. F. Edgerton, Poona, 1944 (Antioch, Rome, and the City of the Greeks — Alexandria — are mentioned in 28, 49); and Ptolemy 7, 1, 63.

51 P. L. Gupta, "Who Ruled in Saurâṣṭra after the Western Kshatrapas?" *Bhâratîya Vidyâ*, 1958, *18*: 83-89.

52 D. Pingree, "A Greek Linear Planetary Text in India," *JAOS*, 1959, *79*: 282-284.

and Neugebauer have shown that the Babylonian linear system lies behind the so-called Egyptian Eternal Tables [53] and appears in an astrological text ascribed to the late fifth-century author Heliodorus [54] (van der Waerden's claim to have found more than a few Babylonian parameters in the Thesauri of the sixth-century astrologer Rhetorius of Egypt [55] cannot be accepted). We may conclude, therefore, that at least some Greek astrologers ignored the epicyclic and eccentric theories developed by Apollonius, Hipparchus, and Ptolemy, and adhered to the Babylonian methods; and the Greek who wrote the original of the Yavanajâtaka in Alexandria between 100 and 150 A. D. was one such astrologer.

Sphujidhvaja mentions the work of the sage Vasiṣṭha; and it is likely that from the Vasiṣṭhasiddhânta are derived the first fifty-six verses of the eighteenth book of Varâhamihira's Pañcasiddhântikâ, which contain another Sanskrit version of Babylonian linear planetary theory. The second book of the Pañcasiddhântikâ contains a summary of the solar and lunar theories of the Vasiṣṭhasamâsasiddhânta.[56] The lunar theory is based on two well-known Babylonian period relations, which also occur in Greek papyri of the second and third centuries A. D.[57] — the equivalence of nine anomalistic months to 248 days and that of 110 anomalistic months to 3031 days. These same two period relations are found in the Pauliśasiddhânta [58] and in the thirteenth-century Candravâkyas of Vararuci,[59] while the second appears in

[53] For the Eternal Tables themselves, see O. Neugebauer, *Egyptian Planetary Texts, Trans. Amer. Phil. Soc.*, 1942, *NS 32*, pt. 2; for their use of Babylonian parameters, see B. L. van der Waerden, "Egyptian 'Eternal Tables,'" *Proc. Sect. Sciences, Kon. Ned. Akad. Wet.*, 1947, *50*: 536-547, 782-788; "Babylonische Planetenrechnung in Ägypten und Indien," *Bibl. Or.*, 1956, *13*: 108-110; *Centaurus*, 1958, *5*: 177; and "Babylonische Methoden in ägyptischen Planetentafeln," *Vjschr. Naturf. Ges. Zürich*, 1960, *105*: 97-144. See also a Demotic text edited by R. A. Parker, "Two Demotic Astronomical Papyri in the Carlsberg Collection," *Acta Orientalia*, 1962, *26*: 143-147 (P. Carlsberg 32 uses Babylonian methods); a Greek papyrus of a similar nature is to be published in *Papiri Greci e Latini* 15.

[54] O. Neugebauer, "On a Fragment of Heliodorus (?) on Planetary Motion," *Sudhoffs Archiv*, 1958, *42*: 237-244. E. Boer, in her edition of Paulus Alexandrinus, Leipzig, 1958, p. 28, says that this passage is not by Heliodorus, and she does not include it in her edition of Heliodorus (?), Leipzig, 1962, pp. 11-12.

[55] *Vjschr. Naturf. Ges. Zürich*, 1955, *100*: 165; Rhetorius' period (1,753,005 years) and planetary parameters are different from the Indian parallels adduced by van der Waerden, and there is no evidence to support the hypothesis that a yuga-system of astronomy was ever known in Greece, much less in Babylon. Professor Neugebauer, who has examined a photograph of the manuscript, informs me that one should read 1,753,200 rather than 1,753,005 (i. e., a c for an ϵ).

[56] For a complete explanation of these obscure verses, see T. S. Kuppanna Sastri, "The Vâsiṣṭha Sun and Moon in Varâhamihira's Pañcasiddhântikâ," *J. Orient. Res., Madras*, 1957, *25*: 19-41. This work is different from the Vasiṣṭhasiddhânta ed. by V. P. Śarmâ, Benares, 1907, and the Vṛddhavasiṣṭhasiddhânta ed. by V. P. Dvivedi in his *Jyautiṣasiddhânta-sañgraha*, Benares Sanskrit Series, fasc. 2, Benares, 1912, pt. 2.

[57] E. J. Knudtzon and O. Neugebauer, "Zwei astronomische Texte," *Bull. Soc. roy. lettres de Lund*, 1946-1947, pp. 77-78; O. Neugebauer, "The Astronomical Treatise P. Ryl. 27," *Det Kgl. Danske Vid. Selskab., hist.-fil. medd.*, 1949, *32*: 2, København; O. Neugebauer, *Exact Sciences*, pp. 161-167, 185-187; and B. L. van der Waerden, "The Astronomical Papyrus Ryland 27," *Centaurus*, 1958, *5*: 177-191.

[58] Pañcasiddhântikâ 3, 4.

[59] Ed. C. Kunhan Raja, Madras, 1948, and in the Haricarita of Parameśvara Bhaṭṭa, ed. V Krishnamacharya, Adyar Library Series 63, Madras, 1948. Neither of these publications was known to the scholars who recognized Babylonian parameters in the vâkya-system; O. Neugebauer, "Tamil Astronomy," *Osiris*, 1952,

the Romakasiddhânta [60] and in the Uttarakhaṇḍa of Brahmagupta's Khaṇḍakhâdyaka.[61] The Vasiṣṭhasamâsasiddhânta computes the true longitude of the moon according to a Babylonian linear zigzag system; and a linear system is the basis of Vasiṣṭha's solar theory, as it also is of Pauliśa's. Therefore, it is apparent that the earliest form of astronomy introduced into India by the Greeks was entirely Babylonian in origin.

Greek epicyclic theory soon followed, however, and probably under the patronage of the same Śaka dynasty of Ujjain. If the last verses of the last chapter of the Pañcasiddhântikâ are in fact based on the Pauliśasiddhânta, then Puliśa — who, despite Bîrûnî,[62] has nothing to do with the fourth-century astrologer Paulus of Alexandria [63] — gives the same values for the mean synodic arcs of the planets as appear in cuneiform tablets.[64] But Puliśa computed solar longitude according to the epicyclic theory,[65] and he included in his siddhânta all of the trigonometry necessary for the solution of problems in epicyclic astronomy, including a table of sines derived from the Greek table of chords.[66]

Later in the fourth century, probably not long after the Pauliśasiddhânta, was written the Romakasiddhânta,[67] whose name betrays its origin. That

10: 252-276; B. L. van der Waerden, "Die Bewegung der Sonne nach griechischen und indischen Tafeln," *Sitzungsber. Bayer. Akad. Wiss., math.-naturwiss. Kl.*, 1952, Nr. 18; I. V. M. Krishna Rav, "The Motion of the Moon in Tamil Astronomy," *Centaurus*, 1955-1956, *4*: 198-220; and B. L. van der Waerden, "Tamil Astronomy," *ibid.*, 221-234.

60 Pañcasiddhântikâ 8, 5.

61 Khaṇḍakhâdyaka 9, 5 in the edition with the commentary of Pṛthûdakasvâmin by P. C. Sengupta, Calcutta, 1941, and in his English translation, Calcutta, 1934; this chapter is not included in the edition with the commentary of Âmarâja by Babua Miśra, Calcutta, 1925.

62 *Alberuni's India*, trans. E. C. Sachau, vol. 1, London, 1914, p. 153.

63 There are at least three Pauliśa- (or Puliśa) siddhântas. The first is the original work, probably of the fourth century; it is lost. The second is the edition of Lâṭadeva, written around 505 A.D.; it also is lost, but a summary is preserved in the Pañcasiddhântikâ. The third is a later work modeled on a standard Indian siddhânta; it is lost except for quotations and references in the commentaries of Pṛthûdaka and Bhaṭṭotpala and in Bîrûnî's works. As Bîrûnî knows only this third Puliśa, his statement probably has no relevance for the first and makes no sense if applied to the text he had before him. Moreover, in the one detail in which the second Puliśa (Pañcasiddhântikâ 3, 17) and Paulus Alexandrinus (28) overlap — the values of the mean daily motion of the sun for the different months in a year — they do not agree. However, the name Pauliśa might still be a Sanskrit version of Paulus; the Pauliśasiddhânta did feel it necessary to give corrections for the differences in longitude between Yavanapura (Alexandria) and Avanti (Ujjain) and Vârâṇasî (Benares); see Pañcasiddhântikâ 3, 13.

64 Neugebauer, *Exact Sciences*, pp. 172-173, 189.

65 Pañcasiddhântikâ 3, 1-3.

66 *Ibid.* 4, 1-15.

67 There are apparently five Sanskrit works of this title. The history of the first two is the same as that of the first two versions of the Pauliśasiddhânta. The third is a revision by Śrîṣeṇa mentioned by Brahmagupta in Brâhmasphuṭasiddhânta 11, 50-51, and elsewhere. The fourth, which is also called Śrîṣavâyaṇa, is preserved in at least twenty-two manuscripts: MSS 34 of 1870/1871, 106 of 1873/1874 and 411 of 1884/1886 of the Bhandarkar Oriental Research Institute, Poona; MSS 3279, 9329, 9376, 13333, and 13421 (i) of the Oriental Institute, Baroda; MS 2790 of the Raghunatha Temple Library of the Mahârâja of Jammu and Kashmir; MSS 4643, 4777, 5068, and 5069 of the Anup Sanskrit Library, Bikaner; MS Or. fol. 981b of Berlin, now at Tübingen; MS 1805 of the India Office Library, London; MS Add. 14,365o of the British Museum, London; MS Wilson 157d of the Bodleian Library, Oxford; MS 8 of the Sanskrit College Library, Benares; MSS 378, 379, and 460 of the Library of the University of Bombay; and MS 259 of the Library of the Asiatic Society of Bombay. This text is obviously based on an Arabic work; see T. Aufrecht, *Catalogi Codicum Manuscriptorum Bibliothecae Bodleianae pars octava, Codices Sanscriticos com-*

origin is also revealed by the fact that the Romaka gives a luni-solar cycle of 2850 years,[68] which is equivalent to the Metonic cycle of 19 years multiplied by 150 so that its tropical year may be equal to Hipparchus', or 365 days plus 1/4 minus 1/300. And whereas Puliśa seems to have used an epicyclic model only for the sun, the Romaka applies this method to both luminaries.[69] Nothing is known of its planetary theory.

Both of these texts were probably based on translations from the Greek made under the patronage of the Kṣatrapas of Ujjain. That dynasty, however, became greatly weakened towards the end of the fourth century, and a new nationalistic state, the Gupta Empire, gained the hegemony of North India. It is an attractive hypothesis to suppose that the court of Samudragupta or of Candragupta II [70] encouraged an important development in Indian astronomy, the merging of two concepts of foreign origin into a new theory of planetary motion. To understand what was achieved we must briefly consider the kalpa.

A kalpa is a period of 4,320,000,000 years; 72,000 of these kalpas or 311,040,000,000,000 years constitute the life of Brahma. Each kalpa is divided into 1000 equal parts called mahâyugas, which are 4,320,000 years apiece, and each mahâyuga contains four smaller yugas which are in the ratios to each other of 4:3, 3:2, and 2:1. The last yuga, then, the kaliyuga, is 1/10 mahâyuga, or 432,000 years. This is a Babylonian number: sexagesimally it would be written 2,0,0,0. It is the span of time given to the Babylonian kingdom before the Flood in the histories of Berossos [71] and Abydenus.[72] It seems likely that it should have become known as a significant number in India at the time when other Babylonian influences were being felt, that is, during the Achemenid occupation of the Indus Valley. In fact, the kalpa appears with an eschatological connotation in the fourth and fifth Rock Edicts of Aśoka [73] and in the Dîghanikâya.[74] However, the first text to describe the kalpa precisely as I have just done was a pre-second-century work which was the common source [75] of a passage occurring in the twelfth book of the Mahâbhârata [76] and in the first book of the Manusmṛti.[77]

plectens, Oxford, 1864, pp. 338-340. The fifth Romaka- (Romaśa) siddhânta is contained in MS 377 of the Library of the University of Bombay. Versions two, three, and perhaps five are dealt with, though not thoroughly, in S. B. Dikshit, "The Romaka Siddhantas," *Indian Antiquary*, 1890, *19*: 133-142; for two, see also J. Burgess, *ibid.*, 284-285.

68 Pañcasiddhântikâ 1, 15; see Thibaut, *op. cit.*, intr. xxvii-xxviii.

69 Pañcasiddhântikâ 8, 1-6.

70 Though the details cannot be accepted, there is little reason to doubt the genuineness of the tradition that Candragupta II was a great patron of learning, including astronomy and astrology, preserved in pseudo-Kâlidâsa's Jyotirvidâbharaṇa 22, 8-12, ed. Sîtarâma Śarmâ, Bombay, 1908.

71 P. Schnabel, *Berossos und die babylonisch-hellenistische Literatur*, Leipzig-Berlin, 1923, fr. 29-30a, pp. 261-263.

72 Fr. 1 in C. Muller, *Fragmenta Historicorum Graecorum*, vol. 4, Paris, 1851, p. 280.

73 E. Hultzsch, *Inscriptions of Asoka, CII* 1, new ed., Oxford, 1925, pp. 189, 191.

74 Dîghanikâya 25, 18 and 28, 16, ed. J. E. Carpenter, *Pali Text Society*, vol. 3, London, 1911, pp. 51, 111; Añguttaranikâya, Catukkanipâta, 156, ed. R. Morris, *Pali Text Society*, pt. 2, London, 1888, p. 142.

75 See G. Bühler, *The Laws of Manu, Sacred Books of the East*, vol. 25, Oxford, 1886, pp. lxxxii-xc (hereafter *SBE*).

76 Mahâbhârata, Śântiparvan 224, 12-30, ed. S. K. Belvalkar, Poona, 1951.

77 Manusmṛti 1, 64-74, ed. with the commentary of Kullûka by Nârâyaṇ Râm Âchârya, 10th ed., Bombay, 1946.

This kalpa of ultimately Babylonian origin was combined by Indian astronomers of the late fourth or early fifth centuries with Greek epicyclic theory. The mean motions of the planets can be described in terms of an integer number of revolutions within a given period as long as that period is fairly long; obviously the kalpa and the mahâyuga were ideally suited for such a use. But a shorter period, 1/24 mahâyuga, or 180,000 years, was employed also. This was the yuga which seems to have been the basis of the system of the original Old Sûryasiddhânta,[78] a work known to us now only through Varâhamihira's summary of the recension made by Lâṭadeva in 505 A. D.[79] The beginning of the yuga was taken to be a mean conjunction of all the planets at Aries 0° at midnight between 17 and 18 February -3101.

The period of 180,000 years, however, was not long enough to permit the use of very accurate parameters. The mahâyuga was somewhat better, though not as good as might be expected. For the mean conjunction in -3101 was taken to mark the beginning of the last and smallest yuga, the kaliyuga of only 432,000 years. Since the mahâyuga itself had to begin with a similar conjunction, one was forced to fit the parameters into a period only 1/10 as long as the whole yuga in order that the conjunction of the beginning of the kaliyuga might in fact take place. Practically speaking, then, one had a period only 2 2/5 as long as 180,000 years rather than one 24 times as long. Âryabhaṭa in 499 solved this problem in part by making the four yugas within the mahâyuga equal;[80] this gave him an effective period of 1,080,000 years. But this defiance of tradition was not welcomed by many in India.

However, even before Âryabhaṭa astronomers had realized the advantages of working with a kalpa,[81] despite the enormous numbers with which one had to compute. That the parameters could be further refined was perhaps not as decisive a factor in their choice of the longer period as the fact that its length allowed one to begin the system with a true conjunction rather than with the mean conjunction which the original Old Sûryasiddhânta and Âryabhaṭa had to accept. For the kalpa gave one enough time to pull the apogees back to the beginning of the zodiac and to endow them with such a slow motion that they would have reached their proper positions in the fifth century A. D.

The Indians of the Gupta Age, therefore, seem to have been the originators of the yuga-system of astronomy and to have developed variants em-

78 Pañcasiddhântikâ 1, 14.

79 I have recently found fragments of Lâṭadeva's work in Pṛthûdaka's commentary on Brahmagupta's Brâhmasphuṭasiddhânta preserved in MS 1781 of the Viśveśvarânanda Vedic Research Institute, Hoshiarpur; MS 339 of 1879/1880 of the Bhandarkar Oriental Research Institute, Poona; and MS 1304 of the India Office Library, London. These should make Lâṭa's role in the development of jyotiḥśâstra much clearer than it is at present.

80 Âryabhaṭîya, Daśagîtikâ 3, ed. with the commentary of Parameśvara by H. Kern, Leyden, 1874; cf. Brâhmasphuṭasiddhânta 1, 9.

81 The siddhânta of Svayambhû (Brahman) to which Âryabhaṭa refers in Golâdhyâya 50 most probably was based on the kalpa-system; see the Paitâmaha- (Brahma) siddhânta, ed. by V. P. Dvivedi in *Jyautiṣasiddhântasaṅgraha*, fasc. 2, pt. 1, and the Brâhmasphuṭasiddhânta (corrected Brahmasiddhânta) of Brahmagupta.

ploying periods of 180,000, 4,320,000, and 4,320,000,000 years. The elements which they used were admittedly of Greek and Babylonian derivation, but only they had the necessary theoretical knowledge and the inspiration.

Our subsequent remarks will indicate the absence of any evidence for much knowledge or inspiration in Sasanian astronomy. But first it is necessary to mention an intriguing community in India which was, I believe, the only post-Achemenid group of Iranians who were in a position historically to influence the development of Indian astronomy and astrology before the Muslim invasions. The Pahlavas, who had established kingdoms in northwestern India in the first century B. C., left descendants who became integrated into Hindu society as a special class of Brâhmaṇas, the Maga Brâhmaṇas; [82] the great sixth-century astrologer Varâhamihira was one of their number.[83] They are known from various reports, all of which portray them as good Hindus whose only idiosyncrasy was an inordinate devotion to the sun. But the important fact about these Magi is that they seem to have had no contact with Iran after the first century A. D., and no one would suggest that the Pahlavas knew of yuga astronomy.[84]

In fact, virtually nothing is known of the astronomy and astrology of pre-Sasanian Iran. There was indeed a Greek astrological text of the second century B. C. ascribed to Zoroaster of which fragments are preserved by Proclus and the Geoponica; [85] the material with which it deals is overwhelmingly Babylonian.[86] But there is reason to believe that it is the product of the Magusaeans of Asia Minor and in no way reflects scientific knowledge in Iran. However, we have seen before that certain Babylonian astronomical and astrological theories were transmitted to India during the Achemenid occupation of the Indus Valley; it is difficult to believe that the Persians were not exposed to the same influences as their remote vassals in India. In fact, a linear shadow text which may be an echo of this influence is preserved in the ninth-century Pahlavi Shâyast Lâ-shâyast.[87]

We have previously mentioned the fact that Buddhists introduced nakṣatra astrology into Iran and Central Asia. The Śârdûlakarṇâvadâna, which contains a thorough exposition of this system, was extremely popular in this area. It was summarized in Chinese by the Parthian prince An Shih-kao

[82] On the Maga Brâhmaṇas, see the most recent, R. C. Hazra, *Studies in the Upapurâṇas*, vol. 1, Calcutta Sanskrit College Res. Series 11, Calcutta, 1958, pp. 29-108.

[83] D. K. Biswas, "The Maga Ancestry of Varâhamihira," *IHQ*, 1949, *25*: 175-183.

[84] The Sasanians, of course, occupied at least some parts of the Indus Valley where the Kuṣâṇas once had reigned, but the extent of their rule was quite limited; see E. Honigmann and A. Maricq, *Recherches sur les Res Gestae Divi Saporis*, Mem. Acad. Roy. de Belgique, Cl. des Lettres 47, 4, Bruxelles, 1953, pp. 94-100, 107 fn. 6. Their cultural influence seems to have been nil. Save for his acceptance of Herzfeld's interpretation of the Paikuli inscription, now disproved by Maricq, the account in A. S. Altekar, "The Extent of the Sassanian Political Domination in India," *M. P. Khareghat Memorial Volume*, vol. 1, Bombay, 1953, pp. 213-220 is a sober summary of what little information is available on this subject.

[85] The fragments (074-097) are collected in J. Bidez and F. Cumont, *Les Mages hellénisés*, vol. 2, Paris, 1938, pp. 207-242; see also fr. 012-052, pp. 158-197.

[86] Pingree, *Materials*, pp. 39, 43, 51-52, 55-62.

[87] Chap. 21, trans. E. W. West, *SBE*, vol. 5, Oxford, 1880, pp. 397-400.

in the second century A. D.[88] and fully translated twice in the third.[89] A long fragment of the Sanskrit text written in about 500 A. D. was among the Weber manuscripts found south of Yarkand,[90] and fragments of fifth-century manuscripts of the Mahâmâyûrîvidyârâjñî, which also deals with nakṣatra astrology to some extent, are preserved among the Bower and Petrovski manuscripts from Kashgar.[91] To reach these places the texts most probably passed through Buddhist communities in the eastern provinces of the Sasanian Empire; and one finds the remains of this Buddhist influence in the second chapter of the Bundahishn, where the twenty-eight nakṣatras are listed with Persian names.[92] Also connected with this type of astrology perhaps is the theory that the moon is the bestower of all benefits upon mankind, which is mentioned in the ninth-century Dâdistân-î Dînîk.[93] Perhaps, if Tabarî's story is not entirely fictitious, the " goodness of birth " which the astrologers observed for the first Ardashîr [94] was the presence of the moon in an auspicious nakṣatra. One might also suggest that Firdôsi's frequent references to the good or bad achrat or constellation of an individual [95] are to be interpreted in a similar way.

However, trustworthy knowledge of Iranian astronomy and astrology is non-existent before the reign of Shâpûr I (240-270). He encouraged the spread of Greek and Indian science within his realm.[96] The hexameters of the first-century astrologer Dorotheus of Sidon, preserved only in fragments in Greek, were translated into Pahlavi under Shâpûr; we now have a late eighth-century Arabic translation of this Pahlavi version made by 'Umar ibn Farrukhân.[97] From this it is clear that the original third-century Pahlavi version was revised at the end of the fourth or the beginning of the fifth century, and some Indian theories were added – in particular, that of the navâṃśas or ninths of a sign. This mixture of Greek and Indian material is characteristic of the Sasanians; it is found also in the fragments of an Arabic translation of the Pahlavi version of the Anthologies of the second-century astrologer Vettius Valens, fragments recently identified by Pro-

[88] Trans. pp. 213-217 Mukhopadhyaya; on An Shih-kao, see E. Zürcher, *The Buddhist Conquest of China*, vol. 1, Leiden, 1959, pp. 32-34.

[89] Pp. xii-xiii Mukhopadhyaya.

[90] Ed. A. F. R. Hoernle, *J. Asiatic Soc. of Bengal*, 1893, *62*: 9-17.

[91] For the Mahâmâyûrî, see S. Oldenburg, *Zapiski Vostočnago otdyeleniya Imp. Russk. Arkheol. Obstchestva* 11, 1897-1898, St. Petersburg, 1899, pp. 218-261; A. F. R. Hoernle, *The Bower Manuscript*, Arch. Surv. India, New Imperial Series 22, Calcutta, 1893-1912, pts. 6-7, pp. 222-240e and pl. xlix-liv; and S. Lévi, " Le catalogue géographique des Yakṣa dans la Mahâmâyûrî," *J. asiatique*, **11e série, 1915, *5*: 19-**138.

[92] Bundahishn 2, 2, trans. E. W. West, *SBE*, vol. 5, Oxford, 1880, p. 11; see W. B. Henning, " An Astronomical Chapter of the Bundahishn," *JRAS*, 1942, 229-248.

[93] Dâdistân-î Dînîk 71, 2, trans. E. W. West, *SBE*, vol. 18, Oxford, 1882, p. 215.

[94] *Annales quos scripsit . . . at-Tabarî*, ed. M. J. de Goeje, prima series, vol. 2, ed. J. Barth and T. Nöldeke, Leiden, 1881-1882, p. 815.

[95] The references are collected by J. Scheftelowitz in *Zeitschr. für Indologie und Iranistik*, 1926, *4*: 326-328.

[96] See the Dênkart quoted by R. C. Zaehner in *Zurvan: A Zoroastrian Dilemma*, Oxford, 1955, p. 8.

[97] See *Kitâb al-Fihrist*, ed. G. Flügel, vol. 1, Leipzig, 1871, p. 239; C. A. Nallino in *A Volume of Oriental Studies Presented to Professor E. G. Browne*, Cambridge, 1922, pp. 262-263, reprinted in his *Raccolta di scritti editi e inediti*, vol. 6, Roma, 1948, pp. 302-303. See also fn. 10 above.

fessor Kennedy in a manuscript in the British Museum.[98] Bîrûnî says that this Pahlavi work is one of the main sources of Abû Ma'shar's Magnus Introductorius.[99] The same British Museum manuscript preserves parts of an Arabic version of the Pahlavi translation of Teucer of Babylon.

We do not know so much of the study of astronomy under Shâpûr, though it does seem that Ptolemy's Syntaxis was one of the Greek works translated in this period.[100] It has sometimes been supposed that Shâpûr's foundation at Jundi-Shâpûr included an observatory, but no observations are known to have been made there in pre-Islamic times.[101]

At some point in early Sasanian history, however, an official astronomical work was compiled, the Zîj-i Shâh.[102] This, as we know from Bîrûnî, was revised under Khusrau I (531-579),[103] another ruler who encouraged Greek or Greco-Syrian and Indian scholars in Iran. At present the various versions are known very imperfectly in the citations of the early Islamic astronomers and astrologers; but from these fragments Professor Kennedy has been able to demonstrate that this zîj contains parameters from the ârddharâtrika or Midnight system of Âryabhaṭa, which is the same as that of the Old Sûryasiddhânta of Lâṭadeva and the Khaṇḍakhâdyaka of Brahmagupta. It has also been possible to show that a series of horoscopes of the vernal equinoxes of the first regnal years of the Sasanian kings, probably due to the ninth-century astrologer from Balkh, Abû Ma'shar, was computed by means of the planetary theory of the Old Sûryasiddhânta.[104] Another possible influence which this text may have had is on a Zoroastrian doctrine preserved in the Greater Bundahishn and in the Shikandgûmânîk vijâr. These works state that the planets are bound by chords to the chariot of the sun.[105] Bîrûnî, in his book On Transits, also attributes this theory to the Persians.[106] A similar idea appears in a Manichaean text from Turfan[107] and in the seventh- or eighth-century Mandaean Ginzâ.[108] The modern Sûryasiddhânta, which is known to preserve many of the theories of its similarly named predecessors, explains the anomalies in planetary motion by the activities of demons stationed at the sun, the apogees, and the nodes, who pull the planets along by chords of wind.[109] The Sasanian concept would appear to be a reflection of the Indian. Therefore, either the Old Sûryasiddhânta itself or a very similar text must have been translated into Pahlavi, perhaps under Khusrau.

Sanskrit astrological works were also popular in Iran. The early Islamic

[98] MS Add. 23,400 of the British Museum.

[99] *Al-Bîrûnî On Transits*, trans. M. Saffouri and A. Ifram with a commentary by E. S. Kennedy, Beirut, 1959, 3: 15 and 5: 10-11.

[100] Zaehner, *Zurvan*, p. 139.

[101] A. Sayili, *The Observatory in Islam*, Publ. Turkish Historical Soc. 7, 38, Ankara, 1960, pp. 50-51, 357-358.

[102] See E. S. Kennedy, "The Sasanian Astronomical Handbook *Zîj-i Shâh* and the Astrological Doctrine of 'Transit' (Mamarr)," *JAOS*, 1958, *78*: 246-262.

[103] E. S. Kennedy, *A Survey of Islamic Astronomical Tables*, *Trans. Amer. Phil. Soc.*, 1956, *NS 46*, pt. 2, p. 130.

[104] See D. Pingree, "Historical Horoscopes," to appear in *JAOS*.

[105] Zaehner, *Zurvan*, pp. 164, 416-417. The authors of the Pahlavi books appear not to have realized the function of these chords.

[106] Bîrûnî On Transits 15: 13-16: 9.

[107] Zaehner, *Zurvan*, p. 161 fn. 3.

[108] *Ibid.*, p. 160 fn. 1.

[109] Sûryasiddhânta 2, 2, ed. K. Chaudhary, Kâśî Sanskrit Series 144, Benares, 1946.

astrologers — many of whom were Persians — incorporated numerous Indian theories into their books, and most of these must have reached them through Pahlavi texts. Of course, there were direct translations from Sanskrit into Arabic made in the eighth century, but these seem to have been mainly of astronomical works, such as the Brâhmasphuṭasiddhânta and the Khaṇḍakhâdyaka of Brahmagupta; Bîrûnî says that in his time no Sanskrit astrological treatises had been translated into Arabic.[110] The Indo-Iranian astrology of these early Islamic authors reached Byzantium at the end of the eighth century in the works of pseudo-Stephanus of Alexandria and of Theophilus of Edessa; more was translated into Greek in the Comnenan period, the late eleventh and twelfth centuries.[111] It arrived in the Latin West in the twelfth and following centuries. These translations are useful because of their preservation of texts which sometimes have been lost in the original Arabic.

The most important of the transmitters of Indo-Iranian astrology was Abû Ma'shar. In his Book of the Thousands, epitomized by al-Sijzî in the late tenth century,[112] he gave a yuga-system of astronomy which he called the Thousands of the Persians. The mean motions of the planets in this system are preserved in Bîrûnî's Book of Instruction in the Elements of the Art of Astrology.[113] The period used is 360,000 years, in the middle of which — on 17 February -3101 — occurred the mean conjunction of the planets at Aries 0°, which, for the Indians, marks the beginning of kaliyuga; Abû Ma'shar interprets it as the indicator of the Flood.

To date the Flood in -3101 is rather strange. But one is not at a loss to explain it. In his Book of Conjunctions Abû Ma'shar says that this date was proposed by someone whose name, corruptly preserved in Arabic, may be Abydenus.[114] Abydenus, it may be remembered, was one of those Greek historians who placed the Babylonian kingdom of 432,000 years' duration before the Flood; and this 432,000 years is the length of the kaliyuga which begins in -3101. Someone aware of both Abydenus' Flood story and the astronomical date of the beginning of kaliyuga has rather sloppily combined the two traditions. As Bîrûnî remarks, the Persians did not usually believe in the Flood; but there were some who did accept it, confining its effectiveness to western Asia.[115] It is surely these Persians whom one must suspect of dating the Flood in -3101, for they occupied the ground, quite literally, between the two ideas which were synthesized. This interpretation agrees with Bîrûnî's statement in the India that Abû Ma'shar's date for the Flood was derived from the Hindu kalpa-theory.[116]

110 India, vol. 2, p. 211.

111 For pseudo-Stephanus, see my paper cited in fn. 104. Theophilus, Apomasar (Abû Ma'shar), and Achmat, whose works are filled with Indian material, are in the process of being edited.

112 The manuscript was discovered by Professor Kennedy, to whom I owe much of what I know of Arabic astronomy and astrology.

113 Ed. and trans. R. R. Wright, London, 1934, pp. 113-114.

114 This information is from a translation by E. S. Kennedy; the identification with Abydenus was originally suggested by Professor A. J. Sachs of Brown University.

115 *The Chronology of Ancient Nations*, trans. C. E. Sachau, London, 1879, p. 27.

116 India, vol. 1, p. 325.

In any case, -3101 cannot be a Greek date for the Flood. The only known astrological Flood-theory in Greece is that derived from Berossos' Babyloniaka, according to which a conjunction of all the planets in Cancer produces an ecpyrosis, or conflagration, whereas a conjunction in Capricorn causes a kataklysmos, or flood.[117] The choice of Cancer and Capricorn is clearly due to a desire to connect the world-year with the summer and winter solstices. In this tradition the Aries conjunction of -3101 is meaningless.

But it is also contrary to astrological theory. The zodiac is divided into four triplicities, which are connected with the four elements. The first consists of Aries, Leo, and Sagittarius, and is fiery; the second of Taurus, Virgo, and Capricorn, and is earthy; the third of Gemini, Libra, and Aquarius, and is airy; and the last of Cancer, Scorpio, and Pisces, and is watery. The conjunction of -3101 occurs in a fiery triplicity and astrologically must indicate, if anything, a conflagration, not a flood. The latter can take place only when there is a conjunction in a watery triplicity. This was recognized by Abû Ma'shar's predecessor, Mâshâ'allâh, who dated the Flood in -3300[118] because in that year occurred a Saturn-Jupiter conjunction in Cancer, the first sign of the watery triplicity; and Mâshâ'allâh expressly states that he is using the Zîj-i Shâh. Cancer is also connected with the Flood in the Pahlavi Bundahishn.[119]

As has been said before, Abû Ma'shar's yuga is 360,000 years split in half by the Flood; in other words, the 180,000 years of the original Old Sûryasiddhânta. In fact, the parameters which Abû Ma'shar gives for the moon, Mars, Venus, and Mercury — if one corrects the last by one — are exactly one-twelfth of those in the Old Sûryasiddhânta of Lâṭadeva. But the parameter for Saturn is one-twelfth of that in the Somasiddhânta, the Brahmasiddhântaof the Śâkalyasaṃhitâ, the Vṛddhavasiṣṭhasiddhânta, and the modern Sûryasiddhânta; and that for Jupiter is one-twelfth of what appears in the Âryabhaṭîya.[120] Therefore, the so-called Thousands of the Persians is really an eclectic Indian system. Abû Ma'shar, of course, probably found it in some Pahlavi text or Arabic translation thereof; but I think its ultimate origin is now clear. In the On Transits, Bîrûnî recognizes that the equations in Abû Ma'shar's zîj were taken from the Zîj-i Shâh, which had got them from an Indian source.[121]

That Abû Ma'shar is extremely unreliable in what he reports, moreover, can easily be shown from a statement of his which Bîrûnî has preserved in his Chronology of the Ancient Nations.[122] There he asserts that, using the system of the Persians, he has found that the planets are not in mean conjunction at Aries 0° at the time of the vernal equinox of -3101, but are scattered between Pisces 27° and Aries 1°. Abû Ma'shar is convicted of

[117] Fr. 37, pp. 266-267 Schnabel.

[118] See my paper cited in fn. 104.

[119] Bundahishn 7, 1-5, pp. 25-26 West.

[120]

	Old Sûryasiddhânta	Âryabhaṭîya	Brahmasiddhânta	Abû Ma'shar
Saturn	146,564	146,564	*146,568*	12,214
Jupiter	364,220	*364,224*	364,220	30,352
Mars	*2,296,824*	*2,296,824*	2,296,832	191,402
Venus	*7,022,388*	*7,022,388*	7,022,376	585,199
Mercury	*17,937,000*	17,937,020	17,937,060	1,494,751 (1,494,750)
Moon	*57,753,336*	*57,753,336*	*57,753,336*	4,812,778

[121] On Transits 30: 10-16.

[122] Chronology, p. 29.

lying by one glance at the parameters of his Thousands of the Persians. Those for the superior planets and the sun are all divisible by two, and as the conjunction of -3101 took place exactly in the middle of the yuga of 360,000 years, they must all be at Aries 0° at that date; the mean positions of the inferior planets, of course, are identical with the mean sun. So Abû Ma'shar's statement is nonsense when referred to his Persian system; but its source, fortunately, is known. In the kalpa-system of the Brâhmasphuṭa-siddhânta the mean planets are precisely between the limits set by Abû Ma'shar at the beginning of kaliyuga. This fact was not unnoticed by the Indians; it is recorded in the Siddhântaśekhara of the eleventh-century Śrîpati.[123] It must have been part of the polemic addressed by the partisans of the mahâyuga against those who preferred the kalpa. Abû Ma'shar has stupidly used it as a criticism of a system for which it is totally irrelevant.

So far nothing original has turned up in Sasanian astronomy and astrology, save for the fact that they synthesized Greek and Indian theories. However, we do know of one concept which seems definitely to be an Iranian innovation. This is the theory that history is the unfolding of the influences of periodically recurring Saturn-Jupiter conjunctions.[124]

Roughly, the idea behind astrological history is this. A Saturn-Jupiter conjunction takes place about every 20 years; a series will occur in the signs of one triplicity for about 240 years, that is twelve conjunctions; and they will have passed through the four triplicities and begin the cycle again after about 960 years. When they shift from one triplicity to another, they indicate events on the order of dynastic changes. The completion of a cycle of 960 years, which is mixed up with various millennial theories, causes revolutionary events such as the appearance of a major prophet. The ordinary course of politics is dependent on the horoscopes of the vernal equinoxes of the years in which the minor conjunctions within a triplicity take place.

The tenth-century astrologer Ibn Hibintâ preserves fragments of an astrological history written on this principle by Mâshâ'allâh, and a Parisian manuscript of a compilation by al-Sijzî contains the horoscopes, but not the interpretations, for such a history written under Hârûn al-Rashîd. Al-Kindî and Abû Ma'shar also wrote on these conjunctions, as did pseudo-Stephanus of Alexandria in Greek.

A non-Greek doctrine such as that of the Saturn-Jupiter conjunctions, occurring in the works of the early Islamic astrologers, may be assumed to have an Iranian origin. But there is more substantial evidence of the theory's Sasanian background; Ibn Khaldun says that Khusrau's famous minister, Buzurjmihr, was familiar with the method, and he cites from Jirâsh as an authority on the subject an astrologer with the obviously Iranian name of Hurmuzdâfrîd.

Astrological history by Saturn-Jupiter conjunctions, then, is the main,

123 Siddhântaśekhara 2, 52-53, ed. Babuâji Miśra, 2 pts., Calcutta, 1932-1947.

124 See my paper cited in fn. 104, where the subject is dealt with more extensively and the appropriate references will be found.

if not the only, Sasanian contribution to astronomy or astrology. Since, through the conjunction of -3101, it is closely linked with the yuga-system of astronomy, one would expect that any astronomer who learned the latter from the Sasanians would also have learned the former. But there is not a trace of the knowledge of these conjunctions in India. This I take to be fairly conclusive evidence that Indian yuga-astronomy could not have been borrowed from Iran, but instead profoundly influenced Sasanian science.

Appendix 1

Annual Administration Report of the Archaeological Department, Gwalior State for the Year 1938-39, Gwalior, 1940, p. 18 and pl. xxviii (b) (not seen).

Krishna Deva, "Coin Devices on Râjghâṭ Seals," *J. Numismatic Soc. India*, 1941, *3*: 77 and pl. v, figs. 17-19 (not seen).

A. S. Gadre, *Archaeology in Baroda (1934-1947)*, Baroda, 1947, pp. 5, 28.

B. B. Lal in *Ancient India*, 1949, *5*: 101-103.

M. D. Desai, "Some Roman Antiquities from Akota near Baroda," *Bull. Baroda Mus.*, 1949-1950, *7*: 21-23.

B. Subbarao, *Baroda Through the Ages (Being a Report of an Excavation Conducted in the Baroda Area 1951-52)*, M. S. Univ. Archaeology Series 1, Baroda, 1953, pp. 6, 32-34, 56-64, 87.

M. Wheeler, *Rome Beyond the Imperial Frontiers*, London, 1954, pp. 151-153.

H. D. Sankalia and S. B. Deo, *Report on the Excavations of Maheshwar and Navdatoli 1952-53*, Deccan College Res. Inst. and M. S. Univ. Publ. 1, Poona-Baroda, 1958, pp. 159-162.

R. C. Agrawala in *J. Bihar Res. Soc.*, 1955, *41*: 296-298.

H. D. Sankalia, B. Subbarao, and S. B. Deo, *The Excavations of Maheshwar and Navdatoli 1952-53*, Deccan College Res. Inst. and M. S. Univ. Publ. 1, Poona-Baroda, 1958, pp. 159-62.

B. Subbarao, *Personality of India*, 2nd ed., Baroda, 1958, pp. 46-47.

M. N. Deshpande in *Ancient India*, 1959, *15*: 90.

H. Rydh, *Rang Mahal*, Acta Archaeologica Lundensia, Series in 4° No. 3, Lund-Bonn-Bombay, 1959, pp. 147-149.

Historical and Cultural Chronology of Gujarat, ed. M. R. Majmudar, Baroda, 1960, pp. 95-96, 96-97.

H. D. Sankalia, S. B. Deo, Z. D. Ansari, and S. Ehrhardt, *From History to Pre-history at Nevasa (1954-56)*, Dept. of Archaeology and Ancient Indian History, Deccan College, Univ. of Poona, Publ. No. 1, Poona, 1960, pp. 69-70, 200-201, 279-281, 307-315, 421, 446.

Indian Archaeology 1960-61 — A Review, New Delhi, 1961, pp. 7-9, 18, 21.

S. N. Chowdhary, "A Bronze Statuette of Atlas from Śâmaḷâjî," *J. Oriental Inst., Baroda*, 1962, *11*: 309-315.

U. P. Shah and R. N. Mehta, "Nagarâ," *ibid.*, 403-406.

PRECESSION AND TREPIDATION IN INDIAN ASTRONOMY BEFORE A.D. 1200

DAVID PINGREE, Brown University

The process by which various non-Ptolemaic elements of the Greek astronomical tradition were transmitted to India and were there transformed into the astronomy of the *siddhāntas* is a subject of complexity and of obscurity. Its elucidation, however, is of great historical importance, both for the understanding it will afford us of the motivation for particular Indian solutions of problems in mathematical astronomy, and for the insight we will obtain from it into those areas of Hellenistic astronomy that, being almost totally eclipsed in Greek by the brilliance of Ptolemy's *Almagest*, can be discerned, though dimly, in the poetry of *jyotiḥśāstra*. The present paper contains an investigation into one aspect of this process, that in which the ideas both of the precession and of the trepidation of the equinoxes were introduced into India and there interpreted in terms of an older Indian tradition of the position of the solstices relative to the *nakṣatras* and in other ways. This example, like that of the planetary model previously discussed in this journal (ii (1971), 80–85), beautifully illustrates the failure of the Greeks to communicate and of the Indians to grasp the full significance of the concepts transmitted.

The *Jyotiṣavedāṅga*[1] of Lagadha (5/4th century B.C.?) states (Ārca 6 = Yājuṣa 7): "The Sun and the Moon begin their northern [course] at the beginning of Śraviṣṭhā [Dhaniṣṭhā]; the southern [course] of the Sun [begins] in the middle of Sarpa [Āśleṣā]. [The beginnings of these two courses occur] always in [the months] Māgha and Śrāvaṇa [respectively]." One also finds this scheme in, for example, the *Parāśaratantra* cited by Utpala (A.D. 966) on the *Bṛhatsaṃhitā*[2] of Varāhamihira (*ca* A.D. 550).

By the time of Varāhamihira a fixed sidereal zodiac was in use in India.[3] In this zodiac the beginning of Aries was identified with the beginning of the nakṣatra Aśvinī; in the fifth and sixth centuries the beginning of Aries was further said to be the point of the vernal equinox. Varāhamihira recognized the discrepancy with the statement of Lagadha (*Bṛhatsaṃhitā* 3, 1–2):

> Once, according to what is said in ancient treatises, the southern ayana of the Sun was from the middle of Āśleṣā, and the northern began with Dhaniṣṭhā. Now [one] ayana of the Sun begins at the beginning of Cancer, the other at the beginning of Capricorn. This is a negation of what was said; the difference is made manifest by direct observations.

In his *Pañcasiddhāntikā*[4] (3, 20–2) Varāhamihira explains this change by a theory of trepidation over an arc of 46;40°—23;20° (identified with the Sun's maximum declination) to either side of the equinox:

> When the sum [of the longitudes] of the Sun and Moon is a revolution, it is called Vaidhṛta [yoga]; but if it is a revolution plus 10 nakṣatras [133;20°],

> Vyatipāta. The time is to be ascertained by means of the degrees attained [by the luminaries]. When the return of the Sun was from the middle of Āśleṣā [at 113;20°], then the ayana [-correction] was positive; now the ayana is from Punarvasu [at 90°]. When the falling away [from the mean position] of the ayana is reversed, then the correction [kṣepa] for the Sun and Moon [equals] the degrees of the maximum declination [kāṣṭhā] of the Sun [23;20°]. There is Vyatipāta if the sum [of the longitudes] of the Sun and the Moon is 180°.

This is the earliest datable reference to a theory of trepidation or precession in India; unfortunately no rate is given. A theory of trepidation was known to Theon of Alexandria (A.D. 361) and to Proclus (A.D. 410–485), and a theory of precession to Hipparchus (*ca* −126); Hipparchus's length for a tropical year was used by Sphujidhvaja (A.D. 269/70) in his *Yavanajātaka* (79, 34) and in the *Romakasiddhānta* summarized by Varāhamihira in his *Pañcasiddhāntikā* (1, 15 and 8, 1). It is not unreasonable to suppose that the *idea* of trepidation or precession was introduced into India by the Greeks, though the *parameters* chosen by the Indians are their own, and that the arguments presented in favour of the hypothesis of a motion of the colures are derived from a particular interpretation of the *Vedāṅgajyotiṣa*. This hypothesis of a Greek influence is strengthened by a passage quoted by Bhāskara (A.D. 629) in his commentary[5] on the *Āryabhaṭīya* of Āryabhaṭa (A.D. 499):

> Here the Romakas [*i.e.*, Romans], who do not know the ultimate purpose, read: 'The sages say that an ayana [begins] from the beginning of Vasudeva [Dhaniṣṭhā] [and another] from the middle of Sarpa [Āśleṣā], whereas it is observed [that they begin] from the beginnings of Capricorn and Cancer. How can this be without a motion [of the solstices]?'

Several of these sources and some new ones, including Maṇindha,[6] are cited by Govindasvāmin (*ca* A.D. 850) in his commentary on the Uttarakhaṇḍa (2, 21b–25a) of Parāśara's *Horāśāstra*[7] (*ca* A.D. 650/750):

> The motion of the ayana has been described by former teachers. In this [matter] Maṇindha [says]: 'Hence the planets move "up" from the prime vertical [= equinoctial colure] 27° in 1800 years'. [It is said] by Sūrya [*Sūryasiddhānta* 3, 9a–b]: 'The circle of the constellations lags to the east 600 [times] in a [Mahā-]yuga'. On the other hand Bhāskarācārya, who completely adheres to the opinion of Āryabhaṭa, says [as shown above, he attributes this to the Romakas]: 'The sages say that an ayana [begins] from the beginning of Vasudeva [Dhaniṣṭhā] [and another] from the middle of Sarpa [Āśleṣā]'. And Varāhamihira [*Bṛhatsaṃhitā* 3, 1a–b]: 'The southern ayana of the Sun was from the middle of Āśleṣā, and the northern began with Dhaniṣṭhā'. In this [matter] Haridatta[8] [A.D. 684] [says]: 'They are 24°; the planets move from that'. The meaning of 'from that' is 'from the ecliptic'.

Govindasvāmin goes on to explain these obscure passages in terms of two schools advocating different rates and arcs of trepidation:

> This is what was said. There is one opinion of Āryabhaṭācārya and so on, another opinion of Maṇindha, Sūrya, and so on. Here Āryabhaṭa and so on [say] that the fact that the [vernal] equinox was at the beginning of Aries

in Śaka 444 [A.D. 522] is established by such instruments as the gnomon.[9] They determined that after this it moves [at the rate of] 0;1° in one year. So it moves 'down' 24°. Then one should add the minutes of trepidation. Afterwards it moves 'up'. Then, subtracting 0;1° a year from 24°, one should add [the remainder].

The opinion of the *Sūryasiddhānta* and so on [is this]. It moves 'down' 27° in 1800 years. Then it moves 'up'. In this [theory] this is the mathematical operation. In 1800 years 27° are obtained. One should operate with the given [years]. Divide the lapsed years of the Kaliyuga by 1800, multiply the remainder by 3, and divide [the product] by 200. [The result] is degrees, and the remainder is minutes. Or else divide the lapsed years of the Kaliyuga by 1800 and multiply the remainder by 54: these are seconds. It is to be known that in this school 0;0,54° represents the amount of trepidation of one year. It was shown by the previous commentator, Bhagadatta [*ca* A.D. 800], that this school is the better; this also is our tradition. Here it is to be understood that, if the quotient of the division by 1800 is an even number, the motion [of trepidation] is 'down', and if it is odd, 'up'.

This opinion of Maṇindha or Manetho seems the older theory of trepidation. The arc of 27° approximates the 26;40° in the difference of two equal nakṣatras between the Vedic lists (of unequal nakṣatras) beginning with Kṛttikā and the fixed sidereal zodiac beginning with Aśvinī; the 26;40° is probably rounded off to 27° to reflect the number of nakṣatras and to provide nice parameters. For the vernal point is assumed to have coincided with the beginning of Aśvinī and of Aries at the beginning of the Kaliyuga in −3101 and again 3600 years later in A.D. 499; the second part of this assumption is very nearly correct. Then the Vedic nakṣatra-lists beginning with Kṛttikā are explained by a progressive motion of the vernal point to fixed Aries 27° (= fixed Kṛttikā 0;20°) over half of this period, and a subsequent retrograde motion back to fixed Aries 0° over the second half; the retrograde motion continues at the same rate after A.D. 499 so that the trepidation will reach its limit at fixed Pisces 3° in A.D. 2299. The rate of trepidation, as Govindasvāmin correctly remarks, is:

$$\frac{27^\circ}{1800^y} = \frac{3^\circ}{200^y} = 0;0,54^\circ/^y.$$

This, of course, is the rate of precession "established" by Yaḥyā ibn Abī Manṣūr on the basis of an observation of the autumnal equinox that he made on 19 September 830.[10]

Govindasvāmin correctly refers to the *Sūryasiddhānta* as adhering to the opinion of Maṇindha; in the modern versions the relevant verses are (3, 9–10b)[11]:

> The circle of the constellations lags 600 [times] to the east in a [Mahā-]yuga. Multiply this by the ahargaṇa and divide [the product] by the civil days [in a Mahāyuga]. Multiply the resulting arc by three and divide [the product] by ten; the [resulting] degrees are to be known as the [amount of] trepidation.

A total oscillation of the vernal point, including both the progressive and the retrograde motions, occurs in 7200 years over an arc of 108°. In a Mahāyuga of 4 320 000 years there are 600 such oscillations; and 108° is $\frac{3}{10}$ of 360°.

A number of later texts follow this original version of Maṇindha's trepidation: see, for example, *Vṛddhavasiṣṭhasiddhānta* (2, 36) and the *Śākalyasaṃhitā's Brahmasiddhānta* (1, 192 and 194).[12] Some scholars, however, did advocate a modification of the parameters in the direction of a slowing down of the velocity of the trepidational motion. Thus Āmarāja (*ca* A.D. 1200) cites in his commentary[13] on the Khaṇḍakhādyaka (3, 11) of Brahmagupta (A.D. 665) a verse which he attributes to the *Uttara*,[14] which is an appendix to the *Khaṇḍakhādyaka* expounding non-ārdharātrika views: "Add 3179 to the Śaka-years and divide [the sum] by 7380; having added or subtracted [the quotient] from the beginning of the [fixed] zodiac, the degrees of declination north or south [are to be computed]".

The addition of 3179 to the epoch of the Śaka era results in the epoch of the Kaliyuga, −3101, when the current oscillation is assumed to have begun. But division by 7380 instead of 7200 decreases the rate of trepidation by 180/7200 or 1/40. Another result of such a division is to place the end of half an oscillation not at Kali 3600 (A.D. 499), but at Kali 3690 (A.D. 589). Āmarāja's further discussion shows that he understands this much; he states that there is no trepidation in Śaka 511 (A.D. 589), but he assumes that the limit of trepidation is 24° from fixed Aries 0°, a value which is connected with the second opinion of Govindasvāmin.

Āmarāja goes on to explain wrongly several other passages which in fact utilize the 24°-limit and the rate of 0;1° per year. Thus, on a passage of the *Karaṇakutūhala*[15] in which Bhāskara (A.D. 1183) states that the trepidation in Śaka 1105 was 11°, he remarks: "Because it makes a 'revolution' in 7380 years, there are 585 336 'revolutions' of the vernal point in a Kalpa [of 4 320 000 000 years]". Indeed, 585 366 × 7380 = 4 320 001 080, so that this number is nearly correct. Āmarāja further explains *Bṛhatsaṃhitā* (3, 1–2) on the grounds that Varāhamihira died in Śaka 509 (A.D. 587), when the trepidation was nil.

In the later Āryabhaṭa's (between *ca* A.D. 950 and 1100) *Mahāsiddhānta*[16] (1, 11–12) the 'revolutions' are numbered 578 159; this implies an oscillation in very nearly 7472 years. In his summary of the *Parāśarasiddhānta* (2, 9), this Āryabhaṭa gives the number of 'revolutions' as 581 709; this implies an oscillation in about 7426½ years.

The second opinion Govindasvāmin claims to belong to the school of Āryabhaṭa[17] because of the verse cited from Bhāskara; but we have seen that Bhāskara quotes this with disapproval from the Romakas. Nevertheless, the opinion does belong to the school of Āryabhaṭa, though the connection is through the date Śaka 444 (A.D. 522) at which the vernal point was, according to Govindasvāmin, demonstrated by means of a gnomon to be at fixed Aries 0°. The date seems rather to have been derived from a misunderstanding of a verse in the *Āryabhaṭīya* (Kālakriyā 10): "When three quarters of a [Mahā-]yuga and 3600 years had passed, then 23 years from my birth here had passed". This was quite early interpreted to mean that Āryabhaṭa was born in Kali 3600 and wrote his *Āryabhaṭīya* in Kail 3623 or Śaka 444.[18]

Another parameter, however, is derived from the tradition preserved by Varāhamihira in the verses of the *Pañcasiddhāntikā* quoted above, and apparently belongs to the school of Pauliśa, which was certainly strongly influenced by Greek methods. Pauliśa, noting the difference of 23;20° between the middle of fixed Āśleṣā and fixed Cancer 0°, concluded that the limit of the arc of trepidation from fixed Aries 0° was 23;20°, which he identified with the maximum declination of the Sun. The usual obliquity of the ecliptic noted in Indian texts is 24°, and it is this amount that the second system takes as the limit of trepidation.

In this system, then, a coincidence of the vernal point and the fixed Aries 0° occurs in A.D. 522, the limit of the trepidation is 24° to either side of the fixed Aries 0°, and the rate of trepidational motion is 0;1° a year or 1° every 60 years. A period of oscillation, then, is 5760 years, and there are 750 such oscillations in a Mahāyuga; but the vernal point and the fixed Aries 0° did not coincide at the beginning of either the Mahāyuga or the Kaliyuga.

The earliest reference to this system seems to be in the Pūrvakhaṇḍa (*ca* A.D. 600/700) of the *Horāśāstra*[19] of Parāśara (3, 31a–b): "Subtract 444 from the Śaka-year and divide [the remainder] by 60; [the result is] the degrees of trepidation". Thereafter it is often repeated. Āmarāja (on *Khaṇḍakhādyaka* 3, 11) says that Pṛthūdakasvāmin (A.D. 864) claimed that the trepidation in Śaka 800 (A.D. 878) would be 6;30°;[20] this should be corrected to 6° as there are 356 (60 × 5;56) years between Śaka 444 and Śaka 800. Muñjāla (A.D. 932) is quoted by several sources[21] as giving a trepidation of 6;50° for Śaka 854, which is 410 (60 × 6;50) years after Śaka 444. Śrīpati (A.D. 1056), in his *Dhruvamānasa*,[22] gives the same rule as appears in Parāśara's *Horāśāstra*. Śatānanda (1099), in his *Bhāsvatī* (8, 1),[23] gives the rate of precession as 1° in 60 years, but makes the year of coincidence Śaka 450 (A.D. 538). And Bhāskara (A.D. 1183) in his *Karaṇakutūhala* (2, 17)[24] gives a trepidation of 11° for Śaka 1105, which is 661 (60 × 11;1) years after Śaka 444.

But there are some modifications of these parameters which occur; motivations for them are not apparent. Thus, Āmarāja cites two verses from a *Karaṇottama* (on *Khaṇḍakhādyaka* 3, 11). The first states that the trepidation for the epoch of this work is 600 minutes; at the usual rate of trepidation this would date the text in Śaka 1044 (A.D. 1122). The next verse is somewhat corrupt, but the most reasonable interpretation that occurs to me is that one should subtract 4217 from the lapsed years (since the beginning of Kaliyuga) and divide (the remainder) by 61. Kali 4217 corresponds to A.D. 1116—just 6 years before the date toward which the first verse pointed. The rate of trepidation—1° every 61 years—means that the traversal of an arc of 24° takes 1464 years, a complete oscillation of 96° 5856 years. There is not an integral number of complete oscillations in a Mahāyuga.[25]

The second variant is quoted by Parameśvara (*ca* A.D. 1380–1460) in his commentary on the *Āryabhaṭīya* (Kālakriyā 10): "Subtract 3600 from the Kali-year and divide [the remainder] by 5808; divide the remainder by 1152. Multiply the remaining years by 2 and divide [the product] by 121. The remainder is the equation in degrees and so on". Here Kali 3600 is substituted for Kali 3623 as the year in which the coincidence of the vernal point and the fixed Aries 0° occurred; in this Parameśvara follows Maṇindha. The rate of trepidation also is changed from 1° in 60 years to 2° in 121 years (1° in 60½ years). As a complete oscillation is 96° (24 × 4), the period is 5808 years (96 × 60½) and the time required for the traversal of an arc of 24° is 1152 years

$(24 \times 60\frac{1}{2})$. There is not an integral number of oscillations in a Mahāyuga.

Finally, Āmarāja (on *Khaṇḍakhādyaka* 3, 11) ascribed to "others" the following line: "Subtract 441 from the Śaka-year, multiply [the remainder] by 2, and divide by 119; the result is in degrees". Śaka 441 or Kali 3620 becomes the year of coincidence, and the rate is changed to 2° in 119 years or 1° in $59\frac{1}{2}$ years. The traversal of an arc of 24°, then, takes 1428 years, a complete oscillation of 96° 5712 years. Again there is not an integral number of oscillations in a Mahāyuga.

So far we have been examining theories of trepidation; but the theory of precession was not unknown in the period we have been investigating. The earliest reference is probably the corrupt number preserved by Bhāskara in his commentary on the *Āryabhaṭīya* and said to be the number of revolutions of the vernal point in a yuga according to the Romakas: viyadrudrakṛtanavadhṛtī or 1 894 110. If dhṛti (18) is emended to a word signifying 1, the resulting 194 110 could be the revolutions in a Kalpa; for the annual motion then would be $\frac{194\,110 \times 360°}{4\,320\,000\,000}$ or approximately 0;0,58°. In fact, for a motion of 0;1° a year or 1° in 60 years there should be 200 000 revolutions of the vernal point in a Kalpa. The deviations from 200 000 that occur in this Romaka-passage and in others shortly to be discussed are presumably due to efforts to make the parameters yield a desirable coincidence of the vernal point and the fixed Aries 0° in about A.D. 500. Without knowing the interval each system would place between the beginning of the Kalpa and A.D. 500 we cannot reconstruct their computations.

The next such parameter of precession was given by Viṣṇucandra (*ca* A.D. 550/600) whose verse is cited by Pṛthūdakasvāmin (on *Brāhmasphuṭasiddhānta* 11, 54):[26] "The yuga of the ayana is said to be 189 411 [revolutions]; this was formerly the opinion of Brahmā, the Sun [Sūrya], and so on". It scarcely seems useful to compute the annual motion resulting from this parameter. The versions of the *Brahmasiddhānta* and the *Sūryasiddhānta* to which Viṣṇucandra refers are no longer available to us.

Finally, Bhāskara (A.D. 1150) has a very difficult passage (*Siddhāntaśiromaṇi*, Golādhyāya 7, 17–18):[27]

> The intersection of the equinoctial and the declinational circles is 'the node of the declination'. Its retrograde revolutions are said by the Saura [*Sūryasiddhānta*] to be 30 000 in a Kalpa. But the precessional motion proclaimed by Muñjāla and so on is correct; in this school its revolutions in a Kalpa are 199 699.

Bhāskara's reference here to the *Sūryasiddhānta* has caused much learned comment as it clearly conflicts with the verse from that work (3, 9) cited as early as Govindasvāmin;[28] it also conflicts with the statement regarding Sūrya's theory of precession made by Viṣṇucandra. The expression of 600 in the *Sūryasiddhānta* is "triṃśatkṛtyo" (thirty twenties); in some versions the word for twenties, "kṛtyo", is corrupt. Perhaps Bhāskara's statement reflects someone's misunderstanding of the *Sūryasiddhānta*, which was interpreted as saying that there are thirty trepidations in a Mahāyuga only, and therefore 30 000 in a Kalpa.

Bhāskara's statement regarding Muñjāla is not without difficulties either, though the number of revolutions in a Kalpa is closer to 200 000 than any other that is attested. Muñjāla, as was shown above, is known to have followed Govindasvāmin's second theory. Unless he wrote a third work besides the *Bṛhanmānasa* and the *Laghumānasa*, Bhāskara's reference must be wrong.

In conclusion, it has been established that before the twelfth century there existed three systems of accounting for the shift in the point at which the ecliptic crosses the equator which became evident when, in the fifth and sixth centuries, it was assumed that the nakṣatras in the Vedic literature and the *Vedāṅgajyotiṣa* referred to the equal nakṣatras of the fixed Indian zodiac. Two solutions involved trepidation over arcs whose limits were determined by an interpretation of the indigenous Indian traditions, the third simple precession; the rates were either 0;0,54° or 0;1° per year, with various modifications chosen to fit into particular contexts which, by and large, cannot now be reconstructed. Both trepidation and precession appeared in Greek before they appear in Sanskrit, and the origins of all three Indian systems are expressly connected with texts dependent on Greek sources: Maṇindha or Maṇittha, Pauliśa, and Romaka. The supposition is quite strong, then, that the theories of trepidation and precession were transmitted to India along with other astronomical theories between the second and fifth centuries A.D., although the specific parameters employed arise out of the adaptation of Greek astronomy to the Indian tradition.

REFERENCES

1. I have used the edition of both the Yājuṣa (with Somākara Śeṣa's *Bhāṣya*) and the Ārca recensions, prepared by Sudhākara Dvivedin, *The Pandit*, NS xxix (1907, repr. Benares, 1907); and the edition by R. Shamasastry (Mysore, 1936). The standard article on precession and trepidation in India remains that written by H. T. Colebrooke a century and a half ago ("On the Notion of the Hindu Astronomers concerning the Precession of the Equinoxes and Motions of the Planets", *Asiatic researches*, xii (1816), 209–50, reprinted in his *Miscellaneous essays*, ii (London, 1837), 374–416). This was used by T.-H. Martin, *Mémoire sur cette question: La précession des équinoxes a-t-elle été connue des égyptiens ou de quelque autre peuple avant Hipparque?* (Paris, 1869), 179–88; Martin is misunderstood by P. Duhem, *Le système du monde*, ii (Paris, 1914), 212–14 and 223–6. Later articles on the subject do not really advance our knowledge of the history of precession and trepidation in India; they represent a continuation of the misunderstanding of the implications of Vedic literature that, as this paper attempts to demonstrate, originated in about the fifth century A.D. Nonetheless, it may not be useless to mention the more significant of these: D. N. Mookerjee, "Notes on Indian Astronomy", *Journal of the Department of Letters, University of Calcutta*, v (1921), 277–302; J. Ghatak, "The Conception of the Indian Astronomers Concerning the Precession of the Equinoxes", *Journal of the Asiatic Society of Bengal*, NS xix (1923), 311–21; S. K. Das, "Precession and Libration of the Equinoxes in Hindu Astronomy", *Journal of the Asiatic Society of Bengal*, NS xxiii (1927), 403–13; R. Krishnamurthy, "Precession or Ayanamsa", *The mathematics student*, xiii (1945), 77–81; and K. V. Abhyankar, "The Precession of the Equinoxes and its Discovery in India", *Acharya Dhruva Smaraka Grantha* (Ahmadabad, 1946), iii, 155–64.
2. I have used the edition of the *Bṛhatsaṃhitā* with the *Vivṛti* of Utpala prepared by Sudhākara Dvivedin, *Vizianagram Sanskrit series*, xii (Benares, 1895–97).
3. It is to this zodiac, whose beginning lies in the vicinity of the star ζ Piscium, that I hereafter refer with the adjective "fixed"; in it the nakṣatras are equal arcs of 13;20° each. In the zodiac of the *Vedāṅgajyotiṣa* the nakṣatras are also equal arcs of 13;20° each, but the initial point of the system is not known to us. In the Vedic texts the nakṣatras are individual stars or groups of stars whose identity is not certain. Much confusion is introduced into the history of Indian astronomy by reading all texts as though they refer to the fixed zodiac; the specific parameters used for the limits of trepidation in Indian astronomy also result from this misunderstanding on the part of Indian astronomers in the fifth and sixth centuries A.D.

4. I have used the edition of O. Neugebauer and D. Pingree, *Det Kongelige Danske Videnskabernes Selskab*, Hist.-Filos. Skr., vi, 1 (Copenhagen, 1970).

5. Unfortunately still unpublished, though an edition is promised by K. S. Śukla of Lucknow. I cite this quotation from T. S. Kuppanna Sastri's edition of Bhāskara's *Mahābhāskarīya*, (with the *Bhāṣya* of Govindasvāmin and the *Siddhāntadīpikā* of Parameśvara), *Madras Government oriental series*, cxxx (Madras, 1957), p. XXVI.

6. Maṇindha is the prevalent South Indian orthography for Maṇittha, which represents the Greek Manethōn. Maṇittha's work was known to Varāhamihira, and so must have been written in the fourth or fifth century. The opinion is also ascribed to Maṇittha by Nīlakaṇṭha (on *Āryabhaṭīya*, Kālakriyā 10; see below, ref. 17).

7. I have used transcripts of MS 3166 in the Mysore Government Oriental Library and of MS D 11498 in the Sarasvati Mahal Library in Tanjore.

8. This verse is not found in his *Grahacāranibandhana* edited by K. V. Sarma, *Journal of oriental research, Madras*, xxiii (1953/54, repr. Madras, 1954).

9. For this method see, for instance, Parameśvara (A.D. 1443) in his second *Goladīpikā* (4, 85–90), ed. K. V. Sarma, *Brahmavidyā*, xx (1956), 119–86 and xxi (1957), 87–144; repr. as *Adyar library series* Paper 32 (Madras, 1957).

10. For the possible connection see *Dumbarton Oaks papers*, xviii (1964), 138.

11. I have used the editions by F.-E. Hall and Bāpū Deva Śāstrin with the *Gūḍhārthaprakāśaka* of Raṅganātha, *Bibliotheca Indica*, xxv (Calcutta, 1854–58); by Sudhākara Dvivedin, *Bibliotheca Indica*, clxxiii (Calcutta, 1910–11, repr. Calcutta, 1925); and by K. S. Śukla (with the *Vivaraṇa* of Parameśvara) (Lucknow, 1957). An apparent misunderstanding of this passage is found in Bhāskara's *Siddhāntaśiromaṇi* (see ref. 27), and Sumatiharṣa (A.D. 1619) in his commentary on Bhāskara's *Karaṇakutūhala* (2, 17; see below, ref. 24), ignoring 3, 9c–d, interprets *Sūryasiddhānta* 3, 9a–b to refer to a trepidation over an arc 60° (30° on either side of Aries 0°) at a rate of 0;1° per year. One complete oscillation, then, takes 7200 years as it should, but the rate of precession is that of Govindasvāmin's second school. Sumatiharṣa ascribes this interpretation to a ṭīkā on the *Sūryasiddhānta*. See also below, ref. 25.

12. Both of these texts are edited by V. P. Dvidevin, *Jyautiṣasiddhāntasaṅgraha*, *Benares Sanskrit series*, 2 fasc. (Benares, 1912).

13. I have used the edition of the *Khaṇḍakhādyaka* (with the *Vāsanābhāṣya* of Āmarāja) prepared by Babua Miśra (Calcutta, 1925).

14. The *Uttarakhaṇḍakhādyaka* is apparently by Brahmagupta himself. In his *Brāhmasphuṭasiddhānta* (11, 54; I have used the edition of Sudhākara Dvivedin (Benares, 1902)), written in A.D. 628, he denies a motion of the solstices; see also Bhāskara's *Vāsanābhāṣya* (on *Siddhāntaśiromaṇi*, Golādhyāya 7, 17–18; see below, ref. 27). But the *Uttarakhaṇḍakhādyaka* was written almost 40 years later.

15. See below, ref. 24.

16. I have used the edition of Sudhākara Dvivedin, *Benares Sanskrit series*, cxlviii–cl, 3 fasc. (Benares, 1910), and the edition of the Pūrvagaṇita by S. R. Sarma (2 vols., Marburg, 1966). Munīśvara (b. A.D. 1603) in his *Marīcī* on the *Siddhāntaśiromaṇi* (Golādhyāya 7, 17–19) refers to this Āryabhaṭa, not to the author of the *Āryabhaṭīya* as imagined by Colebrooke.

17. There is no mention of a motion of the equinoxes in the *Āryabhaṭīya*; I have used the editions of H. Kern (with the *Bhaṭadīpikā* of Parameśvara) (Leiden, 1874), and of K. Sāmbaśiva Śāstrī (1 and 2) and Suranad Kunjan Pillai (3) (with the *Bhāṣya* of Nīlakaṇṭha), *Trivandrum Sanskrit series*, ci, cx and clxxxv (3 vols., Trivandrum, 1930, 1931, and 1957).

18. See, *e.g.*, the texts collected by K. V. Sarma in his edition of Haridatta's *Grahacāranibandhana* (see above, ref. 8), pp. V–VI.

19. I have used the edition of G. L. Śarman and G. Śarman (Bambaī Saṃ. 2008, Śaka 1873 (A.D. 1951)).

20. I do not find this in Pṛthūdakasvāmin's *Vivaraṇa*, edited with the *Khaṇḍakhādyaka* by P. C. Sengupta (Calcutta, 1941); but only the beginning of his commentary on the *Uttarakhaṇḍakhādyaka* is preserved.

21. Apparently in his lost *Bṛhanmānasa*; it is quoted by the commentators on his extant *Laghumānasa* (1, 2), *e.g.*, by Praśastadhara (A.D. 958) as quoted in Majumdar's edition, p. 5, and by Parameśvara, and was noted by al-Bīrūnī (*India* (ed. Hyderabad, 1958), 308; trans. E. C. Sachau (London, 1910), i, 366–7) as occurring in the *Laghumānasa*. For the *Laghumānasa* I have used the editions of N. K. Majumdar (Calcutta, 1951), and of B. D. Āpaṭe, *Ānandāśrama Sanskrit series*, cxxiii (Poona, 1952). For another opinion attributed to Muñjāla by Bhāskara see below, ref. 27.

22. Quoted by Babuāji Miśra in his edition of Śrīpati's *Siddhāntaśekhara* (2 vols., Calcutta, 1932–47), i, 12.
23. I have used the edition of Ṭīkārāma Dhanañjaya (Benares, n.d.).
24. I have used the edition (with the *Gaṇakakumudakaumudī* of Sumatiharṣa) (Bambaī Saṃ. 1958, Śaka 1823 (A.D. 1901)).
25. Another possible interpretation would be: one should add 4217 to the lapsed years (since the epoch of the work) and divide (the result) by 61. One would then assume that the initial year of the Kaliyuga (−3101) was a year of coincidence, that the annual rate of trepidational motion is 0;1,1°, and that the limit of trepidation is 30° from fixed Aries 0° as in the case of Sumatiharṣa's interpretation of the *Sūryasiddhānta* (see above, ref. 11). Then the motion in 4217 years would be about 70°, so that the actual position of the vernal point relative to fixed Aries 0° would be −10°.
26. I have used MS Sanskrit 2769 of the India Office Library, London; see also the *Pañcasiddhāntikā*, i, 8.
27. I have used the edition of Dattātreya Āpaṭe (with Bhāskara's own *Vāsanābhāṣya* and Munīśvara's *Marīcī*), *Ānandāśrama Sanskrit series*, cxxii (2 vols., Poona, 1943–52).
28. See Munīśvara's *Marīcī*, *ad loc.*, and Colebrooke's summary thereof.

Note added in proof:

Since this article was written, the appearance of the complete text of the *Khaṇḍakhādyaka* in the edition by B. Chatterjee (2 vols, Calcutta, 1970) has made it clear that Āmarāja means by *Uttara* (see ref. 14) not Brahmagupta's *Uttarkhaṇḍakhādyaka*, in which the cited verse does not appear, but the *Khaṇḍakhādyakottara* of his (Āmarāja's) teacher Trivikrama (A.D. 1180).

THE MESOPOTAMIAN ORIGIN OF EARLY INDIAN MATHEMATICAL ASTRONOMY

DAVID PINGREE, Brown University

In this paper I intend to advance and offer evidence in support of an hypothesis concerning the dependence of the mathematical astronomy of the *Jyotiṣavedāṅga* on Mesopotamian science of the Achaemenid period.[1] I believe that the evidence in support of the theory that some elements of early Indian astronomy are derived from Mesopotamia is overwhelming, and that the evidence for the rest of my hypothetical reconstruction is persuasive. But I must enter a cautionary note with regard to that portion which relates to the Indian intercalation-cycle: the evidence in both the cuneiform and the Sanskrit sources is so fragmentary that no hypothetical reconstruction of the development or of the interrelation of their respective intercalation-cycles is more than a reasonable guess. I hope that the reader will find my guess more plausible than those of my predecessors.

Though the Vedas and Brāhmaṇas provide us with some crude elements of observational astronomy, such as the standard list of 27 or 28 nakṣatras or constellations associated with the Moon's course through the sky, and some rough parameters, such as the twelve months and 360 nychthemera of a year, mathematical astronomy begins in India with a group of related texts which I intend to explain in this paper. The basic text of this group is the *Jyotiṣavedāṅga*,[2] one of the six aṅgas or "limbs" studied by Vedic priests; its purpose was to provide them with a means of computing the times for which the performances of sacrifices are prescribed, primarily new and full moons. This brief work has come down to us in two recensions: a shorter one of 36 verses associated with the *Ṛgveda*, and a longer one of 43 verses associated with the *Yajurveda*, which latter incorporates 29 verses of the Ṛk-recension. That Ṛk-recension was composed by one Lagadha, who is otherwise unknown, or, according to another interpretation, by Śuci on the basis of Lagadha's teachings; the Yajur-recension names no author, but has the dubious benefit of a bhāṣya or commentary by one Somākara. It is the Yajur-recension that has generally been used by modern scholars also, as it, in two of its additional verses, attempts to adjust the older system of the Ṛk-recension to the familiar terms of medieval Indian astronomy. In this paper the shorter and surely older Ṛk-recension will be used.

We are justified in asserting the originality of the Ṛk-recension not only by its shortness, but also by its parallelism to other pre-medieval Sanskrit texts. In particular we must discuss here the following seven works in addition to the two recensions of the *Jyotiṣavedāṅga*:

1. The *Arthaśāstra* of Kauṭilya[3] is an ancient work on political science. Many scholars have identified the author with the minister of Candragupta Maurya, who established the Mauryan Empire in northern India shortly before 300 B.C., though it seems fairly secure that our recension of Book Two of the *Arthaśāstra* does not antedate the second century A.D.[4] The twentieth chapter of the second book of the *Arthaśāstra* prescribes the duties of the Mānādhyakṣa or Super-

intendant of Measurements, among which is included the duty of supervising the measurements of time. These time-measurements are closely related to those of the *Jyotiṣavedāṅga.*

2. The *Śārdūlakarṇāvadāna*[5] is now the thirty-third story in a Buddhist collection of tales about Bodhisattvas, the *Divyāvadāna.* Originally it was an anti-caste tract in which a king of the Mātaṅgas (that is, Caṇḍālas), Triśaṅku, asks a Brāhmaṇa, Puṣkarasārin, to give his daughter, Prakṛti, to the outcaste's son, Śārdūlakarṇa. Upon the Brāhmaṇa's refusal of this unorthodox request, Triśaṅku proves his status as a Brāhmaṇa by displaying his knowledge of astral divination and astronomy. Our present Sanskrit version is full of interpolations, both in prose and in poetry; but the history of the basic core of the text can be traced back to translations into Chinese by the Parthian prince An Shih-kao, who settled in Loyang in A.D. 148,[6] by Chu Chiang-yen and Chih Ch'ien between *ca* A.D. 220 and 252,[7] and again by Dharmarakṣa between A.D. 266 and 317.[8] The passages which are of interest to us as they reflect the astronomy of the *Jyotiṣavedāṅga* occur in the Chinese translations of the third century.

3. The twelfth chapter of the *Pañcasiddhāntikā*,[9] composed by Varāhamihira at Ujjayinī towards the middle of the sixth century A.D., summarizes a *Paitāmahasiddhānta* whose epoch is 11 January of A.D. 80. This is our most important witness for establishing the priority of the luni-solar period-relation of the Ṛk-recension over that of the Yajur-recension; for, with the exception of an alteration in the position of the winter solstice and Varāhamihira's introduction of the concept of omitted tithis, it is in complete agreement with the former.

4. The *Vasiṣṭhasiddhānta* summarized in the second chapter and in verses 1–60 of the seventeenth chapter of the *Pañcasiddhāntikā* of Varāhamihira represents an Indian adaptation of a Greek version of Babylonian astronomy intermingled with elements of the older Indian borrowings from Babylonian astronomy which appear in the *Jyotiṣavedāṅga* and in the other texts of which we have been speaking. Though the version of the *Vasiṣṭhasiddhānta* available to Varāhamihira apparently used 3 December 499 as its epoch, the reference to Vasiṣṭha as an astronomical authority in the *Yavanajātaka*[10] written by Sphujidhvaja in A.D. 269/270 indicates that the original was probably composed in the second or early third century A.D.

5. The *Yavanajātaka* itself is primarily a versification of a Sanskrit prose translation of a lost Greek astrological text; the Greek original was composed in Egypt, and probably in Alexandria, in the first half of the second century A.D., while the prose Sanskrit translation was made by Yavaneśvara in western India in A.D. 149/150. In the seventy-ninth chapter of the *Yavanajātaka*, Sphujidhvaja describes a Greek adaptation of the Mesopotamian linear astronomy of the Seleucid period, but he intersperses in this material various elements derived from the tradition of the *Jyotiṣavedāṅga.*

6. The fifth, sixth, and seventh upāṅgas of the Jaina canon of sacred literature are in many respects virtually identical; they are entitled respectively the *Sūriyapannatti*, the *Jaṃbuddīvapannatti*, and the *Caṃdapannatti.*[11] The recension of the Jaina canon is traditionally assigned to the Council of Valabhī in Saurāṣṭra, which met during the reign of the Maitraka monarch Dhruvasena I

from *ca* A.D. 519 to 549. Though these texts contain much that is certainly far older than the early sixth century A.D., their intercalation-cycle seems to have been adapted to that of the Yajur-recension as they interpret the 366 days of a solar year as civil rather than sidereal.[12]

7. Many verses ascribed to Garga are cited by Somākara in his bhāṣya on the Yajur-recension, with whose system Garga is in accord as he knows most of that recension's additional verses. Usually this Garga has been identified with the author of the oldest form of the *Gargasaṃhitā*, which may be as old as the first century A.D.[13]; but the ślokas cited by Somākara are not found in that work. Rather one must date this Garga to a considerably later period as he demonstrates a knowledge of the astronomy of the early medieval period—for example, of the four categories of time-measurements (solar, lunar, civil, and nākṣatra) first enunciated in the *Paitāmahasiddhānta* of the *Viṣṇudharmottarapurāṇa*, a work written in all probability in the early fifth century A.D., though they are already hinted at by Sphujidhvaja.[14]

Though the Yajur-recension is thus seen most likely to be a product of the third to fifth centuries A.D., the *Jyotiṣavedāṅga* has often been dated to the twelfth century B.C. on the basis of the position among the nakṣatras that it assigns to the winter solstice. Though this argument, as we shall see, is not worth much, the system of the Ṛk-recension must indeed be much earlier than that of the Yajur-recension since we find it in one text definitely dated in A.D. 80 and in another which in part depends on sources that may go back to the third century B.C. I hope now to show that in fact the astronomy of the Ṛk-recension was formulated in the fifth or fourth century B.C. on the basis of information about originally-Mesopotamian methods and parameters transmitted to India during the Achaemenid occupation of the Indus Valley between *ca* 513 and 326 B.C. This date well suits the Ṛk-recension, whose language is definitely post-Vedic, but not yet classical.

The time-measurements of this earliest period of Indian mathematical astronomy were made by means of two instruments; both of them were invented in Mesopotamia. Furthermore, the mathematical device used for interpreting them—a linear zig-zag function—is also Mesopotamian. The first of these two instruments is an out-flowing water-clock. Its usage is described in the first three pādas of verse 7 of the Ṛk-recension:

> The increase in daylight and the decrease in night-time in the northern course (of the Sun) is a prastha of water, in the southern course it is the reverse.

This means that one employs a linear zig-zag function with a constant difference of one prastha a day to determine the amount of water to be poured into the water-clock for each period of daylight, with a minimum amount at the beginning of the Sun's northern course, that is at the winter solstice, and the maximum amount at the summer solstice.

The escape-hole (nāḍikā) of the water-clock (ghaṭikā or kumbha) is described in the *Śārdūlakarṇāvadāna*[15] as a round tube of gold four digits in length with a rectangular opening at the end. A similar description is given in the *Arthaśāstra*.[16]

> a nāḷikā [*i.e.*, 24 minutes] is measured by an āḍhaka of water flowing through a hole which is four digits long [extending] from a pot [kumbha]. [It is composed of] four māṣakas of gold.

That there is this relationship between the liquid measure, āḍhaka, and the temporal measure, kumbhikā, is also stated in verse 17 of the Ṛk-recension. This out-flowing water-clock is also mentioned by Sphujidhvaja[17] and by Varāhamihira.[18]

In cuneiform sources we find a reference to the use of an out-flowing water-clock in mul Apin,[19] a series which was probably compiled in about 700 B.C.; the earliest copy is dated 687 B.C. A number of other contemporary or even older texts contain similar references, in particular the fourteenth tablet of *Enūma Anu Enlil*. One mathematical text indicates that the shape of the clepsydra was a cylinder as seems to be true of the Indian equivalent. The water-clock itself, then, appears to be a Babylonian invention of the first few centuries of the last millennium B.C. What of its manner of use in India?

The remainder of verse 7 in the Ṛk-recension is also connected with Mesopotamia. It states simply: "Six muhūrtas in an ayana." A muhūrta is a thirtieth of a nychthemeron, or two nāḍikās, and an ayana is the time between two consecutive solstices. So the statement tells us that the magnitude of difference between the longest and shortest daylights is six muhūrtas or a fifth of a nychthemeron. This, of course, means that the ratio of the longest to the shortest day is 3:2—a ratio inappropriate to all parts of India save the extreme north-west, but one that is well-attested in cuneiform texts.

In fact the earliest ratio of longest to shortest daylight in Mesopotamia is 2:1, found in the so-called Astrolabe P, in the fourteenth tablet of *Enūma Anu Enlil* in mul Apin, and elsewhere in tablets dated before the seventh century B.C.[20] The ratio 3:2 used by the Indians, however, was commonly utilized in all Babylonian astronomical texts after *ca* 700 B.C. This tradition must surely be the source of the Sanskrit texts under discussion, and provides us with a *terminus post quem* for those texts.

But not only do our texts use the Babylonian ratio; they also employ a linear zig-zag function to determine the lengths of daylight in intermediate months. Thus the *Arthaśāstra*[21] gives the length of daylight as 15 muhūrtas at the equinoxes, with an increase or a decrease of one muhūrta a month as one proceeds from an equinox toward respectively the summer or the winter solstice; thereby the longest day becomes 18 muhūrtas, the shortest 12. The winter solstice is placed by the *Arthaśāstra* here in Pauṣya, which is the month preceding Māgha wherein the winter solstice occurs according to verse 6 of the Ṛk-recension; this is probably due to an attempt already discernible in the *Paitāmahasiddhānta* of A.D. 80 to have the vernal equinox fall in Caitra, the third month after Pauṣya. Elsewhere[22] the *Arthaśāstra* places the beginnings of the two ayanas in the same months as does the *Jyotiṣavedāṅga*.

A linear zig-zag function for obtaining lengths of daylight identical to the *Arthaśāstra*'s is found in the *Śārdūlakarṇāvadāna*,[23] in the *Yavanajātaka*,[24] and in the *Vasiṣṭhasiddhānta*.[25] This methodology is, of course, undeniably Babylonian; one need only multiply the Babylonian bēru or double-hours by 2;30 to produce the Indian table of muhūrtas.

The same zig-zag function is expressed in verse 22 of the Ṛk-recension, though at intervals of days rather than months.

> Whatever has elapsed of the northern ayana or whatever remains of the southern ayana, multiply it by 2, divide [the product] by 61, and add [to the quotient] 12; [the sum] is the length of daylight [expressed in muhūrtas].

This formula is based on three parameters—12 muhūrtas as the length of daylight at the winter solstice, 183 "days" as the interval between two consecutive solstices, and 6 muhūrtas as the difference between the longest and shortest lengths of daylight. Then, if the time from the winter solstice in "days" is denoted x and the length of daylight on that "day" as $d(x)$, it is assumed that

$$d(x) = 12 + \frac{6x}{183} = 12 + \frac{2x}{61}.$$

The second instrument used for measuring time in this period, though not mentioned in the *Jyotiṣavedāṅga*, is the gnomon. In later Sanskrit texts it is called śaṅku, but in the earliest specimens the gnomon-shadow was named chāyāpauruṣa, so the gnomon itself must have been called puruṣa or man; its length was normally defined as 12 aṅgulas or digits. One of the oldest gnomon-texts is that preserved in the *Arthaśāstra*.[26] Here it is assumed that the noon-shadow is 0 in Āṣāḍha—that is, at the summer solstice if the vernal equinox falls in Caitra as the preceding section of the *Arthaśāstra* prescribes. This indicates a terrestrial latitude of about 24°—that is, approximately the latitude of Ujjayinī—whereas the use of 3 : 2 as the ratio of longest to shortest length of daylight indicates a terrestrial latitude of about 35°—that is, somewhat north of the latitude of Babylon. But though the length of the noon-shadow for the summer solstice has been adjusted to Indian conditions, the *Arthaśāstra* still employs a Babylonian linear zig-zag function to obtain the lengths of the noon-shadows in other months of the year. It is assumed that the length of the noon-shadow at the winter solstice is 12 digits, so that the monthly increment or decrease in the length of the noon-shadow is 2 digits and its length at the equinoxes is 6 digits. Precisely this same linear zig-zag function is found in the *Vasiṣṭhasiddhānta* summarized by Varāhamihira,[27] though with the months replaced by the zodiacal signs occupied by the Sun.

In this system the noon-shadow at the winter solstice measures one gnomon-length. Exactly the same noon-shadow at the winter solstice is given in mul Apin,[28] though that text does not yield a value for the length of the noon-shadow at the summer solstice. Rather it forms a linear zig-zag function for the seasonal hours after sunrise at which the length of the shadow equals one gnomon-length; these values are 3 bēru or 90 uš at the winter solstice, 2 bēru or 60 uš at the summer solstice, and 2;30 bēru or 75 uš at the equinoxes. This implies a monthly difference of 5 uš.

The remainder of the gnomon-text in the *Arthaśāstra* gives shadow-lengths at the passing of various fractions of the day of the summer solstice; presumably simple proportion is to be used to find the shadow-lengths on other days. A similar, though corrupt and fragmentary, table of shadow-lengths at the passing of each muhūrta of the day of the summer stolstice is found in the *Śārdūlakarṇāvadāna*,[29] and another in the Pahlavī *Shāyast nē shāyast*[30]; and mul Apin also preserves such tables in which a number of seasonal-hours is associated with each shadow-length expressed in integer gnomon-lengths. Though the methods of computation of these several tables differ, it is noteworthy that the shadow is assumed in mul Apin to equal 8 gnomon-lengths after $\frac{1}{2}$ seasonal hour on the day of the summer solstice or 45 minutes of an equinoctial hour, whereas both the *Arthaśāstra* and the *Śārdūlakarnāvadāna* state that the shadow equals 8 gnomon-lengths after 1 muhūrta or 48 minutes of an equinoctial hour. It seems plausible, then, that the Sanskrit texts represent an adaptation of a lost Mesopotamian scheme reduced to expression in Indian units of time-measurement, and with further modifications to produce a shadow-length of 0 at noon.

The *Śārdūlakarṇāvadāna*[31] preserves a fragment of another early gnomon-text in which the gnomon measures 16 digits and the longest length of daylight, 18 muhūrtas, occurs in Śrāvaṇa as in the *Jyotiṣavedāṅga*. This scheme may well antedate the one discussed above. It gave the lengths of the noon-shadows for each month in a linear zig-zag system almost identical with that of the *Arthaśāstra*, though the latter used a 12-digit gnomon; the one variation in the *Śārdūlakarṇāvadāna* is that the noon-shadow at the summer solstice is half a digit rather than zero. This shows that in this scheme it was *not* assumed that the situation at Ujjayinī can be generalized for all of India. But it would be unrealistic to think that one could compute the precise latitude for which this shadow-table would work; it should suffice to say that it must have originated in northern India.

From this discussion it should be clear that the mathematical astronomy of this group of texts centering about the *Jyotiṣavedāṅga* has borrowed much from Mesopotamian astronomy of the seventh and sixth centuries B.C., and is especially close to the compilation mul Apin, though, of course, the Indians have altered certain elements to suit their own situation. We now must look at the more important and troublesome part of the *Jyotiṣavedāṅga*, its intercalation-cycle.

In the Ṛk-recension the elements of this yuga or cycle are not clearly stated, but can easily be inferred. It is stated in verse 32 that the yuga consists of five years—a choice perhaps determined by the lustrum mentioned in some Vedic texts, though periods of two, three, four, and six years are also attested in the literature.[32] These five years are solar, and therefore contain 60 solar months. That they also contain 2 intercalary months to make a total of 62 synodic months is clear from verse 4, which gives the following rule for finding the parvan or syzygy.

> Diminish [the current year-number] by 1, multiply [the remainder] by 12 [to get the solar months], multiply [the product] by 2 [to get the "syzygies" of the solar months], add [to the product] the lapsed [syzygies of the present year], and add [to that sum] 2 for each 60; [the result] is called the sum of the syzygies.

The addition of 2 syzygies for each 60 "syzygies" of the solar months is, of course, equivalent to adding 2 intercalary months for each 60 solar months. This same intercalation-rule is found in the *Arthaśāstra*,[33] the *Śārdūlakarṇāvadāna*,[34] and the *Paitāmahasiddhānta* summarized by Varāhamihira.[35]

The number of "days" within this five-year yuga can be computed from verse 18 of the Ṛk-recension:

> The Moon travels through a nakṣatra in 1 day and 7 kalās, the Sun in 13 and 5/9 days.

The Moon's total travel in 5 solar years containing 62 synodic months is 67 sidereal rotations; since in each sidereal rotation it travels by definition through 27 nakṣatras, in a yuga it travels through 1809 nakṣatras. From verse 16 we learn that each day contains 603 kalās, so that to find the number of days, D, in a yuga we form the equation:

$$D = 1809 \times 1\frac{7}{603} = 1830.$$

Similarly, the Sun travels through 135 nakṣatras in 5 years at the rate of $13\frac{5}{9}$ days per nakṣatra. And therefore

$$D = 135 \times 13\frac{5}{9} = 1830.$$

A cruder parameter for solar motion—$13\frac{2}{3}$ days per nakṣatra—is found in verse 12; it yields an impossible 1845 days in a five-year yuga.

If there are 1830 "days" in 5 solar years, each one contains 366 "days" and each ayana contains 183 "days", a parameter confirmed by the rule for the daily increase in the length of daylight discussed above. Furthermore, in 62 months there must be 1860 tithis or thirtieths of a month; this again is confirmed by verse 8 of the Ṛk-recension.

> They say that the first [tithis] in [each] ayana [are] the first, seventh, and thirteenth [in the śuklapakṣa and] the fourth and tenth [in the kṛṣṇapakṣa]; [the series is taken] twice.

Since 1860 tithis contain 10 ayanas, each ayana is 186 tithis or 6 synodic months plus 6 tithis; and the constant difference in the above verse is precisely 6 tithis. The same scheme is used for determining the tithi-numbers of the equinoxes in verse 12.

Precisely these figures are known to lie behind the computations of the *Paitāmahasiddhānta* of A.D. 80. Neugebauer and the present author have shown that they are to be interpreted to mean that each year contains 366 sidereal days, which are equivalent to 365 civil days. In other words, the fundamental relation of the five-year yuga is:

5 years = 1830 sidereal days = 62 synodic months.

This manner of expression is quite consistent with the normal practice of the earliest texts of medieval Indian astronomy, for both the *Paitāmahasiddhānta* of the *Viṣṇudharmottarapurāna*[36] and the *Āryabhaṭīya*[37] composed by Āryabhaṭa in A.D. 499 express the length of their yugas in nākṣatra or sidereal days, from which the civil days can be derived by subtracting the number of revolutions of the Sun in a yuga; and Sphujidhvaja stated the correct relation already in the third century.[38] One of the principal problems that modern scholars have had with the *Jyotiṣavedāṅga* is due to their understanding the text to mean that a solar year equals 366 civil days, which is a parameter approximately three times as inaccurate as that of the Ṛk-recension.

Their justification for understanding the text in this way is to be found in verses 28–29 of the Yajur-recension—two verses not to be found in the Ṛk-recension. These verses state:

> A year [consists of] 366 days, 6 ṛtus [or seasons], 2 ayanas, and 12 solar months; a yuga is this multiplied by 5. The sum of the days plus 5 [equals] the risings of [the nakṣatra] Vāsava [or Dhaniṣṭhā]; [the sum of days] diminished by 62 [equals the risings] of the Moon; [and the sum of days diminished] by 21 [equals the Moon's transits] of the nakṣatras.

Since we know that the Moon travels through 1809 nakṣatras in a five-year yuga, "the sum of days" in these verses is 1830, that is, the number of sidereal days in our interpretation. But these verses also state that the number of risings of a nakṣatra—that is, the number of sidereal days—equals "the sum of days plus 5" or 1830 plus the number of revolutions of the Sun. Clearly, then, the author of these two verses assumed that 1830 is the number of civil days in a five-year yuga, and both medieval and modern scholars have followed him. But I would suggest that the redactor of the Yajur-recension, in perhaps the fourth or fifth century A.D., added these verses in order to provide a statement about the fundamental period-relations of the *Jyotiṣavedāṅga* including that of sidereal to civil days; unfortunately, he did not correctly apprehend what that relation was. As we have shown, the Ṛk-recension of the *Jyotiṣavedāṅga* and the *Paitāmahasiddhānta* of A.D. 80 do not impose this false interpretation on us.

Given this yuga of 5 years, of which each contains 365 civil days, and during which period 2 intercalations are made, we must necessarily ask whether the Indians developed it themselves or adopted it from some external source. Though longitudes in this cycle are measured in Indian units—the 27 nakṣatras now for the first time interpreted as equal arcs of 13;20° each along the ecliptic—there is nothing in the earlier Indian texts to suggest any prior stages of development that could have culminated in the *Jyotiṣavedāṅga*. On the contrary, not only are the time-measuring instruments, the water-clock and the gnomon, along with the linear zig-zag functions utilized in reading them, derived from Babylonia; some of the elements of the intercalation-cycle itself are also, beginning with the division of the mean synodic month into thirty equal units called in Sanskrit tithis (these units appear in a cuneiform tablet of about 600 B.C. which deals with the dates of solstices, equinoxes, and the heliacal risings and settings of Sirius[39]), the thirty muhūrtas of a nychthemeron being an obvious imitation of the thirty tithis of a synodic month. Furthermore, a year of 365 days, though Egyptian in origin, may have been adapted in Iran as early as the second half of the fifth century B.C.[40] This possibility provides us with a hint regarding the intermediary which transmitted so much of Mesopotamian astronomy to the banks of the Indus and Ganges.

If we look in cuneiform texts for a five-year intercalation-cycle, we are frustrated. But a knowledge of the basic period-relations of lunar and solar motion are implicit in the Babylonian eighteen-year eclipse cycle, whose antecedents go back to the eighth century B.C.[41] And, as I have learned recently from Dr Herman Hunger of the Oriental Institute, the second tablet of mul Apin describes a crude intercalation-cycle in which one month is intercalated every three years. The epact in this text is 10 days; somewhat later texts of the Achaemenid period use 10;30 days. Moreover, if legend as preserved by Censorinus is correct, Cleostratus of Tenedos towards the end of the sixth century B.C. had learned—inevitably from a Mesopotamian source—of an eight-year cycle with three intercalary months,[42] and some claim, though on the basis of not completely convincing evidence, that this octaëteris was used in the Babylonian civil calendar from 527 to 503 B.C.[43] In 432 B.C. Meton in Athens proposed the nineteen-year cycle with seven intercalations, which is based on very fine parameters indeed; and Meton may have had Babylonian sources. The Mesopotamian evidence consists of the actual intercalations in the civil calendar, which are regular with respect to the year of intercalation after 483 B.C. with the exception of one anomalous intercalation in 384 B.C.,[44] and it must be remembered that the civil authorities may well have lagged far behind the astronomers in adopting this intercalation-scheme. In Athens itself the Metonic cycle was never employed in the civil calendar. It is not clear what relationship the Egyptian 25-year cycle, introduced in the middle of the fourth century B.C. during Achaemenid rule, might have to the Indian. For the Egyptians 25 years of 365 days each contain 309 synodic months,[45] whereas for the Indians 25 such years contain 310 synodic months. The contemporaneity of these developments in areas linked by the Achaemenid empire suggests a common origin.

It is clear, then, that Babylonian astronomers were capable of devising intercalation-cycles in the seventh, sixth, and fifth centuries B.C., and there is evidence both in the Greek and in the cuneiform sources that they actually did so; and by the early fourth century B.C. they had certainly adopted the quite-accurate nineteen-year cycle. It is my suggestion that some knowledge of these attempts reached India, along with the specific astronomical material we have already discussed, in the fifth or fourth century B.C. through Iranian intermediaries, whose influence is probably discernible in the year-length selected by Lagadha for the *Jyotiṣavedāṅga.* But the actual length of the yuga, five years, was presumably accepted by Lagadha because of its identity with a Vedic lustrum. Not having access to a series of extensive observations such as were available to the Babylonians, he probably was not completely aware of the crudeness of his system. And the acceptance of this cycle by Indians for a period of six or seven centuries or even more demonstrates among other things that they were not interested in performing the simplest acts of observational astronomy.

This leads us directly back to the question of the date of the *Jyotiṣavedāṅga* which we briefly touched upon at the beginning of this paper. Verses 5–6 of the Ṛk-recension may be translated as follows.

> When the Moon and Sun travel in the heavens together with Vāsava [*i.e.*, the nakṣatra Dhaniṣṭhā], then is the beginning of the yuga, the śuklapakṣa of Māgha, and the northern ayana. The Sun and Moon proceed northwards at the beginning of Śraviṣṭhā [*i.e.*, Dhaniṣṭhā], and southwards in the middle of Sārpa [*i.e.*, Āśleṣā]; [these two events occur] always in [the month] Māgha and Śrāvaṇa respectively.

Since the tropical longitude of the beginning of Dhaniṣṭhā was 293;20° in about A.D. 500, that point in the sphere of the fixed stars coincided with the winter solstice in about 1180 B.C. But, though this computation is correct, its relevance to the date of the *Jyotiṣavedāṅga* is not evident. We simply do not know where Lagadha would have placed the beginning of the equal nakṣatra Dhaniṣṭhā with respect to the fixed stars, nor do we know the accuracy with which he could have determined the sidereal longitude of the Sun at the winter solstice. Since a displacement of the beginning of the equal nakṣatra by some 10°, or an error of 10 days in computing the date of the winter solstice, or some combination of these two effects is all that is required to bring the date from the twelfth century to the fifth century B.C., we should not lend much weight to this chronological argument.[46] Furthermore, we must consider that an intercalation-cycle based on such a crude parameter for the length of a year as 365 days does not inspire much confidence in the accuracy with which its author was able or wished to endow the positions of the solstices. The scheme could never pretend to provide more than a very rough approximation to reality, for a few years; it would require constant readjustment if it were really to work. The true significance lies in the fact that with it the Indians accepted in theory the possibility of describing the periodicity of celestial motions in mathematical terms, and that they learned of this possibility from the Babylonians through their Iranian neighbours.

But there is one further question that we must raise before accepting this hypothesis of transmission. Was this an isolated phenomena, or part of a general Iranian influence on Indian culture in the fifth and fourth centuries B.C.? Unfortunately, our answer to that question is rather clouded by the scarcity of literary or archaeological data from the period in question.[47] We do not know how far into India the Achaemenids penetrated, but probably their control did not extend beyond the western parts of the Panjab as Alexander met numerous small and apparently independent states in the Indus valley. But Iranian influence in the early fifth century was sufficiently strong to make possible the safe completion of Scylax's exploratory voyage down the Indus, and Takṣaśila, in the region where Pāṇini seems to have worked, was certainly a city where cross-cultural contacts were frequent. And it is arguable that the enormous and often-studied Iranian influence discerned in Mauryan polity, architecture, sculpture, epigraphy, and the like in the third century B.C. was an inheritance from the pre-Mauryan Nandas' rather than from the post-Alexandrian Greeks' adaptations of Achaemenid forms. And parallel to the suggested Mesopotamian-Iranian influence on Indian mathematical astronomy is the influence of the same cultural complex on Indian omens, which first are mentioned in the Upaniṣads and Buddhist canonical texts of the pre-Mauryan period, though the oldest codifications of these omens available to us were compiled in the early centuries A.D. It is reasonable, then, or at least so I believe, to see the origins of mathematical astronomy in India as just one element in a general transmission of Mesopotamian-Iranian cultural forms to northern India during the two centuries that antedated Alexander's conquest of the Achaemenid empire.

REFERENCES

1. This relationship I originally proposed in *Isis*, liv (1963), 231–3.
2. Both recensions along with Somākara's commentary were first edited by A. Weber, *Über den Vedakalender, namens Jyotisham* (Berlin, 1862) (the text of one recension had been published at Bombay in 1833); the text of the Yajur-recension along with the non-Yajur verses of the Ṛk-recension were reprinted by G. Thibaut, "Contributions to the Explanation of the Jyotisha-Vedánga", *Journal of the Asiatic Society of Bengal*, xlvi (1877), 411–37, esp. 413–6; the Ṛk-recension was reprinted by J. B. Moḍaka of Thana in 1885; one recension was included in *Vedasya ṣaḍaṅga* (Bombay, 1892); both recensions were edited and commented on by Bārhaspatya (Lāla Choṭelāl) in the *Hindustan review* for 1907 (reprinted Allahabad, 1960) and again, with Somākara's bhāṣya, by S. Dvivedin (Benares, 1908). The most recent attempt at an edition and at an understanding of both texts is by R. Shamasastry (Mysore, 1936). The more recent efforts by B. R. Kulkarni, *The Lagna System of the Vedanga Jyotisha* (Dhulia, 1943); by G. Prasad, "The Astronomy of the Vedāṅga Jyotiṣa", *Journal of the Ganganath Jha Research Institute*, iv (1946–47), 239–48; and by A. K. Chakravarty, "The Working Principle of the Vedāṅga Jyotiṣa Calendar", *Indian studies past and present*, x (1968–69), 31–42, as well as the older work by B. G. Tilak, *Vedic chronology and Vedanga Jyotisha* (Poona, 1925), do not advance our understanding of this text.
3. I have used the recent edition with translation and commentary by R. P. Kangle (3 vols, Bombay, 1960–65).
4. T. R. Trautmann, *Kauṭilya and the Arthaśāstra* (Leiden, 1971), 176–84.
5. I have used the edition by S. Mukhopadhyaya (Santiniketan, 1954); this is reprinted in P. L. Vaidya's edition of the *Divyāvadāna, Buddhist Sanskrit texts*, xx (Darbhanga, 1959), 314–425.
6. E. Zürcher, *The Buddhist conquest of China* (2 vols, Leiden, 1959), i, 32–4.
7. Zürcher, *op. cit.*, i, 47–51.
8. *Ibid.*, 65–71.
9. Edited, translated, and commented on by O. Neugebauer and D. Pingree (2 vols, Copenhagen, 1970–71).
10. Edition, translation, and commentary by D. Pingree to appear in the *Harvard oriental series*; the reference is to 79,3.
11. These works have been printed several times by the Jainas in India; I have used the edition of the first two by J. F. Kohl (Stuttgart, 1937). See also G. F. Thibaut, "On the Sūrya-prajñapti", *Journal of the Asiatic Society of Bengal*, xlix (1880), 107–27 and 181–206.
12. *Sūriyapannatti*, I, 3, and XX, 59.
13. On the various Gargas see D. Pingree, *Census of the exact sciences in Sanskrit*, Series A, ii (Philadelphia, 1971), 115b–126a.
14. For the *Paitāmahasiddhānta*, see D. Pingree, "The *Paitāmahasiddhānta* of the *Viṣṇudharmottara-purāṇa*", *Brahmavidyā*, xxxi–xxxii (1967–68), 472–510. Compare Garga as cited by Somākara on Yajur-recension 11 and *Paitāmahasiddhānta*, III, 1; *cf. Yavanajātaka*, 79, 6.
15. Pp. 57–8.
16. II, 20, 35.
17. *Yavanajātaka*, 79, 27; this could be interpreted as either an out-flowing or an in-flowing clock.
18. *Pañcasiddhāntikā*, 14, 31–2.
19. See O. Neugebauer, "The Water Clock in Babylonian Astronomy", *Isis*, xxxvii (1947), 37–43.
20. See, *e.g.*, B. L. van der Waerden, "The Earliest Astronomical Computations", *Journal of Near Eastern studies*, x (1951), 20–34.
21. II, 20, 37–8.
22. II, 20, 61–64.
23. Pp. 53 and 100–8.
24. 79, 26 and 31.
25. *Pañcasiddhāntikā*, 2, 8.
26. II, 20, 39–42.
27. *Pañcasiddhāntikā*, 2, 9–10.
28. E. F. Weidner, "Ein babylonisches Kompendium der Himmelskunde", *American journal of Semitic languages and literatures*, xl (1924), 186–208, esp. 198.
29. Pp. 54–5.
30. Translated by E. W. West, *Sacred books of the East*, v (Oxford, 1880), 397–400.

31. Pp. 100–3.
32. A. A. Macdonell and A. B. Keith, *Vedic index* (2 vols, London, 1912), ii, 412; for a fifth-century reference to the five-year cycle see *Paitāmahasiddhānta*, IV, 4.
33. II, 20, 65–6.
34. Pp. 103–4.
35. *Pañcasiddhāntikā*, 12, 1.
36. III, 32.
37. *Daśagītikā*, 1.
38. *Yavanajātaka*, 79, 8.
39. Published by O. Neugebauer and A. Sachs, "Some Atypical Astronomical Cuneiform Texts. I", *Journal of cuneiform studies*, xxi (1967), 183–218, esp. 183–90; for the use of tithis in this text see 189–90.
40. So, for instance, argues S. H. Taqizadeh, *Old Iranian calendars* (London, 1938), on not entirely convincing evidence.
41. A. J. Sachs, *Late Babylonian astronomical and related texts* (Providence, 1955), p. xxxi. A crude relation

$$223 \text{ months} \approx 18 \text{ solar rotations} + 10;30°$$

is found in a "saros"-text of the early fifth century B.C.; see A. Aaboe and A. Sachs, "Two Lunar Texts of the Achaemenid Period from Babylon", *Centaurus*, xiv (1969), 1–22, esp. 18. In a later "saros"-tablet there occurs a column *X* which registers the difference between a year and 12 months; see A. Aaboe, *Some lunar auxiliary tables and related texts from the Late Babylonian Period* (Copenhagen, 1968), 28–30.
42. *De die natali* 18,5; H. Diels and W. Kranz, *Die Fragmente der Vorsokratiker*, I⁷ (Berlin, 1954), 42.
43. B. L. van der Waerden, *Die Anfänge der Astronomie* (Groningen [ND]), 112.
44. See R. A. Parker and W. H. Dubberstein, *Babylonian chronology* (Providence, 1956), 6.
45. R. A. Parker, *The calendars of Ancient Egypt* (Chicago, 1950).
46. On the difficulties involved in accepting the precision of the statements in the *Jyotiṣavedāṅga* required for their use as chronological indicators see W. D. Whitney, "On the Jyotisha Observation of the Place of the Colures, and the Date derivable from it", *Journal of the Royal Asiatic Society*, n.s, i (1865), 316–31; cf. also D. Pingree, "Precession and Trepidation in Indian Astronomy before A.D. 1200", *Journal for the history of astronomy*, iii (1972), 27–35.
47. Most of the literary evidence is discussed in S. Chattopadhyaya, *The Achaemenids in India* (Calcutta, 1950); *cf.* also R. A. Jairazbhoy, *Foreign influence in Ancient India* (London, 1963), 38–47. For the archaeological evidence see R. E. M. Wheeler, "Iran and India in Pre-Islamic Times", *Ancient India*, iv (1948), 85–103, esp. 92 sqq.

CONCENTRIC WITH EQUANT

DAVID PINGREE, PROVIDENCE

In 1952[1] van der Waerden suggested that the motion of the Sun according to the vâkyas of South India (which belong to the Parahita system founded by Haridatta in 683 on the basis of the Âryapakṣa of Âryabhaṭa[2]) could be explained on the hypothesis of a concentric with equant. Four years later[3] Krishna Rav showed that the same model would account for the lunar motion in these tables. But, in that same year[4], van der Waerden abandoned his hypothesis because he knew of no Greek or Indian text in which such a model is described. One purpose of the present article is to point out that a procedure for solving a concentric with equant is described in a text belonging to the Âryapakṣa, the *Mahâbhâskarîya* of that Bhâskara who wrote a commentary on the *Âryabhaṭîya* in 629; its second purpose is to suggest a pre-Ptolemaic, Peripatetic origin of the model, and therefore of the equant as well.

In *Mahâbhâskarîya* IV 19–21[5] is found a method of computing the effect of a concentric with equant by means of an eccentric with varying eccentricity. In the accompanying figure, E is the equant point at a fixed distance EO from the center of the earth, while D is the center of an eccentric with radius $DS' = OS = R$. D slides up and down the apsidal line DEO in such a fashion that DS' is always parallel to ES. With Bhâskara's rule one finds approximately DO and OS', from which one can find the elongation from the apogee, $DOS' = EOS$. The text states:

> IV 19. Subtract (the Sine of) the final equation from (the Sine of) the koṭi or again add it, depending on the quadrant; the square-root of the sum of the square of that and the square of the bâhu is the hypotenuse.
>
> IV 20. Multiply (the Sine of) the final equation by the hypotenuse (and) divide (the product) by the Radius; add (the quotient) to or subract it from the previous koṭi (repeatedly) until the hypotenuse is equal (to the hypotenuse obtained in the immediately preceding computation).

[1] B. L. van der Waerden, 'Die Bewegung der Sonne nach griechischen und indischen Tafeln,' *SBAW*, Math.-naturwiss. Kl. (1952), 219–232. I wish here to thank my colleagues, Profs. Neugebauer and Toomer, for their suggestions, and Prof. Neugebauer in particular for having drawn the figure which accompanies this paper.

[2] Haridatta's *Grahacâranibandhana* was edited by K. V. Sarma (Madras, 1954); see also C. Kunhan Raja, *Candravâkyas of Vararuci* (Madras, 1948).

[3] I. V. M. Krishna Rav, 'The motion of the Moon in Tamil astronomy,' *Centaurus* IV (1955–56), 198–220.

[4] B. L. van der Waerden, 'Tamil Astronomy,' *ibid.*, 221–234.

[5] The *Mahâbhâskarîya* has been edited three times: with the *Karmadîpikâ* of Parameśvara by B. D. Âpaṭe, *ASS* 126 (Poona, 1945); with the *Bhâṣya* of Govindasvâmin and the *Siddhântadîpikâ* of Parameśvara by T. S. Kuppanna Sastri, *Madras GOS* 130 (Madras, 1957); and with an English translation and commentary by K. S. Shukla (Lucknow, 1960). The three texts are virtually identical; I translate Shukla's, but read koṭyâḥ with the other editions in 19a in place of koṭyâm.

[6] See D. Pingree, 'On the Greek origin of the Indian planetary model employing a double epicycle,' *JHA* 2 (1971), 80–85.

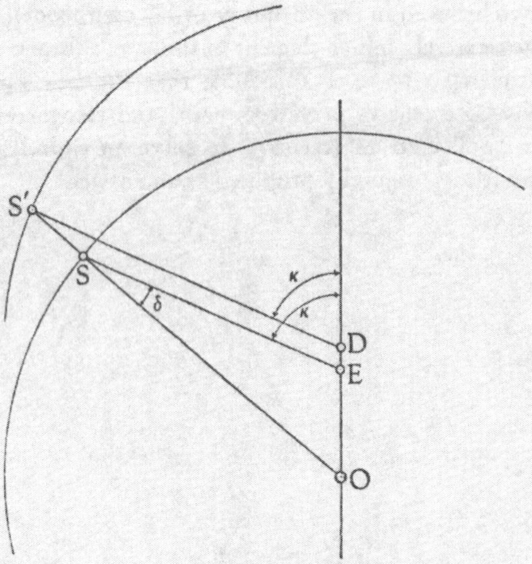

Fig. 1

IV 21. Multiply the Radius by the Sine of the bâhu (and) divide (the product) by the (final) hypotenuse. Add the arc (corresponding to that quotient) to (the longitude of) the apogee according to the quadrant of the argument.

In these verses, the Sine of the final equation is the eccentricity (one begins the computation by using its mean value, *EO*, the distance of the equant from the center of the earth); the Sine of the koṭi is Cos$\varkappa$; and the Sine of the bâhu is Sin$\varkappa$. In IV 9–12 Bhâskara gives an equivalent solution employing an epicycle of varying radius.

These and similar rules appear elsewhere in Sanskrit astronomical texts, though Bhâskara's exposition is the earliest that survives. There can be little doubt that they are typically Indian methods for solving the concentric with equant; and it would be difficult not to conclude that, like the Indian epicyclic and eccentric models of planetary motion, the concentric with equant originated among the Greeks[6]. More specifically, it seems likely that such a model would have originated among Peripatetics attempting to preserve the concentricity of the orbits of at least the two luminaries while allowing their motions as seen from the earth to be anomalous. Evidence exists that such efforts were made in the first and second centuries A. D.[7]

If this reconstruction is valid, it follows that Ptolemy did not invent the equant – nor does he claim to have done so[8]. It is virtually inconceivable that the inventor of the concentric with equant could have studied the planetary models in the

[7] *Ibid.*
[8] *Almagest* IX 5–6.

Almagest and have believed in the adequacy of his own model, based, as it would have been, on one relatively minor element in Ptolemy's theory of the star-planets. On the other hand, it is perfectly plausible that Ptolemy knew of the equant employed in some Aristotelian's primitive model and recognized the usefulness of the device, when applied to an eccentric, to solve an entirely different problem from that to which it was originally proposed as an answer.

INDIAN ASTRONOMY

DAVID PINGREE

Professor of the History of Mathematics, Brown University

(*Read April 21, 1977*)

THE STUDY of astronomy as pursued in India from antiquity until the introduction of the modern science in the nineteenth century does not fit into a Western historian's normal conception of a scientific enterprise.[1] In particular, Western astronomy from Hipparchus till Kepler consistently emphasized the need to combine carefully planned and executed observations with theory in order to determine and perfect geometrical models that generate predictions corresponding as closely as possible to the observed phenomena. The successful pursuit of such programs of astronomical research was certainly rare, but the attempts were many. In India astronomers paid little attention either to observation or to theory. It is my purpose in this paper to describe, without going into technical details, the Indians' approach to the celestial science, to discuss at least some of the causes of their peculiar attitudes, to indicate in what areas their unorthodox methodology led to real contributions to world astronomy, and, finally, to point to one of the effects of their failure to develop a science that could compete with that of the West.

Mathematical astronomy was introduced into India, along with the "sciences" of divination from celestial omens and of astrology, from the West. The earliest infusion was of simple arithmetical schemes for establishing an intercalation cycle and for determining the length of daylight, derived from Akkadian sources, transmitted through Achemenid Iran, and embodied in a text called the *Jyotiṣavedāṅga* that was composed in northwestern India by Lagadha in, probably, the late fifth or early fourth century B.C.[2] This was followed by the translation of various Greek texts on astrology and astronomy into Sanskrit in western India in the second, third and fourth centuries A.D. These texts expounded several different astronomical systems: one represented Greek adaptations of the lunar, solar, and planetary theories developed in Mesopotamia during the Seleucid and Parthian periods;[3] another seems to have been closely related to what little we know of Hellenistic astronomy, and specifically to the theories of Hipparchus; and a third preserved planetary models devised by Peripatetic philosophers to solve the problem of the anomalous motions of the planets, while salvaging the Aristotelian principle of concentricity—that is, retaining a fixed distance for each of the planets from the center of the universe.[4] The preservation of this Greek material is one of India's greatest contributions to the history of science.[5] But each of these astronomical systems was transformed in India into something somewhat different from what one expects in a cuneiform or a Greek text; such transformations are only to be expected when an intellectual system is introduced into an alien culture. Fortunately, the nature of an advanced astronomical theory is such that it cannot be altered beyond indisputable recognition; such elements as the theory of epicycles and eccentrics can only have been developed within the specific traditions of Greek science, and astronomical parameters which are the result of complex interrelationships of observation and theory cannot have been independently arrived at in Mesopotamia, Greece, and India.

The derivation of their mathematical models (arithmetical from the Babylonians, geometric from the Greeks) and their basic parameters from external sources, of course, explains the lack of concern with observational astronomy in India. By this I do not mean that there was no interest in looking at the stars and planets, but that there was no tradition of systematic observations designed to test and improve theory nor any tradition of in-

[1] For a technical account of Indian astronomy see D. Pingree, "Essay on the History of Indian Astronomy," *Dictionary of Scientific Biography* **15** (New York, 1978), pp. 533–633. Biographical and bibliographic information on individual astronomers and their works can be found in D. Pingree, *Census of the Exact Sciences in Sanskrit* (henceforth *CESS*), of which there have so far appeared in Series A vols. 1–3, *Memoirs Amer. Philos. Soc.* **81**, **86**, and **111** (Philadelphia, 1970, 1971, and 1976). Vol. 4 is now in the process of being prepared for publication.

[2] D. Pingree, "The Mesopotamian Origin of Early Indian Mathematical Astronomy," *Jour. History of Astronomy* (henceforth *JHA*) **4** (1973): pp. 1–12.

[3] Represented by the *Yavanajātaka* of Sphujidhvaja, ed. D. Pingree, *Harvard Oriental Series* **48** (Cambridge, Mass., 1978), and by the siddhāntas of Vasiṣṭha, Romaka, and Pauliśa summarized by Varāhamihira in his *Pañcasiddhāntikā*, ed. O. Neugebauer and D. Pingree (2 v., København, 1970–1971).

[4] D. Pingree, "On the Greek Origin of the Indian Planetary Model Employing a Double Epicycle," *JHA* **2** (1971): pp. 80–85; and "Concentric with Equant," *Archives Internationales d'Histoire des Sciences* **24** (1974): pp. 26–29.

[5] D. Pingree, "The Recovery of Early Greek Astronomy from India," *JHA* **7** (1976): pp. 109–123.

PROCEEDINGS OF THE AMERICAN PHILOSOPHICAL SOCIETY, VOL. 122, NO. 6, DECEMBER 1978

strumentation useful for astronomy. Early Indian astronomers, such as Brahmagupta[6] in his *Brāhmasphuṭasiddhānta*, written in 628 A.D., do assert the superiority of their systems over others in producing a coincidence between predictions and observations; but these claims, in the light of the erroneous parameters for planetary mean motions and equations and other astronomical constants utilized in all Sanskrit texts, are not credible. An investigation of Brahmagupta's coordinates of the fixed stars, for instance, in relation to which observations of the planets would have been made, reveals that there is no set of stars visible in the sky that would fit the positions he gives. In fact, there is no accurate star-catalog to be found in the Sanskrit texts before Mahendra Sūri[7] translated a Persian version of an excerpt from Ptolemy's star-catalog in 1370.

Mahendra's work, the *Yantrarāja*, was the first to describe a useful observational instrument, the astrolabe, in Sanskrit. The early Indian treatises contain elaborate explanations of mechanical toys (water-driven revolving globes and elaborate clepsydrae in various animal, bird, and human forms), and of impractical wooden rings and hemispheres, but the basic observational instrument used by Indian astronomers was the simple gnomon for measuring the sun's shadow, from which various problems of local time or latitude could be solved. The only description of an observation of a heavenly body given by Brahmagupta, and also by his contemporary Bhāskara,[8] involves the extravagant construction, through the use of vertical staffs of appropriate length erected on a level platform, of a physical model of the computed sines of the altitudes of the sun and the moon or of two other planets at conjunction. The execution of such a physical model would not be very difficult, but the astronomical theory behind it did not possess the required accuracy in the prediction of the time of the conjunction such that an observation made at that time would succeed. In any case, both Brahmagupta and Bhāskara expressly state that the purpose of these observations was to impress the king or other patron of the astronomer, not to improve the astronomy. One hopes that they had at least occasional success.

Historically one knows that the attempts by Indian astronomers to predict correctly those lunisolar conjunctions and oppositions that result in eclipses met with frequent failure. The importance of such predictions lay in the fact that eclipses were regarded as omens boding ill for the kingdom. From the late sixth century on, beginning in western India, but spreading thence to other areas of central and northern India, this potential ill was averted by the pious donation of land to Brāhmaṇas. Fortunately, these donations were recorded on copper plates, of which many have survived.[9] It was the practice to issue these plates in advance, but to date them on the day of the eclipse. In a large number of cases no eclipse was in fact visible in the locality of the grant on the given date. This, of course, was no argument against the accuracy of the prediction made by the court astronomer, but rather evidence of the efficacy of the donation of land to Brāhmaṇas, which averted not only the evil potentially indicated by an eclipse, but the eclipse itself. In one case, however, we learn from a copper-plate grant issued in 1128 that the astronomer Padmanābha,[10] at the court of the Kalacūri ruler of Ratnapura, Ratnadeva II, actually managed to predict a total lunar eclipse correctly. The rarity of such successes is indicated by his being rewarded with the grant of a village called Ciñcātalāī. The only other records that have come down to us of Indian observations are those made by Parameśvara.[11] He observed five predicted lunar and eight predicted solar eclipses at Gokarṇa and elsewhere on the Malabar Coast between 1398 and 1432. He also claims to have utilized these and other observations in establishing the parameters of his *dṛggaṇita* (observation-computation) system, but in fact those parameters seem mainly to be derived from the already existing *Sūryasiddhānta*; Parameśvara simply expressed them in an unusual form.

If we accept, then, that observation played almost no role in Indian astronomy, what did Indian astronomers do? Most of them, like their contemporaries in other countries, merely copied in various combinations the works of their predecessors. The few who tried to do something new turned to the mathematics utilized by astronomy. This approach is already evident in the earliest period of Indian adaptations of foreign astronomical systems, and undoubtedly reflects the strength of the Indian tradition of mathematics applied, for example, to the geometry of the bricks utilized in building gigantic animal- and bird-shaped altars for Vedic sacrifices. The texts that give the rules for shaping and combining these bricks, the *Śulbasūtras*, are certainly earlier than the Sanskrit translations of Greek astronomical works, and represent but one aspect of an early Indian mathematical tradition.[12] When the Greek astronomical

[6] *CESS* A4.
[7] *CESS* A4.
[8] *CESS* A4.

[9] An as yet incomplete search of the epigraphical literature has turned up more than fifty examples.
[10] *CESS* A4.
[11] *CESS* A4.
[12] See, e.g., B. Datta, *The Science of the Śulba* (Calcutta, 1932).

texts were translated, then, one of the ways in which they were transformed was the alteration of their mathematics. The Hipparchan trigonometrical function, the chord, was replaced by the sine, cosine, and versine functions; the analemmata that Hipparchus apparently had used for solving certain problems in spherical trigonometry were exploited most effectively by the Indians in application to gnomon-problems; the Euclidean theory of continuous fractions was used to solve indeterminate equations which arose from the problem of ascertaining the numbers of rotations of the planets in a *Kalpa* or period of 4,320,000,000 years; and various methods of approximation were developed for the determination of the equations of the planets. Indeed, the predilection for approximations is one of the characteristics of Indian astronomy, though it is combined with a delight in complexity. An example of the latter trend is an extraordinarily involved and lengthy solution of the problem of the altitude of the sun in the special case when it is 45° to the east or west of the meridian. Very much simpler general solutions of the problem of solar altitude, which work perfectly well for these special cases, were also known, but the elaborate solution was obviously regarded with great pride. Some of the work on the mathematics applicable to astronomy undertaken by Indian scientists was, however, of a very high order. One especially admires the investigations of the applications of various series undertaken first by Mādhava of Saṅgamagrāma[13] in Kerala in about 1400; he discovered equivalents of the Taylor series for finding sines and cosines, of the Gregory series for the arc of tangent x, and of the Leibnitz series for $\pi/4$, and he developed approximations to Newton's power series for sines and cosines.[14] Indian astronomers also devoted much effort, especially in Gujarat and Rajasthan in the eleventh through the eighteenth centuries, to devising tables of various functions and forms for the convenient solution of a number of astronomical problems.[15] Some of these innovations in mathematics influenced Sasanian and early Islamic astronomy,[16] and, through Arabic intermediaries, the medieval astronomy of Byzantium and Western Europe.[17] The most lasting Indian contribution in this area was the use of modern trigonometrical functions, but the full understanding of these functions and especially their use in spherical astronomy were due to Islamic scientists of the tenth century and later, and were transmitted to India only in the seventeenth century.[18]

In the course of developing the mathematics of astronomy, the Indians usually devised many alternative solutions to the same problem, some of which were precise, while others were approximations of various levels of accuracy. It is a characteristic of the Indian scientific (or rather *śāstraic*) tradition that learning tends to be cumulative at the expense of being self-consistent. This characteristic is expressed in astronomical treatises or *siddhāntas* by the collocation of many solutions of varying validity for each problem without any attempt, usually, to distinguish the better from the worse. This results obviously in great unevenness and a lack of consistency in these treatises, and reflects the absence of a rigid methodology that helped to inhibit the growth of as fruitful a mathematics in India as existed in Greece or in Islam.

But inconsistency is only one difficulty that confronts the reader of a Sanskrit astronomical treatise. These *siddhāntas*, which were used for the instruction of students, were written in verse, ostensibly so that they could be readily memorized, but in part at least to display the versatility of their authors. The exigencies of the meter and the traditions of the genre prescribed the exclusion from the text both of the rationales for the computational rules and of essential parts of those rules themselves. The practical result of these exclusions was that the *siddhānta* could not be understood without the help of a *guru* or teacher. Of course, commentaries were composed on many of the *siddhāntas*, but typically they merely fill in the missing parts of the versified rules and give worked examples; they never address the real problems of astronomical theory. The student would, of course, not learn from the texts anything about the observational basis of the astronomical system it expounds; this, as we have seen, is because that basis was developed outside of India and never transmitted to India. But he would also not learn from the *siddhāntas* the mathematical bases of the algorithms it prescribed. It is obvious, however, from the continued vitality of the mathematical aspects of astronomy in India that some instruction was available. This

[13] *CESS* A4.

[14] R. C. Gupta, "The Mādhava-Gregory Series," *Mathematics Education* **7** (1973): pp. 67–70; and "Mādhava's and Other Medieval Indian Values of Pi," *Mathematics Education* **9** (1975): pp. 45–48; and A. K. Bag, "Mādhava's Sine and Cosine Series," *Indian Jour. History of Science* **11** (1976): pp. 54–57.

[15] D. Pingree, "Sanskrit Astronomical Tables in the United States," *Trans. Amer. Philos. Soc.* **58**, 3 (1968); and *Sanskrit Astronomical Tables in England* (Madras, 1973).

[16] D. Pingree, "Indian Influence on Sasanian and Early Islamic Astronomy and Astrology," *Jour. Oriental Research, Madras* **34–35**, 1964–1966 (1973): pp. 118–126; and "The Greek Influence on Early Islamic Mathematical Astronomy," *Jour. Amer. Orient. Soc.* **93** (1973): pp. 32–43.

[17] D. Pingree, "The Indian and Pseudo-Indian Passages in Greek and Latin Astronomical and Astrological Texts," *Viator* **7** (1976): pp. 141–195.

[18] D. Pingree, "Islamic Astronomy in Sanskrit," to appear in *Jour. History of Arabic Science.*

was through an oral tradition, a *guruparamparā*, often maintained within a family in which father taught sons, and all utilized a common library of astronomical manuscripts. We know, for example, of one such family that flourished from the tenth century till the thirteenth. One of its early members, Bhāskara Bhaṭṭa, was awarded the title Vidyāpati for his proficiency by the royal astronomer and polymath, Bhojarāja of Dhārā, in the early eleventh century. In the fifth generation from him was the most famous of Indian astronomers, also named Bhāskara;[19] and this second Bhāskara's grandson, Caṅgadeva,[20] and grandnephew, Anantadeva,[21] founded a *maṭha* or college for the study of Bhāskara's works in 1207. Other families of equally long or longer duration are known from later times. But there are also cases of such families of astronomers sending their sons to extra-familial *gurus* to be trained. The most famous of these *gurus* was Gaṇeśa,[22] who lived in a town named Nandipadra near the mouth of the Narmadā in the first half of the sixteenth century. He trained not only his own sons and nephews, but several sons of astronomical families from Vidarbha and elsewhere. These Vidarbhans later settled in Kāśī (that is, Benares), where their descendants a century and more later would still proudly proclaim that their ancestors had studied with Gaṇeśa.

Gaṇeśa developed his own astronomical system, which was an amalgam of elements from several existing systems chosen, he claimed, for their superiority as determined by observation. The observations, however, only confirmed existing parameters; they did not determine new ones. In general, the system of education within families ensured conservatism since most astronomers would never become aware of any system not taught by their fathers and not represented in their family libraries. For this reason each of the five *pakṣas* or schools of astronomy that developed in India from the two systems derived from Greek sources in the fifth century had its own territory in which it was almost absolutely predominant. These provincial schools very rarely communicated with each other; and so, for instance, the brilliant mathematical investigations into power series made in Kerala were not known in any other part of India, and the adaptation of an advanced trigonometry from the *zīj* of Ulugh Beg by the astronomers of Kāśī and Delhi in the seventeenth century had little influence outside of those cities. The spread of Ptolemaic astronomy, begun in the seventeenth century and continued by the *paṇḍitas* of Jayasiṃha[23] in the eighteenth, was fiercely and successfully resisted. And when a Western educational system was introduced into India in the nineteenth century, while other *śāstras* like astrology and medicine were able to survive, Indian astronomy and mathematics were unable to complete and rapidly vanished. But the traditional astronomer, despite the limitations of his science that are apparent to us, had performed two very useful functions in Indian society. He computed the annual calendars or *pañcāṅgas* which gave the dates both of extraordinary celestial events and, more importantly, of the festivals which depend on the complex computations of the beginnings of local *tithis* or lunar days. Many of the Vedic liturgical performances were regulated by the *tithi*, as were also more popular and local religious observances. And second, the astronomers computed the tables used by astrologers in casting horoscopes, and often acted as astrologers themselves. The number of such experts that were needed in India was vast. One consequence of this, and of the diffusion of instruction through families rather than through centralized educational institutions, was the creation of a vast literature in the field and the production of innumerable manuscript copies. There still exist today, despite the final demise of the tradition of composing and copying these astronomical texts in about 1900, approximately 10,000 separate works that deal with some aspect of the exact sciences in Sanskrit, and probably 100,000 manuscripts. The remarks I have made in this paper are based on the reading of only a small portion of this enormous production: the hundred or so texts that have been published, and a couple of thousand manuscripts to which I have had access. It is possible that in the remainder is hidden material that would require a reassessment, radical or otherwise, of Indian astronomy, though that is unlikely. It is certain, however, that much material of great historical interest remains to be discovered, even if we often cannot be overly enthusiastic about the level of astronomical competence displayed in these texts. But it is equally certain that time will rapidly destroy a sizeable percentage of the 100,000 manuscripts before they are exposed to competent eyes. Manuscripts in India in general do not survive the climate and the insects for more than three hundred years; they need to be copied or to be properly preserved in suitably equipped libraries. Such modern libraries are scarce, and the family and scribal traditions that formerly replenished the old *pustakālayas* have died with the usefulness of the *śāstra*. One of the many prices that India and the rest of the world are paying for its Westernization is the loss of this part as much else of its cultural heritage.

[19] *CESS* A4.
[20] *CESS* A3, 39b–40a.
[21] *CESS* A1, 41a–41b.
[22] *CESS* A2, 94a–106b; A3, 27b–28a; and A4.
[23] *CESS* A3, 63a–64b; and A4.

THE PURĀṆAS AND JYOTIḤŚĀSTRA: ASTRONOMY

DAVID PINGREE
BROWN UNIVERSITY

This article examines the origins of the purāṇic and jyotiṣa cosmologies, showing which of the elements in each were influenced by Babylonian and Greek ideas, and how the jyotiṣīs adapted to their own system what purāṇic ideas they could while rejecting all others. The key jyotiṣa text, the *Paitāmahasiddhānta*, is paradoxically preserved in an upapurāṇa, the *Viṣṇudharmottarapurāṇa*. It is further shown that a movement to reconcile the cosmology of the astronomers with that of the purāṇas began in the late seventeenth century, perhaps in an attempt among Indian intellectuals to close ranks against the perceived threat to their traditions posed by Islamic and European astronomy.

THERE EXIST IN A NUMBER of purāṇas, as Kirfel[1] has demonstrated, two descriptions of the universe having a common source. In this common source the earth, *pṛthivī*, with its seven concentric pairs of continents and oceans,[2] is a horizontal disk in the center of a vertical universe enclosed in the *brahmāṇḍa*. That universe contains seven *loka*s above[3] and seven *pātāla*s below.[4] The first three of the upper seven constitute the Vedic triad—the *bhūrloka* being the surface of the earth, the *bhuvarloka* the region between the earth and the sun, and the *svarloka* the region between the Sun and Dhruva, the pole-star. From the center of the earth rises mount Meru,[5] which acts somewhat as does the Vedic *akṣa* or axle that connects heaven and earth (which occurs only as a simile for Visnu!),[6] though the name Meru (or rather, Mahāmeru) appears first in Vedic literature only in the *Taittirīyāraṇyaka* (I.7.1.2); for Meru in the purāṇic text is the axle around which the wheels carrying the celestial bodies rotate. It also serves the function, as do Anaximenes' "higher parts of the earth," of explaining the disappearances of the Sun, the Moon, and the *nakṣatras*. Above these circle the Saptarṣis[7]—Ursa Maior—presumably because that constellation, as was noted in the Babylonian omen series, *Enūma Anu Enlil*, never disappears from the night sky. The *cakra*s of these *jyotīṃṣi* are rotated by chords of wind that bind them to Dhruva, which is located on the tail of the starry Śiśumāra or Dolphin.[8] Dhruva is also a late concept; it first appears in the prescriptions for the marriage ceremony given in the gṛhyasūtras,[9] though there only as an unmoved star, not as one pole of the axis about which the other celestial bodies revolve.

The concepts of Meru and Dhruva serve to date this cosmology to the middle of the last millennium B.C. at the earliest. Indeed, the early Pāli texts of the Buddhists refer to Himavat as the center of the world (Meru is substituted for Himavat only in the *Mahāvastu*),[10] and state that the cause of day and night is the circling by the Sun and the Moon about Sineru (Sumeru).[11] A firmer terminus post quem for the purāṇic text is found in a passage that occurs only in the *Viṣṇupurāṇa* among representatives of version I,[12] but is in all of the bearers of version II of Kirfel's text;[13] this passage refers to the five year *yuga* of Lagadha's *Jyotirvedāṅga* with some of its characteris-

[1] W. Kirfel, *Das Purāṇa vom Weltgebäude* (Bonn, 1954), I 7–12 and II 9–13 (henceforth cited as Kirfel, *Purāṇa*).

[2] Kirfel, *Purāṇa*, I 4; II 1, 6–36; and II 6. See also W. Kirfel, *Die Kosmographie der Inder* (Bonn and Leipzig, 1920), 56–127 (henceforth cited as Kirfel, *Kosmographie*).

[3] Kirfel, *Purāṇa*, I 7, 7–14; II 6, 134–35; and II 9, 20. See also Kirfel, *Kosmographie*, 128–30. For the Vedic *loka*s see J. Gonda, *Loka* (Amsterdam, 1966).

[4] Kirfel, *Purāṇa*, I 5 and II 7. See also Kirfel, *Kosmographie*, 143–47.

[5] Kirfel, *Purāṇa*, I 2, 3–12, and II 2, 15–47, *etc*.

[6] Kirfel, *Kosmographie*, 7.

[7] Kirfel, *Purāṇa*, I 7, 2–4 and 5 c–d, and II 13, 106 and 108 c–d.

[8] Kirfel, *Purāṇa*, I 9; I 12, 24–29; II 10; and II 12.

[9] A. A. Macdonell and A. B. Keith, *Vedic Index*, 2 vols. (rpt. Varanasi, 1958), 1: 405–6.

[10] *Aṅguttaranikāya*, etc., cited by Kirfel, *Kosmographie*, 182; cf. p. 184.

[11] *Jātaka* cited by Kirfel, *Kosmographie*, 190.

[12] Kirfel, *Purāṇa*, I 8, 29–122.

[13] Kirfel, *Purāṇa*, II 9, 85–111; cf. II 13, 126–30.

tic parameters, such as the ratio, 3:2, of the longest to the shortest length of daylight in the year; the astronomy of the *Jyotirvedāṅga* was greatly influenced by ideas introduced into India from Mesopotamia, presumably through Iranian intermediaries, in about 400 B.C.[14] Another Mesopotamian theory that appears in this passage is that of the three paths of heaven to the north, the middle, and the south, which the Babylonians had associated with their gods Enlil, Anu, and Ea respectively. These paths and their elaborations appear in Sanskrit texts on celestial omens which are dependent on the Babylonian series *Enūma Anu Enlil* as three *vīthi*s or paths;[15] in the purāṇic texts they are the *nāgavīthi* to the north, a *madhyamamārga* in the middle, and the *ajavīthi* to the south.[16] But the same passage of the purāṇic text refers to the twelve zodiacal signs, and places the equinoxes and solstices at the beginnings (the ends, mistakenly, in version II) of Meṣa and Tulā and of Makara and Karkaṭa. This points to a time after the introduction into India of Greek astrology in the second century A.D.[17]

The same late date—after about 150 A.D.—for the whole purāṇic text is indicated by its placing the five planets between the *nakṣatramaṇḍala* and the *saptarṣimaṇḍala* in the order established by the Greeks in the third century B.C., but introduced into India, again with Greek astrology, in the second century A.D.;[18] this ascending order is: Mercury, Venus, Mars, Jupiter, and Saturn, above which the purāṇic text adds the Vedic Svarbhānu or Rāhu with his non-Vedic tail, Ketu, to account for the occurrences of solar and lunar eclipses.[19] Not only is the text forced by its theory of the Sun and the Moon (the Moon is above the Sun) to introduce external forces—Rāhu and Ketu—to act as the causes of eclipses; it also cannot ascribe the waxing and the waning of the Moon to their true physical causes, and therefore introduces the idea that Soma, the Moon, is drunk alternately by the Devas and the Pitṛs.[20] The Purāṇa text also devotes many verses to enumerating the beings who accompany the Sun's chariot in each of the twelve months;[21] the first verse of this section in the version of the *Viṣṇupurāṇa* (2.10.1) gives the number of tracks on the Sun's wheel as 183, corresponding to the 183 sidereal days in an *ayana* according to Lagadha's *Jyotirvedāṅga*.

The terminus ante quem for the purāṇic text is provided by the cosmographical aṅgas of the Śvetāmbara Jainas—the *Jambūdvīpaprajñapti*, the *Sūryaprajñapti*, the *Candraprajñapti*, and the *Jīvābhigamasūtra*—which were put in their present form in the early sixth century A.D.; for they provide a greatly elaborated version of the purāṇic tradition.[22] Some of the purāṇas which contain one or the other of the versions of the purāṇic cosmological text are probably as old as the fifth century, so that a date sometime in the third or fourth century A.D. for their common source seems quite likely; we shall see that this cosmology was known to Indian astronomers in about 400 A.D.

But another purāṇic tradition was also adopted by the astronomers—that of the *kalpas*, *manvantaras*, *caturyugas*, and *yugas*, which appear as the chronological framework in the *pañcalakṣaṇa* core of the purāṇas[23] as well as in the *Śāntiparvan* of the *Mahābhārata* (12.224) and in the *Manusmṛti* (1.64–86). A date of about the second century A.D. for this chronological system cannot be very far from the truth; the *kappa*, of course, occurs much earlier in the Buddhist tradition, including the Aśokan inscriptions, but its duration is not in these early references specified.[24]

The basic parameter of the *yuga* system is the *kaliyuga* of 432,000 years; this is a number derived from Babylon, where it was regarded as the period of the kings who reigned before the Flood. In the Babylonian sexagesimal system of writing numbers, 432,000 is 2,0,0,0 (i.e., 2×60^3). In India this number was combined with the decimal system, with the idea

[14] D. Pingree, "The Mesopotamian Origin of Early Indian Mathematical Astronomy," *JHA* 4 (1973): 1–12.

[15] D. Pingree, "Venus Omens in India and Babylon," in *Language, Literature, and History*, AOS 67 (New Haven, 1987), 293–315.

[16] See also Kirfel, *Purāṇa*, II 9, 55–57.

[17] For this introduction of Greek astrology into India see D. Pingree, *The Yavanajātaka of Sphujidhvaja*, 2 vols. (Cambridge, Mass., 1978).

[18] This order lies behind that of the planetary week-days mentioned first in India in *Yavanajātaka* 77. 2–8. At about the same time as Yavaneśvara made the prose translation of the Greek astrological text that is the basis for Sphujidhvaja's work, the Śātavāhana queen, Bālaśrī, in an inscription at Nāsik (*EI* 8 (1905–6): 60–65) gives the order Moon, Sun, *nakṣatras*, and planets, in which only the sequence of the two luminaries differs from that of the purāṇic text.

[19] Kirfel, *Purāṇa*, I 7, 5 a–b; I 12, 16–23; II 12, 2–4; and II 13, 107–9.

[20] Kirfel, *Purāṇa*, I 12, 1–15, and II 11, 58–76.

[21] Kirfel, *Purāṇa*, I 10, 1–21, and II 11, 1–36.

[22] Kirfel, *Kosmographie*, 214–61 and 278–91.

[23] W. Kirfel, *Das Purāṇa Pañcalakṣaṇa* (Leiden, 1927), 12–14.

[24] D. Pingree, "Astronomy and Astrology in India and Iran," *Isis* 54 (1963): 229–46, esp. p. 224.

of four ages, and with a theory of those ages' proportional decline, so as to produce the *caturyuga* of 4,320,000 years in which the four *yuga*s are in the ratio 4:3:2:1. The *kalpa* was defined as 1,000 *caturyuga*s or 4,320,000,000 years; and this period was divided, rather unsatisfactorily, among the 14 Manus. All of these numbers could be expressed in "divine years," of which each equals 360 human years; in this transformation the *kaliyuga* is 1,200 divine years, the *caturyuga* 12,000, and the *kalpa* 12,000,000.

From Lagadha's *Jyotirvedāṅga*, Sphujidhvaja's *Yavanajātaka*, and Varāhamihira's *Pañcasiddhāntikā* we know that other *yuga*s were used for astronomical purposes in India. Thus, the *Jyotirvedāṅga*, as we have noted above, uses a Vedic five-year *yuga* as a rough intercalation cycle; its parameters reappear in the first *Paitāmahasiddhānta*, whose epoch is 80 A.D. and which is summarized by Varāhamihira.[25] A luni-solar *yuga* of 165 years is found in the *Yavanajātaka* (79.2–10 and 14–20), whose date is 269/270 A.D. And the *Romakasiddhānta* summarized by Varāhamihira[26] has a *yuga* of 2850 years that is the smallest number of Hipparchan years of 6,5;14,48 days each that is also a multiple of the Babylonian nineteen year cycle (usually called "Metonic" in the West) and contains an integer number of days. In Greece earlier attempts had been made to find a *magnus annus* in which all the planets make an integer number of rotations; these efforts were ultimately inspired by a passage in Plato's *Timaeus* (39 B–D).

In about 400 A.D. someone in India who had access to Greek astronomical texts which were in many respects based on the work of Hipparchus and other Hellenistic astronomers,[27] but whose planetary models in part represented Peripatetic reactions to the tradition of Greek mathematical astronomers since Apollonius,[28] attempted to combine these Greek traditions with the cosmology and chronology of the purāṇas. His efforts are embodied in the second *Paitāmahasiddhānta*,[29] which is fortuitously preserved for us in a purāṇa—or rather, despite its early age and incredible bulk, in an upapurāṇa—the *Viṣṇudharmottarapurāṇa* (II.166–74).

The *Viṣṇudharmottarapurāṇa* is a compilation of many diverse elements, put together, it would seem, in Kāśmīra in the sixth or seventh century,[30] perhaps during the reign of one of the early Kārkoṭa kings in the last three quarters of the seventh century when Kāśmīra was powerful and prosperous. The first Kārkoṭa, Durlabhavardhana, as we know from Kalhāṇa's *Rājataraṅgiṇī* (4.4 and 6), was a devout worshipper of Viṣṇu. It brings together many separate treatises on technical śāstras, such as grammar and lexicography, alaṅkāra and prosody, nāṭaka and nṛtta, gīta and instrumental music, śilpa and citra, āyurveda and pākaśāstra, as well as more traditional purāṇic fare. The text on astronomy that the compiler chose to incorporate was the *Paitāmahasiddhānta*, which is a prose work in the form of a dialogue between Bhṛgu and Brahmā. Its first chapter deals with military astrology in the form of omens similar to those developed in the last few centuries B.C. from the Babylonian astral and terrestrial omens that had been introduced into India during the Achemenid period. The second chapter gives some basic information concerning the Greek astrology that had come to India through such translations as Yavaneśvara's *Yavanajātaka* of 149/150 A.D. The remaining seven chapters are devoted to the new astronomy in which Greek theories are modified to fit in with some Indian traditions. This new astronomy of the *Paitāmahasiddhānta* is the direct ancestor of the premier *pakṣa* of Indian astronomy, the Brāhma, and was clearly known to and modified by Āryabhaṭa, the author in the late fifth century (he was born in 476) of the two other early *pakṣa*s, the Ārya and the Ārdharātrika. All later Indian *pakṣa*s descended from these three, though the infusion of elements of Islamic astronomy beginning in the tenth century led to modifications of some of them in limited areas.

---

[25] *Pañcasiddhāntikā* 12. For the history of jyotiṣa literature see D. Pingree, *Jyotiḥśāstra* (Wiesbaden, 1981); *Census of the Exact Sciences in Sanskrit*, series A, vols. 1–4 (vol. 5 in preparation) (Philadelphia, 1970–81); and "History of Mathematical Astronomy in India," *Dictionary of Scientific Biography*, vol. 15 (New York, 1978), 533–633. In these three works will be found supporting documentation for most of the statements made in the remainder of this paper.

[26] *Pañcasiddhāntikā* 1. 15.

[27] D. Pingree, "The Recovery of Early Greek Astronomy from India," *JHA* 7 (1976): 109–23.

[28] D. Pingree, "On the Greek Origin of the Indian Planetary Model Employing a Double Epicycle," *JHA* 2 (1971): 80–85.

[29] D. Pingree, "The *Paitāmahasiddhānta* of the *Viṣṇudharmottarapurāṇa*," *Brahmavidyā* 31/32 (1967–68): 472–510.

[30] A lengthy discussion of the *Viṣṇudharmottarapurāṇa* together with a list of its contents is given by R. C. Hazra, *Studies in the Upapurāṇas*, vol. 1 (Calcutta, 1958), 155–218. Hazra dates the work between 400 and 500 A.D.; Priyabala Shah, *Viṣṇudharmottarapurāṇa*, vol. 1 (Baroda, 1958), xxvi, places its date between 450 and 650. But both of these dates are based on those of the texts incorporated in the upapurāṇa, and therefore are only termini post quos.

The date of the *Paitāmahasiddhānta* is pushed to the beginning of the fifth century not only by the fact that Āryabhaṭa at the end of that century was able to draw upon it, but more importantly by the fact that in about 450 A.D. one of its characteristic parameters, the longitude of the apogee of the Sun computed for that date, was known at the Sasanian court at Ctesiphon. This means, as we shall see, that a text going further than the *Paitāmaha* itself in expounding the new astronomy, already existed and had been transmitted to Iran by the middle of the fifth century; the early decades of that century, then, constitute the last possible date of the *Paitāmaha*, which may well, then, have originated in the same intellectual ferment of the Gupta Empire that produced Kālidāsa and so many others.

The author of the *Paitāmaha*, perhaps following the lead of the Greek computations of the *magnus annus*, adapts the purāṇic cosmological system to the problems both of finding the mean longitudes of the planets and of determining the longitudes of the planets' apogees and nodes at any particular time. For a *kalpa* of 4,320,000,000 years is long enough that each of these elements can be positioned at Aries 0° at its beginning and endowed with an integer number of rotations during its course such that, for the Sun, the Moon, and the planets, their mean velocities are essentially correct, and, for the apogees and nodes, their longitudes within historical time are correct. Following its pretence to be based on a revelation made at the beginning of the *kalpa*, however, the *Paitāmaha* does not say how much time had elapsed from that beginning till the time when it was composed. Later texts inform us that the interval between the beginning of the *kalpa* and that of the current *kaliyuga* was a period of 432,000 years multiplied by 4567, or 1,972,944,000 years.

The mathematical problem faced by the author of the *Paitāmaha* in fitting the *kalpa* to the mean motions of the planets was simple. Given certain period relations which tell one that a planet makes x sidereal revolutions in y years, such as were common in both Babylonian and Greek astronomy and had already appeared in India in the *Vasiṣṭhasiddhānta* and the *Pauliśasiddhānta* summarized in Varāhamihira's *Pañcasiddhāntikā*, he had to find by proportion how many rotations each planet makes in 4,320,000,000 years. He further faced the problem of securing an approximation to a conjunction of all the planets at the beginning of the current *kaliyuga*; and so had to find the number of rotations in a *kalpa* that had to be added to or subtracted from his initial values in order that each planet would make close to an integer number of rotations in 1,972,944,000 years. This problem can be expressed as an indeterminate equation of the first degree, and was brilliantly solved by the application of the algorithm of continuous fractions associated with the name of Euclid—the so-called *kuṭṭaka* or pulveriser. The result is that the *Paitāmaha* can list the number of rotations that each planet makes in a *kalpa*—4,320,000,000 for the Sun; 57,753,300,000 for the Moon; 2,296,828,522 for Mars; 17,936,998,984 for Mercury's *śīghra*; 364,266,455 for Jupiter; 7,022,389,492 for Venus' *śīghra*; 146,567,298 for Saturn; 488,105,858 for the Moon's apogee; and 232,311,168 for the Moon's node—and be certain that these will produce reasonably accurate mean longitudes of the planets within his own time and for centuries thereafter. And he didn't have to make a single observation. For the apogees and nodes his task was even simpler. He had only to know where they should be in his own time, and to endow them with small integer numbers of rotations in a *kalpa* chosen so that they would have arrived at their proper positions at the beginning of the current *kaliyuga* but be moving so slowly that they would still be there three or four thousand years later.

You may have noticed that the numbers of rotations in the cases of the two inferior planets, Venus and Mercury, are those of their *śīghras*—that is, of their conjunctions with the Sun. This is because of cosmological considerations. The author of the *Paitāmaha*, of course, was adopting a Greek astronomical system that was based on the conception of the earth as a sphere within a *pañjara* or cage of internesting planetary and stellar spheres. He was rejecting the purāṇic flat earth cosmology, though, as we shall see, preserving what purāṇic elements he could. One thing that the purāṇas did have correctly was the ascending order of the five star planets from the center of the earth, though they misplaced the Sun, the Moon, and the *nakṣatras* below them in following an older cosmology of the first millennium B.C. In that exemplar of Greek astronomy or astrology whence the purāṇas had derived this order it was important to keep Venus below the Sun for, despite the actual order of the cakras, the purāṇic text groups the Moon, Mercury, and Venus together as moving fast, Mars, Jupiter, and Saturn together as moving slowly;[31] and in Indian astronomy, as in its Greek prototypes, the geocentric distances of the planets were regarded as inversely proportional to their velocities. Thus, the Moon which travels 13;10,35° per day is the closest to the earth,

[31] Kirfel, *Purāṇa*, II 13, 97–98.

and Saturn which travels 0;2° per day is the furthest away. The mean velocities of the two inferior planets is the same as that of the Sun, so that to use that parameter would place all three at the same distance from the earth. And Venus' mean daily motion on its epicycle is 0;36° while the mean daily motions of the Sun and of Mars are respectively 0;59° and 0;31°, so that the use of this parameter would place Venus between the Sun and Mars—which, indeed, some Greek cosmologists did. But one can keep Venus below the Sun and above Mercury by using the *śīghra* motion, which for each of the inferior planets is the sum of the mean motion of the Sun and its proper motion on its *śīghra* epicycle; for the *śīghra* velocity of Mercury comes to be 4;5,32° per day, that of Venus 1;36° per day, while the mean velocity of the Sun remains 0;59° per day. This was the solution adopted by the Indians; it also appears earlier in the Greek Keskinto inscription.

Given the theory of the inverse proportionality of velocity to distance and the number of rotations that each planet makes in a *kalpa*, it is easy to compute the actual distance of each of the planetary spheres from the center of the earth as the author of the *Paitāmaha* instructs one to do, though he does not himself carry out the computations; one arrives at distances far different from those given in the purāṇic text. The *Paitāmaha*'s author makes two preliminary assumptions: one is that a minute in the orbit of the Moon equals 15 *yojanas* (this results from the estimate that the Moon's horizontal parallax, which is the earth's radius seen at the Moon's distance, is 0;53°), and that each planet travels an equal number of *yojana*s in a *kalpa*—a number that is equal to the orbit of heaven. This last assumption, of course, is just the common sense deduction that the velocity of each planet measured in *yojanas* per time unit is a constant for all the planets. Since there are 21,600 minutes in a circle, there are 21,600 × 15 = 324,000 yojanas in the orbit of the Moon. The Moon circles this orbit 57,753,300,000 times in a *kalpa*, so that in that period it travels 57,753,300,000 × 324,000 = 18,712,069,200,000,000 *yojanas*, which is the number given in the *Paitāmaha* for the circumference of the outermost sphere. If one divides this number by the number of rotations of any other planet in a *kalpa*, one will find the circumference of that planet's orbit measured in *yojanas*; and then it is a simple matter to compute the radius of that circle, which is the distance of that orbit from the center of the earth. The circumference of the earth, as is implied by the computation from the horizontal parallax of the number of *yojanas* in a minute of the Moon's orbit, is, as is also stated in the *Paitāmaha*, 5,000 *yojanas*.

The celestial bodies in the *Paitāmaha* and all subsequent Indian astronomical texts, then, are arranged in a fashion different from the cosmology of the purāṇic text in every respect save the order of the five star planets. But the author of the *Paitāmaha* did not wish to depart altogether from the purāṇic cosmology. Therefore, in order to preserve something from that tradition, he turned to geography and to celestial mechanics. In geography he made mount Meru that point on the earth's surface through which the axis connecting Dhruva, the north pole, to its counterpart in the south passed, and asserted the existence of the city Laṅkā on the equator to the south of Meru such that the prime meridian passes over both. Later astronomers assert that the prime meridian passes also over Ujjayinī, and add three cities on the equator at quadrants from Laṅkā—Romaka to the west, Siddhapura to the north, and Yamakoṭī to the east. Varāhamihira, in his *Pañcasiddhāntikā* (15, 22–23), written in the middle of the sixth century, was the first to identify these four cities explicitly with the four cardinal peoples of the purāṇic text—the Bhāratas, the Ketumālas, the Kurus, and the Bhadrāśvas.[32]

On the subject of celestial mechanics the author of the *Paitāmaha* says nothing, but later astronomers generally attribute the daily rotations of the heavenly spheres to the force of the purāṇic *pravaha* wind which is wrapped around the axis extending from the south pole through Vaḍavāmukha and Meru to Dhruva; and the motions of the planets on their epicycles are explained as being caused by Asuras or Demons stationed at their *manda* and *śīghra* apogees tugging on chords of wind attached to the planetary chariots.

One further gesture that the author of the *Paitāmaha* makes toward Indian tradition is to justify the study of astronomy by quoting at the end of his work the final verse of Lagadha's *Jyotirvedāṅga*:

vedā hi yajñārtham abhipravṛttāḥ
kālānupūrvā vihitāś ca yajñāḥ
tasmād idaṃ kālavidhānaśāstraṃ
yo jyotiṣaṃ veda sa veda sarvam

The Vedas went forth for the sake of the sacrifices; the sacrifices were established as proceeding regularly in time. Therefore, he who knows *jyotiṣa*, this science of time, knows all.

The same justification for the study of astronomy, that it is necessary for the proper performance of the

[32] Kirfel, *Purāṇa*, I 2, 33, and II 2, 47.

Vedic rituals, stayed alive in the tradition of the Brāhmapakṣa, and was popular again in the seventeenth century when astronomers, searching for a basis of their science in the writings attributed to *devas* or to *ṛṣis*, rediscovered the *Paitāmahasiddhānta* in the *Viṣṇudharmottarapurāṇa*. Thus, the earliest manuscript of the *Paitāmaha* as an excerpt from the upapurāṇa was copied in ŚAKA 1563 = A.D. 1641; it is now known only from a nineteenth-century copy. Bhāskara in his *Siddhāntaśiromaṇi* (*Grahagaṇita* 1.1.9) had already summarized Lagadha's verse; Nṛsiṃha in his *Vāsanāvārttika*, composed in Kāśī in A.D. 1621, quotes Lagadha in his comments on this verse, though perhaps directly from the *Jyotirvedāṅga* rather than through the *Paitāmaha*. But elsewhere (on *Grahagaṇita* 2.34–35) Nṛsiṃha states that Bhāskara obtained his planetary parameters from the *Brahmasiddhānta* preserved at the end of the second *kāṇḍa* of the *Viṣṇudharmottara*, and again (on *Grahagaṇita* 1.2.1–6) attempts to explain the disagreement between Pitāmaha or Brahmā in the *Viṣṇudharmottara* (which, of course, belongs to the Brāhmapakṣa) and Brahmā in the *Śākalyasaṃhitā* (which belongs to the Saurapakṣa) as due to scribal and other corruptions. In both cases Nṛsiṃha is clearly discussing these texts because they were regarded as authoritative because of their ascriptions to a deity. Similarly, Kamalākara in the *Siddhāntatattvaviveka* (1.62) that he completed in Kāśī in 1658, in defending the validity of the Saurapakṣa to which he owed his allegiance, says:

> ced viṣṇudharmottaram eva mūlaṃ
> brāhmaṃ purāṇaṃ vadasīha tat tu
> atantrikair nāśitam eva pūrvaṃ
> sandṛśyate sarvajanaprasiddham

> If you should say in this matter that the Brāhma's source is the *Viṣṇudharmottarapurāṇa*, yet this is seen, as is well known to all people, to have been destroyed previously by non-tantrikas.

Others as well from among the astronomers of seventeenth century Kāśī recognized the historical importance of the *Paitāmahasiddhānta*. But rather than relating their statements, which add little to what I have already said, I turn to consider what the jyotiṣīs have to say directly concerning the purāṇic text that I discussed at the beginning of this paper. The first astronomer to attempt to deal exhaustively with the purāṇic cosmological tradition was Lalla, who wrote his *Śiṣyadhīvṛddhidatantra* in the middle of the eighth century. He devoted *adhikāra* 19 (*bhuvanakośa*) of that work and *adhikāra* 20 (*mithyājñāna*) to this topic. In the *bhuvanakośa* he incorporates the flat earth into the spherical cosmology by inserting the seven *pātālas* into the interior of the earth; the seven oceans and the six *dvīpas* beyond Jambūdvīpa in the southern hemisphere (the names that he gives to the oceans and *dvīpas* are closest to those found in the *Varāhapurāṇa*); the mountain ranges, peoples, and rivers of Jambūdvīpa over the northern hemisphere rather than on a flat surface. Moreover, he turns the three Vedic *lokas*—*bhū*, *bhuvar*, and *svar*—into, respectively, the inside of the earth together with its southern hemisphere, the northern hemisphere (that is, Jambūdvīpa), and Sumeru; and the four upper *lokas* in the purāṇic order—*mahas*, *jana*, *tapas*, and *satya*—into spheres filling the space between the *nakṣatras* and heaven, the Brahmāṇḍagola. That furthest sphere is surrounded by three great circles, the equator, the prime meridian, and the ecliptic. The diurnal rotation of these celestial spheres is still powered by the *pravaha* wind (*Śiṣyadhīvṛddhidatantra* 18.3). This general solution to the problem of the incongruence of the purāṇic and the jyotiṣa cosmologies is repeated in many siddhāntas subsequent to Lalla's.

In his twentieth *adhikāra*, however, Lalla systematically refutes with physical arguments the many unacceptable doctrines of purāṇic astronomy. These false notions include the ideas that solar and lunar eclipses are caused by Rāhu, that the Moon is above the Sun, that Meru causes the darkness of night, that in the *kṛṣṇapakṣa* the Moon is being drunk by the Gods, that the earth is flat, and that it is supported by a tortoise, elephants, or some other physical supports. Lalla's arguments against these false beliefs are based on inferences from observed phenomena, and are essentially correct; he lapses into an argument from authority only in the case of Rāhu where, after proving that Rāhu cannot be the cause of an eclipse, he states that it may be a concomitant because Brahmā by his power causes the Sun to be near Rāhu at the time of an eclipse. It is because of this concomitance, he believes, that the smṛtis and the Vedasaṃhitās claim that an eclipse is caused by Rāhu.

Lalla's is indeed a powerful refutation of the purāṇic errors in astronomy, and his arguments were often repeated—e.g., by Śrīpati in the eleventh century in his *Siddhāntaśekhara*, by Bhāskara in the twelfth in his *Siddhāntaśiromaṇi*, and by Jñānarāja in the early sixteenth in his *Siddhāntasundara*. As far as the astronomers were concerned, that is where the matter rested. Even in the sometimes heated exchanges that took place in the seventeenth century between partisans of traditional Indian astronomy and those who sought to change it by introducing elements, including cosmological concepts, derived from the Islamic interpretation and transformation of Ptolemaic astronomy, the astronomers did not retreat from their rejection of those purāṇic beliefs that Lalla had refuted. The opposition of some to the Islamic system was indeed largely based on an appeal to revealed texts; but they turned to the *Paitāmaha* and to the *Sūryasiddhānta* for their authorities, and ignored the purāṇic text.

However, in the late seventeenth and early eighteenth centuries two texts were written on the relationship between the purāṇic cosmology and jyotiḥśāstra that seem to reflect some deepening awareness of the threat to the traditional Indian sciences. One was composed by Kevalarāma, the Jyotiṣarāya of the court of Jayasiṃha at Jayapura from about 1730 on. He had written such works as a *Brahmapakṣanirāsa* in which he apparently refuted the school of astronomy that was started by the *Paitāmahasiddhānta*, and a *Dṛkpakṣasāraṇī* based on the lunar theory of the seventeenth century French astronomer, de la Hire. But he also composed a *Bhāgavatajyautiṣayor bhūgolavirodhaparihāra* or *Removal of the Disagreement between the Bhāgavatapurāṇa and Astronomy Concerning the Sphere of the Earth.* Unfortunately, I have not as yet been able to secure a copy of this work, and so cannot say how Kevalarāma, who was among the first in India to study modern astronomy, sought to resolve the ancient conflict between the paurāṇikas and the jyotiṣīs; but it will be possible in the future, I hope, to examine his arguments.

Some decades before this jyotiṣī wrote, a paurāṇika had written a short work in 18 verses defending the purāṇic cosmology against that of the astronomers, claiming that the latter is not true but simply a useful tool for computations. This is the same attitude that late Greek Neoplatonists had adopted towards Ptolemy's *Almagest*. The author of the Indian version of this denigration of mathematical astronomy, the *Saurapaurāṇikamatasamarthana* or *Reconciliation of the Opinions of the Sūryasiddhānta and the Purāṇas*, was Nīlakaṇṭha Caturdhara, the famous commentator on the *Mahābhārata*. What inspired him to write this work, and how representative his views were among Indian intellectuals of his time, are topics also awaiting further research.

1 NACIMIENTO Y DIFUSIÓN DE LA CIENCIA: MOVIMIENTOS E INTERCAMBIOS INTELECTUALES

NAISSANCE ET DIFFUSION DE LA SCIENCE: MOUVEMENTS ET ÉCHANGES INTELLECTUELS

THE BIRTH AND SPREAD OF SCIENCE AND THE MOVEMENT OF INTELLECTUALS (THE INVISIBLE UNIVERSITY)

Coordinadores: **Eloy BENITO RUANO** *(Espagne)*
S. C. HUMPHREYS *(U.S.A.)*

INNOVATION AND STAGNATION IN MEDIEVAL INDIAN ASTRONOMY

David PINGREE *(USA)*

The largest pre-modern literature on astronomy still extant today — precariously preserved on hundreds of thousands of rapidly deteriorating manuscripts — is that composed in India, overwhelmingly in the Sanskrit language, in the fourteen centuries between about 400 A.D. and 1800 A.D. But while India was far ahead of Europe, the Islamic countries, and China in the quantity of its publications in this field, its record in quality as judged by modern Western standards is ambiguous. For, despite India's great contributions to the world's sciences of astronomy, the fundamental breakthroughs that led to the almost universally received modern form of astronomy occurred elsewhere.

The basic elements of Indian mathematical astronomy came from the West in several stages. The first of these was the transmission to India in about the eighth century B.C. of Mesopotamian descriptions of the motion of the Sun's rising-point along the Eastern horizon, their recognition of the usefulness of intercalation, their use of New Moon as the beginning of the month, and various elements of their star-lore. The second was the transmission in the late fifth or early fourth century B.C. of Babylonian arithmetical methods of computing the passage of time and the progress of the Sun, the Moon, and the planets; the third was the introduction of Greek adaptations of Babylonian planetary theory in the second century A.D.; and the last an influx of pre-Ptolemaic, Hellenistic astronomy in the third and fourth centuries A.D. This last transmission included geometrical models of planetary motion (eccentric deferents, epicycles, and equants), the parameters of these models, eclipse theory, and some basic mathematical tools for solving problems in plane and spherical trigonometry, though without Menelaus' theorem for dealing directly with angles on the surface of a sphere. Indeed, the Sanskrit texts are now one of our primary sources for the reconstitution of Hellenistic astronomical theory and practice. Indian astronomers made great advances in the mathematics of the Greek systems that they had adopted; for instance, already in the fifth century A.D. they had applied the Euclidean algorithm of continued fractions to the solution of indeterminate equations, a crucial step toward the development of their characteristic mode of expressing the mean motions of the planets in terms of their integer numbers of revolutions in vast periods of time — in mahāyugas of 4,320,000 years and in kalpas of 4,320,000,000 years; they had derived from the Chords of Hipparchus the Sine, Cosine, and Versine funcions that are the foundations of modern trigonometry; and they had fully exploited the possibilities of applying the rudimentary analemmas of Hellenistic astronomy to the solution of numerous problems in spherics and in time-keeping.

Their prestige, as a result of this brilliance, was so great that in the middle of the sixth century the astronomers of the Sasanian Shāh of Iran, Khusrau Anūshirwān, chose to follow the Indian *Zīj al-Arkand* in composing the *Royal Astronomical Tables* rather than Ptolemy's *Almagest*; and in 718 parts of Varāhamihira's *Pañcasiddhāntikā* served as the basis for the *Chiu-chih li* that Gu-tan Hsi-ta composed at the T'ang court. And in trigonometry and analemmas the Indians were the teachers of the Arabs, and ultimately of the West. Indeed, the first work on mathematical astronomy in Arabic that we know of is an adaptation of Brahmagupta's *Khaṇḍakhādyaka* made in Sind in 735, while the first serious work in this field available in Western Europe was the Latin translation, made by Adelard of Bath in Spain in 1126, of al-Majrīṭī's revision of the Arabic *Zīj al-Sindhind* of al-Khwārizmī; al-Khwārizmī's main sources were Arabic versions of a *Mahāsiddhānta* based on Brahmagupta's *Brāhmasphuṭasiddhānta*. As a result, for much of the medieval period many European astronomers computed the positions of the planets, predicted eclipses, and cast horoscopes with methods and tables derived from or influenced by Indian astronomy, and they regarded an Indian city, Ujjayinī, as lying on the world's prime meridian.

The Indian contemporaries of these medieval European astronomers did not cease being innovative in mathematics. From the eleventh century through the eighteenth, for instance, they devised an amazingly rich variety of astronomical tables designed to facilitate the computation of planetary positions, of the occurrences of eclipses, both solar and lunar, and of the characteristic elements of Indian pañcāngas or calendar — tithis, nakṣatras, and yogas. Some of the more ingenious of these tables were invented by Mahādeva in 1316, by Makaranda in 1478, and by Haridatta in 1638. The mathematics of indeterminate equations of the second degree had been studied with partial success by Brahmagupta in the *Brāhmasphuṭasiddhānta* that he published in 628, and the cyclic solution discovered by Jayadeva was described by Udayadivākara as early as 1073; Jayadeva's solution was independently discovered in the West only in the seventeenth century by Pell and Fermat. But the Indians' most brilliant achievement was their discovery, without a knowledge of the calculus, of the misnamed «Gregory's Series» for π and of the power series for Sines, Cosines, and Tangents, all due to the incredible work of Mādhava of Sangamagrāma in Kerala in about 1400 A.D. In Europe these series were rediscovered by the pioneers in calculus — Newton, Leibniz, and Gregory — some two hundred and fifty years after Mādhava and a full century after they had been virtually forgotten in India.

It was not in India, however, despite this extraordinary ability in mathematics, that modern astronomy was born. Its beginnings lie rather in the Muslim world where, in the early eleventh century, Ibn al-Haytham criticized Ptolemy's planetary and lunar models for their contravention of certain principles of physical astronomy based on Aristotle. A similar criticism had been inconclusively leveled against Ptolemy in antiquity by, for instance, Proclus in the fifth century, but astronomers in the Arabic tradition persisted in their efforts to devise physical models of internesting spheres that at least in some important respects obeyed the laws that Aristotle had laid down for the motions of celestial bodies. This work culminated in the successful application to this problem of the so-called «Ṭūsī couple» for converting circular into linear motion by Mu'ayyad al-Dīn al- 'Urḍī and by Naṣīr al-Dīn al-Ṭusī in the middle of the thirteenth century, and the complete resolution of the remaining problems by Ibn al-Shāṭir of Damascus toward the middle of the next century. It was these Arabic mathematical models that Copernicus employed in the *De revolutionibus*, though, of course, he went further toward modern Western astronomy than had the Muslims when, completely abandoning the Aristotelian physics that had motivated the models in the first place, he imagined the Sun to be at the center of the planetary spheres.

We have seen that Indian astronomy began its brilliant career by receiving and improving scientific ideas that entered Bhārata from the outside. This receptivity continued throughout the medieval period, though eventually it encountered, as we shall see, a counter-reaction in the seventeenth century. Thus, Muñjāla in 932 provided a trigonometrical formula that reproduces the effect of the crank-mechanism in Ptolemy's model of the Moon — that is, essentially, the Moon's evection. Śrīpati and Bhojarāja in the middle of the eleventh century had a complete theory of the equation of time, probably also derived from Arabic sources. Mahendra Sūri published the first Sanskrit treatise on the construction and use of the plane astrolabe in about 1370; he based it on an Arabic (or perhaps Persian) text. In the seventeenth century Sanskrit versions of the zīj of Ulugh Beg, of al-Qūshjī's *Risālah dar hay'at*, and of other Persian astronomical works circulated in Delhi, Agra, and Benares, where they influenced such scholars as Nityānanda, Kamalākara, and Ranganātha, and aroused the derision of others like Nṛsiṃha, Munīśvara, and Gadādhara, who protested that some tenets of Persian astronomy contradicted the teachings of the ancient ṛṣis. Finally, in the 1720's and 1730's, under the patronage of Savāī Jayasiṃha, Euclid's *Elements*, Theodosius' *Spherics*, Ptolemy's *Almagest*, al-Zarqālī's *Universal Astrolabe*, and that chapter of Naṣīr al-Dīn al-Ṭūsī's *Tadhkira* together with al-Birjandī's commentary thereon that describes the new mathematical devises that ended up as the mathematical foundation of the Copernican revolution were all translated from Arabic into Sanskrit, along with de La Hire's theory of the Moon and Napier's *Logarithms* from European languages.

These translations were not just too late to rejuvenate Indian astronomy; they were almost entirely without effect because they were generally ignored by Indian astronomers — even including, to a surprising extent, their sponsor the Mahārāja Jayasiṃha. Most professional astronomers in India staunchly resisted the new science; in fact, they continued to produce quite traditional treatises throughout the eighteenth century. Though an abortive effort, inspired by Launcelot Wilkinson, an idealistic English civil servant in Sihore, was made in the 1830's and 1840's to introduce the concepts of modern astronomy to Indian students through Sanskrit, and even to encourage those students to enter the field by first reading the *Sūryasiddhānta*, Bhāskara's *Siddhāntaśiromaṇi*, and Gaṇeśa's *Grahalāghava*, the real training of some Indians in the modern Western form of the science began only in the observatories, colleges, and universities founded on the European model in the middle of the last century and, for the most part, operated in complete ignorance of the medieval Indian tradition. Clearly the scientific work on astronomy that is done in India today is a development from those nineteenth century beginnings, and is not at all dependent on nor are its practitioners aware of the medieval astronomy; from which, indeed, modern Western scientists have nothing to learn about their astronomy, though much to learn about human nature and about the various attitudes towards the preceptible world that may give rise to systems of knowledge reasonably labeled «science».

It is, of course, no more noteworthy that Indian astronomers have discarded medieval Indian astronomy than that European and Middle Eastern astronomers have rejected Ptolemy and Ibn al-Shāṭir. It is important, however, that historians study these superceded scientific systems in order that we all may better understand the intellectual and social factors that shape our various explanations of natural phenomena. One question that we may ask of that history — the one that I wish briefly and tentatively to address in the remainder of this paper — is: what factors in traditional Indian astronomy contributed to the production of a Mādhava but prevented the appearance of a Naṣīr al-Dīn or a Copernicus? For modern Western astronomy, I am convinced, could not have developed in India, though obviously other developments, especially in mathematics, might well have occurred had traditional Indian science been able to survive the nineteenth century as a viable alternative intellectual system capable of attracting competent and imaginative students — if, in short, it had continued to perform its traditional functions in Indian society.

The most obvious reason that Naṣīr al-Dīn could not have been an Indian is that the problem that he set out to solve could not have arisen within the Indian intellectual tradition. This problem was the incompatiability of a physics, requiring that celestial bodies rotate with uniform motion about a center, with the Ptolemaic models, imposed by observational data, in which they rotate with non-uniform motion on eccentric orbits. For the Greeks and their Arabic successors the heavenly spheres participate in this circular motion by their very nature; for the Indians, when they remark at all on the matter, the spheres are rotated by the pravaha wind, but the planets' deviations from uniform motion about the center are caused by arbitrary forces, called asuras or demons, which operate upon them from moving positions within the model of each. The effects of these arbitrary forces are subject to mathematical computation, but their nature was not regarded as relevant or interesting. Moreover, Greek and Arabic astronomy had a firm tradition of employing observations systematically to check and to refine the models and their parameters; the Indians were bereft of any such tradition, though naturally they rejoiced when, usually by sheer luck, computation and observation coincided. The absence of a concern for the mechanics of the solar system and the lack of a theory of observational astronomy lead us to the consideration of some basic characteristics of the traditional Indian concept of the origins of scientific knowledge and to the question of the organization and the teaching of the śāstras or sciences.

Almost universally the Indian śāstras — medicine, grammar, the various schools of philosophy, astrology, and astronomy, among others — claim that the basis of the knowledge that they teach is divine revelation. Only in mathematics, once in a while, does one find statements indicating that this science is a product of human activity. This does not mean, of course, that religion dominated science or that innovation was precluded; I have already given ample evidence that great advances were made. What it did mean was that each science tended to be regarded as a self-sufficient entity independent of all others since independently created by the divine. It was from this perception of the matter that they were taught; one person might be learned in many śāstras, but almost never would he use one to illuminate another. In contrast, the Greek educational tradition, inherited in its Neoplatonic form by the Arabs, strove to integrate the sciences within an encyclopedic framework in which their interconnections were strongly emphasized. Therefore, even if Indian philosophers or physicists had developed concepts of celestial mechanics different from the ideas of the pravaha wind and the asuras, these would not necessarily have been seen as in contradiction with the geometrical models of the astronomers.

Another aspect of the belief that the origin of knowledge lies in revelation is the feeling that the theory underlying the text so received must be correct; it is only the practical details that can be improved upon. As we have seen, the innovations made by Indian astronomers were entirely in the area of mathematics, a field in which they were most ingenious and inventive, but they never questioned the basic assumptions. The debates between the advocates of the old and the new astronomies — be it the new Islamic as in the Mughal period or the modern Western as in the British Raj — turned precisely on the point that the changes proposed were not just in mathematics, but involved the structure of the universe as it had been prescribed by the ṛṣis under divine inspiration. Thus Munīśvara, incensed by the theory that the zero-point of the ecliptic actually moves with respect to the fixed stars (in traditional Indian astronomy precession and trepidation were simply mathematical models used in the computation of the Sun's declination and were not permitted to affect planetary longitudes), criticizes the Persians, their Greek predecessors, and their Indian followers for their arrogance in adhering to a doctrine justified by their own intellectual efforts and in trusting observations even when the conclusions reached are contrary to the opinions of the ṛsis. An extreme example of this obscurantist line of reasoning is Gadādhara's statement, in defence of the theory that the unmoved highest sphere is made of metal rather than of crystal, that, since the god Pitāmaha — that is, Brahmā — does not refute the theory of a metallic sphere, therefore an opponent of that theory is an opponent both of truth and of Brahmā. It must, however, be noted in Gadādhra's defence that he does advance more rational arguments in support of his thesis, such as the blueness of the sky that must be due to the reflection of the Sun's light from a metallic rather than a crystalline surface, and the tremendous weight of the stellar and planetary spheres that a metal sphere could bear while a crystalline sphere would be shattered.

This mixture of arguments based on blind faith with those based on practical considerations is not unusual, in India as in the West. Practical matters were the main concern of Indian scientists. Fundamentally, Indian astronomy was neither mystical nor other-worldly; it was astonishingly playful and pragmatic. It saw its function to be the development of methods of solving the problems encountered in preparing the annual calendar, in predicting eclipses, and in casting horoscopes. It was successful in satisfying the society for which it performed these functions, and conceived of no other questions as ones it ought to attempt to answer. Therefore, the mathematics by which these practival computations were effected were always subject to improvement; even the parameters of the system — the mean motions, the dimensions of the epicycles, or the longitudes of the apogees and nodes — could be changed from what the ṛsis had taught on the justification that, as the yuga ages, the universe and the motions occurring within it degenerate. But there was no incentive to doubt the foundations, of the astronomy, and no perceived conflict between it and any other facet of Indian intellectual and social beliefs.

But even within this highly pragmatic framework changes in parameters took place with great infrequency. This was due to the lack of a tradition of using observations to check and to correct those parameters — a lack which we have previously noted. The earliest observations prescribed in the Sanskrit texts were designed to impress the astronomer's patron with his powers of prediction rather than to provide data for revising his models; thus, in the late 620's both the first Bhāskara and Brahmagupta instructed their students to construct a wooden replica of the trigonometrical constructions used in predicting an eclipse, and to fit it with a sighting tube so that, at the properly determined moment, the patron might see the earth's shadow cross the Moon's surface. The first deliberate and systematic effort to use observations to gather data that we know of was made by Mādhava's student, Parameśvara's, who observed a ong series of solar and lunar eclipses in Kerala between 1393 and 1432. Unfortunately, he does not seem to have been able to use the data to effect any changes in the traditional theory of eclipses or in those of the Sun and the Moon. The strongest statement made in favor of observational astronomy in medieval India is that of Nīlakaṇtha, the pupil of Parameśvara's son, Dāmodara, in the late fifteenth century: «Whatever astronomical system does not disagree with observation, that is to be followed. The one which agrees with observation is to be recognized by contemporaries passing judgment (on the accuracy of different systems) on the occasions of such phenomena as eclipses. But those who, when an older astronomical system disagrees (with observations), produce a new one after investigating the parameters of (the models of) the plantets and learning (their correct values) — they are not, because of this, to be laughed at in this world or to be punished in the next». One

letter throughout Europe; and within a few decades the calculus had transformed European mathematics.

If we briefly turn our attention from the mathematics embedded in astronomical literature to the mathematical literature itself, we must immediately note one striking fact: that is, that though, as has often been stated in this paper, the truly innovative advances made in Indian astronomy were in the mathematics applied to it, the number of texts devoted to mathematics as an independent subject is astonishingly small, and that a surprising number of these have been preserved in only a handful of copies, often in just one. It is true that of one work on arithmetic, Bhāskara's *Līlāvatī*, there exist over five hundred manuscripts in addition to numerous translations into many vernaculars and more than thirty medieval commentaries; moreover, since 1832 it has been printed more than thirty times. Clearly, ever since it was composed in the 1140's the *Līlāvatī* has dominated the teaching of arithmetic throughout the Indian subcontinent. Its companion piece, Bhāskara's textbook on algebra, the Bījagaṇita, was nowhere near so popular; but it is preserved in about two hundred manuscript copies, and was explained by five medieval commentators. It was the text normally chosen for instructing pupils in algebra; but algebra was not as frequently taught as was arithmetic.

In contrast to this marked success of Bhāskara's two works, just one incomplete manuscript has preserved each of the two less popular of the four treatises composed by Bhāskara's own favorite predecessor, the eighth century mathematician Śrīdhara, while the latter's best known opus is found in only a half dozen copies and the fourth is lost altogether. The *Gaṇitatilaka* written by Śrīpati in the middle of the eleventh century survives in one incomplete manuscript, as does the untitled work in the Bakhshālī Manuscript; Nārāyaṇa's *Gaṇitakaumudī*, composed in 1356, is complete in only one copy, though half a dozen others preserve just the last two chapters, brilliant investigations of permutations and combinations and of magic squares, and two manuscripts contain his algebraic *Bījāvataṃsa*, though not in its entirety. In fact, the only work that was anywhere near as popular as Bhāskara's was the *Gaṇitasārasangraha* composed by the Jaina Mahāvīra in the middle of the ninth century. Almost all of the two hundred or so manuscripts of this work were kept in the monasteries of his co-religionists, whose students thereby learned their arithmetic from this leading scientist of their own sect.

The ten mathematical treatises that I have just mentioned are the only ones written in Sanskrit before the sixteenth century that still exist; we know about as many more that were written then but have since been lost. And only half a dozen works, either derivatives of these or simple business artihmetics, were written after 1500. Mathematics, then, in which the Indians had truly excelled from the first millennium B.C., when the *Śulvasūtras* recorded advanced solutions to the problems of the geometry of altar building, till Mādhava some two thousand years later, was not a subject on which many scholars would choose to write a book — there were neither a professional class of mathematicians to use them, nor a market for new mathematical textbooks. Few saw any need to go beyond the standard instructional manuals — the *Līlāvatī*, the *Bījagaṇita*, and the *Gaṇitasārasangraha*.

Like all other sciences in medieval India, mathematics was almost always applied rather than theoretical. It contributed to many other disciplines — plane and solid geometry to altar building and to architecture, combinations and permutations to medicine, to music, and to prosody, the summation of series to Jaina cosmology, and many varieties of mathematics to astronomy; and it was often the most creative element in each. But it had, for most intellectuals, no intrinsic value; it existed simply to serve the purposes of the other śāstras.

The consequent lack of a coherent tradition was accompanied by the lack of a coherent methodology. The power of a rigorous and relentless pursuit of the logical consequences of an axiomatic system was never felt by Indian mathematicians. In fact, the processes by which problems were analyzed and solutions obtained are almost universally left undescribed, so that the most brilliant achievements appear as revelations rather than as the results of rational thought. The one exception of which I am aware is the one which I discussed before, in the School of Mādhava, whose last intellectual descendent, Śaṅkara, in about 1550 wrote a careful and complete exposition of the steps by which his great predecessor had derived all of the series for trigonometrical functions. This is a beautiful piece of work, clear and logical — and is ideal for instructing students in the proper way to do mathematics, though it does not start from axioms or offer proofs. So far as we know, Śaṅkara never had any students till his work was finally printed just fifteen years ago, in 1975.

Medieval Indian astronomers and mathematicians, then, served their society well, but the fields that they cultivated as defined by Western tradition were not well served by the expectations and attitudes of that society. What they did and the ways in which they did it ceased to be attractive once India decided to adopt the values and the methods of modern Western science and its most persuasive by-product, technology. For the historian of science and of intellectual history in general the Sanskrit scientific texts provide a marvellously rich field to explore; for the modern Western scientist they should serve as a reminder of the need for open-mindedness, imagination, and communication.

INDIAN RECEPTION OF MUSLIM VERSIONS OF PTOLEMAIC ASTRONOMY

David Pingree

The history of astronomy, as of all other aspects of human thought, is extraordinarily complex. Much of that complexity is reflected in the subject I am about to address. For both Indian and Muslim conceptions of the forms of the heavens and earth and of the mathematics by which their several motions may be described originated in Hellenistic astronomy, but each descended through various different cultural milieus to become transformed into models, parameters, and mechanisms barely recognizable to each other. In this paper, expanding on what I have previously written on this subject[1] but striving not to repeat excessively what has already been said, I intend to examine how some Indians attempted to make the Muslim interpretation of Ptolemy palatable to their fellows, who frequently dismissed it as foreign rubbish,[2] while others tried to use elements of it simply to buttress the, for them, naturally declining system revealed at the beginning of the yuga by the divine knowledge of the Sun.

For in the tradition of the astronomies of India it had become, by the sixteenth century in Northern India, important to many scientists to emphasize the origin of one's pakṣa or school in a revelation granted by either a divinity or an ṛṣi.[3] The main rivals in the resulting wars of revelations were Brahmā, the creator and recreator of this universe, from whose *Paitāmahasiddhānta*[4] both the Āryapakṣa of

[1] See D. Pingree, "Islamic Astronomy in Sanskrit," *Journal for the History of Arabic Science* 2 (1978): 315–330 [henceforth "Islamic Astronomy"]; and id., "History of Mathematical Astronomy in India," in *Dictionary of Scientific Biography*, vol. 15 (New York, 1978), pp. 533–633, esp. pp. 625–629 [henceforth "History"].

[2] See, e.g., Munīśvara and Gadāhara cited in "Islamic Astronomy," pp. 321–322.

[3] See, e.g., Nṛsiṃha and Kamalākara cited in D. Pingree, "The Purāṇas and Jyotiḥśāstra: Astronomy," *Journal of the American Oriental Society* 110 (1990): 274–280 at 279 [henceforth "The Purāṇas"].

[4] See D. Pingree, *Census of the Exact Sciences in Sanskrit*, Series A, vols. 1–5 (Philadelphia, 1970–1994) [henceforth *CESS*], A4, 259a; and "History," pp. 555-565.

Āryabhaṭa[5] and the Brāhmapakṣa of Brahmagupta[6] truthfully claimed descent, and Sūrya, the Sun god, whose *Sūryasiddhānta*,[7] which had been updated in the sixteenth or seventeenth century,[8] was in fact principally based on the Ārdharātrikapakṣa of Āryabhaṭa,[9] itself a modification of his Āryapakṣa. There are several other divinities and ṛṣis who are quoted as authorities; but, in fact, all of the siddhāntic tradition of cosmology, geography, and mathematical astronomy goes back to Indian adaptations in the fifth century of Greek models and parameters altered to fit existing Indian theories expressed in the Purāṇas.[10]

The cosmology (khagola) of the siddhāntas conceived of the universe (insofar as we can perceive or deduce it) to consist of nine internested spheres, one for each of the seven planets in the Hellenistic order, the eighth bearing the nakṣatras, and the ninth being the sphere of heaven. The distances of the planetary spheres from each other is based on the theory, adumbrated by Plato, that the distance of each from the center of the earth is inversely proportional to its planet's mean velocity. While the inner eight spheres are rotated daily by the pravaha wind which is wrapped around their common axis, each planet moves independently on its own concentric orbit in a motion that is irregular because of the pulls exerted on the planet by demons stationed at the uccas on its manda and śīghra epicycles. The forces that move the celestial bodies, then, are material beings, whose execution of their self-appointed tasks is certainly not eternal, since this cosmos within the Brahmāṇḍa is destroyed and recreated to the rhythm of a Kalpa of 4,320,000,000 years; nor is it necessarily constant, since the world declines drastically over the course of a Mahāyuga of 4,320,000 years. These ideas allowed Indian astronomers, if they so chose, to justify the introduction of foreign models

---

[5] See *CESS* A1, 53a–53b; A2, 15b; A3, 16a; A4, 27b; and A5, 14a–15a; "History," pp. 590–593; and D. Pingree, "Āryabhaṭa, the Paitāmahasiddhānta, and Greek Astronomy," *Studies in History of Medicine and Science* 12, no. 1–2, NS (1993): 69–79.

[6] *CESS* A4, 254b–255b, and A5, 237b; and "History," pp. 565–580.

[7] A full bibliography of the *Sūryasiddhānta* will appear in *CESS* A6; for now see D. Pingrée, *Jyotiḥśāstra*, (Wiesbaden, 1981), pp. 23–24; and "History," pp. 608-610.

[8] "History," pp. 617–618.

[9] "History," pp. 602–608.

[10] See "The Purāṇas" and D. Pingree, "The Recovery of Early Greek Astronomy from India," *Journal for the History of Astronomy* 7 (1976): 109–123.

and parameters, which could be regarded as representing the degeneration of what had existed and been described by a god or an ṛṣi at the beginning of the yuga. But this cosmology precluded most from embracing Aristotelian concepts of natural motion. Though the Indians like Aristotle had five elements (mahābhūtas), they—earth, water, air, fire, and space—all permeate the entire cosmos; in this the Indians were closer to Plato than to his pupil. Muslim astronomers, of course, being devotees of the Stagirite, believed in a radical difference between the sublunar world of naturally linear motion and celestial spheres of naturally circular motion, which ought as well, at least in principle, to be uniform and concentric.

The Indian sphere of the earth (bhūgola) was dominated by the enormous Mt. Meru at the North Pole around which spread over all of the Northern hemisphere the inner continent called Jambūdvīpa, the island of the rose-apple tree, with Bhāratavarṣa or India and Laṅkā to the South (the city which Muslim astronomers of the ninth century had already learned to call the qubbat al-arḍ or Cupola of the Earth), Romaka or Rome to the West, the Siddhas to the North, and Yamakoṭi, the Castle of Yama, to the East. Around Jambūdvīpa flows the ocean of salt water; and the Southern hemisphere is covered by alternating rings consisting of the six remaining continents and six remaining oceans of the Purāṇas, while at the South Pole lies the Vaḍavāmukha, the Mare's Mouth.

As I indicated with respect to the qubbat al-arḍ, this Indian cosmology and geography were familiar to Muslims in the late eighth and the ninth centuries through Arabic translations of both Pahlavī and Sanskrit astronomical texts.[11] But even in this earliest period of Islamic astronomy the Ptolemaic system was also known—for instance, in the work of Māshā'allāh preserved for us in a Latin translation by Gerard of Cremona as the *De elementis et orbibus cœlestibus*, though he also describes Indian models for the planets.[12] This is the earliest treatise in Arabic that can be said to belong to the class of texts called ᶜilm al-hay'a. Later members of this class include the *Kitāb al-tadhkira fī ᶜilm al-hay'a* composed in Arabic by

[11] See D. Pingree, "The Greek Influence on Early Islamic Mathematical Astronomy," *Journal of the American Oriental Society* 93 (1973): 32–43.

[12] See D. Pingree, "Māshā'allāh: Some Sasanian and Syriac Sources," in *Essays on Islamic Philosophy and Science*, ed. G. F. Hourani (Albany, N.Y., 1975), pp. 5–14, esp. pp. 9–12.

Naṣīr al-Dīn al-Ṭūsī in 1261[13] and the *Risālah dar hay'ah* written in Persian by ᶜAlī al-Qūshjī in the 1450s or 1460s.[14] Both were important for the transmission of Islamic astronomy to India; and both contain in general the same sort of information. The *Tadhkira* is divided into four books containing, respectively, mathematical and physical principles; the configuration of the celestial spheres; geography; and the sizes and distances of the earth and the celestial bodies. The *Risālah dar hay'ah* in an introduction and two books treats in far simpler form the same material (in the way that al-Qūshjī conceives of it) as is contained in the first three books of the *Tadhkira*. The readers of these two works, then, are introduced to basic definitions of Euclidean geometry; the five elements and the Aristotelian principles of their motions (Aristotelian principles are omitted by al-Qūshjī); the arrangement of the celestial spheres, of which there are nine as in the Indian tradition, and the spheres of the four sublunar elements; the great and smaller circles on the celestial spheres that are used in astronomy; the solar, lunar, and planetary models and parameters; geography insofar as it is related to astronomy (e.g., the seven climata and terrestrial longitudes and latitudes); and, at least in the *Tadhkira*, the sizes of the earth and of the celestial bodies, and the distances of the latter from the earth. Clearly the main elements of the Islamic tradition that Indian astronomers would find difficulty in receiving are the Aristotelian notions of the causes of physical motions and those features of the arrangement of the spheres and of the planetary models that depend on the Islamic interpretation of how Aristotelian philosophy and physics impose changes in Ptolemy's system.

The translation of Arabic or Persian astronomical texts into Sanskrit presupposes the existence of bilingual individuals and, at least ideally, of technical dictionaries. The oldest Persian-Sanskrit dictionary that we have constitutes the first prakaraṇa of the *Pārasīprakāśa* composed by Kṛṣṇadāsa in the late sixteenth century for the Emperor Akbar.[15] While it contains a number of words that occur in astronomical texts—for example, the names of the planets

[13] Edited by F. J. Ragep, *Naṣīr al-Dīn al-Ṭūsī's* Memoir on Astronomy, 2 vols. (New York, 1993).

[14] An edition of the Persian together with an anonymous Sanskrit translation, the *Hayatagrantha*, is being prepared by D. Pingree and K. Plofker.

[15] See *CESS* A2, 57a–57b; A4, 61b; and A5, 49a.

and words designating measures of time—it is bereft of all of the technical vocabulary of mathematical astronomy. More detailed information concerning the Persian calendar, and, after an excursus on the technical terms of astrology, the Persian words used in arithmetic, trigonometry, and astronomy can be found in a second *Pārasīprakāśa*, that written by Mālajit in 1643, a work for which the title Vedāṅgarāya was bestowed on him by Shāh Jahān.[16] However, in the manuscripts that I have examined,[17] this unpublished text contains only the vocabulary contained in the *Risālah dar hay'ah* up to the end of the third bāb of the first maqāla; the next bāb takes up the planetary models, whose technical vocabulary Mālajit does not discuss.

The date of the Sanskrit translation, entitled *Hayatagrantha*, of al-Qūshjī's *Risālah dar hay'ah* and the name of the translator are both unknown.[18] The earliest dated manuscript that has been discovered, though it is now lost, was copied in 1694, during the long and unpleasant reign of Aurangzeb. I suspect that the translation was made earlier in the seventeenth century, under Jahāngīr or Shāh Jahān, but that dating can not as yet be either confirmed or denied. I should only report that transliterations found in the *Hayatagrantha's* prose are not always identical with those in the Vedāṅgarāya's poetry, but that this fact obviously can not be used to substantiate a claim that the Vedāṅgarāya did not use the *Hayatagrantha*.

The *Risālah dar hay'ah* is not a philosophically oriented text, so that there are in it no direct statements about the laws of Aristotelian physics as there are in the *Tadhkira*. Therefore, the anonymous translator had no difficulty beyond the linguistic in converting it into Sanskrit. He was clearly helped by a collaborator who was versed in Persian and Islamic astronomy, at least at a level sufficient for understanding the *Risālah*; this is clear from the existence of small additions meant to explain the Persian text to the Sanskrit reader. Moreover, a few phrases in the Persian text, mostly of a pious nature, which were felt to be inappropriate for a Hindu audience, were omitted. Otherwise, everything is straightforward, including the planetary models with all their normal Muslim solid spheres sur-

---

[16] See *CESS* A4, 421a–421b; and A5, 305b–306a.

[17] MS British Library Add. 14,357b; MS London, India Office Library 2114d; and MS Sarasvatī Bhavana, Benares 35337 (with his own ṭīkā, the *Pañjikā*).

[18] See "Islamic Astronomy," pp. 326–328; and *CESS* A4, 57a–57b.

rounding eccentrics and equants, their epicycles, and long explanations of how they function to produce the apparent motions of the Sun, the Moon, and the planets. The collaborator is even learned enough to add what is not in the Persian, the information that ᶜallāma Qūshjī, the son of Ulugh Beg's teacher, determined the obliquity of the ecliptic to be 23;30,17°.

However, sometime in the eighteenth century an unknown astronomer interfered with the original translation by adding material, some from the *Sūryasiddhānta* and some from his own wit. So, after al-Qūshjī reports that he observed the Sun's apogee to be at Cancer 2;26° in Muḥarram of 841 AH (July 1437 A.D.), the interpolator computes that, starting from the *Sūryasiddhānta's* solar apogee at Gemini 18° and using al-Qūshjī's rate of precession, 1° in 70 years, the Sun's apogee was in Cancer 2° in 1178 A.H. (which began on 1 July 1764).[19] If the interpolator's date can be fixed from this as about 1765,[20] his location is indicated by his several references to Kāśī—i.e., Benares. In general, the interpolator is someone familiar with both Persian astronomy and the *Sūryasiddhānta*; his interventions are intended to provide comparisons between the two systems.

I do not wish to discuss here the controversy that took place in Kāśī during the reign of Shāh Jahān between two rival families of astronomers, one of which incorporated elements of Islamic astronomy into their otherwise Indian siddhāntas, and the other of which often but not consistently vigorously opposed such practices. I have written at some length—though by no means exhaustively—about this conflict elsewhere.[21] Rather, I would like to examine part of the *Sarvasiddhāntarāja* composed by Nityānanda at Delhi in 1639.[22] Previously, Nityānanda had translated Farīd al-Dīn Ibrāhīm al-Dihlawī's *Zīj-i Shāh Jahān* into Sanskrit for Āsaf Khān;[23] the epoch of that work was in the year in which Shāh Jahān began to reign, 1628. Since the *Zīj-i Shāh Jahān* was based on the *Zīj* of

---

[19] *Hayatagrantha*, ed. V. Bhaṭṭācārya, (Vārāṇasī, 1967), p. 69.

[20] Note that the one Benares manuscript that does not contain the interpolator's remarks, Sarasvatī Bhavana 36934, was copied by Nāgeśa on 16 September 1765.

[21] "Islamic Astronomy," pp. 320–323.

[22] "Islamic Astronomy," pp. 323–326, and *CESS* A3, 173b–174a; A4, 141b; and A5, 182a. I have used the manuscript, now γ 550, in the Wellcome Institute in London.

[23] See *CESS* A3, 173b; A4, 141a–141b; and A5, 182a.

Ulugh Beg as was al-Qūshjī's *Risālah*, Nityānanda's models and parameters in the *Sarvasiddhāntarāja* generally agree with those in the *Hayatagrantha*. But whereas the latter makes no attempt to disguise its foreign origin—all technical terms are first transliterated from the Persian, and then explained in Sanskrit—Nityānanda has felt it necessary, or at least useful, to adopt several interesting stratagems to seduce his readers into believing that he is impeccably orthodox. It is this aspect of his work that I wish at present to review.

After an unexceptional verse in which he pays his homage to Brahmā, he cleverly associates astronomical systems described by gods and by ṛṣis with that proclaimed by Romaka—by which appellation he means both the author of the third-century *Romakasiddhānta* summarized by Varāhamihira in his *Pañcasiddhāntikā*[24] and the Romaka or Muslim whom, as we shall see, he pretends to be the ultimate human authority for Ulugh Beg's astronomy:

> śrīsūryasomaparameṣṭhivasiṣṭhagargā-
> cāryātriromakapulastyaparāśarādyaiḥ||
> tantrāṇi yāni gaditāni jayanti tāni
> sphurjaddhiyā gaṇitagolasphuṭāni||

> Those treatises are victorious which are accurate in
> mathematics and spherics because of (their
> authors') flashing intelligence (and) which were
> proclaimed by (the gods) Sūrya, Soma, and Brahmā
> (and by the ṛṣis) Vasiṣṭha, Garga, Atri, Romaka,
> Pulastya, and Parāśara.

Having thus inserted Romaka among the ṛṣis, he in the next verse praises the human Bhāskara, who like Maya, Āryabhaṭa, and Brahmagupta, followed the pakṣa of Brahmā. And next he claims that in general his efforts have been directed toward investigating Bhāskara's treatise without the modifications added through their own intelligence by others who are far distant from the siddhāntas composed by divinities such as are the Sun and the Moon, siddhāntas filled with good applications of arithmetic, the solution of indeterminate equations, algebra, and the arrangement of the spheres.

---

[24] See *CESS* A5, 562a–562b.

After several more verses proclaiming his intent to produce a simpler text while staying within the Indian tradition, Nityānanda states that he is writing the *Siddhāntarāja* after investigating what he calls the *Romakasiddhānta* (that is, Ulugh Beg's *Zīj*), the *Sūryasiddhānta*, and the *Brāhmasphuṭasiddhānta* of Brahmagupta, of which trio he claims that the Romaka agrees best with observation, though men know that the *Sūryasiddhānta* is like a Veda and that the *Brāhmasphuṭasiddhānta* contains useful methods. He then plunges into the story originally employed in the introduction to an early recension of the *Jñānabhāskara* cast in the form of a dialogue between Sūrya and his charioteer, Aruṇa—a dialogue to which Nityānanda directly refers. According to this story, the Sun, because of the curse of Brahmā, became a Yavana (in the seventeenth century Yavana meant Muslim) in the city of Romaka and was known as Romaka. After the curse was lifted, he became the Sun again, and wrote the *Romakasiddhānta* "which has the form of revelation (śrutirūpam)." It is this work in its entirety that Nityānanda claims now to be repeating.

Our author, then, has substituted the Sun god, Sūrya, for Farīd al-Dīn Ibrāhīm, and he has elevated the science of the Mlecchas to the level of a Veda. This theme is expanded by his later assertion that, though the siddhāntas produced by the gods and ṛṣis are phrased differently, the astronomy is always the same; they all follow the *Sūryasiddhānta*. But human authors can and do err from the divine path. However, the gods and the munis present an astronomy designed to be applicable to dharmaśāstra—i.e., to the determination of the proper times for performing rituals and for observing festivals—and applicable to astrology, while humans deduce certain parameters from observation. Nityānanda's solution to the dilemma of having contradictory theories is typical in India: one is to use both, each in the situations appropriate to it. There exist a plurality of truths, each of which has its proper application. Nityānanda expresses his general criterion of truth in the following verse:

> yad yad uktam ṛṣibhiḥ kila devais
> tat tad atra sakalaṃ saphalaṃ hi||
> pūruṣair aviditāgamatattvaiḥ
> kṣiptam ūharahitaṃ tad asatyam||

> Whatever was said by the ṛṣis and the gods, all of that is here; it is indeed fruitful. That which is added by men who do not know the truths of the sacred books, that which is lacking investigation, is false.

Nityānanda, then, accepts the siddhāntas composed by gods or ṛṣis as true regardless of what they say, while asserting that everything they say is in agreement with the *Sūryasiddhānta*; and he will accept what men say in addition only if it results from investigation, i.e., is properly inferred in accordance with the rules of traditional Indian philosophy, which include perception as one of the bases for valid knowledge.

In the first chapter, then, Nityānanda has established the orthodoxy of his composition of a siddhānta expounding Islamic astronomy by placing Romaka, his stand-in for Farīd al-Dīn, among the ṛṣis; by making him an incarnation of the Sun; by proclaiming the peaceful coexistence of mutually contradictory truths; and by taking the normal Indian position that, while the gods and ṛṣis always speak the truth, man through his own intelligence may discover additional truths that have their proper applications. In the second chapter he endeavors further to justify what he is doing by employing the traditional Indian divisions of time, including the theory of Kalpas, Mahāyugas, and Yugas, as a framework within which to transform Islamic into Indian astronomy. For he converts the Muslim perpetual mean motions of the planets and the longitudes of their apogees and nodes into the standard Indian form of integer numbers of rotations in a Kalpa; though, because they do not properly fit, he is obliged to add bījas or corrections at the end, another traditional Indian device. These bījas, he claims, were determined by observation. To emphasize how close the mean longitudes computed according to the *Zīj-i Shāh Jahān* are to those computed with Indian methods, he compares them with results from the *Sūryasiddhānta* and the Brāhmapakṣa; indeed, they are not very different, as is not very surprising.

In the next chapter Nityānanda compares the parameters of the solar, lunar, and planetary models in the so-called *Romakasiddhānta* with those in the *Sūryasiddhānta* and those in Brahmagupta's *Brāhmasphuṭasiddhānta*. The numbers are simply juxtaposed, without comment. This is followed by an elaborate and lengthy explanation of the finding of the Sine of an arc, the sum or the difference of the Sines of two given arcs, and so on, culminating in the construction of a table of the Sines for every minute of arc between 0° and 90°. All this, of course, represents Ulugh Beg's and al-Kāshī's amplification of the Sine function originally introduced by Indian astronomers in the fourth or fifth century A.D.

Nityānanda proceeds to instruct his reader on how to compute a longitude or a latitude with the Romaka's models. These are indeed the models of the Islamic Ptolemaic tradition with the crank-mechanism and the prosneusis of the Moon, the double eccentricity and the equant of the superior planets and Venus, and the triple eccentricity and crank-mechanism of Mercury. But the geometrical solution for finding the planet's longitude according to each of these models is simply a modification of the methods employed in the siddhāntas for single or double epicycle models.

Having made his points that the Romaka is of impeccably divine origin, that its results are close to those obtained by an already internally divided Indian tradition, and that its innovations are useful, Nityānanda introduces a miniature ᶜilm al-hay'a text discussing geometrical principles, the configuration of the four elemental and nine celestial spheres, and the models of the Sun, the Moon, and the planets. This is followed by elaborate directions for drawing diagrams of these models on a wall in order to instruct one's students. These are, of course, the diagrams familiar to us from both Arabic and Persian manuscripts and their imitations in medieval Latin codices and fifteenth- and sixteenth-century printed books, though the three manuscripts of the *Sarvasiddhāntarāja* that I have been able to examine are not illustrated,[25] though spaces are left for some diagrams in the Wellcome's copy.

A text which does have such diagrams, at least in the two manuscript copies that I have consulted,[26] is the *Jyotiḥsiddhāntasāra* composed by a Mālavīya Brāhmaṇa named Mathurānātha Śukla.[27] He was teaching astronomy to school-children in Kāśī when he was requested by a Rāja, Ḍālacandra, to write a book on the subject. He completed the present work in 1782 and added a commentary which

[25] Besides the Wellcome manuscript, I inspected manuscript 206 of A1883/84 at the Bhandarkar Oriental Research Institute, Poona, and manuscript 2619 at the Rajasthan Oriental Research Institute at Alwar.

[26] These are R.15.124 and R.15.125 at Trinity College, Cambridge.

[27] See *CESS* A4, 349a–350a; and A5, 272b.

presumably helped his students as much as it does us to understand the full details of his adaptation of an as yet unidentified work on hay'a, though its parameters in general are again those of Ulugh Beg.

Following the standard pattern of an ᶜilm al-hay'a text, Mathurānātha begins his khagolavicāra with geometry and the Aristotelian physical principles of motion. There follow descriptions of the celestial spheres, a Ptolemaic-style catalogue of 1025 stars arranged in 48 constellations (missing but alluded to in both the manuscripts available to me), descriptions with parameters of the Islamic Ptolemaic models for the Sun, the Moon, and the planets, latitude theory, the heliacal risings and settings of the planets, and solar and lunar eclipses. The bhūgolavicāra describes the seven climata with their maximum daylights and median terrestrial latitudes, the coordinates used in Islamic mathematical geography, particulars of the seven regions, sunrise and twilight (he knows Ibn Muᶜādh's estimation that it begins or ends when the Sun is 18° below the horizon), some elements of the Phārisī, Rūmī, and Mālikī calendars with remarks on the year of the Phiraṅgis, methods for determining the time of day, the establishing of the distance between two localities by means of simultaneous observations of a lunar eclipse, and the dimensions of the universe measured in farsangs. Though Mathurānātha when necessary uses a Persian word, overwhelmingly his vocabulary is taken from the Sanskrit siddhāntas; and from time to time he inserts, especially into his commentary, information about traditional Indian views concerning the subject that is being discussed in his Persian source. It is not entirely clear what his (or the Rāja's) purpose was in instructing the students, who must have been Brāhmaṇas, in the basic elements of Persian astronomy. It is difficult to believe that they wished the younger generation of Hindus to become more tolerant of Muslims through a knowledge of their astronomy, but other possible motives do not immediately present themselves.

The motivation of Mathurānātha's predecessor, Jayasiṃha,[28] in studying Islamic astronomy is much clearer. Despite all the enthusiasm that he continues to arouse as the man who introduced "modern" science into India, he was in fact a very devout and pious Hindu who believed firmly that the siddhāntas attributed to the gods

[28] See *CESS* A3, 63a–64b; A4, 97b; and A5, 115b–116a.

and the ṛṣis, but particularly the *Sūryasiddhānta*, are true. It is for this reason that he had his pandits write a *Sūryasiddhāntavyākhyā* describing and defending that work's cosmology and planetary models while explaining away any observed defect,[29] and it is for this reason that he had his Jyotiṣarāya, Kevalarāma,[30] write the *Brahmapakṣanirāsa*[31] in which he attempts to show that the tradition of the *Paitāmahasiddhānta* of the *Viṣṇudharmottarapurāṇa*, the *Brāhmasphuṭasiddhānta* of Brahmagupta, and the *Siddhāntaśiromaṇi* of Bhāskara does not have a solid foundation since they are not based on the words of either a god or an ṛṣi. Only the *Sūryasiddhānta* has such a basis, and so do those other works by gods and ṛṣis which are the occasional expressions of the theories of Sūrya in varying verbiage. Clearly this is a view close to that voiced by Nityānanda. And we must conclude that like Nityānanda Jayasiṃha allowed the epithet "true" to be accorded to valid inferences from observed phenomena. Unlike his seventeenth-century predecessor, however, Jayasiṃha did not adapt in Sanskrit the Islamic Ptolemaic models of the universe or of the mechanisms producing the motions of the Sun, the Moon, and the planets. The opening verses of Kevalarāma's *Brahmapakṣanirāsa* correctly describe his basic attitude:

> vedāḥ sampāditā yena teṣāṃ vistāritāḥ kriyāḥ||
> varṇāśramavibhāgaś ca viluptaḥ sthāpitaḥ punaḥ||
> pātitāḥ puṣkare tīrthe paurohityāt padodbhavāḥ||
> śrīsavāīnareśas tu śrautasmārtārthasiddhaye||
> anarham ity asau jñātvā brahmapakṣaṃ nirāsyati||

> He who caused the Vedas to succeed has spread abroad their rituals; the division of the varṇas and the stages of life had slipped, but are restored again; because of his being a purohita the lotuses are caused to fall in the pilgrimage site at Puṣkara lake. In order to achieve the meaning of śruti and smṛti (the Vedas and dharmaśāstra) the king, Savāī

---

[29] I have used manuscript 29498 of the Rajasthan Oriental Research Institute at Jodhpur.

[30] See *CESS* A2, 63a–63b; A4, 63b; and, most importantly, A5, 52a–52b.

[31] I have used manuscript 28628 of the Rajasthan Oriental Research Institute at Jodhpur.

> (Jayasiṃha), recognizing that the Brahmapakṣa is unworthy, annihilates it.

There is not a hint in the *Brahmapakṣanirāsa* of the existence of Islamic, European, or any other Mleccha astronomy.

But Jayasiṃha did, for his own purposes of correcting parameters in this decaying universe,[32] construct observatories and sponsor the translations into Sanskrit of Theodosius' *Spherics*, Euclid's *Elements*, Ptolemy's *Almagest*, and at least parts of some contemporary European works on astronomy. He also had translated, by Nayanasukha with the help of a Persian assistant, Muḥammad Ābidda, the Arabic commentary by al-Birjandī on the eleventh chapter of the second book of Naṣīr al-Dīn's *Tadhkira*[33]—the chapter in which al-Ṭūsī discusses his new models employing the Ṭūsī-couple—while al-Birjandī, in his sharḥ, reports at length on the criticisms of Ptolemy issued by Ibn al-Haytham and on some of the work done by Persian astronomers after Naṣīr al-Dīn, notably by Quṭb al-Dīn al-Shīrāzī in his *Al-tuḥfa al-shāhiyya* and in his *Nihāyat al-idrāk*. The translator and his Persian assistant obviously discussed the meaning of many passages in this dense and difficult book; in some cases they decided that an expansion was necessary to render the text meaningful, and a few times they despaired of rendering a translation and simply omitted a difficult passage. But in general they performed their task well; highly technical Arabic terms are retained in their Persian forms in Sanskrit transliteration, but usually with an explanation of their meaning when they are introduced. An intelligent reader could certainly have made sense of this text; but, so far as we know, it had no readers at all. The unique manuscript was copied by one of Jayasiṃha's scribes, Kṛpārāma, in 1729.[34] This was probably the first copy of Nayanasukha's draft, and, as far as we know, no other copy was ever made. Jayasiṃha received this most important document of the Marāgha School into his library in 1730, but never into his astronomy.

---

[32] Concerning Jayasiṃha's attitude, see D. Pingree, "Indian and Islamic Astronomy at Jayasiṃha's Court," in *From Deferent to Equant* (New York, 1987), pp. 313–328, esp. pp. 315–318.

[33] An edition of the Arabic original of al-Birjandī's sharḥ on *Tadhkira* 2,11 together with Nayanasukha's Sanskrit translation is being prepared by T. Kusuba.

[34] This is manuscript 46 in the collection of the Palace Museum at Jaipur.

Our two most intelligent and informed authors, Nityānanda and Jayasiṃha, have unwittingly exposed a set of basic differences between the Hindu and the Muslim scientific views of astronomy. The Muslim believes, as a Greek would, in the uniformity of nature over distances in time and space, while the Hindu believes that the universe decays over time and that the planets may move differently over Bhāratavarṣa and over the Mlecchas. Moreover, the heavenly spheres perform social functions so that one theory of their behavior is valid for one human purpose, another for another. These philosophical differences over the nature of nature itself made it impossible for a good Brāhmaṇa such as Nityānanda or a good Kṣatriya such as Jayasiṃha to receive more of the Muslim interpretation of Ptolemy than some new parameters and the planetary models bereft of the forces that make them move, but regarded purely, as Proclus viewed those of the *Almagest*, as a means to produce mathematically correct predictions of celestial phenomena. In this they were in agreement with the warmest proponents of Islamic astronomy in seventeenth-century Benares. The mere translation of texts, such as is represented by the *Hayatagrantha* or the *Jyotiḥsiddhāntasāra* of Mathurānātha, was not sufficient to produce an Indian Ulugh Beg.

Yet, in a practical sense, some Indians learned to follow the Muslims in their computations. I offer in evidence of this two manuscripts. One, in Poona, discusses in detail the computation of square-roots, sines, and a gnomon's shadow according to Nityānanda's translation of the *Zīj-i Shāh Jahān*.[35] The second, at Berlin, gives detailed computations of the longitudes and latitudes of the Moon and the planets and the longitude of the Sun at noon in Jayapura on Monday 7 March 1718 according to the *Zīj-i Muḥammad Shāh*.[36] Clearly this was computed some years later, after the *Zīj* had been written. That date of computation is approximated by the computations at the end of the manuscript of a lunar eclipse on Sunday 28 May 1732 and of a solar eclipse visible at Jayapura on Monday 20 May 1734. The whole is accompanied by elaborate diagrams and provides ample evidence that its author is technically competent even if not a convert to the philosophical bases of Islamic astronomy.

---

35 Manuscript 579 of 1895/1902 at the Bhandarkar Oriental Research Institute.

36 Manuscript or. fol. 2973 at the Städtbibliothek.

Bibliography

Bhaṭṭācārya, V., ed. *Hayatagrantha.* Vārāṇasī, 1967.

Pingree, David. *Census of the Exact Sciences in Sanskrit.* Series A, vols. 1–5. Philadelphia: American Philosophical Society, 1970–1994.

——. "The Greek Influence on Early Islamic Mathematical Astronomy." *Journal of the American Oriental Society* 93 (1973): 32–43.

——. "Māshā'allāh: Some Sasanian and Syriac Sources." In *Essays on Islamic Philosophy and Science,* edited by G. F. Hourani, pp. 5–14. Albany, N.Y.: State University of New York Press, 1975.

——. "The Recovery of Early Greek Astronomy from India." *Journal for the History of Astronomy* 7 (1976): 109–123.

——. "History of Mathematical Astronomy in India." In *Dictionary of Scientific Biography,* vol. 15 (1978), pp. 533–633. 18 vols. New York: Charles Scribner's Sons, 1970–90.

——. "Islamic Astronomy in Sanskrit." *Journal for the History of Arabic Science* 2 (1978): 315–330.

——. *Jyotiḥśāstra: Astral and Mathematical Literature.* Wiesbaden: O. Harrassowitz, 1981.

——. "Indian and Islamic Astronomy at Jayasiṃha's Court." In *From Deferent to Equant: A Volume of Studies in the History of Science in the Ancient and Medieval Near East in Honor of E. S. Kennedy.* Annals of the New York Academy of Sciences, vol. 500, edited by D. A. King and G. Saliba, pp. 313–328. New York, 1987.

——. "The Purāṇas and Jyotiḥśāstra: Astronomy." *Journal of the American Oriental Society* 110 (1990): 274–280.

——. "Āryabhaṭa, the Paitāmahasiddhānta, and Greek Astronomy." *Studies in History of Medicine and Science* 12, no. 1–2, NS (1993): 69–79.

Ragep, F. Jamil. *Naṣīr al-Dīn al-Ṭūsī's* Memoir on Astronomy *(al-Tadhkira fī ᶜilm al-hay'a).* 2 vols. New York: Springer-Verlag, 1993.

INDIAN ASTRONOMY IN MEDIEVAL SPAIN

David Pingree

Indian astronomy entered early Islam through several routes[1]. The *ārdharātrikapakṣa* that had been initiated by Āryabhaṭa of Kusumapura in about 500 influenced the *Zīk-i Shāhriyārān* composed in Pahlavī at the Sasanian court in Ctesiphon under Khusrō Anūshirwān in 556. This is known to us now through the astrological computations made by Māshā'allāh ibn Atharī, a Persian Jew from Baṣra, for the several astrological histories that he wrote in the decades before and after 800[2]. From the works of Māshā'allāh extant in Arabic, Greek, and Latin, and from the discussions of them by al-Hāshimī in his *Kitāb ʿilal al-zījāt* and by al-Bīrūnī in his *Kitāb tamhīd al-mustaqirr li-maʿnā al-mamarr*[3] we know that Khusrō's *zīk* utilized the *ārdharātrika*'s apogees of Saturn, Jupiter, and the Sun; and its equations of the center of Jupiter and, within one minute, Mars and the Sun. This *Zīk-i Shāhriyārān* already employed the Persian calendar, and was probably influenced by Ptolemy[4].

The *zīk* of Khusrō strongly influenced that produced under the last of the Sasanian Shāh-i Shāhs, Yazdijird III, which was translated into Arabic as the *Zīj al-Shāh* by al-Tamīmī in about 800. This also is now lost, but much concerning it may be learned from the *Zīj al-mumtaḥan* (*Tabulae Probatae*), the *zīj*es of Ḥabash al-Ḥāsib, and again, al-Hāshimī and al-Bīrūnī[5]. While some modifications were made in the parameters — *e.g.*, in Mars' equation of the center — the structure and calendar of

---

[1]. For a general survey of the process by which Indian astronomy was introduced into Islam see D. Pingree, "The Greek Influence on Early Islamic Mathematical Astronomy", *JAOS* 93, 1973, 32-43.

[2]. E. S. Kennedy and D. Pingree, *The Astrological History of Māshā'allāh*, Cambridge 1971, pp. 69-88.

[3]. F. I. Haddad, E. S. Kennedy, and D. Pingree, *The Book of the Reasons behind Astronomical Tables*, Delmar, N. Y. 1981, pp. 212-216, 262-264, and 283-284.

[4]. D. Pingree, "Indian Influence on Sassanian and Early Islamic Astronomy and Astrology", *JOR Madras* 34-35, 1964-66 (1973), 118-126.

[5]. Haddad, Kennedy, and Pingree, *op. cit.*, pp. 213-216, 220, and 262-263.

Khusrō's *zīk* were retained, and the influence of Ptolemy may even have been expanded.

In the early eighth century the earliest Arabic *zījes* were composed on the basis of Indian methods: the *Zīj al-Jāmi*ᶜ and the *Zīj al-Hazūr* at Qandahar, the *Zīj al-Arkand* at Sind in 735 largely on the basis of the *ārdharātrikapakṣa* as expounded in Brahmagupta's *Khaṇḍakhādyaka*, and the *Zīj al-Harqan* on the basis of the *āryapakṣa* of Āryabhaṭa's *Āryabhaṭīya* in 742[6]. None of these was wildly popular outside of Sind and Afghanistan; we know of them only through al-Hāshimī and al-Bīrūnī. More significant for the transmission of Indian astronomy to the Islamic world was the translation of a Sanskrit text, apparently entitled *Mahāsiddhānta*, into Arabic at the court of al-Manṣūr in Baghdād in the early 770's[7]. Two of those involved in the popularization of this translation, the *Zīj al-Sindhind*, were Muḥammad ibn Ibrāhīm al-Fazārī[8], who, according to one story, was involved in the translation itself, and Yaᶜqūb ibn Ṭāriq[9].

Al-Fazārī, in about 775, wrote on the basis of the *Zīj al-Sindhind*, the *Zīj al-Shāh*, and the Ptolemaic tradition (probably as represented by the *Handy Tables* in their Pahlavī version)[10], a *Zīj al-Sindhind al-kabīr*. Basically, al-Fazārī's mean motions of the planets (including the Sun and the Moon), their nodes, and their apogees were derived from the *brāhmapakṣa* through Brahmagupta's *Brāhmasphuṭasiddhānta*, the equations of the center and of the anomaly from the *Zīj al-Shāh*, and the table of declinations of the Sun from the *Handy Tables*; other computations, involving three different values for the radius, R, in a

---

6. Haddad, Kennedy, and Pingree, *op. cit.*, pp. 207-211.

7. Haddad, Kennedy, and Pingree, *op. cit.*, pp. 216-223.

8. D. Pingree, "The Fragments of the Works of al-Fazārī", *JNES* 29, 1970, 103-123.

9. D. Pingree, "The Fragments of the Works of Yaᶜqūb ibn Ṭāriq", *JNES* 27, 1968, 97-125, and E. S. Kennedy, "The Lunar Visibility Theory of Yaᶜqūb ibn Ṭāriq", *JNES* 27, 1968, 126-132.

10. For the use of the *Handy Tables* in the computation of the prorogation in a Sasanian horoscope see *Dorothei Sidonii Carmen astrologicum*, ed. D. Pingree, Leipzig 1976, III 1, 30-65.

sine-table — 150 from the *Khaṇḍakhādyaka*, 3270 from the *Brāhmasphuṭasiddhānta*, and 3438 from the *Paitāmahasiddhānta* and the *Āryabhaṭīya* — were derived from various Indian and Sasanian sources. Al-Fazārī in some instances — *e.g.*, in the mean motions of Saturn and of the lunar node, and in some of the equations — differs from his identifiable sources. Some of his tables, as had been Ptolemy's for Chords and declinations, were entered with arguments of 0;30°, while the equation tables imitated the Indian practice of intervals of 90°/24 or 3;45° (and its multiples 7;30° and 15°). Though the *brāhmapakṣa* does not allow for precession, al-Fazārī used Ptolemy's value of 1° per 100 years. It appears that al-Fazārī's *Zīj al-Sindhind al-kabīr* followed the *Zīj al-Shāh* in using the epoch of Yazdijird III (16 June 632) and the Persian calendar, though in about 788 he published a *Zīj ᶜalā sinī al-ᶜArab* which utilized the epoch of the Hijra (16 July 622) and the Muslim calendar. In this latter *zīj* the mean motions were given for 1 to 6 *saura* days (a *saura* day is a 360th of a solar year), 6 to 6,0 *saura* days, 1 to 6,0 solar years, etc. In the *Zīj al-Sindhind al-kabīr* the epoch was dawn at Vār-i Yamkart (Bārah), which is noon at Ujjayn (or the Cupola of the Earth).

Al-Fazārī's associate, Yaᶜqūb ibn Ṭāriq, wrote a *zīj*, also in about 775, using the epoch of Yazdijird III, the Persian calendar, and dawn at Bārah; in his table of mean motions the collected years are 10. Unlike al-Fazārī, Yaᶜqūb takes his equations of the center entirely from the *Zīj al-Shāh*. His equations of the anomaly come from the same source except for those of Jupiter and Venus; the latter is from the *ārdharātrikapakṣa*. And his declination table was based on the Indian value of the obliquity, 24°, instead of on Ptolemy's value, as was al-Fazārī's.

Yaᶜqūb also wrote an important work on cosmology, the *Kitāb tarkīb al-aflāk*, apparently in 777/778, and a *Kitāb al-ᶜilal*, in both of which he expounded many Indian astronomical concepts and computations. Neither was, to our knowledge, ever heard of in al-Andalus, but each informs us of elements of the *Sindhind* tradition that were transmitted to Spain.

That transmission was effected primarily through the *Zīj al-Sindhind* composed by Muḥammad ibn Mūsā al-Khwārizmī under al-Ma'mūn in the 820's[11] and the commentary on it written by Aḥmad ibn al-Muthannā' in the tenth century. From a comparison of the extant versions of al-Khwārizmī's *zīj* (the few Arabic fragments; the Samaritan adaptation; and the Latin versions) with the commentaries by Ibn Masrūr[12] and by Ibn al-Muthannā' (this last in Hebrew[13] and Latin[14] versions only) as well as from Abraham ibn Ezra's *De rationibus tabularum*[15] it is possible to arrive at a fairly accurate understanding of the original form of the *Zīj al-Sindhind*[16]. From this analysis it is clear that al-Khwārizmī, while basically following the earlier works of al-Fazārī and of Yaᶜqūb ibn Ṭāriq, furthered the intrusion of Ptolemaic and other material into the tradition. Thus, he began with a discussion of the three calendars used in the Islamic world — that of the Hijra, that of Yazdijird III, and the Seleucid, called the Rūmī or that of Alexander — and gave rules and tables for converting dates in one into the corresponding dates in the other two. However, like al-Fazārī and Yaᶜqūb, al-Khwārizmī used the calendar of Yazdijird III in his *zīj* and also the mean motions of the *Sindhind* tradition. His equations of the center and of the anomaly are virtually identical with those of the *Zīj al-Shāh*, while al-Fazārī had deserted that *zīj* in his equations of the center for the Sun and the Moon, and Yaᶜqūb had in the equations of the anomaly for Jupiter and Venus.

---

11. On al-Khwārizmī see, *e.g.*, F. Sezgin, *Geschichte des arabischen Schrifttums* (henceforth *GAS*), vol. 6, Leiden 1978, pp.140-143.

12. The commentary by Ibn Masrūr, a pupil of the ninth century astrologer, Abū Maᶜshar, is preserved in a manuscript Taymūrīya, Math. 99, in the Dār al-Kutub al-Miṣrīya in Cairo; I am grateful to E. S. Kennedy for supplying me with a film of this text.

13. Edited, translated, and commented on by B. R. Goldstein, *Ibn al-Muthannā's Commentary on the Astronomical Tables of al-Khwārizmī*, New Haven 1967.

14. Edited by E. Millàs Vendrell, *El comentario de Ibn al-Muthannā' a las Tablas Astronómicas de al-Jwārizmī*, Madrid-Barcelona 1963.

15. Edited by J. M. Millàs Vallicrosa, *El libro de los fundamentos de las Tablas astronómicas de R. Abraham ibn ᶜEzra*, Madrid-Barcelona 1947.

16. For its relation to the Indian tradition see D. Pingree, "The Indian and Pseudo-Indian Passages in Greek and Latin Astronomical and Astrological Texts", *Viator* 7, 1976, 141-195, esp. 151-169.

Al-Khwārizmī's computation of the equations and his integration of them follows faithfully an Indian model, though the commentators Ibn Masrūr and Ibn al-Muthannā' attempt to explain his procedure in terms of a Greek eccenter with epicycle model.

In the computation of the equations the Indian Sines were used. The value of the radius, R, used by al-Khwārizmī was that of Brahmagupta's *Khaṇḍakhādyaka*, namely 150. According to Ibn al-Muthannā' he gave two tables of declination, one using the Indian value of the obliquity, 24°, the other the value given by Ptolemy in the *Handy Tables*, 23;51°. For finding right ascensions al-Khwārizmī uses Menelaus' theorem as expounded in the *Almagest*. For the latitudes of the planets and of the Moon as for the first visibility of the lunar crescent al-Khwārizmī follows the *Sindhind* faithfully, while the anomalies tabulated for the planets' first stations are both in fact and according to Ibn Masrūr Ptolemaic — from the *Handy Tables*.

This suffices to show that the process of the mixture of systems had proceeded already quite far in al-Khwārizmī's *zīj*. It contained many parameters and algorithms of Indian origin, derived both from the original *Zīj al-Sindhind* and from the *Zīj al-Shāh* of Yazdijird III, but also contributions from the Ptolemaic tradition, coming both from the *Almagest* and from the *Handy Tables*. The Indian material, however, was largely in the form of unjustified rules, whose bases — *e.g.*, for the computation and integration of the two equations of the planets — had no explanation, and so were necessarily interpreted in the light of Ptolemaic astronomy. In so far as they remained useful, these Indian elements survived; the most lasting contribution of the *Sindhind* tradition, of course, was the trigonometry of Sines, Versines, and Cosines, and some applications of analemmas. But Indian astronomy in its Islamic context was incapable of further development or elaboration since its foundations were not understood; and it virtually disappeared from the mainstream of astronomy in Eastern Islam after al-Battānī, though al-Khwārizmī's *zīj* remained accessible to the unknown founders of Samaritan astronomy[17] and to the

[17]. See D. Pingree, "Al-Khwārizmī in Samaria", *AIHS* 33, 1983, 15-21.

equally obscure compiler of the text which is the basis of a Byzantine treatise of the late eleventh century[18].

The *Zīj al-Sindhind* was brought to al-Andalus not very long after its composition, during the reign of ᶜAbd al-Raḥmān II (821-852)[19]. According to the account of Ibn Saᶜīd, this ruler sent ᶜAbbās ibn Nāṣiḥ to Iraq to acquire books, among which was a copy of the *Sindhind*. In the ninth century a large number of poet astrologers were attached to the court at Cordova, including ᶜAbbās ibn Firnās and Ibn al-Shamir as well as ᶜAbbās ibn Nāṣiḥ. Undoubtedly other astronomical tables besides that of al-Khwārizmī were available to them, including others belonging to the *Sindhind* tradition. Samsó conjectures that this earlier Khwārizmī zīj was a "minor" recension, which was the one adapted by Maslama al-Majrīṭī and Ibn al-Ṣaffār, while a "major" recension, which had demonstrations, was the basis of the commentary of Ibn al-Muthannā'[20]. However, it seems to me that there would have been only one *Zīj al-Sindhind* of al-Khwārizmī, containing canons and tables, and that this was the basis of the commentaries of al-Farghānī, Ibn Masrūr, and Ibn al-Muthannā' (who provided the demonstrations) and of the revision(s) by Maslama and the members of his "school".

The version of the *Zīj al-Sindhind* produced by Maslama al-Majrīṭī at Cordova in the late tenth century is now lost in its Arabic form, though it survives in the Latin translation made by Adelard of Bath in or soon after 1126[21] (there also survives a manuscript containing a revision of

---

18. See A. Jones, *An Eleventh-Century Manual of Arabo-Byzantine Astronomy*, *CAB* 3, Amsterdam 1987.

19. For this account of the introduction of al-Khwārizmī into Spain I rely upon the masterful account by J. Samsó in his *Las ciencias de los antiguos en al-Andalus*, Madrid 1992, pp. 49-56.

20. Samsó, *op.cit.*, pp. 85-87.

21. Edited by A. Björnbo, R. Besthorn, and H. Suter, *Die astronomischen Tafeln des Muḥammed ibn Mūsā al-Khwārizmī*, Copenhagen 1912; translated and commented on by O. Neugebauer, *The Astronomical Tables of al-Khwārizmī*, Copenhagen 1962.

Adelard's translation by Robert of Chester[22]) and a related version apparently due to Petrus Alfonsi[23]. Mercier describes a fragment of a third Latin version of the *zīj*, probably of its form before Maslama's revision[24]; Samsó suggests that this may be the translation, otherwise lost, by Hermann of Carinthia[25]. There were also Arabic recensions produced by two of Maslama's students, Ibn al-Ṣaffār[26] and Ibn al-Samḥ[27]. The astronomical tables of both, following the *Sindhind* tradition, are referred to by Ṣāʿid al-Andalusī[28], as is that of Ibn al-Ṣaffār by Abraham ibn Ezra[29]; fragments of the tables of Ibn al-Ṣaffār survive and are soon to be published[30].

The effects of Maslama's intervention were substantial, but did not suffice to obliterate the traces of Indian astronomy in the *Sindhind* tradition. He changed the calendar and the epoch from those of Yazdijird III to those of Islam, with the result that the collected years for the mean motion tables equal thirty; recalculated various tables for the geographical coordinates of Cordova; and introduced a Sine-table in which the radius,

---

[22]. See most recently R. Mercier, "Astronomical Tables in the Twelfth Century", in C. Burnett, ed., *Adelard of Bath*, London 1987, pp. 87-118, esp. 96-97.

[23]. Mercier, *op.cit.*, pp. 95-96.

[24]. Mercier, *op.cit.*, p. 101.

[25]. Samsó, *op.cit.*, p. 85, fn. 121.

[26]. *GAS*, vol. 6, p. 250.

[27]. *GAS*, vol. 6, p. 249.

[28]. In his *Ṭabaqāt al-umam*. I have used the English tanslation of S. I. Salem and A. Kumar, published under the title *Science in the Medieval World*, Austin 1991; for Ibn al-Ṣaffār see p. 65, for Ibn al-Samḥ p. 64. For the *Ṭabaqāt* see L. Richter-Bernburg, "Ṣāʿid, the *Toledan Tables*, and Andalusī Science", *From Deferent to Equant*, *ANYAS* 500, New York 1987, pp. 373-401.

[29]. *Op.cit.*, (n. 15), pp. 75 and 110.

[30]. Samsó, *op.cit.*, p. 85.

as in Ptolemy's table of Chords, is 60, even though he retained various Indian procedures that presuppose a radius of 150.

After the work of Maslama and his pupils on al-Khwārizmī's *Zīj al-Sindhind* the Indian tradition was next taken up by Ṣāᶜid al-Andalusī in Toledo[31]. Not only did Ṣāᶜid in his *Ṭabaqāt al-umam* of 1068 mention a number of *Sindhind* treatises of the ninth and tenth centuries from the East — Ḥabash's earliest *zīj*[32], al-Nayrīzī's *zīj*[33], Ibn al-Ṣabbāh's *zīj*[34], Ibn al-Muthannā's *Taᶜdīl* of al-Khwārizmī's *zīj*[35], and Ibn al-Ādamī's *Naẓm al-ᶜiqd*[36] (which influenced Ṣāᶜid's uderstanding of trepidation[37]) — but he composed an *Iṣlāḥ ḥarakāt al-kawākib wa al-taᶜrīf bi-khaṭā al-rāṣidīn* in which he defended the *Sindhind* against the assault of his older contemporary, ᶜAbdallāh ibn Aḥmad al-Saraqusṭī[38]. Moreover, Ṣāᶜid was most likely a part of the group of astronomers, including his student al-Zarqālla, who put together the *Toledan Tables*[39] which, in the two Latin versions (the first probably by John of Seville, the second by Gerard of Cremona), greatly influenced Western European astronomy in the late twelfth, the thirteenth, and the early fourteenth centuries. These *Toledan Tables*[40] retain from the Indian material of the *Sindhind* a table of Sines in which the radius is 150 (table 12 of Toomer), the longitudes

---

31. Samsó, *op.cit.*, pp. 144-150.

32. Ṣāᶜid, *op.cit.*, (n. 28), p. 51.

33. *Ibid.*, p. 52.

34. *Ibid.*, p. 52.

35. *Ibid.*, p. 53.

36. *Ibid.*, pp. 46 and 53.

37. Samsó, *op.cit.*, p. 221; and Richter-Bernburg, *op.cit.*, 386.

38. Ṣāᶜid, *op.cit.*, p. 67; see also Richter-Bernburg, *op.cit.*, 377 and 385.

39. Samsó, *op.cit.*, pp. 147-152, and Mercier, *op.cit.*, pp. 104-107; and Richter-Bernburg, 375-376 and 386-390.

40. G. Toomer, "A survey of the Toledan Tables", *Osiris* 15, 1968, 5-174.

of the nodes, the tables of the latitudes of the planets (tables 45-46 Toomer), and formulae relating the apparent diameters of the Sun and the Moon to their velocities. Clearly at this stage trigonometry was the most impressive contribution of Indian science to the West.

But the *Sindhind* was not yet dead. Abū ᶜAbdallāh Muḥammad ibn Ibrāhīm ibn Muḥammad ibn Muᶜādh, who was not known to Ṣāᶜid and who died in 1093, must have composed his zīj at Jaén in about 1080[41]; it is known to us in a Latin translation by Gerard of Cremona, entitled *De diversarum gentium eris, annis ac mensibus, et de reliquis astronomiae principiis*, but commonly called the *Tabulae Jahen*[42]. This work recasts Maslama's version of al-Khwārizmī's *Zīj al-Sindhind* for the geographical coordinates of Jaén, though Ibn Muᶜādh also cites Ibn al-Samḥ[43]. Since the tables of the *Tabulae Jahen* are lost we cannot, in most cases, judge the extent to which Ibn Muᶜādh retained Indian parameters. However, in the surviving canons he states that he has computed the mean longitudes of the planets at epoch (midnight at Jaén on 15/16 July 622) "secundum intentionem Indorum"; that Jaén is 62° West of Arim (Ujjayinī); and that the longitude of the solar apogee is Gemini 17;55°. Moreover, not surprisingly, many of the rules for computing are clearly based on al-Khwārizmī's understanding of the Indian models.

After the year 1100, however, the Arabic tradition of the *Sindhind* in Spain disappears from sight, and we must instead examine the translation of some texts containing Indian astronomical ideas into Latin, Hebrew, and Spanish — a process that continued from the early twelfth century till the late thirteenth. We have already mentioned Adelard of Bath, Hermann of Carinthia, Robert of Chester, and Petrus Alfonsi as translators and transmitters of the *Zīj al-Sindhind* of al-Khwārizmī; and John of Seville and Gerard of Cremona as translators of the *Toledan Tables*. To these may be added Hugo of Santalla for his translation into Latin of Ibn al-Muthannā's commentary on the *Zīj al-Sindhind*, and Abraham ibn Ezra for his translation of the same work into Hebrew as

---

41. Samsó, *op.cit.*, pp. 152-166; see also Richter-Bernburg, *op.cit.*, 381-383.

42. Edited by I. Heller, *De elementis et orbibus coelestibus*, Nürnberg 1549, ff. N i r--Z i r.

43. Ff. N iii r and Y ii r.

well as for his important book on the causes behind the *zījes*, known in Latin as *De rationibus tabularum*. Moreover, many Indian astronomical ideas are scattered through the Latin translations of Arabic astrological texts made in the twelfth century, and of Hebrew ones in the thirteenth. In this way such material gained a wide audience in Western Europe in the late Middle Ages and in the Renaissance while among professional astronomers the *Alfonsine Tables* eclipsed the *Toledan Tables*, which in turn had displaced the *Zīj al-Sindhind*. There are indeed some traces of Indian astronomical computations to be found at the court of Alfonso el Sabio; but what this tradition had offered to Islam that was useful in the context of a Ptolemaic planetary theory operating precariously in a Aristotelian universe had been thoroughly absorbed in Spain by the late thirteenth century, and the *Sindhind* had ceased to function as viable alternative to the Islamic modifications of Ptolemy.

DAVID PINGREE

NĪLAKAṆṬHA'S PLANETARY MODELS

One of the special qualities of Dan Ingalls' humanism was that he appreciated investigations into all aspects of human intellectual activity, that he recognized that science was an important part of Indian culture as it is of ours. I hope that his magnanimity is shared by the readers of this journal.

This paper deals with one occasion when Indian philosophy and astronomy interacted. Philosophy taught an astronomer that perception is a more powerful guide to the real nature of planetary motions than is tradition, even that which claims to be divine revelation. The astronomer, relying on observations, changed both the traditional planetary models and their parameters. However, he failed to justify these changes by showing how his observations made them necessary, and, in the end, he was not radical enough to break completely out of the tradition of geocentrism. Obviously, to notice this limitation is not to criticize the astronomer, but simply to locate him within a particular intellectual tradition wherein what he actually did do was extremely radical.

The only Sanskrit work that has so far been found in which are discussed extensively and carefully the role of observations (pratyakṣa, parīkṣaṇa) and inference (anumāna) as fundamental to the proper practice of astronomy is the *Jyotirmīmāṃsā*[1] of the important Kerala astronomer, Nīlakaṇṭha Somayājin.[2] His view, contrary to the frequent assertion that the fundamental siddhāntas expressing the eternal rules of jyotiḥśāstra are those alleged to have been composed by deities such as Sūrya, is that the astronomers must continually make observations and draw logical conclusions from them so that the computed phenomena may agree as closely as possible with contemporary observations. This may be a continuous necessity because models and parameters are not eternally fixed, but change, or because longer periods of observation lead to more accurate determinations of the models and parameters, or because improved techniques of observing and of interpreting the results may lead to superior solutions.

As part of his demonstration of the disagreement between the various siddhāntas, he computed according to several texts the mean longitudes of each of the seven planets and the lunar apogee for day 1,682,112

Journal of Indian Philosophy **29:** 187–195, 2001.

since the beginning of the Kaliyuga.[3] The calendar equivalent of this ahargaṇa will vary because of the difference in the epoch of the day (sunrise at Laṅkā) and because of the differences in the lengths of the years in the various pakṣas. The epoch of the Kaliyuga in the ardharātrika system was midnight at Laṅkā on 17/18 February in −3101 Julian. The length of a Julian year is 365 1/4 days, so that the ahargaṇa of 1,682,112 corresponds to 4605 years and 135 days. The resulting date in the Julian calendar is 1 August 1504. This date is undoubtedly near the date at which Nīlakaṇṭha completed the *Jyotirmīmāṃsā*.

In the *Jyotirmīmāṃsā*, Nīlakaṇṭha refers to his own *Āryabhaṭīyabhāṣya* which he had finished after 28 July 1501;[4] and in the *Āryabhaṭīyabhāṣya* he refers to both his *Siddhāntadarpaṇa*[5] and his *Tantrasaṅgraha*.[6] According to the chronograms in the first and the last verses of the *Tantrasaṅgraha* the ahargaṇa from the beginning of the Kaliyuga till Nīlakaṇṭha's beginning to write this book was 1,680,548 (= 20 April 1500), and the ahargaṇa till his completing it was 1,680,553 (= 25 April 1500). In part then, the *Jyotirmīmāṃsā* can be considered a justification for the innovations in siddhāantic astronomy that Nīlakaṇṭha introduced in the *Tantrasaṅgraha* and *Siddhāntadarpaṇa*.

And Nīlakaṇṭha did change many parameters. As examples I will present his mean motions for the planets and his sizes for the epicycles in the *Siddhāntadarpaṇa* and in the *Tantrasaṅgraha*.[7] I compare the *Siddhāntadarpaṇa*'s parameters with those of the Brāhmapakṣa.

	Siddhāntadarpaṇa[8] (Rotations in a Kalpa)	*Brāhmapakṣa*[9] (Rotations in a Kalpa)
Saturn	146,571,016	146,567,298
Jupiter	364,160,611	364,226,455
Mars	2,296,862,137	2,296,828,522
Sun	4,320,000,000	4,320,000,000
Venus' śīghra	7,022,270,552	7,022,389,492
Mercury's śīghra	17,937,120,175	17,936,998,984
Moon	57,753,332,321	57,753,300,000
Lunar Apogee	488,123,318	488,105,858
Lunar Node	−232,296,745	−232,311,168

There is no simple relationship between these two columns of numbers. Moreover, although the number of civil days in a Kalpa according to the *Siddhāntadarpaṇa* − 1,577,917,839,500 − is greater than is that of the *Brāhmapakṣa* – 1,577,916,450,000, the numbers of rotations are not uniformly more so that the mean daily motions might be reasonably close to each other; instead, a widely differing pattern of changes in mean daily motions occurs.

	Siddhāntadarpaṇa	*Brāhmapakṣa*
Saturn	0;2,0,23,2,20,40, . . .°	0;2,0,22,51,43,54, . . .°
Jupiter	0;4,59,5,51,55,24, . . .°	0;4,59,9,8,37,23, . . .°
Mars	0;31,26,29,40,12,31, . . .°	0;31,26,28,6,47,12, . . .°
Sun	0;59,8,10,10,18,37, . . .°	0;59,8,10,21,33,30, . . .°
Venus' śīghra	1;36,7,38,25,20,25, . . .°	1;36,7,44,35,16,45, . . .°
Mercury's śīghra	4;5,32,23,39,27,21, . . .°	4;5,32,18,27,45,33, . . .°
Moon	13;10,34,51,51,41,46, . . .°	13;10,34,52,46,30,13, . . .°
Lunar Apogee	0;6,40,55,22,23,6, . . .°	0;6,40,53,56,32,54, . . .°
Lunar Node	–0;3,10,47,58,0,6, . . .°	–0;3,10,48,20,6,41, . . .°

The *Siddhāntadarpaṇa*'s parameters diverge either positively or negatively from the *Brāhmapakṣa*'s in the third sexegesimal place; this will affect the degrees of different planets' longitudes in about 15 to 60 years. Unfortunately, Nīlakaṇṭha neither describes how he arrived at these numbers from observations nor how he revised the length of a year.[10]

The next table compares the rotations of the planets in a Mahāyuga according to the *Tantrasaṅgraha*, the *Āryabhaṭīya*,[11] and the *Sūryasiddhānta*[12]

	Tantrasaṅgraha (Rotations in a Mahāyuga)	Difference	*Āryabhaṭīya* (Rotations in a Mahāyuga)	Difference	*Sūryasiddhānta* (Rotations in a Mahāyuga)
Saturn	146,612	52 = 13 × 4	146,564	4 = 1 × 4	146,568
Jupiter	364,180	44 = 11 × 4	364,224	4 = 1 × 4	364,220
Mars	2,296,864	40 = 10 × 4	2,296,824	8 = 2 × 4	2,296,832
Sun	4,320,000	0	4,320,000	0	4,320,000
Venus' śīghra	7,022,268	120 = 30 × 4	7,022,388	12 = 3 × 4	7,022,376
Mercury's śīghra	17,937,048	28 = 7 × 4	17,937,020	40 = 10 × 4	17,937,060
Moon	57,753,320	16 = 4 × 4	57,753,336	0	57,753,336
Lunar Apogee	488,122	3	488,219	16 = 4 × 4	488,203
Lunar Node	–232,300	74	–232,226	12 = 3 × 4	–232,238

In this case the differences between each pair of columns are always multiples of four rotations except in the cases of the lunar apogee and lunar node according to the *Tantrasaṅgraha*. Again, however, the parameters in column 1 are sometimes greater and sometimes less than those in columns 2 and 3. But the number of civil days in a Mahāyuga in the *Tantrasaṅgraha* is the same as in the *Āryabhaṭīya* – 1,577,917,500. Therefore, the *Tantrasaṅgraha* mean daily motions are sometimes more and sometimes less than the *Āryabhaṭīya*'s

	Tantrasaṅgraha	*Āryabhaṭīya*
Saturn	0;2,0,25,3,37,9, . . .°	0;2,0,22,41,41,32, . . .°
Jupiter	0;4,59,6,50,32,53, . . .°	0;4,59,9,0,38,51, . . .°
Mars	0;31,26,29,47,10,42, . . .°	0;31,26,27,48,54,22, . . .°
Sun	0;59,8,10,13,3,31, . . .°	0;59,8,10,13,3,31, . . .°
Venus' śīghra	1;36,7,38,22,15,43, . . .°	1;36,7,44,17,4,45, . . .°
Mercury's śīghra	4;5,32,49,51,28,57, . . .°	4;5,32,18,54,36,24, . . .°
Moon	13;10,34,51,25,23,43, . . .°	13;10,34,52,39,18,56, . . .°
Lunar Apogee	0;6,40,54,43,19,0, . . .°	0;6,40,59,30,7,38, . . .°
Lunar Node	–0;3,10,47,46,37,57, . . .°	–0;3,10,44,7,49,44, . . .°

Again, the deviations occur in the third sexegesimal place. More disturbing is the fact that, on comparing the mean daily motions of the *Siddhāntadarpaṇa* with those of the *Tantrasaṅgraha*, one notices divergences in the third sexegesimal places of Saturn, Jupiter, Mercury's śīghra, the Lunar Apogee, and the Lunar Node. Are these the results of new computations, or simply of his accepting Āryabhaṭa's year length and the restriction that his new numbers of rotations must (except in the cases of the Lunar Apogee and Node) differ from Āryabhaṭa's by a multiple of four? I see no way to answer this question from the scanty facts that we are given.

Nīlakaṇṭha's apogees and nodes according to the *Siddhāntadarpaṇa* in Kali 4800 (= A.D. 1699)[13] and permanently according to the *Tantrasaṅgraha*[14] are all very close to their values in the *Saurapaksa*;[15] the differences would have only a minute effect on their true longitudes. Similarly, Nīlakaṇṭha's circumferences of the śīghra-epicycles,[16] measured in multiples of eightieths of 360° = 4;30° as also had Āryabhaṭa's been, are identical for both their maximum and minimum sizes with his;[17] the circumference's of the manda-epicycles of the Sun and the Moon are identical with Āryabhaṭa's, that of Mars is the mean of Āryabhaṭa's maximum and minimum, and the rest are independent.

	Nīlakaṇṭha	*Āryabhaṭīya*
Saturn	45°	58;30° – 40;30°
Jupiter	36°	36° – 31;30°
Mars	72°	81° – 63°
Venus	13;30°	18° – 9°
Mercury	63°	31;30° – 22;30°

Only Mercury's epicycle is wildly different. Therefore, the parameters of Nīlakaṇṭha's models[18] that enable one to compute true longitudes from mean and, in the case of observations, mean longitudes from true, have not been radically changed from the traditional parameters except in the case of Mercury's manda-epicycle.

But it has recently been noted that Nīlakaṇṭha radically reformed the traditional Indian planetary models. The claim is that for both superior and inferior planets the śīghravṛtta, whose center is the center of the earth, is the concentric circle on which the mean Sun moves; that the mean Sun on this circle is the center of the mandavṛtta, of designated size for each planet, on which "moves" the mandocca; and that the mandocca is the center of the circle on which the planet moves at its mean velocity (the grahabhramaṇavṛtta). For the superior planets the radius of the śīghravṛtta is r_σ (which is derived as a fraction of the Radius, R, by the formula:

$$r_\sigma = \frac{c_\sigma}{360°}$$

where c_σ is the circumference of the śīghra epicycle given in the text), and the radius of the circle on which the planet moves is R; for inferior planets the radius of the śīghravṛtta is R, that of the circle on which the planet moves is r_σ. This would make Nīlakaṇṭha's system very close to being Tychonic if the mean Sun, whose longitude is measured on the śīghravṛttas, were at the true solar distance. But this interpretation is, I believe, contrary both to Nīlakaṇṭha's own description of his planetary models and to his rules for computing the geocentric distances of the planets, both of which remain for the most part faithful to the theories of traditional siddhāntic astronomy.

In the traditional view[19] the earth is at the center of the concentric deferents bearing the center of the manda epicycles of the Sun and the Moon and the centers of both the manda and the śīghra epicycles of each of the five planets. On each epicycle, whose dimensions are designated in the siddhānta or karaṇa, moves an apogee which "pulls" the mean luminary or planet at its center towards itself. The application of the manda equation to the mean longitudes of the Sun and the Moon produces the true longitudes of those bodies; the true longitudes of the five planets are computed by merging the influences of the two equations for each. The difference between the models for the superior planets and those for the inferior planets is that, while all the centers of the epicycles travel on their concentric deferents at a rate of motion for each yojana that is common to all seven celestial bodies, the centers of the epicycles of the superior planets are the planets themselves, but the centers of the epicycles of the inferior planets have the longitude of the mean Sun. Since they all travel an identical number of yojanas in a yuga, each on its own deferent, the radii of these deferents are inversely proportional to each planets' number of revolutions in a yuga. Thus the increasing number of revolutions in a Kalpa or Mahāyuga as one descends from Saturn to the Moon indicates that the distances of the celestial bodies from the earth increase from that of the Moon to that of Saturn.

Nīlakaṇṭha describes his system most clearly in *Siddhāntadarpaṇa*[21] 19–21b; in this translation I am guided by Nīlakaṇṭha's own commentary.

19 "(The centers of) the circles on which the planets move (grahabhramaṇavṛttāni) move on the manda circle with the velocities of the (manda) apogees.[22] For the Sun and the Moon that (manda circle) has its center at the center of the solid earth.

20. For the other (planets) the center of that (manda circle) moves on the śīghra circle (whose radius is r_σ) with the velocity of the mean Sun. Their śaighra (śīghra circle), which is at the middle of the spheres, is not deflected (in latitude) from the circle of the (ecliptic) constellations.

21. Their own (circles) (svavṛtta = grahabhramaṇavṛtta) of Mercury and Venus, because they are their śaighras (the circles on which they move with their tabulated śīghra velocities),[23] are measured by their (the śīghra-epicycles') degrees."[24]

These are indeed significant changes in the traditional siddhāntic planetary models, though not outside of some of the basic concepts of that tradition: that the system is geocentric, that the single recognized anomalies for each of the two luminaries is accounted for by a single manda epicycle (Nīlakaṇṭha does not take into consideration the second lunar equation found already in Muñjāla in the tenth century), that the two anomalies of each of the five planets are accounted for by two epicycles, the manda and the śīghra. But for the planets he has rearranged the deferent circles and the two epicycles, apparently in an effort to improve the latitude theory. It is a pity that he nowhere discussed in detail the process that led him to this result.

In one other crucial aspect he adheres to tradition. This is in regard to the distances of the orbits of the celestial bodies from the central earth. His rules for computing these distances, or rather the yojanas that measure the circumferences of those orbits and that, when divided by 2π, give the distances in yojanas, are found in the *Tantrasaṅgraha*.

4,8c–9b: "The radius (of the Moon's orbit, that is, $\frac{21{,}600'}{2\pi}$) multiplied by 10 is (the number of) yojanas in the radius of the orbit of the Moon; that multiplied by the rotations of the Moon (in a Mahāyuga) and divided by (the number of) its own rotations (is the number of yojanas in the radius of the orbit) of the Sun."

More succinctly, the mean distance of the Moon from the earth is $\frac{216{,}000}{2\pi}$ yojanas, and that of the Sun $\frac{216{,}000}{2\pi} \times \frac{57{,}753{,}320}{4{,}320{,}000}$ yojanas. The mean circumference of the orbit of the Moon is 216,000 yojanas, that of the orbit of the Sun 2,887,666 yojanas.

8,36c–d: "The (orbit) of (each of) the other (planets) is to be derived from the orbit of the Moon as (was that of) the Sun."

More succinctly, the mean circumference of the orbit of one of the five planets is 216,000 yojanas multiplied by 57,753,320 and divided by the number of its rotations in a Mahāyuga. In this rule, the divisor for Venus is 7,022,268, and the mean circumference of Venus' orbit is 1,776,451 yojanas; the divisor for Mercury's orbit is 17,937,048, and the mean circumference of Mercury's orbit is 695,472 yojanas.

Thus the mean Sun that is the center of the manda circle of the inferior planets is *not* at the distance from the earth occupied by the Sun itself; nor are the centers of the manda epicycles of the superior planets, which are much further away from the earth.

Elsewhere in the *Tantrasaṅgraha* Nīlakaṇṭha makes clear that he wants us to understand his system in this way. For he gives a rule for computing the momentary orbit in yojanas of an inferior planet:

8, 37c–38b: "The accurate orbit of Venus and Mercury is (obtained) from (its mean) orbit multiplied by its śīghra hypotenuse (the distance measured in parts of R from the center of the earth to the planet) and divided by the radius (R) of the (deferent) circle."

The maximum and minimum distances of the planets are R + r_μ + r_σ, and R − (r_μ + $r\sigma$). For Venus the limits are $\frac{1{,}776{,}451}{2\pi} \times \left(1 + \frac{62}{80}\right)$ and $\frac{1{,}776{,}451}{2\pi} \times \frac{18}{80}$ yojanas, for Mercury $\frac{695{,}472}{2\pi} \times \left(1 + \frac{45}{80}\right)$ and $\frac{695{,}472}{2\pi} \times \frac{35}{80}$.

Nīlakaṇṭha made significant changes in Indian planetary models; he did not anticipate Tycho Brahe.

NOTES

[1] *Jyotirmīmāṃsā* ed. K.V. Sarma, *Panjab University Indological Series* 11, Hoshiapur 1977.

[2] D. Pingree, *Census of the Exact Sciences in Sanskrit*, Series A, vols. 1–5, Philadelphia 1971–1994, vol. 3, 175b–177b; vol. 4, 142a–142b; and vol. 5, 186a.

[3] *Jyotirmīmāṃsā*, pp. 27–31.

[4] Ibid., p. 35.

[5] *Siddhāntadarpaṇa*, ed. K.V. Sarma, *Panjab University Indological Series* 7, Hoshiapur 1976; this edition contains what remains of Nīlakaṇṭha's own commentary.

[6] *Tantrasaṅgraha* ed. K.V. Sarma, *Panjab University Indological Series* 10, Hoshiapur 1977; this edition includes the *Yuktidīpikā* and the *Laghuvṛtti*, both by Śaṅkara, a pupil of Nīlakaṇṭha.

[7] See D. Pingree, "History of Mathematical Astronomy in India", in C.C. Gillispie, ed., *Dictionary of Scientific Biography*, vol. 15, New York, 1978, pp. 533–633, esp. pp. 621–623.

[8] *Siddhāntadarpaṇa* 2–7.

[9] Pingree, "History" (see note 7), p. 556.

[10] The only observations of his own that Nīlakaṇṭha refers to in the works associated with his new models are used to show that the trepidational motion affects all the fixed stars, not just the ecliptic. These are listed in his *Siddhāntadarpaṇavyākhyā* (p. 16):

1. The Moon occulted Citrā when the ahargaṇa of Kali was 1,677,647 (= 12 April 1492). The longitude of the Moon at sunset on 11 April 1492 was 197°, its latitude −2°. Citrā is α Virgo, whose longitude in 1492 was 197;20° and whose latitude was −1;56°.
2. Mercury was north of Citrā when the ahargaṇa was 1,678,524;30 (= sunset of 6 September 1494). The longitude of Mercury was 196;48°, its latitude −1;45°.
3. Rohiṇī was seen above the Moon when the ahargaṇa was 1,679,003 (= 28 December 1495); the longitude of the Moon at sunset was 64°, its latitude −4°. Rohiṇī is α Tauri, whose longitude in 1495 was 63;14° and latitude 5;37°.

These observations may also have been used in correcting the mean motions of the Moon and the Mercury, but Nīlakaṇṭha does not say that they were nor does he describe how they would have been analysed to provide a correction.

[11] Pingree, "History", p. 590.

[12] Ibid., p. 608.

[13] Ibid., p. 623.

[14] Ibid., p. 622.

[15] Ibid., p. 610.

[16] Ibid., p. 623.

[17] Ibid., p. 592.

[18] K. Ramasubramanian, M.D. Srinivas, and M.S. Sriram, "Modification of the Earlier Indian Planetary Theory by the Kerala Astronomers (c. 1500 A.D.) and the Implied Heliocentric Picture of Planetary Motion", *Current Science* 66, 1994, 784–790. A seminar on the *Tantrasaṅgraha* and this interpretation of it was held in Madras on 11–13 March 2000; the proceedings are not yet available.

[19] Pingree, "History", p. 558.

[20] Ibid., pp. 556, 591, 608, and 609.

[21] These verses are rather awkwardly translated by Subramanian et al., "Modification", pp. 788–789.

[22] In *Siddhāntadarpaṇa* 22 and its commentary the grahabhramaṇavṛtta is called the pratimaṇḍalam, which is in turn called the jñātabhogagraha, "(the circle) on which

is the planet whose motion is known"; its radius is said to be R (this applies only to the three superior planets). Another circle with equal radius has its circumference touched by a line from the planet to the center (of the earth) at the point where its (geocentric) motion (in longitude) is to be known; this is what Āryabhaṭa called the orbital circle (kakṣyāvṛtta), the computation of whose radius in yojanas will be described below.

[23] The rotations of the śīghrocca of Mercury are given in *Siddhāntadarpaṇa* 4 as being "on its own circle", i.e., on its grahabhramaṇavṛtta; the same must hold for the motion of Venus' śīghrocca in agreement with *Tantrasaṅgraha* I 16–18.

[24] This means that, for the inferior planets, the radius of the svavṛttas are to R, the radius of the concentric deferent, as c_σ is to 360°. The svavṛtta, as Nīlakaṇṭha explains in his commentary, still has 360° in its circumference, but its size is smaller.

Brown University, Providence, Rhode Island

David Pingree

The logic of non-Western science: mathematical discoveries in medieval India

One of the most significant things one learns from the study of the exact sciences as practiced in a number of ancient and medieval societies is that, while science has always traveled from one culture to another, each culture before the modern period approached the sciences it received in its own unique way and transformed them into forms compatible with its own modes of thought. Science is a product of culture; it is not a single, unified entity. Therefore, a historian of premodern scientific texts – whether they be written in Akkadian, Arabic, Chinese, Egyptian, Greek, Hebrew, Latin, Persian, Sanskrit, or any other linguistic bearer of a distinct culture – must avoid the temptation to conceive of these sciences as more or less clumsy attempts to express modern scientific ideas. They must be understood and appreciated as what their practitioners believed them to be. The historian is interested in the truthfulness of his own understanding of the various sciences, not in the truth or falsehood of the science itself.

In order to illustrate the individuality of the sciences as practiced in the older non-Western societies, and their differences from early modern Western science (for contemporary science is, in general, interested in explaining quite different phenomena than those that attracted the attention of earlier scientists), I propose to describe briefly some of the characteristics of the medieval Indian *śāstra* of *jyotiṣa*. This discipline concerned matters included in such Western areas of inquiry as astronomy, mathematics, divination, and astrology. In fact, the *jyotiṣīs*, the Indian experts in *jyotiṣa*, produced more literature in these areas – and made more mathematical discoveries – than scholars in any other culture prior to the advent of printing. In order to explain how they managed to make such discoveries – and why their discoveries remain largely unknown – I will also need to describe briefly the general social and economic position of the *jyotiṣīs*.

David Pingree, a Fellow of the American Academy since 1971, is University Professor in the department of the history of mathematics at Brown University. He teaches about the transmission of science between cultures, and his publications include many editions of astronomical, astrological, and magical works in Akkadian, Arabic, Greek, Latin, and Sanskrit. Most recently he has written "Arabic Astronomy in Sanskrit" (with T. Kusuba, 2002), "Astral Sciences in Mesopotamia" (with H. Hunger, 1999), and "Babylonian Planetary Omens" (with E. Reiner, 1998).

'*Śāstra*' ('teaching') is the word in Sanskrit closest in meaning to the Greek 'ἐπιστήμη' and the Latin 'scientia.' The teachings are often attributed to gods or considered to have been composed by divine *ṛṣis*; but since there were many of both kinds of superhuman beings, there were many competing varieties of each *śāstra*. Sometimes, however, a school within a *śāstra* was founded by a human; scientists were free to modify their *śāstras* as they saw fit. No one was constrained to follow a system taught by a god.

Jyotiḥ is a Sanskrit word meaning 'light,' and then 'star'; so that *jyotiḥśāstra* means 'teaching about the stars.' This *śāstra* was conventionally divided into three subteachings: *gaṇita* (mathematical astronomy and mathematics itself), *saṃhitā* (divination, including by means of celestial omens), and *horā* (astrology). A number of *jyotiṣīs* (students of the stars) followed all three branches, a larger number just two (usually *saṃhitā* and *horā*), and the largest number just one (*horā*).

The principal writings in *jyotiḥśāstra*, as in all Indian *śāstras*, were normally in verse, though the numerous commentaries on them were almost always in prose. The verse form with its metrical demands, while it aided memorization, led to greater obscurity of expression than prose composition would have entailed. The demands of the poetic meter meant that there could be no stable technical vocabulary; many words with different metrical patterns had to be devised to express the same mathematical procedure or geometrical concept, and mathematical formulae had frequently to be left partially incomplete. Moreover, numbers had to be expressible in metrical forms (the two major systems used for numbers, the *bhūtasaṅkhyā* and the *kaṭapayādi*, will be explained and exemplified below), and the consequent ambiguity of these expressions encouraged the natural inclination of Sanskrit *paṇḍits* to test playfully their readers' acumen. It takes some practice to achieve sureness in discerning the technical meanings of such texts.

But in this opaque style the *jyotiṣīs* produced an abundant literature. It is estimated that about three million manuscripts on these subjects in Sanskrit and in other Indian languages still exist. Regrettably, only a relatively small number of these has been subjected to modern analysis, and virtually the whole ensemble is rapidly decaying. And because there is only a small number of scholars trained to read and understand these texts, most of them will have disappeared before anyone will be able to describe correctly their contents.

In order to make my argument clearer, I will restrict my remarks to the first branch of *jyotiḥśāstra* – *gaṇita*. Geometry, and its branch trigonometry, was the mathematics Indian astronomers used most frequently. In fact, the Indian astronomers in the third or fourth century, using a pre-Ptolemaic Greek table of chords,[1] produced tables of sines and versines, from which it was trivial to derive cosines. This new system of trigonometry, produced in India, was transmitted to the Arabs in the late eighth century and by them, in an expanded form, to the Latin West and the Byzantine East in the twelfth century. But, despite this sort of practical innovation, the Indians practiced geometry without the type of proofs taught by Euclid, in

1 For a description of the table of chords, cyclic quadrilaterals, two-point iteration, fixed-point iteration, and several other mathematical terms mentioned in this essay, please see Victor J. Katz, *A History of Mathematics: An Introduction* (New York: HarperCollins, 1993).

which all solutions to geometrical problems are derived from a small body of arbitrary axioms. The Indians provided demonstrations that showed that their solutions were consistent with certain assumptions (such as the equivalence of the angles in a pair of similar triangles or the Pythagorean theorem) and whose validity they based on the measurement of several examples. In their less rigorous approach they were quite willing to be satisfied with approximations, such as the substitution of a sine wave for almost any curve connecting two points. Some of their approximations, like those devised by Āryabhaṭa in about 500 for the volumes of a sphere and a pyramid, were simply wrong. But many were surprisingly useful.

Not having a set of axioms from which to derive abstract geometrical relationships, the Indians in general restricted their geometry to the solution of practical problems. However, Brahmagupta in 628 presented formulae for solving a dozen problems involving cyclic quadrilaterals that were not solved in the West before the Renaissance. He provides no rationales and does not even bother to inform his readers that these solutions only work if the quadrilaterals are circumscribed by a circle (his commentator, Pṛthūdakasvāmin, writing in about 864, follows him on both counts). In this case, and clearly in many others, there was no written or oral tradition that preserved the author's reasoning for later generations of students. Such disdain for revealing the methodology by which mathematics could advance made it difficult for all but the most talented students to create new mathematics. It is amazing to see, given this situation, how many Indian mathematicians did advance their field.

I will at this point mention as examples only the solution of indeterminate equations of the first degree, described already by Āryabhaṭa; the partial solution of indeterminate equations of the second degree, due to Brahmagupta; and the cyclic solution of the latter type of indeterminate equations, achieved by Jayadeva and described by Udayadivākara in 1073 (the cyclic solution was rediscovered in the West by Bell and Fermat in the seventeenth century). Interpolation into tables using second-order differences was introduced by Brahmagupta in his *Khaṇḍakhādyaka* of 665. The use of two-point iteration occurs first in the *Pañcasiddhāntikā* composed by Varāhamihira in the middle of the sixth century, and fixed-point iteration in the commentary on the *Mahābhāskarīya* written by Govindasvāmin in the middle of the ninth century. The study of combinatorics, including the so-called Pascal's triangle, began in India near the beginning of the current era in the *Chandaḥsūtras*, a work on prosody composed by Piṅgala, and culminated in chapter 13 of the *Gaṇitakaumudī* completed by Nārāyaṇa Paṇḍita in 1350. The fourteenth and final chapter of Nārāyaṇa's work is an exhaustive mathematical treatment of magic squares, whose study in India can be traced back to the *Bṛhatsaṃhitā* of Varāhamihira.

In short, it is clear that Indian mathematicians were not at all hindered in solving significant problems of many sorts by what might appear to a non-Indian to be formidable obstacles in the conception and expression of mathematical ideas.

Nor were they hindered by the restrictions of 'caste,' by the lack of societal support, or by the general absence of monetary rewards. It is true that the overwhelming majority of the Indian mathematicians whose works we know were Brāhmaṇas, but there are exceptions (e.g., among Jainas, non-Brahmānical scribes, and craftsmen). Indian society was far from open, but it was not absolutely rigid; and talented mathematicians, whatever their origins, were not ignored by their colleagues. However, astrologers (who frequently were not Brāhmaṇas) and the makers of calendars were the only *jyotiṣīs* normally valued by the societies in which they lived. The attraction of the former group is easily understood, and their enormous popularity continues today. The calendar-makers were important because their job was to indicate the times at which rituals could or must be performed. The Indian calendar is itself intricate; for instance, the day begins at local sunrise and is numbered after the tithi that is then current, with the tithis being bounded by the moments, beginning from the last previous true conjunction of the Sun and the Moon, at which the elongation between the two luminaries had increased by twelve degrees. Essentially, each village needed its own calendar to determine the times for performing public and private religious rites of all kinds in its locality.

By contrast, those who worked in the various forms of *gaṇita* usually enjoyed no public patronage – even though they provided the mathematics used by architects, musicians, poets, surveyors, and merchants, as well as the astronomical theories and tables employed by astrologers and calendar-makers. Sometimes a lucky mathematical astronomer was supported by a Mahārāja whom he served as a royal astrologer and in whose name his work would have been published. For example, the popular *Rājamṛgāṅka* is attributed, along with dozens of other works in many *śāstras*, to Bhojadeva, the Mahārāja of Dhārā in the first half of the eleventh century. Other *jyotiṣīs* substituted the names of divinities or ancient holy men for their own as authors of their treatises. Authorship often brought no rewards; one's ideas were often more widely accepted if they were presented as those of a divine being, a category that in many men's minds included kings.

One way in which a *jyotiṣī* could make a living was by teaching mathematics, astronomy, or astrology to others. Most frequently this instruction took place in the family home, and, because of the caste system, the male members of a *jyotiṣī*'s family were all expected to follow the same profession. A senior *jyotiṣī*, therefore, would train his sons and often his nephews in their ancestral craft. For this the family maintained a library of appropriate texts that included the compositions of family members, which were copied as desired by the younger members. In this way a text might be preserved within a family over many generations without ever being seen by persons outside the family. In some cases, however, an expert became well enough known that aspirants came from far and wide to his house to study. In such cases these students would carry off copies of the manuscripts in the teacher's collection to other family libraries in other locales.

The teaching of *jyotiḥśāstra* also occurred in some Hindu, Jaina, and Buddhist monasteries, as well as in local schools. In these situations certain standard texts were normally taught, and the status of these texts can be established by the number of copies that still exist, by their geographical distribution, and by the number of commentaries that were written on them.

Thus, in *gaṇita* the principal texts used in teaching mathematics in schools were clearly the *Līlavatī* on arithmetic and the *Bījagaṇita* on algebra, both written by Bhāskara in around 1150, and, among Jainas, the *Gaṇitasārasaṅgraha* composed in about 850 by their coreligionist, Mahāvīra. In astronomy there came to be five *pakṣas* (schools): the *Brāhmapakṣa*, whose principal text was the *Siddhānta-*

śiromaṇi of the Bhāskara mentioned above; the *Āryapakṣa*, based on the *Āryabhaṭīya* written by Āryabhaṭa in about 500; the *Ārdharātrikapakṣa*, whose principal text was the *Khaṇḍakhādyaka* completed by Brahmagupta in 665; the *Saurapakṣa*, based on the *Sūryasiddhānta* composed by an unknown author in about 800; and the *Gaṇeśapakṣa*, whose principal text was the *Grahalāghava* authored by Gaṇeśa in 1520. Each region of India favored one of these *pakṣas*, though the principal texts of all of them enjoyed national circulation. The commentaries on these often contain the most innovative advances in mathematics and mathematical astronomy found in Sanskrit literature. By far the most popular authority, however, was Bhāskara; a special college for the study of his numerous works was established in 1222 by the grandson of his younger brother. No other Indian *jyotiṣī* was ever so honored.

Occasionally, indeed, an informal school inspired by one man's work would spring up. The most noteworthy, composed of followers of Mādhava of Saṅgamagrāma in Kerala in the extreme south of India, lasted for over four hundred years without any formal structure – simply a long succession of enthusiasts who enjoyed and sometimes expanded on the marvelous discoveries of Mādhava.

Mādhava (c. 1360 – 1420), an Emprāntiri Brāhmaṇa, apparently lived all his life on his family's estate, Ilaññipaḷḷi, in Saṅgamagrāma (Irinjālakhuḍa) near Cochin. His most momentous achievement was the creation of methods to compute accurate values for trigonometric functions by generating infinite series. In order to demonstrate the character of his solutions and expressions of them, I will translate a few of his verses and quote some Sanskrit.

He began by considering an octant of a circle inscribed in a square, and, after some calculation, gave the rule (I translate quite literally two verses):

> Multiply the diameter (of the circle) by 4 and divide by 1. Then apply to this separately with negative and positive signs alternately the product of the diameter and 4 divided by the odd numbers 3, 5, and so on .... The result is the accurate circumference; it is extremely accurate if the division is carried out many times.

This describes the infinite series:

$$C = \frac{4D}{1} - \frac{4D}{3} + \frac{4D}{5} - \frac{4D}{7} + \frac{4D}{9} \ldots.$$

That in turn is equivalent to the infinite series for π that we attribute to Leibniz:

$$\frac{\pi}{4} = 1 - \frac{1}{3} + \frac{1}{5} - \frac{1}{7} + \frac{1}{9} \ldots.$$

Mādhava expressed the results of this formula in a verse employing the *bhūtasaṅkhyā* system, in which numbers are represented by words denoting objects that conventionally occur in the world in fixed quantities:

> vibudhanetragajāhihutāśanatriguṇavedabhavāraṇabāhavaḥ |
> navanikharvamite vṛtivistare
> paridhimānam idaṃ jagadur budhāḥ ||

A literal translation is:

> Gods [33], eyes [2], elephants [8], snakes [8], fires [3], three [3], qualities [3], Vedas [4], nakṣatras [27], elephants [8], and arms [2] – the wise say that this is the measure of the circumference when the diameter of a circle is nine hundred billion.

The *bhūtasaṅkhyā* numbers are taken in reverse order, so that the formula is:

$$\pi = \frac{2827433388233}{900000000000}$$

(= 3.14159265359, which is correct to the eleventh decimal place).

Another extraordinary verse written by Mādhava employs the *kaṭapayādi* system in which the numbers 1, 2, 3, 4, 5, 6, 7, 8, 9, and 0 are represented by the consonants that are immediately followed by a vowel; this allows the mathematician to create a verse with both a transparent meaning due to the words and an unrelated numerical meaning due to the consonants in those words. Mādhava's verse is:

vidvāṃs tunnabalaḥ kavīśanicayaḥ sarvārthaśīlasthiro
nirviddhāṅganarendraruṅ

The verbal meaning is: "The ruler whose army has been struck down gathers together the best of advisors and remains firm in his conduct in all matters; then he shatters the (rival) king whose army has not been destroyed."

The numerical meaning is five sexagesimal numbers:

0;0,44
0;33,6
16;5,41
273;57,47
2220;39,40.

These five numbers equal, with R = 3437;44,48 (where R is the radius):

$$\frac{5400^{11}}{R^{10}11!}$$

$$\frac{5400^{9}}{R^{8}9!}$$

$$\frac{5400^{7}}{R^{6}7!}$$

$$\frac{5400^{5}}{R^{4}5!}$$

$$\frac{5400^{3}}{R^{2}3!}$$

These numbers are to be employed in the formula:

$$sin\theta = \theta-\left(\frac{\theta}{5400}\right)^3\left[\frac{5400^3}{R^2 3!}-\left(\frac{\theta}{5400}\right)^2\left[\frac{5400^5}{R^4 5!}-\left(\frac{\theta}{5400}\right)^2\left[\frac{5400^7}{R^6 7!}-\left(\frac{\theta}{5400}\right)^2\left[\frac{5400^9}{R^8 9!}-\left(\frac{\theta}{5400}\right)^2\left[\frac{5400^{11}}{R^{10}11!}\right]\right]\right]\right]\right]$$

and this formula is a simple transformation of the first six terms in the infinite power series for $sin\theta$ found independently by Newton in 1660:

$$sin\theta = \theta-\frac{\theta^3}{R^2 3!}+\frac{\theta^5}{R^4 5!}-\frac{\theta^7}{R^6 7!}+\frac{\theta^9}{R^8 9!}-\frac{\theta^{11}}{R^{10}11!}$$

Not surprisingly, Mādhava also discovered the infinite power series for the cosine and the tangent that we usually attribute to Gregory.

The European mathematicians of the seventeenth century derived their trigonometrical series from the application of the calculus; Mādhava in about 1400 relied on a clever combination of geometry, algebra, and a feeling for mathematical possibilities. I cannot here go through his whole argument, which has fortunately been preserved by several of his successors; but I should mention some of his techniques.

He invented an algebraic expansion formula that keeps pushing an unknown quantity to successive terms that are alternately positive and negative; the series must be expanded to infinity to get rid of this unknown quantity. Also, because of the multiplications, as the terms increase, the powers of the individual factors also increase. One of these factors in the octant is one of a series of integers beginning with 1 and ending with 3438 – the number of parts in the radius of the circle that is also the tangent of 45°, the angle of the octant; this means that there are 3438 infinite series that must be summed to yield the final infinite series of the trigonometrical function.

It had long been known in India that the sum of a series of integers beginning with 1 and ending with n is: $n\left(\frac{(n-1)}{2}+1\right)$, that is, $\frac{n^2}{2}-\frac{n}{2}+n$. Since n here equals 3438, Mādhava decided that $n-\frac{n}{2}$, which equals $\frac{3438}{2}$, is negligible with respect to $\frac{3438^2}{2}$. Therefore, an approximation to the sum of the series of n integers is $\frac{n^2}{2}$. Similarly, the sums of the squares of a series of n integers beginning with 1 was known to be $\frac{\frac{n}{2}[2(n+1)^2-(n+1)]}{3}$. If n is large, this is approximately equal to $\frac{n(n+1)^2}{3}$ since $\frac{-(n+1)}{6}$ is negligible. But, with $n = 3438$, $\frac{3438 \times 3439^2}{3}$ is little different from $\frac{3438^3}{3}$. Therefore, as an approximation, the sum of the series of the squares of 3438 integers beginning with 1 is $\frac{n^3}{3}$. Finally, it was known that the sum of the cubes of a series of n numbers beginning with 1 is: $\left(\frac{n}{2}\right)^2(n+1)^2$ or $\frac{n^2(n+1)^2}{4}$. If n is 3438, there is little difference between $\frac{3438^2 \times 3439^2}{4}$ and $\frac{3438^4}{4}$. Therefore, the expression $\frac{n^4}{4}$ is a close approximation to the sum of the cubes of a series of n numbers beginning with 1. From these three examples Mādhava guessed at the general rule that the sum of n numbers in an arithmetical series beginning with 1 all raised to the same power, p, is approximately equal to $\frac{n^{p+1}}{p+1}$.

It had also been realized in India since the fifth century – from examining the sine table in which the radius of the circle, R, is $\frac{21{,}600}{2\pi}$ (which was approximated by 3438) and in which there are 24 sines in a quadrant of 90°, so that the length of each arc whose sine is tabulated is 225′ – that the sine of any tabulated angle θ is equal to θ minus the sum of the sums of the second differences of the sines of the preceding tabulated angles. Mādhava discovered, by some very clever geometry, that the sum of the sums of the second differences approximately equals $\frac{\theta^3}{R^2 3!}$ and that the versine of θ is approximately equal to $\frac{\theta^2}{2R}$. Since $sin^2\theta = R^2 - cos^2\theta$ and $vers\theta = R - cos\theta$, Mādhava could correct the approximation to the versine by the approximation to the sum of the sums of the second differences of tabulated sines; then he could correct the approximation to the sum of the sums of the second differences by the corrected approximation to the versine; and he could continue building up the two parallel series by applying alternating corrections to them. He finally arrives at two infinite power series, equivalent, if $R = 1$, to:

$$sin\theta = \theta - \frac{\theta^3}{3!} + \frac{\theta^5}{5!} - \frac{\theta^7}{7!} + \frac{\theta^9}{9!} \ldots,$$

and

$$cos\theta = 1 - \frac{\theta^2}{2!} + \frac{\theta^4}{4!} - \frac{\theta^6}{6!} + \frac{\theta^8}{8!} \ldots.$$

Subsequent members of the 'school' of Mādhava did remarkable work as well, in both geometry (including trigonometry) and astronomy. This is not the occasion to recite their accomplishments, but I should remark here that, among these members, Indian astronomers attempted especially to use observations to correct astronomical models and their parameters.

This began with Mādhava's principal pupil, a Nampūtiri Brāhmaṇa named Parameśvara, whose family's illam was Vaṭaśreṇi in Aśvatthagrāma, a village about thirty-five miles northeast of Saṅgamagrāma. He observed eighteen lunar and solar eclipses between 1393 and 1432 in an attempt to correct traditional Indian eclipse theory. One pupil of Parameśvara's son, Dāmodara, was

Nīlakaṇṭha – another Nampūtiri Brāhmaṇa who was born in 1444 in the Kelallūr illam located at Kuṇḍapura, which is about fifty miles northwest of Aśvatthagrāma.

Nīlakaṇṭha made a number of observations of planetary and lunar positions and of eclipses between 1467 and 1517. Nīlakaṇṭha presented several different sets of planetary parameters and significantly different planetary models, which, however, remained geocentric. He never indicates how he arrived at these new parameters and models, but he appears to have based them at least in large part on his own observations. For he proclaims in his *Jyotirmīmāṃsā* – contrary to the frequent assertion made by Indian astronomers that the fundamental *siddhāntas* expressing the eternal rules of *jyotiḥśāstra* are those alleged to have been composed by deities such as Sūrya – that astronomers must continually make observations so that the computed phenomena may agree as closely as possible with contemporary observations. Nīlakaṇṭha says that this may be a continuous necessity because models and parameters are not fixed, because longer periods of observation lead to more accurate models and parameters, and because improved techniques of observing and interpreting results may lead to superior solutions. This affirmation is almost unique in the history of Indian *jyotiṣa*; *jyotiṣīs* generally seem to have merely corrected the parameters of one *pakṣa* to make them closely corresponded to those of another.

The discoveries of the successive generations of Mādhava's 'school' continued to be studied in Kerala within a small geographical area centered on Saṅgamagrāma. The manuscripts of the school's Sanskrit and Malayālam treatises, all copied in the Malayālam script, never traveled to another region of India; the furthest they got was Kaṭattanāt in northern Kerala, about one hundred miles north of Saṅgamagrāma, where the Rājakumāra Śaṅkara Varman repeated Mādhava's trigonometrical series in a work entitled *Sadratnamālā* in 1823. This was soon picked up by a British civil servant, Charles M. Whish, who published an article entitled "On the Hindú Quadrature of the Circle and the Infinite Series of the Proportion of the Circumference to the Diameter in the Four Sástras, the Tantra Sangraham, Yocti Bháshá, Carana Paddhati and Sadratnamála" in *Transactions of the Royal Asiatic Society* in 1830.[2] While Whish was convinced that the Indians (he did not know of Mādhava) had discovered calculus – a conclusion that is not true even though they successfully found the infinite series for trigonometrical functions whose derivation was closely linked with the discovery of calculus in Europe in the seventeenth century – other Europeans scoffed at the notion that the Indians could have achieved such a startling success. The proper assessment of Mādhava's work began only with K. Mukunda Marar and C. T. Rajagopal's "On the Hindu Quadrature of the Circle," published in the *Journal of the Bombay Branch of the Royal Asiatic Society* in 1944.

So while the discoveries of Newton, Leibniz, and Gregory revolutionized European mathematics immediately upon their publication, those of Mādhava, Parameśvara, and Nīlakaṇṭha, made between the late fourteenth and early sixteenth centuries, became known to a handful of scholars outside of Kerala in

2 Note that the *Tantrasaṅgraha* was written by the Nīlakaṇṭha whom we have already mentioned, the *Yuktibhāṣa* by his colleague and fellow pupil of Dāmodara, Jyeṣṭhadeva, and the *Karaṇapaddhati* by a resident of the Putumana illam in Śivapura in 1723.

India, Europe, America, and Japan only in the latter half of the twentieth century. This was not due to the inability of Indian *jyotiṣīs* to understand the mathematics, but to the social, economic, and intellectual milieux in which they worked. The isolation of brilliant minds was not uncommon in premodern India. The exploration of the millions of surviving Sanskrit and vernacular manuscripts copied in a dozen different scripts would probably reveal a number of other Mādhavas whose work deserves the attention of historians and philosophers of science. Unfortunately, few scholars have been trained to undertake the task, and the majority of the manuscripts will have crumbled in just another century or two, before those few can rescue them from oblivion.

Islam

INDIAN INFLUENCE ON SASANIAN AND EARLY ISLAMIC ASTRONOMY AND ASTROLOGY

By

David Pingree, Chicago

The study of the history of astronomy and astrology in Iran during the Sasanian period (226-652)[1] is rendered difficult by the fact that none of the contemporary Persian works on these subjects written in the Pahlavi language have been preserved in their original form. But there are numerous passages in other texts of the Sasanian period, and especially in the apologetic literature of the ninth century[2], which give us some inkling of what those works were like. Thus, the *Bundahishn* devotes its second chapter to a discussion of the stars[3], and its fifth chapter is concerned with the horoscope of the creation of Gayômart, the first man, and with other astrological details[4]. The *Dênkart* informs us of the traditional Iranian view of the transmission of the science[5]; this account is repeated and supplemented by a Persian astrologer whom Hârûn al-Rashîd (786-809) placed in charge of his Khizânat al-ḥikma, Abû Sahl al-Faḍl ibn Nawbakht[6], in his *Kitâb al-Nahmaṭân*[7]. But it is mainly through the contemporaries of Ibn Nawbakht, of his father Nawbakht[8], and of his grandson

1. A brief survey from one point of view is given in my "Astronomy and Astrology in India and Iran", *Isis* 44, 1963, 229-246, esp. 240 sqq ; the present paper is intended to provide a more complete (though still non-technical) survey. Note that all dates, unless otherwise specified, are in the Christian Era.

2. The latest complete surveys of Pahlavî literature are J.C.Tavadia, *Die Mitte persische Sprache und Literature der Zarathustrier*, Leipzig, 1956, and J. Duchesne-Guillemin, *La religion de l Iran ancien*, Paris, 1962, pp. 52-63. A new history by J.P. de Menasce will appear in the volume of the *Cambridge History of Iran* devoted to the Sasanian period.

3. Translated by W.B. Henning, " An astronomical Chapter of the Bundahishn ", *JRAS*, 1942, 229-248.

4. The latest translation is by D.N. MacKenzie, " Zoroastrian Astrology in the *Bundahišn* ", *BSOAS* 27, 1964, 511-529.

5. Translated by R.C. Zaehner, *Zurvan : A Zoroastrian Dilemma*, Oxford, 1955, pp. 7-9.

6. See the *Fihrist* of Ibn al-Nadîm, ed. G. Flügel, Leipzig, 1871, p. 274; the *Kitâb ṭabaqât, al-umam* of Ṣâ<id al-Andalusî, ed. L. Cheikho, Beyrouth, 1912, p. 60; trans. R. Blachere, Paris, 1935, p.117; and the *Ta>rîkh a -ḥukamâ'* of Ibn al-Qifṭî, ed. J. Lippert, Leipzig, 1903. pp. 255 and 409.

7. Quoted by Ibn al-Nadîm, pp. 238-239 ; translated in my *The Thousands of Abû Ma>shar*, which is in press at the Warburg Institute, London.

8. Nawbakht, the astrologer of al-Manṣûr (754-775), assisted-Mâshâ>allâh in casting the horoscope of Baghdâd in Rabî I 141 A.H. (12 July-11 August 758) according to al-Ya'qûbî's *Kitâb al-Buldân*, ed. M.J. de Goeje, *BGA* 7, 2nd ed., Leiden, 1892, p. 238 (Cf. p. 241); but al-Bîrûnî in his *Al-Athâr al-bâqiya 'an al qurûn al-khlâiya*, ed. C.E. Sachau, repr. Leipzig, 1923, pp. 270-271 ; trans. C.E. Sachau, London 1879, pp. 262-263, dates the horoscope 23 Tammûz 1074 of the Era of Alexander (24 July 762). See C.A. Nallino, *Raccolta di scritti*

al-Ḥasan[9] that we are able to learn something of the knowledge of the stars in Sasanian Iran. These Arabic sources will be described more fully later in this paper.

Sasanian astronomy, as is characteristic of Sasanian thought in most fields of science and philosophy, was syncretistic—a blend of concepts and methods derived not only from Iran's indigenous traditions, but also from those of her neighbours, and especially of India and the Hellenistic world (this latter influence was felt both directly through Greek and indirectly through Syriac). It is primarily the Indian influence that will be investigated here.

Our sources inform us that the first two Sasanian emperors—Ardashîr I (226-240) and Shâpûr I (240-270)—were dedicated to the expansion of the Iranian intellectual tradition, and supported the translations of Greek and Sanskrit books into Pahlavî[10]. Thus we know that versions were made of Ptolemy's *Syntaxis mathêmatikē*, which the *Dēnkart* calls M.g.st.yk. (Megistê, whence the name *al-Majisṭî*[11]; the ninth century scholar Manushchihr mentions Ptalamayus (Ptolemaios) in connection with Indian and Iranian astronomical tables *Zîk î Hindûk* and the *Zîk î Shahriyârân*)[12]. Also translated into Pahlavî from Greek were the hexameters of the *Pentateuch*, an astrological poem written in the first century

editie inediti, vol. 5, Roma 1944, pp. 200-201. In fact, the horoscope quoted as Nawbhakt's by al-Bîrûnî can be dated 30th July 762 :

	al-Bîrûnî	Computation
Saturn	Aries 26 ; 40 retr.	Taurus 1
Jupiter	Sagittarius	Sagittarius 9 retr.
Mars	Gemini 2 ; 50	Gemini 6
Sun	Leo 8 ; 10	Leo 10
Venus	Gemini 29 ; 0	Cancer 2
Mercury	Cancer 25 ; 7	Cancer 26
Moon	Libra 19 ; 10	c. Libra 16

For the date 24 July 762 to be correct, the Moon would have to be in Cancer 19.

An anecdote involving Nawbakht, his son Abû Sahl, and the Caliph al-Manṣûr is recorded by Abû al-Faraj (Bar Hebraeus) in his *Ta'rîkh mukhtasar al-duwal*, 2nd. ed., Bayrut, 1958, p. 125. In the Nuruosmaniye Mosque in Istanbul MS 2951 ff. vv. 137-138 contain a *Kitâb fîhi sarâ' ir min aḥkâm al-nujûm* which is ascribed to Nawbakht, the Wise.

9. Al-Ḥasan ibn Sahl ibn Nawbakht is mentioned by Ibn al-Nadîm, p. 527, and by Ibn al-Qifṭî, p. 165, besides being quoted by many astrologers. An extremely inaccurate prediction which he made for al-Wâthiq in Dhû al-ḥijja 232 A.H. (19 July - 16 August 847) is recorded by Abû al-Faraj, p. 141. He is perhaps identical with the al-Ḥasan ibn Sahl who represented al-Ma'mûn (813-833) in Irâq ; see, e.g., al-Ya'qûbî's *Ta'rîkh*, ed. Bayrût, 1960, vol. 2, pp. 445 sqq.

10. See the passage from the Dênkart cited in fn. 5, and that from Abû Sahl Ibn Nawbakht cited in fn. 7; *Cf.* also H.W. Bailey, *Zoroastrian Problems in the Ninth century Books*, Oxford, 1943, pp. 80 sqq., and C.A. Nalline, "Tracce di opere greche giunte agli Arabi per, trafila pehlevica ", *A Volume of Oriental Studies Presented to Professor E.G. Browne*, Cambridge, 1922, pp. 345-363, repr. in his *Raccolta*, vol. 6, Roma, 1948, pp. 285-303.

11. Bailey, p. 86.

12. Bailey, p. 80. *Cf.* the history of Anûshirwân's *Zîj al Shâh* given below.

by Dorotheus of Sidon[13]., as well as an unknown text attributed to one Cedrus of Athens, the *Paranatellonta* of Teucer of Babylon, and the *Anthologiae* of Vettius Valens. Ibn Nawbakht tells us that there was also translated a work by an Indian named Faramâsb, which Justi conjectures to be Paramâśva[14].

Other Indian astronomical and astrological ideas were spread to Sasanian Iran through the translations of Buddhist texts which were made in the Eastern provinces of the empire. The *Śārdūlakarṇāvadāna*[15] was certainly known among the faithful of this area since the Parthian prince An Shi-kâo[16] translated the introductory story into Chinese in 148, and a long fragment of the Sanskrit text copied in c. 500 was found south of Yarkand[17]; this work contains a summary of the Babylonian-influenced astronomy and astrology which was current in India between c. 500 B.C. and 100 A.D.[18]. Another such text is the *Mahāmāyūrīmañjarī*[19], which deals in part with nakṣatra-astrology ; fragments of fifth century manuscripts are preserved among the Bower and Petrovski manuscripts from Kashgar. From such sources as these are probably derived the Iranian references to the nakṣatras[20], to Rāhu (who is called Gocihr)[21], and to shadow-tables[22].

The earliest attempt to compose a set of astronomical tables, the *Zīk-ī Shahriyârân* (*Zīj al-Shâh* in Arabic)—however, was apparently composed in 450, during the reign of Yazdijird II (438-457). A reference is preserved from this work by Ibn Yûnis in his *Zīj al-Ḥākimī*[23]: the longitude of the apogee of the Sun at Gemini 17 ; 55°. This parameter is derived from

13. See my forthcoming edition of the fragments of the Greek original and of the Arabic translation of the Pahlavî made by 'Umar ibn al-Farrukhân in c. 800 ; a horoscope in the fourth book indicates that the Pahlavî version was revised in the early fifth century.

14. F. Justi, *Iranisches Namenbuch*, Marburg, 1895, repr. Hildesheim, 1963, p. 90.

15. Edited by S. Mukhopadhyaya, Santiniketan, 1954; repr. in *Divyâvadâna*, ed. P.L. Vaidya, *BST* 20, Darbhanga, 1959, pp. 314-425.

16. E. Zürcher, *The Buddhist Conquest of China*, Leiden, 1959, vol. 1, pp. 32-34.

17. Edited by A.F.R. Hoernle, *JAS Bengal* 62, 1893, 9 - 17.

18. *Cf. Isis* 54, 1963, 231-233 and 240-241.

19. S. Oldenburg, *Zapiski Vostocnago otdyeleniya Imp. Russk. Arkheol. Obstchestva* 11, 1897-98, St. Petersburg, 1899, pp. 218-261 ; A.F.R. Hoernle, *The Bower Manuscript*, *ASI*, New Imp. Ser. 22, pts. 6-7, Calcutta, 1893-1912, pp. 220-240a and pls. xlix-liv; and S. Levi. " Le catalogue geographique des Yakṣa dans la Mahâmâyûrî ", *JA* 11 *e* ser., 5, 1915, 19-138.

20. *JRAS* 1942, 242-246. Henning's date (c. 500) is reasonable, though his method of arriving at it is open to question.

21. E.g., in the horoscope of Gaymôart in *BSOAS* 27, 1964, 513-517, and especially 515-516 where he is specifically compared to a serpent whose head and tail are separated by six zodiacal signs.

22. *Isis* 54, 1963, 240. On shadow-tables in Greek, see now O.Neugebauer, " Uber griechische Wetterzeichen und Schattentafeln", *Sitz. Österreich. Akad. Wiss.*, Phil. - Hist. Kl. 240, 2, Wien 1962, 29-44.

23. E.S. Kennedy and B.L. van der Waerden, *JAOS* 83, 1963, 321 and 323.

the tradition of the *Paitāmahasiddhānta* of the *Viṣṇudharmottarapurāṇa*[24]. This *Paitāmahasiddhānta* is probably what Āryabhaṭa I (fl. 499) intended to refer to when he mentioned the *Svāyambhuva*[25]; it is extensively used by Brahmagupta in his *Brāhmasphuṭasiddhānta* (628)[26], and evidently dates from the early fifth century. Perhaps it was composed during the reign of Candragupta II (c. 376-413), whom tradition asserts to have been a patron of learning, including Jyotiḥśāstra[27]. Some knowledge of it, then, either in a complete translation or in a freer summary, could easily have reached Iran by c. 450.

Āryabhaṭa I, besides writing his well-known *Āryabhaṭīya* to which reference has been made above, composed a second work in which he expounded his ārdharātrika or Midnight System. Though the original text of this work is now lost, the system is reported to us by Lāṭadeva (fl. 505) in his *Old Sūryasiddhānta*[28], by Bhāskara I (fl. c. 600) in his *Mahābhāskarīya*[29], and by Brahmagupta in his *Khaṇḍakhādyaka* (665)[30]. The evidence seems to indicate that one version of the ârdharâtrika system—probably Lāṭadeva's *Old Sūryasiddhānta*—was translated into Pahlavî before 550 and given a title like *Ẓīk ī Arkand*; Arkand appears to be an attempt to render the Sanskrit 'ahargaṇa'. Ali Ibn Sulayman al-Hâshimî in his *Kitâb ilal al-zîjât*[31], which was written in c. 875, quotes from Mâshâ'allâh[32], a Persian Jew from Baṣra (*c.* 750-815),

24. *Viṣṇudharmottarapurāṇa* 2, 166-174 in the Veṅkaṭeśwara ed., Bombay, 1912; this *Paitāmahasiddhānta* was published separately from MS 36938 of the Sarasvatī Bhavan Library Benares, by V.P. Dvivedi in his *Jyautiṣasiddhāntasaṃgraha*, *BSS* 39, fasc. 2, Benares, 1912, pt. 1. See my "The Persian 'Observation' of the Solar Apogee in *ca.* A.D. 450", *JNES* 24, 1965, 334-336.

25. *Āryabhaṭīya* with the *Bhāṣya* of Nīlakaṇṭha Somasutvan, *TSS* 101, 110, and 185, Trivandrum, 1930-1957, *Golapāda* 50; See also Svayambhû mentioned by Garga in the verses quoted by Nīlakaṇṭha *ad loc.* : vol. 3, p. 162.

26. Edited by Sudhâkara Dvivedin, Benares, 1902. I am preparing a new edition to be accompanied by the valuable commentary of Pṛthûdakasvâmin (fl. 864).

27. See pseudo-Kâlidâsa, *Jyotirvidābharaṇa*, ed. Sītârâma Śarmâ, Bombay, 1908, 22, 8-12.

28. This is summarised by Varâhamihira (fl. c. 550) in his *Pañcasiddhāntika*, ed. G. Thibaut and S. Dvivedin, Benares, 1889, chapters 16 and 17; I am preparing a new edition of this work also, based on a number of manuscripts not known to the first editors. For Lâṭadeva's authorship of the *Old Sūryasiddhānta*, see al-Bīrûnī's *Kitāb fī tahqīq mā li 'l-Hind*, ed. Hyderabad, 1958, p. 118 ; trans. E.C. Sachau, London, 1910, vol. 1, p. 153.

29. *Mahābhāskarīya*, ed. K.S. Shukla, Lucknow, 1960, 7, 21-35.

30. The recension of Pṛthûdakasvâmin was published in an inadequate edition based only on the Berlin manuscript by P.C. Sengupta, Calcutta, 1941, and translated by the same scholar, Calcutta 1934 ; the later recension of Āmarâja, who apparently flourished in the thirteenth century, was published by Babua Misra, Calcutta, 1925. I have discovered two manuscripts of the important commentary by Utpala (fl.966), which was also known to al-Bīrûnī.

31. An edition with translation and commentary of this work is being prepared by E.S. Kennedy and myself ; the present reference is to 95 : 15sqq.

32. The collaborator with Nawbakht in casting the horoscope of Bâghdad in 762, Mâshâ' allâh was one of the foremost astrologers of the early 'Abbâsid priod ; see, e.g., Ibn al-Nīdum, pp. 273-274, Ibn al-Qufṭī, p. 327, and Abû al-Faraj, p. 136. See also Kennedy and Pingree, *The Astrological History of Mashā'allāh*, which is shortly to appear.

the story that the Sasanian emperor Khusro Anûshirwân called together his astronomers and astrologers to compare Ptolemy (the Pahlavî version of the *Syntaxis*) with the *Zîj al-Arkand* the Pahlavî presentation of the ārdharātrika system). They found the *Zîj al-Arkand* to be preferable both astronomically and astrologically, and so wrote for Anûshirwân a *Zîj al-Shâh* (*Zîk î Shahriyârân*) based upon it and using four kardajas[33]. This story is substantiated and added to by a passage in the *Al-Qânûn al-Mas ûdî* of al-Bîrûnî[34], who informs us that the convocation of astrologers took place in the twenty-fifth year of Anûshirwân or 555/6.

Mâshâ'allâh used this sixth century version of the *Zîk î Shahriyârân* in computing horoscopes, especially those for his astrological history of the world. A summary of this history has fortunately been preserved by a Christian astrologer of Baghdâd, Ibn Hibintâ (fl. c. 950), in his *Kitâb al-Mughnî* ; and from the sixteen horoscopes that this summary contains it has been possible to extract a fair amount of information about this *zîk's* theory of Saturn and Jupiter[35]. Its relation to the ārdharātrika system is, in fact, not very close, being most evident in its choice of values for the maximum equations and in its use of the method of sines for computing the *mandaphala*.

Mâshâ 'allâh, in the same passage from al-Hâshimî that has been mentioned above, refers to yet another version of the *Zîj al-Shâh,* composed under the last of the Sasanian emperors, Yazdijird III (632-652) ; this was called "The Triple" because it utilized only three kardajas. It is apparently this text which was translated into Arabic by Abû al-Ḥasan 'Alî ibn Ziyâd al-Tamîmî[36]. and which is quoted extensively by al-Hâshimî and by al-Bîrûnî[37]. It seems clear that al-Bîrûnî (whom we shall discuss more fully below) used Yazdijird III's version of the *Zîj-al-Shâh* when he composed his *Zîj al-Sindhind* (*Siddhānta*) in 770/1 or 772/3[38], but it is not certain whether he used the Pahlavî original or was already able to refer to al-Tamîmî's translation.

This brief survey has demonstrated that, on a professional level, Sasanian astronomy was influenced by two Indian traditions : that of the *Paitāmaha-siddhānta* of the *Viṣṇudharmottarapurāṇa* and that of the ārdharātrika system of Āryabhaṭa I. It might not be out of place here to record also the theory found in the fifth chapter of the *Bundahishn*[39] and repeated in numerous

33. Kardaja is an attempt to render the Sanskrit *Kramajyā* ; for its significance see the commentary of al-Hâshimî, *passim*.

34. Ed. Hyderabad, 3 vols., 1954-56 : vol. 3, pp. 1473-74.

35. See the forthcoming publication cited in fn. 32.

36. Ibn al-Nadîm, p. 244.

37. See, for example, E.S. Kennedy, "The Sasanian Astronomical Handbook *Zîj-i Shâh* and the Astrological Doctrine of 'Transit' (*mamarr*)", *JAOS* 78, 1958, 246-262.

38. The details are discussed in my article, "The Fragments of the Works of Ya 'qûb, ibn Ṭâriq".

39. *BSOAS* 27, 1964, 516-517 and 519.

other Iranian sources[40] that explains the *śīghraphalas* by means of chords linking the planets to the chariot of the Sun; these chords are surely the *vāta-raśmis* of the *Modern Sūryasiddhānta*[41].

Equally impressive is the Indian influence on Sasanian astrology. The fifth century revision of the Pahlavî translation of Dorotheus of Sidon, for instance, mentioned and used the navâṃśas. The horoscope of Gayômart in the fifth chapter of the *Bundahishn* is not the Hellenistic *thema mundi*[42], but the Indian horoscope for the birth of the highest type of *mahāpuruṣa* in which all the planets are in their exaltations[43]. And the Islamic astrologers of the early Abbâsid period quote many Persian (*i.e.*, Sasanian) authors who have used Indian sources ; we mention here only Buzurjmihr and Andarz-ghar[44].

The first Arabic astronomical text, however, based directly on a Sanskrit source was a *Zîj al-Arkand* written in Sind in A.H. 117, which equals 103 of the Era of Yazdijird (735)[45]; since this work employs as its base-date the year Śaka 587, it is clear that it was influenced by Brahmagupta's *Khaṇḍakhādyaka*. Al-Bîrûnî revised this *Zîj- al-Arkand* because its language was so atrocious[46]. According to al-Hâshimî[47] two abridgements of the *Zîj al-Arkand* were made in Qandahar: a *Zîj al-Arkand* and a *Zîj al-Hazûr*. Furthermore, a series of horoscopes of the two equinoxes of the years in which the Sasanian emperors began their reigns was computed by means of the ārdharātrika system (the *Zîj al-Arkand* ?) towards the end of the eighth century[48].

A second late 'Ummayid work influenced by Indian astronomy is the *Zîj al-Harqan* cited by al-Bîrûnî[49]; *harqan* is clearly another attempt to transliterate *ahargaṇa*. The epoch of this *zîj* is Sunday 21 Daymâh 110 of the Era of Yazdijird, or 11 March 742. Little else is known of it.

---

40. *Isis* 54, 1963, 242.

41. Ed. K. Chaudhary, *KSS* 144, Benares 1946, 2, 2.

42. See, e.g., A. Bouche-Leclercq, *L'Astrologie grecque*, Paris, 1899, repr. Bruxelles, 1963, pp. 185-188.

43. See *Yavanajātaka* 8, 3-5 and 9, 1 ; my edition is in the press, and will appear in the *Harvard Oriental Series.* Note that in the *Bhudahishn* the Sun is in the first nakṣatra; this allows it to be in its exaltation according to Indian astrologers (Aries 10°), but not according to Greek astrologers (Aries 19°).

44. Some of the major Arabic sources are : pseudo-Mâshâ' allâh in MS 2122 of the Laleli Mosque, Istanbul; the *Kitâb a -Masâ'il* of Abû Yûsuf Ya'qûb ibn Alî al-Qaṣrânî (fl. c. 810); the *Kitâb aṣl al-uṣûl* of Abû al- 'Anbas Aḥmad ibn Muḥammad al-Ṣaymarî (died 888/9); the various works of Sahl ibn Bishr and of Abû Ma'shar (787-886) ; the *Kitâb al-Mughnî* of Ibn Hibintâ; and the *Majmâ aqâwîl al-ḥukamâ al-munajjimîn* of al-Dâmaghânî (fl. 1113/4)

45. Al-Bîrûnî, *India*, ed. pp. 383-384, trans., vol. 2, pp. 48-49.

46. D.J. Boilot, " L'oeuvre d'al-Beruni ; essai bibliographique " *M-DEO* 2, 1955, 161 256, RG 6.

47. *Kitâb 'ilal al-zîjât* 94: 2sqq.

48. D. Pingree, " Historical Horoscopes ", *JAOS* 82, 1962, 487-502 ; and now, *The Thousands of Abû Ma'shar,* section X B.

49. *India*, ed. p. 387, trans., vol. 2, pp. 52-53,

But the most influential translation was that of a work apparently called the *Mahāsiddhānta*[50]. The basic planetary parameters of this work belonged to the tradition of the *Paitāmahasiddhānta* of the *Viṣṇudharmottarapurāṇa* and the *Brāhmasphuṭasiddhānta* of Brahmagupta, though it also included elements derived from the *Āryabhaṭīya*. According to the *Nizâm al-iqd* of Ibn al-Adamî (c. 920), which was completed by al-Qâsim ibn Muhammad al-Madâ'inî in 949/50, a man from India brought this text (which is associated with Fiyaghra, *i.e.*, Vyāghramukha, the Câpa prince under whom Brahmagupta wrote) to Baghdad in 772/3, and the Caliph al-Manṣûr ordered Muhammad ibn Ibrahîm al-Fazârî to translate it ; al-Bîrûnî, who says that the man from India was a member of a delegation from Sind, dates his coming to Baghdad in 770/1. Al-Fazârî obeyed the Caliph, and the result was his *Zîj pal-Sindhind al-Kabîr*; but he was influenced in his choice of values for the maximum equations of the planets more by the *Zîj al-Shâh* than by the *Mahāsiddhānta*. At a later date al-Fazârî wrote another *zîj* based by the same parameters, but with the planetary mean motions tabulated for saura days and using the Hijra calendar instead of that of the Yazdijird III.

Descended from the *Zîj al-Sindhind al-Kabîr* are a large number of texts; the earliest were due to Yaqûb ibn Ṭâriq, who composed a *zîj*, an interesting book entitled *Tarkîb al-aflâk* (written in 777/8), and *Kitâb al-ilal* all of which drew heavily upon the Indian and Sasanian sources mentioned above. The authorship of the *Zîj al-Sindhind al-Ṣaghîr* remains obscure, though its writing must have been approximately contemporaneous with al-Fazârî and Ya'qûb ibn Ṭâriq; and from al-Hâshimî we learn some details of yet a third *Zîj al-Sindhind* which was apparently written in 792/3.

But the most famous *Zîj al-Sindhind* is that of Muhammad ibn Mûsâ al-Khwârizmî (fl. 828), who is known to have written it for the Caliph al-Mamûn (809-833)[51]. Of the Arabic original only fragments, survive[52]; but a recension of it was made in Spain by Maslama ibn Aḥmad al-Majrîṭî (died 1007/8) or by his pupil Ibn al-Ṣaffâr (died 1035). This recension, lost to us in its original Arabic, is preserved in a Latin translation made by Adelhard of Bath in 1126[53]. Besides this translation a number of important commentaries are

50. Al-Bîrûnî in the *India*, ed. p. 356, calls it the *Sidhând al-kabîr*. The evidence supporting the statements made in the next two paragraphs will be found assembled in my article, "The Fragments of the Works of Ya'qûb ibn Ṭâriq", and in Kennedy and Pingree's edition of al-Hâshimî ; it is not necessary to repeat it all here.

51. See the passage from Ibn al-Adamî translated in my "The Fragments of the Works of Ya'qûb ibn Ṭâriq"; on al-Khwârizmî himself see, inter al., Ibn al-Nadîm, p. 274, and Ibn al-Qifṭî, p. 286.

52. Besides the numerous quotations and references in al-Hâshimî and al-Bîrûnî and the lommata in the commentary of Ibn Masrûr, chapters of the *Zîj al-Sindhind* are found in Ibn Hibintâ's *Kitâb al-Mughnî* (München Arab 852 ff. vv. 33-37) and in Nuruosmaniye 2795 ff. vv. 65-66.

53. Edited by A. Bjornbo R. Besthorn, and H. Suter Kobenhavn 1914 ; English translation and commenttary by O. Neugebauer, Kobenhan 1962. *See also* G. Toomer in *Centaurus* 10, 1964, 23-212.

known. That by Ahmad ibn Muhammad ibn Kathîr al-Farghânî[54], who flourished under al-Mamûn and his immediate successors, is extensively quoted by al-Hâshimî and al-Bîrûnî[55]. The *Kitâb ilal al-zîjât* preserved in a unique manuscript in Cairo and attributed to Ibn Masrûr, the pupil of Abû Ma'shar (787-886), is, in fact, another commentary on al-Khwârizmî's *Zîj al-Sindhind.* The Arabic original of the commentary written by Ahmad ibn al-Muthanna' in the tenth or eleventh century is now lost , but we do have of it a Latin translation[56] made by Hugo of Sanctalla in northern Spain for Michael, the Bishop of Tarazona from 1119 to 1151, and a Hebrew translation[57] made by Abraham ibn Ezra of Tudela in c. 1160. Three other commentators namely, Abû Ṭalḥa, Abû al-Ḥasan al-Ahwâzî[58], and Muḥammad ibn 'Abd al-'Azîz al-Hâshimî[59]—are recorded by al-Bîrûnî[60]. Other authors of *zîjes* based in whole or in part on the *Zîj al-Sindhind* are Abû Ma'shar[61], Ḥabash al-Ḥâsib (c. 850)[62], al-Ḥasan ibn al-Ṣabbâḥ (c. 870)[63], al-Faḍl ibn Ḥâtim al-Nayrîzî (c. 900)[64] the sons of Amâjûr (c. 910)[65], and Ibn al-Adamî[66]; all of these lived and worked in the eastern regions of the Islam world. The Spanish tradition of the *Sindhind* is represented, as well as by those authors mentioned previously in connection with al-Khwârizmî, by Ibn al-Samḥ (c. 1010)[67], al-Jahânî (1079), whose *zîj* survives in a Latin translation by Gerard of Cremona (1114-1187)[68], and al-Zarqâlla (c. 1100)[69].

Besides the *Zîj al-Arkand* and the *Zîj- al-Sindhind,* the only other Indian astronomical text mentioned in early Muslim sources is the *Zîi al-Arjabhar* (*Āryabhaṭīya*). This, however, was evidently very imperfectly known, the only Abbâsid astronomer who had any real knowledge of its planetary theory seems to have been the commentator on al-Khwârizmî, Abû al-Ḥasan al-Ahwâzî[70]. Others, however, are said to have studied in India ; Muhammad

---

54. See, e.g., Ibn al-Nadîm, p. 279, and Ibn al-Qifṭî, pp. 78 and 286.

55. See Kennedy and Pingree's edition of al-Hâshimî.

56. A very poor edition was published by E. Milla's Vendrell, Madrid—Barcelona, 1963.

57. See the forthcoming edition by B. Goldstein in the Yale Univrsity Press. Abraham ibn Ezra also has valuable information on the *Sindhind* tradition in his *De rationibus tabularum,* ed. J.M. Milla's Vallicrosa, Madrid-Barcelona, 1947.

58. Boilot, RG. 2 and 3.

59. *Rasâlil a -Bîrûnî,* ed. Hyderabad, 1948, pt. 1, p. 118.

60. See also his own work, Boilot, RG 1.

61. For his *Zîjal-Hazârât* see my *The Thousands of Abû Ma'shar.*

62. E. S. Kennedy, " A Survey of Islamic Astronomical Tables ",*TAPhS, NS.* 46, 1956, 123-177, nos. 15 and 16.

63. Kennedy, " Survey ", No. 31.

64. Kennedy, " Survey ", Nos. 46 and 75.

65. Kennedy, " Survey ", No. 90.

66. Kennedy, " Survey ", No. 18.

67. Kennedy, " Survey ", No. 26.

68. Ed. J. Heller, Nuremburg, 1549.

69. Kennedy, " Survey ", No. 24.

70. Al-Bîrûnî, *India,* ed. p. 357, trans., vol. 2. 19.

ibn Ismâîl al-Tanûkhî, for instance, is supposed to have brought back curious notions regarding trepidation from India[71]. In fact, the value of precision arrived at by Yaḥyâ ibn abî Manṣûr after he observed the equinox of 19 September 830 is that which pertains to the *Old Sūryasiddhānta* of Lāṭadeva and to various other Sanskrit texts[72].

Besides the astronomer in the delegation sent from Sind in the early 770's a number of Indian astrologers visited Baghdad in the Abbâsid period. The most notable of these was Kanaka, who served Harûn al-Rashîd ; he is perhaps identical with the homonymous astrologer cited by Kalyâṇavarman in his *Sārāvalī*[73]. Other Sanskrit names are corruptedly recorded by Ibn al-Nadîm[74]: J.w.d.r, S.n.j.h.l, N.h.q, Râḥ. h, Ṣ.k.h, Dâh.r, An. k.w, Z.n.k.l, Ar.y.k.l, J.b.h.r. An.d.y, and J.bâr.y. The fourth of these names should undoubtedly be read Râja ; he is an astrologer frequently quoted by such Arabic compilers as al-Qaṣrânî and al-Ṣaymarî. Through such works as these the theories of Indian astronomy and astrology came to permeate the scientific literature of the Muslim and Christian worlds long before Vasco da Gama sailed boldly onto the Indian Ocean.

---

71. Ṣâ'id al-Andalusî ed. p. 56, trans. p. 112, and Ibn al-Qifṭî, p. 281.

72. *Dumbarton Oaks Papers* 18, 1964, 138.

73. For a discussion of Kanaka, see my *The Thousands of Abū Ma'shar* and " The Fragments of the Works of Ya'qûb ibn Ṭâriq. "

74. *Fihrist*, p. 271.

THE FRAGMENTS OF THE WORKS OF YAᶜQŪB IBN ṬĀRIQ[1]

DAVID PINGREE, University of Chicago

THE following collection of fragments and the succeeding article by E. S. Kennedy present all the material so far discovered relevant to one of the earliest of ᶜAbbāsid astronomers and attempt to interpret that material historically and scientifically. Many absurd assertions have been made concerning early Islamic science by historians who have not had the time or ambition to read the original sources but who are content to continue the historiographic tradition begun in Spain in the twelfth century. These two articles and similar collections of other early Muslim astronomers and astrologers will attempt to provide a different basis for assessing the formative period of science in Baghdād.

The *Fihrist* of Ibn al-Nadīm (p. 278), which is copied in a rather inaccurate manner by Ibn al-Qifṭī (p. 378), tells us of Yaᶜqūb very little indeed.

> Yaᶜqūb ibn Ṭāriq, one of the best astronomers. Among his books are: *Kitāb taqṭī kardajāt al-jayb; Kitāb mā irtafaᶜa min qaws niṣf al-nahār*; and *Kitāb al-zīj maḥlūl fī al-Sindhind li-daraja daraja*, which is in two books; the first is on the science of the sphere, and the second on the science of the dynasty (*duwal*).

The first of these works must have described the method of converting a table of sines whose argument is expressed in intervals of 3;45° (a normal Indian table of *kramajyā*'s, from which, apparently through Pahlavī, comes the Arabic *kardaja*) to one whose argument is expressed in intervals of 1°. It was either a part of, or was used in writing, the *Kitāb al-zīj*. The second work apparently deals with the problem of determining the altitude of the Sun from the day-circle. It may have been extracted from the *zīj*, or from the *Kitāb al-ᶜilal* which Ibn al-Nadīm neglects to mention. The fragments of the third work are discussed below. The subject of the second book of this *zīj*, *ᶜilm al-duwal*, seems extremely peculiar; perhaps one should amend the text of both Ibn al-Nadīm and Ibn al-Qifṭī(!) to *ᶜilm al-dawr*, "science of revolution(s)."

The *Sindhind* upon which Yaᶜqūb's *zīj* was based was, of course, that translated by al-Fazārī[2] from a Sanskrit work allied to the *Paitāmahasiddhānta* of the *Viṣṇudharmottarapurāṇa*[3] and the *Brāhmasphuṭasiddhānta* of Brahmagupta.[4] The Sanskrit work was brought to Baghdād by a member of an embassy sent from Sind to the court of al-Manṣūr (754–775);[5] the date is variously reported as being A.H. 154 (24 Dec. 770–12

[1] The best previous discussion of Yaᶜqūb is by Nallino, *Raccolta*, 5, 215 ff.

[2] On al-Fazārī see D. Pingree, "The Fragments of the Works of al-Fazārī," to appear in a future issue of *JNES*, and the commentary to the forthcoming edition of al-Hāshimī's *Kitāb ᶜilal al-zījāt* by E. S. Kennedy and D. Pingree.

[3] → *JNES*, 24 (1965), 334–36.

[4] Brahmagupta wrote the *Brāhmasphuṭasiddhānta* at Bhillamāla (Bhinmal) in southern Rājasthān in Śaka 550 (A.D. 628) during the reign of Vyāghramukha of the Cāpa family (24, 7). Ibn al-Adamī, according to Ṣāᶜid al-Andalusī, ed. p. 50, trans. p. 102, refers to Q.b.gh.r. (*var.* F.y.gh.r) the Indian king; according to Ibn al-Qifṭī (p. 270) he refers to F.y.gh.r. This latter pointing is correct, and Fyaghra is Vyāghra(mukha)..

[5] See Ibn al-Adamī in the passages cited above in n. 4; al-Bīrūnī, *India*, ed. p. 351, trans. Vol. 2, p. 15; and al-Hāshimī, *Kitāb ᶜilal al-zījāt* 95 v: 3 ff.

Dec. 771)[6] and A.H. 156 (2 Dec. 772–20 Nov. 773).[7] Al-Bīrūnī[8] gives another date for the embassy, A.H. 161 (9 Oct. 777–27 Sept. 778) which, falling outside of al-Manṣūr's reign, is probably instead the date of Yaᶜqūb ibn Ṭāriq's *Tarkīb al-aflāk*. A fourth date, 1,972,947,868 years from the beginning of the *kalpa* (A.D. 767), can be discerned in al-Hāshimī,[9] but it is not clear to what it refers. The member of the embassy from Sind who is associated with the *Sindhind* was later identified, mainly by Andalusian scholars,[10] with Kanaka, the astrologer of Hārūn al-Rashīd;[11] this identification has no basis.

Besides the three works mentioned by Ibn al-Nadīm, which are probably really one (the *zīj*), we know from a number of sources that Yaᶜqūb wrote a book entitled *Tarkīb al-aflāk*. This was a treatise, as its title indicates, on the arrangement of the heavens; as was indicated above, it was probably composed in 777/8. A third work was his *Kitāb al-ᶜilal*, which explained the rationale for the mathematical procedures followed by astronomers; a long series of books of this sort was written by subsequent writers.[12] Below are given the fragments with some brief comments of Yaᶜqūb's *Z(īj)*, *T(arkīb al-aflāk)*, and *K(itāb al-ᶜilal)*.

ZĪJ

Z 1. Al-Hāshimī, *Kitāb ᶜilal al-zījāt*, 95 v: 12–17.

As for Yaᶜqūb ibn Ṭāriq, he composed his *zīj* for Bārah, and its revolutions are in agreement with (those of) the *Sindhind* as to the four hundredth part, but he made its collected (years) (*majmūᶜ*) ten years according to years of the Persians and their months, up to noon of the day after your day. Its apogees and nodes agree with the *Sindhind*. However, as for the equation(s), some of them are from the sayings of the Persians. (As for) the division of its equation(s) and its other operations, some of them are according to the dictum of the Indians and some according to the Persians.

Commentary

Bārah, as we shall see from Z 3 and Z 4, is the same as the Tārah which al-Bīrūnī locates with Yamakoṭi 90° East of the meridian of Ujjayinī-Laṅkā and on the equator. Yamakoṭi is one of the four cities on the equator 90° apart mentioned by Āryabhaṭa (*Āryabhaṭīya, Gola* 13) and Varāhamihira (*Pañcasiddhāntikā* 13, 17; see Pingree, *The Thousands of Abū Maᶜshar*, p. 45); the four are, proceeding Eastwards, Laṅkā, Yamakoṭi, Siddhapura, and Romaka. Yamakoṭi, "the top of Yama's (land)," was identified at some point with the Var-i-Yamkart which the *Mēnok-i-χrat* (27, 27–31) states was built by Yamšēt to save creatures from a flood. *Vara* of Var-i-Yamkart is clearly Bārah, which al-Fazārī takes for the prime meridian of the *Sindhind*; Yaᶜqūb here follows al-Fazārī, who chose Bārah as prime meridian to eliminate the six hour

[6] Al-Bīrūnī, *India*, ed. p. 351, trans. Vol. 2, p. 15.

[7] Ibn al-Adamī in the passages cited above in n. 4.

[8] *India*, ed. p. 397, trans. Vol. 2, pp. 67–68.

[9] *Kitāb ᶜilal al-zījāt*, 104 v: 1.

[10] Especially by Abū Muᶜādh al-Jahānī (fl. 1079), f. Zi, and by Abraham ben Ezra (*ca.* 1090–1167) in fr. Z2 below and in *Liber de rationibus tabularum*, p. 92; cf. *The Thousands of Abū Maᶜshar*, p. 16. The fantasies of pseudo-al-Majrīṭī's *Ghāyat al-ḥakīm wa aḥaqq al-natījatayn bi al-taqdīm*, ed. pp. 278 ff., trans. pp. 285 ff., make "Kanka the Indian" a wise man and king who built Memphis and performed many other fabulous feats. The name, then, signified for the Spaniards a quite different personality from the Kanka whose rather ordinary astrological theories are cited by al-Qaṣrānī, Ibn Hibintā, and al-Bīrūnī.

[11] See *al-Bīrūnī, Chronology*, ed. p. 132, trans. p. 129.

[12] Among them may be named ᶜUmar ibn al-Farrukhān al-Ṭabarī, Muḥammad ibn Kathīr al-Farghānī, ᶜAbdallāh ibn Masrūr, ᶜAlī ibn Sulaymān al-Hāshimī, Muḥammad ibn ᶜAbd al-ᶜAzīz al-Hāshimī, al-Bīrūnī, and Abraham ben Ezra.

difference between the sunrise epoch of the *Paitāmahasiddhānta-Brāhmasphuṭasiddhānta* and his own noon epoch.

Though the *Mēnok-i-χrat* (27, 57–58) says that Syāvaχš, not Yamšēt, built Kangdiz, the latter seems to have been identified with Var-i-Yamkart, and its longitude is the prime meridian of Abū Maʿshar's *Zīj al-hazārāt* (*The Thousands of Abū Maʿshar*, pp. 41–45.)

The 400th part is the fraction of an hour which, when added to 365¼ days plus ⅕ hour, results in the *Sindhind* parameter for the length of a solar year—namely, 6,5;15, 30,22,30 days. This year length was used by al-Fazārī (see al-Hāshimī, 95 v: 18), and by the *Paitāmahasiddhānta* of the *Viṣṇudharmottarapurāṇa* and by Brahmagupta in his *Brāhmasphuṭasiddhānta*; in the two Sanskrit works it is referred to as the occurrence of 4,320,000,000 revolutions of the Sun in 1,577,916,450,000 days. It is certain that Yaʿqūb's other parameters for mean motions were taken from the *Sindhind*, as were those of the Sun and of the apogees and nodes specifically referred to by al-Hāshimī (see Z 5).

	Paitāmahasiddhānta-Brāhmasphuṭasiddhānta	al-Fazārī (acc. to al-Bīrūnī, *India*, ed. pp. 352–53, trans. Vol. 2, p. 16; cf. al-Hāshimī, 104: 16 ff.)
Saturn	146,567,298	146,569,284
Saturn's apogee	41	41
Saturn's node	584	584
Jupiter	364,226,455	364,226,455
Jupiter's apogee	855	855
Jupiter's node	63	63
Mars	2,296,828,522	2,296,828,522
Mars' apogee	292	292
Mars' node	267	267
Sun	4,320,000,000	4,320,000,000
Sun's apogee	480	480
Venus' conjunction	7,022,389,492	7,022,389,492
Venus' apogee	653	653
Venus' node	893	893
Mercury's conjunction	17,936,998,984	17,936,998,984
Mercury's apogee	332	332
Mercury's node	521	521
Moon	57,753,300,000	57,753,300,000
Moon's apogee	488,105,858	488,105,858
Moon's node	232,311,168	232,312,138
Fixed stars	omitted	120,000
Civil days	1,577,916,450,000	1,577,916,450,000

These parameters yield the following mean daily motions of the planets.

	Paitāmahasiddhānta-Brāhmasphuṭasiddhānta	al-Fazārī
Saturn	0;2,0,22,51,45,45°	0;2,0,22,57,36,16°
Jupiter	0;4,59,9,8,37,23	0;4,59,9,8,37,23
Mars	0;31,26,28,6,47,45	0;31,26,28,6,47,45
Sun	0;59,8,10,21,33,30	0;59,8,10,21,33,30
Venus' conjunction	1;36,7,44,35,18,26	1;36,7,44,35,18,26
[Venus' anomaly	0;36,59,34,13,44,56	0;36,59,34,13,44,56]

	Paitāmahasiddhānta-Brāhmasphuṭasiddhānta	al-Fazārī
Mercury's conjunction	4;5,32,18,27,45,30	4;5,32,18,27,45,30
[Mercury's anomaly	3;6,24,8,6,12,0	3;6,24,8,6,12,0]
Moon	13;10,34,52,46,30,13	13;10,34,52,46,30,13
Moon's apogee	0;6,40,53,56,32,55	0;6,40,53,56,32,55
[Moon's anomaly	13;3,53,58,49,57,18	13;3,53,58,49,57,18]
Moon's node	– 0;3,10,48,20,6,41	– 0;3,10,48,22,58,46

The calendar utilized by Yaᶜqūb ibn Ṭāriq was the Persian, consisting of 365 days divided into 12 months of 30 days each and five epagomenal days added after the eighth month:

1. Farwardīn	1–30	7. Mihr	181–210
2. Urdībihisht	31–60	8. Ābān	211–240
3. Khardādh	61–90	Epagomenal days	241–245
4. Tīr	91–120	9. Ādhar	246–275
5. Murdādh	121–150	10. Day	276–305
6. Shahrīwar	151–180	11. Bahman	306–335
		12. Isfandārmudh	336–365

The epoch—16 June 632—falls in the first year of the reign of Yazdijird III. The use of noon epoch differs from the tradition of the *Paitāmahasiddhānta* and *Brāhmasphuṭasiddhānta*, which use dawn; it is rather the usage of Ptolemy. Al-Fazārī's father, Abū Isḥāq Ibrāhīm ibn Ḥabīb al-Fazārī, (or rather al- Fazārī himself) is supposed already to have written a *zīj* employing the Arab calendar (*Fihrist*, p. 273); the calendar used by the son, Muḥammad ibn Ibrāhīm al-Fazārī (*Fihrist*, p. 79), in his *Sindhind al-kabīr* was probably, like Yaᶜqūb's, the Persian calendar of Yazdijird III. Al-Khwārizmī, who, according to Ibn al-Adamī (in the passage cited in n. 4), retained the mean motions of the *Sindhind al-kabīr* of the younger al-Fazārī, used the Yazdijird calendar (so Ibn Masrūr and Ibn al-Muthannā; see also Abraham ben Ezra, *De rationibus tabularum*, p. 74).

As the *ahargaṇa* to dawn of Tuesday 16 June 632 is 720,635,806,312 days and we can assume six hours difference between dawn and noon which disappears in the difference between Lañkā and Bārah, we can compute the following epoch positions of the planets, their apogees, and their nodes (note that al-Hāshimī specifies "up to noon of the day after your day"; this indicates that the epoch positions were given for day 1 rather than day 0).

Noon of Tuesday 16 June 632 at Bārah

	mean longitude	apogee	node
Saturn	♏29;11,7°	♐20;55°	♋13;12°
Jupiter	♑1;29,25	♍22;32	♊22;1
Mars	♒9;51,35	♌8;24	♈21;54
Sun	♊25;59,48	♊17;55	
Venus (conj.)	♎26;49,38	♊21;15	♉29;47
Venus (anom.)	120;49,50		
Mercury (conj.)	♐17;2,12	♏14;54	♈21;11
Mercury (anom.)	171;2,24		
Moon	♓28;44,56	♉26;48,35	♑24;5,7
Moon's anomaly	301;56,21		
Regulus	♌7;14,4		

Given this data, it is only a question of time to compute Yaᶜqūb's mean motion tables for 10-year periods (collected years), single years (from 1 to 10), 30 days (Persian months), and for each of the 365 days of a Persian year.

Yaᶜqūb, like al-Fazārī, does indeed derive most of his equations from the *Zīj al-Shāh*; see Z 7 and Z 8. The "division of its equation(s)" perhaps signifies the interval which he employs in the column of the argument; Ibn al-Nadīm indicates that this was 1°, which certainly differs from ancient Indian practice. The "other operations" must include his method of combining the equation of the center with the equation of the anomaly; we can say nothing about the system he used.

Z 2. Abraham ben Ezra, Preface to his translation of Ibn al-Muthannā's *Fī ᶜilal zīj al-Khwārizmī* (pp. 147–48 Goldstein).

So he (the king al-Ṣaffāh [750–54]) gave great wealth to the Jew who had translated the above-mentioned book (from Sanskrit), so that he might go to the city of Arin (i.e., Ujjain), which lies on the equator at the latitude of Aries and Libra, where on all the days of the year the length of daylight is equal to the length of night, neither longer nor shorter; he was especially to bring their wise men to the king. The Jew went and made use of many tricks, until a wise man of Arin decided to come to the king for a large sum (of money) after the Jew promised him that he would stay for only one year, and then could return home. Then the wise man, whose name was Kankah, was brought to the king, and taught the Arabs the basis of number, which lies in nine characters. Then a learned man named Yaᶜqūb ibn Sharah (Ṭāriq) translated from the language of this learned (Indian), through the medium of the Jew who translated in Arabic, a book of tables of the seven planets, all operations relating to the earth, rising-times, declination, the ascendent, the determination of the beginnings of the astrological houses, the knowledge of the superior planets(?), and solar and lunar eclipses. However, in this book he did not note down the causes of all these things, but mentioned only the fact according to (the Indian's) tradition. In this (book) the mean motions of the planets are according to the calculation of the Indians, who call their cycle *hāzarwān*(?); this cycle comprehends 432,000 (read 4,320,000,000) years.

COMMENTARY

The fable of Kankah, the Jewish intermediary, and Yaᶜqūb is clearly based on contemporary Spanish techniques of translating Arabic into Latin, and has no historical foundation. The subjects dealt with in the translation, according to Abraham, are normally treated in Sanskrit *siddhānta*'s in the first five or six chapters; in the *Brāhmasphuṭasiddhānta*, for instance, the *Daśādhyāyī* consists of:

1. *Madhyamādhikāra.* Mean motions.
2. *Spaṣṭādhikāra.* True longitudes.
3. *Tripraśnādhikāra.* On the gnomon shadow, time, and the ascendent.
4. *Candragrahaṇādhikāra.* Lunar eclipses.
5. *Sūryagrahaṇādhikāra.* Solar eclipses.
6. *Udayāstādhikāra.* Heliacal risings and settings.
7. *Candrasṛṅgonnatyadhikāra.* Lunar crescent.
8. *Candracchāyādhikāra.* Shadow of the Moon.
9. *Grahayutyadhikāra.* Planetary conjunctions.
10. *Bhagrahayutyadhikāra.* Planetary latitude and transits.

For an embassy from Sind to the court of al- Ṣaffāh in A.H. 136 (753/4), see al-Yaᶜqūbī, Vol. 2, p. 361.

Z 3. Al-Bīrūnī, *Al-Qānūn al-Masᶜūdī*, vol. 2, p. 547.

Localities along the equator, possessing no latitude.

1. The island Lañkā (*lanka*), called in books the Cupola of the earth. Long. 100;50° Lat. 0;0°

2. Tārah (read Bārah), which al-Fazārī and Yaᶜqūb ibn Ṭāriq mention. Long. 190;50° Lat. 0;0°

3. Yamakoṭi (jamkūt) at the Eastern extremity, which is Yamkard (*jamākard*) according to the Persians; there is no building beyond it according to the Indians. Long. 190;0° Lat. 0;0°

COMMENTARY

The 90° difference between the longitudes of Lañkā and Bārah has already been discussed and explained above; but that al-Bīrūnī here gives different longitudes to Bārah and Yamakoṭi is most surprising. Al-Bīrūnī's longitudes are measured eastwards from the Isles of the Blessed.

Z 4. Al-Bīrūnī, *India*, ed. p. 259, trans. Vol. 1, pp. 303–304.

Yamakoṭi is, according to Yaᶜqūb and al-Fazārī, the country where is the city Tārah (read Bārah) within a sea. I have not found the slightest trace of this name (i.e., Bārah) in Indian literature. As *koṭi* means "castle" and Yama is the angel of death, the word reminds me of Kangdiz, which, according to the Persians, had been built by Kaykāʾūs or Jam in the most remote East, behind the sea.

COMMENTARY

Koṭi does not mean "castle" in Sanskrit, but "tip," "top," "eminence"; the word *koṭa*, however, does signify "castle." Yama is indeed, in Indian mythology, the guide of the dead.

Z 5. Al-Bīrūnī, *India*, ed. pp. 351–52, trans. Vol. 2, p. 15.

It is one of the conditions of a *kalpa* that in it the planets, with their apogees and nodes, must unite in 0° of Aries, *i.e.* in the point of the vernal equinox. Therefore, each planet makes within a *kalpa* a certain number of complete revolutions or cycles. These revolutions of the planets, as known through the *zīj*(es) of al-Fazārī and Yaᶜqūb ibn Ṭāriq, were derived from an Indian who came to Baghdād as a member of the political mission which Sind sent to the Caliph, al-Manṣūr, in A.H. 154 (24 Dec. 770–12 Dec. 771). If we compare these secondary statements with the primary statements of the Indians, we discover discrepancies, the cause of which is not known to me. Is their origin due to the translation of al-Fazārī and Yaᶜqūb? or to the dictation of that Indian? or to the fact that afterwards these computations have been corrected by Brahmagupta, or someone else?

COMMENTARY

The revolutions of the planets, their apogees, and their nodes in a *kalpa* have already been given in the commentary to Z 1; and the date of the embassy from Sind has been discussed in the introduction. Al-Bīrūnī's question regarding the origin of the different parameters for Saturn and the lunar node in the *Brāhmasphuṭasiddhānta* and in the *Sindhind* cannot be answered as yet.

Z 6. Al-Bīrūnī, *India*, ed. p. 356, trans. Vol. 2, pp. 18–19.

We meet in this context (i.e., the difference between a *caturyuga* of 4,320,000 years and a *kalpa* of 4,320,000,000 years) with a curious circumstance. Evidently al-Fazārī and Yaʿqūb sometimes heard from their Indian master to this effect, that his calculation of the revolutions of the planets was that of the great *Siddhānta*, while Āryabhaṭa reckoned with a one-thousandth part of it. They apparently did not understand him properly, and imagined that *āryabhaṭa* (*arjabhad*) means "a thousandth part."

COMMENTARY

This hypothetical reference of the unknown Indian to the "*great Siddhānta*," combined with the title of al-Fazārī's *zīj*, the *Sindhind al-kabīr*, permits one to surmise that the original Sanskrit text was entitled *Mahāsiddhānta*.

It may be noted that Āryabhaṭa in fact signifies "noble warrior," but the meaning "a thousandth" is found again in al-Hāshimī (93 v: 22). For the Arabic comparisons between the *Sindhind*, the *Arjabhar*, and the *Arkand*, see *The Thousands of Abū Maʿshar*, p. 16.

Z 7. Al-Bīrūnī, *On Transits*, 30:10–16.

As to the Indians and the Persians, they have a common opinion (regarding the maximum equation of the center); and so the *zīj*'es of the *Shāh*, of Abū Maʿshar, and of Yaʿqūb ibn Ṭāriq contain nothing on which they differ except only one thing, the difference of which does not exceed one minute . . .. And they have for Saturn 8;37°, and for Jupiter 5;6°, and for Mars 11;12°, and for Venus 2;13°, and for Mercury 4°.

COMMENTARY

In Sanskrit *siddhānta*'s, not the maximum equation of the center is given but the *mandaparidhi*. But, by using the formula

$$\mathrm{Sin}(E_{max}) = r$$

where r is the radius of the *manda*-circle, and the different values of R (R = 3438′ in the *Paitāmahasiddhānta*; R = 3270′ in the *Brāhmasphuṭasiddhānta*), we can compute the appropriate maximum equations of the center.

	Paitāmahasiddhānta		*Brāhmasphuṭasiddhānta*	
	mandaparidhi	E_{max}	*mandaparidhi*	E_{max}
Saturn	30°	4;46,46°	30°	4;46,47°
Jupiter	33	5;15,33	33	5;15,35
Mars	70	11;12,28	70	11;12,41
Sun	13;40	2;10,31	13;20	2;7,20
			14	2;13,42
Venus	11	1;45,3	9	1;25,57
			11	1;45,3
Mercury	38	6;3,31	38	6;3,33
Moon	31;36	5;2,7	30;44	4;53,50
			32;28	5;10,28

Clearly these are far from Yaʿqūb's values.

A Sanskrit *karaṇa*, however, will sometimes list maximum equations; and so, whereas the *Old Sūryasiddhānta* of Lāṭadeva and the summary in the *Mahābhāskarīya* give only the *mandaparidhi*'s of the *ārdharātrika* system, Brahmagupta's *Khaṇḍakhādyaka* informs us of that system's maximum equations of the center.

	Ārdharātrika system		*Zīj al-Shāh* of
	mandaparidhi	E_{max}	Yazdijird III.
Saturn	60°	9;36°	8;37°
Jupiter	32	5;6	5;6
Mars	70	11;10	11;12
Sun	14	2;14	2;14
Venus	14	2;14	2;13
Mercury	28	4;28	4;0
Moon	31	4;56	4;56

Al-Fazārī, as can be seen from the values preserved by al-Bīrūnī (*On Transits*, 24:17–25:5 and 30:19–31:8), in one *zīj* in which R = 150′ mixes maximum equations from the *Zīj al-Shāh* and the *ārdharātrika* system with other, new ones. (For different formulas of his wherein R = 3270′, see *On Transits*, 25:9–26:3.)

	al-Fazārī's E_{max} (R = 150′)	
Saturn	8;37,30°	(cf. *Zīj al-Shāh*)
Jupiter	5;6	*Zīj al-Shāh*
Mars	11;10	*ārdharātrika* system
Sun	2;11,15	
Venus	2;15	
Mercury	4;0	*Zīj al-Shāh*
Moon	5;0	

Yaᶜqūb, as indicated by al-Bīrūnī, here deserts al-Fazārī to return completely to the values of the *Zīj al-Shāh*.

Z 8. Al-Bīrūnī, *On Transits*, 54:15–17

And it must be that Yaᶜqūb ibn Ṭāriq is in agreement (concerning the maximum equations of the anomaly) with the two of them (*i.e.*, al-Fazārī and al-Khwārizmī); but what is in his *zīj* for Jupiter is decreased by 0;22°, and for Venus decreased by 0;55°.

COMMENTARY

Since we know the maximum equations of the anomaly for al-Khwārizmī, as well as those in the *Zīj al-Shāh*, which was al-Fazārī's source (see *On Transits*, 54:5–14), we can easily restore Yaᶜqūb's values.

	Zīj al-Shāh, al-Fazārī, and al-Khwārizmī	Yaᶜqūb ibn Ṭāriq
Saturn	5;44°	5;44°
Jupiter	10;52	10;30
Mars	40;31	40;31
Venus	47;11	46;16
Mercury	21;30	21;30

Using the same method as indicated above, we can find the maximum equations of the anomaly for the *Paitāmahasiddhānta* and the *Brāhmasphuṭasiddhānta*.

	Paitāmahasiddhānta		*Brāhmasphuṭasiddhānta*	
	śīghraparidhi	E_{max}	*śīghraparidhi*	E_{max}
Saturn	40°	6;22,9°	35°	5;34,46°
Jupiter	68	10;53,6	68	10;53,19
Mars	243	42;28,36	243;40	42;37,39
Venus	258	45;48,4	258	45;48,12
			263	46;57,43
Mercury	132	21;31,43	132	21;31,30

For the *ārdharātrika* system one has values closer to those of the *Zīj al-Shāh* and of Ya^cqūb, but still not identical with them.

	ārdharātrika system			
	śīghraparidhi	E_{max}	*Zīj al-Shāh*	Ya^cqūb
Saturn	40°	6;20°	5;44°	5;44°
Jupiter	72	11;30	10;52	10;30
Mars	234	40;30	40;31	40;31
Venus	260	46;15	47;11	46;16
Mercury	132	21;30	21;30	21;30

Ya^cqūb's 46;16° for Venus indicates that he had access to the *ārdharātrika* system independently of the *Zīj al-Shāh*.

Z 9. Abraham ben Ezra, *De rationibus tabularum*, p. 92.

The Indians said that the (maximum) declination of the Sun is 24°, as Iacob Abentaric transmitted (to us) from the words of Chenche (*i.e.*, Kanaka), the most learned of the Indians.

COMMENTARY

The value 24° is the standard Indian parameter for the obliquity of the ecliptic. For Kanaka, see our remarks in the introduction.

TARKĪB AL-AFLĀK

T 1. Al-Bīrūnī, *India*, ed. pp. 397–400, trans. Vol. 2, pp. 67–68.

The only Indian traditions we have regarding the distances of the planets are those mentioned by Ya^cqūb ibn Ṭāriq in his book, *Tarkīb al-aflāk*; and he had drawn his information from the well-known scholar who, in A.H. 161 (9 Oct. 777–27 Sept. 778), accompanied an embassy to Baghdād. First, he gives a metrological statement: "A finger is equal to six barleycorns which are put one by the side of the other. An arm is equal to twenty-four fingers. A *farsakh* is equal to 16,000 arms."

Here, however, we must observe that the Indians do not know the *farsakh*, that it is, as we have already explained (Vol. 1, p. 167), equal to one half a *yojana*.

Further, Ya^cqūb says: "The diameter of the earth is 2,100 *farsakh*, its circumference 6,596 $\frac{9}{25}$ (read 6597 $\frac{9}{25}$) *farsakh*."

On this basis he has computed the distances of the planets as we exhibit them in the following table.

However, this statement regarding the size of the earth is by no means generally agreed to by all the Indians. So, for example, Puliśa reckons its diameter as 1,600 *yojana*'s, and its circumference as 5,026 $\frac{14}{15}$ *yojana*'s; while Brahmagupta reckons the former as 1,581 *yojana*'s, and the latter as 5,000 *yojana*'s.

If we double these numbers, they ought to be equal to the numbers of Ya^cqūb; but this is not the case. Now, the arm and the mile are respectively identical according to the measurement both of us and of the Indians. According to our computation, the radius of the earth is 3,184 miles. Reckoning, according to the custom of our country, 1 *farsakh* = 3 miles, we get 6,728 *farsakh* (for the circumference); and reckoning 1 *farsakh* = 16,000 arms, as is mentioned by Yaᶜqūb, we get 5,046 *farsakh*. Reckoning 1 *yojana* = 32,000 arms, we get 2,523 *yojana*'s.

The following table is borrowed from the book of Yaᶜqūb ibn Ṭāriq.

Planets	Distances	*Farsakhs*	Radii of the Earth
	Radius of the earth	1,050	1
Moon	Smallest distance	37,500	35 $\frac{5}{7}$
	Middle distance	48,500	46 $\frac{4}{21}$
		(read 48,250)	(read 45 $\frac{20}{21}$)
	Greatest distance	59,000	56 $\frac{4}{21}$
	Diameter	5,000	4 $\frac{16}{21}$
Mercury	Smallest distance	64,000	60 $\frac{20}{21}$
	Middle distance	164,000	156 $\frac{4}{21}$
	Greatest distance	264,000	251 $\frac{3}{7}$
	Diameter	5,000	4 $\frac{16}{21}$
Venus	Smallest distance	269,000	256 $\frac{4}{21}$
	Middle distance	709,500	675 $\frac{5}{7}$
	Greatest distance	1,150,000	1,095 $\frac{5}{21}$
	Diameter	20,000	19 $\frac{1}{21}$
Sun	Smallest distance	1,170,000	1,114 $\frac{2}{7}$
	Middle distance	1,690,000	1,609 $\frac{11}{21}$
	Greatest distance	2,210,000	2,104 $\frac{16}{21}$
	Diameter	20,000	19 $\frac{1}{21}$
Mars	Smallest distance	2,230,000	2,123 $\frac{17}{21}$
	Middle distance	5,315,000	5,061 $\frac{19}{21}$
	Greatest distance	8,400,000	8,000
	Diameter	20,000	19 $\frac{1}{21}$
Jupiter	Smallest distance	8,420,000	8,019 $\frac{1}{21}$
	Middle distance	11,410,000	10,866 $\frac{2}{3}$
	Greatest distance	14,400,000	13,714 $\frac{2}{7}$
	Diameter	20,000	19 $\frac{1}{21}$
Saturn	Smallest distance	14,420,000	13,733 $\frac{1}{3}$
	Middle distance	16,220,000	15,447 $\frac{13}{21}$
	Greatest distance	18,020,000	17,161 $\frac{19}{21}$
	Diameter	20,000	19 $\frac{1}{21}$
Zodiac	Radius of outside	20,000,000	19,047 $\frac{13}{21}$
	Radius of inside	19,962,000	1,866 $\frac{2}{3}$
		(read 19,600,000)	(read 18,666 $\frac{2}{3}$)
	Circumference from outside	125,664,000	

COMMENTARY

In this curious passage on distances al-Bīrūnī quite unequivocally states that his source is Yaᶜqūb's *Tarkīb al-aflāk*; but, in an earlier work, the *Taḥdīd*, he ascribed similar parameters regarding the size of the earth and of the zodiacal circle to other sources. Thus, once (*Taḥdīd*, 211:21–212:1) he says that al-Fazārī in his *zīj* stated that the circumference of the earth is 6,600 *farsakh*, where a *farsakh* is 16,000 cubits ("arms"). Later on (*Taḥdīd*, 228:10–229:19), in discussing an Indian book whose title he gives as *Kitāb taḥdīd al-arḍ wa al-falak* (*Bhūgolādhyāya*?), he attributes to it the following parameters:

circumference of the earth	6,597 $\frac{9}{25}$ *farsakh*
diameter of the earth	2,100 *farsakh*
circumference of the zodiac	125,664,400 *farsakh* (read 125,664,000)
diameter of the zodiac	40,000,000 *farsakh*

He then proceeds (*Taḥdīd*, 229:20–230:3): "As the authors of the *Sindhind al-ṣaghīr* have dropped the zeroes from the beginning (right) of the number of days given in the *Sindhind al-kabīr* (of al-Fazārī), and have dropped in it an equal number of zeroes from the number of the Sun's rotations, they have done so in this case. They have made the ratio of the diameter to the circumference equal to the ratio of 40,000 to 125,664, which was mentioned by al-Khwārizmī in his *Zīj*, and in his *Kitāb al-jabr wa al-muqābala*, after dividing each of those two numbers by two."

Thus it appears that the parameters for the diameters and circumferences of the earth and the zodiac (which, of course, are really values of π) were already given by al-Fazārī in his *Sindhind al-kabīr*, probably taken from the *Bhūgolādhyāya* (? of the *Mahāsiddhānta*?) (for their derivation, see T 2.) Yaᶜqūb ibn Ṭāriq, using some unknown system of computation, gave as well the smallest, middle, and greatest distances of the planets and their diameters. It is to be observed that the middle distance is midway between the smallest and the greatest, and that the greatest distance of any planet plus its diameter equals the smallest distance of the next planet in the series. This means that the distances are to the nearest points on circumferences of the planets, not to their centers; and that each planet can be tangent to the next planet above or below it.

If we measure in earth-radii from the center of the earth to the center of each planet at mean distance and multiply by 21, we get the following table:

	earth-radii × 21	differences
Moon	1,015	2,315
Mercury	3,330	11,060
Venus	14,390	19,610
Sun	34,000	72,500
Mars	106,500	121,900
Jupiter	228,400	96,200
Saturn	324,600	75,400
Zodiac (outside)	400,000	

These numbers bear no relation to the sidereal periods of the planets, nor is there any apparent connection between the maximum and minimum distances and the planets' maximum and minimum equations.

The distances of the planets are referred to by Brahmagupta in his *Brāhmasphuṭasiddhānta* in verses 11–12 of Chapter 21, which is entitled *Golādhyāya* (cf. *India*, Vol. 2, pp. 70–71). Bragmagupta states the following things (cf. also *Āryabhaṭīya*, *Daśagītikā* 4):

(a) The circumference of the lunar orbit is 324,000 *yojana*'s;
(b) all of the planets in a *kalpa* travel an equal number of *yojana*'s;

(c) that number of *yojana*'s is the circumference of heaven;
(d) the circumference of the zodiac is 60 times the circumference of the solar orbit.

Using this information and the revolutions of the planets in a *kalpa* one can construct the following table of circumferences of orbits; for comparison I give also the circumferences according to the *Bhūgolādhyāya* of the modern *Sūryasiddhānta* (12, 185–90), where the numbers of revolutions of the planets are different from those in the *Brāhmasphuṭasiddhānta* and where the *yojana*'s of circumference have been rounded off. The radii of these orbits, which are the distances of the planets from the center of the earth, depend on one's value for π.

	Brāhmasphuṭasiddhānta	*Sūryasiddhānta*
Moon	324,000	324,000
Mercury's conj.	$1{,}043{,}210\ \frac{1561237670}{2242124873}$	1,043,209
Venus' conj.	$2{,}664{,}629\ \frac{1627580383}{1755597373}$	2,664,637
Sun	$4{,}331{,}497\ \frac{1}{2}$	4,331,500
Mars	$8{,}146{,}916\ \frac{82430924}{1148414261}$	8,146,909
Jupiter	$51{,}374{,}821\ \frac{54182089}{72845291}$	51,375,764
Saturn	$127{,}668{,}787\ \frac{8412079}{24427883}$	127,668,255
Zodiac	259,889,850	259,890,012
Heaven	18,712,069,200,000,000	18,712,080,864,000,000

We know from elsewhere in the *Brāhmasphuṭasiddhānta* that Brahmagupta assumed the circumference of the earth to be 5,000 *yojana*'s (1, 36) and its diameter 1,581 *yojana's* (21, 32). By simple proportion we find that the radius of the lunar orbit is 51,204.4 *yojana*'s or 64.76 earth radii. Ptolemy (*Syntaxis* 5, 13) finds the mean distance of the Moon at syzygies to be 59 earth-radii, at quadrature 38;43 earth-radii, and the radius of the Moon's epicycle to be 5;10 earth-radii; so the maximum distance of the Moon from the center of the earth would be 64.16 earth-radii, which is practically identical with Brahmagupta's value. Nor is it very far from Yaᶜqūb's maximum distance.

In order to compare the diameters of the planets according to Yaᶜqūb with various Indian values, I tabulate below the figures in *yojanas* given by Brahmagupta in the *Brāhmasphuṭasiddhānta* (1, 36 and 21, 32), by Āryabhaṭa in the *Āryabhaṭīya* (*Daśagītikā* 5), and by the modern *Sūryasiddhānta* (1, 59; 4, 1; and 7, 13–14).

diameters	Brahmagupta	Āryabhaṭa	*Sūryasiddhānta*
earth	1581	1050	1600
(earth's circum.)	5000	⟨3298.68⟩	⟨5059.64⟩
Moon	480	315	480
Mercury		21	45
Venus		63	60
Sun	6522	4410	6500
Mars		$12\frac{3}{5}$	30
Jupiter		$31\frac{1}{2}$	$52\frac{1}{2}$
Saturn		$15\frac{3}{4}$	$37\frac{1}{2}$

The only parameter among these which at all corresponds to Yaᶜqūb's is Āryabhaṭa's diameter of the earth, which is precisely that of the *Tarkīb al-aflāk*; the latter's circumference of the earth is 3,298 17/25 *yojana*'s, which is also precisely Āryabhaṭa's circumference using Āryabhaṭa's value of π. As was noted above, these measurements of the earth had been given by al-Fazārī in his *Sindhind al-kabīr* and in the *Bhūgolādhyāya* (? of the *Mahāsiddhānta*?) before Yaᶜqūb used them.

It will be noticed that Puliśa (i.e., the latest *Pauliśasiddhānta*), as quoted by al-Bīrūnī, agrees with the modern *Sūryasiddhānta* in making the diameter of the earth 1,600 *yojana*'s, but disagrees as to the circumference, which he takes to be $5{,}026\frac{14}{15}$ *yojana*'s. The difference reflects differing values of π. The *Sūryasiddhānta* uses $\sqrt{10}$ for π; Āryabhaṭa (*Gaṇitapāda* 10) the ratio $\frac{62832}{20000}$; and Brahmagupta and the latest Pauliśa the stated ratios of the circumference to the diameter of the earth.

The Sanskrit metrological units to which Yaᶜqūb refers are the following:

6 *yava*'s (barley corns)	= 1 *aṅgula* (finger)
24 *aṅgula*'s	= 1 *hasta* (hand)
32,000 *hasta*'s	= 1 *yojana*.

T 2. Al-Bīrūnī, *India*, ed. p. 132, trans. Vol. 1, p. 169.

The same relation (i.e., $1{:}3\frac{177}{1250}$) is derived from the old theory, which Yaᶜqūb ibn Ṭāriq mentions in his book, *Tarkīb al-aflāk*, on the authority of his Indian informant, namely that the circumference of the zodiac is 1,256,640,000 *yojana*'s (read 125,664,000 *farsakh*'s), and that its diameter is 400,000;000 *yojana*'s (read 40,000,000 *farsakh*'s).

These numbers presuppose the relation between circumference and diameter to be as $1{:}3\frac{56640000}{400000000}$ (read $\frac{5664000}{40000000}$). These two numbers may be reduced by the common divisor of 360,000 (read 32,000). Thereby we get 177 as the numerator and 1,250 as the denominator; and this is the fraction which Puliśa had adopted.

Commentary

The dimensions of the zodiac have been mentioned above in T 1. The value of π, $\frac{125664000}{40000000}$, is precisely equal to Āryabhaṭa's $\frac{62832}{20000}$, since numerator and denominator have both been multiplied by 2 (the 2 converts *yojana*'s to *farsakh*'s). And Āryabhaṭa's value of π, as we have seen, gives earth-dimensions of $3298\frac{17}{25}$ *yojana*'s and 1050 *yojana*'s, or $6597\frac{9}{25}$ *farsakh*'s and 2100 *farsakh*'s. Therefore, it is clear that the dimensions of the zodiac in the *Bhūgolādhyāya* (? of the *Mahāsiddhānta*?) as those of the earth are directly derived from the *Āryabhaṭīya*.

That the latest *Pauliśasiddhānta* used $\pi = 3\frac{177}{1250}$, however, does not agree with its earth-dimensions as given in T 1, namely $5{,}026\frac{14}{15}$ *yojana*'s and 1,600 *yojana*'s. These dimensions give a value of π equal to $3\frac{851}{8000}$.

Note that Āryabhaṭa's expression of π, $\frac{62832}{20000}$, is simply a fractional way of expressing the decimal 3.1416 which, as al-Biruni notes, was used by al-Khwārizmī in his *Kitāb al-mukhtaṣar fī ḥisāb al-jabr wa al-muqābala* (ed. p. 51, trans. pp. 71–72).

T 3. Al-Bīrūnī, *India*, ed. p. 266, trans. Vol. 1, p. 312.

If the circumference of the earth is 4,800 *yojana*'s, the diameter is nearly 1,527; but Puliśa reckons it as 1,600, Brahmagupta as 1,581 *yojana*'s, each of which is equal to eight miles. The same value is given in the *Zīj al-Arkand* as 1,050. This number, however, is, according to ibn Ṭāriq, the radius, whilst the diameter is 2,100 *yojana*'s, each *yojana* being reckoned as equal to four miles, and the circumference is stated as 6,596$\frac{9}{25}$ (read 6,597$\frac{9}{25}$) *yojana*'s.

COMMENTARY

As explained in the next paragraph by al-Bīrūnī, the earth-circumference of 4,800 *yojana*'s was given by Brahmagupta in his *Khaṇḍakhādyaka* (1, 15); he further indicates that in the *Uttarakhaṇḍakhādyaka* Brahmagupta corrected this circumference, but the appropriate verse has not yet been located. The ratio 4,800 to 1,527 gives a value of π equal to 3$\frac{73}{509}$.

A *yojana* is here taken to equal eight miles; in T 1 it was assumed to equal six miles. What are called *yojana*'s of ibn Ṭāriq here, being half an ordinary *yojana* (of eight miles), are, in fact, the *farsakh*'s of T 1.

This fragment informs us that the *Zīj al-Arkand*, like the *Bhūgolādhyāya* (? of the *Mahāsiddhānta*?), al-Fazārī, and Yaᶜqūb, derived its figure for the diameter of the earth from the *Āryabhaṭīya*; in most other matters we know that it rather depended on Āryabhaṭa's *ārdharātrika* system.

T 4. Al-Bīrūnī, *India*, ed. p. 269, trans. Vol. 1, p. 316.

Yaᶜqūb ibn Ṭāriq says in his *Tarkīb al-aflāk* that the latitude of Ūjjayn is four parts (*ajzā*ᵓ) and three-fifths, but he does not say whether it lies in the north or the south. Besides, he states it, on the authority of *al-Arkand*, to be four parts and two-fifths of a part.

COMMENTARY

For Yaᶜqūb, *juz*ᵓ seems here to signify a digit of the noon equatorial shadow. A noon equatorial shadow of 4;36 digits implies a latitude of 22° (see K 5); the latitude of Ūjjain, of course, is northern. The *Zīj al-Arkand* to which al-Bīrūnī refers in the next sentence after the passage quoted above, as it mentions al-Manṣūra, is evidently that composed in Sind in 735, for which see the commentary on al-Hāshimī 93v:23–94r:2; according to this *zīj* the latitude of Ūjjain is 22;29°, which implies a noon equatorial shadow of almost 5 digits. The 4;24 digits cited by Yaᶜqūb, then, must be from another version of the *Zīj al-Arkand*—presumably that which existed in Sasanian times.

As fragments T 5 through T 11 deal with the computation of the *ahargaṇa*, it may be well here to summarize what the *Paitāmahasiddhānta* and the *Brāhmasphuṭasiddhānta* have to say on the subject.

A. Parameters.

a. 1 *Kalpa* = 4,320,000,000 years = 1,000 *Mahāyuga*'s.
b. 1 *Kṛtayuga* = 1,728,000 years (432,000 × 4).
c. 1 *Tretāyuga* = 1,296,000 years (432,000 × 3).
d. 1 *Dvāparayuga* = 864,000 years (432,000 × 2).
e. 1 *Kaliyuga* = 432,000 years (432,000 × 1).
f. 1 *Mahāyuga* = 4,320,000 years (432,000 × 10).
g. 1 *Kalpa* = 14 *Manvantara*'s + 15 *sandhi*'s (*Kṛtayuga*'s).
h. 1 *Manvantara* = 71 *Mahāyuga*'s = 306,720,000 years.
i. 14 *Manvantara*'s = 4,294,080,000 years.
j. 15 *sandhi*'s = 25,920,000 years.
k. 1 *Kalpa* = 4,320,000,000 years.

l. In a *kalpa* there are 1,582,236,450,000 revolutions of the fixed stars.
m. Civil days = revolutions of the fixed stars—revolutions of the Sun = 1,577,916,450,000.
n. *Saura* months = revolutions of the Sun × 12 = 51,840,000,000.
o. *Saura* days = *saura* months × 30 = 1,555,200,000,000.
p. In a *kalpa* there are 57,753,300,000 revolutions of the Moon.
q. *Nakṣatra* days = revolutions of the Moon × 27 = 1,559,239,100,000.
r. Synodic months = revolutions of the Moon—revolutions of the Sun = 53,433,300,000.
s. *Tithi*'s = synodic months × 30 = 1,602,999,000,000.
t. *Adhimāsa*'s = synodic months—*saura* months = 1,539,300,000.
u. *Avama*'s = *tithi*'s—civil days = 25,082,550,000.

B. Elapsed time between the beginning of the *kalpa* and the epoch of the Śaka era (A.D. 78), according to the *Brāhmasphuṭasiddhānta*.

6 *Manvantara*'s	=	1,840,320,000 years
7 *sandhi*'s	=	12,096,000 years
27 *Mahāyuga*'s	=	116,640,000 years
1 *Kṛtayuga*	=	1,728,000 years
1 *Tretāyuga*	=	1,296,000 years
1 *Dvāparayuga*	=	864,000 years
to beginning of *Kaliyuga*		1,972,944,000 years (432,000 × 4,567).

Of the present *Kaliyuga* (epoch—3101) there elapsed 3,179 years to the epoch of the Śaka era; thus, the total number of elapsed years from the beginning of the *kalpa* to the epoch of the Śaka era is 1,972,947,179.

C. The calculation of the *ahargaṇa* according to the *Paitāmahasiddhānta*.
a. Multiply the elapsed *Manvantara*'s by 71.
b. Multiply the product by the number of years in a *Mahāyuga*.
c. Add the product of the number of elapsed *Manvantara*'s increased by one multiplied by a *Kṛtayuga*.
d. Add the number of years that have elapsed of the present *Mahāyuga*; the sum is the number of elapsed years from the beginning of the *kalpa* to the beginning of Caitra of the present year.
e. Multiply the sum of elapsed years by 12, and add to the product the number of months that have elapsed of the present year; the sum is the number of elapsed *saura* months. (This is inaccurate.)
f. Multiply the elapsed *saura* months by 30, and add to the product the number of *tithi*'s that have elapsed of the present month; the sum is the number of elapsed *saura* days. (This is inaccurate.)
f. Find the number of elapsed *adhimāsa*'s from the proportion:

$$\frac{\text{elapsed } saura \text{ days}}{\text{elapsed } adhim\bar{a}sa\text{'s}} = \frac{saura \text{ days in a } kalpa}{adhim\bar{a}sa\text{'s in a } kalpa} = \frac{1555200000000}{1593300000}.$$

h. Multiply the number of elapsed *adhimāsa*'s by 30, and add to the product the number of elapsed *saura* days; the sum is the number of elapsed *tithi*'s.

i. Find the number of elapsed *avama*'s from the proportion:

$$\frac{\text{elapsed } tithi\text{'s}}{\text{elapsed } avama\text{'s}} = \frac{tithi\text{'s in a } kalpa}{avama\text{'s in a } kalpa} = \frac{1602999000000}{25082550000}.$$

j. The number of elapsed civil days (the *ahargaṇa*) equals the number of elapsed *tithi*'s minus the number of elapsed *avama*'s.

D. The calculation of the *ahargaṇa* in the *Brāhmasphuṭasiddhānta*: eight rules.

I (1, 29–30). (Cf. Kennedy, Engle, and Wamstad, in *JNES*, 24 (1965), 274–84).

a. Multiply the elapsed years by 12, and add to the product the elapsed synodic months of the present year; the sum is the number of elapsed *saura* months. (This is inaccurate; see *Paitāmahasiddhānta* a–e).

b. Find the elapsed *adhimāsa*'s from the proportion:

$$\frac{\text{elapsed } saura \text{ months}}{\text{elapsed } adhimāsa\text{'s}} = \frac{saura \text{ months in a } kalpa}{adhimāsa\text{'s in a } kalpa}.$$

(Cf. *Paitāmahasiddhānta* g.)

c. The elapsed synodic months are the sum of the elapsed *saura* months and the elapsed *adhimāsa*'s. (Cf. *Paitāmahasiddhānta* h.)

d. Multiply the elapsed synodic months by 30, and add the *tithi*'s that have elapsed of the present month; the sum is the number of elapsed *tithi*'s. (Thus Brahmagupta avoids adding *tithi*'s to *saura* days.)

e. Find the elapsed *avama*'s from the proportion:

$$\frac{\text{elapsed } tithi\text{'s}}{\text{elapsed } avama\text{'s}} = \frac{tithi\text{'s in a } kalpa}{avama\text{'s in a } kalpa}.$$

(See *Paitāmahasiddhānta* i.)

f. The elapsed civil days (the *ahargaṇa*) equal the elapsed *tithi*'s minus the elapsed *avama*'s. (See *Paitāmahasiddhānta* j.)

g. Divide the *ahargaṇa* by 7; the remainder indicates the week-day, as the first day of the *kalpa* was a Sunday.

II (13, 11). Elapsed *saura* days from elapsed *adhimāsa*'s.

We employ the following symbols:

A capital letter indicates units in a *kalpa*, a lower-case letter elapsed units; a bar over the letter indicates that it is an integer number—e.g., $\bar{a}$ is the largest integer contained in a.

A = *adhimāsa*'s
C = civil days
M = synodic months
N = *saura* months
R = remainder (in division)
S = *saura* days
T = *tithi*'s
U = *avama*'s

Given the proportion: $\frac{a}{A} = \frac{s}{S}$, we can say that

$$a = \frac{A \times s}{S} = \bar{a} + \frac{R}{S}.$$

From this it follows that

$$s = \frac{\left(\bar{a} + \frac{R}{S}\right) \times S}{A} = \frac{(\bar{a} \times S) + R}{A}.$$

This is the formula Brahmagupta gives: the elapsed *saura* days equal the product of the integer elapsed *adhimāsa*'s multiplied by the *saura* days in a *kalpa*, first increased by the remainder, and then divided by the *adhimāsa*'s in a *kalpa*.

III (13, 12–13). Elapsed *saura* days from elapsed *avama*'s.

First one finds the elapsed *tithi*'s by a formula similar to that used in II:

$$t = \frac{(\bar{u} \times T) + R}{U}.$$

Then one finds the elapsed *adhimāsa*'s from the proportion:

$$\frac{a}{A} = \frac{t}{T}.$$

The final step is to convert the elapsed *adhimāsa*'s into elapsed *adhimāsa*-days, and to subtract these from the elapsed *tithi*'s:

$$s = t - 30a.$$

IV (13, 14). Elapsed *saura* days from elapsed civil days.

The first step is to find the elapsed *tithi*'s from the proportion:

$$\frac{c}{C} = \frac{t}{T}.$$

The rest of the procedure is a repetition of III.

V (13, 15–16). Elapsed civil days from elapsed *adhimāsa*'s and elapsed *avama*'s.

Following a procedure similar to that of II, and given that

$$\frac{c}{C} = \frac{u}{U},$$

one derives

$$u = \frac{c \times U}{C} = \bar{u} + \frac{R}{C}$$

and thence

$$\bar{u} = \frac{(c \times U) - R}{C}.$$

But, since

$$\bar{u} + c = t,$$

it follows that

$$t = \frac{(c \times U) - R}{C} + c = \frac{[c \times (U + C)] - R}{C}.$$

But, since

$$U + C = T,$$

it follows that

$$t = \frac{(c \times T) - R}{C}.$$

Similarly, given that

$$\frac{t}{T} = \frac{a}{A},$$

one derives

$$\bar{a} = \frac{(t \times A) - R'}{T}.$$

Substituting $\frac{(c \times T) - R}{C}$ for t, one gets

$$\bar{a} = \frac{(c \times T \times A) - (R \times A)}{C \times T} - \frac{R'}{T},$$

and this reduces to

$$\bar{a} = \frac{c \times A}{C} - \frac{(R \times A) + (R' \times C)}{C \times T}$$

or

$$c \times A = (\bar{a} \times C) + \frac{(R \times A) + (R' \times C)}{T}.$$

If we call $\frac{(R \times A) + (R' \times C)}{T}$ the "corrected remainder" or R_c, then it follows that

$$c = \frac{(\bar{a} \times C) + R_c}{A}.$$

And Brahmagupta states that the elapsed civil days equal the product of the integer elapsed *adhimāsa*'s multiplied by the civil days in a *kalpa*, first increased by the "corrected remainder," and then divided by the *adhimāsa*'s in a *kalpa*.

VI (13, 17). Elapsed civil days from elapsed *avama*'s.

Using the proportion

$$\frac{c}{C} = \frac{u}{U}$$

one proceeds as in II to get

$$c = \frac{(\bar{u} \times C) + R}{U}.$$

VII (13, 18). Elapsed civil days from elapsed *saura* months.

First one finds elapsed synodic months from the proportion:

$$\frac{m}{M} = \frac{n}{N}.$$

From this it follows that, letting t_c represent the elapsed *tithi*'s of the current month:

$$t = 30m + t_c.$$

The elapsed civil days are found from the proportion:

$$\frac{c}{C} = \frac{t}{T}.$$

VIII (13, 14). Elapsed civil days from elapsed *saura* days.

First one finds the elapsed *adhimāsa*'s from the proportion:

$$\frac{s}{S} = \frac{a}{A}.$$

Then it follows that

$$t = s + 30a.$$

The elapsed *avama*'s are found from the proportion:

$$\frac{t}{T} = \frac{u}{U}.$$

And, of course,

$$c = t - u.$$

T 5. Al-Bīrūnī, *India*, ed. p. 297, trans. Vol. 1, p. 353.

Māna (*mān*) and *pramāṇa* (*paramān*) mean measure. The four kinds of measure are mentioned by Yaʿqūb ibn Ṭāriq in *Tarkīb al-aflāk*, but he did not know them thoroughly; and, besides, the names are misspelled, if this is not the fault of the copyists. They are:

sauramāna (*sawramān*), which is the solar measure;
sāvanamāna (*sābanamān*), which is the (Sun-)rise measure;
candramāna (*čandramān*), which is the lunar measure;
nakṣatramāna (*nakshatramān*), which is the lunar mansion measure.

COMMENTARY

These are the four standard measurements of time in Sanskrit texts. They are listed by the *Paitāmahasiddhānta* (p. 1), and discussed in detail in Chapter 23 (*Mānādhyāya*) of the *Brāhmasphuṭasiddhānta*. Al-Bīrūnī's further discussion gives the lengths of the days in each of these measurements, based on the parameters of the *Sindhind* (and of Yaʿqūb); cf. also *Chronology*, ed. p. 13, trans. p. 15.

1. *Sauramāna*. A *saura* day is $\frac{1}{360}$ of a sidereal year. Al-Bīrūnī says that 1 *saura* day equals $1\frac{5609}{384000}$ civil days. This he derives from the fact that a sidereal year has $365\frac{827}{3200}$ civil days (i.e., 6, 5;15, 30, 22, 30 days).

2. *Sāvanamāna*. A civil day is simply 24 hours (approximately the time between two successive sunrises).

3. *Candramāna*. A lunar day or *tithi* is $\frac{1}{30}$ of a mean synodic month. Al-Bīrūnī says that 1 *tithi* equals $\frac{5016051}{31558329}$ civil days. Sachau corrects this to $\frac{10519443}{10686660}$. As there are 1,577,916,450,000 civil days and 1,602,999,000,000 *tithi*'s in a *kalpa*, the correct fraction is $\frac{3506481}{3562220}$; see *India*, ed. p. 372, trans. Vol. 2, p. 35. The denominator of al-Bīrūnī's fraction, 31,558,329, is $\frac{1}{5000}$ of the number of civil days in a *kalpa*; if we multiply the

numerator, 5,016,051, by 5,000 we get 25,082,550,000, which is the number of *avama*'s in a *kalpa*. So al-Bīrūnī's fraction represents the part of an *avama* that equals a civil day, *not* the part of a civil day that equals a *tithi*,

4. *Nakṣatramāna*. A *nakṣatra* day is the time it takes the mean Moon to travel through one *nakṣatra*, where a *nakṣatra* is $\frac{1}{27}$ of a circle or 13;20°. Al-Bīrūnī states that 1 *nakṣatra*-day equals $1\frac{417}{35002}$ civil days; he derives this from a sidereal month of $27\frac{11259}{35002}$ days. As the Moon makes 57,753,300,000 revolutions in the 1,577,916,450,000 civil days in a *kalpa*, this parameter for the length of a sidereal month is correct.

T 6. Al-Bīrūnī, *India*, ed. pp. 360–61, trans. Vol. 2, p. 23.

As regards *adhimāsa* (*ādimāsah*), the word means "the first month," for *ādi* means "beginning." In the books of Yaᶜqūb ibn Ṭāriq and al-Fazārī this name is written *padhamāsah* (read *malamāsah*, for Sanskrit *malamāsa*). *Padh* means "end," and it is possible that the Indians call the leap month by both names; but the reader must be aware that these two authors frequently misspell or disfigure the Indian words, and that there is no reliance on their tradition.

COMMENTARY

The leap months (i.e., the difference between the synodic months and the *saura* months in a *kalpa*) are called *adhimāsa*'s by both the *Paitāmahasiddhānta* and the *Brāhmasphuṭasiddhānta*. Evidently al-Fazārī's source (the *Mahāsiddhānta*?) called them *malamāsa*'s; بذ is a misreading (apparently old) of مل.

Al-Bīrūnī's interpretations demonstrate his lack of real familiarity with Sanskrit. *Ādi* does indeed mean "beginning"; but, had he looked at any Sanskrit astronomical text, he would know that the correct form is *adhimāsa*, meaning "additional month." And, though he elsewhere states correctly that *mala* means "dirt" (*India*, ed. p. 358, trans. Vol. 2, p. 20), he is wrong to think that *pada* means "end"; its true significance is "step."

T 7. Al-Bīrūnī, *India*, ed. p. 364, trans. Vol. 2, p. 26.

Yaᶜqūb ibn Ṭāriq has made a mistake in the computation of the solar days; for he maintains that you get them by subtracting the solar cycles of a *kalpa* from the civil days of a *kalpa*, i.e. the universal civil days. But this is not the case. We get the solar days by multiplying the solar cycles of a *kalpa* by 12 in order to reduce them to months, and the product by 30 in order to reduce them to days, or by multiplying the number of cycles by 360.

In the computation of the *tithi*'s he has first taken the right course, multiplying the lunar months of a *kalpa* by 30, but afterwards he again falls into a mistake in the computation of the *avama*'s (*ayyām al-nuqṣān*). For he maintains that you get them by subtracting the solar days from the *tithi*'s, while the correct thing is to subtract the civil days from the *tithi*'s.

COMMENTARY

Clearly, as al-Bīrūnī reports Yaᶜqūb's rule for computing the *saura* days, it is totally wrong. One might suggest that the original had the rule given in Am: civil days = revolutions of the fixed stars − revolutions of the Sun.

Ya^{c}qūb's second rule, for computing *tithi*'s, is given in As. The third rule should, of course, be u = t − c, not u = t − s; but it has already appeared in the first rule that Ya^{c}qūb's expression for civil days was confused with that for *saura* days.

T 8. Al-Bīrūnī, India, ed. pp. 370–71, trans. Vol. 2, pp. 33–34.

We have already pointed out (T 7) a mistake of Ya^{c}qūb ibn Ṭāriq in the calculation of the universal solar days and *avama*'s. As he translated from the Indian language a calculation the reason of which he did not understand, it would have been his duty to examine it and to check the various numbers of it one by the other. He mentions in his book also the method of *ahargaṇa*, i.e. the resolution of years, but his description is not correct; for he says:

"Multiply the months of the given number of years by the number of *adhimāsa*'s which have elapsed up to the time in question, according to the well-known rules of *adhimāsa*. Divide the product by the solar months. The quotient is the number of integer *adhimāsa*'s plus its fractions which have elapsed up to the date in question."

The mistake is here so evident that even a copyist would notice it; how much more a mathematician who makes a computation according to this method. For he multiplies, by the partial instead of the universal *adhimāsa*'s.

Besides, Ya^{c}qūb mentions in his book another and perfectly correct method of resolution, which is this: "When you have found the number of months of the years, multiply them by the number of the synodic months, and divide the product by the solar months. The quotient is the number of *adhimāsa*'s together with the number of the months of the years in question.

"This number you multiply by 30, and you add to the product the days which have elapsed of the current month. The sum represents the *tithi*'s.

"If, instead of this, the first number of months were multiplied by 30, and the past portion of the month were added to the product, the sum would represent the partial solar days; and if this number were further computed according to the preceding method, we should get the *adhimāsa*-days together with the solar days."

COMMENTARY

Ya^{c}qūb's first rule is to find the elapsed *adhimāsa*'s from the proportion:

$$\frac{a}{A} = \frac{n}{N}.$$

As al-Bīrūnī delights in pointing out, he mistakenly substitutes a for A. Essentially the same proportion was used by Brahmagupta (VIII):

$$\frac{a}{A} = \frac{s}{S}.$$

This form avoids fractions in the numerator (n is almost always an integer plus a fraction).

Ya^{c}qūb's method of computing the elapsed *tithi*'s from the elapsed *saura* months is derived from Bragmagupta (VII):

$$\frac{m}{M} = \frac{n}{N}$$

$$t = 30m + t_c.$$

But his second rule for doing the same thing combines elements of the *Paitāmahasiddhānta* with those of the *Brāhmasphuṭasiddhānta*:

$$s = 30n + t_c \text{ (\textit{Paitāmahasiddhānta} f.)}$$

$$\frac{a}{A} = \frac{s}{S} \text{ (Brahmagupta VIII)}$$

$$t = s + 30a \text{ (Brahmagupta VIII).}$$

T 9. Al-Bīrūnī, India, ed. pp. 374–75, trans. Vol. 2, p. 38.

Because the majority of the Indians, in reckoning their years, require the *adhimāsa*, they give preference to this method, and are particularly painstaking in describing the methods for the computation of the *adhimāsa*'s, disregarding the methods for the computation of the *avama*'s and the sum of the days (*ahargaṇa*). One of their methods of finding the *adhimāsa* for the years of a *kalpa* or *caturyuga* or *kaliyuga* is this:

They write down the years in three different places. They multiply the upper number by 10, the middle by 2,481, and the lower by 7,739. Then they divide the middle and lower numbers by 9,600, and the quotients are days for the middle number and *avama*'s for the lower number.

The sum of these two quotients is added to the number in the upper place. The sum represents the number of the complete *adhimāsa*-days which have elapsed, and the sum of that which remains in the other two places is the fraction of the current *adhimāsa*. Dividing the days by 30, they get months.

Yaʿqūb ibn Ṭāriq states this method quite correctly.

COMMENTARY

This procedure, of course, is based on the difference between a solar year and 12 synodic months, that is, on the epact. The reasoning behind this peculiar expression of the epact can be demonstrated as follows; in order to express units in a sidereal year, I add a subscript y. Y represents the years in a *kalpa*; y represents elapsed years.

$$c_y = \frac{C}{Y} = \frac{1577916450000}{4320000000} = 360 + 5 + \frac{2481}{9600}$$

$$u_y = \frac{U}{Y} = \frac{25082550000}{4320000000} = 5 + \frac{7739}{9600}$$

$$s = 360y.$$

Since $5 + \frac{2481}{9600}$ is the difference between the civil and *saura* days in a year, it follows that

$$c - s = y \times \left(5 + \frac{2481}{9600}\right);$$

and, of course,

$$u = y \times \left(5 + \frac{7739}{9600}\right) = t - c.$$

Therefore

$$y \times \left(10 + \frac{2481}{9600} + \frac{7739}{9600}\right) = t - c + c - s = t - s = 30a.$$

The amount in parentheses is the epact in *tithi*'s; it amounts to $11\frac{31}{480}$, which is a little high.

T 10. Al-Bīrūnī, *India*, ed. p. 380, trans. Vol. 2, p. 44.

Yaᶜqūb ibn Ṭāriq has a note to the same effect: "Multiply the given civil days by the universal *tithi*'s and divide the product by the universal civil days. Write down the quotient in two different places. In the one place multiply the number by the universal *adhimāsa*-days (read *adhimāsa*'s) and divide the product by the universal *tithi*'s. The quotient gives the *adhimāsa*'s. Multiply them by 30 and subtract the product from the number in the other place. The remainder is the number of partial solar days. You further reduce them to months and years."

COMMENTARY

This rule is taken from Brahmagupta (IV):

$$\frac{t}{T} = \frac{c}{C} \text{ (solve for t)}$$

$$\frac{a}{A} = \frac{t}{T} \text{ (solve for a)}$$

$$s = t - 30a.$$

T 11. Al-Bīrūnī, *India*, ed. pp. 380–81, trans. Vol. 2, p. 45.

The following rule of Yaᶜqūb for the computation of the partial *avama*'s by means of the partial *adhimāsa*'s is found in all the manuscripts of his book:

"The past *adhimāsa*'s, together with the fractions of the current *adhimāsa*, are multiplied by the universal *avama*'s, and the product is divided by the universal solar months. The quotient is added to the *adhimāsa*'s. The sum is the number of the past *avama*'s."

This rule does not, as I think, show that its author knew the subject thoroughly, nor that he had much confidence either in analogy or experiment.

COMMENTARY

Unless we can emend this strange rule, al-Bīrūnī's assessment of it is correct. The present text (assuming that by "fractions of the current *adhimāsa*" Yaᶜqūb means only R, the numerator) is equivalent to:

$$u = \frac{(\bar{a} + R) \times U}{N}.$$

It must be emended to:

$$u = \frac{(\bar{a} \times U) + R}{A}.$$

If this emendation is correct, the rule is based on the procedure used by Brahmagupta (II).

T 12. Al-Hāshimī, *Kitāb ᶜilal al-zījāt*, 114 r: 10–12.

Al-Farghānī (in his commentary on al-Khwārizmī's *Zīj al-Sindhind*) related concerning Yaᶜqūb ibn Ṭāriq that he claimed that it (i.e., the apogee) is an alidade with which God rectifies the heavens, and he relates the same concerning Māshāʾallāh. This saying about Māshāʾallāh is better established than that concerning Yaᶜqūb ibn Ṭāriq.

COMMENTARY

As in determining the position of a star a man arranges in one straight line of vision the two holes in the alidade of his astrolabe and the celestial object being sighted, so God arranges in one straight line the earth, the center of the deferent circle, the apogee, and the planet. It is unclear why al-Hāshimī states that this saying is better established about Māshāʾallāh than about Yaʿqūb, but cf. al-Masʿūdī, *Kitāb al-tanbīh wa al-ishrāf*, p. 222.

T 13. Ibn Masrūr, *Kitāb ʿilal al-zījāt*, ff. 46–47.

Verily Ptolemy claimed that, as knowledge of the *Almagest* comes before beginning the practice of determining planetary motions, speed and slowness have two parts for each circle. In the circle with eccentricity, there are the speed and slowness which are due to the apogee-orb in accordance with what was depicted for us in the book *Tarkīb al-aflāk*. The second (speed and slowness) are due to the difference of the center of the epicycle with respect to the zodiacal circle, as we explained in the beginning (?).

COMMENTARY

This fragment is included here because, aside from the work of ʿUṭārid ibn Muḥammad (*Fihrist*, p. 278), Yaʿqūb's is the only book entitled *Tarkīb al-aflāk* known to us. The reference is probably to the same section of the work that T 12 is derived from.

KITĀB AL-ʿILAL

K 1. Al-Bīrūnī, *On Shadows*, 51:9–52:19.

And when one puts together the operations of the workers in this field about that, the routes they travelled and the numbers they put will not remain hidden, as is the case with Muḥammad ibn Ibrāhīm al-Fazārī, Yaʿqūb ibn Ṭāriq, Muḥammad ibn Mūsā al-Khwārizmī, Ḥabash al-Ḥāsib, Abū Maʿshar al-Balkhī, al-Faḍl ibn Nadīm al-Nayrīzī, Muḥammad ibn Jābir al-Battānī, and Abū al-Wafāʾ al-Buzjānī. All of these explained in their *zīj*'es that, if the shadow is squared and the gnomon is squared and the (square) root of their sum is taken, it will be the cosecant. . . . And when the cosecant is determined for all of them, some of them proceed to the cosine of the altitude, and some to the sine of the altitude itselt. . . . However, those who proceed toward the sine of the altitude itself divide by the cosecant the product of the gnomon and the total sine. . . . However, al-Fazārī, al-Khwārizmī, Yaʿqūb ibn Ṭāriq, Abū Maʿshar, and the author of the *Zīj al-Shāh* prescribe the division of 1,800 by the cosecant, and it is the product of 150 multiplied by 12.

COMMENTARY

If ATE is on the plane of the observer's horizon, and if B is the zenith, H the Sun, EL the gnomon, KL the shadow, and EK the cosecant, then

$$\mathrm{EK} = \sqrt{\mathrm{EL}^2 + \mathrm{KL}^2},$$

and HT, the sine of the Sun's altitude, can be derived by

$$\mathrm{HT} = \frac{\mathrm{EL} \times \mathrm{EH}}{\mathrm{EK}}.$$

This rule is also found in the *Khaṇḍakhādyaka* (3, 10–11). The inclusion of the *Zīj al-Shāh* among the texts exhibiting the form substituting 1,800 for R × the gnomon indicates that this rule is derived from the *Zīj al-Arkand,* to whose tradition the *Khaṇḍakhādyaka* also belongs. From the *Zīj al-Shāh* it has been adopted by al-Fazārī despite its use of R = 150′ instead of R = 3270′.

K 2. Al-Bīrūnī, *On Shadows*, 53:6–14.

For the inverse of this, if the altitude is assumed known and the shadow of the gnomon is wanted for that time, the ratio of the sine of the altitude to its cosine is the ratio of the gnomon to its shadow; and from this the gnomon is multiplied by the cosine of the altitude, and the result is divided by the sine of the altitude, and the shadow results. And this operation in the *Zīj al-Shāh,* Yaᶜqūb, al-Khwārizmī, Ḥabash, Abū Maᶜshar, al-Nayrīzī, and al-Battānī does not differ except (to the extent) that the above operation differs. I mean that some of them omit mentioning (the units of) the gnomon when it is multiplied by itself, while others specify its parts according to what has been assumed in their *zīj*'es.

COMMENTARY

From the situation as previously given it is obvious that indeed

$$\frac{HT}{ET} = \frac{EL}{KL}.$$

K 3. Al-Bīrūnī, *On Shadows*, 147:15–18.

And Yaᶜqūb ibn Ṭāriq . . . in his saying: "Divide by the hypotenuse of the shadow at the time 1,800, and multiply what comes out by 150. Divide what results by the sine of the noon altitude, and there comes out a sine. We find its (corresponding) arc and take of it for each 15° one equal hour."

COMMENTARY

The first part of this rule is identical with that in K 1; the result is HT, the sine of the Sun's altitude. The remaining part of the rule gives a rough approximation to the time elapsed since sunrise based on the two following considerations:

1. At noon, HT equals the sine of the noon altitude, and the arc between the intersection of the ecliptic and the horizon and that of the meridian and the ecliptic on an equinoctial day is 90°; the Sine of 90° is R = 150′.
2. At sunrise, HT is 0, and the arc also is 0.

Yaᶜqūb's rule makes all intermediate points proportional. It will only begin to be accurate when the Sun is on the equator.

K 4. Al-Bīrūnī, *On Shadows*, 84:2–6.

And it is possible that the need for the determination of the time is so urgent that it allows no time for adjusting the instrument. (Let us suppose) the gnomon set up on a plane inclined to the plane of the horizon, (but) parallel to (one) standing vertically. So we mark on that plane at the head of the shadow a mark so as to retain the desired (thing, and) we correct (it) afterwards. And that is what Yaᶜqūb ibn Ṭāriq mentioned of its computation in his *Kitāb fī al-ᶜilal.*

COMMENTARY

This technique of immediately recording a shadow and later adjusting the angles so as to tell what time it was when the shadow was cast would be particularly valuable to astrologers, who need to record a particular instant but have not beforehand prepared a gnomon.

K 5. Al-Bīrūnī, *On Shadows*, 94:19–95:4.

That is why al-Nayrīzī, and Yaʿqūb ibn Ṭāriq said concerning its (the equinoctial shadow's) determination: "Multiply the sine of the latitude of the locality by the gnomon, and divide the result by the cosine of the local latitude; there results the equatorial shadow." There is some doubt as to the words of Yaʿqūb, because he calls the sine a straight chord (*watar mustaqīm*).

COMMENTARY

The rationale of this common rule is easily seen from the following: O is the observer, OH the plane of his horizon, Z his zenith, S the Sun on the equator at noon, and AB the gnomon. Then CS = OH is sine ϕ, and SH is cos ϕ. Clearly the equatorial shadow, OB, is found by the proportion:

$$\frac{OB}{OH} = \frac{AB}{SH}.$$

This rule is given, *inter alia*, in *Brāhmasphuṭasiddhānta* 3, 28. For the term "straight chord" see K6.

K 6. Al-Bīrūnī, *On Shadows*, 127:16–129:1.

This (i.e., the sine of the equation of daylight) is exactly what Yaʿqūb ibn Ṭāriq explained in his *Kitāb al-ʿilal*. For he said: "Take the reversed chord of the distances of the end of each (zodiacal) sign, and subtract it from 3,438; there will remain the chord of the noon of the sign's daily circle. Multiply the straight chord of this distance by the equatorial shadow, and divide the result by the digits of the gnomon. Multiply what results by 3,438, and divide the resulting amount by the chord of the hoop (*ṭawq*) of the sign's daily circle. That which results make into an arc; and it is the excess for Aries, and the deficiency for Virgo."

This is because the straight chord for him is the ordinary sine, and the reversed (chord) is the versed (sine); this (above-)mentioned number is the minutes of the total sine according to Āryabhaṭa (*Āryabhaṭīya, Daśagītikā* 10); the chord of the hoop of the sign's daily path is the cosine of its declination, I mean half the diameter of its daily path; the distance of the sign is its declination; and the (above-)mentioned excess and deficiency refer to the differences between a right ascension and an (oblique) ascension for the locality. Since the arc-sine comes out for him in minutes, he claims that the result is in *prāṇas* (*barān*), that is "respirations," because, according to the Indians, an adjusted (respiration) is equal to the revolution of one minute of an equatorial *zaman*. Each six *azmān* make a *ghaṭikā* (*kahrī* ms.), I mean one of the minutes of the day, whose seconds are called *vināḍī* (*banārī*); but the common people among them call them *jasha* (ms.; *ḥabasha* ed.) and *iṣājaka* (ms.; *ḍājaka* ed.).

COMMENTARY

This rule employs a value of R different from that used in K 1 and K 3, namely 3438′ instead of 150′. As al-Bīrūnī remarks, the former value is found in the *Āryabhaṭīya*; it occurs many other places as well, e.g. in the *Paitāmahasiddhānta*. The terminology of "straight chord" and "reversed chord" is a direct translation of the Sanskrit *kramajyā* and *utkramajyā* used, e.g., in the *Brāhmasphuṭasiddhānta*. The time measurements are:

6 *prāṇa*'s = 1 *vināḍī* (or *caṣaka*)
60 *vināḍī*'s = 1 *nāḍī* or *ghaṭikā*
60 *ghaṭikā*'s = 1 nychthemeron.

The phrasing of the first part of the rule indicates that "distance of the end of the sign" is not δ but $90^\circ - \delta$; for

$$\text{Vers. } \theta = \text{R} - \text{Sin }(90^\circ - \theta).$$

Then,

$$R - \text{Vers }(90^\circ - \delta) = \text{Sin } \delta.$$

The rule as given, *e.g.* in *Brāhmasphuṭasiddhānta* 2, 57–58 is:

$$\text{Sin } \gamma = \frac{\text{Sin } \delta \times \text{equatorial shadow}}{12} \times \frac{\text{R}}{\text{Cos } \delta}$$

Cos δ is, of course, the Sine of the Sun's day-circle; see K 5. A proof of this theorem can be found in O. Neugebauer, *The Astronomical Tables of al-Khwārizmī*, "Hist. Filos. Skr. Dan. Vid. Selsk." 4, 2 (København, 1962), pp. 50–51.

K 7. Al Bīrūnī, *On Shadows*, 131:1–7.

And in the *Kitāb al-ʿilal* of Yaʿqūb ibn Ṭāriq (it says): "Multiply the equatorial shadow by the chord of the increment for Aries, namely the sine of its right ascension, and divide what results by the noon shadow in the position of the maximum ascension of the signs at the equator (it is 26;58 digits), and we mean by that the equatorial shadow for the position whose latitude equals the inclination of the ecliptic. So there comes out the chord of the deficiency of the zone of Aries, and the increment of the zone of Virgo; make it an arc, and it will be the decrease for Aries and the increase for Virgo."

COMMENTARY

Yaʿqūb here uses three elements: the sine of right ascension (Sin α), the noon equinoctial shadow at latitude ϕ (Cot ϕ), and what he calls "the noon shadow in the position of the maximum ascension of the signs at the equator, . . . and we mean by that the equatorial shadow for the position whose latitude equals the inclination of the ecliptic." This seems to refer to Cot $\bar{\epsilon}$, but the value he gives, 26;58 digits, is equal to Cot ϵ. Therefore, the formula is:

$$\text{Sin } \gamma = \text{Sin } \alpha \times \frac{\text{Cot } \phi}{\text{Cot } \epsilon}.$$

This is correct; for details, see Kennedy's commentary on the *On Shadows*.

SPURIOUS FRAGMENT

Ṣāʿid al-Andalusī, *Kitāb ṭabaqāt al-umam*, ed. p. 60, trans. p. 117.

Among those who have thoroughly studied astrology and who followed in the path of foreigners—Persians, Greeks, and so on—Yaʿqūb ibn Ṭāriq is outstanding; he is the author of the *Kitāb al-maqālāt*, which deals with the horoscopes of the Caliphs and Kings and of the fate of those for whom one does not know the time of the birth.

COMMENTARY

As no other reference to Yaʿqūb involves astrology, and as Ṣāʿid is an extremely unreliable source, it seems best to regard the *Kitāb al-maqālāt* as spurious.

BIBLIOGRAPHY

Primary Sources

Abraham ben Ezra. *De rationibus tabularum*. Published as *El libro de los fundamentos de las Tablas astronómicas*. Ed. J. M. MILLÁS VALLICROSA. Madrid–Barcelona, 1947.

Abraham ben Ezra. Hebrew translation of Ibn al-Muthannā's *Fī ʿilal zīj al-Khwārizmī*. The prologue is published by M. Steinschneider in *ZDMG*, 24 (1870), 325–92 (see esp. pp. 353–54); cf. C. A. Nallino, *Raccolta*, 5, 218. The text has been edited by B. Goldstein, *Ibn al-Muthannâ's Commentary on the Astronomical Tables of al-Khwârizmî*, New Haven and London, Yale University Press, 1967.

Āryabhaṭa. *Āryabhaṭīya*. Ed. H. Kern, Leyden, 1874; ed. *TSS*, 101, 110 and 185. Trivandrum, 1930–1957. Trans. W. E. Clark. Chicago, 1930.

Al-Bīrūnī. *Chronology*. *Al-āthār al-bāqiya*. Ed. C. E. Sachau. Leipzig, 1878. English translation by C. E. Sachau. London, 1879.

———. *India*. *Fī taḥqīq mā li ʾl-Hind*. Ed. Hyderabad, 1958. English translation by E. C. Sachau. 2 vols. London, 1910.

———. *On Shadows*. *Ifrād al-maqāl fī amr al-ẓilāl*. Ed. as part 2 of *Rasāʾil al-Bīrūnī*. Hyderabad, 1948. Unpublished English translation by E. S. Kennedy.

———. *On Transits*. *Tamhīd al-mustaqarr li taḥqīq maʿna al-mamarr*. Ed. as part 3 of *Rasāʾil al-Bīrūnī*. Hyderabad, 1948. English translation by M. Saffouri and A. Ifram with a commentary by E. S. Kennedy. Beirut, 1959.

———. *Al-Qānūn al-Masʿūdī*. 3 vols. Ed. Hyderabad, 1954–1956.

———. *Taḥdīd*. *Taḥdīd nihāyāt al-amākin li taṣḥīḥ masāfāt al-masākin*. Ed. P. Būljākūf. Cairo, 1964. English translation by J. Ali, *The Determination of the Coordinates of Positions for the Correction of Distances between Cities*, American University of Beirut Centennial Publications. Beirut, 1967.

Brahmagupta. *Brāhmasphuṭasiddhānta*. Ed. S. Dvivedin. Benares, 1902.

———. *Khaṇḍakhādyaka*. Ed. B. Miśra, Calcutta, 1925, and P. C. Sengupta, Calcutta, 1941. English translation by P. C. Sengupta. Calcutta, 1934.

Al-Hāshimī. *Kitāb ʿilal al-zījāt*. M.S. Bodl. Seld. A 11 ff. 96v–137. Unpublished English translation by Fuad I. Haddad and E. S. Kennedy.

Ibn Masrūr. *Kitāb ʿilal al-zījāt*. MS. Taymur, Math. 99. Unpublished translation by E. S. Kennedy.

Ibn al-Nadīm. *Fihrist*. Ed. G. Flügel. 2 vols. Leipzig, 1871–1872.

Ibn al-Qifṭī. *Taʾrīkh al-ḥukamāʾ*. Ed. J. Lippert. Leipzig, 1903.

Al-Jahānī. Latin translation by Gerard of Cremona. Ed. J. Heller with the *De elementis et orbibus coelestibus* of Māshāʾallāh. Noribergae 1549, ff. Ni-Zii.

Al-Khwārizmī. *Kitāb mukhtaṣar fī ḥisāb al-jabr wa al-muqābala*. Ed. F. Rosen. London, 1831.

———. *Zīj al-Sindhind*. Latin translation by Adelard of Bath of the revision of al-Majrīṭī. Ed. A. Bjørnbo, R. Besthorn, and H. Suter. København, 1914. English translation and commentary by O. Neugebauer. København, 1962.

Al-Majrīṭī, Pseudo-. *Picatrix*. *Ghāyat al-ḥakīm*. Ed. H. Ritter. Hamburg, 1933. German translation by H. Ritter and M. Plessner. London, 1962.

Al-Masʿūdī, *Kitāb al-tanbīh wa al-ishrāf*, ed. M. J. de Goeje, *BGA*, 8, Leiden, 1894.

Mēnok-i-χrat. For edition of Pahlavī text and of French translation cited in this article see M. Molé, *Culte, mythe et cosmologie dans l'Iran ancien*. Paris, 1963. Pp. 429–32.

Paitāmahasiddhānta of the *Viṣṇudharmottarapurāṇa*. Ed. as *Khaṇḍa* 2, *adhyāyas* 166–74 of the *Viṣṇudharmottarapurāṇa*. Bombay, 1912. Also ed. in fasc. 2 of the *Jyautiṣasaṅgraha* by V. P. Dvivedin, *BSS* 39. Benares, 1912.

Ptolemy. *Syntaxis mathēmatikē*. Ed. J. L. Heiberg. 2 vols. Leipzig. 1898–1903. German translation by K. Manitius. 2 vols. Leipzig, 1912–1913.

Ṣāʿid al-Andalusī. *Kitāb ṭabaqāt al-umam*. Ed. L. Cheikho. Beyrouth, 1912. French translation by R. Blachère. Paris, 1935.

Sūryasiddhānta (modern). Ed. S. Dvivedin. Calcutta, 1909–1911. English translation by E. Burgess in *JAOS*, 6 (1860); reprinted Calcutta, 1929.

Varāhamihira. *Pañcasiddhāntikā*. Ed. with a Sanskrit commentary and an English translation by G. Thibaut and S. Dvivedin. Benares, 1889; reprinted Lahore, 1930.

Al-Yaᶜqūbī, *Taʾrīkh al-Yaᶜqūbī*, 2 vols., Bayrūt, 1960.

Secondary Sources

C. A. Nallino, *Raccolta di scritti editi e inediti*. Vol. 5. Roma, 1944.

→ E. S. Kennedy, S. Engle, and J. Wamstad, "The Hindu Calendar as Described in al-Bīrūnī's Masudic Canon," *JNES*, 24 (1965), 274–84.

→ D. Pingree, "The Persian 'Observation' of the Solar Apogee in *ca.* A.D. 450," *JNES*, 24 (1965), 334–36.

D. Pingree, *The Thousands of Abū Maᶜshar*, "Studies of the Warburg Institute," 30. London, 1968.

THE FRAGMENTS OF THE WORKS OF AL-FAZĀRĪ

DAVID PINGREE, University of Chicago

AL-FAZĀRĪ is the name of the person most directly connected with the transmission of a Sanskrit work on astronomy—the *Mahāsiddhānta*(?) belonging to what later became known as the *Brahmapakṣa*—to the Arabs in the early part of the eighth decade of the eighth century of the Christian era. This was not the first infusion of Indian astronomical theories into Islām; prior to it was the mediating influence of the *Zīj al-Shāh* in the versions of Khusrau Anūshirwān and of Yazdijird III, the composition of the *Zīj al-Arkand* and its derivatives at Qandahār in 735 and after, and that of the *Zīj al-Harqān* in 742. The versions of the *Zīj al-Shāh* and the *Zīj al-Arkand* largely depend on the *ārdharātrika* system developed by Āryabhaṭa. But, despite al-Fazārī's importance, great confusion exists concerning both his and his father's names and their works.

In his valuable discussion of al-Fazārī, Nallino (pp. 209–15) concludes that his name was probably Ibrāhīm ibn Ḥabīb, and that he made an astrolabe and wrote a *zīj* based on the *Sindhind*; Nallino is followed by Brockelmann (*Suppl.* I, 391) and Kennedy (No. 2). Suter, on the other hand, speaks of two personalities: an Ibrāhīm ibn Ḥabīb (p. 3) who wrote various works and constructed an astrolabe, and his son Muḥammad (pp. 4–5) who was involved in the translation of the *Sindhind*. The confusion is due to the inaccuracies of the Islāmic biographers and bibliographers and to the existence of several contemporaries bearing the name al-Fazārī. There is, for instance, an Abū ʿAbdallāh Muḥammad ibn Ibrāhīm ibn Ḥabīb ibn Sulaymān ibn Samura ibn Jundab al-Fazārī (*Fihrist*, p. 79), who was reputed as an authority on geomancy, and a Muḥammad ibn Ibrāhīm al-Fazārī (*Fihrist*, p. 164), who was a "slave poet"; but elsewhere (*Fihrist*, p. 273; cf. al-Masʿūdī, *Murūj al-dhahab*, 8, 290–91, and Ibn al-Qifṭī, p. 57) Ibn al-Nadīm writes of an Abū Isḥāq Ibrāhīm ibn Ḥabīb al-Fazārī of the family of Samura ibn Jundab, who was the first in Islām to make a plane astrolabe and who wrote a *Kitāb al-qaṣīda fī ʿilm al-nujūm* (*Poem on the Science of the Stars*), a *Kitāb al-miqyās li'l-zawāl* (*Measurement of Noon*), a *Kitāb al-zīj ʿalā sinī al-ʿArab* (*Astronomical Tables According to the Years of the Arabs*), a *Kitāb al-ʿamal bi 'l-asṭurlāb wa huwa dhāt al-ḥalaq* (*Use of the Armillary Sphere*), and a *Kitāb al-ʿamal bi'l-asṭurlāb al-musaṭṭaḥ* (*Use of the Plane Astrolabe*). These seem to be son and father respectively.

But Ibn al-Qifṭī (p. 270) says of Muḥammad ibn Ibrāhīm al-Fazārī that he was outstanding in the science of the stars, a speaker about future events, experienced in the *tasyīr* of the planets, and the first in the Islāmic religion, at the beginning of the ʿAbbāsid dynasty, who was concerned with this subject; al-Fazārī's work on *tasyīr* is referred to by Abū Maʿshar as cited by Shādhān in his *Mudhākarāt* (see *CCAG*, 5,1;148). And the astronomer whose fragments are here collected is called Muḥammad ibn Ibrāhīm al-Fazārī by Ibn al-Adamī (Frag. Z 1), by Ṣāʿid al-Andalusī (ed. p. 13, trans. p. 46, and ed. p. 60, trans. p. 117), and by al-Bīrūnī (Frag. Z 15, 18, and 25 and Frag. Q 2). A reasonable solution to this problem of al-Fazārī's name would perhaps be to assume with Nallino that only one person is referred to in the passages cited above but that, in the entry on the astronomer al-Fazārī, Ibn al-Nadīm or his manuscript tradition has omitted the "Muḥammad ibn" with which the entry should have begun.

This suggestion is strengthened by the fact that the later biographical tradition refers to him as Muḥammad. Thus Yāqūt (*Irshād*, p. 268) gives a long genealogy of twenty-seven generations beginning with Muḥammad ibn Ibrāhīm ibn Ḥabīb ibn Samura ibn Jundab. Yāqūt further quotes al-Marzubānī to the effect that Muḥammad ibn Ibrāhīm al-Fazārī al-Kūfī was an expert on the stars; and Yaḥyā ibn Khālid al-Barmakī's saying that four men were without equal in their specialties: al-Khalīl ibn Aḥmad, Ibn al-Muqaffaᶜ, Abū Ḥanīfa, and al-Fazārī; and Jaᶜfar ibn Yaḥyā's saying that he did not see more exceptional men in their fields than al-Kisāʾī in grammar, al-Aṣmaᶜī in poetry, al-Fazārī in the stars, and Zulzul in playing the lute. Yāqūt adds that al-Fazārī wrote "a *qaṣīda* which raised up the abode of the *zīj*ʾes of the astrologers; it is doubled and long, and fills with its commentary ten volumes (*ajlād*)." He then quotes the first lines. Al-Ṣafadī (1, 336–37), who copies some of Yāqūt's note, quotes more of the *qaṣīda* (Frag. Q 1):

"Glory to Allāh, the high and mighty, to whom are superiority and great grandeur, the most noble, the unique one, the magnanimous benefactor, the creator of the highest seven in order. The sun's light brightens the darkness, and the full Moon's light spreads to the horizon. The sphere, revolving in its course for the greatest of affairs, moves in one of the seas; on it the stars, all of them, are agents; and some of them are permanent, some transitory, and some rise, while others set."

Though this passage is not very illuminating with respect to the contents of al-Fazārī's works, we do learn from it that he was of an old Arab family from Kūfa; that his name was Muḥammad; and that his astronomical or astrological poem filled ten volumes. His dates can be determined by the statements of his biographers that he was the first astronomer in Islām, flourishing at the beginning of the ᶜAbbāsid dynasty (under al-Manṣūr according to al-Masᶜūdī), and by Ibn al-Adamī's account (Frag. Z 1) of the embassy from Sind in 773 (but cf. Frag. Z 3); also, Frag. Z 23 can be dated *ca.* 790.

However, there is one other datable story involving al-Fazārī, that incidentally fits in with his renown as an astrologer. Al-Yaᶜqūbī (*Kitāb al-buldān*, p. 238) recounts that the time for beginning the construction of Baghdād was selected by Nawbakht *al-munajjim* and Māshāʾallāh ibn Sāriya; and al-Bīrūnī (*Chronology*, ed. pp. 270–71, trans. pp. 262–63) gives the horoscope drawn up by Nawbakht, which can be dated 30 July 762:

	Horoscope	Computation (30 July 762)
Saturn	Aries 26;40° retr.	Taurus 1°
Jupiter	Sagittarius	Sagittarius 9
Mars	Gemini 2;50	Gemini 6
Sun	Leo 8;10	Leo 10
Venus	Gemini 29;0	Cancer 2
Mercury	Cancer 25;7	Cancer 27
Moon	Libra 19;10	Libra 18

Al-Yaᶜqūbī goes on to say that the work was done "in the presence of Nawbakht, Ibrāhīm ibn Muḥammad al-Fazārī (*sic!*), and (ᶜUmar ibn al-Farrukhān) al-Ṭabarī, the *munajjimīn*, the masters of calculation."

Of the works ascribed to al-Fazārī, aside from the verses of the *qaṣīda* translated above and Frag. Q 2, we apparently have fragments only of his *Zīj al-Sindhind al-kabīr* and of his *Zīj ᶜalā sinī al-ᶜArab*. It is not easy to assign these fragments to one *zīj* or the other; but we are clearly told that Frag. Z 6 and 15 come from the former, and more information about this *zīj* is contained in Frag. Z 1. It is also certainly true that Frag. Z 5 comes from the same source, and most probably all of the rest save for Frag. Z 10 (perhaps) and 23. These last two, then, and a part of Frag. Z 2 provide all the information available to us about the *Zīj ᶜalā sinī al-ᶜArab*.

The *Zīj al-Sindhind al-kabīr* was, of course, closely connected to the *Zīj al-Sindhind* translated from Sanskrit at the court of al-Manṣūr—according to Ibn al-Adamī, by al-Fazārī himself. But it was an extremely eclectic work, as were most early Islāmic *zīj*ʾes. Though the mean motions were indeed derived from the *Sindhind* (Frag. Z 4 and 5; cf. Frag. S 1), as were also the rules for computing the equations (Frag. Z 15 and 16; cf. Frag. S 2), and much of the geographical material (Frag. Z 17, 19–21, and 24; cf. Frag. S 3), the equations of the planets were taken from other sources—most notably, the *Zīj al-Shāh* (Frag. Z 11–14); a Persian source disguised as Hermes is discerned also (Frag. Z 17 and 18). And there is great inconsistency in the text. Three values of R are used; 3,270 (*Brāhmasphuṭasiddhānta*) in Frag. Z 12 and 16; 150 (*Zīj al-Shāh* and *Zīj al-Arkand*) in Frag. Z 11, 13, 15, and 25; and 3,438 (*Paitāmahasiddhānta* and *Āryabhaṭīya*) in Frag. Z 12 and S 3. Moreover, the maximum equation of the Sun according to the various formulas given by al-Fazārī can be: 2;11,15° in Frag. Z 11 and 12, or 2;14° (*Zīj al-Shāh*) in Frag. Z 16.

The fragments collected here are those to which al-Fazārī's name is attached, and a few in which the *Sindhind* only is mentioned, but which present material closely connected with one or several of al-Fazārī's genuine fragments. The commentary contains references to those other passages, in so far as they have been discovered, which attribute similar theories or methods to the *Sindhind* or to the Indians. But there are many other quotations from various *Zījāt al-Sindhind*; these have been omitted here as they may refer to the original translation from the Sanskrit (cf. Frag. S 1 and 2), to the *Sindhind al-ṣaghīr* (cf. Frag. S 3), to the *Zīj al-Sindhind* of al-Khwārizmī, or to other works of the same title. It is, of course, almost inevitable that al-Fazārī's *zīj* contained material on the Greek-letter phenomena, on the New Moon, and on solar and lunar eclipses, and there is much information about these problems in *Sindhind* sources; but they have been omitted here as it is not absolutely certain in what form they appeared in al-Fazārī's work.

Z 1. Ibn al-Adamī, *Naẓm al-ᶜiqd*, quoted by Ibn al-Qifṭī, pp. 270–71 (cf. p. 266), and by Ṣāᶜid al-Andalusī, ed. pp. 49–50, trans. pp. 102–103.

Al-Ḥusayn ibn Muḥammad ibn Ḥamīd, who is known as Ibn al-Adamī, in his large *zīj* called *Naẓm al-ᶜiqd* (*Niẓām al-ᶜiqd* in Ṣāᶜid), says that, in the year 156 (2 Dec. 772–20 Nov. 773), there came to the Caliph al-Manṣūr (755–775) a man from India, an expert in the calculation (*ḥisāb*) called *al-Sindhind*, concerning the motions of the planets (*nujūm*). He also had with him, in a book consisting of 12 ("a number of" in Ibn al-Qifṭī) chapters, equations (*taᶜādīl*) made according to *kardaja*'s calculated for each half degree, together with various operations on the sphere for the two eclipses (i.e. solar and lunar), the ascensions of the signs, and other things. He said that he had abridged the *kardaja*'s attributed to the king among kings (i.e. Mahārāja) of the Indians named Fiyaghra ("Qabaghra" in Ṣāᶜid; i.e. Vyāghramukha); they were calculated to the minute.

Al-Manṣūr ordered the translation of this book into Arabic, and that there should be written from it a book which the Arabs might use as a basis for the motions of the planets (*kuwākib*). Muḥammad ibn Ibrāhīm al-Fazārī was put in charge of this; and he made of it a book which the astronomers called *al-Sindhind al-kabīr*. By *sindhind* they mean in the Indian language (i.e. Sanskrit) "eternity" (*al-dahr al-dāhir*). Most of the people of that time used it until the days of the Caliph al-Maʾmūn (813–833). Abū Jaᶜfar Muḥammad (om. Ṣāᶜid) ibn Mūsā al-Khwārizmī abridged it for him, and made of it his *zīj* which is famous in the cities of Islām. In it he relied on the mean (motions) of *al-Sindhind*, but differed from it in the equations and the declination. He made the equations according to the opinion of the Persians, and the declination of the Sun according to the opinion of Ptolemy.

Commentary

This story of the transmission of the *Zīj al-Sindhind* to Baghdād has been discussed in "The Fragments of the Works of Yaᶜqūb ibn Ṭāriq," *JNES*, 26 (1968), pp. 97–125 (for the date of the delegation cf. also al-Hāshimī 95v:3). In that article it has been established that the Sanskrit work in 12 (?) chapters, while based on the *Brāhmasphuṭa-siddhānta* written by Brahmagupta in 628 at Bhillamāla under the Cāpa monarch Vyāghramukha, was a later text probably entitled *Mahāsiddhānta* which incorporated certain modifications of the *Brāhmasphuṭasiddhānta*. The calculation of tables in which the interval between entries in the column of arguments (*kardaja*'s) is 0;30° (see also Frag. Z 2) is redolent more of Ptolemy's table of chords (*Syntaxis* 1,11) than of any of the Indian *siddhāntas*, in which the *kardaja* is normally 3;45° or a multiple thereof (though Brahmagupta's *Dhyānagrahopadeśādhyāya* (47–56) uses a *nakṣatra* or 13;20°); Al-Hāshimī (115r:2) states that the *Sindhind* has only one *kardaja*. The table in which the *kardaja*'s were calculated to the minute might possibly be that in *Brāhmasphuṭas-iddhānta* 2,2–9, wherein R = 3,270 minutes; otherwise it makes little sense.

Sindhind (i.e. *siddhānta*) is also translated thus by al-Yaᶜqūbī (*Taʾrīkh*, 1, 84) among others, and it is rendered by al-Hāshimī (95v:7) as meaning "forever and ever" (*abad al-abad*), while al-Bīrūnī (*India*, ed. p. 118, trans. Vol. 1, p. 153) says that it signifies "that which is straight and is not crooked or changing"; further references will be found in Nallino (p. 204). In fact, it means "the perfected (book)."

For the identity of al-Khwārizmī's equations with those of the *Zīj al-Shāh* of Yazdijird III, see the commentary on al-Hāshimī (95r:21–95v:3); but al-Fazārī had already started this (see Frag. Z 2 and 11–14). For his use of Ptolemy's (really Theon's) value for the obliquity of the ecliptic, see Neugebauer, pp. 96–97.

Z 2. Al-Hāshimī 95v:17–96r:14.

As for al-Fazārī, he took the cycles which were mentioned by the Indians, (using) a four-hundredth part (of an hour). He mentioned it, and put it at the beginning of his book, (*Zīj al-Sindhind*) *al-kabīr*; then he mentioned the statement (concerning) the one two-hundredth part.

But when he saw the *Sindhind*, and the people's need for it, and the length of its operations in multiplication and division, and the tedious nature of the computations, he abridged it in a chapter (*bāb*), in which no one had preceded him. He expressed the cycles in sexagesimals, and so organized it in (sexagesimal) places (*bāb*). He extracted from them the motions of the planets in (*saura*) days, from (one) day up to 60 days. He extracted the mean planetary positions from these days, because he divided the cycles of each planet by 60 and set the remainder which did not contain 60 in another (sexagesimal) place until he arrived finally by this method at his objective. He said: it is imagined that each planet rotates in the heaven one (integer number of) rotation(s) during all the days of the (world-)year; and the number of (sexagesimal) places which came out for him from his division—such as degrees, minutes, and seconds—is named for the last place among the digits of this division, and its place according to the name of the fraction which is at its place.

Then he performed operations with the cycles, operations which are related to those which come out of these operations; and no one preceded him in this subject. He composed a treatise for each section in astronomical calculation, explaining in it these difficulties. Then (he wrote) a treatise about the operation; and it is said that no one in his time was his equal in (the domain of) Islām. This book was the last book he composed, as is said, because he classified many *zīj*'es, Arabic and Persian.

He put in his *zīj* shortened tables; and he divided (the arguments of) the sine, the declination, and the equation (tables) according to half-degrees, according to their magnitudes among the Persians, as a matter of convention, not for the sake of amelioration. He set up the equation (tables) of his *zīj* according to the Persian division in their equation (tables), but (he put) the apogees and nodes according to the *Sindhind*.

Commentary

With this whole passage compare the discussion of the *Sindhind* in al-Masᶜūdī (*Tanbīh*, ed. pp. 220–23, trans. pp. 293–96). For the year-length—6,5;15,30,22,30 days—see Yaᶜqūb Frag. Z 1, and al-Hāshimī (95v:4–5 and 109v:14–16), who ascribes it to the *Sindhind*; cf. also the table on f. 9 of Esc. Ar. 927. It is another expression of the fact that 4,320,000,000 sidereal years equal 1,577,916,450,000 days, for which see Frag. Z 5 and S 1, al-Yaᶜqūbī (*Taʾrīkh*, 1, 86–87), and *The Thousands of Abū Maᶜshar* (pp. 30–34). The fraction 1/200 in place of 1/400 gives a year of 6,5;15,30,45 days, or 1,577,916,900,000 days in a *kalpa*; see also al-Hāshimī (95v:7–9). Abraham ibn Ezra (p. 75) erroneously ascribes to the *Sindhind* a year of 365¼ days and ⅛ hour (= 6,5;15,30 days).

The second paragraph refers to a work in which al-Fazārī gave tables of the mean motions of the planets for 1 to 60 *saura* days; 1,0 to 6,0 *saura* days (6,0 *saura* days equal one year); 1 to 60 years; and 60 to some unknown multiple of 60 years. For the details of his computation, see the commentary on al-Hāshimī (95v:19–96r:14). Assuming that this second *zīj* is the *Zīj ᶜalā sinī al-Arab*, it must have contained tables for converting *Sindhind* dates into Hijra dates; see Frag. Z 9, al-Hāshimī 134r:18–134v:14, and Ṣāᶜid al-Andalusī (ed. p. 60, trans. p. 117). Al-Masᶜūdī (*Tanbīh*, ed. p. 199, trans. pp. 266–68) lists al-Fazārī with al-Farghānī, Yaḥyā ibn abī Manṣūr, al-Khwārizmī, Ḥabash, Māshāʾallāh, Muḥammad ibn Khālid al-Marwarrūdhī, Abū Maᶜshar Jaᶜfar ibn Muḥammad al-Balkhī, Ibn al-Farrukhān al-Ṭabarī, al-Ḥasan ibn al-Khaṣīb, Muḥammad ibn Jābir al-Battānī, al-Nayrīzī, "and others" as having referred to the eras of Nabonassar, Philip, the Hijra, and Yazdijird.

On the use of intervals of 0;30° in the column of arguments, see Frag. Z 1. For al-Fazārī's use of the equations of the *Zīj al-Shāh*, see Frag. Z 11–14; but for his adherence to the *Sindhind* positions of the apogees and nodes, see Frag. Z 5.

Z 3. Al-Bīrūnī, *India*, ed. pp. 351–52, trans. Vol. 2, p. 15.

It is one of the conditions of a *kalpa* that in it the planets, with their apogees and nodes, must unite in 0° of Aries, i.e. in the point of the vernal equinox. Therefore, each planet makes within a *kalpa* a certain number of complete revolutions or cycles. These revolutions of the planets, as known through the *Zīj*(es) of al-Fazārī and Yaᶜqūb ibn Ṭāriq, were derived from an Indian who came to Baghdād as a member of the political mission which Sind sent to the Caliph, al-Manṣūr, in A.H.154 (24 Dec. 770–12 Dec. 771). If we compare these secondary statements with the primary statements of the Indians, we discover discrepancies, the cause of which is not known to me. Is their origin due to the translation of al-Fazārī and Yaᶜqūb? or to the dictation of that Indian? or to the fact that afterwards these computations have been corrected by Brahmagupta, or someone else?

Commentary

This is Frag. Z 5 of Yaᶜqūb; cf. also al-Hāshimī (93v:14–18), al-Yaᶜqūbī (*Taʾrīkh*, 1, 86), Abraham ibn Ezra (pp. 88–89), and *The Thousands of Abū Maᶜshar* (pp. 33–34).

Z 4. Al-Bīrūnī, *India*, ed. p. 356, trans. Vol. 2, pp. 18–19.

We meet in this context (i.e. the difference between a *caturyuga* of 4,320,000 years and a *kalpa* of 4,320,000,000 years) with a curious circumstance. Evidently al-Fazārī and Yaᶜqūb sometimes heard from their Indian master to this effect, that his calculation of the revolutions of the planets was that of the great *Siddhānta*, while Āryabhaṭa reckoned with a one-thousandth part of it. They apparently did not understand him properly, and imagined that *āryabhaṭa* (*arjabhad*) means "a thousandth part."

Commentary

This if Frag. Z 6 of Yaᶜqūb. The *Zīj al-Sindhind al-ṣaghīr* seems to have used a *caturyuga* instead of a *kalpa*; see Frag. S 3.

Z 5. Al-Bīrūnī, *India*, ed. pp. 352–53, trans. Vol. 2, p. 16 (cf. ed. pp. 354–55, trans. Vol. 2, pp. 17–18). (I have rearranged the table.)

Commentary

See the commentary on Frag. Z 1 of Yaᶜqūb; these parameters are ascribed to the *Sindhind* by al-Hāshimī (104r:16–105v:8; cf. 107v:23–108r:23). See also *The Thousands of Abū Maᶜshar* (pp. 31–37). For the motions of the apogees and nodes in the *Sindhind*, see al-Hāshimī (114v:5–7; see also 117r:23–117v:2), al-Bīrūnī (Chronology, ed. p. 9, trans. p. 11), and Abraham ibn Ezra (pp. 77–78 and 109). The motion of the fixed stars equals the Ptolemaic value of precession—1° every 100 years.

Planets	Brahmagupta	al-Fazārī
Sun	4,320,000,000	4,320,000,000
Its apogee	480	480
Moon	57,753,300,000	57,753,300,000
Its apogee	488,105,858	488,105,858
Its node	232,311,168	232,312,138
Mars	2,296,828,522	2,296,828,522
Its apogee	292	292
Its node	267	267
Mercury	17,936,998,984	17,936,998,984
Its apogee	332	332
Its node	521	521
Jupiter	364,226,455	364,226,455
Its apogee	855	855
Its node	63	63
Venus	7,022,389,492	7,022,389,492
Its apogee	653	653
Its node	893	893
Saturn	146,567,298	146,569,284
Its apogee	41	41
Its node	584	584
Fixed stars	———	120,000

Z 6. Ibn Masrūr f. 47.

The mean (motions) are extracted from their (the planets') revolutions in the world-years, elevated to years after years, as was done by al-Fazārī in the *Sindhind al-kabīr*.

COMMENTARY

This fragment indicates that al Fazārī, in his *Zīj al Sindhind al-kabīr* as well as in his *Zīj ᶜalā sinī al-ᶜArab* (see Frag. Z 2), tabulated the mean motions of the planets year by year, and therefore presumably month by month and day by day. What number of collected years he may have used can only be guessed at; Yaᶜqūb (Frag. Z 1) used 10.

Z 7. Al Hāshimī 97r:1–3.

He (Abū Maᶜshar) (also) put in it (the *Zīj al-hazārāt*) a chapter on the altitude and depression of the heaven(s), but it does not agree with what was mentioned about it by Theon and Yaḥyā ibn abī Manṣūr, nor what was done about it by al-Fazārī and Ḥabash, nor what is mentioned in the *Arkand*.

COMMENTARY

This probably refers to the matter of planetary distances, which is dealt with by Yaᶜqūb (Frag. T 1–3); see also Frag. S 3 below and *The Thousands of Abū Maᶜshar* (p. 56).

Z 8. Al-Bīrūnī, *India*, ed. pp. 360–61, trans. Vol. 2, p. 23.

As regards *adhimāsa* (*adimāsah*), the word means "the first month," for *ādi* means "beginning." In the books of Yaᶜqūb ibn Ṭāriq and al-Fazārī this name is written *padamāsah* (read *malamāsah*, for Sanskrit *malamāsa*). *Pad* means "end," and it is possible that the Indians call the leap month by both names; but the reader must be aware that these two authors frequently misspell or disfigure the Indian words, and that there is no reliance on their tradition.

COMMENTARY

This is Frag. T 5 of Yaᶜqūb.

Z 9. Al-Bīrūnī, *India*, ed. p. 128, trans. Vol. 1, p. 165.

Al-Fazārī in his *zīj* uses the word *pala* in place of a minute of the days; I have not found this meaning for it in the books of the nation (of Indians), but they use it in the sense of "correction" (*taᶜdīl*).

COMMENTARY

The word *pala* in Sanskrit astronomical texts is one of the normal terms for a minute (= sixtieth) of a *ghaṭikā*; the latter is a sixtieth of a nychthemeron. The Sanskrit word signifying "correction" or "equation" is *phala*. Al-Bīrūnī is not displaying an advanced knowledge of *jyotiḥśāstra* in this passage.

Z 10. Al-Hāshimī 100r:4–6.

For anyone who wants that determination (I mean of the base), and what I have presented, verily the best of tables for this purpose is the *Mujarrad* Table, made up for 210 years. Indeed, al-Fazārī and Ḥabash and Abū Maᶜshar put it in their *zīj*'es.

COMMENTARY

Al-Fazārī's use of the *Mujarrad* Table for determining the week-day in the Arabic calendar supports the conjecture that he, and not his father, was the author of the *Zīj ᶜalā sinī al-ᶜArab*. The *Mujarrad* Table is found in both of the *zīj*'es of Ḥabash (Yeni Cami 784 f. 86v and Berlin 5750 f. 16; cf. al-Bīrūnī, *Chronology*, ed. p. 198, trans. p. 179), and in the *Zīj al-mumtaḥan* of Yaḥyā ibn abī Manṣūr (Esc. Ar. 927 f. 18). The table is, of course, simply the normal Arabic 30 year cycle of 10,631 days multiplied by 7. But Yaḥya intercalates in the second, fifth, seventh, tenth, thirteenth, sixteenth, eighteenth, twenty-first, twenty-fourth, twenty-sixth, and twenty-ninth years of every 30-year period, while Ḥabash intercalates in the third, sixth, eighth, eleventh, thirteenth, sixteenth, nineteenth, twenty-first, twenty-fourth, twenty-seventh, and thirtieth. It is Yaḥyā's pattern which is followed by al-Khwārizmī (see NO 2795 f. 65v); therefore, I assume that it was the pattern of al-Fazārī, and transcribe Yaḥyā's *Mujarrad* Table (see facing page).

Z 11. Al-Bīrūnī, *On Transits*, 24:17–25:3.

As for the amount ascribed to the *Sindhind*, with Yasᶜā al-Maʾmūnī's addition to it, it (the maximum equation of the Sun) is 2;11°. It is this that al-Fazārī made use of in subtracting from the sine of the argument of the Sun ⟨an eighth⟩ of it, and in doubling the sine of the argument of the Moon to obtain their equations. And thus the maximum equation of the Sun comes out equal to 2;11,15°, and that of the Moon equal to 5°; and that is as though the total sine were 150 minutes.

COMMENTARY

Yasᶜā al-Maʾmūnī seems to be an error; but the procedure ascribed to al-Fazārī is all right, as:

$$150 \times \tfrac{7}{8} = 131\tfrac{1}{4} = 2;11,15°,$$

and

$$150 \times 2 = 300 = 5°.$$

Y	D	Y	D	Y	D	Y	D	Y	D	Y	D	Y	D	M	D
1	6	31	4	61	2	91	7	121	5	151	3	181	1	al-Muḥarram	0
2	3	32	1	62	6	92	4	122	2	152	7[c]	182	5	Ṣafar	2
3	1	33	6	63	4	93	2	123	7	153	5	183	3	Rabīᶜ al-awwal	3
4	5	34	3	64	1	94	6	124	4	154	2	184	7[d]	Rabīᶜ al-ākhar	5
5	2	35	7	65	5	95	3	125	1	155	6	185	4[e]	Jumādā al-ūlā	6
6	7	36	5	66	3	96	1	126	6	156	4	186	2	Jumādā al-ukhrā	1
7	4	37	2	67	7	97	5	127	3	157	1	187	6	Rajab	2
8	2	38	7	68	5	98	3	128	1	158	6	188	4[e]	Shaᶜbān	4
9	6	39	4	69	2	99	7	129	5	159	3	189	1	Ramaḍān	5
10	3	40	1	70	6	100	4	130	2	160	7	190	5	Shawwāl	7
11	1	41	6	71	4	101	2	131	7	161	5	191	3[f]	Dhū al-Qaᶜda	1
12	5	42	3	72	1	102	6	132	4	162	2	192	7	Dhū al-Ḥijja	3
13	2	43	7	73	5	103	3	133	1	163	6	193	4		
14	7	44	5	74	3	104	1	134	6	164	4	194	2		
15	4	45	2	75	7	105	5	135	3	165	1	195	6		
16	1	46	6	76	4	106	2	136	7	166	5	196	3		
17	6	47	4	77	2	107	7	137	5	167	3	197	1		
18	3	48	1	78	6	108	4	138	2	168	7	198	5		
19	1	49	6	79	4	109	2	139	7	169	5	199	3[f]		
20	5	50	3	80	1	110	6	140	4	170	2	200	7		
21	2	51	7[a]	81	5	111	3	141	1	171	6	201	4		
22	7	52	5	82	3	112	1	142	6	172	4	202	2		
23	4	53	2	83	7	113	5	143	3	173	1	203	6		
24	1	54	6	84	4	114	2	144	7	174	5	204	3		
25	6	55	4	85	2	115	7	145	5	175	3	205	1		
26	3	56	1	86	6	116	4	146	2	176	7	206	5		
27	1	57	6	87	4	117	2	147	7	177	5	207	3[f]		
28	5	58	3	88	1	118	6	148	4	178	2	208	7		
29	2	59	7[b]	89	5	119	3	149	1	179	6	209	4		
30	7	60	5	90	3	120	1	150	6	180	4	210	2		

a 1 ms. b 2 ms. c 1 ms. d 5 ms. e 3 ms. f 2 ms.

Cf. Frag. Z 12, 13, and 15. Al-Jayhānī, quoted by al-Bīrūnī (*On Transits* 23:17–24:1), says that the maximum equation of the Sun in the *Zīj al-Sindhind* is 2;10,46,40°; cf. the 2;10,31° of the *Paitāmahasiddhānta* (see the commentary on Frag. Z 7 of Yaᶜqūb).

Z 12. Al-Bīrūnī, *On Transits* 25:9–17.

And there was mentioned in some of the books a story about al-Fazārī regarding the equation of the Sun, where he multiplies the sine of its argument in the *kardaja*'s of the *Sindhind* by 105 and divides the product by 2,616; and that in the *kardaja*'s of Āryabhaṭa by 7 and divides the product by 180. In the case of the equation of the Moon he multiplies the sine of its argument in the *kardaja*'s of the *Sindhind* by 10 and divides the product by 109 (read *tisᶜa* for the text's *sabᶜa*); and that in the *kardaja*'s of Āryabhaṭa by 10 and divides the product by 116 (ed. has 117).

Commentary

R for the *Sindhind*, as is noted by al-Bīrūnī (*On Transits* 27:1–3), is the same as in Brahmagupta's *Brāhmasphuṭasiddhānta* (2,5)—namely, 3,270 (see Frag. Z 16); then its maximum equations of the Sun and Moon, according to al-Fazārī's rule, will be: for the Sun:

$$3270 \times \frac{105}{2616} = 131\tfrac{1}{4} = 2;11,15°,$$

and for the Moon:

$$3270 \times \frac{10}{109} = 300 = 5°.$$

These two values are precisely equivalent to those in Frag. Z 11. Al-Fazārī's formula for computing the equation of the Sun is also given precisely in the form it has here in al-Bīrūnī's *Rasāʿil* 1, p. 133, with an example in which the sine of the Sun's argument is 1,635; 1,635 is the eighth sine in Brahmagupta's table, corresponding to an arc of 30°. The values 2;11,15° and 5° are not the values of the *Paitāmahasiddhānta*, the *Brāhmasphuṭasiddhānta*, or the *Zīj al-Shāh*; see the commentary on Frag. Z 7 of Yaʿqūb.

The second part of these rules indicates that al-Fazārī had some knowledge of a work of Āryabhaṭa; and in the *Āryabhaṭīya* R = 3,438. Then, for the Sun:

$$3438 \times \tfrac{7}{180} = 133\tfrac{7}{10} = 2;13,42°;$$

and for the Moon:

$$3438 \times \tfrac{10}{116} = 296\tfrac{11}{29} \approx 4;56,23°$$

In all cases, since it is assumed that

$$E_{max} = R \times \frac{\text{circumference of the epicycle}}{360°}$$

we can easily compute the underlying values of the circumferences of the epicycles. These are:

	Al-Fazārī	Āryabhaṭa
Sun	$14\tfrac{49}{109} \approx 14;27°$	14°
Moon	$33\tfrac{3}{109} \approx 33;1,50°$	$31\tfrac{1}{29} \approx 31;2°$

In fact, the circumferences of the epicycles of the Sun and Moon according to the *Āryabhaṭīya* (*Daśagītikā* 8) are 13:30° and 31;30°; the first of these would have yielded the formula:

$$E_{max} = 3438 \times \tfrac{3}{80},$$

and the second:

$$E_{max} = 3438 \times \tfrac{7}{80}.$$

Since al-Fazārī does not use the fractions $\tfrac{3}{80}$ and $\tfrac{7}{80}$, it is clear that he did not use the *Āryabhaṭīya*; rather, he started with the values for the circumferences of the epicycles 14° and 31° (the fraction $\tfrac{10}{116}$ represents an approximation). Precisely these values are found in the *ārdharātrika* system of astronomy which Āryabhaṭa developed in a lost work, and it is from Āryabhaṭa's lost work, then, that al-Fazārī has derived his rules. However, it is not absolutely necessary to assume that he also found the value R = 3,438 in this lost work of Āryabhaṭa.

Z 13. Al-Bīrūnī, *On Transits*, 30:19–31:5.

And the rule of al-Fazārī is proportional to these quantities. He suggests in the case of Saturn to multiply the sum of the sine (of the argument) and its tenth and (one third of its tenth and) one-sixth of its tenth by 3; and for Jupiter to double the sun of the sine and one-fifth of its tenth; and for Mars to multiply the sum of the sine and its tenth and a sixth of its tenth by 4; and for Venus to diminish from the sine one-tenth of it; and for Mercury to add to the sine three-fifths of it.

COMMENTARY

These rules for obtaining the equations of the center, like those in Frag. Z 11 and 15, employ the value R = 150. The maximum equations of the planets then are:

Saturn	$(150 + \frac{150}{10} + \frac{150}{30} + \frac{150}{60}) \times 3 = 8;37,30°$
Jupiter	$(150 + \frac{150}{50}) \times 2 = 5;6°$
Mars	$(150 + \frac{150}{10} + \frac{150}{60}) \times 4 = 11;10°$
Venus	$(150 - \frac{150}{10}) = 2;15°$
Mercury	$(150 + \frac{450}{5}) = 4°$

For these values of the maximum equations, see the commentary on Frag. Z 7 of Yaᶜqūb. The rule for Venus is applied to the Sun by al-Hāshimī (119r:12–14), and the resulting equation (E_{max} = 2;15°) ascribed to the *Zīj al-Shāh* and the *Zīj al-Sindhind*; cf. Frag. Z 16.

Z 14. Al-Bīrūnī, *On Transits* 54:5–14.

In the *Zīj al-Shāh* it (the maximum equation of the anomaly) is 5;44° for Saturn, . . . for Jupiter 10;52°, . . . for Mars 40;31° (ed. has 41;30°), . . . for Venus 47;11° . . . and for Mercury 21;30°. . . . But al-Fazārī and al-Khwārizmī have them like what is in the *Zīj al-Shāh,* since it is the Indian way.

COMMENTARY

See the commentary on Frag. Z 8 of Yaᶜqūb.

Z 15. Al-Bīrūnī, *Rasāʾil* 1, p. 120.

On the explanation of the equation by computation, Muḥammad ibn Ibrāhīm al-Fazārī in the *Zīj al-Sindhind al-kabīr* (*al-kathīr* in the ed.).

He says: "We multiply the sine of the argument by $\frac{2}{5}$ of the base (*aṣl*) and divide the product by 60; there results the side (*ḍilᶜ*). We multiply the cosine of the argument by $\frac{2}{5}$ of the base (*aṣl*) and divide the product by 60; we add the result to R if the argument is less than a quadrant, but subtract it from R (if the argument is greater than a quadrant). This sine of addition or subtraction is a side (*ḍilᶜ*) for it (the triangle)." The operation proceeds as we have described according to the opinion of al-Bustānī and al-Hāshimī; it is exactly identical and does not differ at all. Therefore we substitute for the remainder what was said here.

COMMENTARY

The method of the *Sindhind* according to al-Bustānī (al-Battānī?) and al-Hāshimī will be found in Frag. S 2; see Kennedy and Muruwwa (p. 116). The only difference between that method and this one is that al-Fazārī substitutes $\frac{2}{5} \times \frac{1}{60} = \frac{1}{150}$ for 1/R; see al-Bīrūnī (*Rasāʾil* 1, pp. 156–57). This demonstrates that R = 150 in at least some passages of the *Zīj al-Sindhind al-kabīr*; cf. Frag. Z 11, 13, and 25.

Z 16. Al-Bīrūnī, *Rasāʾil* 1, pp. 177–78.

As for what we said on the authority of al-Fazārī, R in the *kardaja*'s of *al-Sindhind* is 3,270, and its ratio to 134 which are the minutes of the maximum equation (of the Sun) is like the ratio of 1,635 to 67. According to this ratio is set down the ratio of the sine of the argument to its equation in principle, not in fact as his words indicate. If it were, one would instruct one to multiply the sine of the argument in these *kardaja*'s by 134 or by 67 and to divide the product by 3,270 or by 1,635, so that, according to this principle and basis, the equation should result.

COMMENTARY

It appears from this that al-Fazārī used the common rule:

$$\text{equation} = \text{sine of argument} \times \frac{\text{maximum equation}}{\text{R}}$$

at least in computing the solar equation. The formula (the "method of sines") and the value for the maximum equation (2;14°) occur in the *ārdharātrika* system and in the *Zīj al-Shāh*; al-Fazārī has simply thrown in the *Brāhmasphuṭasiddhānta*'s value of R. This passage, of course, contradicts Frag. Z 11 and Z 12. But, that the maximum equation of the Sun equals 2;14° according to the Indians is stated by Abraham ibn Ezra (p. 78); cf. also al-Hāshimī cited in the commentary on Frag. Z 13, who takes it to be 2;15°.

Z 17. Al-Bīrūnī, *Taḥdīd nihāyat al-amākin*, ed. pp. 211–12, trans. p. 177.

However al-Fazārī has mentioned in his *zīj* that the Indians consider the circumference of the earth to be 6,600 *farsakh*'s, where a *farsakh* is 16,000 cubits. He mentioned also that Hermes considered it to be 9,000 *farsakh*'s, where a *farsakh* is 12,000 cubits. So the share of one degree out of 360°, according to the Indians, is 18⅓ *farsakh*'s; if each of them is three miles, then one degree would be equal to 55 miles, where each mile is 5,333⅓ cubits. But, according to Hermes, it would be 25 *farsakh*'s which are equal to 75 miles, where each mile is 4,000 cubits. Then al-Fazārī claimed that some sages estimated each degree to be equivalent to 100 miles, and hence their estimate of the circumference of the earth would be 12,000 *farsakh*'s.

COMMENTARY

That the circumference of the earth is 6,600 *farsakh*'s (which equal 3,300 *yojana*'s) is stated (in *yojana*'s) by Lalla (*Śiṣyadhīvṛddhidatantra* I 1,56; see also Frag. S 3); Āryabhaṭa (*Daśagītikā* 5) gives it as $3{,}298\frac{17}{25}$ *yojana*'s. The latter value has been traced through the *Zīj al-Sindhind al-kabīr* and the *Zīj al-Sindhind al-ṣaghīr* (see Frag. S 3) to the *Bhūgolādhyāya* (?) of the *Mahāsiddhānta* (?) (see Frag. T 1–2 of Yaᶜqūb). The value 3,300 (= 6,600) here attested for al-Fazārī, then, is probably a rounding; but al-Yaᶜqūbī (*Taʾrīkh*, 1, 85) says that, according to the Indians, the diameter of the earth is 2,100 *farsakh*'s, its circumference 6,300 (*sic!*) *farsakh*'s of which each one equals 16,000 cubits.

Hermes' 9,000 *farsakh*'s equal 108,000,000 cubits or 270,000 stadia. I do not know of another occurrence of exactly this value, but it lies between Archimedes' 300,000 stadia and Eratosthenes' 250,000 (see Bunbury, 1, 620–21). On al-Fazārī's use of Hermes, see Frag. Z 18.

Z 18. Yāqūt, *Muᶜjam*, ed. vol. 1, p. 27, trans. p. 42.

Abū al-Rayḥān (al-Bīrūnī) said: "Hermes upheld this (Persian) division (of the inhabited world into 7 *kishwars*), according to that which Muḥammad ibn Ibrāhīm al-Fazārī attributed to him in his *zīj*. Since Hermes was one of the ancients, it appears that in his time no other division was used; otherwise, matters pertaining to mathematics and astronomy were more particularly within the province of Hermes." Abū al-Rayḥān continued: "al-Fazārī added, 'each *kishwar* is 700 *farsakh*'s by 700 *farsakh*'s.'"

COMMENTARY

This quotation from al-Bīrūnī, like Frag. Z 17, demonstrates al-Fazārī's use of a Persian source which passed under the name of Hermes. There are many descriptions of the *kishwar*'s; for brevity's sake I refer only to al-Bīrūnī (*Tafhīm*, p. 142, and *Taḥdīd*, ed. pp. 135–36, trans. pp. 101–102). It should be noted that Ibn al-Fakīh (p. 7) reports: "Hermes asserts that the length of each clime is 700 *farsakh*'s by 700"; the same dimensions for each clime are attributed to the Indians by al-Yaᶜqūbī (*Taʾrīkh*, 1, 85).

Z 19. Al-Bīrūnī, *Taḥdīd nihāyāt al-amākin*, ed. p. 157, trans. p. 121.

When a comparison was made between the two systems (i.e., eastern and western), it was found that the longitude of one and the same place, as determined by the easterners, exceeds the supplement of its longitude, as determined by the westerners, by 10°. Al-Fazārī assumes in his *zīj* that the difference is 13;30°.

COMMENTARY

According to Frag. Z 20, al-Fazārī places the longitude of Bārah at 190;50° East of the Western Isles; the difference, then, between his longitudes and those of Ptolemy should be 10;50° rather than 13;30°. Evidently, the manuscript of either the *Taḥdīd* or the *Qānūn* is erroneous, or else al-Bīrūnī's knowledge of al-Fazārī's *zīj* has improved between the composition of the *Taḥdīd* in 1025 and that of the *Qānūn* in 1031. The latter explanation seems the better, as al-Bīrūnī only displays an extensive acquaintance with al-Fazārī in his later works.

Z 20. Al-Bīrūnī, *Al-Qānūn al-Masᶜūdī*, 2, 547.

Localities along the equator, possessing no latitude.

1. The island Lañkā (*lanka*), called in books the Cupola of the earth.

Long. 100;50° Lat. 0;0°

2. Tārah (read Bārah), which al-Fazārī and Yaᶜqūb ibn Ṭāriq mention.

Long. 190;50° Lat. 0;0°

3. Yamakoṭi (*jamkūt*) at the Eastern extremity, which is called Yamkard (*jamakard*) according to the Persians; there is no building beyond it according to the Indians.

Long. 190;0° Lat. 0;0°

COMMENTARY

This is Frag. Z 3 of Yaᶜqūb. The *Sindhind*, according to al-Hāshimī (93v:8–14), seems to have placed Bārah 90° East of the Cupola of the earth (on the meridian of Ujjain), which in turn lay 90° East of Thule (*bula*) in the Green Sea; cf. also *The Thousands of Abū Maᶜshar* (p. 45).

Z 21. Al-Bīrūnī, *India*, ed. p. 259, trans. Vol. 1, pp. 303–304.

Yamakoṭi is, according to Yaᶜqūb and al-Fazārī, the country where is the city Tārah (read Bārah) within a sea. I have not found the slightest trace of this name (i.e. Bārah) in Indian literature. As *koṭi* means "castle" and Yama is the angel of death, the word reminds me of Kangdiz, which, according to the Persians, had been built by Kaykaᶜus or Jam in the most remote East, behind the sea.

COMMENTARY

This is Frag. Z 4 of Yaᶜqūb.

Z 22. Al-Hamdānī, 1, 45.

The latitude of Makka according to al-Fazārī is $23\frac{1}{3}$ degrees, . . . and its longitude according to al-Fazārī is 116° from the East. . . . Al-Fazārī says that its (Madīna's) latitude is 30° except for a fraction, and this is not found. And he says that the longitude of Bayt al-Maqdis (Jerusalem) is 127°, and its latitude $31\frac{5}{6}$ degrees.

COMMENTARY

This passage along with Frag. Z 20 indicates that al-Fazārī included a table of the longitudes of famous cities in one of his *zījes*.

Z 23. Al-Masᶜūdī, *Murūj al-dhahab*, 2nd ed., 2, 376–78.

We think that we should finish this chapter with a summary of the extents of the kingdoms and of how close to or far from each other they are according to the calculation of al-Fazārī, the author of a *Kitāb al-zīj* and of a *Qaṣīda fī hayʾāt al-nujūm wa al-falak* (*Poem on the Forms of the Stars and of the Sphere*). Strength is with Allāh!

Al-Fazārī said that the administration of the Commander of the Faithful (extends) from Farghāna and the furthest part of Khurāsān to Ṭanja (Tangier) in the West ⟨4⟩3,700 *farsakh*'s, and from the Bāb al-abwāb (Darband) to Judda (Jidda) 700 *farsakh*'s, and from the Bāb to Baghdād 300 *farsakh*'s, and from Makka to Judda 38 miles.

The administration of Ṣīn (China) in the East 31,000 *farsakh*'s by 11,000 *farsakh*'s.

The administration of Hind (India) in the East 11,000 *farsakh*'s by 7,000 *farsakh*'s.

The administration of Tabbat (Tibet) 500 *farsakh*'s by 230 *farsakh*'s.

The administration of the Kābul Shāh 400 *farsakh*'s by 60 *farsakh*'s.

The administration of the Tughuzghuz (Toghuzghuz or Toquz-Oghuz) among the Turks 1,000 *farsakh*'s by 500 *farsakh*'s.

The administration of the Turks by the Khāqān 700 *farsakh*'s by 500 *farsakh*'s.

The administration of the Khazars and Alans 700 *farsakh*'s by 300 *farsakh*'s.

The administration of Burjān (the Bulgars) 1,500 *farsakh*'s by 300 *farsakh*'s.

The administration of Ṣaqāliba (the Slavs) 3,500 *farsakh*'s by 700 *farsakh*'s.

The administration of the Romans in Qusṭanṭiniya (Constantinople) 5,000 *farsakh*'s by 420 *farsakh*'s.

The administration of Rūmiya (Rome) of the Romans 3,000 *farsakh*'s by 700 *farsakh*'s.

The administration of Andalus by ᶜAbd al-Raḥman ibn Muᶜāwiya 300 *farsakh*'s by 80 *farsakh*'s.

The administration of Idrīs al-Fāṭimī 1,200 *farsakh*'s by 120 *farsakh*'s.

The administration of the coast of Sijilmāsa by the Banū al-Muntaṣir 400 *farsakh*'s by 80 *farsakh*'s.

The administration of Anbiya 2,500 *farsakh*'s by 600 *farsakh*'s.
The administration of Ghāna, the land of gold, 1,000 *farsakh*'s by 80 *farsakh*'s.
The administration of Warām 200 *farsakh*'s by 80 *farsakh*'s.
The administration of Nakhla 120 *farsakh*'s by 60 *farsakh*'s.
The administration of Wāḥ 60 *farsakh*'s by 40 *farsakh*'s.
The administration of Buja 200 *farsakh*'s by 80 *farsakh*'s.
The administration of Nūba (Nubia) by al-Najāshī (the Ethiopian king) in the West 1,500 *farsakh*'s by 400 *farsakh*'s.
The administration of Zanj in the East 7,600 *farsakh*'s by 500 *farsakh*'s.
The administration of Isṭūlā by Aḥmad ibn al-Muntaṣir 400 *farzakh*'s by 250 *farsakh*'s.
In longitude (the inhabited world extends) 78,480 *farsakh*'s, in latitude 25,250 *farsakh*'s.

COMMENTARY

This is one of the earliest lists of countries in an Arabic source and clearly reflects the political situation of around 788. For ᶜAbd al-Raḥman ibn Muᶜāwiya ruled Andalusia from 756 to 788; the Banū Muntaṣir, whom Nallino (p. 212) identifies with the Banū Midrār, began to reign at Sijilmāsa in *ca.* 786; and Idrīs al-Fāṭimī ruled in Morocco from 788 to 793. Moreover, the Kābul Shāhs were driven from Kābul some time between 775 and 809; and, if the Toquz-Oghuz are the Sha-tᶜo Turks as suggested by Barthold (pp. 89–90) rather than the Uighurs (see Dunlop, p. 39), they were suppressed by the Tibetans in 790 (Pelliot, pp. 60 and 121). Therefore, if this fragment comes from the *Zīj al-Sindhind al-kabīr*, that work was written after Yaᶜqūb's *Tarkīb al-aflāk*, which was probably composed in 777/78. Hence I conclude that the fragment most likely comes from al-Fazārī's other *zīj*. This conjecture is supported by the fact that the dimensions given here outrageously contradict the circumference of the earth in Frag. Z 17 and 24.

Z 24. Al-Bīrūnī, *India*, ed. pp. 267–68, trans. Vol. 1, pp. 314–15.

As for the calculation of the *deśāntara* (*dīshantar*: longitudinal difference) from the latitudes of two cities, al-Fazārī teaches in his *zīj*: "Add together the squares of the sines of the latitudes of the two cities, and find the root of the sum; this is the portion (*ḥiṣṣa*). Then square the difference between these two sines and add it to the portion. Multiply the sum by 8 and divide the product by 377; there results the 'lofty' (*jalīla*) distance between the two (cities). Then multiply the difference between the two latitudes by the *yojana*'s of the circumference of the earth, and divide the product by 360." It is known that this is to convert the difference between the two latitudes measured in degrees and minutes into that measured in *yojana*'s. He says: "Subtract the square of the quotient from the square of the 'lofty' distance, and find the (square) root of the remainder; these are the straight *yojana*'s." It is evident that this (latter) is the difference between the meridians of the two cities on the circle of latitude; from this one knows that the "lofty" distance is the distance between the two cities.

COMMENTARY

As al-Bīrūnī remarks, this procedure is erroneous, and I do not find such a method in Sanskrit texts. If the sine of the latitude of one city be denominated a, that of the other b, and the longitudinal difference c, the formula amounts to:

$$c^2 = \left((\sqrt{a^2 + b^2} + a^2 - b^2) \times \frac{8}{377}\right)^2 - \left((a - b) \times \frac{\text{circumference}}{360}\right)^2$$

Since the term—(a − b) × circumference/360—represents the *yojana*'s of the distance between the points where the two parallels of latitude intersect a parallel of longitude, and since c is the *yojana*'s of distance between the points where the two parallels of longitude intersect a parallel of latitude, the first term—$(\sqrt{a^2 + b^2} + a^2 - b^2) \times \frac{8}{377}$ —must be the *yojana*'s of direct distance between the two cities. But of course this direct distance has no relationship with the respective latitudes of the two cities, and the constant $\frac{8}{377}$ (= 1:47$\frac{1}{8}$) cannot be explained. The circumference of the earth, according to al-Fazārī, is 6,597$\frac{9}{25}$ *farsakh*'s; see Frag. Z 17 and S 3.

In *Brāhmasphuṭasiddhānta* 1, 36–37 (cf. also *Mahābhāskarīya* 2, 3–4 and Frag. S 3) is found the following rule: "The circumference of the earth is 5,000 (*yojana*'s). Multiply this by the difference between the latitudes, and divide (the product) by 360; the square of the longitudinal difference (*deśāntara*) (from the prime meridian) is diminished by the square of that quotient. The daily progress (of the planet) multiplied by the square-root of the remainder and divided by the circumference of the earth is taken in minutes of arc; the result is negative to the east of the north-south line running through Ujjayinī, positive to the west." Here the first value computed is parallel to the $(a - b) \times \frac{\text{circumference}}{360}$ of al-Fazārī, and the final value is the c of the previous formula converted into minutes of arc; the *deśāntara*, however, is computed from the difference in *ghaṭikā*'s between the time of an eclipse computed for the prime meridian and the time of the same eclipse observed at one's locality.

Z 25. Al-Bīrūnī, *On Shadows*, 51:9–52:19.

And when one puts together the operations of the workers in this field about that, the routes they traveled and the numbers they put will not remain hidden, as is the case with Muḥammad ibn Ibrahīm al-Fazārī, Yaᶜqūb ibn Ṭāriq, Muḥammad ibn Mūsā al-Khwārizmī, Habash al-Ḥāsib, Abū Maᶜshar al-Balkhī, al-Faḍl ibn Nadīm al-Nayrīzī, Muḥammad ibn Jābir al-Battānī, and Abū al-Wafāʾ al-Buzjānī. All of these explained in their *zīj*'es that, if the shadow is squared and the gnomon is squared and the (square) root of their sum is taken, it will be the cosecant. . . . And when the cosecant is determined for all of them, some of them proceed to the cosine of the altitude, and some to the sine of the altitude itself. . . . However, those who proceed toward the sine of the altitude itself divide by the cosecant the product of the gnomon and the total sine. . . . However, al-Fazārī, al-Khwārizmī, Yaᶜqūb ibn Ṭāriq, Abū Maᶜshar, and the author of the *Zīj al-Shāh* prescribe the division of 1,800 by the cosecant, and it is the product of 150 multiplied by 12.

Commentary

This is Frag. K 1 of Yaᶜqūb.

S 1. Al-Bīrūnī, *India*, ed. pp. 309–10, trans. Vol. 1, p. 368.

And our authorities (*aṣḥāb*) call them (the days of a *kalpa*) "the days of the *Sindhind*" or "the days of the world"; they are 1,577,916,450,000—in solar years 4,320,000,000; in lunar years 4,452,775,;000 in years of which each consists of 360 civil days 4,383,101,250; and in *divya* (*diba*) years 12,000,000.

COMMENTARY

The number of synodic months in a *kalpa* is 53,433,300,000 (cf. Frag. Z 5); a twelfth of this is 4,452,775,000, which then is the number of lunar years. Similarly, 1,577,916,450,000 divided by 360 equals 4,383,101,250. A *divya* day is a revolution of the Sun (*Brāhmasphuṭasiddhānta* 1, 25); so a *divya* year is 360 solar years, and there are 12,000,000 *divya* years in a *kalpa*.

For the days in a *kalpa*, see also al-Ya^cqūbī (*Ta^ɔrīkh*, 1, 86), Ibn al-Muthannā (p. 152 Hebrew, pp. 105–106 Latin), Abraham ibn Ezra (pp. 88–89), and *The Thousands of Abū Ma^cshar* (p. 31); the intervals in days from the beginning of the *kalpa* to various epochs will be found in a text published in *The Thousands of Abū Ma^cshar* (p. 34, fn.) and in al-Hāshimī (108v:4–7).

S 2. Al-Bīrūnī, *Rasā^ɔil* 1, pp. 118–19.

On the explanation of the equation by computation, Muḥammad ibn Jābir al-Bustānī (al-Battānī?) mentioned this in his *zīj*, and Muḥammad ibn ^cAbd al-^cAzīz al-Hāshimī also explained it in two places in his *Kitāb ta^clīl li-Zīj al-Khwārizmī* without any proofs; in one of these two (places) he declared: "Computation of the equation according to the opinion of *al-Sindhind*.

"We multiply the sine of the argument by the base (*aṣl*) and divide the product by R; there results then a side (of the triangle) (*ḍil^c*). We multiply the cosine of the argument by the base (*aṣl*) and divide the product by R; we add the result to R if the argument is less than 90°, or we subtract it from R if the argument is greater than 90°. Then we multiply whatever results from the addition or subtraction by itself, and we add to the product of the multiplication of the side (*ḍil^c*) by itself. We take the square root of the sum, and this is the hypotenuse (*quṭr*). Then we multiply the side (*ḍil^c*) by R and divide the product by the hypotenuse (*quṭr*); there results the sine of the equation.

"As for what concerns the second part, we add the square of the base (*aṣl*) to R^2 and take the square root of the sum; this is the hypotenuse (*quṭr*). Then we multiply the base (*aṣl*) by R and divide the product by the hypotenuse (*quṭr*); there results the sine of the equation."

COMMENTARY

In terms of Indian astronomy, the base (*aṣl*) is the radius of the epicycle (for $\frac{aṣl}{R}$ the Indians normally substitute the circumference of the epicycle divided by 360); the side (*ḍil^c*) is the *bhujaphala*; R ± the cosine of the argument is the *koṭiphala*; and the hypotenuse (*quṭr*) is the *karṇa*. Then the rule in the first paragraph corresponds to that in, e.g. *Paitāmahasiddhānta* 4,10 and *Brāhmasphuṭasiddhānta* 2,12–15; the rule in the second paragraph covers the special case when the argument equals 90°, so that the *bhujaphala* is the *aṣl* and the *koṭiphala* is R. See Frag. Z 15.

S 3. Al-Bīrūnī, *Taḥdīd nihāyāt al-amākin*, ed. pp. 228–30, trans. pp. 193–95.

The Indians have a book on this subject, known as *Kitāb taḥdīd al-arḍ wa al-falak* (*Bhūgolādhyāya*?). First its author derives half the town's parallel (*ṭawq al-madār*). He multiplies the versed sine of the town's latitude by the number of *farsakh*'s in the earth's semi-circumference, which is considered to be 3,298$\frac{17}{25}$ *farsakh*'s. Then he divides the product by 3,438 minutes, and subtracts the quotient from half a rotation, which is 180°; there remains half the parallel of that town. If the latitudes of the two towns are equal, he multiplies the difference in their longitudes by half the parallel, and then divides the product by 180; the quotient obtained thereby is expressed in grand *farsakh*'s. Then he adds to the quotient one-sixth of it, and he claims that the sum obtained is a measure of the distance, on the road used by humans and animals. If the two longitudes are equal, he multiplies the difference between the two latitudes by a quarter of the earth's circumference, which is 1,649$\frac{17}{50}$ *farsakh*'s; then he divides the product by 90, and there results grand *farsakh*'s. He adds a quarter of it to get the measure of the road; that is what the author has claimed. If the two longitudes as well as the two latitudes are different, he obtains the displacement (between the two towns) by taking the difference between the two latitudes; then he squares the difference and retains the product. Further, he multiplies the longitude of each of the two towns by the respective half of the town's parallel, and divides each product by 180. He takes the difference between the two quotients and squares it; then he adds the squared difference to the retained number, and extracts the (square-)root of the sum; the unit in the root is a grand *farsakh*. He adds to the *farsakh*'s of the root one-third of their amount to get the measure of the road.

The object of that operation is to evaluate the *ṭawq al-madār*, which is half the circumference of the parallel of latitude, in the *farsakh* unit for a great circle, whose amount is 6,597$\frac{9}{25}$ *farsakh*'s. Since the diameter of the earth, according to their tradition, is 2,100 *farsakh*'s, its circumference would be 6,600 *farsakh*'s, which is 3$\frac{1}{7}$ times the diameter, according to the ratio which was derived by Archimedes.

But, in India, this ratio (of circumference to diameter) is the ratio of 3,927 to 1,250, because it was communicated to them, by divine revelation and angelic disclosure, that the circumference of the circle of the stars, that is, the circle of the zodiac, is 125,664,400 (read 125,664,000) *farsakh*'s, and that its diameter is 40,000,000 *farsakh*'s. So, according to this ratio, if the diameter of the earth is 2,100 *farsakh*'s according to their hearsay evidence, the circumference of the earth would be 6,597$\frac{9}{25}$ *farsakh*'s. As the authors of *al-Sindhind al-ṣaghīr* have dropped the zeroes from the beginning of the number of the days given in *al-Sindhind al-kabīr* and have dropped in it an equal number of zeroes from the number of the Sun's rotations, they have done so in this case. They have made the ratio of the diameter to the circumference equal to the ratio of 40,000 to 125,664, which was mentioned by al-Khwārizmī in his *zīj*, and in the *Kibāb al-jabr wa al-muqābala*, after dividing each of these two numbers by two. But those two numbers have a common factor of 32; therefore their ratio is what we stated above.

And I say: The ratio of one circumference to another circumference is equal to the ratio of the diameter of the former to the diameter of the latter. Similarly, for the semi-circumferences, the ratio of the radius of the parallel of latitude to the radius of the sphere is equal to the ratio of the semi-circumference of the parallel to the semi-circumference of a great circle. But if we regard the circumference as made up of 360 parts, the authors of the two *Sindhind*'s regard it as made up of 114;36. Now, half this number is 57;18; and if it is expressed in minutes, its amount is 3,438. Therefore, in their tables, thay have put R equal to this amount, and they have evaluated the other sines accordingly.

COMMENTARY

This whole passage is exhaustively discussed by al-Bīrūnī (*Taḥdīd*, ed. pp. 230–31, trans. pp. 196–97 for case one; ed. p. 232, trans. p. 197 for case two; ed. pp. 232–33, trans. pp. 197–99 for case three; and ed. p. 234, trans. pp. 199–200 for the increments). Clearly the rules for determining the distance between towns on the same parallels of longitude or latitude are similar to al-Fazārī's (Frag. Z 23):

$$(a - b) \times \frac{\text{circumference}}{360};$$

the main oddity is the addition of $\frac{1}{6}$ to the longitudes and $\frac{1}{4}$ to the latitudes. The third problem differs from that in Frag. Z 23 in that here the longitudinal difference (*deśāntara*) is known; but again the addition of $\frac{1}{3}$ is puzzling. Al-Bīrūnī severely criticizes the solution of problems one and three and the increments; the reader is referred to his discussion for details.

For the circumference of 6,600 *farsakh*'s, see Frag. Z 17 and the commentary thereon. In his *Κύκλου μέτρησις*, Archimedes proved that π lies between $3\frac{10}{71}$ and $3\frac{1}{7}$. The ratio $6{,}597\frac{9}{25}$:2,100 = 125,664,000:40,000,000 is derived from the *Āryabhaṭīya* (see Frag. T 1–3 of Yaʿqūb); Al-Yaʿqūbī (*Taʾrīkh*, 1, 85) says that, according to the Indians, the orbit of the Moon is 125,664 *farsakh*'s, and its diameter 40,000 *farsakh*'s. The reference to al-Khwārizmī's *Algebra* is: ed. p. 51, trans. pp. 71–72.

The *Zīj al-Sindhind al-kabīr* sounds like al-Fazārī's work, but it seems to have used R = 3,438 (the value in, e.g., *Paitāmahasiddhānta* 3,12 and *Āryabhaṭīya*, *Daśagītikā* 10); this value of R reflects a value of π equal to $3\frac{27}{191}$. In Frag. Z 12 and 16 we find that al-Fazārī uses R = 3,270, and in Frag. Z 11, 13, 15, and 25 R = 150. Clearly he made no effort to be consistent, but incorporated any numbers he found into his *zīj*.

Q 2 (for Frag. Q 1 see p. 104). Al-Bīrūnī, *On Shadows* 142:7–144:9.

Because the *zīj*'es of the Indians are composed in the meter which they call *shalūk* (*śloka*), one of the authors of the *zījāt al-Sindhind* composed his *zīj* in meter for that reason. . . . Similarly Muḥammad ibn Ibrāhīm al-Fazārī wrote in poetry in his *Al-Qaṣīda al-najūmīya*; he said in it concerning (the computation of) the remainder of daylight:

"If you desire (to know) what remains of daylight or what has passed of it by means of the most suitable calculation, then—may Allāh guide you with kindness!—made a rod whose measure is (a criterion) for the perfection of measure, six and six (digits); see to its durability, and make it as long as a span. Set it upright in a level place. Then look at the shadow, where it ends, and measure it with the rod (lacuna of one? word). Whatever results, that is from counting and from the calculation of your determined shadow. Add to it the equivalent of the shadow(!) of the rod, and subtract from it the shadow on the noon of your day; all of this has a share in your concern, and in it lies the accomplishment of your objective. Whatever is left, divide two and seventy by it until it is exhausted; this, upon my life!, is clear in meaning. When you divide, note the sexagesimal place (*bāb*) of the quotient. These are the complete hours from the computation of the straight path. If the day is before (you), they pass one by one until they cross noon entirely and completely; if the day is behind (you), they are left one by one in reverse order till the setting of the Sun, until you no (longer) see."

COMMENTARY

Though the full meaning of this passage eludes us, yet the general procedure which is described remains clear. If we denote the shadow at any given time s, that at noon s_n, and the hours since sunrise or before sunset x, al-Fazārī's formula is equivalent to:

$$\frac{12}{(s + 12) - s_n} = \frac{x}{6}.$$

Then, if $s = s_n$, six seasonal hours have passed since sunrise; if $s = s_n + 2\frac{2}{5}$, five hours have passed; if $s = s_n + 6$, four hours have passed; if $s = s_n + 12$, three hours have passed; if $s = s_n + 24$, two hours have passed; and if $s = s_n + 60$, one hour has passed. It will be necessary for s to be infinitely greater than s_n for x to equal 0. The underlying scheme assumes a geometrical progression in the differences between s and s_n for the second, third, and fourth seasonal hours of the day, with faster increases as one approaches sunrise or sunset; presumably the $2\frac{2}{5}$ for the fifth hour is an approximation of 3, which one expects to complete the geometrical progression. The primitiveness of this scheme makes on suspect a much older source.

BIBLIOGRAPHY

Arabic, Hebrew, and Latin sources

Abraham ibn Ezra. *El libro de los fundamentos de las Tablas astronómicas*. Ed. J. M. Millás Vallicrosa. Madrid-Barcelona, 1947.

Al-Bīrūnī. *Chronology*. Ed. C. E. Sachau. Repr. Leipzig, 1923. Translation by C. E. Sachau. London, 1879.

Al-Bīrūnī. *India*. Ed. Hyderabad-Decan, 1958. Translation by C. E. Sachau, 2 vols. Repr. London, 1914.

Al-Bīrūnī. *On Shadows*. Ed. *Rasāʾil*, Pt. 2.

Al-Bīrūnī. *On Transits*. Ed. *Rasāʾil*, Pt. 3. Translation by M. Saffouri and A. Ifram. *AUBOS* 32, Beirut, 1959.

Al-Bīrūnī. *Al-Qānūn al-Masᶜūdī*. 3 vols. Hyderabad–Deccan, 1954–56.

Al-Bīrūnī. *Rasāʾil*. Hyderabad–Deccan, 1948.

Al-Bīrūnī. *Taḥdīd nihāyāt al-amākin*. Ed. P. Bolgakoff, Cairo, 1962. Translation by J. Ali, Beirut, 1967.

Al-Hamdānī. *Ṣifat jazīrat al-ᶜArab*. Ed. D. H. Müller. 2 vols. Leiden, 1884–91.

Al-Hāshimī. *Kitāb ᶜilal al-zījāt*. Ed., trans., and commented on by F. Haddad, E. S. Kennedy, and D. Pingree.

Ibn al-Fakīh. *Mukhtaṣar kitāb al-buldān*. Ed. M. J. de Goeje, *BGA*, 5. Leiden, 1885.

Ibn Masrūr. *Kitāb ᶜilal al-zījāt*. MS. Taymūr, Math. 99.

Ibn al-Muthannā. *Commentary on the Astronomical Tables of al-Khwārizmī*. Hebrew translation ed. by B. R. Goldstein. New Haven, 1967. Latin translation ed. by E. Millás Vendrell. Madrid–Barcelona, 1963.

Ibn al-Nadīm. *Kitāb al-Fihrist*. Ed. G. Flügel. 2 vols. Leipzig, 1871–72.

Ibn al-Qifṭī. *Taʾrīkh al-ḥukamāʾ*. Ed. J. Lippert. Leipzig, 1903.

Al-Masᶜūdī. *Kitāb al-tanbīh wa al-ishrāf*. Ed. M. J. de Goeje, *BGA*, 8. Leiden, 1894. Translated by B. Carra de Vaux. Paris, 1897.

Al-Masᶜūdī, *Murūj al-dhahab*. Ed. B. de Meynard and P. de Courteille. 9 vols. Paris, 1861–1917. 2d ed. (Vols. 1 and 2 only). Rev. C. Pellat, Beyrouth, 1962–66.

Al-Ṣafadī. *Kitāb al-wāfī bu ʾl-wafayāt*. Ed. H. Ritter, BI 6a. Vol. 1, 2d ed. Wiesbaden, 1962.
Ṣāʿid al-Andalusī. *Kitāb ṭabaqāt al-umam*. Ed. L. Cheikho. Beyrouth, 1912. Translated by R. Blachère. *PIHEM*, 28. Paris, 1935.
Al-Yaʿqūbī. *Kitāb al-buldān*. Ed. M. J. de Goeje. *BGA*, 7. 2d ed. Leiden, 1892.
Al-Yaʿqūbī. *Taʾrīkh*. 2 vols. Bayrūt, 1960.
Yāqūt. *Irshād al-arīb ilā maʿrifat al-adīb*. Ed. D. S. Margoliouth. "E. J. W. Gibb Mem. Ser.," 6. Vol. 6. 2d ed. London, 1931.
Yāqūt. *Muʿjam al-buldān*. Ed. F. Wüstenfeld. 6 vols. Leipzig, 1866–72. Introductory chapters translated by W. Jwaideh. Leiden, 1959.

Sanskrit sources

Āryabhaṭīya. Ed. H. Kern. Leiden, 1874.
Brāhmasphuṭasiddhānta. Ed. S. Dvivedin. Benares, 1902.
Mahābhāskarīya. Ed. K. S. Shukla. Lucknow, 1960.
Paitāmahasiddhānta. Translated by D. Pingree. *Brahmavidyā* 31–32 (1967–68), 472–510.
Śiṣyadhīvṛddhidatantra. Ed. S. Dvivedin. Benares, 1886.

Secondary sources

Barthold, V. V. *A History of the Turkman People*. In *Four Studies on the History of Central Asia*. Translated by V. and T. Minorsky. Vol. 3. Leiden, 1962. Pp. 73–170.
Brockelmann, C. *Geschichte der arabischen Literatur*. Vols. I²–II². Leiden, 1933–49 and *Suppl*. I–III. Leiden, 1937–42.
Bunbury, E. H. *A History of Ancient Geography*. 2 vols. Repr. New York, 1959.
Catalogus Codicum Astrologorum Graecorum. Vol. 5, Pt. 1. Ed. F. Cumont and F. Boll. Bruxelles, 1904.
Dunlop, D. M. *The History of the Jewish Khazars*. *POS*, 16. Princeton, 1954.
Kennedy, E. S. *A Survey of Islamic Astronomical Tables*. *TAPhS*, NS 40, 2. Philadelphia, 1956.
Kennedy, E. S., and A. Muruwwa. "Bīrūnī on the Solar Equation." *JNES*, 17 (1958), 112–21.
Nallino, C. A. *Raccolta di scritti editi e inediti*. Vol. 5. Roma, 1944.
Neugebauer, O. *The Astronomical Tables of al-Khwārizmī*. Hist. Filos. Skr. Dan. Vid. Selsk. 4, 2. København, 1962.
Pelliot, P. *Histoire ancienne du Tibet*. Paris, 1961.
Pingree, D. "The Fragments of the Works of Yaʿqūb ibn Ṭāriq." *JNES*, 26 (1968), 97–125.
Pingree, D. *The Thousands of Abū Maʿshar*. *SWI*, 30. London, 1968.
Suter, H. *Die Mathematiker und Astronomen der Araber und ihre Werke*. *AGMW*, 10. Leipzig, 1900.

THE GREEK INFLUENCE ON EARLY ISLAMIC MATHEMATICAL ASTRONOMY

David Pingree
Brown University

Some concepts of Greek mathematical astronomy reached Islam in the eighth century through translations and adaptations of Sanskrit and Pahlavī texts. These represented largely non-Ptolemaic ideas and methods which had been altered in one way or another in accordance with the traditions of India and Iran. When to this mingling of Greco-Indian and Greco-Iranian astronomy was added the more Ptolemaic Greco-Syrian in the late eighth and early ninth centuries, and the completely Ptolemaic Byzantine tradition during the course of the ninth, the attention of Islamic astronomers was turned to those areas where these several astronomical systems were in conflict. This led to the development in Islam of a mathematical astronomy that was essentially Ptolemaic, but in which new parameters were introduced and new solutions to problems in spherical trigonometry derived from India tended to replace those of the *Almagest*.

The problem of the influence of Greek mathematical astronomy upon the Arabs (and in the following I have generally excluded from consideration the related problems of astronomical instruments and star-catalogues) is immensely complicated by the fact that the Hellenistic astronomical tradition had, together with Mesopotamian linear astronomy of the Achaemenid and Seleucid periods and its Greek adaptations, already influenced the other cultural traditions that contributed to the development of the science of astronomy within the area in which the Arabic language became the dominant means of scientific communication in and after the seventh century A.D. An investigation of this probelm, then, must begin with a review of those centers of astronomical studies in the seventh and eighth centuries which can be demonstrated to have influenced astronomers who wrote in Arabic. This limitation by means of the criterion of demonstrable influence will effectively exclude Armenia, where Ananias of Shirak worked in the seventh century,[1] and China, where older astronomical techniques,[2] some apparently derived ultimately from Mesopotamian sources,[3] were partially replaced by Indian adaptations of Greek and Greco-Babylonian techniques rendered into Chinese at the T'ang court in the early eighth century.[4] But it leaves Byzantium, Syria, Sasanian Iran, and India.

While astronomy had been studied at Athens by Proclus[5] and observations had been made by members of the Neoplatonic Academy in the fifth

[1] F. C. Conybeare, "Ananias of Shirak (A.D. 600-650 c.)," *BZ* 6, 1897, 572-584; R. A. Abramian, *Anania Shirakatsi*, Erevan 1958; and W. Petrie, "Ananija Schirakazi—ein armenischer Kosmograph des 7. Jahrhunderts," *ZDMG* 114, 1964, 269-288.

The reader should note that, in writing this survey, I have disregarded the rather divergent views of B. L. van der Waerden; these have been most recently expounded in his *Das heliozentrische System in der griechischen, persischen und indischen Astronomie*, Zürich 1970.

[2] N. Sivin, *Cosmos and Computation in Early Chinese Mathematical Astronomy*, Leiden 1969; K. Yabuuti, "Astronomical Tables in China, from the Han to the T'ang Dynasties," *History of Chinese Science and Technology in the Middle Ages*, Tokyo (?) 1963, pp. 445-492.

[3] E. S. Kennedy, "The Chinese-Uighur Calendar as Described in Islamic Sources," *Isis* 55, 1964, 435-443, esp. 441.

[4] K. Yabuuti, "The Chiuchih-li, an Indian Astronomical Book in the T'ang Dynasty," *op. cit.*, pp. 493-538; O. Neugebauer and D. Pingree, *The Pañcasiddhântikâ of Varâhamihira*, København 1970/71, vol. 1, p. 16.

[5] Proclus (410-485) was the author of a summary of Ptolemaic astronomy entitled *Hypotypōsis tōn astronomikōn hypotheseōn*, ed. C. Manitius, Leipzig 1909; the summary of Geminus' *Eisagōgē* entitled *Sphaira tou Proklou* is probably not his.

and early sixth centuries,[6] and while Ammonius, Eutocius, Philoponus, and Simplicius had written about astronomical problems at Alexandria in the early sixth century,[7] a hundred years later the tradition was transferred to Constantinople, where Stephanus of Alexandria—perhaps in imitation of the Sasanian *Zīk-i Shahriyārān*—prepared in 617/618 a set of instructions with examples illustrating the use of the *Handy Tables* of Theon for the Emperor Heraclius.[8] Such studies, however, were soon abandoned, not to be revived in Byzantium till the ninth century, when their restoration seems to have been due to the stimulus of the desire to emulate the achievements of the Arabs. Except for the texts of the *Little Astronomy*[9] and some passages reflecting Greco-Babylonian astronomy in pseudo-Heliodorus[10] and Rhe-

[6] A series of seven observations made between 475 and 510 are preserved in several manuscripts, of which the oldest was written in the tenth century; the best edition is by J. L. Heiberg, *Ptolemaei Opera astronomica minora*, Leipzig 1907, pp. xxv-xxvii. The observations are:

I. A conjunction of Mars and Jupiter on 6/7 Pachon 214 Diocletian (1 May 498) observed by Heliodorus (on 1 May 498 Mars and Jupiter were in conjunction at approximately Virgo 1°).

II. A conjunction of Saturn and the Moon observed on 27/28 Mechir 219 (22 February 503) by Heliodorus and his brother (i.e., Ammonius) (on the night of 21/22 February 503 Saturn and the Moon were in conjunction at approximately Cancer 5;34°).

III. A conjunction of Venus and the Moon at Capricorn 13° at an elongation of 48° from the Sun on 21 Athyr 192 (19 November 475) observed at Athens by "the divine" (i.e., Proclus) (on 18 November 475 Venus and the Moon were in conjunction at approximately Capricorn 15°, and the Sun was at Scorpio 28° at an elongation of 47°).

IV. A transit of Regulus by Jupiter on 30 Thoth 225 (27 September 508) (on 27 September 508 Jupiter was approximately at Leo 9°, very close to Regulus).

V. A transit of α Hyadis by the Moon at Taurus 16;30° on 15 Phamenoth 225 (11 March 509) observed by Heliodorus (in the early evening of 11 March 509 the Moon was at Taurus 16;30° close to α Hyadis).

VI. A conjunction of Mars and Jupiter on 9/10 (read 19) Payni 225 (13 June 509) (early on 13 June 509 Mars and Jupiter were in conjunction at approximately Leo 12;40°).

VII. Observations of Venus ahead of Jupiter in 226, but then behind it on the 28th of an unspecified month (Venus was in conjunction with Jupiter at Virgo 6° on 13 October 509 or 15 Paophi 226; they were next in conjunction at Virgo 19° on 21 August 510 or 27 Mesore 226. It must be the latter conjunction that Heliodorus here records as Venus was ahead of Jupiter—i.e., rose before it—on 26 Mesore, and was behind it on the 28th).

[7] Ammonius, a pupil of Proclus, is said to have been the author of a lost work on the astrolabe (it is not "Aegyptius"' *Hermēneia tēs tou astrolabou chrēseōs*, ed. H. Hase, *RhM* 6, 1839, 158-171, and A. Delatte, *Anecdota Atheniensia et alia*, vol. 2, Paris 1939, pp. 254-262). He is also said by a certain philosopher named Stephanus to have published a *kanonion* using the Era of Philip and Egyptian years (*CCAG* 2; 182), and is sometimes identified with the Awmûniyûs who also is said to have used Egyptian years, but whose work is only known in the recension of al-Zarqâlî (J. M. Millás Vallicrosa, *Estudios sobre Azarquiel*, Madrid-Granada 1943-1950, pp. 153-234 and 379-392; M. Boutelle, "The Almanac of Azarquiel," *Centaurus* 12, 1967, 12-19).

Eutocius was evidently the author of a *Prolegomena tēs Ptolemaiou megalēs syntaxeōs*, as yet unpublished (J. Mogenet, *L'Introduction à l'Almageste*, Bruxelles 1956).

Philoponus, a pupil of Ammonius, wrote a well-known treatise *Peri tēs tou astrolabou chrēseōs kai kataskeuēs*, ed. H. Hase, *RhM* 6, 1839, 129-156; see P. Tannery, "Notes critiques sur le Traité de l'astrolabe de Philopon," *RPh* 12, 1888, 60-73 (= *Mémoires scientifiques*, vol. 4, Toulouse-Paris 1920, pp. 241-260); J. Drecker, "Des Johannes Philoponos Schrift über das Astrolab," *Isis* 11, 1928, 15-44. There is an English translation by H. W. Greene in R. T. Gunther, *The Astrolabes of the World*, vol. 1, Oxford 1932, pp. 61-81.

Simplicius, another pupil of Ammonius, composed a commentary on Aristotle's *Peri ouranou*, ed. J. L. Heiberg, *CAG* 7, Berlin 1894.

[8] The fullest description presently available is by H. Usener, *De Stephano Alexandrino*, Bonn 1880, pp. 33-54 (= *Kleine Schriften*, vol. 3, Leipzig-Berlin 1914, pp. 289-319).

[9] On the history of the *Mikros astronomos*, see D. Pingree in *Gnomon* 40, 1968, 15-16.

[10] Heliodorus, who was the principal observer in the series of observations noted in fn. 6, was the author neither of the *Prolegomena* to Ptolemy (probably written by Eutocius; see fn. 7) nor of the *Eis ton Paulon*, edited by E. Boer, Leipzig 1962 (apparently written by Olympiodorus; see L. G. Westerink, "Ein astrologisches Kolleg aus dem Jahre 564," *BZ* 64, 1971, 6-21). He is falsely associated with a short text containing a Greco-Babylonian theory of the planets by F. Boll in *CCAG* 7; 119-122; see also O. Neugebauer, "On a Fragment of

torius of Egypt,[11] Byzantine astronomy in this period was solidly Ptolemaic.

The history of astronomical writings in Syriac before the rise of Islam is difficult to trace. The works of Bar Daiṣan,[12] his pupil Philip,[13] and of George, the Bishop of the Arabs,[14] indicate that sufficient knowledge of the subject must have existed to permit the casting of horoscopes; for this all that is really needed, of course, are tables, and it is certain that a Syriac version of the *Handy Tables* existed.[15] There may also have been a Syriac translation of Ptolemy's *Syntaxis* since some of the Arabic versions are said to have been made from that language.[16] In fact, it has been claimed that Sergius of Rêshʿainâ, the early sixth century translator, was responsible for the Syriac version,[17] and in any case he did write on astrology and on the motion of the Sun.[18]

Heliodorus (?) on Planetary Motion," *Sudhoffs Archiv* 42, 1958, 237-244 (the same text as is found in Vat. gr. 208 also occurs on ff. 200-201 of Laur. 28, 46).

[11] Rhetorius of Egypt, who flourished in the first quarter of the sixth century, includes in his *Peri phuseōs kai kraseōs kai geuseōs tōn z̄ asterōn* a very schematic version of the Greco-Babylonian planetary theory. It was edited by F. Boll in *CCAG* 7; 214-224; see also the excerpts of Isidore of Kiev edited by S. Weinstock in *CCAG* 5, 4; 133-152 and by D. Pingree in *Albumasaris De revolutionibus nativitatum*, Leipzig 1968, pp. 245-273. He also, in chapter 51 of his *Ek tōn Antiochou thēsaurōn* (ed. F. Boll in *CCAG* 1; 163), gives a list of Babylonian period relations for the planets, as do also John Lydus in *Peri mēnōn* III 16 (ed. R. Wuensch, Leipzig 1898, pp. 56-57) and others; for these period relations in Islamic astrology see E. S. Kennedy and D. Pingree, *The Astrological History of Mâshâ'allâh*, Cambridge, Mass. 1971, p. 132, and D. Pingree, *The Thousands of Abû Maʿshar*, London 1968, pp. 64 and 66-68.

[12] Bar Daiṣan (154-222/3) of Edessa represents a peculiar combination of Christianity and Gnosticism with astrology; see H. J. W. Drijvers, *Bar Daiṣan*, Assen 1965, and fns. 13 and 14.

[13] Philip, the pupil of Bar Daiṣan, wrote an exposition of his master's teachings, the *Book of the Laws of Countries*, most recently edited by Drijvers, Assen 1965.

[14] The astronomical letters of George, the Bishop of the Arabs (died 724), are published by V. Ryssel, "Die astronomischen Briefe Georgs des Araberbischofs," *ZA* 8, 1893, 1-55; see also F. Nau, "Notes d'astronomie syrienne," *JA* 10, 16, 1910, 208-228, esp. 208-219, and F. Nau, "La cosmographie au VIIᵉ siècle," *ROC* 15, 1910, 225-254, esp. 229-230, 239-240, and 245. George attributes to Bar Daiṣan (p. 48 Ryssel) a scheme of rising-times of the zodiacal signs equivalent to the Babylonian "System A" (O. Neugebauer, "The Rising Times in Babylonian Astronomy," *JCS* 7, 1953, 100-102), though he derives (p. 49 Ryssel) his longest periods of daylight for the seven climata from Ptolemy's *Syntaxis* 2, 13 (O. Neugebauer, "On Some Astronomical Papyri and Related Problems of Ancient Geography," *TAPhS*, NS 32, 1942, 251-263, esp. 257 fn. 39). George also knows the Ptolemaic value (1° per century) of precession (p. 53 Ryssel), but in general does not display a very profound knowledge of astronomy.

[15] Bar Hebraeus in the thirteenth century (*Chronography*, translated E. A. Wallis Budge, Oxford 1932, vol. I, p. 54) attributes to Theon a work on the use of the astrolabe, an introduction to the *Syntaxis*, and a *Book of the Canon* whose only memorable feature for Bar Hebraeus seems to be the discussion of the theory of trepidation; see also his *Ta'rîkh mukhtaṣar al-duwal*, Bayrût 1958, p. 73. But we know that a version was available to Severus Sebokht already in the seventh century; see F. Nau, "La cosmographie," pp. 237 and 240, F. Nau, "Le traité sur les 'Constellations' écrit en 660 par Sévère Sébokt évêque de Qennesrin," *ROC* 27, 1929-30, 327-410, and 28, 1931-32, 85-100, and O. Neugebauer, "Regula Philippi Arrhidaei," *Isis* 50, 1959, 477-478. Moreover, al-Battânî (C. A. Nallino, *Al-Battânî sive Albatenii Opus Astronomicum*, 3 vols., Milan 1899-1907, vol. 2, pp. 192 and 211) used a Syriac version of Ptolemy's *kanōn basileōn* extending through the reign of Theodosius III (died 717) and a Syriac adaptation of his geographical tables. Ibn al-Nadîm in his *Fihrist* (ed. G. Flügel, 2 vols., Leipzig 1871-72, p. 244) reports that Ayyûb and Simʿân translated the *Zîj Baṭlimûs* (*Handy Tables*) into Arabic (apparently from the Syriac) for Muḥammad ibn Khâlid ibn Yaḥyâ ibn Barmak, who must have flourished in about 800.

[16] Ibn al-Nadîm (*Fihrist*, pp. 267-268; cf. Ibn al-Qifṭî, *Ta'rîkh al-ḥukamâ'*, ed. J. Lippert, Leipzig 1903, pp. 97-98) states that the first translation of the *Almagest* into Arabic was made for Yaḥyâ ibn Khâlid ibn Barmak (died 807), and was "commented on" by Abû al-Ḥasan and Salm, the head of the Bayt al-ḥikma. Nallino (*Al-Battânî*, vol. 2, p. viii; *Raccolta di scritti editi e inediti*, vol. 5, Roma 1944, p. 264) has contended that this ancient version was made from the Syriac and was used by al-Battânî.

[17] Sergius (died 536), the son of Elias, is identified by A. Baumstark (*Lucubrationes Syro-Graecae*, Leipzig 1894, p. 380) with Sarjûn ibn Haliyâ al-Rûmî whose name is recorded at the end of the unique Leiden manuscript of al-Ḥajjâj's Arabic translation of the *Almagest*.

At Qenneshrê, moreover, in the middle of the seventh century lived the monk Severus Sebokht of Nisibis, who was familiar not only with the *Handy Tables*, and the Greek tradition of astrolabes, but also with Indian numerals.[19] This early Indian influence on Syrian science reminds us of the similarities between Ḥarrânian and Indian worship of the planets and of the knowledge of the Sanskrit names of the planets at Ḥarrân evident from the *Ghâyat al-ḥakîm* falsely ascribed to al-Majrîṭî.[20] Also in the late eighth century there came from Edessa the chief astrologer of the Caliph al-Mahdî, Theophilus, a man learned in Greek, in Syriac, and in Arabic.[21] His writings do not reveal what astronomical texts he used, but they do indicate the existence of some sort of astronomy in Syria. They also indicate again an acquaintance with Indian material. This Indian influence on Syrian science probably came through Sasanian Iran; but, despite this contact with the east, Syrian astronomy seems to have been predominantly Ptolemaic.

The Iranians, whatever their astronomy may have been in the Parthian period besides the linear astronomy preserved in cuneiform at Babylon and Uruk,[22] in the very earliest reigns of the Sasanian dynasty became familiar with Ptolemy's *Syntaxis*—*Hē megistē biblos* or, in Pahlavî, *Megistî*—as well as with Greek and Indian astrological texts through the translations sponsored by Ardashîr I and Shâpûr I.[23] In the late fourth century the Era of Diocletian was employed in computing a horoscope inserted into the earlier Pahlavî translation of the Greek astrological poem of Dorotheus of Sidon;[24] this may indicate the existence of a set of astronomical tables in Pahlavî using that era.

Al-Ḥajjâj ibn Maṭar made his translation, perhaps from Sergius' Syriac, for al-Ma'mûn in 829/30; cf. below fn. 96. Some further information on the Arabic versions of Ptolemy will be found in O. J. Tallgren, "Survivances arabo-romane du Catalogue d'étoiles de Ptolémée," *Studia Orientalia* 2, 1928, 202-283.

[18] E. Sachau, *Inedita Syriaca*, Halle 1870, pp. 101-126. Sergius also translated into Syriac the pseudo-Aristotelian *Peri kosmou* edited by P. A. Lagarde, *Analecta Syriaca*, Leipzig 1858, pp. 134-158.

[19] Severus Sebokht of Nisibis, the Bishop of Qenneshrê in the middle of the seventh century, wrote voluminously on astronomy. Some of his works have been published in Nau's articles cited above in fns. 14 and 15, others by Sachau, *op. cit.*, pp. 127-134, and his work on the astrolabe by Nau in *JA* 9, 13, 1899, 56-101 and 238-303 (there is an English translation of this by Mrs. Margoliouth in Gunther, *op. cit.*, pp. 82-103); see also F. Nau, "Un fragment syriaque de l'ouvrage astrologique de Claude Ptolémée intitulé le Livre du fruit," *ROC* 28, 1931-32, 197-202.

[20] The *Ghâya. al-ḥakîm*, falsely attributed to Maslama al-Majrîṭî (died 1007/8), was composed between 954/5 and 959/60 The Arabic original was edited by H. Ritter, Leipzig-Berlin 1933, and translated into German by H. Ritter and M. Plessner, London 1962. The Ḥarrânian adaptations of Indian material are principally located on pp. 202-225 of the text, pp. 213-237 of the translation; they have been freely mingled with material from a variety of other sources.

[21] Theophilus (died 785) wrote several astrological works, now extant mainly in Greek, though there also exist many fragments in Arabic. He was an astrological advisor to the ʿAbbâsid court from the 750s till his death shortly after al-Mahdî's in 785. I am at present preparing an edition of all of his astrological writings, in Greek and in Arabic, which I hope will elucidate further his dependence on Indian sources.

[22] The basic publication of this material is O. Neugebauer, *Astronomical Cuneiform Texts*, 3 vols., London 1955; see also A. Sachs, *Late Babylonian Astronomical and Related Texts*, Providence 1955, and O. Neugebauer and A. Sachs, "Some Atypical Astronomical Cuneiform Texts," *JCS* 21, 1967, 183-214, and 22, 1969, 92-111.

[23] See the texts and references in D. Pingree, *The Thousands*, pp. 7-13; the translation of the *Almagest* by Rabban al-Ṭabarî to which Abû Maʿshar refers (*ibid.*, p. 17) was probably made from the Pahlavî.

[24] This horoscope occurs in Book 3 of the Arabic translation (*ca.* 800) of the Pahlavî version of the astrological poem of Dorotheus of Sidon (*ca.* 75); I am preparing an edition of this work. The horoscope is dated, according to the text, "in the year 96 (read 97) of the years of Darînûs (= Diocletian!) in the month Mihr (= Phamenoth!) on the second day", which corresponds to 26 February 381. The horoscopes and modern computations for that date are:

	text	26 February 381
Saturn	Taurus 4;34°	Taurus 7°
Jupiter	Libra 20;10	Libra 27
Mars	Taurus 24;55	Taurus 26
Sun	Pisces 6;50	Pisces 9
Venus	Pisces 26;50	Pisces 29
Mercury	Pisces 19;55	Pisces 21
Moon	Virgo 19;7	Virgo 20

But, in any case, it appears that in *ca.* 450 a *Royal Canon—Zîk-i Shahriyârân*—was composed; the one element that we know from it, the longitude of the Sun's apogee, is a parameter of the brâhmapakṣa of Indian astronomy.[25] A century later, in 556, Khusrau Anûshirwân ordered his astrologers to compare an Indian text called in Arabic the *Zîj al-Arkand* (*arkand* being a corruption of the Sanskrit ahargaṇa) with Ptolemy's *Syntaxis*; the Indian text, which belonged to the ârdharâtrikapakṣa, was, surprisingly, found to be superior, and a new redaction of the *Zîk-i Shahriyârân* was based upon it;[26] this redaction was still available to Mâshâ'allâh at the beginning of the ninth century.[27] A final version of the *Zîk-i Shahriyârân*, like its predecessor incorporating many ârdharâtrika parameters though employing others of unknown origin as well, was published during the reign of the last Sasanian monarch, Yazdijird III.[28] In its Pahlavî form it was probably the zîj used by the computer of a series of horoscopes illustrating the early history of Islam, computed shortly after 679,[29] and of another series computed during the reign of Hârûn al-Rashîd.[30] It was also known to the authors of a *Zîj al-Jâmi'* and of a *Zîj al-Hazûr*, both written at Qandahâr not very long after 735,[31] and to al-Fazârî and Ya'qûb ibn Ṭâriq in the last quarter of the eighth century.[32] The eclectic nature of these *Royal Canons* is evident from what we know of their parameters; but nothing can now be recovered of any geometric or cinematic models which they may have proposed to account for planetary motions. Mathematically, they favored simplified techniques leading to rough approximations. Even more than the Indians, the Iranians of the Sasanian period, though heirs through both Greece and India of Greek astronomy, concentrated on pragmatic methods and astrologically useful results rather than on theoretical models of the structure of the heavens.

Aside from these official sets of tables in Iran, we know from Neusner's studies that the Jews in Mesopotamia during the Sasanian period were familiar with astrology;[33] the astronomical tables that may have been available to them, however, remain obscure. It also seems likely that astronomical studies flourished in Sasanian Ṭabaristân and Khurâsân because these provinces were the places of origin of several early 'Abbâsid astronomers and astrologers who were acquainted with the Pahlavî tradition, but again all further details are missing.[34]

Among the Indians, who first learned something of Babylonian lunar and solar period relations and linear techniques of solving astronomical problems from the Iranians during the Achaemenid occupation of the Indus Valley, methods of computing planetary positions for casting horoscopes based on Greek adaptations of the Mesopotamian planetary theories of the Seleucid and Parthian periods became known in the middle of the second century A.D. or a little earlier.[35] In a somewhat later period—perhaps in the fourth or early fifth century—Greek texts expounding both a double-

---

25 E. S. Kennedy and B. L. van der Waerden in *JAOS* 83, 1963, 321 and 323; D. Pingree, "The Persian 'Observation' of the Solar Apogee in *ca.* A.D. 450," *JNES* 24, 1965, 334-336. Further evidence for the Indian influence on Sasanian astronomy will be found in D. Pingree, "Astronomy and Astrology in India and Iran," *Isis* 54, 1963, 229-246, and D. Pingree, "Indian Influence on Early Sassanian and Arabic Astronomy," *JORMadras* 33, 1963-64, 1-8.

26 D. Pingree, *The Thousands*, pp. 12-13; F. I. Haddad, E. S. Kennedy, and D. Pingree, *The Book of the Reasons Behind Astronomical Tables* (to appear; hereafter referred to as al-Hâshimî), sec. 7.

27 E. S. Kennedy and D. Pingree, *The Astrological History*.

28 Like other Pahlavî astronomical texts it is primarily known through citations by Arab authors. It was translated into Arabic by an otherwise unknown Abû al-Ḥasan 'Alî ibn Ziyâd al-Tamîmî, and extensively used by al-Hâshimî (e.g.. secs. 8 and 64-66) and by al-Bîrûnî (see in particular the material collected and discussed by E. S. Kennedy, "The Sasanian Astronomical Handbook *Zîj-i Shâh* and the Astrological Doctrine of 'Transit' (*mamarr*)." *JAOS* 78. [1958], 246-262).

29 D. Pingree. *The Thousands*, pp. 114-121.

30 *Ibid.*, pp. 93-114.

31 Al-Hâshimî, secs. 5-6 and 35.

32 See below fns. 48 and 58.

33 J. Neusner. *A History of the Jews in Babylonia*, vol. 4, Leiden 1969, pp. 330-334, and vol. 5, Leiden 1970 pp. 190-193.

34 From Ṭabaristân came Rabban al-Ṭabarî (see above fn. 23), the family of 'Umar ibn al-Farrukhân al-Ṭabarî (*fl. ca.* 760-815), and Yaḥyâ ibn Abî Manṣûr (see below fn. 73); from Khurâsân came the families of Nawbakht and Barmak (see above fns. 15 and 16) as well as Abû Ma'shar (787-886).

35 D. Pingree, "Greco-Babylonian Astronomy and its Indian Derivatives," to appear in *JORMadras*.

epicycle model based partially on the principles of Aristotelian physics and a pre-Ptolemaic eccentric-epicyclic model were translated into Sanskrit.[36] This non-Ptolemaic Greek astronomy was modified by the Indians both in the light of their existing traditions of adaptations of Irano-Babylonian and Greco-Babylonian theories and methods and in that of their own developments in trigonometry and algebra. In the fifth century three separate schools representing different mixtures of these elements came into existence: the brâhmapakṣa, which, as we have seen, already influenced the *Zîk-i Shahriyârân* in *ca.* 450; the ârdharâtrikapakṣa, which influenced the *Zîk-i Shahriyârân* of Anûshirwân and Yazdijird III; and the âryapakṣa, which was in part known in late eighth century Baghdâd,[37] apparently through the *Zîj al-Harqan.*[38]

It is clear, therefore, that Greek influence on early Islamic astronomy came from many sides and in many distorted forms. We must now examine in chronological order what is known of astronomers who wrote in Arabic in order to gauge the extent of the dependence of each on ultimately Greek sources and the route by which this influence reached him.

The earliest astronomical texts in Arabic seem to have been written in Sind and Afghanistân. The first is the *Zîj al-Arkand* (different from the Pahlavî work available to the astrologers of Anûshirwân), which was written in Sind—probably at al-Manṣûra—in 735 on the basis of the *Khaṇḍakhâdyaka* composed by Brahmagupta at Bhillamâla in 665, but displaying some knowledge of the *Zîk-i Shahriyârân* of Yazdijird III.[39] From this zîj are derived two others written at Qandahâr, the *Zîj al-Jâmiʿ* and the *Zîj al-Hazûr,*[40] which we have had occasion to refer to previously. We do not have sufficient information concerning these texts to judge whether or not they described the Greco-Indian double-epicycle model, or possessed any other identifiable Greek element.[41] A similar mixture of Indian and Iranian materials was evidently present in the *Zîj al-Harqan,*[42] whose epoch is the year 110 of Yazdijird or A.D. 742, and which gives, in verse, a rough rule for determining the equations of the center of the two luminaries according to the parameters of the âryapakṣa; the same rule was given later by al-Fazârî.[43] Since this rule is based on the "method of sines," it reflects an Indian simplification of computing with the Greco-Indian double-epicycle model.

The next phase in the history of Arabic astronomical texts is marked by a further infusion of Indian material accompanied by translations of the Ptolemaic *Syntaxis* and *Handy Tables* and of the Iranian *Zîj al-Shâh;* this activity took place at the early ʿAbbâsid court, and particularly under al-Manṣûr and Hârûn al-Rashîd. When the former was planning to build his new capital, Baghdâd, he assembled the astrologers Nawbakht, Mâshâ'allâh, al-Fazârî, and ʿUmar ibn al-Farrukhân al-Ṭabarî to determine an auspicious moment; they chose 30 July 762.[44] It is most likely that they used the Pahlavî original of the *Zîj al-Shâh* of Anûshirwân in casting this horoscope; at least we know that this was used by Mâshâ'allâh for horoscopes he cast during the reign of Hârûn al-Rashîd for his astrological history of the caliphs,[45] for a group of horoscopes he computed between 768 and 794,[46] and for the horoscopes cited in the world history which he wrote in *ca.* 810.[47]

[36] D. Pingree, "On the Greek Origin of the Indian Planetary Model Employing a Double Epicycle," *JHA* 2, 1971, 80-85.

[37] Al-Fazârî (see below fn. 48), frags. Z 4, Z 12, Z 17, and S 3; Yaʿqûb ibn Ṭâriq (see below fn. 58), frags. Z 6, T 1-3, and K 6; D. Pingree, *The Thousands,* pp. 16, 19, and 29-32; and al-Hâshimî, secs. 3 and 25.

[38] See below fns. 42 and 43.

[39] Al-Hâshimî, sec. 4, and the commentary on Yaʿqûb, frag. T 4.

[40] Al-Hâshimî, secs. 5-6 and 35.

[41] As al-Bîrûnî translated the *Khaṇḍakhâdyaka* into Arabic and entitled this translation *Tahdhîb zîj al-Arkand,* it is uncertain to what extent his numerous citations in the *India* and in *Al-Qânûn al-Masʿûdî* of the *Zîj al-Arkand* can legitimately be referred to the "translation" of 735.

[42] E. S. Kennedy, "A Survey of Islamic Astronomical Tables," *TAPhS,* NS 46, 1956, 123-177, no. X206 (p. 137).

[43] Al-Bîrûnî, *Rasâ'il,* Hyderabad-Deccan 1948, III 26, and *On Transits,* translated M. Saffouri and A. Ifram, Beirut 1959, 26: 4-19; al-Fazârî, frag. Z 12.

[44] D. Pingree in *JNES* 29, 1970, 104.

[45] E. S. Kennedy and D. Pingree, *The Astrological History,* pp. 129-143.

[46] *Ibid.,* pp. 175-185.

[47] *Ibid.,* pp. 1-125. If the Latin *De scientia motus orbis,* published by I. Stabius, Nuremberg 1504, and by I. Heller, Nuremberg 1549, is really Māshā'allāh's, he knew something of Ptolemaic astronomy, probably from Ḥarrān.

But another of the astrologers consulted on the propitious moment for the foundation of Baghdâd was Muḥammad ibn Ibrâhîm al-Fazârî, the scion of an ancient Arab family of al-Kûfa.[48] When an embassy sent to the court of al-Manṣûr from Sind in 771 or 773 included an Indian learned in astronomy, the caliph ordered al-Fazârî to translate with his help a Sanskrit text related to the brâhmapakṣa and apparently entitled the *Mahâsiddhânta*; this work seems to have been dependent on the *Brâhmasphuṭasiddhânta* written by Brahmagupta in 628. The result of this collaboration was the *Zîj al-Sindhind al-kabîr*, whose elements, however, are derived not only from the *Mahâsiddhânta*,[49] but also from the âryapakṣa[50] (probably through the *Zîj al-Harqan*), the *Zîj al-Shâh*,[51] Ptolemy[52] (perhaps the Pahlavî version), and a Persian geographical text ascribed to Hermes.[53] While mentioning the rules for computing the equations of the center by means of the "method of sines", al-Fazârî also describes the Indian method of computing this equation by means of an epicycle model.[54] This epicyclic model from a Sanskrit source and the Ptolemaic value of precession apparently from a Pahlavî source are the first specifically Greek elements identified in an Arabic astronomical text.

Later in his life—apparently in *ca.* 790[55]—al-Fazârî wrote the first zîj to contain tables for converting dates according to other calendars into those of the Muslim calendar.[56] He also composed a lengthy poem on astronomy, and various works on the astrolabe and armillary sphere; the last two probably were based on Greek or Syriac texts as we know of no treatises on the astrolabe in Pahlavî, and none in Sanskrit before the fourteenth century.[57] This eclectic and not very profound production by al-Fazârî formed the foundation of the Sindhind tradition which survived in the eastern domains of Islam into the tenth century, but much later in Spain and the Latin west.

Associated with al-Fazârî in his conversations with the Indian astronomer from Sind was one Ya'qûb ibn Ṭâriq.[58] Like al-Fazârî he wrote a zîj (employing as epoch the Era of Yazdijird III) in which some elements from the *Mahâsiddhânta*[59] were mixed with others from the *Zîj al-Shâh*,[60] but he also introduced yet others from an ârdharâtrika source—presumably the *Zîj al-Arkand*[61]—and from Ptolemy.[62] In 777 or 778 he composed a second work, the *Tarkîb al-aflâk*, in which he deals at length with methods of computing the ahargaṇa which he has derived from the *Mahâsiddhânta*,[63] but also with the distances of the planets in a fashion that appears more like that of Ptolemy in the *Planetary Hypotheses*[64] than like that of the Indians. The precise method by which he computed these distances is not known, however, and it is not yet possible definitely to assert a Greek influence on him at this point.

As we have noted previously, the Ptolemaic *Syntaxis* and *Handy Tables* were first translated into Arabic during the reign of Hârûn al-Rashîd, under the patronage of the family of Barmak. Under the same patronage Aḥmad ibn Muḥammad al-Nihâwandî began making observations at Jundishâpûr for use in his *Al-zîj al-mushtamil*, thus beginning the long tradition of the observatory in Islam.[65]

This tradition of observation was continued under al-Ma'mûn, with the instruments invented by the Greeks, at both Baghdâd and Damascus. One purpose of this activity was to solve the problem of contradictory parameters; for, though the three systems known to astronomers of the early 'Ab-

---

[48] D. Pingree, "The Fragments of the Works of al-Fazârî," *JNES* 29, 1970, 103-123.

[49] Al-Fazârî, frags. Z 1-6, Z 8-9, Z 11-12, Z 16, Z 24, S 1, and S 2.

[50] See above fns. 37 and 43.

[51] Al-Fazârî, frags. Z 2, Z 11-15, and Z 25.

[52] Al-Fazârî, frag. Z 5.

[53] Al-Fazârî, frags. Z 17 and Z 18.

[54] Al-Fazârî, frag. S 2.

[55] Al-Fazârî, frag. Z 23.

[56] Al-Fazârî, frags. Z 2 and Z 10.

[57] The earliest Sanskrit text on the astrolabe is the *Yantrarâja* composed by Mahendra in *ca.* 1360.

[58] D. Pingree, "The Fragments of the Works of Ya'qûb ibn Ṭâriq," *JNES* 27, 1968, 97-125; E. S. Kennedy, "The Lunar Visibility Theory of Ya'qûb ibn Ṭâriq," *JNES* 27, 1968, 126-132.

[59] Ya'qûb, frags. Z 1-2, Z 5-6, and Z 9.

[60] Ya'qûb, frags. Z 1 and Z 7.

[61] Ya'qûb, frags. Z 7 and Z 8.

[62] E. S. Kennedy, "Lunar Visibility Theory," 128.

[63] Ya'qûb, frags. T 5-11.

[64] Ya'qûb, frag. T 1; cf. B. R. Goldstein, "The Arabic Version of Ptolemy's *Planetary Hypotheses*," *TAPhS*, NS 57, 1967, 4, pp. 6-8.

[65] A. Sayili, *The Observatory in Islam*, Ankara 1960, pp. 50-51. This important study is an essential source for much of what follows.

bâsid period—the Indian (Sindhind), the Iranian (Shâh), and the Ptolemaic (*Almagest*, newly available in the translation completed by al-Ḥajjâj in 829 or 830)—were all descendents of Hellenistic astronomy, their parameters differed substantially. One rather bizarre effort at reconciliation was made between 840 and 860 by the astrologer Abû Maʿshar of Balkh in his *Zîj al-hazârât*, his *Kitâb al-ulûf*, and his *Kitâb ikhtilâf al-zîjât*. He claimed that the three systems were all aberrant derivatives of a unique ante-Diluvian revelation whose exposition he had discovered in a manuscript buried at Iṣfahân by the legendary Ṭahmûrath. In fact, Abû Maʿshar's system is a curious admixture of elements lifted from the various texts known to him.[66]

But the astronomers of al-Ma'mûn sought a less fanciful solution. They tried to establish the correct parameters on the basis of new observations, though it is evident that some of their revised parameters are in fact confirmations of ones already known. Their new value of precession, for example, is Indian,[67] and their new mean motions of the planets seem largely to be simply Ptolemy's corrected for the new length of a tropical year.[68]

The personnel involved in the observations made under al-Ma'mûn at the Shammâsîya in Baghdâd in 828 and 829 included Sanad ibn ʿAlî,[69] Yaḥyâ ibn Abî Manṣûr,[70] and Muḥammad ibn Mûsâ al-Khwârizmî,[71] and the principal observor at Qâsîyûn near Damascus in 831-833 was Khâlid ibn ʿAbd al-Malik al-Marwrûdhî.[72] Except for Sanad, a converted Jew, all of these astronomers came from Iran; and each wrote at least one zîj. We now possess two of these in imperfect form, the *Zîj al-Mumtaḥan* of Yaḥyâ ibn Abî Manṣûr and others and the *Zîj al-Sindhind* of al-Khwârizmî. Let us now examine these in order to assess the extent of Greek influence, direct and indirect, on each.

A manuscript of the Escorial claims to contain the *Zîj al-Mumtaḥan* of Yaḥyâ, though it is clear that only in part does it reproduce that work.[73] The introduction proclaims that Yaḥyâ is following the *Almagest* with corrections due to his own observations made at the Shammâsîya in Baghdâd during the course of a solar year.[74] Those sections of the Escorial manuscript that have been investigated recently and that can be confidently attributed to Yaḥyâ tend to bear out this claim, though only partially. As was noted previously, the determination of the value of precession—1° in 2/3 of a century—merely confirms an older Indian parameter, and this, then, is used to "correct" the mean motions of the planets according to Ptolemy.[75] The table of solar equations is based on a Ptolemaic eccentric model, but utilizes a new parameter, 1;59°, for the maximum, while the table of lunar equations is equivalent to that in the *Handy Tables* save that an extra column is added allowing the reduction of longitudes on the lunar orbit to longitudes on the ecliptic; in this extra column the Ptolemaic value of the inclination of the lunar orbit, 5°,

---

[66] D. Pingree, *The Thousands*.

[67] D. Pingree, "Gregory Chioniades and Palaeologan Astronomy," *Dumbarton Oaks Papers* 18, 1964, 135-160, esp. 138; D. Pingree, "Precession and Trepidation in Indian Astronomy Before A.D. 1200," *JHA* 3, 1972, 27-35.

[68] This problem is being investigated by my student J. Skold.

[69] The only extant works of Sanad known to me are his *Kitâb al-qawâṭiʿ* (BM Or. 3577 ff. 329v-335; see Ibn al-Nadîm, p. 275), his *Fî ḥaṣr al-qawâṭiʿ* (BM Or. 3577 ff. 335-336), and his *Maʿrifat al-hilâl wa mâ fîhinna al-nûr* (Bodleian Marsh 663 p. 208). See also Ibn Yûnus, *Zîj al-Ḥâkimî*, in Caussin de Perceval, "Le Livre de la grande Table Hakémite," *Notices et extraits des manuscrits* 7, Paris 1803, pp. 16-240, esp. p. 56.

[70] Yaḥyâ ibn Abî Manṣûr, originally coming from Ṭabaristân under the name Bizîst ibn Fîrûzân, was the principal author of the *Zîj al-Mumtaḥan*; see fn. 73 and also Ibn Yûnus, pp. 56, 62-68, 74, 78, 114-116, 146, and 230-236.

[71] Al-Khwârizmî is now the best known scientist of al-Ma'mûn's circle. His *Zîj al-Sindhind* is discussed below.

[72] Khâlid's zîjes are noted by Kennedy, "Survey," no. 97 (p. 136).

[73] Kennedy, "Survey," pp. 145-147; J. Vernet, "Las 'Tabulae Probatae'," *Homenaje a Millás-Vallicrosa*, 2 vols., Barcelona 1956, vol. 2, pp. 501-522; E. S. Kennedy, "Parallax Theory in Islamic Astronomy," *Isis* 47, 1956, 33-53, esp. 44-46; H. Salam and E. S. Kennedy, "Solar and Lunar Tables in Early Islamic Astronomy," *JAOS* 87, 1967, 492-497; and E S. Kennedy, "The Solar Eclipse Technique of Yaḥyâ b. Abî Manṣûr," *JHA* 1, 1970, 20-38.

[74] Vernet, *op. cit.*, p. 508.

[75] See above fn. 68.

is not employed, but rather a new value, 4;46°.[76] The planetary equations are taken from Ptolemy with the exception of the equation of the center of Venus, which is made to be the same as that of the Sun.[77] The tables for computing solar eclipses use 23;51° as the value of the obliquity of the ecliptic, a parameter from the *Handy Tables*, but otherwise are based on procedures related to, but different from, those of Indian astronomers.[78] The latitude theory is also derived from India—evidently through the Sindhind—as is also the parameter for the maximum lunar latitude, 4;30°; for the other planets the maximum latitudes are closer to Ptolemy's than to the Indians'.[79] In general, then, for computing planetary longitudes Yaḥyâ is basically Ptolemaic, but he depends on the Sindhind and perhaps the Shâh for methods of solving other problems.

The *Zîj al-Sindhind* of al-Khwârizmî, on the other hand, is largely derived from the works of al-Fazârî, which it replaced as the main representative of the Sindhind tradition. Though we now possess, along with a few fragments of the original zîj,[80] only a Latin translation made in 1126 by Adelard of Bath of the revision due primarily to Maslama al-Majrîṭî of Cordoba in *ca.* 1000,[81] there is a substantial commentary literature which helps us to reconstruct the early ninth century state of the text; the principal commentators are al-Farghânî in *ca.* 850,[82] Abû Maʿshar's pupil, Ibn al-Masrûr, in *ca.* 875,[83] and Ibn al-Muthannâ in the tenth century.[84] In addition many later Andalusian texts belong to this tradition.[85] An analysis of this material shows that the models and methods underlying the zîj's tabular values are with a few exceptions the Indian or Indo-Iranian permutations and refinements of non-Ptolemaic Greek material as transmitted through al-Fazârî, including a table of sines; the exceptions are the tables of epicyclic anomalies at which each planet has its first station,[86] the tables of right ascension,[87] the table of the equation of time,[88] and various parameters embedded in tables (such as that of the obliquity of the ecliptic),[89] all derived from the *Handy Tables*.

It is clear, then, that astronomy in the time of al-Ma'mûn developed in two ways: one school retained most of the Indian Sindhind tradition but supplemented it from the Iranian Shâh and the Ptolemaic *Handy Tables*, and the other, while accepting the models of Ptolemaic astronomy, adjusted the parameters in accordance with its own observations, and continued to incorporate elements from the Sindhind or Shâh wherever these seemed "easier" than their Ptolemaic coun-

[76] Salam and Kennedy, *op. cit.*; al-Hâshimî, secs. 73-74.

[77] Ibn Yûnus, pp. 230-236.

[78] Kennedy, "Parallax Theory" and "Solar Eclipse Technique." In his table of solar declinations (Vernet, p. 515) Yaḥyâ gives the obliquity as 23;33°.

[79] Kennedy, "Survey," pp. 146 and 173.

[80] D. Pingree in *JNES* 29, 1970, 110-111; al-Hâshimî, secs. 16.2 and 85. Another, brief astronomical text of al-Khwârizmî has been studied by E. S. Kennedy, "Al-Khwârizmî on the Jewish Calendar," *SM* 27, 1961, 55-59.

[81] A. Bjørnbo, R. Besthorn, and H. Suter, *Die astronomischen Tafeln des Muḥammed ibn Mûsâ al-Khwârizmî*, København 1914; O. Neugebauer, *The Astronomical Tables of al-Khwârizmî*, København 1962. See also Kennedy, "Parallax Theory," 47-51; E. S. Kennedy and M. Janjanian, "The Crescent Visiblity Table in al-Khwârizmî's Zîj," *Centaurus* 11, 1965, 73-78; and E. S. Kennedy and W. Ukashah, "Al-Khwârizmî's Planetary Latitude Tables," *Centaurus* 14, 1969, 86-96.

[82] There are fragments in al-Hâshimî, sec. 15, in Ibn al-Muthannâ (see below fn. 84), and in al-Bîrûnî, *Rasâ'il*, I 128 and 168. Al-Farghânî was also the author of a summary of Ptolemaic astronomy published by J. Golius, Amsterdam 1669; see also R. Campani, "Il 'Kitâb al-Farghânî' nel testo arabo e nelle versioni," *RSO* 3, 1910, 205-252.

[83] MS Taymûriyya, Math. 99; see Ibn al-Nadîm, p. 277.

[84] The Latin translation by Hugo of Sanctalla (*fl. ca.* 1130) is edited by E. Millás Vendrell, *El comentario de Ibn al-Mutannâ' a las Tablas Astronómicas de al-Jwârizmi*, Madrid-Barcelona 1963; the two Hebrew versions, of which one was written by Abraham ben Ezra in 1160, by B. R. Goldstein, *Ibn al-Muthannâ's Commentary on the Astronomical Tables of al-Khwârizmî*, New Haven-London 1967.

[85] In particular one thinks of the recent studies of al-Zarqâlî (see above fn. 7 and G. J. Toomer, "The Solar Theory of az-Zarqāl: A History of Errors," *Centaurus* 14, 1969, 306-336), al-Jahânî (H. Hermelink, "Tabulae Jahen," *AHES* 2, 1964, 108-112), the *Toledan Tables* (G. J. Toomer, "A Survey of the Toledan Tables," *Osiris* 15, 1968, 5-174), and Abraham ben Ezra (J. M. Millás Vallicrosa, *El libro de los fundamentos de las Tablas astronómicas de R. Abraham ibn ʿEzra*, Madrid-Barcelona 1947).

[86] Neugebauer, *The Astronomical Tables*, p. 101.

[87] *Ibid.*, pp. 104-105.

[88] *Ibid.*, p. 108.

[89] *Ibid.*, pp. 48 and 95-96.

terparts. The results were eclectic and internally inconsistent astronomical systems, in some respects permitting more accurate computations than would those of al-Fazârî or Yaʿqûb, but still manifesting a basic lack of control over the geometric models that lie behind them. This control was to come only with al-Battânî.

But preceding al-Battânî's zîj was the work of Ḥabash al-Ḥâsib of Marw, who wrote extensively on the observations made under al-Ma'mûn[90] and himself composed several zîjes;[91] he is known to have been engaged in observational activities between 829 and 864.[92] There are two manuscripts extant of zîjes ascribed to Ḥabash, one at Berlin and the other at Istanbul, and many later authors discuss his views on various topics. A cursory examination of this material makes clear that his geometric models are essentially Ptolemaic, though many of his parameters are taken from the *Zîj al-Mumtaḥan* and some of his eclipse-theory from the Sindhind. He further introduces many refinements into his tables facilitating their practicality, including a table of tangents.

At this point in the history of eastern Islamic astronomy, with the existence of Arabic versions of the *Almagest* and of the *Handy Tables* and with the compilation of Yaḥyâ's and Ḥabash's zîjes, the predominance of Ptolemy was virtually secured; the Sindhind and the Shâh became primarily of historical interest. But before the process of Ptolemaicization was completed, however, the preliminary Greek works on celestial mechanics, the so-called *Little Astronomy*, were translated into Arabic in Syria by Isḥâq ibn Ḥunayn,[93] Thâbit ibn Qurra,[94] and Qusṭâ ibn Lûqâ,[95] with the possible participation of al-Kindî.[96] Isḥâq

---

[90] Ibn Yûnus, p. 160.

[91] Kennedy, "Survey," nos. 15 and 16 (pp. 126-127) and pp. 151-154; A. Sayili, "The Introductory Section of Habash's Astronomical Tables Known as the 'Damascene' Zîj," *Ankara Univ. Dilve Tarih-Coğrafya Fakültesi Dergisi* 13, 1955, 133-151; E. S. Kennedy and W. R. Transue, "A Medieval Iterative Algorism," *AMM* 63, 1956, 80-83; Kennedy, "Parallax Theory," 42-43; E. S. Kennedy and M. Agha, "Planetary Visibility Tables in Islamic Astronomy," *Centaurus* 7, 1960, 134-140, esp. 137; Salam and Kennedy, "Solar and Lunar Tables"; and E. S. Kennedy, "An Early Method of Successive Approximations," *Centaurus* 13, 1969, 248-250. For a late eleventh century Greek text which may have been influenced by Ḥabash see O. Neugebauer, *Commentary on the Astronomical Treatise Par. gr. 2425*, Bruxelles 1969. Many Arabic authors also discuss Ḥabash—e.g., Ibn Amâjûr as cited by Ibn Yûnus, pp. 126-142, al-Hâshimî, secs. 13 and 74, Abû Naṣr Manṣûr in *Rasâ'il*, Hyderabad-Deccan 1948, and al-Masʿûdî and al-Bîrûnî in various of their works. All of these will be investigated by J. Skold in his work on Ḥabash.

[92] Ibn Yûnus, pp. 98-100 and 170-174. There are manuscripts of two works of his on Ptolemaic instruments noticed in M. Krause, "Stambuler Handschriften islamischer Mathematiker," *QSt* B 3, 1936, 437-532, esp. 446-447.

[93] Isḥâq (died 910/11) translated Euclid's *Data* (Naṣîr al-Dîn al-Ṭûsî, *Majmûʿ al-rasâ'il*, 2 vols., Hyderabad-Deccan 1939-40, no. 1) and Menelaus' *Spherics* (also connected with al-Māhānī; M. Krause, *Die Sphärik von Menelaos aus Alexandrien*, Göttingen 1936), and collaborated with Thâbit in translating Archimedes' *On the Sphere and the Cylinder* (al-Ṭûsî, no 12). See Ibn al-Nadîm, p. 285, and F. Rosenthal, "Isḥâq b. Ḥunayn's Ta'rîḫ al-aṭibbâ'," *Oriens* 7, 1954, 55-80.

[94] Thâbit of Ḥarrân (834-901) corrected Isḥâq's translations; translated Archimedes' *Lemmata* (al-Ṭûsî, no. 10) and, with Qusṭâ, Theodosius' *Spherics* (al-Ṭûsî, no. 3; translated into Sanskrit by Nayanasukhopâdhyâya in 1730); corrected Autolycus' *On the Moving Sphere* (al-Ṭûsî, no 2.) and *On Risings and Settings* (al-Ṭûsî, no. 13) (see J. Mogenet, *Autolycos de Pitane*, Louvain 1950); and he composed the *Data* (al-Ṭûsî, no. 9). On Thâbit's life see E. Wiedemann, "Über *Ṯâbit ben Qurra*, sein Leben und Wirken," *Sitz. Phys.-Med. Soz. Erlangen* 52-53, 1920-21, 189-219 (= *Aufsätze zur arabischen Wissenschaftsgeschichte*, 2 vols., Hildesheim-New York 1970, vol. 2, pp. 548-578). For some more of his astronomical works see below fn. 100. Some of his mathematical texts have recently been published: translations of two Archimedean treatises in *Rasâ'il Ibn Qurra*, Hyderabad-Deccan 1948, and of Nicomachus, edited by W. Kutsch, Beirut 1958.

[95] Qusṭâ of Baʿalbak (died 912) translated Theodosius' *Spherics* (with Thâbit) and *On Habitations* (al-Ṭûsî, no. 4), and, perhaps, Hypsicles' *Anaphoricus* (al-Ṭûsî, no. 14; see fn. 96). For Qusṭâ's life see G. Gabrieli, "Nota bibliografica su Qusṭa ibn Lûqâ," *Rend. R. Acc. Lincei* 5, 21, 1912-13, 341-382, and W. H. Worrell, "Qusta ibn Luqa on the Use of the Celestial Globe," *Isis* 35, 1944, 285-293.

[96] Al-Kindî (died *ca.* 875) corrected Qusṭâ's translation of Hypsicles' *Anaphoricus* according to al-Ṭûsî, no. 14, and the Arabic biographical literature, though he is not named in the manuscripts of the pre-Ṭûsî

and Thâbit were involved as well in a new translation of the *Almagest* from the Greek,[97] and Thâbit with the translation of Ptolemy's *Planetary Hypotheses*.[98] Thâbit also wrote numerous treatises on astronomy;[99] many of these minor tracts exist in Latin translations, of which a number have recently been published.[100] These works of Thâbit demonstrate that he did indeed understand Ptolemaic astronomy as expounded in the *Almagest*, and also attest to his knowledge of Theon's *Handy Tables* and the *Zîj al-Mumtaḥan*. These various books and others written by the Syrian translators and scholars of the second half of the ninth century provided the Arabs with a curriculum for instruction in the Ptolemaic system, and prepared the way for al-Battânî's *Zîj al-Ṣâbi'*.

For it is al-Battânî[101] who represents the culmination of this phase in the history of Islamic astronomy. He mastered both the Ptolemaic models and solutions and also the new trigonometry developed from the Sindhind tradition by scholars such as Ḥabash. While accepting the Ptolemaic cinematic models, however, he acknowledged the necessity of improving the Greek parameters by means of new observations; he took over many of the values established by the astronomers of the time of al-Ma'mûn, in particular those recorded in the *Zîj al-Mumtaḥan*, and revised others on the basis of his own observations. After al-Battânî the Ptolemaic character of mathematical astronomy in Islam was firmly established, to be challenged only by those who were disturbed by its incongruency with Aristotelian physics; the influence of the non-Ptolemaic systems of India and Iran which, though ultimately of Greek origin, had been largely transformed in the direction of mathematical simplification through approximative methods, waned everywhere save in Spain after contributing little besides trigonometry to the later phases of Islamic astronomy.

Greek astronomy, then, or rather several of its products, reached Islam by various routes, on some of which it was subjected to major modifications. Through India came a double-epicycle model of planetary motion that had been generated by the controversy of the first and second centuries between Greek mathematical astronomers and Peripatetic philosophers. With this also came the rudiments of trigonometry; various mathematical techniques of approximation; relatively crude theories of eclipses, planetary latitudes, geographical distances, and so on; and a host of new parameters, of which those for the mean planetary motions were based on the conception of universal conjunctions occurring at regular intervals over vast periods of time. Through Iran came a Sasanian adaptation of some of this Greco-Indian astronomy fused with various elements of Ptolemaic theory. The mixture of these two traditions in the 'Abbâsid period led to an eclecticism characterized by internal inconsistencies and evident imprecision. In response to this unsatisfactory and confused state al-Ma'mûn inaugurated a program of careful translations of

---

text used by M. Krause in V. de Falco, M. Krause, and O. Neugebauer, *Hypsikles. Die Anfangszeiten der Gestirne*, Göttingen 1966, pp. 61-84. Among other astronomical works of al-Kindî is one on the instrument dhât al-shu'batayn (E. Wiedemann, "Über eine astronomische Schrift von *al-Kindî*," *Sitz. Phys.-Med. Soz. Erlangen* 42, 1910, 294-300 (= *Aufsätze*, vol. 1, pp. 660-666), and an epitome of Ptolemaic astronomy based on al-Ḥajjâj's translation (F. Rosenthal, "Al-Kindî and Ptolemy," *Studi orientalistici in onore di Giorgio Levi della Vida*, 2 vols., Roma 1956, vol. 2, pp. 436-456).

[97] This also was revised by al-Ṭûsî and exists in many manuscript copies. Al-Ṭûsî's recension was translated into Sanskrit by Jagannâtha in 1732 under the title *Siddhântasamrâṭ* (edited by R. Śarman, 3 vols., New Delhi 1967-69).

[98] B. R. Goldstein, "The Arabic Version".

[99] Krause, "Stambuler Handschriften," 453-457.

[100] *De anno Solis* in F. J. Carmody, *The Astronomical Works of Thabit b. Qurra*, Berkeley-Los Angeles 1960, pp. 63-79 (on the grave deficiencies of Carmody's book see O. Neugebauer in *Speculum* 37, 1962, 99-103); O. Neugebauer, "Thâbit ben Qurra 'On the Solar Year' and 'On the Motion of the Eighth Sphere'," *PAPhS* 106, 1962, 264-299, esp. 264-289.

De motu octave spere in Carmody, pp. 102-113; Neugebauer, 290-299.

De hiis que indigent expositione antequam legatur Almagesti in Carmody, pp. 131-139.

De imaginatione spere et circulorum eius in Carmody, pp. 140-144.

De quantitatibus stellarum et planetarum et proportione terre in Carmody, pp. 145-148.

De figura sectore in Carmody, pp. 159-164.

[101] Al-Battânî, a Ḥarrânian whose observations were made at al-Râqqa between 877/8 and 918/9 and who died in 929/30, wrote as his principal astronomical treatise the *Zîj al-Ṣâbi'* edited by Nallino.

Greek and Syriac texts, and in particular of the *Almagest* and of the *Handy Tables*; of a modernization of the Sindhind tradition; and of new observations leading to adjustments of the Ptolemaic parameters. Despite the strange aberrations of Abû Maʿshar and the temporarily continuing vitality of the Sindhind tradition, scholars in the latter half of the ninth century carried on the process of the Ptolemaicization of Islamic astronomy. The process was completed in the great work of al-Battânî, which represents a level of control over observation and theory that the best astronomers in the Muslim world never lost.

Later developments in Islamic astronomy (aside from alterations of parameters, which new observations might at any time justify) were not generated by mathematical, but by philosophical considerations. These developments cannot here be adequately explored, but are interesting within the context of a discussion of Greek influence because they reflect the fundamental incompatibility of Aristotelian physics with Ptolemaic planetary astronomy. Al-Biṭrûjî[102] and others attempted to devise mathematical models that would preserve the concentricity of the planetary orbits, and the School of Marâgha[103] invented geometric devices by which the non-uniform motion of the Ptolemaic deferent about its center was banished from planetary theory. The complaints of the Peripatetics and these Muslim responses were not without influence on the Western astronomers who faced the same problems in the Renaissance.

102 The main study is now B. R. Goldstein. *Al-Biṭrûjî: On the Principles of Astronomy*, 2 vols., New Haven 1971; see also L. Gauthier, "Une réforme du système astronomique de Ptolémée tentée par les philosophes arabes du XIIe siècle," *JA* 10, 14, 1909, 483-510; the edition of the Latin translation (made by Michael Scot in 1217) by F. J. Carmody, Berkeley-Los Angeles 1952; and B. R. Goldstein, "On the Theory of Trepidation," *Centaurus* 10, 1964, 232-247; see also F. J. Carmody, "Regiomontanus' Notes on al-Biṭrûjî's Astronomy," *Isis* 42, 1951, 121-130.

103 V. Roberts, "The Solar and Lunar Theory of Ibn ash-Shâṭir: A Pre-Copernican Copernican Model," *Isis* 48, 1957, 428-432; E. S. Kennedy and V. Roberts, "The Planetary Theory of Ibn al-Shâṭir," *Isis* 50, 1959, 227-235; F. Abbud, "The Planetary Theory of Ibn al-Shâṭir: Reduction of the Geometric Models to Numerical Tables," *Isis* 53, 1962, 492-499; V. Roberts, "The Planetary Theory of Ibn al-Shâṭir: Latitudes of the Planets," *Isis* 57, 1966, 208-219; and E. S. Kennedy, "Late Medieval Planetary Theory," *Isis* 57, 1966, 365-378.

Al-Bīrūnī's Knowledge of Sanskrit Astronomical Texts

David Pingree

The Moslem interpreter of Indian culture best known to the West, Abū 'l-Rayḥān Muḥammad al-Bīrūnī, was carried off to Ghazna by Sulṭān Maḥmūd in 1018 A.D.[1] During the course of the following years he gained access to the Sanskrit literature current in northwestern India at the time, and made much of it available in Arabic by means of a series of books that he published in the late 20s and in the 30s of the eleventh century. In this paper I intend to consider the means by which he studied several Sanskrit texts on astronomy which were among the most important sources for his knowledge of Indian science, and to examine the question of the reliability of his reporting of Indian astronomical and physical theories.

Al-Bīrūnī was neither the first nor the last Moslem to study Indian astronomy. There were three pakṣas (that is, schools of astronomy) whose systems—and, in particular, whose parameters and computational techniques—had been partially familiar to Arabic-reading scholars since the eighth century A.D.[2] The *brāhmapaksa* was initiated by the *Paitāmahasiddhānta* of the *Viṣṇudharmottarapurāṇa*[3] in the early fifth century, and its influence was already felt in Sasanian Iran when the first *Zīk i Shahriyārān* was composed in about 450.[4] But the brāhmapakṣa's most prominent representative was the *Brāhmasphuṭasiddhānta* composed by Brahmagupta at Bhillamāla in southern Rājasthān in 628.[5] Based mainly upon this text, though including some elements from the āryapakṣa, was a work, apparently entitled *Mahāsiddhānta* (*al-Sindhind al-kabīr*), of which a copy was carried to Baghdād in 771 or 773 by a member of a delegation from Sind. This *Mahāsiddhānta* was used by al-Fazārī and of Yaᶜqūb ibn Ṭāriq[7] in their several zījes composed between about 775 and 790, though they also derived some material from the Ptolemaic tradition and from the *Zīk i Shahriyārān* of Yazdijird III. The works of al-Fazārī and of Yaᶜqūb are the bases of the *Sindhind* tradition of Islamic astronomy,[8] which al-Bīrūnī had studied thoroughly and through which he had learned something of Indian astronomy long before his sojourn in the Panjab.

The second Indian pakṣa of which elements reached Islam was the ārdharātrika or midnight-system, initiated by Āryabhaṭa in about 500.[9] This was apparently the foundation of the version of the Pahlavī *Zīk i Shahriyārān* that was published under Khusraw Anūshirwān in 556;[10] much of this zīk was, naturally, repeated in the similar zīk issued under Yazdijird III, probably in the 630s, which was translated into Arabic as the *Zīj al-Shāh* by al-Tamīmī toward the end of the eighth century.[11] The ārdharātrikapakṣa was also followed by Brahmagupta in his *Khaṇḍakhādyaka*[12] written in 665, which, with its commentaries, was often cited by al-Bīrūnī in the works that he wrote in the years immediately on either side of 1030,[13] but whose elements also reached him through the Arabic version made in Sind in 735 and published under the title *Zīj al-Arkand.*[14]

Finally, the third of the Sanskrit pakṣas that we must consider is the āryapakṣa, also initiated by Āryabhaṭa, in the *Āryabhaṭīya*, whose epoch is 499. Elements of this pakṣa, as we have noted above, were included in the *Mahāsiddhānta* utilized by al-Fazārī and by Yaᶜqūb ibn Ṭāriq. There was also a metrical Arabic version of at least a part of the *Āryabhaṭīya*, published as the *Zīj al-Harqān* in 742,[15] and a fuller translation, the *Zīj al-Arjabhar*, was made, apparently toward the end of the eighth century, and was used by one Abū 'l-Ḥasan al-Ahwāzī.[16]

Al-Bīrūnī was familiar with many of the parameters and procedures of the brāhmapakṣa and of both of Āryabhaṭa's pakṣas through the medium of at least some of these early Arabic translations before he went to India; in fact, most of our knowledge of these lost zījes is derived from al-Bīrūnī himself. But the Indian systems were not presented in these texts with any integrity; their authors generally conflated material from various zījes without naming their sources, and thereby offered an extremely distorted and inaccurate picture of Indian astronomy. Al-Bīrūnī, through his access to Sanskrit texts, certainly increased the knowledge of Indian astronomy among readers of Arabic, but, as we shall see, was still unable to avoid many unfortunate distortions of these Sanskrit texts. The best translations appear to have been those made into Persian by scholars patronized by the Tughlūqs in the late fourteenth and by the Mughals in the sixteenth and seventeenth centuries,[17] but they enjoyed a very limited circulation only within India.

As we have mentioned, much of our knowledge of the lost works of the two earliest representatives of the *Sindhind* tradition, al-Fazārī and Yaᶜqūb ibn Ṭāriq, we owe to al-Bīrūnī, who severely (and usually justly) condemns them for their incompetence in astronomy.[18] Occasionally, moreover, he also chastises them for misunderstanding their Sanskrit sources—for instance, for thinking that the proper name Āryabhaṭa (Arjabhar in Arabic) is a technical term meaning "a thousandth part."[19] But we shall see that al-Bīrūnī's own control of the highly technical Sanskrit employed in composing metrical treatises on astronomy was minimal. One instructive example is found in the *India*, written after he had spent several years studying Indian astronomy.[20]

> As regards adhimāsa [spelled adimāsah in Arabic], the word means "the first month," for ādi means "beginning." In the books of Ya^{c}qūb ibn Ṭāriq and al-Fazārī this name is written padamāsah [an early scribal error for the correct malamāsa]. Pad means "end" [in fact, it means "foot." Anta means "end"; al-Bīrūnī read alif, nūn dāl instead of bā', dāl], and it is possible that the Indians call the leap month by both names; but the reader must be aware that these two authors frequently misspell or disfigure the Indian words, and that there is no reliance on their tradition.

In fact, the earlier translators, in this case as in some others, were right, and al-Bīrūnī failed to understand two simple Sanskrit words because he read them in faulty Arabic transliterations rather than in Nāgarī.

And, despite his reputation as a Sanskrit scholar, he surely had no direct acquaintance with the Sanskrit originals of the two siddhāntas which I wish to discuss in some detail; they represent the brāhmapakṣa and the ārdharātrikapaksa. Al-Bīrūnī did have a copy of the *Āryabhaṭīya*, though he did not realize it and used it very little;[21] his Arabic version of the *Karaṇatilaka*, a representative of the saurapakṣa written by Vijayananda at Vārāṇasī in 966, was one of his earliest efforts to have a Sanskrit text translated, apparently undertaken in 1026, and is so full of elementary mistakes that its editor suspects the hand of a student.[22] I would suggest rather that al-Bīrūnī's role in the production of the *Ghurrat al-zījāt* was confined to that of a sponsor, editor, and explicator of some paṇḍita's incompetent effort.

The two siddhāntas to be considered here are the *Brāhmasphuṭasiddhānta* of Brahmagupta, which we have already referred to as the principle representative of the brāhmapakṣa, and the later *Pauliśasiddhānta*,[23] which belongs to the ārdharātrikapakṣa. This *Pauliśasiddhānta* is now lost save for al-Bīrūnī's citations and fragments of it preserved in the commentaries on Brahmagupta's *Khaṇḍakhādyaka* composed by Pṛthūdakasvāmin[24] at Sthāneśvara (or Sthānvīśvara) in the plains northwest of Delhi in 864, by Utpala[25] in Kashmir in 968, and by Āmarāja[26] at Ānandapura in Gujarāt in about 1200. It is clear, therefore, that this siddhānta was available in northwestern India from the ninth through the twelfth century at least. But we know further, from its fragments, that this *Pauliśasiddhānta* also was written at Sthāneśvara, and that is was already quoted by an eighth-century commentator named Balabhadra, of whom we shall say more later. In fact, we can tentatively localize both Balabhadra and the author of the *Pauliśasiddhānta* in the empire of Yaśovarman of Kānyakubja, who ruled over the region around the Yamunā-Ganges doab from about 725 to 750.[27]

But, when al-Bīrūnī mentions the *Pauliśasiddhānta* in the *On Shadows,*[28] he names its author "Pulisa al-Yunānī" [Paulus the Greek]; when he mentions him in the *India,*[29] he calls him "Pulisa al-Yunānī from the city of Sayntra," and he conjectures that Sayntra is a corruption of Alexandria. This is a triply erroneous statement, and a good place to commence our criticism of al-Bīrūnī's misunderstandings of this text.

The statement concerning Pauliśa occurs in a listing of the five zījes of the Indians. These are the five siddhāntas which gave its name to the *Pañcasiddhāntikā* written by Varāhamihira at Ujjayinī in about 550:[30] the *Sūrya*, apparently composed by Lātadeva in 505 on the basis of the ārdharātrikapakṣa; the *Vāsiṣṭha*, written in about 200 on the basis of a Greco-Babylonian tradition; the *Pauliśa* and the *Romaka*, originally fourth or fifth century versions of Greek works, but known to Varāhamihira in Lāṭadeva's recensions; and the *Paitāmaha* or *Brāhma*, composed in A.D. 80 in the tradition of the *Jyotiṣavedāṅga's* adaptation of Babylonian astronomy. Al-Bīrūnī did not have a copy of the *Pañcasiddhāntikā* itself, though he knew of some quotations from it in Balabhadra's commentary on the *Brāhmasphuṭasiddhānta*[31] and in Utpala's commentary on Varāhamihira's *Bṛhatsaṃhitā.*[32]. In the *Bṛhatsamhitā* Varāhamira names the five siddhāntas of the *Pañcasiddhāntikā,*[33] and al-Bīrūnī evidently used that list in constructing his own.

But he had some assistance that was not very expert. He states that the author of the *Vāsiṣṭha* is Viṣṇucandra and that of the *Romaka* Śrīṣen. Sanskrit texts—primarily Brahmagupta, Balabhadra, Pṛthūdaka, and Utpala—inform us that Viṣṇucandra composed a *Vāsisthasiddhānta*[34] and Srīṣeṇa a *Romakasiddhānta*[35] in about 600, that is, some decades after Varāhamihira wrote the *Pañcasiddhāntikā.* Furthermore, al-Bīrūnī identifies Varāhamihira's *Paitāmahasiddhānta*, whose epoch we have seen to be A.D. 80, with Brahmagupta's *Brāhmasphuṭasiddhānta*, composed more than half a millennium later, and Lāṭadeva's recension of the *Pauliśasiddhānta* with the text written more than two centuries later at Sthāneśvara. There can be no question that he was poorly informed concerning the history of Indian astronomy.

So much for the first mistake in al-Bīrūnī's statement. There remain two others to be explained—his calling Puliśa a Yunānī or Greek, and his assertion that he comes from Alexandria. It is not difficult to unravel the process by which these two errors arose. The paṇḍita wrote out his translation of the

Pauliśasiddhānta, naturally, in Arabic characters; when he wrote the Arabic name of Sthāneśvara—that is, Tanīsar—he omitted the diacritical marks. So al-Bīrūnī misread *ta', nūn, yā', sīn, rā'* without points as *sīn, yā', nūn, tā', rā'*—Sayntra. Apparently, he was no longer able to consult the translator about the correctness of this reading, or he chose not to do so when he was composing the *India*; rather, accepting this mysterious reading and combining it with a genuine Indian tradition that the original Pauliśa was a Yavana or foreigner, he called the author of the later *Pauliśasiddhānta* a Greek from Alexandria.

That modern scholars have, on the basis of this triple error, assumed that the Greek astrologer, Paulus of Alexandria, who composed the second edition of his Εισαγωγικα in 378, also wrote the original of the *Pauliśasiddhānta* summarized by Varāhamihira, is one of the less encouraging results of the respect paid to al-Bīrūnī's alleged control of Sanskrit.[36] In a later, and as yet unpublished little treatise, the *Maqāla fī sayr sahmī al-saᶜāda wa al-ghayb*,[37] al-Bīrūnī still calls Puliśa a Greek, but correctly reads the name of his city as Tānīsar. This leads him to suppose either that there was a common source of both the Greek and the Indian astronomical systems (an idea reminiscent of Abū Maᶜshar's strange history of science)[38], or that Puliśa had emigrated from Alexandria to the plains of Panipat.

If we now turn to al-Bīrūnī's quotations from the translation of the later *Pauliśasiddhānta* that he inserted in his *On Transits*, in his *India*, in his *al-Qānūn al-Masᶜūdī*, and in his *On Shadows* and compare them with the original Sanskrit verses that have been preserved by the commentators, we are constantly brought to an awareness of the inadequacies of the translator, of the ineptness of the interpreter's explanation of the text, and of al-Bīrūnī's tendency to understand the Indian text in the terms with which he was familiar—that is to say, those of Ptolemaic astronomy and of Aristotelian physics. Let us now consider a few of the many possible examples.

Almost all of the lengthy computations that al-Bīrūnī in the *India* and in the *Qānūn* reports as being derived from the *Pauliśa* are certainly the interpreter's as they would have no place in a Sanskrit siddhānta. In some cases al-Bīrūnī himself has caught his paṇḍita's errors and unjustly attributed them to the *Pauliśa*; for example, at one point in the *India*[39] there is a computation of the lapsed *adhimāsas* from the lapsed *tithis*, wherein the translator has substituted saura days for tithis, and al-Bīrūnī criticizes the *Pauliśa*, to which such an example could not belong. Other errors, however, al-Bīrūnī does not notice. In the *On Transits*,[40] in discussing the *Pauliśa's* computation of the equations of the planets, al-Bīrūnī constantly assumes that the maximum equation, itself expressed in minutes, is taken to be equal to the radius of the epicycle, whereas the preserved Sanskrit verses of the *Pauliśasiddhānta*[41] correctly, for an Indian text, state that the radius of the epicycle is the Sine of the maximum equation. But at one point[42] al-Bīrūnī found in the translation some obviously mistaken values for the maximum equations of the center of the five "star-planets." This leads him to complain: "And as for what is in the Indian zījes which we have studied, this is at the utmost of confusion to the extent that its measure is not possible. So suspicion turns toward the copies which have come to us and toward the interpreter (mufassir) who dictated to us."

This passage suggests the method which al-Bīrūnī followed in his studies of Sanskrit astronomical texts. As we have seen before, some of his errors are due to his misreadings of an unpointed Arabic manuscript; these indicate that his translator was not present when he read the texts. The absence of the translator is also indicated by a passage in the *On Shadows*[43] wherein the *mutarjim* is chastised for his failings. On the other hand, in the *India*[44] al-Bīrūnī criticizes his *mufassir* for giving him, presumably orally, the wrong explanation of the relation between a muhūrta, which is a thirtieth of a seasonal day, and a ghaṭikā, which is a sixtieth of a mean day. Thus, the *Pauliśa* reached al-Bīrūnī through a double layer of opaqueness—the rough, inaccurate, and interpolated translation of an absent pandita, and the incompetent exegesis of a local *mufassir*. This twofold removal from the original clarifies al-Bīrūnī's curious statement in one passage of the *India*:[45] "Puliśa says in his siddhānta that Puliśa the Greek says . . ."; the first Puliśa is the translator inserting his own comments regarding the text that he has translated.

Al-Bīrūnī's tendency to comprehend this text, thus rather dimly revealed to him, in terms of Greek rather than of Indian science, is evident in his discussion of the *Pauliśa*'s alleged physical theories. In the preserved Sanskrit verses of the *Pauliśa*,[46] the circumference of the orbit of the heavens and its radius are given in yojanas; and we are informed by Balabhadra[47] that the followers of Āryabhaṭa believed that the orbit of the heavens is the limit to which the sun's rays reach. Thus far does the genuine Indian tradition go. But al-Bīrūnī claims[48] that Puliśa believes that the planetary bodies are fiery (that is, self-luminous) while the starry sphere is watery (that is, reflective), and that the weak solar rays reflected at a great distance from this watery eighth sphere produce the blue color of the sky. Such a physical explanation of a color-phenomenon is totally foreign to Sanskrit siddhāntas and could not have been in the *Pauliśa*.

Utpala quotes from the *Pauliśasiddhānta*:[49]

The earth is formed round like a circle, and is motionless in the boundless sky; it consists of the five mahābhūtas. In its middle is Meru, [the home] of the Gods. Above that is the [north] pole in the sky; the circle of the constellations, which is bound to it by chords of wind, revolves in their risings and settings as they are driven by the wind.

These verses represent the normal transformation by Indian astronomers of the Purāṇic cosmology in which Mount Meru rises at the center of a flat, circular earth. When Greek mathematical astronomy was introduced into India in the third and fourth centuries, it became necessary to regard the earth as a sphere suspended in the center of a nesting of planetary and stellar spheres in order to be able to operate with the Greek geometric models. While accepting this spherical cosmology, Indian astronomers retained Mount Meru as the terrestrial north pole and explained the diurnal rotation of the heavens by the Purāṇic mechanism of chords of wind wrapped around the axis of the world and pulled by demons.

But al-Bīrūnī[50] claims that Puliśa could not believe in a material axis extending to the sphere of the fixed stars and holding the earth in place in the center since he must believe that weights fall to the earth from every direction; in other words, he implies that Puliśa subscribes to an Aristotelian argument in favor of a geocentric universe. Further, al-Bīrūnī states that Puliśa "believes . . . that the motion of what is at the periphery is the cause of the non-motion of what is at the center." This type of argument for the immobility of the earth is also Greek rather than Indian.

We shall shortly consider similar misinterpretations in al-Bīrūnī's discussion of passages of Brahmagupta's *Brāhmasphuṭasiddhānta*. But here the situation is even further complicated by the fact that al-Bīrūnī's paṇḍita translated not just the *Brāhmasphuṭasiddhānta*, but also the commentary on it composed by Balabhadra at Kānyakubja in about 750—and by the fact that he usually failed to distinguish in his translation between Brahmagupta's verses and Balabhadra's exegesis. Though we still possess the whole of the *Brāhmasphuṭasiddhānta* in Sanskrit, no manuscripts of Balabhadra's *Bhāṣya* are known to exist. However, there do survive three copies, all incomplete, of the great commentary on the *Brāhmasphuṭasiddhānta* written by Pṛthūdakasvāmin at Sthāneśvara shortly before 864.[51] A close comparison of this commentary with the relevant passages in al-Bīrūnī's *India* and with some sections of Utpala's *Vivṛti* on Varāhamihira's *Bṛhatsaṃhitā* shows that all three have depended on Balabhadra's lost commentary.[52] Al-Bīrūnī thereby is of great assistance in our efforts to reconstruct this lost eighth-century Sanskrit text, which is of great importance because of the almost unique interest that Balabhadra displayed in the physical model of the universe implied by the Indian adaptations of Greek astronomy and in the history of the development of cosmological theories in India. This latter interest is especially manifested in his numerous quotations from earlier Sanskrit texts.

In fact, it is clear that almost all of the knowledge of the astronomical works of Lāṭadeva, Viṣṇucandra, and Varāhamihira that al-Bīrūnī demonstrates in the *India* is derived from Balabhadra; a few quotations seem to come from Utpala's commentary on the *Bṛhatsaṃhitā*. Furthermore, Balabhadra is al-Bīrūnī's main source for information about Aryabhaṭa's *Aryabhaṭīya*. It is true, as has been noted above, that al-Bīrūnī had access to a manuscript of the original *Āryabhaṭīya*, with a late commentary mentioning the later *Pauliśasiddhānta* and Pṛthūdakasvāmin,[53] from which one of his paṇḍitas made some translations, but he unaccountably remained ignorant of the fact that this Āryabhata is identical with the one cited by Brahmagupta and Balabhadra.[54] Therefore, he distinguished them as Āryabhaṭa al-kabīr and Āryabhaṭa of Kusumapura, an error which has led several modern historians of science astray.[55]

But rather than follow that desultory path, let us trace al-Bīrūnī's steps as he gradually discovered the *Brāhmasphuṭasiddhānta*. When he wrote the *Chronology* in 1000, he knew something of it through the zījes of al-Fazārī and of al-Khwārizmī and through the *Kitāb al-ulūf* and *Zīj al-hazārāt* of Abū Maᶜshar.[56] These works were still his sole sources of knowledge of Brahmagupta's work when he composed the *Taḥdīd al-amākin*[57] in 1025 and the *On the Solar Equation*[58] in, apparently, 1027; his first dated work to indicate a more direct acquaintance with the *Brāhmasphuṭasiddhānta* is the *Astrology*,[59] which he wrote in 1029. When he composed the *India* in 1030/1031 his paṇḍita had translated Chapter 21, on the sphere, which was the first in Balabhadra's recension; Chapter 1, on the mean motions of the planets; Chapter 11, criticisms of his predecessors; and a sample page of Chapter 20, on prosody.[60] Little more knowledge is demonstrated in the *Qānūn* of 1031, but the *On Shadows*, which was probably written a little while after the *India*, shows his acquaintance with Chapter 3, on problems of local time, local latitude, and the ascendent on the local horizon, and that part of Chapter 12 which is concerned with gnomons.[61] It also demonstrates that he has acquired a translation of part of the *Bhāṣya* of Pṛthūdakasvāmin;[62] in the *India* he includes Pṛthūdakasvāmin in a list of scholars of whom he only knows the names while remaining ignorant of even the titles of their works.[63] As the citations from the *Brāhmasphuṭasiddhānta* in his other works are relatively meagre, we will restrict our consideration to the *India*, in which dozens of pages allegedly expound the views of Brahmagupta.

These pages are, unfortunately, full of gross misconceptions and wrong translations. As noted above, al-Bīrūnī's paṇḍita did not usually indicate where his translation shifted from Brahmagupta to Balabhadra; therefore, long passages written by the commentator are erroneously attributed to the *Brāhmasphuṭasiddhānta*. Moreover, many passages–especially the verses quoted by Balabhadra from earlier works–are improperly understood by the translator. And lastly, al-Bīrūnī, as he also did with regard to the *Pauliśa*, attributes to the Indians Greek physical theories and methods of geometric proof. This tendency is particularly regrettable in the light of al-Bīrūnī's criticism of Ibn al-Muqaffaᶜ for making tendentious additions to his translation of the *Kalīla wa Dimna*.[64] We shall now consider briefly, as one example, what a few verses of the *Brāhmasphuṭasiddhānta* and their commentary underwent at the hands of al-Bīrūnī and his paṇḍitas.

In verses 4–5 of Chapter 21 Brahmagupta says:

> The sphere of the stars, which is bound at the two poles and is driven by the pravaha wind, moves at the horizon to the right for the Gods, to the left for the Demons. Elsewhere, in every place, (by the amount that) the pole is depressed (toward the horizon, the equator of) the sphere of the constellations is elevated. At Laṅkā (on the terrestrial equator, the equator of) the sphere is the east-west line, and the two poles are on the horizon.

As was the case with the verse of the *Pauliśa* that was cited earlier, this is the normal Indian astronomical cosmology based on an adaptation of the Purāṇic flat earth to a Greek spherical universe, and the usual explanation of the diurnal rotation of the heavens as caused by chords of wind wrapped about the axis of that universe. For Balabhadra, as we know from Pṛthūdaka and Utpala,[65] these verses are the occasion for a lengthy discussion of the nature of diurnal rotation, buttressed by citations from Varāhamihira, Āryabhaṭa, Pauliśa, the *Vāsiṣṭha*, and Lāṭadeva. Much of this discussion is reproduced by al-Bīrūnī in the *India*,[66] though largely attributed to Brahmagupta himself, interspersed with excerpts from other sections of Balabhadra's commentary, and frequently completely misunderstood.

> Brahmagupta says in the *Brahmasiddhānta*: "Some people maintain that the first motion [from east to west] does not lie in the meridian [read celestial equator], but belongs to the earth." But Varāhamihira refutes them by saying: "If that were the case, would not a bird return to its nest as soon as it had flown away from it towards the west?"

So far, al-Bīrūnī is free from error save that the reference to Āryabhaṭa's theory of the diurnal rotation of the earth and Varāhamihira's refutation of it was made by Balabhadra, not by Brahmagupta. Al-Bīrūnī continues: "Brahmagupta says in another place of the same book: 'The followers of Āryabhaṭa maintain that the earth is moving and heaven resting. People have tried to refute them by saying that, if such were the case, stones and trees would fall away from the earth.' " This again is from Balabhadra, not from Brahmagupta, and it is from the same, not from another place. The idea is that the violence of a diurnal rotation of the earth would hurl objects on the earth's surface into space. But al-Bīrūnī proceeds with a curious Aristotelian misinterpretation compounded by a failure to understand Balabhadra's argument:

> But Brahmagupta does not agree with them, and says that that [*i.e.*, that stones and trees would fall away from the earth] would not necessarily follow from their theory, apparently because he thought that all heavy things are attracted towards the center of the earth.

In other words, al-Bīrūnī thinks that Brahmagupta (whom he mistakenly speaks of instead of Balabhadra) argued against the contention that things would fly off the earth if it were moving on the grounds of the Aristotelian concept of the natural motion of earth and water, the "heavy" elements, toward the center of the earth. But, as we learn from reading Pṛthūdaka and Utpala, Balabhadra, though he does mention the fact that clods of dirt thrown up by students return to the same point on the earth's surface as an argument in favor of the earth's stability, did not at all contest the argument put forward by the opponents of Āryabhaṭa against the diurnal rotation of the earth; rather, he added another argument to it in the form of a quotation from Varāhamihira's *Pañcasiddhāntikā*:[67] "Another thing: if there were (a rotation) of the earth (every) day, bees, geese, flags, and so on would always be driven to the west; if it were revolving slowly, how would it revolve (once a day)?

Varāhamihira, as did Indian scientists generally, regards the air surrounding the earth with the objects in it as a stable system which would be turned into an eastward flowing stream by the violence of the earth's rotation. In order to counter the argument that the movement may be nonviolent–*i.e.*, slow–he asks how the earth then could rotate daily. Al-Bīrūnī omits the principle and repeats a garbled version of the secondary argument: "He [that is, Brahmagupta] says: 'On the contrary, if that [diurnal rotation of the earth] were the case, the earth would not vie in keeping an even and uniform pace with the minutes of heaven, the prāṇas of the times.' "

In this paraphrase the words "the minutes of heaven, the prāṇas of the times" are a tautologous explanation of the Sanskrit word "kālāsus," "time-minutes," of which there are by definition 21,600 in a day. I assume this represents a part of the mufassir's attempt to explain the text to al-Bīrūnī; but he remains puzzled, and remarks: "There seems to be some confusion in this chapter, perhaps by the fault of the translator."

Another example of confusion. In his commentary on *Brāhmasphuṭasiddhānta* 21, 3, Balabhadra quotes some verses from Āryabhaṭa, the *Vāsiṣṭha*, and Lāṭadeva, stating that the universe is a sphere consisting of the five mahābhūtas: earth, water, fire, wind, and space.[68] We have already considered a similar verse from the *Pauliśa*, and it repeats a very common Indian concept. But in a passage of the *India*[69] dependent on this section of Balabhadra's commentary, al-Bīrūnī disastrously misinterprets these verses:

> Āryabhaṭa inquires into the nature of the world, and says that it consists of earth, water, fire, and wind, and that each of these elements is spherical. Likewise, Vasiṣṭha and Lāṭa say that the five elements, that is earth, water, fire, wind, and heaven, are spherical.

By transferring the attribute of sphericity from the universe itself to the five mahābhūtas, al-Bīrūnī forcibly converts the Indians to a form of Aristotelianism, to a belief of which they were completely unaware.

Elsewhere in the *India* al-Bīrūnī attempts to convert Brahmagupta himself to the same belief. The latter writes in *Brāhmasphuṭasiddhānta* 21, 2: "The sphere of the earth, which is formed by the good and bad deeds of creatures, is surrounded by the spheres of the Moon, Mercury, Venus, the Sun, Mars, Jupiter, and Saturn, and ending with the sphere of the constellations."

Al-Bīrūnī's paṇḍita misunderstood Brahmagupta's reference to the karma, good and bad, of creatures, which forms the earth, and we find in the *India* a partially Moslem, partially Peripatetic interpretation:[70]

> Brahmagupta says in the first chapter of the *Brahmasiddhānta* where he enumerates the heavens, placing the Moon in the nearest heaven, the other planets in the following ones, and Saturn in the seventh: "The fixed stars are in the eighth heaven, and this has been created spherical in order to last forever, that in it the pious may be rewarded, the wicked be punished, since there is nothing behind it." He indicates in this chapter that the heavens are identical with the spheres, and he gives them in an order which differs from that of the traditional literature of their creed, as we shall show hereafter in the proper place. He indicates, too, that the spherical can be slowly influenced from without. He evinces his knowledge of the Aristotelian notions regarding the spherical form and circular motion, and that there is no body in existence behind the spheres.

Neither the concept of the eternity of the spheres, nor the idea of the teleological cause of the world's creation by a retributive God, nor Aristotle's theory of the nature of circular motion is in fact to be found in the Sanskrit text.

There are numerous other passages in which the translator failed to understand the Sanskrit, or in which al-Bīrūnī gratuitously attributes Peripatetic notions or Greek geometrical proofs to Brahmagupta; an example of the mistaken ascription of a geometrical proof is to be found in the several pages of the *India*[71] devoted to a rule for computing the apparent diameter of the earth's shadow at the Moon's distance, in which al-Bīrūnī even claims to find a lacuna in the manuscript of the *Brāhmasphuṭasiddhānta*, whereas such a proof has no place in an Indian astronomical text, either mūla or bhāṣya.

I hope the evidence so far given is sufficient to convince the reader that, at least in the field of astronomy, al-Bīrūnī was not an accurate reporter of the contents of the Sanskrit texts which he claims to describe. I also hope that two of the reasons for al-Bīrūnī's errors and distortions have been established: his reliance on incompetent translators and explicators necessitated by his own inability to read the highly technical Sanskrit employed in jyotiḥśāstra,[72] and his uncritical assumption that Indian scientists adhered to the doctrines of Aristotelian physics.

But, in closing I must state that the magnitude of his achievement is little lessened by these flaws. From the point of view of his Moslem contemporaries, his knowledge of Indian astronomy was unique. And from our point of view, his accounts of the lost siddhāntas of Vijayananda and of Pauliśa, and of the *Bhāṣya* of Balabhadra, though now we know with what caution we must use them, are invaluable contributions to our knowledge of the Indian astronomical tradition.

NOTES

1. On al-Bīrūnī, see E. S. Kennedy, *Dictionary of Scientific Biography* (hereafter *DSB*), (New York, 1970), vol. 2, pp. 147–158.

2. In general, see D. Pingree, "Indian Influence on Early Sassanian and Arabic Astronomy," *JORMadras*, (1963–1964), 33: 1–8.

3. D. Pingree, "The *Paitāmahasiddhānta* of the *Viṣṇudharmottarapurāṇa*," *Brahmavidyā*, (1967–1968), 31–32; 472–510.

4. D. Pingree, "The Persian 'Observation' of the Solar Apogee in *ca*. A.D. 450," *JNES* (1965), 24: 334–336.

5. S. Dvivedin, ed. (Benares, 1902). On Brahmagupta, see D. Pingree in *DSB*, vol. 2, pp. 416–418.

6. D. Pingree, "The Fragments of the Works of al-Fazārī," *JNES* (1970), 29: 103–123.

7. D. Pingree, "The Fragments of the Works of Ya^{c}qūb ibn Ṭāriq," *JNES* (1968), 27: 97–125; E. S. Kennedy, "The Lunar Visibility Theory of Ya^{c}qūb ibn Tāriq," Ibid., pp. 126–132.

8. D. Pingree, "The Indian and Pseudo-Indian Passages in Greek and Latin Astronomical and Astrological Texts," to appear in *Viator*.

9. On Āryabhaṭa, see D. Pingree, *Census of the Exact Sciences in Sanskrit*, series A vol. 1, (Philadelphia 1970), pp. 50a–53b, and vol. 2, (Philadelphia, 1971), p. 15b.

10. F. Haddad, E. S. Kennedy, and D. Pingree, *The Kitāb ^{c}ilal al-zījāt of al-Hāshimī* (New York, 1975), section 7.

11. D. Pingree, "The Greek Influence on Early Islamic Mathematical Astronomy," *JAOS* (1973), 93: 32–43, esp. 36, n. 28.

12. B. Chatterjee, ed., 2 vols. (Calcutta, 1970).

13. In *On Transits*, edited as part 3 of *Rasā'il al-Bīrūnī* (Hyderabad: 1948); English trans. by M. Saffouri and A. Ifram, with a commentary by E. S. Kennedy (Beirut, 1959), 32:15–33:1; in *India*, (ed. Hyderabad, 1958; English trans. by E. C. Sachau [London, 1879], 2 vols. ed. pp. 120–121, trans. vol. 1, p. 156; ed. p. 266, trans. vol. 1, p. 312; ed. p. 346, trans. vol. 2, p. 7; ed. pp. 381–382, trans. vol. 2, pp. 46–47; ed. pp. 392–393, trans. vol. 2, p. 60; ed. p. 410, trans. vol. 2, p. 79; ed. pp. 413–417, trans. vol. 2, pp. 83–88; ed. pp. 419–420, trans. vol. 2, pp. 90–92; ed. p. 440, trans. vol. 2, p. 116; and ed. pp. 442–443, trans. vol. 2, pp. 119–120 (*cf.* also the commentary on the *Khaṇḍakhādyaka* mistakenly attributed to Balabhadra in ed. pp. 121 and 494, trans. vol. 1, p. 156, and vol. 2, p. 187); *Al-Qānūn al-Mas^{c}ūdī* (Hyderabad, 1954-1956), 3 vols., pp. 180 and 973; and *On Shadows* (ed. as part 2 of *Rasā'il al-Bīrūnī*), pp. 41, 133, and 150. See also Pingree, "Greek Influence," p. 37, n. 41.

14. C. E. Sachau, ed., *Chronology* (repr. Leipzig, 1923); English trans. by C. E. Sachau (London, 1879), ed. p. 25, trans. p. 29; *On Transits* 32:15–33:1; *India*, ed. p. 266, trans. vol. 1, p. 312; ed. pp. 268–269, trans. vol. 1, pp. 315–316; ed. p. 346, trans. vol. 2, p. 7; ed. pp. 383–384, trans. vol. 2, pp. 48–49; *Qānūn*, pp. 174 and 180; *On Shadows*, pp. 33, 35, 133, and 134; see also al-Hāshimī, section 4.

15. *On Transits* 26:4–19; and *India*, ed. p. 387, trans. vol. 2, pp. 52–53.

16. See al-Hāshimī, sections 3 and 25; Pingree, "Greek Influence," p. 37 n. 37.

17. E.g., the anonymous translation of Varāhamihira's *Bṛhatsaṃhitā*, described by C. A. Storey, *Persian Literature* (London, 1958), vol. 2, pt. 1, p. 38, and the translations of Bhāskara's *Līlāvatī* by Fayḍī and of his *Bījaganita* by Rushdī described by Storey, pp. 4–5.

18. Al-Fazārī fr. Z 9 (al-Bīrūnī confuses pala and phala); fr. Z 3 (= Ya^{c}qūb fr. Z 5); frs. Z 24 and S 3; Ya^{c}qūb frs. T 5, T 7, T 8, and T 11.

19. Al-Fazārī fr. Z 4 = Ya^{c}qūb fr. Z 6.

20. Al-Fazārī fr. Z 8 = Ya^{c}qūb fr. T 6.

21. D. Pingree, "Brahmagupta, Balabhadra, Pṛthūdaka, and al-Bīrūnī," to appear in the *Proceedings of the World Sanskrit Congress*, n. 18.

22. S. S. H. Rizvi, "A Unique and Unknown Book of al-Bīrūnī: *Ghurrat-*

uz-Zijat or *Karana Tilaka*," *Islamic Culture* (1963), 37: 112–130, 167–187, and 223–245; (1964), 38: 47–74 and 195–212; and (1965), 39: 1–26 and 137–180, esp. (1963), 37: 114.

23. D. Pingree, "The Later *Pauliśasiddhānta*," *Centaurus* (1969), 14: 172–241; further fragments may be found in al-Bīrūnī, *On Shadows*, pp. 41, 114, 115, 139, and 160.

24. P. C. Sengupta, ed. (Calcutta, 1941).

25. Ed. in the edition of the *Khaṇḍakhādyaka* by B. Chatterjee cited in n. 12, above; see also, D. Pingree, "The Beginning of Utpala's Commentary on the *Khaṇḍakhādyaka*," *JAOS* (1973), 93, 469–481.

26. B. Misra, ed. (Calcutta, 1925); see also, B. Chatterjee, ed. *Khaṇḍakhādyaka*, vol. 1, pp. 170–174. On Āmarāja, see Pingree, *Census*, series A, vol. 1, pp. 50a–50b, and vol. 2, pp. 15a–15b.

27. Pingree, "Brahmagupta, Balabhadra," n. 4.

28. See the references in n. 23.

29. Pauliśa fr. P 1.

30. O. Neugebauer and D. Pingree, eds. (København, 1970–1971), 2 vols.

31. For the Sanskrit fragments, see Pingree, "Brahmagupta, Balabhadra," ns. 13 and 14.

32. S. Dvivedin, ed. (Benares, 1895), 2 vols., (repr. Vārāṇasī: 1968).

33. *Bṛhatsaṃhitā* 2, vol. 1, p. 22.

34. See *Pañcasiddhāntikā*, vol. 1, pp. 10–12.

35. Ibid., p. 12.

36. *E.g.*, in W. and H. G. Gundel, *Astrologumena* (Wiesbaden, 1966), p. 239.

37. Pauliśa fr. P 41.

38. D. Pingree, *The Thousands of Abū Maᶜshar* (London, 1968).

39. Pauliśa fr. P 29.

40. Pauliśa frs. P 32, P 34, P 35, and P 37.

41. Pauliśa fr. P 36.

42. Pauliśa fr. P 34.

43. *On Shadows*, p. 114.

44. Pauliśa fr. P 38.

45. Pauliśa fr. P 53.

46. Pauliśa fr. P 59.

47. *India*, ed. p. 183, trans. vol. 1, p. 225.

48. Pauliśa fr. P 54.

49. Pauliśa frs. P 51 and P 52.

50. Pauliśa fr. P 55.

51. I am preparing an edition of this on the basis of the three known manuscripts: BORI 339 of 1879/1880; IO Sanskrit 2769; and VVRI 1781.

52. Pingree, "Brahmagupta, Balabhadra."

53. Pauliśa fr. P 25.

54. Pauliśa fr. P 8 and n. 21, above.

55. *E.g.*, G. R. Kaye, "The Two Āryabhaṭas," *BM* 3, 10, 1909/1910, 289–292, justly criticized by B. Datta, "Two Āryabhaṭas of al-Biruni," *BCMS* (1926), 17: 59–74.

56. *Chronology*, ed. p. 9, trans. p. 11; ed. p. 13, trans. p. 15 (*cf.* Yaᶜqūb fr. T 5); ed. p. 25, trans. p. 29 (cf. Pingree, *The Thousands*, p. 30 n. 2, p. 35, and p. 39); ed. p. 26, trans. p. 31; ed. p. 52, trans. p. 61; ed. p. 274, trans. p. 266; and in *Documenta Islamica Inedita*, J. Fück, ed. (Berlin: 1952), p. 73 (cf. Pingree, *The Thousands*, p. 38).

57. Al-Fazārī frs. Z 19 and S 3; *cf.* Yaᶜqūb fr. T. 1.

58. Al-Fazārī frs. Z 15, Z 16, and S 2.

59. R. R. Wright, ed. (London, 1934), section 204.

60. Pingree, "Brahmagupta, Balabhadra."

61. All the relevant passages will be discussed in my edition of the *Brāhmasphuṭasiddhānta*.

62. *On Shadows*, p. 141.

63. *India*, ed. p. 123, trans. vol. 1, p. 158.

64. Ibid.

65. Pingree, "Brahmagupta, Balabhadra."

66. *India*, ed. p. 231, trans. vol. 1, pp. 276–277; *cf.* Pingree, "Brahmagupta, Balabhadra."

67. *Pañcasiddhāntikā* 13, 7.

68. Pingree, "Brahmagupta, Balabhadra."

69. *India*, ed. p. 223, trans. vol. 1, p. 268; *cf.* Pingree, "Brahmagupta, Balabhadra."

70. *India*, ed. p. 182, trans. vol. 1, pp. 223–224; *cf.* Pingree, "Brahmagupta, Balabhadra."

71. *India*, ed. pp. 407–409, trans. vol. 2, pp. 75–77; *cf.* Pingree, "Brahmagupta, Balabhadra."

72. This inability makes very implausible his claim to be himself translating Euclid's *Elements* and Ptolemy's *Almagest* into Sanskrit, set forth in the *India*, ed. p. 106, trans. vol. 1, p. 137.

Byzantine, Medieval, and Renaissance Europe

GREGORY CHIONIADES AND PALAEOLOGAN ASTRONOMY

David Pingree

FOR almost seven centuries following the publication of the commentary on the Handy Tables of Theon by Stephanus of Alexandria[1] little interest was shown in mathematical astronomy in Byzantium. It is true that, in the ninth century, under the leadership of Leo the Mathematician,[2] the text of Ptolemy's Almagest was studied and copied,[3] and that scholars in the eleventh and twelfth centuries had learned something of Arabic science. But it seems improbable that many, save perhaps the astrologers, had the motivation or the training necessary for an attempt to understand more than the most elementary principles of the motions of the celestial spheres; and even the astrologers really needed nothing beyond an ability to manipulate tables.

This neglect continued into the thirteenth century, both at Nicaea and in Constantinople after it had been recovered from the Latins. But the beginnings of a revival of astronomical studies can be traced to the early decades of this century when a few scholars sought to sustain Greek learning under the patronage of John III Vatatzes (1222–1254) and Theodore II Lascaris (1254–1258).

Nicephorus Blemmydes,[4] who taught at the Imperial court from 1238 to 1248 and whose pupils included George Acropolites,[5] reawakened an interest in ancient Greek science which had been virtually dead since the time of Michael Psellos[6] in the eleventh century. His *Epitome physica*[7] is a completely unoriginal book, and its treatment of astronomy (chapters 25–30) is pitifully inadequate. He has very little that is sensible to say about planetary theory; but he does demonstrate that he has read Aristotle, Cleomedes, and Euclid

[1] See H. Usener, *De Stephano Alexandrino* (Bonn, 1880), pp. 33–54, reprinted in his *Kleine Schriften*, 3 (Leipzig-Berlin, 1914), pp. 289–319, and l'Abbé Halma, Πτολεμαίου καί Θέωνος Πρόχειροι Κανόνες, 3 (Paris, 1825), pp. 101–112; see also G. Sarton, *Introduction to the History of Science*, 1 (Baltimore, 1927), pp. 472–473.

[2] See J. L. Heiberg, "Der byzantinische Mathematiker Leon," *Bibliotheca Mathematica*, NF 2 (1887), pp. 33–36; Sarton, 1, pp. 554–555; and E. E. Lipšic, "Vizantijskij učenyj Lev Matematik," *Vizantijskij Vremennik*, *N.S.*, 2 (1949), pp. 106–149.

[3] Heiberg in his edition of the Almagest (Leipzig, 1898–1903) lists three ninth-century manuscripts (Par. gr. 2389, Vat. gr. 1291, and Vat. gr. 1594), one of the tenth (Marc. gr. 313), one of the twelfth (Vat. gr. 180), two of the thirteenth (Par. gr. 2390 and Vat. gr. 184), and two of the thirteenth or fourteenth (Marc. gr. 311 and Vat. gr. 1038).

[4] See K. Krumbacher, *Geschichte der byzantinischen Literatur*, 2nd ed. (Munich, 1897), pp. 550–554; Sarton, 2 (1931), p. 971; and H.-G. Beck, *Kirche und theologische Literatur im byzantinischen Reich* (Munich, 1959), pp. 671–673. For an example of his astronomical wisdom, see J. B. Bury, "An Unpublished Poem of Nicephorus Blemmydes," BZ, 10 (1901), pp. 418–424 and P. N. P(apageorgiu), "Zu Nikephoros Blemmydes, B. Z. X 419 (Bury)," *ibid.*, p. 545.

[5] See Krumbacher pp. 286–288; A. Heisenberg, *Georgii Acropolitae opera*, 2 (Leipzig, 1903), pp. III–XIII; Sarton, 2, pp. 1113–1114; and Beck, pp. 674–675.

[6] See C. Zervos, *Un philosophe néoplatonicien du XI*e *siècle, Michel Psellos* (Paris, 1920). On Byzantine education in general, see F. Fuchs, *Die höheren Schulen von Konstantinopel im Mittelalter*, *Byzantinisches Archiv*, 8 (Leipzig-Berlin, 1926); for its condition under Psellos, see L. Bréhier, "L'enseignement supérieur à Constantinople dans la dernière moitié du XIe siècle," *Revue internationale de l'enseignement*, 38 (1899), pp. 97–112. Aristotle, of course, continued to be taught until the fall of the capital in 1204.

[7] Edited in PG, 142, cols. 1005–1302. What is printed as the last chapter (cols. 1303–1320) is, in fact, Nicephorus' commentary on the eighth Psalm; it is different from what is given in col. 1357 ff., which seems to be by Euthymius Zigabenus.

with some comprehension, and he observed at least one lunar eclipse, that of 18 May 1258.[8]

An account[9] of an observation of a solar eclipse by his pupil George Acropolites in the company of the Imperial court on 3 June 1239 reveals the intellectual atmosphere in which Nicephorus was working. The Empress Irene asked Acropolites, then only twenty-one years old, what had caused this phenomenon. He, though he had just begun his studies under Blemmydes, was able to reply correctly that the Moon was interposed between the Earth and the Sun. The court physician, Nicolaus, scoffed at this ridiculous response, and the Empress, trusting her doctor, called Acropolites a fool. She quickly regretted her use of this derogatory term, not because she realized the correctness of Acropolites' explanation, but because she considered it improper to insult one engaged in philosophical studies. Two years later the Empress died; the philosopher seriously suggests that the eclipse was a portent of that unfortunate event, as was also the appearance of a bearded comet. It was Acropolites who, after the capture of Constantinople by Michael VIII Palaeologus in 1261, restored mathematics to the capital; he taught Euclid and Nicomachus to George (later Gregory) of Cyprus and others.[10]

Among his pupils was, apparently, George Pachymeres,[11] a man who progressed much further in astronomical studies than had his teacher. Pachymeres' knowledge of this subject is, naturally, set forth in the fourth book of his Quadrivium.[12] To a large extent this consists of elaborate instructions for the multiplication of sexagesimal numbers, a procedure he regarded as incredibly difficult, a discussion of the risings, settings, and culminations of various constellations, and a number of the fundamental doctrines of astrology, many of which are also found in the *Epitome physica* of his mentor's mentor. He is capable of such improbable statements as: "They say that a yearly revolution of the Sun takes place in 365 degrees (μοίραις for ἡμέραις), 14 minutes, and 48 seconds"; but his planetary theory is far more complete than that of his predecessor, and he himself is far from being confused about everything.

George of Cyprus' friend John Pediasimus[13] continued Blemmydes' study of Cleomedes' Κυκλικὴ θεωρία μετεώρων, on which he wrote a commentary; and

[8] 27, 15 (col. 1265).

[9] George Acropolites, Χρονικὴ συγγραφή, 39 (pp. 67–68, Bonn; 1, pp. 62–64, Heisenberg).

[10] See George of Cyprus' autobiography in W. Lameere, *La tradition manuscrite de la correspondance de Grégoire de Chypre, Études de philologie, d'archéologie et d'histoire anciennes publiées par l'Institut Historique Belge de Rome*, 2 (Brussels-Rome, 1937), p. 185; on George of Cyprus, see Krumbacher pp. 476–478 and Beck pp. 685–686.

[11] See Krumbacher, pp. 288–291; Sarton, 2, pp. 972–973; and Beck, p. 679.

[12] Edited by P. Tannery and E. Stéphanou, Studi e Testi, 94 (Vatican City, 1940). The quotation is from 4, 7 (p. 364).

[13] See Krumbacher, pp. 556–558; Sarton, 3 (1947), pp. 682–683; V. Laurent in *Échos d'orient*, 31 (1932), pp. 327–331; and Beck, pp. 710–711. For his harmonic interpretation of seven- and nine-month births, see V. de Falco, "L'aritmologia pitagorica nei Commenti ad Esiodo," *Rivista indo-greco-italica*, 7, 3/4 (1923), pp. 25–54 and *In Ioannis Pediasimi libellum de partu septemmestri ac novemmestri nondum editum* (Naples, 1923); F. Cumont, "L'opuscule περὶ ἑπταμήνων καὶ ἐννεαμήνων," *Revue belge de philologie*, 2 (1923), pp. 5–21; and J. L. Heiberg in BZ, 25 (1925), pp. 145–146; cf. also Psellos' tract on the same subject edited by S. Weinstock in *Catalogus codicum astrologorum Graecorum*, 9, 1 (Brussels, 1951), pp. 101–103.

other mathematicians of this period were Maximus Planudes,[14] who composed one of the first treatises on Indian numerals in Byzantium[15] and an exegesis of the first two books of Diophantus,[16] and his pupil Manuel Moschopulus, who wrote the first Western treatise on the construction of magic squares.[17] But a scholar more directly interested in the study of mathematical astronomy was Planudes' friend Manuel Bryennius,[18] whose only surviving work is on harmonics, but who is praised by Theodore Metochites[19] as his initiator into the secrets of the heavens in 1314.

Metochites, besides being the most powerful courtier in the empire of Andronicus II, was one of the most intelligent men in Byzantium. With some assistance from Bryennius, but mainly through his own efforts, he succeeded in mastering the Μεγάλη σύνταξις, a feat of which he was justifiably proud and which astounded his contemporaries.[20] His success had two immediate results: the publication of several treatises on Ptolemaic astronomy, of which the most impressive is the Στοιχείωσις and commentary on the Almagest, and the instruction of his brilliant pupil Nicephorus Gregoras.[21] Thus he raised the level of sophistication in Byzantine astronomy to a height it had not attained for centuries.

Gregoras continued Metochites' method of simply trying to understand and explicate the classical texts in his works on eclipses and the astrolabe[22]—subjects in the exposition of which, as in almost everything, he was opposed by

[14] See Krumbacher, pp. 543–546; Sarton, 2, pp. 973–974; C. Wendel, "Planudea," *BZ*, 40 (1940), pp. 406–445; and Beck, pp. 686–687.

[15] It would seem that Indian numerals were known in Byzantium by the twelfth century (see P. Tannery, "Les chiffres arabes dans les manuscrits grecs," *Revue archéologique*, 3rd Ser., 7 [1886], pp. 355–360 reprinted in his *Mémoires scientifiques*, 4 [Toulouse–Paris, 1920], pp. 199–205). But, as there is no evidence that the scholium of the monk Neophytus (P. Tannery, "Le scholie du moine Néophytos sur les chiffres hindous," *Revue archéologique*, 3rd Ser., 5 [1885], pp. 99–102 reprinted in *Mém. Sc.*, 4, pp. 20–26) was written before Planudes' Ψηφοφορία κατ' Ἴνδους (edited by C. I. Gerhardt, *Das Rechenbuch des Maximus Planudes* [Halle, 1865]), the latter must be accepted as the earliest text on the subject to survive in Greek.

[16] Edited by P. Tannery in *Diophanti Alexandrini opera omnia*, 2 (Leipzig, 1895), pp. 125–255.

[17] See Krumbacher, pp. 546–548; P. Tannery, "Le traité de Manuel Moschopoulos sur les carrés magiques," *Annuaire de l'Association pour l'Encouragement des Études Grecques en France*, 20 (1886), pp. 88–118 reprinted in *Mém. sc.*, 4, pp. 27–60; and Sarton, 3, pp. 679–681. The most recent treatment of the history of magic squares is that of S. Cammann, "The Evolution of Magic Squares in China," *Journal of the American Oriental Society*, 80 (1960), pp. 116–124.

[18] See Krumbacher, p. 599 and Sarton, 3, pp. 745–746.

[19] See K. N. Sathas, Μεσαιωνικὴ Βιβλιοθήκη, 1 (Venice, 1872), pp. ι'–ρλη'; Krumbacher, pp. 550–554; R. Guilland, "Les poésies inédites de Théodore Métochite," *Byzantion*, 3 (1926), pp. 265–302 reprinted in his *Études byzantines* (Paris, 1959), pp. 177–205; Sarton, 3, pp. 684–688; I. Ševčenko, "Observations sur les recueils des *discours* et des *poèmes* de Th. Métochite et sur la bibliothèque de Chora à Constantinople," *Scriptorium*, 5 (1951), pp. 279–288; H. Hunger, "Theodoros Metochites als Vorläufer des Humanismus in Byzanz," BZ, 45 (1952), pp. 4–19; H.-G. Beck, *Theodoros Metochites* (Munich, 1952); R. J. Loenertz, "Théodore Métochite et son père," *Archivum Fratrum Praedicatorum*, 23 (1953), pp. 184–194; J. Verpeaux, "Le cursus honorum de Théodore Métochite," *Rev. ét. byz.*, 18, (1960), pp. 195–198; and I. Ševčenko, *Études sur la polémique entre Théodore Métochite et Nicéphore Choumnos*, Corpus Bruxellense Historiae Byzantinae, Subsidia, 3 (Brussells, 1962).

[20] Sathas p. κ', note 3.

[21] Guilland, *Ét. byz.*, pp. 180–186. On Gregoras, see R. Guilland, *Essai sur Nicéphore Grégoras. L'homme et l'oeuvre* (Paris, 1926), and *Correspondance de Nicéphore Grégoras* (Paris, 1927); Sarton, 3, pp. 949–953; and Beck, pp. 719–721.

[22] The two treatises on the astrolabe are edited by A. Delatte, *Anecdota Atheniensia et alia*, 2 (Paris, 1939), pp. 195–235.

Barlaam of Calabria.[23] Only in regard to the calendar did Gregoras claim to have achieved a better result than had his ancient predecessors. Ptolemy, following Hipparchus, had found that the length of a tropical year was 6,5; 14,48 days; Gregoras claims[24] that he himself, by observation, had discovered that it is not quite so long. Isaac Argyrus,[25] in recording this discovery, states that the correction required is obtained by subtracting $\frac{1}{200}$ of a day from the Ptolemaic figure, or about $\frac{1}{100}$ of a day from $365\frac{1}{4}$ days. This amounts to adding $0;0,18^0$ per year to the Ptolemaic value of precession, $0;0,36^0$, a procedure which yields the result $0;0,54^0$ per year. As Isaac remarks, the "Persians" had already arrived at this conclusion. It is the well-known parameter of 1^0 of precession every 66 years which is ascribed to οἱ νεώτεροι by Simeon Seth in the eleventh century.[26] The same parameter is found in a scholium to Pappus' Prolegomena to the Almagest discovered by Tannery;[27] as this scholium refers to ephemerides which began with 1 March 1032 and attributes the parameter in question to οἱ νεώτεροι, its author may well be Simeon. The scholium correctly refers to the observation of the autumnal equinox by Yaḥyâ ibn Abî Manṣûr under the Caliph al-Ma'mûn on 19 September 830, but does not realize that the value of precession arrived at can actually be traced back to Lâṭadeva's version of the Sûryasiddhânta (A.D. 505) and the lost work of Maṇittha (Μανέθων?).[28]

Simeon Seth also refers to a precession of $0;0,54^0$ per year in an unpublished text in a fourteenth-century manuscript, Vat. gr. 1056, which is a copy of a twelfth-century codex.[29] This text occurs among a collection of five star-

[23] See Krumbacher, pp. 102 and 625; Sarton, 3, p. 583; and Beck, pp. 717–719.

[24] Ῥωμαϊκὴ ἱστορία, 8, 13 (1, pp. 364–373, Bonn).

[25] His work on the computus, written in 1373, is edited by D. Petavius, *Uranologion* (Paris, 1630), pp. 359–383; the passage referred to is on page 381 (cf. also p. 382). It is reprinted in PG, 19, cols. 1279–1316. On Isaac, see G. Mercati, *Notizie di Procoro e Demetrio Cidone, Manuele Caleca e Teodoro Meliteniota, ed altri appunti per la storia della teologia e della letteratura bizantina del secolo XIV*, Studi e Testi, 56 (Vatican City, 1931), pp. 233–236 and *passim*; Sarton, 3, pp. 1511–1512; and Beck, pp. 729–730.

[26] In Περὶ χρείας τῶν οὐρανίων σωμάτων, 70, in Delatte, p. 124. For Simeon, see Krumbacher, pp. 615 and 896 and Sarton, 1, p. 771; for a peculiarity in his astronomical terminology, see M. V. Anastos, "Ὑπόγειος, a Byzantine term for perigee, and some Byzantine views of the date of perigee and apogee," *Orientalia Christiana Periodica* 13 (*Miscellania Guillaume de Jerphanion*) (1947), pp. 385–403.

[27] P. Tannery, "Les éphémérides chez les byzantins," *Bulletin des sciences mathématiques*, 2nd Ser., 30 (1906), pp. 59–63 reprinted in *Mém. sc.*, 4, pp. 289–293. The whole scholium has now been edited by J. Mogenet "Une scolie inédite du Vat. gr. 1594 sur les rapports entre l'astronomie arabe et Byzance," *Osiris*, 14 (1962), pp. 198–221; the passage referred to is on p. 209.

[28] The account by T.-H. Martin in his *Mémoire sur cette question*: *La précession des équinoxes a-t-elle été connue des égyptiens ou de quelque autre peuple avant Hipparque?* (Paris, 1869), pp. 179–188, is based entirely on H. T. Colebrooke, "On the Notion of the Hindu Astronomers concerning the Precession of the Equinoxes and Motions of the Planets," *Asiatic Researches*, 12 (1818), pp. 211–252, reprinted in his *Miscellaneous Essays*, 2 (London, 1837), pp. 374–416. The account given by P. Duhem in his *Le système du monde*, 2 (Paris, 1914), pp. 212–214 and 223–226, is a rather confused version of Martin. I have found in a manuscript (3166 of the Oriental Research Institute, Mysore) of an early ninth-century commentary by Govindasvâmin on the Uttarakhaṇḍa of the Bṛhatpârâśarahorâśâstra a passage which considerably clarifies the Indian tradition and raises the possibility that the Indian theories of precession and trepidation are derived from a Greek source.

[29] It gives horoscopes for the coronations of Alexius I Comnenus in 1081 and of Manuel I Comnenus in 1143. On folio 6^v is a horoscopic diagram beneath which is written: Μηνὶ Μαρτίῳ λα′ ἡμέρᾳ δ′ ἰνδικτιῶνος ϛ′ ἔτους ͵ϛχνα′ γέγονεν ἡ ἀναγόρευσις τῆς βασιλείας τοῦ κραταιοῦ καὶ εὐσεβοῦς ἡμῶν βασιλέως τοῦ Πορφυρογεννήτου κυροῦ Μανουῆλ τοῦ Κομνηνοῦ εἰς Μάμισταν. ἡ δὲ τελευτὴ τοῦ τρισμακαρίστου ἀοιδίμου βασιλέως καὶ πατρὸς αὐτοῦ κατὰ τὸν η′ τοῦ Ἀπριλλίου ἡμέρᾳ ε′ τῆς αὐτῆς ἰνδικτιῶνος. At the bottom of the page, in a different hand, is written: Μηνὶ Σεπτεμβρίῳ κδ′ ἡμέρᾳ δ′ ἰνδικτιῶνος δ′ ἔτους ͵ϛχπη′ γέγονεν ἡ τελευτὴ τοῦ

catalogues (fols. 30ᵛ–33) derived from the Zîj al-Ḥâkimî,[30] Kûshyâr bin Labbân,[31] the Egyptians,[32] and Abû Maʿshar,[33] as quoted in the Kitâb al-Mughnî of Ibn Hibintâ.[34] The catalogues are dated respectively in 1156, 1161, probably 1142, 1161, and 1148; the longitudes of the stars are fairly consistently computed by allowing a precession of 1^0 for every 66 years from the time of Ptolemy. Thus it is clear that Gregoras' observations, if he really made any, merely confirmed a parameter which had been known in Byzantium for at least two and a half centuries.

The final figures in this conservative tradition of astronomy are Nicolaus

---

τρισμακαρίστου καὶ ἀοιδίμου βασιλέως ἡμῶν κυροῦ Μανουῆλ τοῦ Κομνηνοῦ τῷ τῶν μοναχῶν μετασχηματισθέντος σχήματι καὶ [μετὰ] μετονομασθέντος. ἐν τῷ μέσῳ δὲ τοῦ λζ′ ἔτους τῆς βασιλείας αὐτοῦ γέγονεν ἡ τούτου τελευτή. I list below the longitudes of the planets as given in the horoscope and those computed for 31 March 1143 at *ca.* 10 A.M.:

	Horoscope	*Computation*
Saturn	Aries 6;0	Aries 3
Jupiter	Aquarius 9;15	Aquarius 8
Mars	Leo 9;36	Leo 5
Sun	Aries 16;52	Aries 16
Venus	Pisces 22;28	Pisces 21
Mercury	Aries 16;8	Aries 4
Moon	Libra 1;4	Libra 0
Ascendent	Gemini 25	c. 10 A.M.

On folio 7, beneath another horoscopic diagram, is the text: Μηνὶ Ἀπριλλίῳ α′ ἡμέρᾳ ε′ ἰνδικτιῶνος δ′ ἔτους ͵ϛφπθ′ εἰσῆλθεν ὁ βασιλεὺς κῦρος Ἀλέξιος εἰς τὸν παλάτιον, καὶ ἀνηγορεύθη. I give below the positions according to the horoscope and according to my computations for *ca.* 10 A.M. of 1 April 1081:

	Horoscope	*Computation*
Saturn	Aquarius 26;40	Aquarius 28
Jupiter	Scorpio 10;50	Scorpio 11
Mars	Leo 19;43	Leo 21
Sun	Aries 18;22	Aries 18
Venus	Aries 27;32	Aries 26
Mercury	Aries 22;34	Aries 24
Moon	Sagittarius 28;53	Sagittarius 18
Ascendent	Gemini 10	c. 10 A.M.

Below this is yet another horoscope with the inscription: τὸ τοιοῦτον θεμάτιόν ἐστι τῆς τοῦ βασιλέως Ἀλεξάνδρου ἐξελεύσεως. I have not been able to date this successfully. The longitudes given in the diagram are as follows:

	Horoscope
Saturn	Virgo 22;16
Jupiter	Leo 12;23
Mars	Capricorn 19;55
Sun	Scorpio 10;30
Venus	Scorpio 9;20
Mercury	Scorpio 13;40
Moon	not given
Ascendent	Scorpio 10;0

[30] The Zîj al-Kabîr al-Ḥâkimî was written by Ibn Yûnis in Cairo *ca.* 990; see E. S. Kennedy, *A Survey of Islamic Astronomical Tables*, Transactions of the American Philosophical Society, N.S., 46, 2 (Philadelphia, 1956), no. 14.

[31] Kûshyâr wrote *ca.* 1010 the Zîj al-Bâligh (Kennedy no. 7) and the Zîj al-Jâmiʿ (Kennedy no. 9 and pp. 156–157).

[32] Perhaps Ibn Yûnis is being referred to again.

[33] Abû Maʿshar (787–886) was the most important astrologer of his time. It was mainly through him that Islam, and thereby Byzantium, learned something of Indian and Sasanian astrology; see my papers "Historical Horoscopes," *JAOS*, 82 (1962), pp. 487–502 and "Astronomy and Astrology in India and Iran," *Isis*, 54 (1963), pp. 229–246.

[34] Ibn Hibintâ's Kitâb al-Mughnî was written in 941; see C. Brockelmann, *Geschichte der arabischen Literatur*, 1 (Leiden, 1943), p. 252.

Cabasilas,[35] the Hesychast mystic, who wrote a long-winded but intelligent commentary on the third book of the Μεγάλη σύνταξις,[36] and Theodore Meliteniotes,[37] perhaps a relative of Metochites, who explains the astrolabe and Ptolemaic astronomy in the first two books of his 'Αστρονομική τρίβιβλος, written before 1368. As far as has been determined at present this school of astronomers added nothing new to their classical heritage except for Gregoras' realization that the "new" parameter for the length of the tropical year necessitated a reform of the calendar. But their explication of texts was carried on at a fairly high level of comprehension.

The spirit of the late thirteenth and early fourteenth century in Byzantium within a certain group of intellectuals was one of immense pride in the great tradition of which the Empire was the heir and which it was her duty to safeguard from barbarization. Even Metochites, for all his zeal to achieve classicism (most evident in the hexameters of his verses), was criticized by Nicephorus Chumnos for the error of contradicting Plato in his astronomical theories.[38] Since he was raised in such an atmosphere, one might guess that one of Metochites' aims in writing the Στοιχείωσις was to demonstrate the superiority of Ptolemaic astronomy over its rivals. In the introduction to that work he has a hypothetical colleague exclaim,[39] "Be a Greek, and shun the theories of the Indians, the Scythians, or the Persians, or any other foreign ideas!" This command certainly reflects the attitude of many Byzantines in the early decades of the fourteenth century toward those who, like Maximus Planudes, were familiar with foreign astronomical texts. The problem now is to identify these texts.

Professor Neugebauer has shown[40] that a treatise preserved on folios 232^v–285^v of Par. gr. 2425, a manuscript of the fifteenth century, is an exposition of methods of solving astronomical problems derived from an Islamic source, perhaps one of the zîjes of Ḥabash al-Ḥâsib.[41] Besides the common Arabic values for the obliquity of the ecliptic (23;35^0) and the maximum latitude of the Moon (4;46^0), it utilizes the value of R (150) which appears in Brahmagupta's Khaṇḍakhâdyaka[42] and many early Islamic zîjes influenced by that work. The Byzantine text, as Neugebauer proves, was written between 1072 and 1086.

But Metochites must have been referring to something more extensive than the text in the Paris manuscript, and to something more immediate. George

35 See Krumbacher, pp. 158–160; Sarton, 3, pp. 1438–1439; and Beck, pp. 780–783.

36 Edited by I. Camerarius, Κλ. Πτολεμαίου Μεγάλης Συντάξεως Βίβλ. ιγ (Basel, 1538), pt. 2, pp. 131–194.

37 See H. Usener, *Ad historiam astronomiae symbola* (Bonn, 1876), pp. 8–21 reprinted in *Kl. Schr.*, 3, pp. 330–349; Krumbacher pp. 623 and 625; Mercati pp. 172–191; Sarton, 3, pp. 1512–1514; and Beck, p. 792.

38 On the conflict between Chumnos and Metochites see, besides Ševčenko's *Études sur la polémique*, his "Le sens et la date du traité 'Anepigraphos' de Nicéphore Chumnos," *Bull. de l'Acad. roy. de Belgique, Classe des Lettres*, 5th Ser., 35 (1949), pp. 473–488, and J. Verpeaux, *Nicéphore Choumnos* (Paris, 1959), pp. 52–62 and 151–170.

39 Sathas, pp. πη'-πθ'.

40 In a paper which has not yet been published.

41 See Kennedy nos. 15, 16, and 39 and pp. 151–154.

42 Khaṇḍakhâdyaka 3, 8 in the edition of Babua Misra (Calcutta, 1925); 1, 30 in that of P. C. Sengupta (Calcutta, 1941).

Chrysococces[43] tells us what it was. In the introduction to his Ἐξήγησις εἰς τὴν σύνταξιν τῶν Περσῶν, written in or shortly after 1347, he says, in summary: "I studied Persian astronomy with a priest from Trebizond named Manuel. He reported that a certain Chioniades, who had been raised in Constantinople, fell in love with mathematics and other sciences. After he had mastered medicine,[44] he wished to study astronomy; he was informed that, in order to satisfy his desire, he would have to go to Persia. He traveled to Trebizond, where he was given some assistance by the Emperor Comnenus, and thence proceeded to Persia itself, where he persuaded yet another Emperor to aid him. He eventually learned all that he wished to know, and returned to Trebizond, bearing away from Persia a number of astronomical texts which he translated into Greek. The best of these texts had no commentary; the present ἐξήγησις fulfills the need for one."

Sixteen of the letters of Gregory Chioniades have been preserved in a manuscript in Vienna.[45] From these we learn that he did indeed travel to Persia—in fact to Tabrîz, the Mongol capital, where he was Orthodox Bishop[46]—and that he received some assistance from the Emperor of Trebizond, Alexius II Comnenus (1297–1330).

In several Greek manuscripts of the early fourteenth century translations of Arabic or Persian astronomical works are found. Vat. gr. 211 (V), which was written before 1308,[47] contains versions of the Zîj as-Sanjarî of ʿAbd ar-Raḥmân al-Khâzinî (*ca.* 1120), the Greek freedman of a judge in Marv,[48] the Zîj al-ʿAlâʾî of ʿAbd al-Karîm ash-Shirwânî al-Fahhad (*ca.* 1150),[49] a short

[43] See I. Bullialdus, *Astronomia philolaica* (Paris, 1645), *Tabulae philolaicae*, pp. 211–232; H. Usener, *Ad historiam*, pp. 23–37 reprinted in *Kl. Schr.*, 3, pp. 350–371; Lampros in Νέος Ἑλληνομνήμων, 15 (1921), pp. 332–336; U. Lampsides, "George Chrysococcis, le médecin, et son oeuvre," *BZ*, 38 (1938), pp. 312–322; and Sarton, 3, p. 688. No account of Chrysococces is reliable.

[44] In Ambr. 693 (Q 94 sup.), a manuscript of the fifteenth or sixteenth century, on folios 336–347, is a text entitled: Ἀντίδοται ἐκ Περσείας κομισθεῖσαι καὶ ἐξελληνισθεῖσαι παρὰ τοῦ φιλοσοφωτάτου καὶ ἰατρικωτάτου κυροῦ Γεωργίου τοῦ Χιονιάδου. George Choniates (Sarton, 3, p. 438), whose name is the *lectio facilior*, must be fictious; but is George Chioniades the same as Gregory?

[45] Edited by Ἰ. Β. Παπαδοπούλου, "Γρηγορίου Χιονιάδου τοῦ ἀστρονόμου Ἐπιστολαί," Ἐπιστημονικὴ Ἐπετηρὶς τῆς Φιλοσοφικῆς Σχολῆς τοῦ Πανεπιστημίου Θεσσαλονικῆς, 1 (1927), pp. 151–205; cf. the notes by Χ. Χαριτωνίδης, *ibid.*, pp. 260–280. On Chioniades, see also Μητροπολίτης Τραπεζοῦντος Χρυσάνθος in Ἀρχεῖον Πόντου, 4–5 (1933), pp. 332–340, and Sarton, 3, p. 438. N. A. Οἰκονομίδης in his excellent article, "'Σημείωμα περὶ τῶν ἐπιστολῶν Γρηγορίου τοῦ Χιονιάδου," Ἀρχεῖον Πόντου, 20 (1955), pp. 40–44, has demonstrated that Gregory's eighth letter, addressed to Lucites, was written in (or shortly after) September 1301, but his contention that Gregory was then in Trebizond is not convincing. It is, however, known from Joseph Lazaropoulos (A. Papadopoulos-Cerameus, *Fontes historiae imperii Trapezuntii* [Petropol, 1897] pp. 65–66; cited by both Metropolitan Chrysanthus and Oeconomides) that, "Γρηγόριος ἐν ἱερεῦσι καὶ μονοτρόποις σεβασμιώτατος... ὁ Χιονιάδης καλούμενος, προστάξει βασιλικῇ καὶ ἀξιώσει, ἅμα πρεπούσῃ δωροφορίᾳ" was in Trebizond on 24 June 1302, presumably on his way to the Mongol Court.

[46] See my article "Δάρας τὸ νῦν λεγόμενον Ταυρές," *Bull. de l'Acad. Roy. de Belgique, Classe des Lettres*, 5th Ser., 48 (1962), pp. 323–326.

[47] The owner has noted dates four times on blank pages or in margins. The first note is on folio 160ᵛ, where the year is mentioned as 1619 of the Romans, 707 of the Arabs, 677 of the Persians, and 6816 of Creation (between 30 December 1307 and 20 June 1308); on folio 174ᵛ is recorded 14 May of the year 6828 of Creation and 1631 of the Romans (14 May 1320); on folio 180ᵛ, in the margin, 13 March of the year 6830 of Creation or 18 Khardâd of the year 681 (read 691) of the Persians (13 March 1322); and on folio 234 1620 of the Romans, now 1633, 6817 of Creation, now 6830, 708 of the Arabs, now 722, and 678 of the Persians, now 691 (A.D. 1309, now 1322).

[48] Kennedy no. 27 and pp. 158–161.

[49] Kennedy no. 84.

zîj of Shams ad-Dîn al-Bukhârî, and some less important texts; the same translations, in somewhat different order, are found in a fifteenth-century codex, Vat. gr. 1058 (v).[50] Identical texts and, in addition, the long zîj of Shams ad-Dîn are preserved in Laur. 28, 17 (L), which was written in 1323; but the tables which appear in the two Vaticani are not present in the Laurentianus. These tables—those belonging to the Zîj as-Sanjarî as well as those belonging to the Zîj al-ʿAlâʾî—are the source of many of the tables upon which Chrysococces wrote his exegesis; this fact proves that the texts in V, v, and L are translations by Gregory Chioniades.

The Zîj as-Sanjarî has survived complete in at least two Arabic manuscripts (BM Or. 6669 and Vat. ar. 761), and extracts from it are found in two others. The Greek version, while preserving the order of the original (see the appendix in which I compare the two), represents a shortened redaction. The calendaric tables in Chioniades' translation are for Arab, Persian, Roman, and Sultanic (or Malikî) years, with a special table of Syrian months added. The tables of the mean motions of the planets are for hours, days, months, years, and thirty years according to the Arabic calendar, the epoch being 421 Hijra or A.D. 1030 at a geographical longitude of 90° E. The tables of equations and of latitudes are straightforward; the methods involved have, for the most part, been discussed by Neugebauer.

The tables of the Zîj al-ʿAlâʾî are far more interesting for several reasons, not the least of which is that the Arabic original is lost. The calendaric material explains Roman, Persian, Arab, Hebrew, and Sultanic years. Epochal longitudes of the planets are given both for the beginning of the Era of Yazdijird (16 June 632) at a geographical longitude of 84° E and for the beginning of the year 541 Yazdijird (1 February 1172). Though parameters are given for the mean motions of the planets for one day, a month of the Persians, a year of the Arabs, a year of the Persians, thirty years of the Arabs, thirty years of the Persians, and 36,0,0,0,0,0,0,0 or 100,776,960,000,000 days to six sexagesimal places, they do not seem to have been used in constructing the tables of mean motions. The last table of parameters—that for 36,0,0,0,0,0,0,0 days—belongs to the Zîj as-Sanjarî rather than the Zîj al-ʿAlâʾî, but some of the other parameters are known from other sources to be al-Fahhad's. Further investigation is required to clarify this situation.

The tables of the mean motions of the planets themselves are arranged substantially as are those in the Zîj as-Sanjarî, except that the elements of the Persian calendar are substituted for those of the Arabic. The tables of correction of the Sun include the epoch longitude of the Sun's apogee, the anomaly, and the equation for each degree of corrected anomaly, so that the true longitude of the Sun minus the motion of its apogee from epoch can be read off directly. The tables of corrections for the other planets are arranged in such a way that the equations are always positive; this is achieved by adding 360° or some other constant equal to or greater than the maximum equation to all entries.

[50] Analyzed by O. Neugebauer in his *Studies in Byzantine Astronomical Terminology*, Transactions of the American Philosophical Society, N.S., 50, 2 (Philadelphia, 1960).

Both the Zîj as-Sanjarî and the Zîj al-ʿAlâʾî contain numerous tables of trigonometrical functions such as are common in Islamic astronomical works.

The Zîj of Shams ad-Dîn al-Bukhârî preserved in L is not accompanied by tables, but a study of the examples which this text gives proves that it is a commentary on the tables of the Zîj al-ʿAlâʾî. Shams ad-Dîn, according to his incomplete horoscope,[51] was born 11 June 1254 in Bukhârâ; the many suggestions concerning his identity are without exception uninformed conjectures.[52] The examples in his zîj are dated between 12 February 1293 and 18 November 1296, though he also discusses the horoscope of one Fakhr ad-Dîn, who was born in Tabrîz on 25 August 1268 and who presumably consulted the astronomer-astrologer when he was in his late twenties. It appears that Shams ad-Dîn was working in Tabrîz before Ghâzân Khân established an observatory there in 1300,[53] but the references to Naṣîr ad-Dîn aṭ-Ṭûsî and the Zîj-i Îlkhânî[54] indicate that he had some contact with the observatory at Marâgha.[55] Chioniades calls him his teacher.

The dates of the examples given in the translations of the Zîj al-ʿAlâʾî and the Zîj as-Sanjarî indicate that Chioniades was in Tabrîz in 1295 and 1296. Soon afterwards he seems to have returned to Constantinople, where he stayed for a few years; the references to 12 March and 30 April 1302 and his seventh letter, one of several to Constantine Lucites, must belong to this period. But he returned to the capital of the Mongols as Bishop, possibly in connection with the attempt made by Andronicus II to form an alliance with Ghâzân Khân in the summer of 1302.[56] To this second voyage is to be ascribed his correspondance with Alexius II Comnenus of Trebizond. His translations seem to belong to the brief period when he was in Constantinople between about 1298 and 1302.

L does not contain any of the tables given in V and v; but it does have another set which was undoubtedly also a part of Chioniades' corpus of Islamic astronomy. This set is found again, at least in part, in Vat. gr. 191, wherein there are also examples for dates ranging from 14 April 1298 to 1 May 1302; apparently all these examples are worked for the geographical coordinates of Constantinople.[57] In one of them the Zîj al-ʿAlâʾî is quoted. These tables also appear in Vat. gr. 185, fols. 1–21.

The epoch of these tables is 1404 of the Romans or A.D. 1093. The tables of the mean motions of the planets are for hours, days, months, years, and twenty

[51] V, fol. 24, v fol. 323^{v}, and L, fol. 207.

[52] J. Gildemeister had suggested Shams ad-Dîn as-Samarqandî (Usener, *Ad historiam*, p. 15 and *Kl. Schr.*, 3, pp. 339–340); this identification is repeated provisionally by Sarton in 2, pp. 1020–1021, but he preferred Shams ad-Dîn Mîrak al-Bukhârî in 3, p. 699.

[53] See A. Sayili, *The Observatory in Islam, Publications of the Turkish Historical Society*, 7th Ser., No. 38 (Ankara, 1960), pp. 226–232.

[54] Kennedy no. 6 and pp. 161–162.

[55] See Sayili, pp. 188–223.

[56] Ghâzân Khân received an embassy from Andronicus II at the end of August 1302; see B. Spuler, *Die Mongolen in Iran, Iranische Forschungen*, 1 (Leipzig, 1939), p. 101 (cf. also pp. 107 and 253).

[57] The tables and texts of Vat. gr. 191 have been studied by Neugebauer in an unpublished paper which he was kind enough to let me read. The references, on folio 319^{v}, to earthquakes on 1 June and 17 July and a lunar eclipse on 18 May 1296 apparently are not to be connected with Chioniades.

years according to the Roman calendar. The equations are tabulated in such a way that the total correction is found immediately. The arguments and anomalies are arranged horizontally and vertically in steps of 12^0 so that the whole table is thirty columns by thirty. The corrections are normed so as to be always positive; the equations themselves appear to be close to those of Ptolemy.

The tables upon which George Chrysococces commented were, as he himself says, translated from the "Persian" by Chioniades. They are constructed for a place which is 72^0 E and 38^0 N; these are the geographical coordinates of Tabrîz according to Chioniades. In fact, as has been remarked, many of Chrysococces' tables are copies or derivatives of those in the Zîj as-Sanjarî and the Zîj al-ʿAlâ'î.

From the former are taken the table of Famous Cities,[58] and one of the star-catalogues,[59] but the latter has contributed much more, including all of the calendaric material.[60] With minor variations, such as the dropping of one sexagesimal place, Chrysococces' tables of the daily and hourly mean motions of the Moon, its node, and the five star-planets are taken directly from the Zîj al-ʿAlâ'î; but his parameters for the annual motions are slightly different from al-Fahhad's, a circumstance which involves some inconsistency, but not of a very serious sort. More damaging is the fact that the tables for the determination of the possibility of eclipses are taken from a work which used Arab years though all the rest of his tables are constructed according to the Persian calendar; this lack of coordination, however, is undoubtedly due to the innate advantage of a lunar calendar in compiling such tables. Other tables connected with eclipses, those for parallax, are taken directly from Ptolemy's Handy Tables. Finally, most of the tables for computing planetary latitudes are borrowed with slight modifications from the Zîj al-ʿAlâ'î.

Chrysococces gives two different sets of tables for the mean yearly motions of the planets. The first shows these motions for 1 to 10, 10 to 100, and 100 to 1000 years. These are the tables which were published by Bullialdus and are usually taken to be Chrysococces'; but there is some evidence to indicate that they are from the Zîj-i Īlkhânî of Naṣîr ad-Dîn aṭ-Ṭûsî.[61] The parameters for one year are exactly $\frac{1}{10}$th of those for 10; they in turn are $\frac{1}{10}$th of those for 100; and these $\frac{1}{10}$th of those for 1000. Therefore, though the lower numbers are given with an apparent accuracy of four sexagesimal places (five in the case of the Sun), the only significant parameters are those for 1000 years, which are carried out to only two sexagesimal places (three in the case of the Sun).

The second set of tables of the mean annual motions of the planets in Chrysococces' work gives the mean longitudes at the beginning of every

[58] Vind. phil. gr. 190 (W), fol. 155 = V, fol. 128ᵛ. This was edited by Bullialdus pp. 230–232.

[59] W, fol. 222 = V, fols. 153ᵛ–154. The second catalogue, in Vind. phil. gr. 87, fol. 33, is dated in the year 6854 of Creation or 715 Yazdijird (A.D. 1346).

[60] W, fol. 150 = V, fol. 161ᵛ; W, fols. 150ᵛ–151 = V, fols. 163–163ᵛ; and W, fols. 151ᵛ–152 = V, fols. 154ᵛ–155.

[61] Nasîr ad-Dîn (Kennedy, p. 161) gives tables arranged in the same way. The only parameter recorded by Kennedy, that for the yearly motion of the apogee, is identical with that in Chrysococces' tables.

Persian year from 710 to about 765, that is, from A.D. 1340 to A.D. 1395. The parameters used in constructing these tables, as has been remarked, are close to but not identical with those which appear in the Zîj al-ʿAlâ'î. The entries in the tables of corrections are consistently positive for all the planets, a constant equal to or greater than the maximum equations being added to all numbers in each case; the same procedure had been followed previously by al-Fahhad. Again, the equations themselves are virtually identical with the Ptolemaic values.

In the manuscripts of the Ἐξήγησις four tables which give the motion of the Moon in minutes and hours for various daily motions are superscribed τοῦ Χρυσοκόκκου.[62] These seem to be the only tables in the whole set which were actually constructed by that astronomer. They are remarkably simple; it was necessary only to divide the daily motions by 24 and the resultant hourly motions by 60. There is, then, nothing whatsoever to indicate that Chrysococces was in any way original.

Another Greek text which utilizes Islamic materials is the Παράδοσις εἰς τοὺς Περσικοὺς προχείρους κανόνας, which has been mistakenly ascribed to Isaac Argyrus. It was written shortly after 25 December 1352, the date for which planetary longitudes are computed in one of its examples. Mercati[63] has suggested that it is a first draft of the third book of Theodore Meliteniotes' Ἀστρονομικὴ τρίβιβλος; this, in fact, is what it seems to be. The two are identical even to the point of sharing an obvious error in converting a Persian into a Christian date. In his preface to this book Meliteniotes has derived his information concerning Islamic astronomers from the first chapter of Chioniades' translation of the zîj of Shams ad-Dîn,[64] but the tables of which he explains the use are Chrysococces' version of Chioniades'. There is nothing new here.

I have not discussed some other Palaeologan adaptions of foreign astronomical tables—for instance, the Greek version of the Alfonsine Tables made on Cyprus by John the Astrologer in 1340,[65] Demetrius Chrysoloras' Latin Tables, whose epoch is 1377,[66] Michael Chrysococces' commentary of 1435 on the Hexapterygon of Immanuel ben Jacob Bonfils,[67] originally written in Hebrew in 1365, or Marcus Eugenicus' exegesis of 1444 on the Latin translation of the tables of Jacob ben David ben Yom-tob,[68] who wrote in 1361; Isaac Argyrus' New Tables, whose epoch is 1 September 1367, are adaptations of Ptolemy's. There is also evidence, as is shown in an appendix, for the existence of Tables of Palaeologus, perhaps written in Nicaea in or shortly after 1436. But enough has been investigated to enable us now to be much clearer about

[62] See W, fols. 170v–172v and Vind. phil. gr. 87, fols. 7v–8v.

[63] Mercati, p. 175.

[64] This was not realized by L. H. Gray, "Zu den byzantinischen Angaben über den altiranischen Kalender," *BZ*, 11 (1902), pp. 468–472.

[65] In Vat. gr. 212, fols. 26–104v.

[66] See Krumbacher, p. 110; Beck, p. 751; and Vat. gr. 1059, fols. 482–489. For Demetrius Chrysoloras, Michael Chrysococces, and Marcus Eugenicus, see also G. Mercati, *Scritti d'Isidoro il cardinale Ruteno*, Studi e Testi, 46 (Rome, 1926), pp. 40–50.

[67] His commentary, which is preserved in a number of manuscripts, is being studied by Neugebauer.

[68] See Krumbacher, pp. 115–117 and 496–497, and Beck, pp. 755–758.

the situation of astronomy in Byzantium in the early fourteenth century than was hitherto possible. It is especially apparent that the role of Trebizond as an astronomical center has been vastly overrated.[69] It is true that men like Constantine Lucites[70] and Andreas Libadenus[71] showed an interest in this science and communicated with Chioniades and Gregoras about it; but it is not at all certain that there ever was an Academy of Astronomical Studies or an observatory in Trebizond. Certainly Chioniades, to whose activities as a translator nearly all of the Islamic influence on Palaeologan astronomy can now be traced, seems merely to have passed through Trebizond on his journeys between Istanbul and Tabrîz. It is significant that in all the texts associated with him one finds examples worked for Constantinople and Tabrîz; Trebizond is mentioned only in the preface to Chrysococces' Ἐξήγησις.

Appendix I

parameters from chioniades' works

It should be noted that those parameters which are labeled "approximate" have been sqeezed from the tables as they appear in the manuscripts. They cannot be accepted as definitive until they are confirmed by a thorough study of the textual tradition and an exhaustive investigation into the structure of the tables. But the deviations from the correct figures are probably not very great. For the inferior planets the motion of the anomaly is given.

Yearly mean motions according to the Zîj as-Sanjarî (one year equals 354 days) (approximate)

Saturn	11;52
Jupiter	29;26
Mars	3,5;31
Sun	5,48;55,13
Venus	3,8(sic for 18);15
Mercury	19;46
Moon	5,44;27
Apogee	0;0,52,55

Daily mean motion according to the Zîj as-Sanjarî (approximate)

Saturn	0;2
Jupiter	0;5

[69] Especially by I. B. Papadopoulos. His attempt to create a second Ananias of Shirak and to identify Tychicus with Constantine Lucites is particularly open to criticism. Ananias' autobiography and computus were translated by F. C. Conybeare, "Ananias of Shirak (A.D. 600–650 c.)," *BZ*, 6 (1897), pp. 572–584, and all his works, including those on astronomy, are edited by R. A. Abramian, *Anania Shirakatsi* (Erevan 1958). There seems to be no question that he was a contemporary of Stephanus of Alexandria; see also H. Thorossian, *Histoire de la littérature arménienne* (Paris, 1951), pp. 106–107.

[70] Guilland, *Correspondance*, p. 347, and Beck, pp. 793–794.

[71] See especially his predictions for the year 1336, edited by F. Boll, in *Catal. Cod. Astrol. Graec.*, 7 (Brussels, 1908), pp. 152–160; also Krumbacher, p. 422; N. Banescu, "Quelques morceaux inédits d'Andréas Libadénus," Βυζαντίς, 2 (1911–1912), pp. 358–395; and Beck, p. 794.

Mars	0;31,20
Sun	0;59,8,20
Venus	0;37
Mercury	3;6,20
Moon	13;10,30

Maximum equations according to the Zîj as-Sanjarî (approximate)

Saturn	6;31 and 6;13
Jupiter	5;15 and 11;3
Mars	11;25 and 41;9
Sun	2;12,23
Venus	2;23 and 45;59
Mercury	3;2 and 22;2
Moon	5;1

Yearly mean motion according to the Zîj al-'Alâ'î (one year equals 365 days) (approximate)

Saturn	12;13
Jupiter	30;20
Mars	3,11;16
Sun	5,59;44,51
Venus	3,45;2
Mercury	53;57
Moon	2,9;23

Maximum equations according to the Zîj al-'Alâ'î (approximate)

Saturn	6;31 and 6;25
Jupiter	5;15 and 11;19,30
Mars	11;25 and 38;57,30
Sun	1;58,56
Venus	1;59 and 46;37
Mercury	3;2 and 22;31
Moon	

Period relations from the Zîj as-Sanjarî

PLANETS	REVOLUTIONS	CONJUNCTIONS	DAYS	PERSIAN YEARS and DAYS	MEAN DAILY MOTION (not in text)
Saturn	$1+7;20^0$	29	10965	30y 15d	$0;2,0,35,47^0$
	$2+1;58^0$	57	21551	59y 16d	$0;2,0,36,4^0$
Jupiter	$1+4;48^0$	11	4388	12y 8d	$0;4,59,17,20^0$
	$7+0;8^0$	76	30315	83y 20d	$0;4,59,16,25^0$
Mars	$17+11;6^0$	15	11699	32y 19d	$0;31,26,39,12^0$
	$42+3;9^0$	37	28857	79y 22d	$0;31,26,39,35^0$

10*

PANETS	REVOLU-TIONS	CONJUNC-TIONS	DAYS	PERSIAN YEARS and DAYS	MEAN DAILY MOTION (not in text)
Sun	$1-0;15^0$	0	365	1y 0d	$0;59,8,13,9^0$
	$4+0;2^0$	0	1461	4y 1d	$0;59,8,22,40^0$
	$8+0;4^0$	0	2922	8y 2d	$0;59,8,22,40^0$
	$25-0;1^0$	0	9131	25y 6d	$0;59,8,20,42^0$
Venus	$8-1;54^0$	5	2920	8y 0d	$0;36,59,10,41^0$
	$16-4;47^0$	10	5839	15y 364d	$0;36,59,33,29^0$
	$24-6;49^0$	15	8759	23y 364d	$0;36,59,25,33^0$
Mercury	$13+2;50^0$	41	4751	13y 6d	$3;6,24,10,18^0$
	$46+0;16^0$	145	16802	46y 12d	$3;6,24,22,58^0$
	79	249	28854	79y 19d	$3;6,24,1,47^0$
Moon	$45+2,18;43^0$	45	1240	3y 145d	$13;10,34,59,1^0$
	$271+1,43;28^0$	269	7412	20y 112d	$13;10,35,2,6^0$
	$572+5,3;48^0$	568	15651	42y 321d	$13;10,35,1,53^0$

Tables of mean motions, apparently from the Zîj al-ʿAlâʾî
For a Persian year (365 days)

Saturn	12;13,39,27,44,5,45 (daily motion of 0;2,0,36,4,33,33)
Jupiter	30;20,30,14[a],24,25 (daily motion of 0;4,59,15,39,41)
Mars	3,11;17[b],12,27,22,45 (daily motion of 0;31,26,39,51,21)
Sun	5,59;45,45,43[c],13[d],45[e] (daily motion of 0;59,8,20,35,25)
Venus	3,45;1,49,41,44,56[f],10[g] (daily motion of 0;36,59,28,43,1,38)
Mercury	53;58,14,38,17[h],15[i],0[j] (daily motion of 3;6,24,22,7,59)
Moon	2,9;23,6,42,49,40 (daily motion of 13;10,35,1,55,32)

a 16MS b 14MS c 55MS d 27MS e 5MS f 53MS g 0MS h 33MS i 55MS j 58MS

For an Arab year (354 days)

Saturn	11;51,32,50,53,56,42 (daily motion of 0;2,0,36,4,33,33)
Jupiter	29;25,38,24,7,54 (daily motion of 0;4,59,15,39,41)
Mars	3,5;31,19,8,57,54 (daily motion of 0;31,26,39,51,21)
Sun	5,48;55[a],13[b],28,57,30 (daily motion of 0;59,8,20,35,25)
Venus	3,18;14,55,25,51,38[c],12[d] (daily motion of 0;36,59,28,43,1,38)
Mercury	19;47,46,35,6,6 (daily motion of 3;6,24,22,7,59)
Moon	5,44;26,41,21,38[e],48[f] (daily motion of 13;10,35,1,55,32)

a 59MS b 43MS c 36MS d 0MS e 37MS f 36MS

For 30 Persian years (mistaken parameters for yearly motion used)

Saturn	6;49,43,52,2,52,30 (30 × 12;13,39,27,44,5,45)
Jupiter	3,10;15,8,12,12,30 (30 × 30;20,30,*16*,24,25)
Mars	5,38;36,13,41,22,30[a] (30 × 3,11;17,12,27,22,45)
Sun	5,52;52,37,53,32,30 (30 × 5,59;45,45,*15*,27,5)

Venus	4,30;54,50,52,26,30[b] (30×3,45;1,49,41,44,*53,0*)
Mercury	2,59;7,19[c],16,57[d],59[e] (30×53;58,14,38,*33,55,58*)
Moon	4,41;33,21,25 (30×2,9;23,6,42,*50*)

a 32MS b 59MS c 49MS d 17MS e 30MS

For days		For a Persian month (30 days)
Saturn	0;2,0,36,4,33,33	1;0,18,2,16,46,30
Jupiter	0;4,59,15,39,41	2;29,37,49,50,30
Mars	0;31,26,39,51,21	15;43,19,55,40,30
Sun	0;59,8,20,35,25[a]	29;34,10,17,42,30
Venus	0;36,59,28,43,1,38	18;29,44,21,30,48,49
Mercury	3;6,24,22,7,59	1,33;12,11,3,59,30
Moon	13;10,35,1[b],55,32	35;17,30,57,45,54

a 35MS b 30MS

For 30 Arab years (10631 days)

Saturn	56;8,32,3,48,30,3 (10631×0;2,0,36,4,33,33)
Jupiter	2,43;44,3,56,13,31 (10631×0;4,59,15,39,41)
Mars	2,51;25,27[a],47,21,51 (10631×0;31,26,39,51,21)
Sun	38;27,16,15,14,35[b] (10631×0;59,8,20,35,25)
Venus	1,14;14,37,11,42[c],23[d],58 (10631×0;36,59,28,43,1,38)
Mercury	4,28;3,45,36,30,49 (10631×3;6,24,22,7,59)
Moon	38;17,6,9,58,3,0 (should be 38;17,6,10,34,52)

a 26MS b 34MS c 41MS d 47MS

Revolutions in 36,0,0,0,0,0,0,0 days

Saturn	12,3,36,27,21,18 (1 rev. in 2,59,5;5,22,7,40,48 − days)
Jupiter	29,55,33,58,6,0 (1 rev. in 1,12,10;40,3,8,38,24 + days)
Mars	3,8,39,59,8,6,0 (1 rev. in 11,26;55,36,0,43,12 − days)
Sun	5,54,50,3,32,30,0 (1 rev. in 6,5;14,27,10,4,48 + days)
Venus	3,41,56,52,18,9,46 (1 rev. in 9,43;55,15,18,43,12 + days)
Mercury	18,38,26,12,40,54,0 (1 rev. in 1,55;52,4,50,9,36 + days)
Moon	1,19,3,30,11,33,12,0 (1 rev. in 27;19,17,41,17,48 − days)
Apogee	53,45,36,38 (1 rev. in 40,10,42,23 + days)

Yearly mean motions according to the tables in L (one year equals 365 days) (approximate)

Saturn	12;13,20
Jupiter	30;20,30
Mars	3,11;17
Sun	5,59;45,40
Venus	
Mercury	
Moon	2,9;23

Combined maximum equations according to L (approximate)

Saturn	12;47
Jupiter	16;34
Mars	53;58
Sun	2;0,30
Venus	47;56,30
Mercury	26;49
Moon	7;38,30

Yearly mean motions according to Chrysococces (one year equals 365 days) (approximate)

Saturn	12;12,48
Jupiter	30;19,45
Mars	3,11;16,30
Sun	5,59;44,49
Venus	3,45;1,40
Mercury	53;57
Moon	2,9;23,2
Apogee	0;0,50,55

Maximum equations according to Chrysococces (approximate)

Saturn	6;32 and 6;13
Jupiter	5;15 and 11;3
Mars	11;25 and 42;12
Sun	2;0,30
Venus	1;59 and 45;59
Mercury	3;2 and 22;1
Moon	6;20,18,30

Yearly mean motions according to the 1000-year tables of Chrysococces (one year equals 365 days)

Saturn	12;12,48,2
Jupiter	30;19,20,25,48
Mars	3,11;16,19,25,12
Sun	5,59;44,48,38,1,48
Venus	3,45;1,46,39
Mercury	53;58,14,38,24
Moon	2,9;23,4,58,59
Apogee	0;0,51,25,42,36

Mean motion for 1000 years according to the 1000-year tables of Chrysococces

Saturn	5,33;20,30
Jupiter	1,22;20,30
Mars	1,52;3,40
Sun	1,46;50,33,50

Venus	29;37,30
Mercury	5,30;44,0
Moon	2,23;53,3
Apogee	14;17,8,35

Appendix II

A comparison of the contents of the Zīj as-Sanjarī and its Byzantine translation. The numbers refer to folia, the sigla to the following manuscripts: L = Laurentianus 28, 17. LA = Or. 6669 of the British Museum, London. V = Vaticanus graecus 211. VA = Vaticanus arabus 761. v = Vaticanus graecus 1058.

Preface VA 1^v.

Chapter 1. On fundamentals. VA 5^v.
Chapter 2. On the starting-point and the general approach. VA 7, LA 57.
Chapter 3. On the manner of approaching the question of the Sun. VA 10, LA 59^v.
Chapter 4. On the manner of approaching the question of the Moon. VA 12^v, LA 61.
Chapter 5. On the manner of approaching the question of the superior planets. VA 14, LA 62^v.
Chapter 6. On the manner of approaching the question of the inferior planets. VA 15^v, LA 7^v.

Index of treatises and sections. VA 18, LA 1: L 81^v, V 38, v 273^v.

Treatise 1. On calendars. VA 20: (Part 1) L 82, V 38^v, v 274.
- Section 1. On their bases. VA 20, LA 3^v: (Chapter 1) L 82^v, V 39, v 274^v.
  - Chapter 1. On days, months, and years as components of calendars. VA 20^v, LA 3^v.
  - Chapter 2. On the epochs of calendars. VA 21^v, LA 4^v.
  - Chapter 3. On the differences between calendars. VA 22^v, LA 5.
- Section 2. On calendars in use. VA 23, LA 5^v: (Chapters 2–4) L 83^v, V 39^v, v 275.
  - Chapter 1. On months and years in detail. VA 23, LA 5^v.
  - Chapter 2. On the week-days which begin years and months. VA 24.
  - Chapter 3. On the years of a calendar in days and on the elevation of the days into years and months. VA 25.
  - Chapter 4. On transforming dates from one calendar into another by computation. VA 26.
  - Chapter 5. On transforming dates from one calendar into another by the use of tables. VA 26^v, LA 9.
- Section 3. On festivals. VA 27^v, LA 10: (Chapter 5) L 95^v, V 46^v, v 280.
  - Chapter 1. On what pertains to the days of a month. VA 28, LA 10.
  - Chapter 2. On what pertains to the days of a month and a week VA 28^v, LA 11.

Chapter 3. On what pertains to solar and lunar years and to the days of the week. VA 29, LA 11.
Chapter 4. On the years of the Hebrews and their festivals. VA 30, LA 12.

Treatise 2. On fundamentals, on sines, and on shadows. VA 31, LA 13: (Part 2) L 100^{v}, V 50, v 282^{v}.

Section 1. On fundamentals. VA 31^{v}, LA 13.
Chapter 1. On interpolation. VA 31^{v}, LA 13: (Section 1) L 100^{v}, V 50, v 282^{v}.
Chapter 2. On multiplication and division. VA 32^{v}, LA 13^{v}.
Chapter 3. On the value of π. VA 32^{v}, LA 14.

Section 2. On sines, versines, and arcs. VA 33, LA 14: (Section 2) L 102, V 51, v 283.
Chapter 1. On sines. VA 33^{v}, LA 14^{v}.
Chapter 2. On versines. VA 34, LA 15.
Chapter 3. On arcs. VA 34^{v}, LA 15^{v}.

Section 3. On the three shadows. VA 34^{v}, LA 15^{v}: (Section 3) L 103^{v}, V 52^{v}, v 283^{v}.
Chapter 1. On the first shadow. VA 35, LA 16.
Chapter 2. On the second shadow. VA 35^{v}, LA 16.
Chapter 3. On transforming one shadow into the other VA 36, LA 16^{v}.

Treatise 3. On ascendants. VA 36, [LA 16^{v}].

Section 1. On declination, geographical latitude, altitude, and rising-times for sphaera recta. VA 36, LA 16^{v}: (Part 3) L 103^{v}, V 52^{v}, v 283^{v}.
Chapter 1. On declination. VA 36^{v}, LA 16^{v}: (Chapter 1) L 103^{v}, V 52^{v}, v 284.
Chapter 2. On geographical latitude. VA 37, LA 17: (Chapter 2) L 104, V 53, v 284.
Chapter 3. On the maximum altitudes of the Sun and the stars on the meridian. VA 37, LA 17^{v}: (Chapter 3) L 104^{v}, V 53^{v}, v 284^{v}.
Chapter 4. On knowing the declination from the geographical latitude and the maximum altitude. VA 37^{v}, LA 17^{v}.
Chapter 5. On the rising-times of the signs for sphaera recta VA 37^{v}, LA 17^{v}: (Chapter 4) L 105, V 53^{v}, v 284^{v}.

Section 2. On the azimuth of the point of sunrise and the equation of daylight. VA 38, LA 18: (Part 4) L 105^{v}, V 54, v 284^{v}.
Chapter 1. On the azimuth of the point of sunrise. VA 38, LA 18: (Chapter 1) L 106, V 54, v 285.
Chapter 2. On the equation of daylight. VA 38, LA 18: (Chapter 2) L 106^{v}, V 54^{v}, v 285.

Chapter 3. On the versine of the day-(arc) VA 38v, LA 18v.
Chapter 4. On the arcs of day and night and their hours. VA 38v, LA 18v: (Chapter 3) L 107v, V 55, v 285v.
Chapter 5. On oblique ascensions. VA 39v, LA 19: (Chapter 4) L 108, V 55v, v 286.
Section 3. On the situations of the fixed stars. VA 40, LA 19v: (Part 5) L 108, V 55v, v 286.
Chapter 1. On the correction of their longitudes. VA 40, LA 19v: (Chapter 1) L 108v, V 56, v 286.
Chapter 2. On their distances from the equator and their maximum altitudes. VA 40v, LA 20: (Chapter 2) L 109, V 56, v 286v.
Chapter 3. On the transits of the stars. VA 41, LA 20v: (Chapter 3) L 109v, V 57, v 287.
Chapter 4. On simultaneously rising and setting points. VA 41, LA 20v: (Chapter 4) L 110, V 57v, v 287.
Chapter 5. On the times of risings and settings of the stars. VA 41v, LA 21: (Chapter 5) L 110v, V 57v, v 287v.
Section 4. On what has passed. VA 41v, LA 21: (Part 6, Chapter 1) L 111, V 58, v 287v.
Chapter 1. On what has passed of the day. VA 42, LA 21.
Chapter 2. On what has passed of the night. VA 42, LA 21v.
Chapter 3. On seasonal hours. VA 43, LA 22.
Section 5. On the rising-times and equalizations of the houses. VA 43v, LA 22v.
Chapter 1. On the rising-times. VA 43v, LA 22v: (Chapter 2) L 113, V 59v, v 288v.
Chapter 2. On the hours from the rising-times. VA 44, LA 23: (Chapter 3) L 113v, V 60, v 289.
Chapter 3. On the equilization of the houses. VA 44, LA 23: (Chapter 4) L 114, V 60, v 289.
Section 6. On altitude, the qibla, and the times of prayer. VA 45, LA 23v.
Chapter 1. On altitude. VA 45, LA 24: (Chapter 5) L 115, V 61v, v 289v.
Chapter 2. On the declination-circle of the Sun. VA 46, LA 24: (Chapter 6) L 116, V 62, v 290.
Chapter 3. On the qibla. VA 46, LA 24v: (Chapter 7) L 116v, V 62v, v 290v.
Chapter 4. On the times of prayers. VA 47v, LA 25v.

Treatise 4. On the mean motions of the planets. VA 48v, LA 26v: (Part 7) L 118, V 63v, v 291.
Section 1. On corrections of the revolutions. VA 51, LA 28.
Chapter 1. On the correction of the bases of motions. VA 51, LA 28.
Chapter 2. On the advantages and disadvantages of the revolutions. VA 52, LA 28v.

Section 2. On deriving the mean motions of the planets from computation and from tables. VA 52^v, LA 29: (Chapter 1) L 118, V 64, v 291^v.

Chapter 1. On mean motions from computation. VA 52^v, LA 29.

Chapter 2. On mean motions from tables. VA 53^v, LA 30.

Section 3. On corrections of the mean motions. VA 54^v, LA 30^v: (Chapter 2) L 120, V 65^v, v 292^v.

Chapter 1. On their corrections with regard to the remainder of the difference between the two longitudes. VA 54^v, LA 30^v.

Chapter 2. On the corrections of the mean motions with regard to the equation of time. VA 55, LA 31.

Section 4. On the mean motions with regard to special computations. VA 55^v, LA 31^v.

Chapter 1. On the mean motions for a geographical longitude of 90^0 using the Arabic calendar. VA 56, LA 31^v.

Chapter 2. On the correction of the mean motions for special operations. VA 56^v, LA 32^v.

Section 5. Introduction to world-years and Sultanic years from the three well-known calendars. VA 57, LA 32^v: (Chapter 3) L 121, V 66, v 293.

Chapter 1. On the differences between the world-years and the calendars. VA 57^v, LA 33.

Chapter 2. Introduction to the world-years or Sultanic years. VA 58, LA 33^v.

Chapter 3. On establishing the rules for computing true longitudes. VA 58^v, LA 34: (Chapter 4) L 122, V 67, v 293^v.

Treatise 5. On computing true longitudes and latitudes. VA 59, [LA 34^v]: (Part 8) L 123^v, V 68, v 294.

Section 1. On computing true longitudes. VA 59, LA 34^v: (Chapter 1) L 124, V 69, v 294^v.

Chapter 1. On computing the true longitude of the Sun. VA 59, LA 34^v.

Chapter 2. On computing the true longitude of the Moon. VA 59^v, LA 34^v.

Chapter 3. On computing the true longitudes of the lunar nodes. VA 60, LA 35.

Chapter 4. On computing the true longitudes of the five star-planets. VA 60, LA 35.

Chapter 5. On the exactitude of computations of true longitudes. VA 60^v, LA 35^v.

Section 2. On retrograde and direct motion. VA 61, LA 35^v: (Chapter 2) L 127, V 71, v 296.

Chapter 1. (no title). VA 61, LA 35^v.

Chapter 2. Fundamentals of retrograde and direct motion. VA 61, LA 35^v.

Section 3. On the latitudes of the planets. VA 61^{v}, LA 36: (Chapter 3) L 128^{v}, V 72, v 296^{v}.

Chapter 1. On the latitude of the Moon. VA 61^{v}, LA 36.

Chapter 2. On the latitudes of the superior planets. VA 61^{v}, LA 36.

Chapter 3. On the latitudes of the inferior planets. VA 62, LA 36^{v}.

Section 4. On the velocities and diameters of the Sun and Moon. VA 63, LA 37: (Chapter 4) L 132, V 75, v 298^{v}.

Chapter 1. On daily motion. VA 63, LA 37.

Chapter 2. On the diameter of the Sun. VA 63, LA 37^{v}.

Chapter 3. On the diameter of the Moon. VA 63, LA 37^{v}.

Chapter 4. On the diameter of the shadow. VA 63^{v}, LA 37^{v}.

Chapter 5. On daily motions and diameters in the tables. VA 63^{v}, LA 38.

Treatise 6. On parallax. VA 64, LA 38: (Part 9) L 133, V 76, v 299.

Section 1. Fundamentals necessary for parallax. VA 64, LA 38: (Chapter 1) L 133, V 76, v 299.

Chapter 1. On general matters. VA 64, LA 38.

Chapter 2. On the distance of the Moon from the meridian. VA 64^{v}, LA 38^{v}.

Chapter 3. On the latitude of the clime. VA 64^{v}, LA 38^{v}.

Chapter 4. On the altitude of any degree in the zodiacal circle. VA 64^{v}, LA 38^{v}.

Chapter 5. On the altitude of the Moon. VA 65, LA 38^{v}.

Chapter 6. On the distance of the Moon from the earth. VA 65, LA 39.

Chapter 7. On the three angles necessary in eclipses. VA 66, LA 39^{v}.

Section 2. On parallax by computation. VA 66^{v}, LA 40: (Chapter 2) L 135, V 78, v 300.

Chapter 1. On parallax of the Sun and Moon in the circle of altitude. VA 66^{v}, LA 40.

Chapter 2. On longitudinal and latitudinal parallax of the Moon. VA 67, LA 40^{v}.

Chapter 3. On the correction of the place of the Moon. VA 67, LA 40^{v}.

Section 3. On parallax. VA 67^{v}, LA 40^{v}: (Chapter 3) L 137^{v}, V 79^{v}, v 301.

Chapter 1. On longitudinal and latitudinal parallax. VA 67^{v}, LA 41.

Chapter 2. On the correction of the place of the Moon in longitude and latitude. VA 68^{v}, LA 41^{v}.

Section 4. On the priority of what is observed. VA 69, LA 41^{v}.

Chapter 1. On the priority of what is observed. VA 69, LA 42.

Chapter 2. On the diversity of conditions for the priority of what is observed. VA 69^{v}, LA 42^{v}.

Chapter 3. On the rule of priority in every table for every eclipse. VA 70, LA 42^{v}.

Treatise 7. On conjunctions and oppositions. VA 70, LA 42v: (Part 10, Section 1) L 138, V 80v, v 301v.
  Section 1. On knowing this from computations of the equations. VA 70v, LA 43.
    Chapter 1. On the hours of conjunction and opposition by means of the equations. VA 70v, LA 43.
    Chapter 2. On their fractions. VA 71, LA 43v.
    Chapter 3. On their risings. VA 71v, LA 43v.
  Section 2. On operations regarding conjunctions and oppositions. VA 71v, LA 43v.
    Chapter 1. On the mean motions of the Sun and Moon. VA 72, LA 44.
    Chapter 2. On the equations of the Sun and Moon. VA 72, LA 44.
    Chapter 3. On the hours of the equations. VA 72v, LA 44v.
    Chapter 4. On the fraction of conjunction and opposition. VA 73, LA 45.
    Chapter 5. On its transformation into nighttime and daytime. VA 73, LA 45.

Treatise 8. On eclipses. VA 73v, LA 45.
  Section 1. On lunar eclipses. VA 73v, LA 45: (Section 2) L 140, V 82, v 302v.
    Chapter 1. Fundamentals for lunar eclipses. VA 73v, LA 45v.
    Chapter 2. On the prerequisites for an eclipse. VA 74, LA 45v: (Chapter 1) L 140, V 82, v 302v.
    Chapter 3. On the times of an eclipse. VA 75, LA 46v.
    Chapter 4. On its duration. VA 75v.
    Chapter 5. On eclipses from the tables. VA 76: (Chapter 2) L 141v, V 84, v 303.
    Chapter 6. On the magnitude of a lunar eclipse. VA 77.
    Chapter 7. On the projection of a lunar eclipse. VA 77v.
    Chapter 8. On the conclusion of an eclipse. VA 79, LA 47.
  Section 2. On solar eclipses. VA 79, LA 47v: (Section 3) L 142v, V 84v, v 303v.
    Chapter 1. Fundamentals for solar eclipses. VA 79v, LA 47v: (Chapter 1) L 142v, V 84v, v 303v.
    Chapter 2. On the correction of hours for the middle of an eclipse. VA 79v, LA 48.
    Chapter 3. On determining the fact of an eclipse and its magnitude. VA 81v, LA 49v: (Chapter 2) L 143v, V 85v, v 304.
    Chapter 4. On the times of eclipses. VA 82, LA 49v.
    Chapter 5. On knowing the eclipse by means of a table. VA 83v, LA 50v.
    Chapter 6. On the corona and the duration of an eclipse. VA 84, LA 51.

Chapter 7. On the magnitude of an eclipse. VA 84^{v}, LA 51^{v}.
Chapter 8. On the darkness of an eclipse and its projection. VA 85, LA 52.

Treatise 9. On first visibilities. VA 85^{v}, LA 52^{v}: (Part 11) L 147^{v}, V 89, v 306.
Section 1. On first visibility of the lunar crescent. VA 85^{v}, LA 52^{v}.
Chapter 1. Fundamentals for the first visibility of the lunar crescent. VA 86, LA 53: (Chapter 1) L 148^{v}, V 89^{v}, v 306^{v}.
Chapter 2. On using simple arcs. VA 87, LA 53^{v}: (Chapter 2) L 151^{v}, V 92^{v}, v 308.
Chapter 3. On correcting the simple arcs according to al-Khāzinī. VA 87^{v}.
Chapter 4. On using the arcs according to al-Battānī. VA 88, LA 54.
Chapter 5. On using the arcs according to Thābit ibn Qurra al-Ḥarrānī. VA 89.
Chapter 6. On what al-Khāzinī looks for in first visibilities. VA 89^{v}.
Chapter 7. On the azimuth of the crescent. VA 90^{v}, LA 55: (Chapters 3, 4, and 6) L 152^{v}, 154, 155^{v}; V 93, 94^{v}, 95^{v}; v 308, 309, 310.
Section 2. On heliacal risings and settings. VA 91, LA 55^{v}: (Chapter 5) L 154^{v}, V 95, v 309.
Chapter 1. Fundamentals for the superior planets. VA 91, LA 55^{v}.
Chapter 2. On appearances and disappearances according to the tables. VA 92.

Treatise 10. On the transfers of years. VA 92^{v}: (Part 12) L 156^{v}, V 97, v 310^{v}.
Section 1. On the transfers of world-years. VA 92^{v}: (Chapter 1) L 157, V 97, v 310^{v}.
Chapter 1. On the times of the transfers of years. VA 93.
Chapter 2. On the mean transfer. VA 93^{v}.
Chapter 3. On the excess of a year. VA 94.
Chapter 4. On the ascendant of the transfer. VA 94^{v}.
Section 2. On the location of the rays of the planets. VA 95: (Chapter 2) L 159, V 98^{v}, v 311^{v}.
Chapter 1. Fundamentals for the location of the rays and their motions. VA 95^{v}.
Chapter 2. On the location of the rays. VA 96.
Chapter 3. On the location of the rays through computation of the horizon and the planets' points of rising. VA 96^{v}.
Chapter 4. On the location of the rays through computation of half of the day-arc. VA 97.
Chapter 5. On the location of the rays according to the opinion of Ptolemy. VA 97^{v}.
Section 3. On the motion of the Haylāj. VA 98^{v}: (Chapter 3) L 162^{v}, V 102, v 313^{v}.

Chapter 1. On the motion of the Haylāj VA 98v.
Chapter 2. On the position of the division. VA 99v.
Section 4. On the intihā' and its motion. VA 100: (Chapter 4) L 165v, V 105, v 315v.
Chapter 1. On the intihā' and its motion. VA 100.
Chapter 2. On the motion of the ascendant-degree of the year-transfer. VA 100v.
Chapter 3. On the transfers of months. VA 100v.
Chapter 4. On the motion of the ascendant of the year-transfer. VA 101.

The Separate Treatise. VA 101.
Section 1. On the ascendant from the altitude of the Moon. VA 101.
Chapter 1. On the hours of the arrivals of the planets at the meridian and the degrees of the transits. VA 101v, LA 56.
Chapter 2. On the hours through estimation. VA 102v, LA 56v.
Chapter 3. On the transformation of the observed altitude of the Moon into its true altitude. VA 103.
Chapter 4. On the correction of the hours and on the ascendant. VA 103v.
Section 2. On the correction of geographical longitude. VA 104.
Chapter 1. (no title). VA 104.
Chapter 2. On geographical longitude from the altitude of the Sun. VA 104v.
Section 3. On changing the computation of true longitudes from one place to another. LA 63.
Chapter 1. On changing the position of the planet. LA 63.
Chapter 2. On changing the hours in conjunctions, oppositions, and lunar eclipses. LA 63.
Chapter 3. On changing the ascendants. LA 63v.
Chapter 4. On altitudes and hours. LA 63v.
Chapter 5. On the visibility of the lunar crescent and solar eclipses. LA 63v.
Section 4. On the true daily motions of the planets. LA 64.
Section 5. On the conjunctions of the planets. LA 65v.
Chapter 1. On the times of the conjunctions. LA 65v.
Chapter 2. On the motion from one conjunction to the next. LA 66.
Section 6. On spheres and transits. LA 66v.
Chapter 1. On the bases of the spheres and their measurements. LA 66v.
Chapter 2. On the minutes of ascent and descent, and on the transits of the planets in their conjunctions. LA 67v.

Appendix III

THE TABLES OF PALAEOLOGUS (BRITISH MUSEUM, MS BURNEY 91, FOL. 8)

Χρὴ γινώσκειν διὰ μεθόδου εὑρεῖν τὰ ἔτη ἃ ἐμετεβλήθησαν παρ' ἡμῶν ἀπὸ τοὺς μῆνας τῶν Περσῶν εἰς τοὺς μῆνας τῶν 'Ρωμαίων. ἀρχὴν τήρησον τὰ Ἑλληνικὰ ἔτη πόσα εἰσὶν, καὶ ἐξ αὐτῶν τῶν πεπληρωμένω ἐτῶν ἄφελον ἔτη ͵ϛϡμγ (6943 A.M. began 1 Sept. 1435 A.D.). τὰ λοιπὰ ἕξωμεν τὰ ῥηθέντα ἔτη. καὶ ὡς ἐν ὑποδείγματος (corrected to ὑποδείγματι) ἔστω Ἑλληνικὰ ἔτη πεπληρομένα εἰς τὴν α' τοῦ Μαρτίου ͵ϛϡαν (A.D. 1 March 1443). ἐξ αὐτῶν ἄφελον ͵ϛϡμγ. τὰ λοιπὰ ζ̄ ἕξομεν ἔτη πεπληρομένα. τὸ Ἑλληνικὸν ἔτος ἄρχεται ἀπὸ τὴν α' τοῦ Σεπτευρίου, καὶ τὸ ἡμέτερον ἀπὸ τὴν α' τοῦ Μαρτίου. καὶ πρόσχες ἵνα μὴ συνάρξῃς αὐτά, καὶ ἐπακολουθήσῃ τι ἄτοπον. ἀλλὰ τὸ μὲν Σεπτέβριον πρὸ μηνῶν ϛ̄, τὸ δὲ τὸν Μάρτιον μετὰ μῆνας ϛ̄. καὶ ἐκ περιουσίας τοῦτο ἄφελε ἀπὸ τὰ πεπληρομένα ἔτη τὰ Περσικὰ ω̄ε̄ (805 Yazdijird began A.D. 28 Nov. 1435) ἢ καὶ ἀπὸ τὸ Ἰουδαϊκὸν Ἑξαπτέριγον (the Hexapterygon of Michael Chrysococces) ιθσογ καὶ ἁπλᾶ ἔτη η̄ (273 × 19 + 8 = 5195 years) καὶ τὰ λοιπά εἰσι τὰ ἡμέτερα ἔτη.

Γίνωσκαι πῶς εἰς τὰ ͵ϛϡμγ ἔτη πεπληρομένα Μαρτίου α' ἦν Περσικὰ ἔτη πεπληρομένα ω̄ε̄ Τυρμᾶ ε' (A.D. 1 March 1436). ἐν τούτοις ἦν ὁ "Ηλιος εἰς τὴν μέσην κίνησιν Τοξότῃ ῑϛ λη' νδ'' μετὴν μέθοδον τῶν ἐτῶν τοῦ Παλαιολόγου, μετὰ γοῦν τὴν μέθοδον τῶν κανονίων τοῦ Χρισοκκούκη (George Chrysococces) εἰς Τοξότην ῑϛ λη' ιζ''. τὸ ὕψωμα αὐτοῦ μετοῦ Παλαιολόγου β̄ κ̄θ ιθ' κζ'', μετοῦ Χρισοκκούκη β̄ κ̄θ ιθ' κθ''.

"Ἔτι τῆς Σελήνης ἡ μέση κίνησις μετοῦ Παλαιολόγου δ̄ κ̄ ιβ' νε'', ἡ ἰδία κίνησις δ̄ ῑβ ιζ' κθ'', τὸ κέντρον ῑ ῑθ ν' η''. μετοῦ Χρισοκκούκη ἡ μέση ὁδὸς δ̄ κ̄ ιβ' νϛ'', ἡ ἰδία κίνησις δ̄ ῑβ ιζ' κζ'', τὸ κέντρον ῑ ῑθ μη' κζ''. αἱ ἐποχαὶ αὗται ἐγένοντο εἰς τὸ μῆκος μοιρῶν ν̄ζ.

Chrysococces gives the mean motion of the Sun, measured from its apogee, as $2,54;3,11^0$ at the beginning of 765 Yazdijird (A.D. 8 Dec. 1395), and the longitude of the solar apogee on the same date as $1,28;44,27^0$. The mean Sun and the apogee have respectively yearly motions of $5,59;44,49,20^0$ and $0;0,50,55^0$. At the beginning of 769 Yazdijird (A.D. 7 Dec. 1399) Chrysococces says that the mean motion of the Moon is $3;15,55^0$, its anomalistic motion $29;42,1^0$, and its double elongation $3,34;7,16^0$; their yearly motions are respectively $2,9;23,2^0$, $1,28;43,7^0$, and $4,19;14,44^0$. For these functions, without the solar apogee, the motions for the 95 days from Farvardîn 1 to Tîr 5 are respectively $1,32;38,50^0$, $2,38;34,52^0$, $2,28;6,30^0$, and $2,11;51,37^0$. From these data one gets the following values for 5 Tîr 805 Yazdijird (A.D. 1 March 1436).

	CHRYSOCOCCES	TEXT	TEXT—CHRYSOCOCCES
mean Sun	$4,16;34,54^0$	$4,16;38,17^0$	$+ 0;3,23^0$
solar apogee	$1,29;18,24^0$	$1,29;19,29^0$	$+ 0;1,5^0$
mean Moon	$2,19;39,59^0$	$2,20;12,56^0$	$+ 0;32,57^0$
lunar anomaly	$2,11;40,43^0$	$2,12;17,27^0$	$+ 0;36,44^0$
double elongation	$5,18;49,17^0$	$5,19;48,27^0$	$+ 0;59,10^0$

Chrysococces' figures are for the longitude of Tybênê (Tabrîz), which is 72^0 E; the text's are for a longitude of 57^0 E—i.e., on a parallel running close to Nicaea. The difference between the two is 15^0 or one hour. Therefore, to multiply the entries in the column headed Text—Chrysococces by 24 should result in mean daily motions:

	(TEXT—CHRYSOCOCCES) × 24
mean Sun	1;21,12°
solar apogee	0;26°
mean Moon	13;10,48°
lunar anomaly	14;41,36°
double elongation	23;40°

Clearly, though these numbers, with the exception of that for the solar apogee, are all in the vicinity of what they should be, there was a serious lack of accuracy in the text's computations.

The Palaeologan Tables give results very close to those obtained by using Chrysococces' work. It appears that it is the purpose of the text preserved in Burney 91 to give the epoch values of these Palaeologan Tables; thus, it seems probable that the Palaeologan Tables were written shortly after 1 March 1436 in or near Nicaea, and that the structure of its tables for determining the longitudes of the Sun and the Moon was modelled on Chrysococces' tables.

Additional Note: After this paper had already reached final proof, the author had the opportunity of examining MS 859 of the Hamidiye Collection in the Süleymaniye Library in Istanbul, and found that the Arabic text of a shortened version of the Zîj as-Sanjarî which it contains is the same as that translated into Greek by Chioniades.

THE INDIAN AND PSEUDO-INDIAN PASSAGES IN GREEK AND LATIN ASTRONOMICAL AND ASTROLOGICAL TEXTS

•

by David Pingree

CONTENTS

I. Introduction
II. Kalyāṇa (parapegma)
III. Curtius Rufus (new moon months; pakṣas)
IV. Philostratus (planetary weekdays)
V. ‘Abd al-Bāri‘
VI. Sasanians (nakṣatras; planetary chords; decans; *Zīk i Shahriyārān*).
VII. Severus Sebokht (lunar nodes)
VIII. Theophilus of Edessa (military astrology; zodiacal topothesia)
IX. Māshā'allāh ("Era of the Flood"; planetary chords; cosmic magnet; navāṃśas)
X. *Zīj al-Sindhind* (Kalpa; Caturyuga; mean motions; year-length; sidereal zodiac; trepidation; longitudes of apogees and nodes; ahargaṇa; mean longitudes of planets; longitudinal difference; accumulated epact; trigonometric functions; equation of center; obliquity of ecliptic and method of declinations; equation of anomaly; combined effect of equations; time to first or second station; ascensional difference; terrestrial latitude; gnomon-shadows; lunar latitude; apparent diameters of sun, moon, and earth's shadow; eclipse-limit; totality of eclipse; duration of eclipse and of totality; color of eclipse; longitudinal parallax; latitudinal parallax; latitudes of planets; value of π)
XI. Ṛṣi (interrogations)
XII. Bhūridāsa and Buzurjmihr (Jovian dodecaeteris; theft)
XIII. Abū Ma‘shar (nativity of Ceylonese prince; childbirth; ketu; lunar nodes; terms; decans; revolution of years of nativities; place of sun in nativity; astrological places; fulfillment of interrogations)
XIV. Ja‘far al-Hindī (order of orbits of planets and fixed stars; benefic and malefic planets; quarters of a month; nakṣatras)
XV. Al-Qabīṣī (karaṇas)
XVI. Simeon Seth (precession; star-catalog)
XVII. Vaticanus graecus 1056 (interrogations)

XVIII. Parisinus graecus 2506 (lordships of months of pregnancy)
XIX. *Picatrix* (names of planets)
XX. Shams al-Dīn al-Bukhārī (year-beginning; adhimāsas; Sin η; śaṅkutala and bāhu)
XXI. Conclusions

I. Introduction

Astronomy and astrology in India[1] are not indigenous sciences, but are local adaptations and developments of Mesopotamian,[2] Greco-Babylonian,[3] and Greek[4] texts; and, at an early stage of their developments, parts of the Indian traditions had influenced Sasanian[5] and Syriac science before the rise of Islam. There existed, therefore, a more or less common understanding of astronomy and astrology in those regions of the world where Latin, Greek, Syriac, Pahlavī, and Sanskrit were used, though each culture had its particular idiosyncrasies and its special areas of sophistication. Islam was the heir to all of these traditions,[6] which it was able to synthesize precisely because of their common features. The object of this paper is to attempt to isolate as many as possible of those elements of the Islamic adaptations of Indian astronomy and astrology that were included in the massive influx of translations of

[1] A survey of the Indian material relevant to these and related sciences, including mathematics, will be found in D. Pingree, *Census of the Exact Sciences in Sanskrit* (hereafter *CESS*), of which volumes 1 and 2 of Series A have been published: A1 as vol. 81 of the Memoirs of the American Philosophical Society (Philadelphia 1970), A2 as vol. 86 (Philadelphia 1971); A3 is in press. Indian influences on European mathematics will be the subject of a separate paper.

[2] See especially D. Pingree, "Astronomy and Astrology in India and Iran," *Isis* 54 (1963) 229-246, and "The Mesopotamian Origin of Early Indian Mathematical Astronomy," *Journal of the History of Astronomy* (hereafter *JHA*) 4 (1973) 1-12.

[3] See especially D. Pingree, "A Greek Linear Planetary Text in India," *Journal of the American Oriental Society* (hereafter *JAOS*) 79 (1959) 282-284; chap. 79 of D. Pingree, ed., *The Yavanajātaka of Sphujidhvaja*, to appear in the Harvard Oriental Series; and O. Neugebauer and D. Pingree, eds. *The Pañcasiddhāntikā of Varāhamihira,* Danske videnskabernes Selskab, Hist.-filos. Skrifter 6. 1, 2 parts (Copenhagen 1970-1971).

[4] See D. Pingree, "On the Greek Origin of the Indian Planetary Model Employing a Double Epicycle," *JHA* 2 (1971) 80-85; "Precession and Trepidation in Indian Astronomy before A.D. 1200," *JHA* 3 (1972) 27-35; and "Concentric with Equant," *Archives internationales d'histoire des sciences* 24 (1974) 26-28; for astrology see the commentary to the *Yavanajātaka,* and D. Pingree, ed., *The Vṛddhayavanajātaka of Mīnarāja,* to appear in the Gaekwad Oriental Series; and D. Pingree, "The Indian Iconography of the Decans and Horās," *Journal of the Warburg and Courtauld Institutes* 26 (1963) 223-254, and "Representation of the Planets in Indian Astrology," *Indo-Iranian Journal* 8 (1965) 249-267.

[5] See D. Pingree, "Indian Influence on Sassanian and Early Islamic Astronomy and Astrology," *Journal of Oriental Research, Madras* 34-35 (1964-1966) 118-126, and F. Haddad, E. S. Kennedy, and D. Pingree, eds., *The Kitāb 'ilal al-zījāt of al-Hāshimī,* to appear in the series of the Society for the Study of Islamic Philosophy and Science.

[6] See D. Pingree, "The Greek Influence on Early Islamic Mathematical Astronomy," *JAOS* 93 (1973) 32-43.

Arabic science into Byzantine Greek and into Latin, as well as in the subsequent translations of this material from Greek into Latin and from Latin into Greek; I omit the immense quantity of Hebrew and vernacular texts, and also those Greek and Latin works whose information concerning Indian science is secondary within each culture. Though I have utilized as many printed and manuscript sources as could reasonably be obtained, I am aware that much that is relevant must have escaped my notice, and can only hope that others more versed especially in the Latin sources will continue this work. I have tried, where possible, to add to the citation of the Greek or Latin text one of a Sanskrit passage of appropriate antiquity expressing the same or a similar idea; but, in the interest of avoiding excessive length, I have refrained from citing the Arabic intermediaries and from translating any passage.

The period within which Arabic scientific texts were translated into Greek extended from the ninth to the fourteenth century, into Latin from the twelfth to the thirteenth only, though many more Latin translations were made in two centuries than Greek in six. The reason for the disparity, of course, lies in part in the fact that the Greeks generally possessed scientific texts superior or equal to those of the Arabs whereas the Latins did not. Moreover, in astronomy at least, the Latins received an Arab form more imbued with Indian elements than did the Byzantines; for the latter learned primarily from the astronomers of Marāgha, who had essentially anticipated Copernicus's mathematically significant reform of the Ptolemaic planetary models and had progressed far beyond the level attained by Indian astronomers, whereas the former came into contact first with a provincial Arab culture which, astronomically, had not passed through the process of Ptolemaicization that had occurred in the East. Thus, as we shall see, the earliest set of astronomical tables in Latin is the only surviving representative of the major Arabic adaptation of Indian astronomy; the Eastern Arabs did not regard such a primitive work as worthy of preservation.

II. KALYĀṆA

There existed extensive knowledge of India, more or less accurate, in the West in the Hellenistic and Roman Imperial periods. Little was known of Indian astronomy, however. Undoubtedly the most astonishing piece of evidence is an inscription of the late second century B.C. found during the excavations of the theater at Miletus.[7] This is a calendar of the heliacal risings and settings of certain fixed stars in which a number of earlier authorities are mentioned; among these is ὁ Ἰνδῶν Καλλανεύς.[8] These citations are as follows:

[7] See H. Diels and A. Rehm, "Parapegmenfragmente aus Milet," *Sitzungsberichte der Preussischen Akademie der Wissenschaften zu Berlin* 23 (1904), Philos.-Hist. Cl., 92-111; and A. Rehm, "Weiteres zu den milesischen Parapegmen," *ibid.* 752-759. I hope to deal separately with the astronomical reports of the followers of Alexander.

[8] See *CESS* A2, 24b.

1. Κατὰ δὲ Ἰνδῶν Καλ[λανέ]α Σκορπίος δύνει μετὰ βρον[τ]ῆς καὶ ἀνέμου.
2. [Ὑάδες ἑσπ]έριαι ἐπιτέλλουσιν [κατὰ Ἰνδῶν Καλ]λανέα.
3. [Κατὰ δὲ Ἰ]νδῶν Καλλανέα [Πλειάδες ἑσπ]έριαι δύνουσιν [καὶ ἐπι]σημαίνει χαλάζηι.
4. Αἴξ ἑσπερία δύνει κατὰ Ἰνδ[ῶν] Καλλανέα.
5. [Ὑάδες ἑῷ]ιαι ἐπιτέλ[λουσιν κατ' Ἰνδῶ]ν Καλλα[νέα].

The name Καλλανεύς is surely a transliteration of the Sanskrit Kalyāṇa or one of its Prakrit equivalents, but it is unclear whether it is a personal name or a generic term for a gymnosophist. Stars in three of the constellations named by Kalyāṇa are used as *yogatārās* in later Indian astronomy; these identifications can safely be made from the polar coordinates given in the *Brāhmasphuṭasiddhānta* composed by Brahmagupta at Bhillamāla in Rajasthan in 628:[9]

1. Σκορπίος. Jyeṣṭhā is α Scorpii.
2. Ὑάδες. Rohiṇī is δ^1 Tauri.
3. Πλειάδες. Kṛttikā is η Tauri.
4. Αἴξ. Capella or α Aurigae is not a *yogatārā.*

The appearance of Capella on the list naturally raises doubts about the possibility of its Indian origin. These are strengthened by the non-occurrence of this method of weather-prediction in India. For heliacal risings of fixed stars were certainly observed at an early period in India, as is proved by a famous passage in the *Taittirīyabrāhmaṇa*:

> yat puṇyaṃ nakṣatram/ tadvat kurvītopavyuṣam/ yadā vai sūrya udeti/
> atha nakṣatraṃ naiti/ yāvati tatra sūryo gacchet/ yatra jaghanyaṃ paśyet/
> tāvati kurvīta yatkārī syāt/ puṇyāha eva kurute/[10]

But it was the moon's conjunctions with the *nakṣatras* that were thought to affect the weather, not their heliacal risings or settings; see, for instance, the *varṣādhyāya* in the *Śārdūlakarṇāvadāna,*[11] a Buddhist anti-caste tract at least as old as the first century A.D. We must conclude, then, that even if Καλλανεύς was an Indian, he followed a Greek rather than an Indian tradition.

III. CURTIUS RUFUS

Further evidence of a knowledge of Indian astronomy or astrology among the ancient Greeks is very scanty indeed. Quintus Curtius Rufus in the middle of the first

[9] See D. Pingree in *Dictionary of Scientific Biography* (henceforth *DSB*) 2 (New York 1970) 416-418; vols. 1-10 have so far appeared. The data concerning the stars is taken from my edition of the *Brāhmasphuṭasiddhānta* now under preparation.

[10] N. S. Goḍabole, ed., *Taittirīyabrāhmaṇa* 1.5.2.1, Ānandāśrama Sanskrit Series 37, 2 vols. (Poona 1898).

[11] S. Mukhopadhyaya, ed., *Śārdūlakarṇāvadāna* (Santiniketan 1954) 68-79.

century A.D. ascribes to the Indians a form of prediction from the stars (astral omens), the use of *amānta* (from new moon) rather than *pūrṇimānta* (from full moon) months, and the division of each synodic month into two *pakṣas,* each containing fifteen "days" (*tithis*).[12] Strabo at the end of the first century B.C. had also noted that the Brāhmaṇas study astronomy.[13] The astronomical system to which these two authors refer would have been that of the *Jyotiṣavedāṅga.*

IV. PHILOSTRATUS

Philostratus, in his life (written in the early third century) of Apollonius of Tyana, a wandering sage of the end of the first century, reports that an Indian philosopher named Iarchas was an expert in astrology, and gave Apollonius seven planetary rings of which the appropriate one was to be worn on each planetary weekday.[14] Though astral divination was introduced into India from Mesopotamia in the fifth or fourth century B.C., the first work on genethlialogy in Sanskrit that we know of is the translation made by Yavaneśvara in A.D. 149/150 from a Greek text written in Egypt (probably Alexandria) in the first half of the second century A.D.; of the translation of Yavaneśvara there survives a versification, the *Yavanajātaka,* made by Sphujidhvaja in A.D. 269/270, in which the planetary weekdays are mentioned.[15] The anecdote, therefore, appears improbable in the context of Apollonius's lifetime, though it has been used as evidence that the Indians knew the planetary weekdays already in the first century A.D.[16] Considering the generally unreliable character of Philostratus's romance, one must certainly repugn this story's authenticity.

V. 'ABD AL-BĀRI'

The only other reference to Indian astronomy or astrology in pre-Islamic Greek and Latin texts that I am aware of occurs in the seventh-century *Chronicon Paschale,*[17] which says that one Ἀνδουβάριος of the race of Arphaxad first composed works on astronomy for the Indians at the time of the Tower of Babel. The tale is obviously a fable, and the name (one should read Ἀβδουβάριος) Semitic.[18] Curtius Rufus's reference to the *amānta* months and the *pakṣas,* then, remains the only genuine trace of a knowledge of Indian astronomy in the West in the classical period.

[12] *De rebus gestis Alexandri Magni* 8.9; ed. H. Barden, 2 vols. (Paris 1947-1948).

[13] Γεωγραφικά 15.70; ed. A. Meineke, 3 vols. (Leipzig 1866); see also Clemens Alexandrinus, Στρωμάτεις 3.194; ed. O. Stählin and L. Früchtel, 2 vols. (Berlin 1960-1970).

[14] Τὰ εἰς τὸν Τυανέα Ἀπολλώνιον 5.41; ed. F. C. Conybeare, 2 vols. (London 1912).

[15] *Yavanajātaka* (n. 3 above) 79.55.

[16] A. Cunningham, "The Probable Indian Origin of the Names of the Weekdays," *Indian Antiquary* 14 (1885) 1-4.

[17] L. Dindorff, ed., *Chronicon Paschale* 1 (Bonn 1832) 64.

[18] *CESS* A1, 43a.

VI. SASANIANS

The Indian influence on Sasanian science has already been referred to. Specifically, with regard to astronomy and astrology, one knows the following data.

1. A book by an Indian named Farmāsb (the name may be an attempt to transliterate Parameśvara)[19] is said by Ibn Nawbakht, one of Hārūn al-Rashīd's astrologers, to have been translated into Pahlavī under Ardashīr I (226-240).[20]

2. References to the *navāṃśas* (called *nō bahr*) were inserted into the Pahlavī translation of the astrological poem of Dorotheus of Sidon in the third and fourth century.[21]

3. The *Bundahišn,* in a passage which must date from after the early fifth century, lists the twenty-seven *nakṣatras* of Indian astronomers, beginning with Aśvinī.[22]

4. The *Bundahišn* in another passage ascribes to Gayomart the horoscope of a *mahāpuruṣa.*[23]

5. This same passage refers to the chords that bind the planets to the chariot of the sun, an idea that occurs elsewhere in Sasanian texts[24] and that seems to be Indian in origin; it reached the Latin West, as we shall see below, through Māshā'-allāh.

6. Indian descriptions of the decans seem to have reached Abū Ma'shar through a Sasanian source.[25]

7. It is probable that a Sasanian astronomical *zīk* was composed in about A.D. 450; if so, the one parameter ascribed to it is derived from the *Brāhmapakṣa* of Indian astronomy.[26]

[19] F. Justi, *Iranisches Namenbuch* (Marburg 1895) 90, conjectures Paramāśva, which seems much less likely as a name in Sanskrit.

[20] Quoted by Ibn al-Nadīm, *Kitāb al-Fihrist* (Cairo n.d.) 345-348; see D. Pingree, *The Thousands of Abū Ma'shar,* Studies of the Warburg Institute 30 (London 1968) 9-10.

[21] D. Pingree, ed., *Carmen astrologicum* 5.5, 26 (Leipzig 1976).

[22] W. B. Henning, "An Astronomical Chapter of the Bundahishn," *Journal of the Royal Asiatic Society* (1942) 229-248, esp. 242-246. Henning's argument about the date is irrelevant since the list is Indian. But the earliest Sanskrit text to place Aśvinī at the beginning of Aries is 3.2 of the *Paitāmahasiddhānta* of the *Viṣṇudharmottarapurāṇa,* a work of the early fifth century; see D. Pingree, "The *Paitāmahasiddhānta* of the *Viṣṇudharmottarapurāṇa,*" *Brahmavidyā* 31-32 (1967-1968) 472-510.

[23] The passage is translated by D. N. Mackenzie, "Zoroastrian Astrology in the Bundahišn," *Bulletin of the School of Oriental and African Studies* 27 (1964) 511-529; for its interpretation see D. Pingree, "Māshā'allāh: Some Sasanian and Syriac Sources," in G. F. Hourani, ed., *Essays on Islamic Philosophy and Science* (Albany 1975) 5-14. There exists a Byzantine version of this horoscope, edited by J. Bidez in *Catalogus codicum astrologorum graecorum* (hereafter *CCAG*) 5.2 (Brussels 1906) 131-137.

[24] Pingree, "Astronomy and Astrology" (n. 2 above) 242.

[25] Pingree, "Indian Iconography" (n. 4 above).

[26] The allegation that the Persians observed the solar apogee in ca. A.D. 450 is made by Ibn Yūnus; see E. S. Kennedy and B. L. van der Waerden, "The World-year of the Persians," *JAOS* 83

8. In 556 Anūshirwān had his astronomers compose a *Zīk i Shahriyārān* which employed parameters from a *Zīj al-Arkand*, which evidently belonged to the *ārdharātrikapakṣa.*[27] These *ārdharātrika* parameters also were used in the *Zīk i Shahriyārān* written under Yazdijird III between 632 and 652,[28] and some were retained by early 'Abbāsid astronomers.

9. A series of horoscopes of the vernal equinoxes of the coronation years of the Sasanian kings was computed in early 'Abbāsid times by means of an Indian astronomical system close to the *ārdharātrikapakṣa,* though it is not clear whether this was available in Pahlavī or was introduced into Islam only in the eighth century.[29]

VII. SEVERUS SEBOKHT

Evidence for an Indian influence on Syriac science is more meager; one can conjecture that it came through Sasanian intermediaries. The earliest detectable reference to Indian science in Syriac is in the works of Severus Sebokht of Nisibis, the bishop of Qenneshrē in the middle of the seventh century.[30] In A.D. 662 he wrote the following (I quote the French translation of Nau):

> "J'omets maintenant de parler de la science des Hindous, qui ne sont même pas Syriens, de leurs découvertes subtiles dans cette science de l'astronomie – (découvertes) qui sont plus ingénieuses que celles des Grecs même et des Babyloniens – et de la méthode diserte de leurs calculs, et de leur comput qui surpasse la parole, je veux dire celui qui (est fait) avec neuf signes. Si ceux qui croient être arrivés seuls à la limite de la science parce qu'ils parlent grec, avaient connu ces choses, ils seraient peut-être convaincus, bien qu'un peu tard, qu'il y en a aussi d'autres qui savent quelque chose, non seulement des Grecs, mais encore des hommes de langue différente."[31]

The problem of the history of the Indian numerals to which Severus here alludes will be dealt with more extensively elsewhere; what interests us now is his reference to Indian discoveries in astronomy. Unfortunately, he is not specific, and the only

---

(1963) 315-327, esp. 323. For the Indian derivation of the parameter see D. Pingree, "The Persian 'Observation' of the Solar Apogee in ca. A.D. 450," *Journal of Near Eastern Studies* (hereafter *JNES*) 24 (1965) 334-336.

[27] See al-Hāshimī, section 7 (n. 5 above), and the commentary thereto. This *zīk* of Anūshirwān was used by Māshā'allāh; see E. S. Kennedy and D. Pingree, *The Astrological History of Māshā'allāh* (Cambridge, Mass. 1971) 69-88.

[28] See E. S. Kennedy, "The Sasanian Astronomical Handbook *Zīj-i Shāh* and the Astrological Doctrine of 'Transit' (*Mamarr*)," *JAOS* 78 (1958) 246-262.

[29] D. Pingree, "Historical Horoscopes," *JAOS* 82 (1962) 487-502, and (n. 20 above) 82-93.

[30] References will be found in Pingree (n. 6 above) 34-35, esp. nn. 14, 15, and 19.

[31] F. Nau, "Notes d'astronomie syrienne," *Journal Asiatique* 10.16 (1910) 209-228, esp. 225-227.

possible indication of an Indian influence that I have discovered in his published writings is his reference to the ascending and descending nodes of the moon as being called the head and the tail respectively.[32] They are ordinarily named the head and tail of a celestial snake called Rāhu in Sanskrit astronomical texts; thus, for example, in Varāhamihira's *Pañcasiddhāntikā* 7.2. Frequently the head is called Rāhu, the tail Ketu; of Ketu more will be said later on. But the concept of a snake or dragon (*gōčihr*) whose head and tail are the lunar nodes is also Sasanian; it is found, for example, in the *Bundahišn.*[33] It seems likely, then, that some lost Pahlavī work was an intermediary between the Indians and Severus Sebokht.

VIII. THEOPHILUS OF EDESSA

The same may be conjectured with respect to the Syrian astrologer, Theophilus of Edessa, who, after serving as advisor to the Caliph al-Mahdī (775-785), died – allegedly at the age of ninety – on 15/16 July 785.[34] Though best known for his lost translations of the Homeric epics into Syriac, Theophilus's three surviving works, all astrological, are preserved more or less intact only in Greek. They are:

1. *Πόνοι περὶ καταρχῶν πολεμικῶν*, in forty-one chapters, addressed to his son Deucalion. This work on military astrology has many similarities to such Sanskrit works as Varāhamihira's *Bṛhadyātrā,* written in about 550;[35] compare, for example, 1.2, which states that the invader is indicated by the ascendant, the besieged by the descendant, with *Bṛhadyātrā* 2.13.

2. [*Ἀποτελεσματικά*], in thirty chapters, also addressed to Deucalion. This is an introductory astrological textbook which also seems to have been influenced by Indian astrology. Compare, for example, 13 (see also Protagoras of Nicaea in Hephaestio 3.47.56),[36] which contains a topothesia of the zodiacal signs, with Sphujidhvaja, *Yavanajātaka* 1.14-25.[37]

This Indian zodiacal topothesia was also used by Abū Ma'shar in his *Kitāb al-madkhal al-kabīr* (6.8; see also Ghulām Zuḥal in London, British Library MS Add. 23,400 fol. 17 and al-Bīrūnī's *Tafhīm*).[38] Thence it was translated before 1000 into Byzantine Greek as *Μυστήρια* 3.22.[39] This appears in a fuller form in the Latin translation made by Hermann of Carinthia in 1140.[40]

[32] *Ibid.* 219-224.

[33] MacKenzie (n. 23 above) 515-516.

[34] I am now preparing an edition of his surviving works in Greek and Arabic from which the question of his sources should be clarified.

[35] Varāhamihira, *Bṛhadyātrā,* ed. D. Pingree (Madras 1972).

[36] Protagoras of Nicaea in *Hephaestio,* ed. D. Pingree, 2 vols. (Leipzig 1973-1974).

[37] These verses occur also in Mīnarāja's *Vṛddhayavanajātaka* (n. 4 above) as 1.4-15.

[38] R. R. Wright, ed., *The Book of Instruction in the Elements of the Art of Astrology* (London 1934) 221.

[39] Ed. I. Heeg, *CCAG* 5.3 (Brussels 1910) 131-132.

3. Περὶ καταρχῶν διαφόρων. This important text on catarchic astrology was originally divided into twelve sections, one for each of the astrological places; most of the first ten sections survive. It is likely that in this work also Theophilus has drawn upon Indian material available in Syriac, but no details can at present be given. It is also highly probable that some of the Indian material in the *Picatrix*, which will be discussed below, came through Syrian intermediaries at Ḥarrān.

IX. MĀSHĀ'ALLĀH

Māshā'allāh, the son of Atharī, a Persian Jew from Baṣra, was active as an astrologer in 'Irāq from 762 till 809; he died in about 815.[41] He wrote a large number of astrological treatises, of which some survive in Arabic, more in Latin translations, and a few fragments in Byzantine Greek. Like Theophilus he had access to Syriac sources, but more influential in his intellectual development was Sasanian science. Of interest to us are the following:

1. The *De elementis et orbibus coelestibus* or *De scientia motus orbis*[42] in twenty-seven chapters is a Latin translation by Gerard of Cremona of a lost Arabic original. Chapter 1 contains the statement that the Indians have said that God is the cause of things just as the sun is the cause of heat, without His knowing that He is the cause. It is not clear to me what Indian sect, if any, adheres to the view that the Creator is ignorant of his own creation, though Brahma (or anyone else) may well be indifferent.

In Chapter 20 Māshā'allāh argues that the heart of Leo and of Scorpio, and the Vulture and the Fish, were in the time of the Flood in certain degrees of the ecliptic, but that now their longitudes are greater. This statement concerning the precession of the equinoxes contains a reference to the "Era of the Flood," a Sasanian date corresponding to -3360, or 259 years before the beginning of the current Kaliyuga, according to Indian astronomers, at midnight of 17/18 February -3101 or at dawn of Friday 18 February -3101. This "Flood" era was used by Māshā'allāh in his *Kitāb fī al-qirānāt wa al-adyān wa al-milal,*[43] wherein he discusses horoscopes cast by means of the *Zīk i Shahriyārān* of Anūshirwān, which, as we have seen, was strongly influenced by Indian astronomy.

Others who cite this "Era of the Flood," or the Jupiter-Saturn conjunction signifying it in -3380,[44] and its difference from the beginning of the Kaliyuga, also

[40] Abū Ma'shar, *Introductorium in astronomiam,* published by E. Ratdolt (Augsburg 1489). I have not used the translation by John of Seville.

[41] D. Pingree in *DSB* 9 (New York 1974) 159-162.

[42] I. Heller, ed., *De elementis et orbibus coelestibus* (Nuremberg 1549); cf. also M. Power, *An Irish Astronomical Text* (London 1914). This work is discussed in some detail in Pingree (n. 23 above).

[43] Kennedy and Pingree (n. 27 above) 41-44, 70, 77-78, and 93-94.

[44] *Ibid.* 40-41, 70, 77, and 90-93.

called the "Era of the Flood," include al-Jahānī in his *De diversarum gentium eris, annis ac mensibus, et de reliquis astronomiae principiis,*[45] which survives only in a Latin translation of Gerard of Cremona; in chapter 27 he cites Kanaka,[46] an Indian who practiced astrology in Baghdād in the late eighth century, on this subject. One may also cite 'Umar ibn al-Farrukhān, a contemporary of Māshā'allāh and Kanaka, as quoted by an anonymous author in Newminster in 1428,[47] and Abū Ma'shar, who discusses the matter in his *Kitāb al-qirānāt* according to the Latin translation by John of Seville.[48]

Elsewhere in the *De orbibus* Māshā'allāh describes simple planetary models employing only eccenters and epicycles which are very similar to those described by Āryabhaṭa in about 499 and by his followers; however, these may be independent reflexions of lost Hellenistic models.[49]

Another element of Indian (and Sasanian) astronomy, the chords of wind that cause the inequalities in planetary motion, is referred to by Māshā'allāh in two fragments quoted together by 'Alī ibn abī al-Rijāl in his *Kitāb al-bāri'*; there is a Latin translation by Aegidius de Thebaldis.[50] The *locus classicus* in Sanskrit astronomy for the chords of wind that cause the inequalities of planetary motion is in the *Sūryasiddhānta.*[51] This theory of the planetary chords is also well established as having been popular among Sasanian astronomers, as has been noted above.

One other quasi-astronomical concept of Māshā'allāh's which may have, ultimately, an Indian origin is found in his *De ratione circuli et stellarum et qualiter operantur in hoc seculo,* also known as the *Epistola Messahalae de rebus eclipsium et de coniunctionibus planetarum in revolutionibus annorum mundi,* which survives in a Latin translation by John of Seville[52] and Hebrew by Abraham ibn Ezra.[53] In chapter 1 Māshā'allāh compares the influence of the planets on the earth to the influence of a magnet on iron. In the sixth century Varāhamihira had stated that the

[45] Al-Jahānī, *De diversarum gentium eris . . .*, ed. I. Heller (Nuremberg 1549); on al-Jahānī or al-Jayyānī, an eleventh-century astronomer from Jaén in Andalusia, see Y. Dold-Samplonius and H. Hermelink in *DSB* 7 (New York 1973) 82-83.

[46] See D. Pingree in *CESS* A2.19a-19b, and in *DSB* 7.222-224.

[47] Kennedy and Pingree (n. 27 above) 191.

[48] Abū Ma'shar, *De magnis coniunctionibus* 1.1; printed for Melchior Sessa by Iacobus Pentius de Leucho (Venice 1515); see also Pingree (n. 20 above) 41-42.

[49] Pingree (n. 23 above).

[50] Ibn abī al-Rijāl, *De iudiciis astrorum* 5.15; ed. A. Stupa (Basel 1551); on Ibn abī al-Rijāl, an eleventh-century astrologer from Qayrawān in Tunisia, see D. Pingree in *Encyclopaedia of Islam,* new ed. 3 (Leiden 1971) 688 (hereafter *EI*).

[51] MM. S. Dvivedī, ed., *Sūryasiddhānta* 2.1-2, Bibliotheca Indica 173 (Calcutta 1925). The position of the *śīghrocca* on its epicycle, it should be noticed, is determined by the longitude of the mean sun measured on the concentric deferent. The theory of chords binding the planets is also referred to by Ibn al-Muthannā, ed. Millás (n. 68 below) 120, ed. Goldstein (n. 69 below) 46 and 172.

[52] Māshā'allah, *De ratione circuli . . .*, ed. I. Heller (Nuremberg 1549), and N. Pruckner in *Iulii Firmici Materni Astronomicω̃n libri VIII* (Basel 1551) pt. 2, 115-118.

[53] B. Goldstein, "The Book on Eclipses of Māshā'allāh," *Physis* 6 (1964) 205-213.

earth is kept at the center of the universe like a piece of iron at the end of a magnet.[54]

Another fragment of Māshā'allāh in Greek describes the Indian theory of the *navāṃśas,* how to find their lords, and their use in interrogations. These *navāṃśas* first appear in the *Yavanajātaka* of Sphujidhvaja;[55] that the Sasanians had knowledge of them is proved by their having been inserted in the Pahlavī version of Dorotheus, as has been noted previously. This text is edited from Vaticanus graecus 1056 (V) fol. 48rv in Appendix 1, together with a related Byzantine text.[56]

A knowledge of methods of finding the lords of the *navāṃśas* also reached the West through the works of Abū Ma'shar. I edit in Appendix 2 *Μυστήρια* 1.140 from Angelicus graecus 29 (E) fols. 34v-35.

Abū Ma'shar also describes methods of finding the lords of the *navāṃśas* in his *Kitāb al-madkhal al-kabīr* 5.14, which was twice translated into Latin as we have seen, but was omitted by the Byzantine compiler of *Μυστήρια* 3, and in his *Kitāb aḥkām sinī al-mawālīd* 3.9, which was translated into Greek[57] before 1000 and from Greek into Latin[58] in the thirteenth century. Another route by which a knowledge of these *navāṃśas* reached the West was through the Latin translation by John of Seville of al-Qabīṣī's *Al-madkhal ilā ṣinā'at aḥkām al-nujūm.*[59]

X. Zīj al-Sindhind

Though some Indian astronomical texts related to the *Āryabhaṭīya* written by Āryabhaṭa[60] (whose epoch is 499) and to the *Khaṇḍakhādyaka* written by Brahmagupta[61] (whose epoch is 665) had influenced Arabic astronomy in the first half of the eighth century, the primary direct infusion of Indian material into Islam occurred when an embassy from Sind to the court of al-Manṣūr in Baghdād in 771 or 773 included an Indian who had the text of a *siddhānta,* probably entitled *Mahāsiddhānta,* which belonged to the Brāhmapakṣa of Indian astronomy.[62] This was

[54] *Pañcasiddhāntikā* (n. 3 above) 13.1.

[55] *Yavanajātaka* (n. 3 above) 1.35, 1.41, 52.9; cf. 59.1-11.

[56] Part of this related text is copied in pseudo-Palchus 114, edited by S. Weinstock in *CCAG* 9.1 (Brussels 1951) 161-168. This same chapter in pseudo-Palchus contains an excerpt from a similar text preserved in the tenth- or eleventh-century Byzantine translation of an astrological compendium ascribed to Aḥmad the Persian, wherein it is 2.26.

[57] Abū Ma'shar, *De revolutionibus nativitatum,* ed. D. Pingree (Leipzig 1968); see also the recension of Isidore of Kiev on pp. 240-244, and 9.7.

[58] Ed. H. Wolf in *Εἰς τὴν τετράβιβλον τοῦ Πτολεμαίου ἐξηγητὴς ἀνώνυμος* (Basel 1559) 207-279.

[59] Al-Qabīṣī, *Libellus ysagogicus,* published by E. Ratdolt (Venice 1482) d 3; on al-Qabīṣī, a tenth-century astrologer from Aleppo, see D. Pingree in *EI* 4.340-341, and in *DSB*.

[60] Al-Hāshimī section 3 (n. 5 above).

[61] *Ibid.* section 4.

[62] For the *pakṣas* see D. Pingree, "On the Classification of Indian Planetary Tables," *JHA* 1 (1970) 95-108; on Brahmagupta see n. 9 above.

rendered into Arabic with additions from other sources by al-Fazārī[63] as the *Zīj al-Sindhind al-kabīr.* Al-Fazārī wrote several other astronomical works related to the *Sindhind,* as did also Ya'qūb ibn Ṭāriq.[64]

This material was used by al-Khwārizmī[65] in his *Zīj al-Sindhind,* composed during the reign of al-Ma'mūn (813-833). Though this work is now lost in Arabic save for some fragments, it was revised in Spain in the tenth century by al-Majrīṭī and Ibn al-Ṣaffār, and in that recension became the first serious text on mathematical astronomy known to the Latin West when Adelard of Bath translated it in 1126.[66] The *Zīj al-Sindhind* was commented on in the ninth century by several authors including al-Farghānī;[67] this commentary was used in the tenth century as a basis for his own exegesis by Ibn al-Muthannā. The Arabic original of Ibn al-Muthannā's work is lost, but was translated into Latin by Hugh of Sanctalla[68] in the twelfth century, and into Hebrew by Abraham ibn Ezra[69] in the same century; Abraham also used it in his *De rationibus tabularum.*[70] Further Arabic texts dependent on the *zīj* of al-Khwārizmī and written in Spain in the eleventh century include the *Tabulae Jahen,* translated into Latin in the twelfth century by Gerard of Cremona,[71] and the *Toledan Tables,* apparently also translated by Gerard.[72] The *Toledan Tables* were translated into Greek on Cyprus in the fourteenth century, apparently by George Lapithes.[73] One final channel through which the *Sindhind* reached the West was through the works of Abū Ma'shar, who used it in his lost *Zīj al-hazārāt.*[74]

In the following discussion of the discernible impact of the *Sindhind* tradition on Byzantine and medieval Latin astronomy and astrology, I follow the order of the *Brāhmasphuṭasiddhānta (BSS).*[75] I will refer for the *ārdharātrikapakṣa* or *Arkand,*[76] which also influenced the *Sindhind,* to Brahmagupta's *Khaṇḍakhādyaka* (*Kh*).[77]

63 D. Pingree, "The Fragments of the Works of al-Fazārī," *JNES* 29 (1970) 103-123.

64 D. Pingree, "The Fragments of the Works of Ya'qūb ibn Ṭāriq," *JNES* 27 (1968) 97-125.

65 G. J. Toomer in *DSB* 7.358-365.

66 A. Bjørnbo, R. Besthorn, and H. Suter, eds., *Die astronomischen Tafeln des Muḥammad ibn Mūsā al-Khwārizmī,* Hist.-filos. Skr. (n. 3 above) 3.1 (Copenhagen 1914); O. Neugebauer, *The Astronomical Tables of al-Khwārizmī,* Hist. Filos. Skr. Dan. Vid. Selsk. 4.2 (Copenhagen 1962). On the two independent examinations of the Arabic material represented in Latin manuscripts see G. J. Toomer's review of Neugebauer's book in *Centaurus* 10 (1964) 203-212, esp. 210-212.

67 See al-Hāshimī sections 11, 15 and 34 (n. 5 above); on al-Farghānī see A. I. Sabra in *DSB* 4 (New York 1971) 541-545.

68 E. Millás Vendrell, ed., *El comentario de Ibn al-Muṯannā' a las Tablas Astronómicas de al-Jwārizmī* (Madrid 1963).

69 B. R. Goldstein, *Ibn al-Muthannā's Commentary on the Astronomical Tables of al-Khwārizmī* (New Haven 1967); this book includes a second Hebrew translation also. On Abraham see the inadequate article by M. Levey in *DSB* 4.502-503.

70 Abraham ibn Ezra, *El libro de los fundamentos de las tablas astronómicas,* ed. J. M. Millás Vallicrosa (Madrid 1947).

71 I. Heller, ed., *Tabulae Jahen* (Nuremberg 1549).

72 See G. J. Toomer, "A Survey of the Toledan Tables," *Osiris* 15 (1968) 5-174.

73 See D. Pingree, "The Byzantine Version of the *Toledan Tables*: The Work of George Lapithes?", to appear in *Dumbarton Oaks Papers* (hereafter *DOP*) 30.

74 See Pingree (n. 20 above).

75 S. Dvivedin, ed., *Brāhmasphuṭasiddhānta* (Benares 1902).

The length of a Caturyuga according to *BSS* 1.7a-b is 4,320,000 years, that of a Kalpa according to *BSS* 1.10 is 4,320,000,000 years. Though these values were well known to Arab astronomers, and corruptly to Abraham ibn Ezra,[78] they do not to my knowledge appear in any Greek or Latin text. Moreover, the planets, their nodes, and their apogees are stated in *BSS* 1.14 to conjoin at the beginning of Aries at the beginning and the end of a Kalpa. This is clearly stated by Ibn al-Muthannā.[79]

The mean motions of the planets in the *Sindhind* and the longitudes of their apogees and nodes are (save in the case of Saturn) directly derived from their rotations in a Kalpa as given by the *BSS,* and their mean daily motions from that and the number of civil days in a Kalpa assumed by the *BSS.* This information is given in *BSS* 1.15-22.

These rotations are precisely identical with those given by al-Fazārī with the exceptions that he gives Saturn 146,569,284 instead of 146,567,298, and that he allows the sphere of the fixed stars to rotate 120,000 times while Brahmagupta denies any motion to the sphere of the fixed stars.[80] *BSS* 1.22 indicates that the civil days in a Kalpa number 1,577,916,450,000; this number is correctly given by Ibn al-Muthannā.[81] Ibn al-Muthannā proceeds to give the correct procedure for determining a planet's mean daily motion. The number is less correctly preserved by Abraham ibn Ezra,[82] who also refers to the conjunctions at the beginning and end of the Kalpa.

If one follows the procedure prescribed by Ibn al-Muthannā, one finds that the mean daily motions of the planets and of the moon's apogee and node are precisely identical in al-Fazārī and in tables 4-20 of al-Khwārizmī's *zīj*;[83] these in turn are identical with the values in the *BSS* save in the case of Saturn:

BSS	*al-Fazārī*
0;2,0,22,51,43,54$^{o/d}$	0;2,0,22,57,36,16$^{o/d}$

One aspect of the mean solar motion in the *BSS* is the fact that a sidereal year contains 6,5;15,30,22,30 days. This parameter appears in table 115 of al-Khwārizmī's

[76] The *Zīj al-Arkand* is referred to by name by Abraham ibn Ezra as contradicting the number of civil days in a Kalpa of the *Sindhind* (n. 70 above) 89. Abraham's statement is true; the *ārdharātrikapakṣa* uses a different year-length from that of the Brāhmapakṣa, and in any case employs a Caturyuga with mean grand conjunctions at Aries 0° instead of a Kalpa with true grand conjunctions.

[77] Brahmagupta, *Khaṇḍakhādyaka,* ed. B. Chatterjee, 2 vols. (Calcutta 1970).

[78] In the preface to his Hebrew translation of Ibn al-Muthannā, Goldstein (n. 69 above) 147-148; this is Pingree (n. 64 above) fr. Z 2.

[79] Millás (n. 68 above) 105; Goldstein (n. 69 above) 26 and 152.

[80] Pingree (n. 63 above) fr. Z 5; cf. Pingree (n. 64 above) fr. Z 1.

[81] Millás (n. 68 above) 105; cf. Goldstein (n. 69 above) 26 and 152.

[82] Abraham ibn Ezra (n. 70 above) 88-89.

[83] See Neugebauer (n. 66 above) 90-94; see also chapters 18-21, and J. J. Burckhardt in *Vierteljahresschrift der naturforschende Gesellschaft, Zürich* 106 (1961) 213.

zīj[84] and is directly expressed by Ibn al-Muthannā[85] in two forms: as days and hours, and as an excess of revolution in time-degrees. For 0;15,30,22,30 days = ¼ day + 0;12,9 hours; and 0;15,30,22,30 days = 93;2,15 time-degrees. Roundings of these parameters are given by Abraham ibn Ezra;[86] for ¼ + 1/120 = 0;15,30, which number of days corresponds to 93 time-degrees.

As the year-length indicates, the zodiac of the *BSS* is fixed with respect to the fixed stars.[87] The result of this situation is obviously that discrepancies in longitude will arise between tables based on Indian parameters and those using a tropical zodiac. This was noticed by Abraham ibn Ezra:

> Nam si quis velit coequare planetas in tabulis Indorum secundum tempus Ptolomei, inveniet Solem intrasse caput Arietis 7 diebus ante illum diem in quo intrare Ptholomeus probavit illum. Si autem velimus equare Solem cum intrat caput Arietis secundum magistros probationum vel cum alibi est in nostro tempore, inveniemus eum intrare caput Arietis 9 diebus ante illum diem in quo secundum tabulas Indorum intrare invenietur, et hoc quidem negari non potest cum possit fieri huius rei manifestatio et per astrolabium et per umbram.[88]

Abraham states that he wrote this work in 1154,[89] that is, about 600 years after the Indian zodiac was fixed; at that time there was a difference of about 3° between the beginnings of Ptolemy's tropical and the Indians' sidereal zodiac. The two coincided in A.D. 837, so that the difference in the date of the beginning of the solar year in 1154 should be only about 3 days instead of 7. Abraham's error arises from comparing the longitudes of Regulus according to Ptolemy, the *Zīj al-Mumtaḥan,* and the Indians:

> Similiter invenimus quod Ptholomeus probavit in suo tempore cor Leonis in 3° gradu Leonis esse fere, hodie vero invenitur probatione in 18° gradu Leonis; Indi vero asserunt cor Leonis semper inmobilem consistere in 25 minuto 10ⁱ gradus.[90]

The 7 presumably is an approximation to 10;25 − 3, the 9 an inaccurate result of 18 − 10;25. But in fact the longitude of Regulus in 137 according to Ptolemy was Leo 2;30°, so that in 1154 it would have been Leo 12;40°. The longitude of Regulus according to *BSS* 10.1-8 is Leo 9°; Abraham's 10;25° comes from applying to that longitude 1;25° of precessional motion, which represents the 142 years between the date of the *BSS,* 628, and the date of al-Fazārī, 770, at the Ptolemaic rate of precession that al-Fazārī followed. The rate of precession in the *Zīj al-Mumtaḥan* is 1° in 66 years, as will be seen below; this results in a longitude for Regulus of Leo 17;54°, which is close enough to Abraham's figure.

[84] Neugebauer 131.
[85] Millás (n. 68 above) 200; Goldstein (n. 69 above) 133-134.
[86] Abraham ibn Ezra (n. 70 above) 75; see also 79.
[87] See Pingree, "Precession and Trepidation" (n. 4 above).
[88] Abraham ibn Ezra (n. 70 above) 81; see also 86.
[89] *Ibid.* 109.

Though Abraham recognizes the fixed nature of the Indian zodiac, yet he refers to an Indian theory of trepidation,[91] employing two small circles whose centers are the beginnings of Aries and Libra respectively, and having an amplitude of 8°. It is certainly true that the Indians had various theories of precession and trepidation derived from the Greeks,[92] though Brahmagupta denies them; but neither the model nor the parameter ascribed to them by Abraham is attested in the Sanskrit texts investigated by me.

Before discussing the longitudes of the apogees and the nodes in al-Khwārizmī's *zīj,* we must note that a knowledge of their motion was a part of the *Sindhind* tradition;[93] in fact, Abū Ma'shar apparently stated that the motion of the sun's apogee in 1000 years is 0;2,14°, when according to the *BSS* it should be 0;2,24°.[94] Abraham ibn Ezra states that the node of Saturn moves less than 0;2° in 1000 years.[95] According to the verses of the *BSS* cited above, Saturn's node rotates 584 times in a Kalpa – that is, it travels 210,240 degrees in 4,320,000,000 years, or 0;2,55,12° in 1000 years. Perhaps one should read *tria* in place of *dua* in Abraham's text.

Elsewhere Abraham apparently describes the nodal motion as retrograde.[96] The apogees of the five star-planets, however, are considered by the Indians to be fixed with respect to the fixed stars according to Abraham.[97] This refers to the sidereally fixed apogees of the five star-planets in the *ārdharātrikapakṣa*; see *Kh* 2.6a-b.

In order to determine the mean longitudes of any of the planets, apogees, or nodes at a particular time, one must first know the interval from the beginning of the Kalpa till that time. Brahmagupta divides the interval into three parts: from the beginning of the Kalpa till the beginning of the present Kaliyuga (which was at sunrise at Laṅkā on Friday 18 February -3101) is 1,972,944,000 years; from the beginning of the Kaliyuga till the beginning of the Śaka era (which was on 16 March 78) is 3179 years; and from the beginning of the Śaka era till the desired time. The first two intervals and their sum are given in *BSS* 1.26-27. The first interval was known to Abū Ma'shar in days, along with the correct intervals in days from the beginning of the Kalpa to various other epochs and historical events;[98] I quote here a long passage from the Latin translation of his *De magnis coniunctionibus*:

> (a) Et estimaverunt Indi quod principium fuit die dominica Sole ascendente, et est inter eos (scilicet inter illum diem et illum diem Diluvii)

[90] *Ibid.* 82; see also 84.
[91] *Ibid.* 77.
[92] Pingree, "Precession and Trepidation" (n. 4 above).
[93] Al-Hāshimī sections 25 and 30 (n. 5 above); Millás (n. 68 above) 106-107; Goldstein (n. 69 above) 27 and 153, Pingree (n. 20 above) 33.
[94] Pingree (n. 20 above) 50 n. 1.
[95] Abraham ibn Ezra (n. 70 above) 109.
[96] *Ibid.* 102.
[97] *Ibid.* 101.
[98] Pingree (n. 20 above) 34-44 and 130.

septingenta milia milium milium et viginti milia milium milium et sexcenta milia milium et triginta quatuor milia milium et quadraginta [*read* quadringenta] milia et quadraginta duo milia et septingenta et quindecim dies, qui erunt anni Persici milies milies milia et nongentesies mille [*read* milies] milia <· · ·> et trecenties mille et quadragesies mille et nongenti et triginta octo anni et trecenti quadraginta quatuor [*read* quinque] dies. Et fuit Diluvium die Veneris vicesimo septimo die mensis Rabe Primi, et est dies 29 ex Cibat, et est dies decimus quartus ex Adristinich.

(d) Fuerunt ergo inter Diluvium et primum diem anni in quo fuit Alhigira 3837 anni et 268 dies, et erunt secundum annos Persarum 3725 anni et 348 dies.

(e) Et inter Diluvium et diem Dezdabir regem [*read* regis] Persarum, ab inicio regni cuius ceperunt Persici eram, scilicet die Martis, scilicet illo die † exarisan, fuerunt 3735 anni et 10 menses et 22 dies secundum annos Persarum. Et similiter interfecerunt Gezdagirth die Martis 22 diebus Rabe Primi anno 11 de Alhigera.

(j) Et inter diem primum anni Alhigere et regum [*read* regem] Gezdagir fuerunt 3634 [*read* 3624] dies; erunt secundum annos Persicos 9 anni et 11 menses et 9 dies.

(c) Et fuerunt inter Diluvium et [inter] tempus Habentis duo cornua secundum annos Romanos 2790 et 126 [*read* 226] dies.

(i) Et inter duo cornua Habente<m> et primum annum de Alhigera 932 anni et 287 dies secundum annos Romanos; et erunt Arabici 974 anni et 294 dies.

(b) Inter Diluvium et Philippum, super <quem> coequantur tabule Ptholomei et cum quo faciunt eram suam Egyptii, 2778 anni et <268 dies secundum annos Romanorum; et erunt 2780 anni et> 232 dies secundum annos Egiptiorum; et sunt convenientes duo cornua Habentia [*read* Habentis] annis.

(f) Et inter annos Philippi et Habentia [*read* Habentis] duo cornua 22 [*read* 11] anni et 316 [*read* 323] dies; et Philippus fuit prior.

(g) Et inter Philippum et Alhigeram 946 [*read* 945] anni et 316 [*read* 116] dies secundum annos Egiptiorum et Persarum; et erunt Arabici 974 [*read* 973] anni et 313 [*read* 243] dies.

(h) Et inter Philippum et Gezdargid 950 [*read* 955] anni et 90 dies fuerunt.[99]

The correct intervals, using Friday 18 February −3101 as the epoch of the Flood, are:

[99] It follows *De magnis coniunctionibus* 4.12 (n. 48 above).
[100] Pingree (n. 20 above) 43-44.
[101] *Ibid.* 38-39.

Interval	*Years and Days*	*Days*
a. Kalpa → Flood	1,972,944,000 (Indian) 1,974,340,938 (Persian) and 345 days	720,634,442,715
b. Flood → Philip	2,780 (Persian) and 232 days 2,778 (Julian) and 268 days	1,014,932
c. Flood → Alexander	2,790 (Julian) and 226 days	1,019,273
d. Flood → Hijra	3,837 (Arab) and 269 days 3,725 (Persian) and 348 days	1,359,973
e. Flood → Yazdijird	3,735 (Persian) and 322 days	1,363,597
f. Philip → Alexander	11 (Julian) and 323 days	4,841
g. Philip → Hijra	945 (Persian) and 116 days 973 (Arab) and 243 days	345,041
h. Philip → Yazdijird	955 (Persian) and 90 days	348,665
i. Alexander → Hijra	932 (Julian) and 287 days 962 (Arab) and 154 days	340,700
j. Hijra → Yazdijird	9 (Persian) and 339 days	3,624

a. Kalpa → Flood. The interval in days is correct, but that in Persian years and days is short by one day – that is, the given interval is that between the beginning of the Kalpa and the Thursday preceding the Flood. This may be due to a misunderstanding of the difference between the midnight and the sunrise epochs, or it may simply be a scribal error. The calendar-dates for the epoch of the Flood are taken from Abū Ma‘shar, whose computations have been discussed elsewhere.[100]

b. Flood → Philip. If one assumes that something has dropped from the text, the computation is correct, and does not contain the erroneous date of Philip embedded in Abū Ma‘shar's computation as preserved by al-Hāshimī.[101]

c. Flood → Alexander. If a minor scribal error is corrected, the computation is correct.

d. Flood → Hijra. The difference between the 269 days of my computation and the 268 days of Abū Ma‘shar's arises from his erroneously rounding off to one the fractional intercalary day involved in 3,837 Arab years. Otherwise his computation is correct.

e. Flood → Yazdijird. The computation is correct. 22 Rabiʿ I of 11 A.H. is 16 June 632, the epoch of the era of Yazdijird.

f. Philip → Alexander. The 316 days here is probably a scribal error influenced by the erroneous 316 in the next interval, g.

g. Philip → Hijra. Both year-numbers are too great by one, and both day-numbers were wrongly copied.

h. Philip → Yazdijird. The year-number was wrongly copied.

i. Alexander → Hijra. The computation of this interval in Arab years and days is totally wrong; the source of the error is not clear to me.

j. Hijra → Yazdijird. The computation is correct.

Table 1 of the *zīj* of al-Khwārizmī gives many of these intervals, but with much greater confusion and using 17 February −3101 as the epoch of the Flood.[102] Some of these intervals also appear in various manuscripts of the *Toledan Tables.*[103] And an elaborate chronological table is included among the Arabic texts translated into Greek by Gregory Chioniades in about 1300;[104] it does not appear in the Arabic manuscripts of the *Zīj al-Sanjarī,* so it probably was taken from the lost *Zīj al-ʿAlāʾī* written by al-Fahhad in the twelfth century. The author uses the scheme set forth by Abū Maʿshar in his *Kitāb al-ulūf* and *Zīj al-hazārāt* wherein a *yuga* contains 360,000 years split in half by the occurrence of the Flood on Friday 18 February −3101.[105] I edit the text from fol. 167 of Vaticanus graecus 211 in Appendix 3.

In this list the epoch of the era of Noah is Friday 18 February −3101; that of the era of Nabuchodonosor is Thursday 26 February −746; that of Alexander (the era of Philip) is Sunday 12 November −323; that of the Romans (the Seleucid era) is Monday 1 October −311; that of the Arabs (the era of the Hijra) is Friday 16 July 622; that of the Persians (the era of Yazdijird) is Tuesday 16 June 632; and that of the Sulṭān Malikshāh is Tuesday 12 March 1079. The table gives intervals in years (Persian in lines 1, 2, 3, and 6; Julian in line 4; and Arab in line 5) and in days, with the latter expressed in both decimals (with Indo-Arabic numerals) and sexagesimals (with Greek numerals). The period from the Flood to the Hijra is too much by one day — that is, the Flood is dated 17 February −3101 — both in the day-interval and the year-interval; however, for computing intervals between other epochs and the Hijra, the correct number of days is used except in the case of the epoch of Malikshāh. The interval in days between the Hijra and Malikshāh is 166,794 or 1526768 – 1359974; the interval should be 166,795 days. In the intervals between the Hijra and the era of Yazdijird expressed in years, an Arab year is assumed to be 354½ days rather than 354 11/30 days; the latter value would give 10 years and 82 days. In the interval between the Hijra and the era of Malikshāh, though the correct

[102] Neugebauer (n. 66 above) 82-84; cf. chap. 1 of al-Jahānī (n. 45 above) N ii – O iii.

[103] Toomer (n. 72 above) 18, and Pingree (n. 73 above).

[104] D. Pingree, "Gregory Chioniades and Palaeologan Astronomy," *DOP* 18 (1964) 133-160.

[105] Pingree (n. 20 above) 27-31.

Arab year-length is used, the erroneous interval of 166,794 days also is utilized. Finally, it should be noted that the period before the Flood is taken to be 180,000 Persian years of 365 days each, whereas Abū Ma'shar's years contain 365;15,32,24 days each.

Given the interval of 1,363,597 days from the beginning of the Kaliyuga to the epoch of Yazdijird and the 720,634,442,715 days from the beginning of the Kalpa to the beginning of the current Kaliyuga, one can determine with the Brāhmapakṣa's rotations in a Kalpa the longitudes of the apogees and nodes on 16 June 632.[106] The computed longitudes are precisely identical with those of al-Khwārizmī,[107] except that Ibn al-Ṣaffār has misread the entry for Saturn – 8^S 20;55° with a *kaf* (20) – as 8^S 4;55° with a *dal* (4). This confusion is noted by Abraham ibn Ezra.[108]

Brahmagupta's basic formula for finding planetary mean longitudes for mean sunrise at the meridian of Laṅkā, on which lies also Ujjayinī, is:

$$r = R \cdot \frac{c}{C} \ ,$$

where R is the planet's revolutions in a Kalpa, C the civil days in a Kalpa, c the lapsed civil days (*ahargaṇa*), and r the lapsed revolutions. The rule is given in *BSS* 1.31; it is repeated by Ibn al-Muthannā.[109] Laṅkā is not mentioned in the *Sindhind* tradition, but Arab astronomers place in the same geographical location (on the equator on the prime meridian) the city of Ujjayinī, which in the Indian tradition is at a latitude of 24° N on the prime meridian. Ujjayinī was originally transliterated Ūzayn,[110] but this was soon corrupted to Arīn. It is often referred to as Arin in the Latin translations of al-Khwārizmī,[111] Ibn al-Muthannā citing al-Farghānī,[112] and al-Jahānī,[113] and in later authors.

Brahmagupta gives a general and correct rule for applying a correction for the longitudinal difference between Laṅkā and one's location in *BSS* 1.34. A rule for computing the longitudinal difference, based on the assumption that the longitude of Arin is 90° E, was given by al-Khwārizmī.[114]

Brahmagupta gives the following formula for finding the accumulated epact, ϵ.

$$\epsilon = \left(10 + \frac{2481}{9600} + \frac{7739}{9600}\right) y$$

[106] See Pingree (n. 64 above) fr. Z 1.

[107] Chaps. 8, 18, and 19, and tables 27-56 (see Neugebauer [n. 66 above] 99 and 103); see also Pingree (n. 26 above). The longitudes of various apogees are also referred to by Abraham ibn Ezra (n. 70 above) 77-78; 121; and 147; cf. chap. 9 of Al-Jahānī (n. 45 above) Q i v-Q ii.

[108] Abraham ibn Ezra (n. 70 above) 109-110.

[109] Millás (n. 68 above) 126; Goldstein (n. 69 above) 26-27 and 152-153.

[110] Pingree (n. 20 above) 45.

[111] Introduction; chaps. 7, 24, and 25; see also Neugebauer (n. 66 above) 86, 151, and 211.

[112] Millás (n. 68 above) 122; Goldstein (n. 69 above) 49.

[113] Al-Jahānī, chap. 8 (n. 45 above) Q 1, and chap. 27, Z i v.

[114] Chap. 7; cf. Millás (n. 68 above) 122; Goldstein (n. 69 above) 49.

where y is the number of years elapsed since the beginning of the Kalpa. The rule is based on the calculation that the epact in the normal sense equals 11;3,52,30 tithis. The rule is found in *BSS* 1.39-40, and appears among the fragments of Ya'qūb ibn Tāriq.[115] It does not occur in the Latin translation of al-Majrīṭī's recension of the *zīj* of al-Khwārizmī, but a part of it appeared in the original, as we know from a lemma of Ibn al-Muthannā.[116] There the fraction 2481/9600 is the fraction of a day by which the length of a year in the *Sindhind* exceeds 365 days; for

$$\frac{2481}{9600} = 0;15,30,22,30.$$

In the second chapter of the *Brāhmasphuṭasiddhānta* Brahmagupta deals with the computation of the true longitudes of the planets. He begins with a versified table of Sines (*BSS* 2.2-5) and Versines (*BSS* 2.6-9); these are the only two trigonometrical functions normally tabulated by Indian astronomers, though they also make frequent use of the Cosines. In the *Brāhmasphuṭasiddhānta* R = 3270, which is a value known to al-Fazārī,[117] and there are 24 entries in a quadrant so that the interval between arguments is 3;45°, the standard Indian value. But in the *Khaṇḍakhādyaka* R = 150 and there are 6 entries in a quadrant, so that the interval between arguments is 15°; see *Kh* 3.6, which is derived from *BSS* 25.16. This value of R was also known to al-Fazārī,[118] who apparently derived it from the Sasanian *Zīj al-Shāh* or the *Zīj al-Arkand.* The tables of Sines in Adelard's Latin translation of al-Majrīṭī's version of al-Khwārizmī's *zīj* use R = 60,[119] but it is known that in the Arabic original R = 150. Therefore we find a diameter of 300 in Ibn al-Muthannā.[120] This value of R is also referred to by Abraham ibn Ezra,[121] though with some confusion as he infers that the Indians made the radius rather than the diameter equal to 300. The value R = 150 is also used in the *Toledan Tables*[122] and in an anonymous Byzantine treatise of the late eleventh century.[123]

It is also clear that in al-Khwārizmī's *zīj* the intervals of argument in the Sine-table were 15° as in the *Khaṇḍakhādyaka*; he and other Arabic astronomers call the interval of argument in any table a *kardaja,* a term derived through a Pahlavī intermediary from the Sanskrit *kramajyā* (Sine; as opposed to *utkramajyā,* Versine). The evidence is found in Ibn al-Muthannā again.[124]

115 Pingree (n. 64 above) fr. T 9.

116 Millás (n. 68 above) 200; Goldstein (n. 69 above) 144.

117 Pingree (n. 63 above) frs. Z 12 and Z 16.

118 *Ibid.* frs. Z 11, Z 13, Z 15, and Z 25.

119 Tables 58 and 58a; but see Neugebauer (n. 66 above) 54 and 104.

120 Millás (n. 68 above) 124; Goldstein (n. 69 above) 51 and 178; cf. Abraham (n. 70 above) 131.

121 Abraham (n. 70 above) 126-127.

122 Toomer (n. 72 above) 27; Pingree (n. 73 above). See also O. Neugebauer and O. Schmidt, "Hindu Astronomy at Newminster in 1428," *Annals of Science* 8 (1952) 221-228.

123 O. Neugebauer, *A Commentary on the Astronomical Treatise Par. gr. 2425,* Mémoires de l'Académie royale de Belgique, Classe des lettres 69.4 (Brussels 1969) 27-33 and 37-38.

The planetary model used by Brahmagupta in both the *Brāhmasphuṭasiddhānta* and the *Khaṇḍakhādyaka* is that employing two epicycles,[125] whereas that employed by al-Khwārizmī is a model employing an eccenter and an epicycle. Such an eccenter-epicycle model was also known to Brahmagupta (*BSS* 14.1-18) and to other Indian astronomers, but does not seem to have been often used in computations. But al-Khwārizmī (following al-Fazārī) probably derived the models as also many of the parameters from the Sasanian *Zīj al-Shāh.*

The equation of the center depends on the planet's distance from its apogee. We have already seen that the longitudes of the apogees according to al-Khwārizmī were computed from the elements of the *Brāhmasphuṭasiddhānta.* It remains to be noted that the Arabic term for apogee, *al-awj,* is derived from the Sanskrit *ucca*; in Latin it appears as *elaug* or *alauge.*[126] The maximum equations of the center according to the *ārdharātrikapakṣa* and the *Zīj al-Shāh,*[127] and according to al-Khwārizmī[128] are:

	Ārdharātrikapakṣa	*Zīj al-Shāh*	*al-Khwārizmī*
Saturn	9;36°	8;37°	8;36°
Jupiter	5;6°	5;6°	5;6°
Mars	11;10°	11;12°	11;13°
Sun	2;14°	2;14°	2;14°
Venus	2;14°	2;13°	2;14°
Mercury	4;28°	4;0°	4;2°
Moon	4;56°	4;56°	4;56°

Close approximations to the *ārdharātrika* values are given in *Kh* 2.6c-7.

In al-Khwārizmī's tables the maximum equations of the center occur at arguments of 90° because they are computed according to the so-called "Method of Declinations" (sun and moon)[129] or "Method of Sines" (five star-planets).[130] Brahmagupta computes all equations of the center according to the "Method of Sines" (*Kh* 2.6c-7 cited above in conjunction with *Kh* 1.16-17). The "Method of Declinations," however, was also known to Brahmagupta, whose declinations in any case form a sine-function (*BSS* 2.55). The rule is given more directly in *BSS* 3.61-62. Incidentally, this verse informs us that the obliquity of the ecliptic is assumed to be 24° by Brahmagupta as by all other Indian astronomers. This was the value originally used by al-Khwārizmī in one table according to Ibn al-Muthannā, though in another he

[124] Millás (n. 68 above) 123; Goldstein (n. 69 above) 49 and 176; cf. Abraham (n. 70 above) 130, and Millás (n. 68 above) 127.

[125] Pingree, "On the Greek Origin" (n. 4 above). The varying velocities of the planets in the different quadrants of the Indian epicycle are referred to by Abū Ma'shar 7.1 (n. 40 above).

[126] Al-Khwārizmī, chaps. 8 and 18; Millás (n. 68 above) 107.

[127] Pingree (n. 64 above) fr. Z 7; cf. Pingree (n. 63 above) frs. Z 12 and Z 13.

[128] Tables 22-56.

[129] Neugebauer (n. 66 above) 95-96.

[130] *Ibid.* 100.

used 23;51°,[131] the value of the *Handy Tables*. The Latin version gives only the latter.[132]

The equation of the anomaly of a planet is computed from an epicycle by Greek, Indian, and Arabic astronomers, though it was the Indians who introduced the use of Sines to solve the angles. Again the *Sindhind* tradition follows the *ārdharātrikapakṣa* and *Zīj al-Shāh* in the maximum values of the equations of anomaly. I tabulate below the maximum equations (and the arguments at which they occur when known) according to the *ārdharātrikapakṣa* (*Kh* 2.8-17), the *Zīj al-Shāh*,[133] and al-Khwārizmī:[134]

	Ārdharātrikapakṣa		*Zīj al-Shāh*	*al-Khwārizmī*	
	equation	*argument*	*equation*	*equation*	*argument*
Saturn	6;20°	96°	5;44°	5;44°	95°-98°
Jupiter	11;30°	108°	10;52°	10;52°	98°-103°
Mars	40;30°	135°	40;31°	40;31°	128°-129°
Venus	46;15°	141°	47;11°	47;11°	135°
Mercury	21;30°	120°	21;30°	21;30°	112°-113°

Ibn al-Muthannā states that in al-Khwārizmī's *zīj* the equations were originally given in tables with different *kardajāt* or intervals of argument for each planet; the smallest *kardaja* was 3;45°, which we have seen to be the interval of arguments in the Sine-table of the *Brāhmasphuṭasiddhānta*. The use of these *kardajāt* placed the maximum equations at the following arguments:

Saturn	97;30°
Jupiter	97;30°
Mars	127;30°
Venus	135°
Mercury	112;30°

Then by linear interpolation between *kardajāt* al-Khwārizmī computed the values for each degree of argument.

The amount of the equation of the anomaly should vary with the distance from the earth of the center of the epicycle, which depends on its position on the eccentric deferent. Indian texts, which normally do not use an eccentric, naturally ignore this variation. This fact confused Abraham ibn Ezra,[135] who thought that the Indians' failure to agree with Ptolemy necessarily led them to commit errors.

[131] Millás (n. 68 above) 130; Goldstein (n. 69 above) 63; cf. Abraham (n. 70 above) 77, 92 (fr. Z 9 of Ya'qūb ibn Ṭāriq), and 143.

[132] Tables 21-26.

[133] Pingree (n. 63 above) fr. Z 14 and (n. 64 above) fr. Z 8.

[134] Tables 27-56; cf. Millás (n. 68 above) 118-119; Goldstein (n. 69 above) 43-44 and 169-170.

[135] Abraham (n. 70 above) 111; see 117, and cf. 85-86 and 89.

In fact, the Indians integrate the two equations by a method of calculating alternating values of each equation for arguments modified by the application of a half or the whole of the other equation. One such procedure is followed by al-Khwārizmī,[136] others by Brahmagupta (*BSS* 2.35 and *Kh* 2.18), and others by others. It is only necessary to note here that the method in the *Sindhind* is indeed Indian.

For computing the time till or since first or second station, Brahmagupta prescribes the division of the difference between the anomaly of the phase and the planet's anomaly by the difference between its *śīghra*-velocity and its corrected *manda*-velocity (*BSS* 2.49; see also *Kh* 2.19). The same rule is in essence given by al-Khwārizmī,[137] and is repeated by Ibn al-Muthannā.[138]

In problems relating to time and locality also the *Sindhind* is indebted to Brahmagupta. He computes the Sine of the ascensional difference (Sin γ) in the following manner. First he finds the radius of the sun's day-circle (r_d):

$$r_d = \sqrt{R^2 - \mathrm{Sin}^2\,\delta}\ ;$$

then the earth-sine (e):

$$e = \frac{\mathrm{Sin}\,\delta \cdot s_o}{12}\ ,$$

where s_o is the noon equinoctial shadow and 12 the height of the gnomon. Finally,

$$\mathrm{Sin}\,\gamma = \frac{e \cdot R}{r_d}\ .$$

This algorithm is given in *BSS* 2.56-58. Neugebauer has shown that this procedure lies behind the rule given in chapter 26 of al-Khwārizmī's *zīj*.[139]

Brahmagupta correctly states the relation between the sun's zenith-distance at noon ($90^\circ - a_n$ where a_n is the sun's altitude at noon) and the local terrestrial latitude (φ):

$$\varphi = 90^\circ - a_n \pm \delta\ ,$$

with the appropriate sign for south or north declination. This is given in *BSS* 3.13 and is also found in al-Khwārizmī's *zīj*.[140]

Among many other rules regarding shadows Brahmagupta gives the following (*BSS*

136 Neugebauer (n. 66 above) 23-30.
137 Chap. 14.
138 Millás (n. 68 above) 122; Goldstein (n. 69 above) 48 and 175.
139 Neugebauer (n. 66 above) 50-53; see also Neugebauer and Schmidt (n. 122 above).
140 Chap. 24; see also Millás (n. 68 above) 133-134; Goldstein (n. 69 above) 66-68.

3.48-49): if a is the altitude of the sun, $\bar{a}$ the coaltitude, s_n the noon shadow, and h_n the noon hypotenuse, then

$$s_n = \frac{\text{Sin}\ \bar{a} \cdot 12}{\text{Sin}\ a}$$

and

$$s_n = \sqrt{h_n^2 - 12^2}\ .$$

The same rules are given by al-Khwārizmī.[141]

In dealing with the computation of lunar and solar eclipses the *Sindhind* tradition is also very dependent on Brahmagupta. The latter in *BSS* 4.5 finds the lunar latitude, β_m, from:

$$\text{Sin}\ \beta_m = \frac{\text{Sin}\ \omega \cdot 270}{R}$$

where ω is the longitude of the moon diminished by that of its node and 270' = 4;30° is the maximum lunar latitude. Virtually the same formula was seen by Neugebauer to underlie the table of lunar latitude in al-Khwārizmī's *zīj*:[142]

$$\sin \beta_m = \sin \omega \cdot \sin i\ ,$$

where i = 4;30°. These two expressions are approximately identical because

$$\text{Sin}_{3270}\ 270' \approx 270.$$

In the next verse, *BSS* 4.6, Brahmagupta gives rules for finding the apparent diameters (d) of the disks of the sun, the moon, and the earth's shadow – in modern notation:

$$d_s = b_s \cdot \frac{11}{20}$$

$$d_m = b_m \cdot \frac{10}{247}$$

$$d_u = \frac{8b_m - 25b_s}{60}\ ,$$

where b is the *bhukti* or daily progress measured in minutes. A false etymology of the word *bhukti* and equivalents of the first two of these formulas are given in

[141] Chap. 28 and 28a; see also Neugebauer (n. 66 above) 105; Millás (n. 68 above) 151-154; Goldstein (n. 69 above) 87-89.

[142] Tables 21-26 and Neugebauer (n. 66 above) 97-98; cf. Millás (n. 68 above) 154-156; Goldstein (n. 69 above) 89-92.

chapters apparently added to the *zīj* of al-Khwārizmī;[143] all three formulas were used in constructing his eclipse-tables,[144] and are specifically repeated in a lemma by Ibn al-Muthannā.[145] These relations are also found in the anonymous Byzantine text in Parisinus graecus 2425.[146]

The eclipse-material in chapters 31 and 33 of the Latin version of al-Khwārizmī's *zīj* has been reworked by al-Majrīṭī, but the original can be recovered from Ibn al-Muthannā and Abraham ibn Ezra and is again remarkably similar to chapter 4 of the *Brāhmasphuṭasiddhānta*. This section in Ibn al-Muthannā begins, however, with a lemma stating that lunar eclipses are possible only if the moon is 13° or less from its node.[147] This limit is not given by Brahmagupta in chapter 4, but is found in the *Pañcasiddhāntikā* (6.2; cf. 7.5).

To determine whether or not a lunar eclipse is total, Brahmagupta subtracts the lunar latitude from half the sum of the apparent diameters of the moon and the earth's shadow; if the remainder is greater than the apparent diameter of the moon, the eclipse is total (*BSS* 4.7). The same method is explained in several lemmata by Ibn al-Muthannā.[148]

Brahmagupta continues with rules for computing the half-duration of the eclipse (Δt_e) and the half-duration of totality (Δt_t) (*BSS* 4.8). With reference to a lunar eclipse, they are:

$$\Delta t_e = \sqrt{(r_u + r_m)^2 - \beta_m^2} \Big/ b_m - b_s$$

$$\Delta t_t = \sqrt{(r_u - r_m)^2 - \beta_m^2} \Big/ b_m - b_s.$$

Brahmagupta converts the resulting times into an elongation of arc ($\Delta\lambda_e$ and $\Delta\lambda_t$) by:

$$\Delta\lambda_e = \Delta t_e \cdot \frac{b_m}{60}$$

$$\Delta\lambda_t = \Delta t_t \cdot \frac{b_m}{60}.$$

This rule is given in *BSS* 4.9.

Brahmagupta assumes that the lunar orbit is parallel to the ecliptic during an

143 Chaps. 29-30a.

144 Tables 61-66; see Neugebauer (n. 66 above) 107.

145 Millás (n. 68 above) 165-169; Goldstein (n. 69 above) 104-109; cf. Abraham ibn Ezra (n. 70 above) 166.

146 Neugebauer (n. 123 above) 33-34.

147 Millás (n. 68 above) 169-170; Goldstein (n. 69 above) 109-110.

148 Millás (n. 68 above) 171-172; cf. Abraham (n. 70 above) 166-167.

eclipse while al-Khwārizmī correctly takes into consideration the fact that it is inclined. Therefore, al-Khwārizmī's computation of those times and arcs as preserved by Ibn al-Muthannā[149] differs from Brahmagupta's though it remains an adaptation rather than a rejection of the Indian method. However, precisely Brahmagupta's method was used by the anonymous Byzantine text in Parisinus graecus 2425.[150]

Finally, Brahmagupta makes the color of a lunar eclipse depend on its magnitude:

1.	beginning and end	smoky
2.	partial	black
3.	more than half	black-coppery
4.	total	tawny

These colors are given in *BSS* 4.19. The colors were evidently in the original of al-Khwārizmī's *zīj*, though all that we have left is a non-specific, philosophical discussion by Ibn al-Muthannā[151] and what appears to be a corrupt – certainly different – version in the fourteenth-century Barcelona tables associated with Pedro IV "el Ceremonioso" of Aragon.[152]

Brahmagupta, like all Indian astronomers, computes a longitudinal ($\pi\lambda$) and a latitudinal ($\pi\beta$) component of parallax. The first depends on the sun's elongation ($\Delta\lambda$) to the east or west of the nonagesimal (V); the maximum is assumed to be 4 *ghaṭikās* = 1/15 day when the sun is on the horizon, the minimum 0 when the sun is at V. The second depends on the zenith-distance of the sun or moon on the altitude-circle of the nonagesimal; if the zenith-distance is 90°, the maximum parallax of 4 *ghaṭikās* is again assumed to occur. Further, since the noon zenith-distance of a body on the ecliptic is $\phi - \delta$ it follows that:

if $\phi = \delta$, the noon zenith-distance is 0 and $\pi_\beta \approx 0$;

if $\phi + \delta = 90^\circ$, then $\pi_\beta \approx 0;4$ days.

These rules are given in *BSS* 5.2-3a, and these assumptions also lie behind the parallax-computations of al-Khwārizmī.[153]

In computing longitudinal parallax Brahmagupta first finds the Sine of the altitude of V, or Sin a (V). To do this he first finds the ascensional difference, ω, in right ascension between the longitude of the ascendant and that of V; from this he obtains δ from:

$$\text{Sin}\ \delta = \text{Sin}\ \omega \cdot \frac{\sin \phi}{\sin \bar{\phi}} \cdot \frac{R}{r},$$

[149] Millás (n. 68 above) 173-179; Goldstein (n. 69 above) 110-119.

[150] Neugebauer (n. 123 above) 39-40.

[151] Millás (n. 68 above) 180-181; Goldstein (n. 69 above) 119-120; cf. Abraham (n. 70 above) 167.

[152] J. M. Millás Vallicrosa, *Las tablas astronómicas del rey Don Pedro el Ceremonioso* (Madrid 1962) 238.

[153] Neugebauer (n. 66 above) 71-72.

where r is the radius of the day-circle (actually, Cos δ). Then, of course,

$$ZV \approx \phi - \delta$$

$$\text{Sin ZV} = \text{Sin } \bar{a}(V)$$

$$\text{Cos ZV} = \text{Sin } a(V).$$

These rules are hinted at in *BSS* 5.3b-d.

The final formula is:

$$\pi_\lambda = 4 \cdot \frac{\text{Sin } a(V) \cdot \text{Sin } \Delta\lambda}{R^2} .$$

Brahmagupta gives it in *BSS* 5.4. These rules are the source of al-Khwārizmī's procedure in chapter 34 and table 77[154] except that he evidently used the value of R in the *Khaṇḍakhādyaka*, 150. Roughly the same procedure for finding π_λ is given in *Kh* 5.1-2 as in *BSS* 5.3-4.

If z is the noon zenith-distance and v the planet's velocity, then Brahmagupta's rule for finding latitudinal parallax is:

$$\pi_\beta = \bar{v} \cdot \frac{\text{Sin z}}{15\text{ R}}$$

This rule is given in *BSS* 5.11. Precisely the same procedure was used in *Khaṇḍakhādyaka* 5.3 to derive the equivalent (with R = 150):

$$\pi_\beta = \frac{13\text{ Sin z}}{40} .$$

This is evidently the form of the rule that al-Khwārizmī[155] and the anonymous Byzantine text in Parisinus graecus 2425[156] used.

The maximum latitudes of the planets in the *Brāhmasphuṭasiddhānta* are those of the *Paitāmahasiddhānta* of the *Viṣṇudharmottarapurāṇa*:[157]

Saturn	2;10°
Jupiter	1;16°
Mars	1;50°
Venus	2;16°
Mercury	2;32°

These values are given in *BSS* 9.1.

[154] *Ibid.* 71 and 123-125; see also Millás (n. 68 above) 160-163; Goldstein (n. 69 above) 121-125.

[155] Neugebauer (n. 66 above) 121-123; see also Millás (n. 68 above) 163-164; Goldstein (n. 69 above) 126-129.

[156] Neugebauer (n. 123 above) 29-32.

[157] *Paitāmahasiddhānta* 3.18.

The maximum latitudes given in the first part of the *Khaṇḍakhādyaka* are those of the *Āryabhaṭīya*:[158]

Saturn	2;0°
Jupiter	1;0°
Mars	1;30°
Venus	2;0°
Mercury	2;0°

These values are found in *Kh* 8.1c-d. But in the *uttara* section of the *Khaṇḍakhādyaka* the maximum latitude of Mercury is stated to be 2;30° while the rest remain as was stated in the first section (*Kh* 14.1c-d). It is this version of the verse that was used in the *Sindhind* tradition. These maximum latitudes were used in al-Khwārizmī's tables, but have now been replaced by erroneous values equal to those of the *uttara* section of the *Khaṇḍakhādyaka* multiplied by 2;30 = 150/60;[159] the original numbers are retained in a lemma by Ibn al-Muthannā.[160]

As we have seen before, the longitudes of the nodes of the planets in the *Sindhind* were derived from the *Brāhmasphuṭasiddhānta* despite the fact that the maximum latitudes come from the *uttara* section of the *Khaṇḍakhādyaka*. The *Brāhmasphuṭasiddhānta* (*BSS* 9.8-10) and, in a less complex form, the *Khaṇḍakhādyaka* (*Kh* 8.5), agree on the computation of a planet's latitude. If ω is the elongation on the deferent between a superior planet and its node, or the elongation on the *śīghra*-epicycle between the *śīghrocca* of an inferior planet and its node, and if ρ is the final *śīghra*-hypotenuse, and if β is the latitude, then

$$\beta = \beta_{\max} \cdot \frac{\text{Sin } \omega}{\rho} .$$

This is precisely the procedure in the lemmata of Ibn al-Muthannā, who gives a specific formula first for finding the final *śīghra*-hypotenuse,[161] and this same procedure, with the erroneous multiplication of $\beta_{\max}$ by 2;30, lies behind chapter 17 and columns 7 and 8 of tables 27 to 56 of the Latin version of al-Khwārizmī's *zīj*.[162]

One further element in the Latin *Sindhind* material that is specified as being Indian and is found in the *Brāhmasphuṭasiddhānta* (12.40 and 21.15) is the approximation $\pi \approx \sqrt{10}$. Many others had used this approximation before Brahmagupta, including the author of the *Paitāmahasiddhānta* of the *Viṣṇudharmottarapuraṇā*[163]

[158] *Āryabhaṭīya*, Daśagītikā 6; ed. H. Kern (Leiden 1875).

[159] Neugebauer (n. 66 above) 39-40 and 103; see also tables 45-46 of the *Toledan Tables* in Toomer (n. 72 above) 69-70, and Pingree (n. 73 above).

[160] Millás (n. 68 above) 157; Goldstein (n. 69 above) 93.

[161] Millás 156-157; Goldstein 92-93.

[162] Neugebauer (n. 66 above) 34-41 and 101-103; cf. chap. 10 of al-Jahānī (n. 45 above) Q ii v-Q iii.

[163] *Paitāmahasiddhānta* 3.6.

and Varāhamihira.[164] In the *Brāhmasphuṭasiddhānta* it occurs in the chapters on mathematics and the sphere; it probably was transmitted to the Arabs with other mathematical material independently of the *Mahāsiddhānta*. This value of π is attributed to the Indians by Abraham ibn Ezra.[165]

XI. Ṛṣi

There appear to have been a number of Indian astrologers at Baghdād in the early ᶜAbbāsid period besides the astronomer from Sind who was involved in the *Sindhind*. We have already mentioned Kanaka while discussing Māshā'allāh; here we must pause to consider a few more. First among them is Irshā al-Hindī, whose name is probably an attempt to transliterate the Sanskrit Ṛṣi[166] or Ārṣya. He is cited often in the Arabic astrological works of the ninth century – for example, in the *Kitāb aṣl al-uṣūl* of al-Ṣaymarī – and reached the West through the translation of Ibn abī al-Rijāl. The first citation deals with an interrogation concerning buried treasure, and uses the ninth degree before the ascendant.[167] I know of no precise parallel to this in Indian texts, though the question is sometimes treated in them.[168]

The next quotation deals with the question of whether or not a woman is pregnant, and uses, among other criteria, the navāṃśa in the ascendant.[169] Again, an exact parallel to this passage in Sanskrit is not known to me, though the subject is dealt with.[170]

The last fragment of Irshā concerns the sex of the unborn child, and uses Jupiter and Mercury.[171] The Indian rules for answering this query also depend on Jupiter and Mercury, as in a verse by Sphujidhvaja.[172]

XII. Bhūridāsa and Buzurjmihr

The only Indian astronomers or astrologers besides Andubarius whom I know to be named in Byzantine texts are Phorēdas (a name which may correspond to Sanskrit Bhūridāsa, though no astronomer or astrologer of that name is known in India) and Porzozomchar (which is probably a misreading of Buzurjmihr, written in Arabic, as Burzujumhar).

164 Varāhamihara 4.1 (n. 35 above).
165 Abraham (n. 70 above) 124.
166 *CESS* A1.59a.
167 'Alī ibn abī al-Rijāl 1.39 (n. 50 above).
168 E.g., Bhojarāja, *Vidvajjanavallabhā* 15.3, ed. D. Pingree (Baroda 1970).
169 Ibn abī al-Rijāl 1.42 (n. 50 above).
170 E.g., Pṛthuyaśas, *Ṣaṭpañcāśikā* 7.5, ed. D. Jhā (Benares 1947); Bhaṭṭotpala, *Praśnajñāna* 33, ed. V. S. Sastri and M. R. Bhat (Bangalore 1949).
171 Ibn abī al-Rijāl 1.46 (n. 50 above).
172 *Yavanajātaka* (n. 3 above) 66.1.

The first occurs in the so-called *Epitome Parisina,*[173] where a cycle of twelve years (*dodecaëteris*) beginning with Jupiter in Sagittarius is attributed to Erimarabus, "whom the Egyptians call a prophet and the discoverer of astronomy," and to Φορηδὰς ὁ Ἰνδός. Though Jovian *dodecaeterides* are well known in Sanskrit,[174] they never begin with Sagittarius. The whole story of these Egyptian and Indian sages appears fabulous, though it remains possible that Φορηδάς represents a real Sanskrit name. The quotation from Buzurjmihr is a much damaged note on the recovery of stolen goods added by a later hand (V^2) in the margin of fols. 81v-82 of Vaticanus graecus 1056; I edit it in Appendix 4. Though the recovery of stolen objects was a matter about which Indian astrologers could expect to be consulted,[175] this quotation is of primary interest for the history of Sasanian astrology and need not be discussed further here.

XIII. Abū Ma'shar

One of the most important transmitters of a knowledge of Indian astrology to the Arabs was Abū Ma'shar, who was born at Balkh on 10 August 787 and died at Wāsiṭ on 9 March 886.[176] An anecdote, found in the *Mudhākarāt* of his pupil Shādhān,[177] demonstrates that he had some direct contact with Ceylon (called India in the Greek); I cite the tenth century Byzantine translation, of which there exists a Latin version.[178]

There are two possible dates in Abū Ma'shar's lifetime for this horoscope. Both have problems, but the later date, 23 December 884, though very close to the date of Abū Ma'shar's death, has fewer and is the more likely to be correct.

	Horoscope	*12 Jan. 826*	*23 Dec. 884*
Saturn	Cancer	Gemini 25°	Gemini 28°
Jupiter	Aries	Aries 22°	Aries 11°
Mars	*Omitted*	Capricorn 16°	Aries 20°
Sun	Capricorn	Capricorn 27°	Capricorn 7°
Venus	*Omitted*	Pisces 12°	Aquarius 18°
Mercury	Aquarius	Capricorn 2°	Capricorn 13°
Moon	Capricorn	Capricorn 29°	Capricorn 29°
Ascendant	Gemini	ca. 3 P.M.	ca. 4 P.M.

[173] F. Cumont in *CCAG* 8.3 (Brussels 1912) 91-92.

[174] Eg., Varāhamihira, *Bṛhatsaṃhitā* 8.3-14, ed. A. V. Tripāṭhī, 2 vols. (Varanasi 1968).

[175] E.g., Sphujidhvaja, *Yavanajātaka* (n. 3 above) 64.7-15, and Bhojarāja (n. 168 above) 16.7-11.

[176] D. Pingree in *DSB* 1 (New York 1970) 32-39.

[177] An edition of the Arabic original of Shādhān's *Mudhākarāt* with the Byzantine translation and its Latin version is being prepared by E. S. Kennedy and D. Pingree; in the following I quote from the preliminary form of the edition of the Greek.

The king of Ceylon in 884 was Sena II (851-885), though his dates are disputed. It appears that three of his sons ruled Ceylon, as Kāśyapa V (913-923), Dappula IV (923), and Dappula V (923-934). It is not clear whether or not the horoscope interpreted by Abū Ma'shar belongs to any one of them. But it is known that Sena II's reign was the period of Ceylon's greatest flourishing in the ninth century, and the most likely to have experienced the influence of Arabic astrology.

One of Abū Ma'shar's arguments in defense of his interpretation of the Ceylonese prince's horoscope was that Saturn rules India. This is iterated in another part of the *Mudhākarāt.*[179] His theory of the possible beneficence of Saturn is not known to me from Sanskrit texts, though the strength of its maleficence is diminished in certain situations. The only other direct reference to India in the *Mudhākarāt* concerns the ease or difficulty of childbirth, where difficulty is indicated by the presence of Saturn in a masculine and Mars in a feminine sign.[180] This also appears not to be a genuine Indian theory.

In one other place, however, Abū Ma'shar speaks of the celestial body *kayd,* a word adapted by the Sasanians from the Sanskrit *ketu.*[181] He reports that ancient scientists give it a motion of 2;30° a year. *Ketu* has two meanings in Sanskrit astrology: the descending node of the moon (Abū Ma'shar's *σύνδεσμος τῶν σφαιρῶν*) and a comet (perhaps his *ἀστὴρ νότιος λαμπρός*). It is never, so far as I am aware, given a retrograde motion of 2;30° (a *dodecatemorion*) per year, but this is the well-known parameter for *kayd* in Islamic *zījes* as demonstrated by Kennedy; it is equivalent to one rotation in 144 years, and retrograde motions of approximately 0;0,24,39,27,7° in a day and 0;12,19,44° in a 30-day month. Neugebauer has published from the manuscripts of Gregory Chioniades's translations one of his tables of the motion of *κάϊτ* and pointed out that the pseudo-planet is also discussed by Theodore Meliteniotes (who used Chioniades's translations); I edit in Appendix 6 another table of Chioniades found on fol. 155 of Vaticanus graecus 211.[182] This table obviously is based on the parameter of −2;30° of motion in a year of 365 days. The rules at the beginning indicate that *κάϊτ* was in Aries 0° in −54 Yazdijird or A.D. 579, and therefore again at the beginning of 91 Yazdijird or A.D. 723 as in the Islamic *zījes.* The epoch of the Sulṭānic years is A.D. 1079, as we have seen before; a period of 144 years began in A.D. 1011 or 69 rather than 66 years before the end of one Sulṭānic year.

[178] *Mudhākarāt* chap. 59, edited in Appendix 5. For the Latin version see the unsatisfactory article by L. Thorndike, "Albumasar in Sadan," *Isis* 45 (1954) 22-32.

[179] *Mudhākarāt* chap. 41; see n. 177 above.

[180] *Ibid.* chap. 138.

[181] *Ibid.* chap. 49. On *kayd* see W. Hartner, "Le problème de la planète kaïd," *Oriens-Occidens* (Hildesheim 1968) 268-286, and "The Pseudoplanetary Nodes of the Moon's Orbit in Hindu and Islamic Iconographies," *ibid.* 349-404; E. S. Kennedy, "Comets in Islamic Astronomy and Astrology," *JNES* 16 (1957) 44-51; and O. Neugebauer, "Notes on al-Kaid," *JAOS* 77 (1957) 211-215.

[182] Concerning this manuscript and Chioniades's translations see Pingree (n. 104 above).

All of this, while employing a transmutation of the Sanskrit *ketu*, is non-Indian; but another text preserved on fols. 175v-176 of Parisinus graecus 2506,[183] pretending to an Indian origin, indicates that the "Chaldaeans" called the ascending and descending nodes of the moon the head and tail of a dragon; in fact, in Sanskrit, as we saw in discussing Severus Sebokht, they are called the head and the tail.

In the discussion of Theophilus of Edessa above it has been indicated that Abū Ma'shar's *Kitāb al-madkhal al-kabīr* contains Indian material; it is, in fact, one of the principle conduits for the transmission of genuine Indian astrological doctrines to the West. In particular we may note here the information it contains about the Indian terms and decans. The terms are given almost correctly in the Latin translation of Hermann;[184] the Greek paraphrase omits this chapter. As representative of the Sanskrit tradition I refer to Sphujidhvaja.[185]

Shortly after this passage in the *Introductorium* Abū Ma'shar correctly describes one of the Indian rules for determining the lords of the decans.[186] The Byzantine paraphrase again has omitted this chapter. For the Sanskrit tradition I refer to Varāhamihira.[187]

The iconography of the decans as described by Abū Ma'shar[188] had a considerable impact on astrological illustrations in the West.[189] It has long been recognized that his descriptions are derived (probably through a Sasanian intermediary) from Varāhamihira's *Bṛhajjātaka*,[190] a work of the sixth century. It has recently been demonstrated that Varāhamihira's descriptions are a mixture of those of the decans and *horās* in the *Yavanajātaka* of Sphujidhvaja, and that these latter are misinterpretations (influenced by Śaivite iconography) of the Greco-Egyptian pictures in a Greek manuscript translated into Sanskrit by Yavaneśvara in A.D. 149/150.[191] The descriptions of the decans in the thirteenth-century Latin translation of pseudo-al-Majrīṭī's *Ghāyat al-ḥakīm*, entitled *Picatrix*, are derived from Abū Ma'shar, but the fields of magical operation over which each presides come from some other, as yet unidentified source.[192]

183 In *CCAG* 7 (Brussels 1908) 125. In line 11 I would restore ⟨δύο το⟩ῦ κύκλου for Boll's ⟨δύο⟩ κύκλων.

184 *Introductorium* 5.10 (n. 40 above).

185 *Yavanajātaka* (n. 3 above) 1.42.

186 *Introductorium* 5.13 (n. 40 above); of *Picatrix* (n. 192 below) 2.12. W. Gundel, *Dekane und Dekansternbilder*, Studien der Bibliothek Warburg 19 (Glückstadt-Hamburg 1936) is not entirely satisfactory.

187 Varāhamihira, *Bṛhajjātaka* 1.11 d, ed. V. Subrahmanya Sastri (Mysore 1929).

188 Abū Ma'shar, *Kitāb al-madkhal al-kabīr* 6.1; the Arabic original was edited by K. Dyroff in F. Boll, *Sphaera* (Leipzig 1903) 482-539; the Greek version by F. Boll in *CCAG* 5 1 (Brussels 1904) 156-169; see also O. Neugebauer, "Variants to the Greek Translation of Abū Ma'shar's Version of the Paranatellonta of Varāhamihira and Teukros." *Bulletin de l'Académie royale de Belgique*, Classe des lettres 5.43 (1957) 133-140; and the Latin in *Introductorium* 6.2 (n. 40 above).

189 A. Warburg in his *Gesammelte Schriften* 2 (Leipzig 1932) 459-482 and 627-644.

190 Varāhamihira, *Bṛhajjātaka* (n. 187 above) chap. 27.

191 Pingree, "Indian Iconography" (n. 4 above).

I have referred above in discussing Māshā'allāh to the doctrine of the *navāṃśas* described by Abū Ma'shar in 3.9 of his *Kitāb aḥkām taḥāwīl sinī al-mawālīd,* which was translated into Greek and thence into Latin. He continues then to claim that, in determining the ruler of a year, the Indians use the lord of the first *navāṃśa* in the sign which the revolution of the years of the nativity has reached.[193] It is true that some Indian astrologers describe a method of continuous horoscopy, called *aṣṭaka-varga,* based on the planets' transits of the places and planetary positions of the nativity; the earliest exposition of this theory is by Sphujidhvaja.[194] It differs, however, from the method ascribed to the Indians by Abū Ma'shar. I suspect that the use of the *navāṃśas* in revolutions was begun by Sasanians familiar with the fourth book of Dorotheus.[195]

Abū Ma'shar frequently in his other works in Arabic refers to Indian theories of one sort or another, but little of this material was translated into either Greek or Latin. The main repository of it in Greek is the Μυστήρια, where are also preserved the Byzantine versions of Shādhān's *Mudhākarāt* and of Abū Ma'shar's *Kitāb al-madkhal al-kabīr.* I edit in Appendix 7, from Angelicus graecus 29 fol. 23, Μυστήρια 1.76, a text in which the indications of the sun in each of the astrological places are listed. For comparison we may refer to various verses in Sphujidhvaja's *Yavanajātaka.*[196] But only in a few places do the Sanskrit verses contain material corresponding to Abū Ma'shar's. His assignment of effects to the sun's being in the several places seems to be related to the aspects of a native's life that these places influence. Abū Ma'shar assumes that the second place influences familial wealth, the third brothers, the fourth parents, the fifth children, the sixth illness, the seventh marriage, and the eighth death; these are all normal Hellenistic associations. But the Indians assign relatives to the fourth and enemies to the sixth; I cite again Sphujidhvaja.[197] It is clear, therefore, again that Abū Ma'shar did not have an Indian source for this passage; and again one suspects a Sasanian origin.

The Hellenistic conception of the spheres of influence of the τόποι also appears in the following chapter of the Μυστήρια,[198] which I edit in Appendix 8 from fol. 38rv of Angelicus graecus 29 and from fol. 97v of Vaticanus graecus 1056, where it has been added by a second and later hand; this chapter also displays a Hellenistic (and specifically Dorothean) methodology in the first paragraph, not Indian.

The final allegedly Indian passage in the Μυστήρια[199] concerns the nature of

[192] *Picatrix* 2.11; the Arabic is in *Pseudo-Mağrīṭī. Das Ziel des Weisen,* ed. H. Ritter (Leipzig 1933) 126-132. I am preparing an edition of the Latin version for the Warburg Institute.

[193] Abū Ma'shar (nn. 57, 58 above) 3.10.

[194] *Yavanajātaka* (n. 3 above) 43.11-51.

[195] Pingree (n. 21 above)

[196] *Yavanajātaka* (n. 3 above) 20.4; 25.1, 5, 13; 26.1, 4, 7, 12, 20, 23, 25, and 27; see also Varāhamihira, *Bṛhajjātaka* (n. 187 above) 20.1-3.

[197] *Yavanajātaka* 1.70-72.

[198] Μυστήρια 1.143.

[199] *Ibid.* 1.78.

good or ill that will befall a querist during the year, and the time at which the prediction will come true. The Indians do have a method for determining the time at which the prediction relating to an interrogation will come true, but it is very different indeed from that ascribed to them in the Byzantine text.[200] I edit that text from fol. 23 of Angelicus graecus 29 in Appendix 9.

XIV. JA'FAR AL-HINDĪ

Ja'far, who may possibly be identical with Abū Ma'shar, is the author of a work on astrology as applied to meteorology translated by Hugh of Sanctalla as the *Liber Gaphar de mutatione temporis.*[201] This is replete with references to Indian theories. In general, Ja'far claims that the Indians attributed changes in the sublunar world to the mixing of stellar with lunar influences, as the moon is the intermediary between the planets and the earth. Further, he seems to state that the constellations lie between the orbits of Mercury and the moon.[202] Though the moon plays a prominent role in Indian astrology, and especially in the *saṃhitā* literature to which meteorological astrology generally belongs, I know of no statement as extreme as Ja'far's, and no Indian believed that the moon transmitted stellar influences to the sublunar world; this is a Greek concept. However, the implication that the fixed stars lie between the orbits of the moon and Mercury is found in the *Purāṇas;* I refer to a composite text of the *Brahmapuraṇa,* the *Kūrmapurāṇa,* and the *Viṣṇupurāṇa.*[203]

Further on Ja'far incorrectly states that Venus and Jupiter are regarded as benefic by the Indians, the rest as malefic.[204] In fact, the Indians count only the sun, Saturn, and Mars as malefic, as in a verse of Sphujidhvaja.[205] On the same page Ja'far erroneously ascribes to the Indians a belief in the significance of the quarters of the month.

However, he then alludes[206] to the Indian system of 28 *nakṣatras* including Abhijit, and the other Indian system of 27 *nakṣatras* of which each contains 13;20°. It has already been noted above that the Sasanians were acquainted with the Indian *nakṣatras.* At some point a text (in Pahlavī or in Arabic) concerning activities to be undertaken or avoided when the moon is in each of the *nakṣatras* (*manāzil al-qamar* in Arabic) was put together from three sources: an Indian text, the fifth book of

[200] See Sphujidhvaja, *Yavanajātaka* 63 (n. 3 above) with the commentary, and Bhojarāja (n. 168 above) 17.16-21.

[201] Ja'far, *Liber Gaphar de mutatione temporis,* published with a similar work ascribed to al-Kindī by P. Liechtenstein (Venice 1507). For an Indian method of meteorological astrology see Bhojarāja 14 (n. 168 above).

[202] Ja'far, fol. c 1r-v.

[203] W. Kirfel, *Das Purāṇa vom Weltgebäude* (Bonn 1954) 48-49.

[204] Ja'far (n. 201 above) fol. c 1v.

[205] *Yavanajātaka* (n. 3 above) 1.109.

[206] Ja'far (n. 201 above) fol. c 2.

Dorotheus, and a Persian tradition. The Arabic version of this text was copied by 'Alī ibn abī al Rijāl, through whom it reached the Latin West;[207] and it was translated into Byzantine Greek;[208] it was also used in part by the author of the *Picatrix*, through which again it appears in Latin.[209] Many parallels to these texts are to be found in the third-century (?) *Tantra* of Parāśara, as quoted in Bhaṭṭotpala's tenth-century *vivṛti* on the *Bṛhatsaṃhitā* of Varāhamihira.[210] Parāśara classifies the twenty-eight *nakṣatras* into seven categories – fixed (*dhruva*), sharp (*dāruṇa*), fierce (*ugra*), swift (*kṣipra*), soft (*mṛdu*), common (*sādharaṇa*), and unstable (*cara*):

1. Aśvinī = al-Naṭḥ or al-Sharaṭān. swift.
2. Bharaṇī = al-Buṭayn. fierce.
3. Kṛttikā = al-Thurayyā. common.
4. Rohiṇī = al-Dabarān. fixed.
5. Mṛgaśiras = al-Haq'a. soft.
6. Ārdrā = al-Han'a. sharp.
7. Punarvasu = al-Dhirā'. unstable.
8. Puṣya = al-Nathra. swift.
9. Āśleṣā = al-Ṭarf. sharp.
10. Maghā = al-Jabha. fierce.
11. Pūrvaphalgunī = al-Zubra. fierce.
12. Uttaraphalgunī = al-Ṣarfa. fixed.
13. Hasta = al-'Awwā. swift.
14. Citrā = al-Simāk al-a'zal. soft.
15. Svāti = al-Ghafr. unstable.
16. Viśākhā = al-Zubānā. common.
17. Anurādhā = al-Iklīl. soft.
18. Jyeṣṭhā = al-Qalb. sharp.
19. Mūla = al-Shawla. sharp.
20. Pūrvāṣāḍhā = al-Na'ā'im. fierce.
21. Uttarāṣāḍhā = al-Balda. fixed.
22. Abhijit = Sa'd al-dhābiḥ. swift.
23. Śravaṇa = Sa'd bula'. unstable.
24. Dhaniṣṭhā = Sa'd al-su'ūd. unstable.
25. Śatabhiṣak = Sa'd al-akhbiyya. unstable.
26. Pūrvabhadrapadā = al-Fargh al-awwal. fierce.
27. Uttarabhadrapadā = al-Fargh al-thānī. fixed.
28. Revatī = Baṭn al-ḥūt. soft.

Parāśara associates certain acts with the *nakṣatras* of each category; the source of the Arabic text has associated these same acts with specific members of each category. The parallels with the Western texts suffice to demonstrate the derivation of much of

[207] Ibn abī al-Rijāl 7.101 (n. 50 above).
[208] Ed. S. Weinstock in *CCAG* 9.1 (Brussels 1951) 138-156.
[209] *Picatrix* 1.4 (n. 192 above)
[210] On 97.6-11.

this material from an Indian source. The Arabic version contains more of such material, as do also the sections ascribed to Dorotheus. The sources of this text on the *nakṣatras*, and of the similar text on the decans discussed under Abū Ma'shar, appear to have been much the same; it seems probable that both originated in Sasanian Iran.

XV. AL-QABĪṢĪ

Previously, in discussing the *navāṃśas* under Māshā'allāh, we have had occasion to refer to al-Qabīṣī; he records also another astrological doctrine which he attributes to the Indians.[211] He calls this albuzic. After a conjunction of the sun and the moon, the first 12 hours are ruled by the sun, the next 12 by Venus, the next 12 by Mercury, and so on until the next conjunction. Each period of 12 hours is divided into three subperiods, each of which contains four hours; these subperiods are ruled by the three lords of the triplicity of the planet which rules the 12-hour period. Essentially the same theory is found in a text preserved in the margin of fol. 208 of Vaticanus graecus 1056, which I edit in Appendix 10.

The periods of 12 hours measured from a conjunction of the sun and the moon are approximately the Indian *karaṇas*, but the Indian method of determining their lords is completely different. Out of the 60 *karaṇas* in a synodic month beginning with a conjunction four are fixed: 58 (*śakuni*), 59 (*catuṣpada*), 60 (*nāga*), and 1 (*kiṃstughna*). The other 56 *karaṇas* form eight series of seven each (*vava, vālava, kaulava, tailika, gara, vaṇija,* and *viṣṭi*). Their lords are:

1. kiṃstughna. Māruta.
2,9,16,23,30,37,44,51. vava. Indra.
3,10,17,24,31,38,45,52. vālava. Brahmā.
4,11,18,25,32,39,46,53. kaulava. Mitra.
5,12,19,26,33,40,47,54. tailika. Āryamān.
6,13,20,27,34,41,48,55. gara. Bhūr.
7,14,21,28,35,42,49,56. vaṇija. Lakṣmī.
8,15,22,29,36,43,50,57. viṣṭi. Yama.
58. śakuni. Kali.
59. catuṣpada. Vṛṣa.
60. nāga. Sarpa.

This system is described by Varahāmihira.[212] The origin of al-Qabīṣī's method of assigning lords according to the descending order of the planets (that is, in accordance with their lordships of the hours) is not apparent; but his use of the lords of the triplicities as lords of the thirds of the *karaṇas* is likely to be due to some astrologer familiar (as were the Sasanians) with Dorotheus of Sidon.

[211] Al-Qabīṣī 4 (n. 59 above) fols. d 4v-d 5.
[212] Varāhamihira (n. 174 above) 99.1-2.

XVI. SIMEON SETH

It is well known that the astronomers of al-Ma'mūn in the early ninth century established that the rate of the precession of the equinoxes is 1° in every 66 years approximately, or about 0;0,54° per year. This is, in fact, an Indian parameter.[213] I have discussed elsewhere its appearance in Byzantine texts of the eleventh and fourteenth centuries.[214] I edit, in Appendix 11, the text of Simeon Seth there referred to, from fol. 32 of Vaticanus graecus 1056.

Simeon Seth's list of stars presents many problems. It bears no direct derivative relationship to Ptolemy's catalog, and two of its nine stars remain unidentified. In the following I give the modern names and Ptolemy's coordinates and magnitudes, as well as the differences between Simeon's and Ptolemy's longitudes and latitudes (always Simeon – Ptolemy).

Stars	λ	$\Delta\lambda$	β	$\Delta\beta$	*Magnitudes*
1. α Andromedae	♓ 17;50	+14;10	+26;0	0	2.3
2. Unidentified					
3. γ Orionis	♉ 24;0	+14;30	−17;30	0	2
4. β Geminorum	♊ 26;40	+15;50	+6;15	−3;0	2
5. α Hydrae	♌ 0;0	+15;30	−20;30	0	2
6. Unidentified					
7. δ Leonis	♌ 14;10	+15;30	+13;40	0	2.3
8. β Leonis	♌ 24;30	+15;10	+11;50	−2;41	1.2
9. β Virginis	♌ 29;0	+15;20	+0;10	––	3

XVII. VATICANUS GRAECUS 1056

There is a curious collection of astrological opinions about interrogations preserved on fols. 151v-154 of Vaticanus graecus 1056;[215] it includes excerpts falsely ascribed to Vettius Valens, to Dorotheus, and to Pythagoras, as well as alleged quotations from the works of the Babylonians, the Egyptians, the Hellenes, the Indians, the Persians, and Abū Ma'shar. I intend to discuss this text more fully in an edition of Valens; here I need only remark that in so far as the "Indian" passages are not commonplaces of interrogational astrology they are also not specifically Indian in character.

XVIII. PARISINUS GRAECUS 2506

This manuscript, on fols. 158v-159v, contains a Christianized treatise on the lordship of the months of a pregnancy, which I edit in Appendix 12. The lordships of the months are determined for the first seven from the ascending order of the planetary

[213] Pingree, "Precession and Trepidation" (n. 4 above).
[214] Pingree (n. 104 above) 138-139.
[215] Ed. I. Heeg in *CCAG* 5.3 (Brussels 1910) 110-121.

orbits beginning with the sun, those of the last two from the descending order beginning with Saturn; it results that every even month is ruled by a malefic or potentially malefic planet:

1.	Sun	2.	Mars
3.	Jupiter	4.	Saturn
5.	Moon	6.	Mercury
7.	Venus	8.	Saturn
9.	Jupiter		

Totally different lords of the months of pregnancy are found in Sanskrit texts. Sphujidhvaja lists Mars, Venus, Jupiter, the sun, the moon, Saturn, the moon, the ascendant, the moon, and the sun.[216] The more common Indian lords – Venus, Mars, Jupiter, the sun, Saturn, the moon, Mercury, the lord of the ascendant, the moon, and the sun – are enumerated first by Mīnarāja.[217]

XIX. PICATRIX

Several passages from the *Picatrix* of an ultimately Indian origin have already been referred to in this article;[218] there are many others, including portions of the Ḥarrānian prayers to the planets.[219] Here I wish only to remark that, through this magical route, the Sanskrit names of the planetary deities (in very corrupt form) appear in Latin. I give in parallel columns the names in each language as they appear in the Arabic and the Latin, with the appropriate Sanskrit names following the Arabic for the Indian names.

	Arabic	*Latin*
Saturn:		
Arabic (Arabicum)	Zuhal	Zohal
Persian (Feniz)	Kaywān	Keyhneri
"Roman" (Romanum)	Qurūnus	Karonoz
Greek (Grecum)	Aqirūnus	Hacoronoz
Indian (Indianum)	Shanashar (Śanaiścara)	Satar
Jupiter:		
Arabic	Mushtarī	Mistery
Persian	Birjīs	Bargiz
Foreign (Romanum)	Hurmuz	Dermiz
Greek	Zāūsh	Raus
Indian	Wihasfaṭi (Bṛhaspati)	Hauzfat

[216] *Yavanajātaka* (n. 3 above) 5.9.

[217] *Vṛddhayavanajātaka* (n. 4 above) 3.9 c-17.

[218] *Picatrix* (n. 192 above); decanic images in 2.11 and lunar mansions in 1.4.

Mars:		
Arabic	Mirrīkh	Marech
Persian	Bahrām	Baharaz
"Roman"	Rays	Rabiz
Greek	Aras	Hahuez
Indian	Anjārā (Aṅgāra)	Bahaze
Sun:		
Arabic	Shams	Iaxems
Persian (Caldeum)	Mihr	Maer
"Roman"	Īliyūs	Aylebuz
Indian	Āras (Āditya ?)	Araz
Venus:		
Arabic	Zuhara	Zohara
Persian	Anāhīd	Ateyhnit
"Roman"	Afrūdīṭī	Affludita
Greek	Ṭiyāniyā	Atruenita
Indian	Surfa (Śukra)	Sarca
Mercury:		
Arabic	'Uṭārid	Hotaric
Persian	Tīr	Tyr
"Roman"	Hārūs	Haruz
Greek	Hurmus	Hermes
Indian	Budha (Budha)	Meda
Moon:		
Arabic	Qamar	Tamar
Persian	Māh	Heme
"Roman"	Sālīnī	Zelec
Greek	Sam'ā'īl	Zamail
Indian	Sūma (Soma)	Teryz

XX. Shams al-Dīn al-Bukhārī

Between the years 1295 and 1302 Gregory Chioniades translated several Arabic *zījes* into Greek including the *Zīj al-'Alā'ī*, the *Zīj al-Sanjarī*, and a *zīj* written by his teacher in Tabrīz, Shams al-Dīn al-Bukhārī.[220] In the introduction to his *zīj* Shams al-Dīn makes several comments on Indian astronomers;[221] his information is correct. The Indian calendar-year begins with the conjunction of the sun and moon of Caitra,

[219] *Ibid.* 3.7.
[220] Pingree (n. 104 above).
[221] Ed. F. Cumont in *CCAG* 1 (Brussels 1898) 85-87.

which is the synodic month in which the sun enters Aries. An intercalary month in Sanskrit is called an *adhimāsa* or *malamāsa*; *padamāsa* is a common Arabic mislection of the latter, of which νατμασάν is evidently a corruption.

The Indian year-beginning is also referred to in the *Zīj al-Sanjarī.*[222] More impressive is the fact that the *zīj* in 4.1 gives the Indian rule for computing the Sine of the rising amplitude of the sun, Sin η:[223]

$$\text{Sin}\ \eta = \text{Sin}\ \delta \cdot \frac{R}{\text{Sin}\ \bar{\phi}}\ .$$

Brahmagupta gives this in *BSS* 3.64. And in 6.5 the *zīj* gives the Indian rule for finding the distance of the perpendicular from the sun to the plane of the horizon, from the east-west line.[224] If we call that distance the *bāhu*, the text states:

$$\text{bāhu} = \text{śaṅkutala} \pm \text{Sin}\ \eta\ ;$$

further,

$$\text{śaṅkutala} = \text{Sin}\ a \cdot \frac{\text{Sin}\ \bar{\phi}}{\text{Sin}\ \phi}\ ,$$

where a is the sun's altitude. This rule is used by Brahmagupta; one will find it expressed, for example, in *Paitāmahasiddhānta* 9.1.[225]

XXI. CONCLUSIONS

From this rapid survey of very scattered and disparate material it is apparent that the most massive transmission of Indian astral science to the West was through the *Sindhind*; but the tables of al-Khwārizmī and the associated commentaries were little and briefly known. The main influence was through the absorption of some Indian methods (primarily trigonometric) and parameters into Islamic Ptolemaic works. The *Sindhind* tradition at its inception in late eighth-century Baghdad was not purely Indian; by the time it reached Adelard of Bath in the early twelfth century many sections had been "modernized"; and the *Toledan Tables*, which appeared in Latin half a century later, but which had been composed in Arabic half a century earlier than Adelard's translation, contain very little indeed of al-Khwārizmī's Indian material. The Byzantine astronomical texts displaying some knowledge of Indian science are interesting, but had virtually no known influence except for the translation of Chioniades.

[222] 1.1 on fol. 83v of Laurentianus 28.17.

[223] See O. Neugebauer, *Studies in Byzantine Astronomical Terminology*, Transactions of the American Philosophical Society n.s. 50.2 (Philadelphia 1960) 29.

[224] *Ibid.* 30.

[225] Trans. Pingree (n. 22 above).

In astrology Indian influence upon the Arabs seems to have been primarily through Sasanian Iran, though some of it was direct. Indian astrology itself was derived from Greece; it acquired more of a Greek coloring as it passed first into Pahlavī, then into Arabic, and finally into Greek and Latin. Thus there is much that is presented in Greek and Latin as being from India that cannot be identified in the few Sanskrit texts on astrology that can be confidently dated before about A.D. 1000. There remain primarily the iconography of the decans, the catarchic astrology of the *nakṣatras,* the *navāṃśas,* the zodiacal topothesia, the *karaṇas,* and elements of the prayers to the planetary deities as Indian contributions, however perverted, to astrology and magic in the West. Their impact on the Byzantine and western European traditions of these two "sciences" and related arts was not negligible, but they were never of central importance. The Indian astral sciences in the second, third and fourth centuries were far more profoundly influenced by the Greek than the latter's spiritual descendants a millennium later by the former.

APPENDIX 1

Vatican Library MS Vat. gr. 1056 (V) fol. 48rv. See above at n. 56.

Περὶ τοῦ γνῶναι τὴν ἐρώτησιν ὡς ὁ σοφώτατος Μασάλα *φησίν, ἀλλὰ καὶ ὡς οἱ* Ἰνδοὶ *ἀπὸ τῆς μεθόδου τοῦ ἐνάτου τοῦ* Περσιστὶ *καλουμένου νουπάχρατ.*

Περὶ *τοῦ τίς ἡ ἐρώτησις ὁ* Μασάλα *φησὶν ὡς οὐχ εὕρομεν ἀκριβέστερον μαθεῖν τὴν ἐρώτησιν ἢ ἀπὸ τοῦ κυρίου τοῦ ὡροσκόπου, ἵνα λάβῃς τοῦτον ὡς σημειωτικὸν ἀστέρα τῆς ὑποθέσεως, ὡσαύτως καὶ τὸν ἀστέρα τὸν ὁρῶντα τὸν ὡροσκόπον καὶ τὸν κύριον αὐτοῦ ἐπεὶ ὁ λογισμὸς τοῦ ἐρωτῶντος ἀπὸ τοῦ κυρίου τοῦ ὡροσκόπου ἐπιγινώσκεται καὶ ἀπὸ τοῦ ὁρῶντος αὐτὸν ἀστέρος καὶ ἀπὸ τῆς μοίρας τοῦ ὡροσκόπου· ἀλλὰ μηδὲ τὸν ἴσως εὑρεθέντα ἀστέρα ἐν τῷ ὡροσκόπῳ καταλήψῃς μήτε μὴν τὸν ὁρῶντα ἀστέρα τὴν μοῖραν τοῦ ὡροσκόπου μοιρικῶς· καὶ οἷον ἂν εὕρῃς δυνατώτερον ἐκ τούτων, ἰδὲ ἐν ποίῳ τόπῳ ἐστί, καὶ ἐκ τούτου ἐπιγνωσθήσεταί σοι ἡ ἐνθύμησις τοῦ ἐρωτῶντος.*

Λέγουσι *δὲ καὶ οἱ* Ἰνδοὶ *ὅτι· ἰδὲ τοὺς ἀστέρας τοὺς ὁρῶντας τὸν ὡροσκόπον καὶ τοὺς ὄντας εἴς τινας τόπους τοῦ ὡροσκόπου, ὡσαύτως καὶ τὸν κύριον τοῦ ἐννάτου ἤτοι τοῦ νουπάχρατ, περὶ οὗ καὶ παρακατιὼν δηλωθήσεται ὅπερ τὸ ἔννατον εὑρήσεις· ἰδὲ δὲ καὶ τὸν ὁριοκράτορα καὶ τριγωνοκράτορα τοῦ ὡροσκόπου· καὶ οἷον εὕρῃς δυνατώτερον καὶ οἰκειότερον τῷ ὡροσκόπῳ τυγχάνοντα, ἰδὲ ἐν ποίῳ τόπῳ ἐστί, καὶ ἐκ τοῦ τόπου ἐπιγνωσθήσεται ἡ ἐρώτησις.*

Τὸ *δὲ ἔννατον ὃ καὶ οἱ* Ἰνδοὶ *νουπάχρατ καλοῦσι γινώσκεται οὕτως. ἰστέον ὅτι τῷ τοιούτῳ ὀνόματι, ἤγουν τὸ ἔννατον, ἔστι διακοσίων λεπτῶν ἤτοι τριῶν μοιρῶν καὶ τρίτου μοίρας ὀρθῆς. ἔσται οὖν ἐν ἑκάστῳ ζῳδίῳ ἐννέα ἔννατα, ὧν ἕκαστον ἔχει ἴδιον ἐπικρατήτορα. καὶ εἰ μὲν τὸ ζῴδιόν ἐστι* Κριὸς *ἢ* Λέων *ἢ* Τοξότης, *ἔσται ἑκάστου πρώτου ἐννάτου ἐν τοῖς τοιούτοις ζῳδίοις κύριος ὁ τοῦ* Κριοῦ *οἰκοδεσπότης*

῞Αρης, τοῦ δὲ δευτέρου κυριεύσει ἡ ᾽Αφροδίτη ἡ κυρία τοῦ Ταύρου, *τοῦ δὲ τρίτου ἐννάτου κυριęύσει ὁ ῾Ερμῆς ὁ κύριος τῶν Διδύμων, τοῦ δὲ δʹ ἐννάτου ἡ Σελήνη ἡ κυρία τοῦ* Καρκίνου, *τοῦ δὲ εʹ ὁ ῞Ηλιος ὁ κύριος τοῦ Λέοντος, τοῦ δὲ ςʹ ὁ ῾Ερμῆς ὁ κύριος τῆς Παρθένου·καὶ οὑτωσὶ ποιῶν εὑρήσεις τὸν κύριον τοῦ ἐννάτου.*

2 Περσιστῇ V ‖ 4 *τοῦ*[1] sup. lin. scr. V ‖ 6 *τοῦ*[2] sup. lin. scr. V ‖ 9 *μοιρηκῶς* V ‖ 21 *τοξότης*]*ταῦρος*V.

With this must be considered a chapter on fols. 123v-124 of Venice, Bibl. Naz. Marciana MS gr. 335 (H).

ρνηʹ Περὶ τοῦ γνῶναι τὴν ἐρώτησιν ἀπὸ τοῦ καλουμένου νουπὰρχ ἤγουν μερισμοῦ.

Στῆσον τὸν ὡροσκόπον, καὶ ἰδὲ τὰς μοίρας αὐτοῦ πόσαι εἰσίν.καὶ μέρισον ταύτας παρὰ τὸν γ̄ ⟨γʹ⟩, καὶ ἀπόλυε ταύτας ἀπὸ τοῦ ὡροσκοποῦντος ζῳδίου, διδοὺς κατὰ γ̄ ⟨γʹ⟩ ἓν ζῴδιον.καὶ εἰς οἷον ἂν καταλήξῃ βλέπε, καὶ ἀπὸ τῆς σημασίας τοῦ τόπου καὶ τοῦ ἐν αὐτῷ ἀστέρος λέγε τὴν ἐρώτησιν.

῾Υποδείγματος χάριν.γέγονεν ἡ ἐρώτησις ὡροσκοπούσης μοίρας ιʹ τοῦ Κριοῦ, καὶ ἦν τρία νουπάρχ.ἀπέλυσα ταῦτα ἀπὸ τοῦ ὡροσκόπου, καὶ κατήντησεν ὁ ἀριθμὸς εἰς τὸν γʹτόπον, ἐν ᾧ ἀνεπόδιζεν ὁ Κρόνος. *καὶ εἶπον διὰ τοῦτο ὅτι ἡ ἐρώτησίς ἐστι περὶ ἀποδήμου, πότε ἥξει. ἦν δὲ ὁ ῾Ερμῆς ὁ κύριος τοῦ νουπὰρχ μεσουρανῶν ἐν τῷ ῾Υδροχόῳ μετὰ ῾Ηλίου, καὶ εἶπον ὡς οὗτος ὁ ἀπόδημος ἐκτήσατο καὶ ἔχει μετ᾽ αὐτοῦ πολλούς, καὶ συμπίλισίς ἐστιν ἐν τῷ οἴκῳ αὐτοῦ διότι ὁ ῞Ηλιος ὁ κρατῶν τοῦ ὑψώματος τοῦ ὡροσκόπου καὶ τοῦ φωτὸς τοῦ κόσμου ἐμεσουράνησεν ἐν τῷ ῾Υδροχόῳ μετὰ τοῦ ῾Ερμοῦ.ὁ δὲ κύριος αὐτοῦ ὁ* Κρόνος *ἐν τοῖς Διδύμοις ἦν τῷ οἴκῳ τοῦ ῾Ερμοῦ, καὶ διὰ τοῦτο ἐτεκμηράμην ὡς ὁ ἀπόδημός ἐστιν ὁ ᾽Αμερουμνῆς. προσελαβόμην δὲ συνεργὸν πρὸς βεβαίωσιν τὸν* Κρόνον, *ὃς κύριος ἔτυχε τοῦ κοσμικοῦ χρόνου, πρὸς δὲ καὶ τοῦ ῾Ηλίου καὶ τοῦ ῾Ερμοῦ, καὶ ἐδήλωσε τὴν ἐρώτησιν εἶναι, πότε ἐπανήξει ὁ ᾽Αμερουμνῆς.*

᾽Ελαβον δὲ καὶ δύο τεκμήρια περιεκτικώτερα ἃ ἐκέκτητο ὁ Μασάλα *καὶ οἱ ἀρχαῖοι.καὶ γὰρ θ̄ εἰσὶν ἐν τῷ ὡροσκόπῳ δηλωτικοί·ὁ κύριος αὐτοῦ καὶ ὁ κύριος τοῦ ὑψώματος καὶ τοῦ τριγώνου καὶ τοῦ ὁρίου καὶ τοῦ θʹ καὶ ὁ ἀστὴρ πρὸς ὃν γίνεται ἡ κόλλησις τῆς μοίρας τοῦ ὡροσκόπου καὶ ὁ ἐπὼν τῷ ὡροσκόπῳ ἀστὴρ καὶ ὁ κλῆρος τῆς τύχης καὶ ὁ κύριος αὐτοῦ, καὶ ὁ διέπων καὶ ὁ κύριος τοῦ ῾Ηλίου ἡμέρας καὶ τῆς* Σελήνης *νυκτός. βλέπε οὖν ποῖος ἐκ τούτων ἔχει πλείονας λόγους, καὶ εἰ ἔστιν ὁ ἐπικρατήτωρ, καὶ ἰδὲ τίς τούτῳ·συνάπτει τῶν ἀστέρων, καὶ τίνος τόπον κυριεύει·καὶ πρὸς ἐκεῖνο ἀποτέλει.ἐπεὶ δὲ ζῳδίου τινὸς μέρος ἐστὶ καθ᾽ ὑπόθεσιν τῆς ἀποδημίας καὶ μέρος τοῦ μεσουρανήματος, λάμβανε τὸν ἔχοντα πλείονας μοίρας, καὶ καταλίμπανε τὸν ἔχοντα τὰς ἐλάττους, καὶ ἀνάγαγε τὸν σκοπὸν εἰς τὰ πλείονα.εἰ μὲν οὖν ἐστιν ὁ ἐπικρατήτωρ ἐν τῷ ὡροσκόπῳ μὴ συνάπτων τινί, ἐρωτᾷ περὶ αὐτοῦ, ἐν δὲ τῷ ηʹπερί τινος ἐξοδιασθέντος παρ᾽ αὐτοῦ.καὶ οὕτως ἀποτέλει κατὰ τὴν φύσιν τῶν ιβ̄ τόπων ταῦτα εἰ μὴ ἐπισυνάπτει ὁ ἐπικρατήτωρ τινί. εἰ δὲ συνάπτει, ἔχε τὸν συνάπτοντα αὐτῷ ἐπικρατήτορα, καὶ ἰδὲ τίνες συνάπτουσιν αὐτῷ, καὶ ἀποφαίνου τὸν σκοπὸν κατὰ τὴν φύσιν τοῦ τόπου οὗ κυριεύει ὁ συνάπτων ἀστήρ.*

Καὶ εἰ ἔστιν ὁ ἐπικρατήτωρ ἐν ἰδίῳ ταπεινώματι ἐρωτᾷ περὶ κλοπῆς ἢ περί τινος πεπτωκότος πράγματος, εἰ δὲ μεταβαίνει ἀπὸ ζῳδίου εἰς ζῴδιον περὶ ἀποδημίας, εἰ δὲ συνάπτει τῷ κυρίῳ τοῦ ηʹ ἢ τοῦ ιβʹκακοποιοῖς οὐδὶ περὶ θανάτου ἢ φόβου, εἰ δὲ

The Indian and Pseudo-Indian Passages in Greek and Latin Astronomical and Astrological Texts

στηρίξει πρὸς τὸ ἀναποδίσαι περὶ πράγματος πότε εὐθυνθήσεται, εἰ δὲ ἀναποδίζει περὶ ἀπορίας πράγματος εἰ δέ ἐστι μετὰ τοῦ Ἀναβιβάζοντος ἐν τῷ οἰκείῳ ὑψώματι ἢ μεσουρανῶν, ἔστιν ἡ ἐρώτησις περὶ θεοῦ ἢ πίστεως ἢ βασιλέως ἢ ἄρχοντος. εἰ δὲ σύνεστι τῇ Ἀφροδίτῃ καὶ ἐφορᾷ τοῦτον ὁ Ἑρμῆς, ἐρωτᾷ περὶ ὑπολήψεως γυναικείας, εἰ δὲ μετὰ τοῦ Καταβιβάζοντος περὶ ἀντιλογίας καὶ διαμάχης, εἰ δὲ
40 ὡροσκοπεῖ ἡ Σελήνη περὶ μάχης ἢ φήμης. εἰ δ' ἔστιν ἐν τῷ δ' ἢ μετὰ τοῦ Ἀναβιβάζοντος ἢ τοῦ Καταβιβάζοντος ἐν τῷ ϛ', ἐρωτᾷ περὶ θησαυροῦ κεχωσμένου· τὸ δ' αὐτὸ καὶ εἰ ὁ κύριος τοῦ β' ἐστὶν ἐν τῷ δ' καὶ ὁ κύριος τοῦ δ' ἐν τῷ ὡροσκόπῳ. εἰ δὲ ἐν τῷ μεσουρανήματί ἐστιν ὁ ἐπικρατήτωρ μετὰ τοῦ Ἄρεως, ἔστι δὲ καὶ τὸ
45 ζῴδιον πυρῶδες, ἐρωτᾷ περὶ χρυσοποιΐας. εἰ δὲ μετὰ τοῦ Καταβιβάζοντός ἐστιν ὁ ἐπικρατήτωρ, ἡ ἐρώτησίς ἐστι περὶ μαγείας· καὶ εἰ ἐπίδοι τοῦτον ὁ Ἑρμῆς, σημαίνει τὴν μαγείαν ἀληθῆ εἶναι. τὸ δ' αὐτὸ λέγε καὶ εἰ ἐπικρατήτωρ ἐστὶν ὁ Κρόνος καὶ σύνεστι τῷ Ἑρμῇ ἢ συσχηματίζεται αὐτῷ. καὶ γὰρ ἡ ἐρώτησις ἔσται περὶ μαγείας. καὶ ὅτε γένηται ἔξαυγος (ἤγουν ἑῷος) ὁ ἐπικρατήτωρ, ἔστιν ἡ ἐρώτησις περὶ
50 δαιμονῶντος, καὶ μάλιστα εἰ σὺν τῇ ἑῴᾳ φάσει συνάπτει κακοποιῷ ἢ ἐν ὁρίοις ὑπάρχει τοῦ Ἑρμοῦ ἢ τοῦ Κρόνου. εἰ δὲ ὡροσκοπεῖ εἷς τῶν οἴκων τοῦ Ἑρμοῦ καὶ οἱ ἐπικρατήτορες ἐπίκεντροι ὦσιν, συνάπτοντος αὐτοῖς τοῦ Ἑρμοῦ, ἡ ἐρώτησίς ἐστι περὶ γραμμάτων.

18 ὅ H ‖ 39 ὑψολήψεως H ‖ 45 ἐρρωτᾷ H ‖ 46 ἐπείδοι H

APPENDIX 2

Rome, Bibl. Angelica MS gr. 29 (E) fols. 34v-35.

Περὶ τοῦ νουποῦχρα τῶν Ἰνδῶν, ἤτοι περὶ τοῦ δασμοῦ τῶν μοιρῶν ἑκάστου ζῳδίου.

Ἀπονέμουσιν οἱ Ἰνδοὶ τοῦ Κριοῦ τὰς πρώτας μοίρας γ̄ καὶ τὸ γ' τῆς μιᾶς μοίρας τῷ Ἄρει, τὰς δὲ ἑξῆς γ̄ μοίρας καὶ τὸ γ' τῇ Ἀφροδίτῃ, τὰς δὲ ἑξῆς γ̄ μοίρας καὶ τὸ
5 γ' τῷ Ἑρμῇ, τὰς δὲ ἑξῆς τῇ Σελήνῃ, τὰς δὲ ἑξῆς τῷ Ἡλίῳ, τὰς δὲ ἑξῆς τῷ Ἑρμῇ, τὰς δὲ ἑξῆς τῇ Ἀφροδίτῃ, τὰς δὲ ἑξῆς πάλιν τῷ Ἄρει, ⟨τὰς δὲ ἑξῆς τῷ Διΐ⟩.
Τοῦ δὲ Ταύρου τὰς α̅ γ̄ μοίρας καὶ τὸ γ' ἀπονέμουσι τῷ Κρόνῳ, τὰς δὲ ἑξῆς τῷ Κρόνῳ, τὰς δὲ ἑξῆς τῷ Διΐ, τὰς δὲ ἑξῆς τῷ Ἄρει, τὰς δὲ ἑξῆς τῇ Ἀφροδίτῃ, τὰς δὲ ἑξῆς τῷ Ἑρμῇ, τὰς δὲ ἑξῆς τῇ Σελήνῃ, τὰς δὲ ἑξῆς τῷ Ἡλίῳ, τὰς δὲ ἑξῆς τῷ
10 Ἑρμῇ.
Τῶν δὲ Διδύμων τὰς α̅ μοίρας γ̄ καὶ τὸ γ' ἀπονέμουσι τῇ Ἀφροδίτῃ, τὰς δὲ ἑξῆς τῷ Ἄρει, τὰς δὲ ἑξῆς τῷ Διΐ, τὰς δὲ ἑξῆς τῷ Κρόνῳ, ⟨τὰς δὲ ἑξῆς τῷ Κρόνῳ⟩, τὰς δὲ ἑξῆς τῷ Διΐ, τὰς δὲ ἑξῆς τῷ Ἄρει, τὰς δὲ ἑξῆς τῇ Ἀφροδίτῃ, καὶ τὰς ἑξῆς τῷ Ἑρμῇ.

Τοῦ δὲ Καρκίνου τὰς α΄ γ̅ μοίρας καὶ τὸ γ΄ ἀνέδωκαν τῇ Σελήνῃ, τὰς δὲ ἑξῆς τῷ Ἡλίῳ, τὰς δὲ ἑξῆς τῷ Ἑρμῇ, τὰς δὲ ἑξῆς τῇ Ἀφροδίτῃ, τὰς δὲ ἑξῆς τῷ Ἄρει, τὰς δὲ ἑξῆς τῷ Διΐ, τὰς δὲ ἑξῆς τῷ Κρόνῳ, ⟨τὰς δὲ ἑξῆς τῷ Κρόνῳ⟩, καὶ τὰς ἑξῆς τῷ Διΐ.

Τοῦ δὲ Λέοντος τὰς α΄ γ̅ μοίρας καὶ τὸ γ΄ ἀπονέμουσι τῷ Ἄρει, τὰς δὲ ἑξῆς τῇ Ἀφροδίτῃ, τὰς δὲ ἑξῆς τῷ Ἑρμῇ, τὰς δὲ ἑξῆς τῇ Σελήνῃ, τὰς δὲ ἑξῆς τῷ Ἡλίῳ, τὰς δὲ ἑξῆς τῷ Ἑρμῇ, τὰς δὲ ἑξῆς τῇ Ἀφροδίτῃ, τὰς δὲ ἑξῆς τῷ Ἄρει, τὰς δὲ ἑξῆς τῷ Διΐ.

Τῆς δὲ Παρθένου τὰς α΄ γ̅ μοίρας καὶ τὸ γ΄ ἀπονέμουσι τῷ Κρόνῳ, καὶ τὰς ἑξῆς τῷ Κρόνῳ, τὰς δὲ ἑξῆς τῷ Διΐ, τὰς δὲ ἑξῆς τῷ Ἄρει, τὰς δὲ ἑξῆς τῇ Ἀφροδίτῃ, τὰς δὲ ἑξῆς τῷ Ἑρμῇ, τὰς δὲ ἑξῆς τῇ Σελήνῃ, τὰς δὲ ἑξῆς τῷ Ἡλίῳ, καὶ τὰς ἑξῆς τῷ Ἑρμῇ.

Τοῦ δὲ Ζυγοῦ τὰς α΄ γ̅ μοίρας καὶ τὸ γ΄ ἀπονέμουσι τῇ Ἀφροδίτῃ, τὰς δὲ ἑξῆς τῷ Ἄρει, τὰς δὲ ἑξῆς τῷ Διΐ, τὰς δὲ ἑξῆς τῷ Κρόνῳ, καὶ τὰς ἑξῆς τῷ Κρόνῳ, ⟨τὰς δὲ ἑξῆς τῷ Διΐ⟩, τὰς δὲ ἑξῆς τῷ Ἄρει, τὰς δὲ ἑξῆς τῇ Ἀφροδίτῃ, ⟨τὰς δὲ ἑξῆς τῷ Ἑρμῇ⟩.

Τοῦ Σκορπίου τὰς α΄ γ̅ μοίρας καὶ τὸ γ΄ ἀπονέμουσι τῇ Σελήνῃ, τὰς δὲ ἑξῆς τῷ Ἡλίῳ, τὰς δὲ ἑξῆς τῷ Ἑρμῇ, τὰς δὲ ἑξῆς τῇ Ἀφροδίτῃ, τὰς δὲ ἑξῆς τῷ Ἄρει, τὰς δὲ ἑξῆς τῷ Διΐ, τὰς δὲ ἑξῆς τῷ Κρόνῳ, καὶ τὰς ἑξῆς τῷ Κρόνῳ, καὶ τὰς ἑξῆς τῷ Διΐ.

Τοῦ δὲ Τοξότου τὰς α΄ γ̅ μοίρας καὶ τὸ γ΄ ἀπονέμουσι τῷ Ἄρει, τὰς δὲ ἑξῆς τῇ Ἀφροδίτῃ, τὰς δὲ ἑξῆς τῷ Ἑρμῇ, τὰς δὲ ἑξῆς τῇ Σελήνῃ, τὰς δὲ ἑξῆς τῷ Ἡλίῳ, τὰς δὲ ἑξῆς τῷ Ἑρμῇ, τὰς δὲ ἑξῆς τῇ Ἀφροδίτῃ, τὰς δὲ ἑξῆς τῷ Ἄρει, καὶ τὰς ἑξῆς τῷ Διΐ.

Τοῦ δὲ Αἰγοκέρωτος τὰς α΄ γ̅ μοίρας καὶ τὸ γ΄ ἀπονέμουσι τῷ Κρόνῳ, καὶ τὰς ἑξῆς τῷ Κρόνῳ, τὰς δὲ ἑξῆς τῷ Διΐ, τὰς δὲ ἑξῆς τῷ Ἄρει, τὰς δὲ ἑξῆς τῇ Ἀφροδίτῃ, τὰς δὲ ἑξῆς τῷ Ἑρμῇ, τὰς δὲ ἑξῆς τῇ Σελήνῃ, τὰς δὲ ἑξῆς τῷ Ἡλίῳ, καὶ τὰς ἑξῆς τῷ Ἑρμῇ.

Τοῦ δὲ Ὑδροχόῳ τὰς α΄ γ̅ μοίρας καὶ τὸ γ΄ ἀπονέμουσι τῇ Ἀφροδίτῃ, τὰς δὲ ἑξῆς τῷ Ἄρει, τὰς δὲ ἑξῆς τῷ Διΐ, τὰς δὲ ἑξῆς τῷ Κρόνῳ, καὶ τὰς ἑξῆς τῷ Κρόνῳ, τὰς δὲ ἑξῆς τῷ Διΐ, τὰς δὲ ἑξῆς τῷ Ἄρει, τὰς δὲ ἑξῆς τῇ Ἀφροδίτῃ, καὶ τὰς ἑξῆς τῷ Ἑρμῇ.

Τῶν δὲ Ἰχθύων τὰς α΄ γ̅ μοίρας καὶ τὸ γ΄ ἀπονέμουσι τῇ Σελήνῃ, τὰς δὲ ἑξῆς τῷ Ἡλίῳ, τὰς δὲ ἑξῆς τῷ Ἑρμῇ, τὰς δὲ ἑξῆς τῇ Ἀφροδίτῃ, τὰς δὲ ἑξῆς τῷ Ἄρει, τὰς δὲ ἑξῆς τῷ Διΐ, τὰς δὲ ἑξῆς τῷ Κρόνῳ, καὶ τὰς ἑξῆς τῷ Κρόνῳ, καὶ τὰς ἑξῆς τῷ Διΐ.

8 *κρόνῳ*] symb. Solis E || 19 *τὸ*] *τῷ* E || 21 post *ἑρμῇ* iter. 20 *τὰς* 2 – 21 *ἑρμῇ* E.

Appendix 3

Vatican Library MS Vat. gr. 211 fol. 167. See above at n. 104.

Χρονογραφία τῶν ἐτῶν καὶ τῶν ἡμερῶν · κανόνια τῆς φανερώσεως τῶν δύο ἐτῶν τῶν κατὰ μῆκος καὶ πλάτος

πλάτος μῆκος	ἔτος Νῶε. ἡ ἀρχὴ τούτου ἡμέρα ϛ΄	ἔτος Ναβουχοδονόσορ. ἡ ἀρχὴ τοῦ ἔτους τούτου ἡμέρα δ΄	ἔτος ᾿Αλεξάνδρου. ἡ ἀρχὴ τοῦ ἔτους τούτου ἡμέρα α΄	ἔτος ῾Ρωμαίων. ἡ ἀρχὴ τοῦ ἔτους τούτου ἡμέρα β΄	ἔτος ᾿Αράβων. ἡ ἀρχὴ τοῦ ἔτους τούτου ἡμέρα ϛ΄	ἔτος Περσῶν. ἡ ἀρχὴ τοῦ ἔτους τούτου ἡμέρα γ΄	ἔτος Σουλτάνου Μαλιξᾶ. ἡ ἀρχὴ τοῦ ἔτους τούτου ἡμέρα γ΄[1]
ἔτος Νῶε ὅμοιον τῶν Περσῶν	ἀπὸ κτίσεως κόσμου χρόνοι τόσοι 180000	χρόνοι 2356 ἡμέραι σλβ	χρόνοι 2780 ἡμέραι σλβ	χρόνοι 2792 ἡμέραι ρϛγ	χρόνοι 3725 ἡμέραι τμθ	χρόνοι 3735 ἡμέραι τκβ	χρόνοι 4182 ἡμέραι τλη
ἔτος Ναβουχοδονόσορ ὅμοιον Περσῶν	860172 γ νη νϛ ιβ	ἀπὸ κτίσεως κόσμου χρόνοι 182356 σλβ ἡμέραι	χρόνοι 424 ἡμέραι 0	χρόνοι 439 ἡμέραι τκϛ	χρόνοι 1369 ἡμέραι ρκα	χρόνοι 1379[1] ἡμέραι ϛ[2]	χρόνοι 1826 ἡμέραι ρϛ
ἔτος ᾿Αλεξάνδρου ὅμοιον Περσῶν	1014932 δ μα νε λβ	154760 ο μβ νθ κ	ἀπὸ κτίσεως κόσμου χρόνοι 182780 σλβ ἡμέραι	χρόνοι 11 ἡμέραι τκϛ	χρόνοι 945 ἡμέραι ριζ	χρόνοι 955 ἡμέραι ϛ	χρόνοι 1402[2] ἡμέραι ρϛ[3]
ἔτος ῾Ρωμαίων ὅμοιον τῶν ῾Ρωμαίων	1019273 δ μγ ζ νγ	159101 ο μδ ια μα	4341 ο α ιβ κα	ἀπὸ κτίσεως κόσμου χρόνοι 182792 ρϛγ ἡμέραι	χρόνοι 932[1] ἡμέραι σπζ	χρόνοι 942 ἡμέραι σνη[3]	χρόνοι 1389[4] ἡμέραι ρξβ[5]

APPENDIX 3 (Continued)

ἔτος Ἀράβων	1359974	499801	345041	340700	ἀπὸ κτίσεως κόσμου	χρόνοι 10	χρόνοι 470[6]
τὰ ἔτη τούτων Ἀρραπικά	ς ιξ μς ιδ	β ιη ν α	α λε ν μα	α λδ λη κ[1]	χρόνοι 183725 τμθ ἡμέραι	ἡμέραι οθ	ἡμέραι σμβ
ἔτος Περσῶν	1363597	503425	348665	344324	3624	ἀπὸ κτίσεως κόσμου	χρόνοι 447
τὰ ἔτη τούτων Περσικά	ς ιη μς λξ	β ιθ ν κε	α λς να ε	α λε λη μδ	ο α ο κδ	χρόνοι 183735 τκβ[4] ἡμέραι	ἡμέραι ις[7]
ἔτος Σουλτάνου	1526768[1]	666596	511836	507495[2]	166794	163171[5]	ἀπὸ κτίσεως κόσμου
Μελιξᾶ τὸ ἔτος τούτου κατὰ τὴν ὁδὸν Ἡλίου	ξ[2] δ ς η	γ ε θ νς	β κβ ι λς	β κ νη[3] ιε	ο μς ιθ νδ	ο με ιθ λα[6]	χρόνοι 184182 ἡμέραι τλη[8]
	[1] 1526767 MS [2] δ MS			[1] κα MS [2] 507695 MS [3] ιη MS	[1] 933 MS	[1] 1359 MS [2] ςς MS [3] σνθ MS [4] τλβ MS [5] 163173 MS [6] λβ MS	[1] ξ′ MS [2] 1826 MS [3] πς MS [4] 1403 MS [5] νη MS [6] 1382 MS [7] ιξβ MS [8] ςλθ MS

APPENDIX 4

Vatican Library MS Vat. gr. 1056 fols. 81v-82. See above at n. 175.

Ἰνδός τις σοφὸς λεγόμενος Πορζὸζόμχαρ εἶπεν· ιβ̄ εἰσὶ σημασίαι τῆς εὑρέσεως τοῦ κλέμματος.

Μία μὲν τὸ ὁρᾶν τὸν κύριον τοῦ δύνοντος τὸν κύριον τοῦ ὡροσκόπου.

Δεύτερον τὸ ὁρᾶν μὲν τὸν κύριον τοῦ δύνοντος τὸν κύριον τοῦ ὡροσκόπου, εἶναι δὲ καὶ τὸν κύριον τοῦ δύνοντος ὕπαυγον· τότε γὰρ φόβῳ τοῦ ἐξουσιαστοῦ ἀντιστρέψει ὁ κλέπτης ἃ ἔκλεψεν.

Γ΄ ὅταν ἀστήρ τις βλέπῃ μὲν τὸν κύριον τοῦ δύνοντος καὶ τὸν Ἥλιον, εἶτα ἀπέρχεται πρὸς τὸν κύριον τοῦ ὡροσκόπου.

Δ΄ ὅτε ἀστήρ τις ἀπὸ σχηματισμοῦ τοῦ Ἡλίου ἐξερχόμενος ἀπέρχεται πρὸς τὸν κύριον τοῦ ὡροσκόπου· γ. . . φως (?) γὰρ τοῦ ἐξουσιαστοῦ ἀντιστρέψει τὸ κλέμμα ὁ κλέπτης.

Ε΄ ὅτε ἀστήρ τις ἀπὸ τοῦ κυρίου τοῦ μεσουρανήματος ἐξερχόμενος ἀπέρχεται πρὸς τὸν κύριον τοῦ ὡροσκόπου.

Ϛ΄ ὅτε ὁρᾷ ὁ κύριος τοῦ ὡροσκόπου τὸν κύριον τοῦ μεσουρανήματος· βοηθήσεσθαι γὰρ τότε παρὰ ἐξουσιαστοῦ πρὸς τὸ λαβεῖν τὸν κλέπτην.

Ζ΄ ὅτε ὁ κύριος τοῦ β΄ τόπου ὡροσκοπεῖ.

Η΄ ὅτε ὁρᾷ ὁ κύριος τοῦ ὡροσκόπου τὸν κύριον τοῦ β΄ καὶ ὁ κύριος τοῦ β΄ τὸν κύριον τοῦ ὡροσκόπου.

Θ΄ ὅτε ὁρᾷ ὁ κύριος τοῦ η΄ τὸν κύριον τοῦ β΄.

Ι΄ ὅτε ὁ κύριος τοῦ β΄ καὶ ὁ κύριος τοῦ ζ΄ εἰσὶν ἐν τοῖς κέντροις.

[fol. 82] σημεῖον γὰρ . . . .

ἐξελθεῖν τὸν κλέπτην ἢ τὸ κλέμμα . . . . .

⟨ΙΒ΄⟩ ὅτε συνάπτει ἡ Σελήνη τῷ κυρίῳ τοῦ ὡροσκόπου.

APPENDIX 5

Shādhān, *Mudhākarāt* chapter 59. See n. 178 above.

Εἶπεν ὁ Ἀποσαΐτ· ἀπεστάλη πρὸς τὸν Ἀπομάσαρ γενέθλιον τοῦ υἱοῦ τοῦ βασιλέως τῶν Ἰνδῶν· εἶχε γὰρ παρ᾽ αὐτῷ φίλον ὅστις ἐπῄνει αὐτὸν διηνεκῶς· ἦν δὲ ὡροσκόπος οἱ Δίδυμοι, καὶ ὁ Ἥλιος ἔτυχεν ἐν τῷ Αἰγοκέρωτι, καὶ ἡ Σελήνη ὁμοίως πλὴν ἤδη ἐχωρίσθη ἐκ τῶν τοῦ Ἡλίου αὐγῶν· ἦν δὲ ὁ Ἑρμῆς ἐν τῷ Ὑδροχόῳ, καὶ ὁ Ζεὺς ἐν τῷ Κριῷ, καὶ ὁ Κρόνος ἐν τῷ Καρκίνῳ, τὴν ἀκμὴν ἔχων τοῦ ἀναποδισμοῦ· ἡ δὲ ἀκμὴ τοῦ ἀναποδισμοῦ τότε γίνεται ὅτε διαμετρήσει ὁ ἀστὴρ τὸν Ἥλιον.

Καὶ ἀπετέλεσεν ὁ Ἀπομάσαρ ὅτι ζήσεται ἔτη ἰσάριθμα τοῖς μέσοις ἔτεσι τοῦ Κρόνου· καὶ ὡς ἤκουσα τὸ τοιοῦτον ἀποτέλεσμα, εἶπον τῷ Ἀπομάσαρ· δόξα σοι ὁ Θεός· τί ἀγαθὸν εἶδες ὅτι δέδωκας αὐτῷ τοσαῦτα ἔτη; ὁ γὰρ ἀνέτης αὐτοῦ ἐν τῇ *ἀκμῇ τοῦ ἀναποδισμοῦ ἐστι καὶ ἀποκεκλικώς· καὶ ἀνάγκη ἐστὶ πολλὴ εἰ δώσω αὐτῷ τὰ ἐλάχιστα αὐτοῦ ἔτη, μᾶλλον δὲ κατὰ τὰς τῶν πολλῶν μεθόδους διὰ τὸν ἀναποδισμὸν ὑφαιροῦμαι ἀπὸ τῶν ἐλαχίστων αὐτοῦ ἐτῶν β̄ πέμπτα· καὶ ἠρξάμην καὶ ἐγὼ καὶ οἱ παρόντες καταμέμφεσθαι τὸν Ἀπομάσαρ καὶ λέγειν αὐτῷ· σὺ μέγα ὄνομα ἔχεις παρὰ τῷ βασιλεῖ τῶν Ἰνδῶν, καὶ εἰσι παρ᾽ αὐτῷ καὶ Ἰνδοὶ ἐπιστήμονες· καὶ ὅτε ἴδωσι τὸ τοιοῦτόν σου ἀποτέλεσμα, ἀντ᾽ οὐδενός σε λογίσονται, καὶ πρὸς τούτοις στερηθήσῃ καὶ τῶν αὐτοῦ δωρεῶν.*

Εἶπεν δὲ ὁ Ἀπομάσαρ πρὸς ἡμᾶς· φρόνιμοι ὄντες πῶς λαλεῖτε εἴ τι δόξει ὑμῖν; οὐκ οἴδατε ὅτι οἱ Ἰνδοὶ πολυζώϊτοί εἰσιν, καὶ οἱ πλείονες ζῶσιν ἕως οὗ καταγηράσωσιν παντελῶς; οὐ γινώσκετε δὲ καί, ὅτε τελευτήσει τις πρὶν ἢ καταλάβῃ τὰ μέσα ἔτη τοῦ Κρόνου, λογίζονται τὸ τοιοῦτον θαῦμα μέγα; οὐ γινώσκετε δὲ καὶ τοῦτο, ὅτι ἡ Ἰνδία τῷ Κρόνῳ ἀνήκει, καὶ διὰ τοῦτο οὐ πάνυ βλάπτει τι ὁ Κρόνος ἐν τοῖς γενεθλίοις αὐτῶν εἰ μή γε ἀποκεκλικὼς εἴη; εἶπον τῷ Ἀπομάσαρ· καὶ ἐν τῷ τοιούτῳ γενεθλίῳ ἀποκεκλικώς ἐστιν. εἶπεν· οὔ. ἐν γὰρ τῷ β′ τόπῳ ἦν, καὶ ὁ β′ τόπος οὔκ ἐστιν ἀπόκλιμα, ἀλλὰ κυρίως ἀποκλίματά εἰσι καὶ φαῦλα, ὅ τε ϛ′ τόπος καὶ ὁ ιβ′, ὁ δὲ β′ τόπος ἔχει μυστήρια μεγάλα. εἶπον δὲ τῷ Ἀπομάσαρ· σὺ ἐδίδαξας ἡμᾶς ὅτι ὁ ἀναποδισμὸς ὑφαιρεῖ ἐκ τῆς δόσεως τῶν ἐτῶν β̄ πέμπτα, καὶ πῶς οὐκ ἀφῆκας ἐν τῷ τοιούτῳ γενεθλίῳ β̄ πέμπτα; εἶπέ μοι δὲ ὅτι· διὰ τοῦτο οὐκ ἀφεῖλον ταῦτα ἀπὸ τῆς τῶν ἐτῶν δόσεως, ὅτι ἦν ὁ Κρόνος ἐν φρέατι· καὶ εἶπον τῷ Ἀπομάσαρ· τί εἰσὶ τὰ λεγόμενα φρέατα; εἶπε δέ· ἀπλανεῖς ἀστέρες ἐνδυναμοῦντες τοὺς κακοποιοὺς καὶ ἀμβλύνοντες τὴν ἐνέργειαν τῶν ἀγαθοποιῶν.

APPENDIX 6

Vatican Library MS Vat. gr. 211 fol. 155. See above at n. 182.

Κανόνια τῆς κινήσεως τοῦ κάϊτ.

Κρατοῦνται οἱ τέλειοι Σουλτανικοὶ χρόνοι καὶ ξ̄ϛ̄ ἑνοῦνται τούτοις, ἢ κρατοῦνται οἱ τετελειωμένοι χρόνοι τῶν Περσῶν καὶ ν̄δ̄ προστίθενται τούτοις, ἢ κρατοῦνται οἱ τετελειωμένοι τῶν Ῥωμαίων πρὸ τοῦ Ἀδὰρ καὶ ρ̄κ̄ προστίθενται τούτοις· ἀφ᾽ ἑκάστου δὲ τούτων ἀνὰ ρ̄μ̄δ̄ ἀφαιροῦνται· εἴ τι καταλειφθῇ, κατ᾽ ἐναντίον ἐκείνου γίνεται εἰσέλευσις εἰς τὰ κανόνια τῶν χρόνων, τῶν μηνῶν καὶ τῶν ἡμερῶν· εἴ τι εὑρεθῇ ἀπὸ ζῳδίων, μοιρῶν, λεπτῶν καὶ δευτέρων λεπτῶν, τοῦτο ἀεὶ ἀφαιρεῖται ἀπὸ τῶν ῑβ̄ ζῳδίων· εἴ τι οὖν καταλειφθῇ, αὐθημερινόν ἐστι τοῦ κάϊτ.

δωδεκαετηρίδες Ἡλίου	κίνησις τοῦ κάϊτ ζῴδια	ἁπλᾶ ἔτη τοῦ Ἡλίου	κίνησις τοῦ κάϊτ μοῖραι	λεπτά
ιβ	α	α	β	λ
κδ	β	β	ε	ο
λϛ	γ	γ	ζ	λ

δωδεκαετηρίδες Ἡλίου	κίνησις τοῦ κάϊτ ζῴδια	ἁπλᾶ ἔτη τοῦ Ἡλίου	κίνησις τοῦ κάϊτ μοῖραι	λεπτά
μη	δ	δ	ι	ο
ξ	ε	ε	ιβ	λ
οβ	ς	ς	ιε	ο
πδ	ζ	ζ	ιζ	λ
ϟς	η	η	κ	ο
ρη	θ	θ	κβ	λ
ρκ	ι	ι	κε	ο
ρλβ	ια	ια	κζ	λ
ρμδ	ιβ	ιβ	λ	ο

	κίνησις τοῦ κάϊτ			κίνησις τοῦ κάϊτ		
μῆνες	μοῖραι	λεπτά	λεπτά β'	ἡμέραι	λεπτά	λεπτά β'
Φαρβαντίν	ο	ιβ	κ	α	ο	κε
Ἀρτεμπίστ	ο	κδ	μ	β	ο	ν
Χορντάτ	ο	λς	νθ	γ	α	ιδ
Τίρ	ο	μθ	ιθ	δ	α	λθ
Μουρντάτ	α	α	λθ	ε	β	γ
Σαριβάρ	α	ιγ	νθ	ς	β	κη
Μμέρ	α	κς	ιη	ζ	β	νγ
Ἀπάν	α	λη	λη	η	γ	ιη
Ἀδδέρ	α	ν	νη	θ	γ	μβ
Ντάϊ	β	γ	ιζ	ι	δ	ζ
Μπαχμάν	β	ιε	λζ	κ	η	ιγ
Ἐσφιταρδμάδ	β	κζ	νζ	λ	ιβ	κ

APPENDIX 7

Rome, Bibl. Angelica MS gr. 29 fol. 23. See above at n. 196.

Περὶ γενεθλίου · μέθοδος Ἰνδικὴ πάνυ θαυμασιωτάτη.

Ὅτε γεννηθῇ τις καὶ τύχῃ ὁ Ἥλιος ἐν τῷ ὡροσκόπῳ ἢ ἐν τῷ ἰδίῳ ὑψώματι, τεύξεται ἐξουσίας μεγάλας · εἰ δὲ ἐν τῷ β΄ τόπῳ, ἔσται τῶν ἀναγκαίων ἐπιδεής · εἰ δὲ ἐν τῷ γ΄, μισεῖται παρὰ τῶν ἀδελφῶν αὐτοῦ · εἰ δὲ ἐν τῷ δ΄, μικροῦ ὄντος τελευτήσουσιν οἱ γονεῖς αὐτοῦ · εἰ δὲ ἐν τῷ ε΄, ἄτεκνος γενήσεται · εἰ δὲ ἐν τῷ ς΄, ὑγιαίνων ἔσται · εἰ δὲ ἐν τῷ ζ΄, συζευχθήσεται γυναικὶ ἀπὸ ξένης · εἰ δὲ ἐν τῷ η΄, τελευτήσει μεταξὺ τῶν συγγενῶν αὐτοῦ · εἰ δὲ ἐν τῷ θ΄, ἐπικτήσεται πλοῦτον ἀπὸ χώρας μικρᾶς · ⟨εἰ δὲ ἐν τῷ ι΄, - - -⟩ · εἰ δὲ ἐν τῷ ια΄, δηλοῖ φίλους, συνεργοὺς καὶ συγγενεῖς · εἰ δὲ ἐν τῷ ιβ΄, δυστυχὴς ἔσται.

APPENDIX 8

Rome, Bibl. Angelica MS gr. 29 (E) fol. 38rv, and Vatican Library MS Vat. gr. 1056 (V^2) fol. 97v. See above at n. 198.

Περὶ τοῦ γνῶναι τὰ περὶ τῶν ἐξουσιαστῶν, πῶς ἀποτελεσθήσονται, κατὰ τὰς Ἰνδικὰς μεθόδους.

Λέγουσιν οἱ Ἰνδοὶ ὅτι ἐπὶ τῆς ἐξουσιαστικῆς καταρχῆς ἰδὲ τοὺς τριγωνοκράτορας τοῦ Ἡλίου καὶ τῆς Σελήνης, ἰδὲ καὶ τοὺς οἰκοδεσπότας τῆς προγενομένης συνόδου ἢ τῆς πανσελήνου· καὶ εἴπερ εὕροις τοὺς τοιούτους ἀγαθυνομένους, λέγε ἀγαθὴν τὴν ἀπόβασιν, εἰ δὲ κακυνομένους εὕροις αὐτούς, κακὴν ἀποφαίνου τὴν ἀπόβασιν· εἰ δὲ εὕροις τὴν Σελήνην ἀγαθυνομένην καὶ ὡροσκοποῦσαν, καλή ἐστιν ἡ καταρχή· εἰ δὲ εὕροις ἐν τῷ ὡροσκόπῳ ἀστέρα τινά, ἐκεῖνός ἐστι σημαντὴρ τῆς ἐξουσίας ἐκείνης· εἰ δέ ἐστιν ἐν τῷ μεσουρανήματι ἀστήρ, ἐκεῖνος σημαίνει τὰ περὶ τῆς ἐξουσίας· εἰ δὲ εὕροις ἀστέρα ἐν τῷ δύνοντι, ἐκεῖνος σημαίνει τὰ τῶν ἐχθρῶν.

Ὁ δὲ δεύτερος τόπος σημαίνει τὰ περὶ τοῦ πλούτου αὐτοῦ, ὁ δὲ τρίτος σημαίνει τὰ περὶ τῶν ἀδελφῶν αὐτοῦ, ὁ δὲ δ΄ τὰ περὶ τῆς φαμιλίας αὐτοῦ, ὁ δὲ ε΄ τὰ περὶ τῶν τέκνων αὐτοῦ, ὁ δὲ ς΄ σημαίνει τὰ περὶ τῶν δούλων αὐτοῦ καὶ ὑποζυγίων, ὁ δὲ ζ΄ τόπος σημαίνει τὰ περὶ τῶν γυναικῶν καὶ δουλίδων ἃς ἔχει εἰς κοίτην, ὁ δὲ η΄ τόπος σημαίνει τὸν θάνατον αὐτοῦ, ἐν ποίῳ τρόπῳ γενήσεται, ὁ δὲ θ΄ τόπος σημαίνει τὴν πίστιν αὐτοῦ, καὶ ἣν ἔχει εὐλάβειαν, καὶ ὅπως ἀναστρέφει περὶ τὴν δικαιοσύνην, ἔτι δὲ καὶ τὰς ἀποδημίας αὐτοῦ, ἄρα γε εὐτυχεῖς ἔσονται ἢ δυστυχεῖς, ὁ δὲ ι΄ τόπος σημαίνει τὰ περὶ τῆς δόξης αὐτοῦ καὶ τοῦ ἀγαθοῦ κηρύγματος ἢ τοῦ ἐναντίου, καὶ ὁποία ἐστὶν ἡ διάθεσις τῶν ὑπηκόων, ὁ δὲ ια΄ τόπος σημαίνει τὰ τῶν φίλων αὐτοῦ, ὅπως διάκεινται πρὸς αὐτόν, ὁ δὲ ιβ΄ σημαίνει τὰ τῶν ἐχθρῶν αὐτοῦ καὶ τῶν βουλευομένων ἀπόστασιν ποιεῖν κατ᾽ αὐτοῦ.

Εἰ δὲ ὡροσκοπεῖ ἀστὴρ ἐν τῇ καταρχῇ καὶ ἔστιν ἀγαθοποιὸς καὶ ὑπάρχει ἐν τῷ ἰδίῳ οἴκῳ ἢ τῷ ὑψώματι ἢ τῷ τριγώνῳ ἢ τῷ ὁρίῳ, λέγε ἀγαθὴν εἶναι τὴν καταρχήν· ἀγωνίσθητι δὲ ποιῆσαι ἐν τῷ καιρῷ τῆς καταρχῆς τὸν Δία ἑῷον καὶ μεσουρανοῦντα, μὴ ποίησον δὲ αὐτὸν ἀποκεκλικότα, ποίησον δὲ τὸν οἰκοδεσπότην τῆς Σελήνης ἐπίκεντρον, καὶ μὴ τύχῃ ἀναποδίζων· εἰ γὰρ τὰ τοιαῦτα δύνῃ κατορθῶσαι, μεγάλη ἔσται ἡ ἐξουσία καὶ ἐπίμονος· παντοίῳ δὲ τρόπῳ σπούδευσον ἵνα οἱ ἀγαθοὶ ὦσιν ἐπίκεντροι· κατὰ πολὺ γὰρ ἐνεργεῖ τὸ ἐπίκεντρον τῶν ἀγαθοποιῶν ἐν τῇ καταρχῇ· μὴ ἔστω δὲ ὁ κύριος τοῦ ὡροσκόπου ἀναποδίζων ἢ ἐν τῷ ἰδίῳ ταπεινώματι· εἰ γὰρ τοιουτοτρόπως ἔσται, ὑποστήσεται ὁ ἐξουσιαστὴς ἀνάγκας ἐν τῇ ἐξουσίᾳ αὐτοῦ.

1-2 tit. om. V^2 || 3 ante *λέγουσιν* add. *περὶ τῆς ἐξουσιαστικῆς καταρχῆς* V^2 | *ὅτι – καταρχῆς* om. V^2 || 6 *κεκακωμένους* V^2 | *εὕροις αὐτούς* om. V^2 | *ἀποφαίνου τὴν ἀπόβασιν* om. V^2 || 7 *καλλή* V^2 || 8 *σημάντωρ* V^2 || 10 *δύνοντι*] *ὡροσκόπω* E || 11 *β΄* E | *γ΄* E | *σημαίνει* om. V^2 || 12 *αὐτοῦ* om. V^2 || 12 – 21 *ὁ – κατ᾽ αὐτοῦ*] *καὶ καθεξῆς ὁμοίως τοῖς σημαινομένοις τοῦ θεματίου τόποις* V^2 || 22 *εἰ δὲ*] *εἶτα ἰδέ, καὶ εἰ* V^2 | *καὶ ἔστιν* om. V^2 || 23 *τῷ*^{2,3} om. E || 24 *τὸν δία* om. V^2 || 25 *μὴ – αὐτὸν*] *καὶ μὴ* V^2 | *ποίησον δὲ τὸν*] *τὸν δὲ* V^2 || 26 *τύχῃ* om. V^2 | *ἀναποδίζοντα* V^2 | *τὰ – κατορθῶσαι*] *κατορθώσεις ταῦτα οὕτως* V^2 || 27 – 29 *παντοίῳ – καταρχῇ*] *ἀεὶ δὲ σπεῦδε ἵνα εἰσὶν οἱ ἀγαθοποιοὶ ἐπίκεντροι ὡς ἔχοντες ἐπὶ τοῖς κέντροις ἐνέργειαν ἐν ταῖς καταρχαῖς* V[2] ‖ 29 *ὁ δὲ κύριος τοῦ ὡροσκόπου μὴ ἔστω* V[2] ‖ 29 – 30 *ἐν – ταπεινώματι*] *τεπεινούμενος* V[2] ‖ 30 *εἰ – ἔσται* om. V[2] | post *ὑποστήσεται* add. *γὰρ τότε* V[2]

APPENDIX 9

Rome, Bibl. Angelica MS gr. 29 fol. 23. See above at n. 199.

Περὶ τοῦ γνῶναι ἀγαθὸν ἢ φαῦλον κατὰ τὸν καιρὸν καθ' ὃν ἡ ἐρώτησις · ἐκ τῶν Ἰνδικῶν ἀποτελεσμάτων.

Εἰ βούλει γνῶναι τί πάθος ἀγαθὸν ἢ φαῦλον ἐν τῷ ἔτει σου, στῆσον τὸν ὡροσκόπον καὶ ἰδὲ τὸν κύριον αὐτοῦ, ἐν ποίῳ τόπῳ ἐστίν · καὶ εἰ μέν ἐστιν ἐν τῷ ὡροσκόπῳ, ἰδὲ ποῖος αὐτὸν ἐφορᾷ ἀστήρ · καὶ εἴπερ εὕροις τὸν ἐφορῶντα αὐτόν, ἰδὲ ποίου τόπου κυριεύει ἐκ τῶν ιβ̅ ζῳδίων · καὶ εἴπερ ἐστὶν ὁ ἀστὴρ ἐκεῖνος ἀγαθοποιὸς καὶ κυριεύει τοῦ β' τόπου, γενήσεται αὐτῷ κέρδος τῷ β' μηνί · εἰ δὲ ἐν τῷ γ' τόπῳ ἐστίν, ὠφεληθήσεται παρὰ ἀδελφοῦ τῷ γ' μηνί · καὶ οὑτωσὶ ἐπὶ τῶν λοιπῶν. ἰδὲ δὲ καὶ τὸν ὡροσκόπον, καὶ ἐὰν ἐφορᾷ τοῦτον ἡ Σελήνη ἢ ὁ κύριος αὐτῆς, γίνωσκε ὅτι ἀγαθόν τι ἀποβήσεται τῷ ἐρωτῶντι περὶ τὴν τοῦ ἔτους ἀρχήν · εἰ δὲ ὁ κύριος τοῦ ὡροσκόπου ἐφορᾷ τὸν ὡροσκόπον, καὶ τοῦτον ἐφορᾷ ἡ Σελήνη, συμβήσεται ἀγαθὸν τῷ ἐρωτῶντι περὶ τὸ μέσον τοῦ ἔτους · εἰ δὲ οὔτε ἡ Σελήνη οὔτε ὁ κύριος τοῦ ὡροσκόπου ἐφορᾷ τὸν ὡροσκόπον, οὔτε ἀγαθὸν οὔτε φαῦλον συμβήσεται τῷ ἐρωτῶντι κατὰ τὸ ἔτος ἐκεῖνο.

APPENDIX 10

Vatican Library MS Vat. gr. 1056 fol. 208.

Λέγουσι δὲ καὶ οἱ Ἰνδοὶ ὅτι μετὰ τὴν σύνοδον Ἡλίου καὶ Σελήνης αἱ ιβ̅ ὧραί εἰσι τοῦ Ἡλίου, αἵτινες καθὼς εἰσιν ἐναντίαι · ἀπὸ γοῦν τῶν τοιούτων ιβ̅ ὡρῶν αἱ πρῶται δ̅ ὧραί εἰσι πάνυ ἐναντίαι, αἱ δὲ δεύτεραι δ̅ ὧραι κουφότεραι, αἱ δὲ ἕτεραι δ̅ ὧραι κουφότεραι · εἶτα αἱ μετ' αὐτὰς ἕτεραι ιβ̅ ὧραί εἰσι τῆς Ἀφροδίτης καὶ ὑπάρχουσιν ἀγαθαί · αἱ ἕτεραι τοῦ Ἑρμοῦ καί εἰσι μέσαι ἤγουν καθ' ἣν ἂν ἔχῃ ψῆφον ὁ Ἑρμῆς · αἱ ἕτεραι τῆς Σελήνης καί εἰσιν ἀγαθαί · αἱ ἕτεραι τοῦ Κρόνου καί εἰσιν ἐναντίαι · σημαίνουσι γὰρ βραδυτῆτα · αἱ ἕτεραι τοῦ Διὸς καί εἰσιν ἀγαθαὶ ἐπὶ πάσης καταρχῆς · αἱ ἕτεραι τοῦ Ἄρεως καί εἰσιν ἐναντίαι πλὴν ἐναντιώτεραί εἰσιν αἱ τοῦ Ἡλίου.

APPENDIX 11

Vatican Library MS Vat. gr. 1056 (V) fol. 32.

Τοῦ Σὴθ ἐκείνου

Περὶ δὲ τῆς τῶν ἀπλανῶν ἀστέρων κινήσεως καὶ τῆς τούτων μερικῆς ἐποχῆς ὡς ἔοικεν οὐκ ἐπέστησε προσοχῇ ἀκριβῇ ὁ ἐπὶ σοφίᾳ περίδοξος Πτολεμαῖος διὰ τὸ τούτων δυσκίνητον ἀμελῶς πρὸς τοῦτο διατεθεὶς καὶ μὴ προσήκουσαν ἐπιδειξάμενος ἐπιμέλειαν · τοιγάρτοι καὶ κατ' ἐνιαυτοὺς ἑκατὸν ἕκαστον τούτων ἔφη μοῖραν ζῳδιακὴν διέρχεσθαι · ἡμεῖς δὲ πολλοῖς ἐντυχόντες περὶ τούτου συγγράμμασιν Αἰγυπτιακοῖς ν̅δ̅ . λεπτὰ δεύτερα κατ' ἐνιαυτὸν ἔρχεσθαι τούτων ἕκαστον διαβεβαιοῦμεν. τινὲς δὲ τούτων κλιμακτῆρας εἰώθασι ποιεῖν ἐν τοῖς γενεθλίοις καί τινες εὐημερίας καὶ εὐτυχίας. οὐ μὴν ἀλλὰ καὶ τῷ Τελμουσῇ περὶ τούτου κοινολογησάμενοι καὶ τοῦτο (συμφωνοῦντες τοῖς Αἰγυπτίοις) εὑρόντες ἀληθὲς τὸ πρᾶγμα προφανῶς ἐπεγνώσαμεν · ὅσους δὲ καὶ ἡμεῖς τῇ συνεχῇ πείρᾳ καὶ τῷ μακρῷ χρόνῳ καλοποιοὺς ἐφεύρομεν καὶ κακοποιοὺς ἀπὸ παρατηρήσεως ἀκριβοῦς, καὶ τούτων τὰς ἐποχὰς καὶ τὰς φύσεις καὶ τὰ πλάτη καὶ τὰ μεγέθη καταγραψόμεθα · ἐξαιρέτως δὲ ἀναιρέτας καὶ καλοποιοὺς ἐφεύρομεν τοὺς ἐλάχιστον πλάτος ἔχοντας καὶ μέχρι τῶν ι̅β̅ μοιρῶν περιλαμβανόμενοι · ἔχουσι δὲ οἱ τοιοῦτοι οὕτως.

Καὶ ἐπειδη ἐν τῇδε τῇ βίβλῳ εἰσὶ καταγεγραμμένοι οἱ τοιοῦτοι ἀπλανεῖς ἀπὸ τῆς βίβλου τοῦ 'Απομάσαρ ἄνευ τινῶν ὀλιγοστῶν, κατὰ τοῦτο περιττὸν ἡμῖν ἔδοξε καταγράψασθαι τούτους καὶ ἐνταῦθα, τοὺς δὲ μὴ ὄντας καταγεγραμμένους καταγραφόμεθα οἵτινες καί εἰσιν.

ὁ κοινὸς Ἵππου καὶ Ἀνδρομέδας	♈ β βο. κς ο	μέγεθος β	κρᾶσις ♂ ☿
ὁ λαμπρὸς τῆς Πλειάδος	♉ ιβ βο. δ λ		κρᾶσις ♂ νεφελοειδής
ὁ ἡγούμενος τοῦ ὤμου τοῦ Ὠρίωνος	Π θ λ νο. ιζ λ	μέγεθος β	κρᾶσις ♂ ☿
ὁ ἐπὶ τῆς κεφαλῆς τοῦ ἐπομένου Διδύμου	♋ ιβ λ βο. γ ιε	μέγεθος β	κρᾶσις ♂
ὁ αὐχὴν τοῦ Ὕδρου	♌ ιε λ νο. κ λ	μέγεθος β	κρᾶσις ♄ καὶ ♀
ὁ ἐν τῇ ῥάχῃ τοῦ Λέοντος	♌ κς λ βο. β λε	μέγεθος β	κρᾶσις ♀
ὁ ἐπὶ τῆς ὀσφύος τοῦ Λέοντος	♌ κθ μ βο. ιγ μ		κρᾶσις ♄ ♀
ὁ ἐν τῇ οὐρᾷ τοῦ Λέοντος	♍ θ μ βο. θ θ	μέγεθος α	κρᾶσις ♄ ♂
ὁ ἐπ' ἄκρας τῆς νοτίου ἀριστερᾶς πτέρυγος τῆς Παρθένου	♍ ιδ κ βο.		κρᾶσις ♂

8 διαβεβαιοῦσι V || 9 Bātlımūs = Ptolemaeus. || 21 κυνὸς V || 25 ὀρίωνος V.

APPENDIX 12

Paris, Bibl. Nat. MS gr. 2506 (B) fols. 158v-159v.

Περὶ τῶν ἐγκυμονουσῶν γυναικῶν, διὰ τί οἱ ἄρτιοι μὲν τῶν μηνῶν τῆς τούτων ἐγκυμονίας εἰσὶ κλιμακτηρικοὶ καὶ ἐπικίνδυνοι, οἱ δὲ περιττοὶ μᾶλλόν εἰσιν ἀγαθοὶ καὶ ἀκίνδυνοι · λύσις Ἰ⟨ν⟩δική.

Θεόπτης Μωϋσῆς μετὰ τῶν ἄλλων τῶν περὶ κοσμογενέσεως καὶ ταῦτά φησιν · καὶ ἐποίησεν ὁ Θεὸς τοὺς ἀστέρας εἰς σημασίαν καὶ εἰς καιροὺς καὶ εἰς ἐνιαυτούς · δῆλον ἐκ τούτου ὡς οὐδὲ ἡ τῶν ἀνθρώπων γένεσις ἢ ἑτέρου οἱουδήτινος τῶν γενέσει καὶ φθορᾷ ὑποκειμένων ἀμέτοχός ἐστιν τῆς τῶν ἀστέρων σημασίας · καὶ ἐπεὶ ὁ Ἥλιος, κἂν τὴν μέσην ἐπέχει ζώνην καὶ οὐχὶ τὴν πρώτην καὶ ἀνωτάτην, πρῶτός ἐστι τῶν ἄλλων καὶ ὥσπερ τις ἀρχηγὸς καὶ ἐξουσιαστής, καὶ οἷον βασιλεὺς τῶν ἄλλων πάντων προτάσσεται, καθότι παρὰ τοῦ πανσόφου ἐτεύχθη δημιουργοῦ εἰς τὸ δι' αὐτοῦ τὸ φῶς παράγεσθαι καὶ τὸ σκότος διώκεσθαι καὶ τὰ κεκρυμμένα φανεροῦσθαι. τούτου γὰρ εἰς τὸν ἀνατολικὸν ὁρίζοντα φθάνοντος, ἡ ἡμέρα ἀπάρχεται, τὸ φῶς ἐφαπλοῦται, καὶ πάντα τὰ ἀφανῆ φανερὰ γίνονται. τὰ φυτά τε τῆς γῆς δι' αὐτοῦ φύονται καὶ ἐκ τοῦ ἀφανοῦς εἰς τὸ φανερὸν ἔρχονται, καὶ τὰ τῶν παντοδαπῶν δὲ ζῴων γένη πρὸς ἃς οὗτος ποιεῖται τροπὰς καὶ συλλαμβάνουσι καὶ τίκτουσιν, καὶ καθολικῷ τῷ λόγῳ ἡ χρονικὴ ἐπὶ τούτοις ἐπικρατεῖ κατάστασις.

Ἀλλὰ καὶ ὅπερ ἡ τῆς γυναικὸς μήτρα ὑποδέχεται καὶ συλλαμβάνει σπέρμα ἐκ τοῦ ἀδήλου καὶ ἀφανοῦς εἰς τὸ δῆλον πάντως καὶ ἤδη φανερὸν ἔρχεται. διὰ τοῦτο προσηκόντως ὁ Ἥλιος κυριεύει τοῦ πρώτου μηνὸς τῆς συλλήψεως, κατὰ κυκλικὴν δὲ περίοδον, καθὼς πραγματικῶς ὑποδειχθήσεται, ὁ τῇ αὐτοῦ ζώνῃ πλησιάζων Ἄρης τοῦ δευτέρου, καὶ κατὰ τὴν αὐτὴν ἀκολουθίαν ὁ Ζεὺς τοῦ τρίτου · τότε γὰρ καὶ τὸ ἔμβρυον σωματοῦσθαι καὶ ἀναζωοῦσθαι ἄρχεται καὶ ἀνεπαίσθητον ποιεῖται τὴν κίνησιν · καὶ εἰ μὲν εἰς τὸν ἐπιμερισμὸν τοῦ Διὸς ἡ κίνησις γένηται ἤτοι εἰς τὴν πρώτην δεκάδα, ζωογονεῖται τὸ βρέφος καὶ ἑπταμηνιαῖον τίκτεται, εἰ δὲ εἰς τὸν ἐπιμερισμὸν τοῦ Κρόνου ἤτοι εἰς τὴν δευτέραν δεκάδα, ὀκταμηνιαῖον τίκτεται καὶ οὐ ζωογονεῖται τὸ βρέφος ἀλλ' οὐδὲ ἡ μήτηρ ἐπὶ τὸ πλεῖστον σῴζεται, εἰ δὲ εἰς τὸν ἐπιμερισμὸν τῆς Σελήνης ἤτοι εἰς τὴν τρίτην δεκάδα ἡ κίνησις γένηται, ἀνεμποδίστως τὸ βρέφος τίκτεται εἰς τὸν νενομισμένον καιρὸν ἤτοι εἰς τὸν θ΄ μῆνα · ὁ Κρόνος πάλιν τοῦ δ΄ μηνός, καὶ ἡ Σελήνη τοῦ ε΄, ἐν ᾗ μηνοκρατορίᾳ φανερὰ καὶ κατάδηλος ἡ τοῦ βρέφους γίνεται κίνησις · καὶ εἰ μὲν εἰς τὸν ἐπιμερισμὸν αὐτῆς τῆς Σελήνης ἡ κίνησις γένηται ἤτοι εἰς τὴν πρώτην δεκάδα, ζωογονηθήσεται καὶ ἀβλαβῶς τὸ βρέφος τεχθήσεται, εἰ δὲ εἰς τὸν ἐπιμερισμὸν τοῦ Ἑρμοῦ ἤτοι εἰς τὴν δευτέραν δεκάδα καὶ τύχῃ τότε κεκακωμένος ὁ Ἑρμῆς, οὐκ ἀνεμποδίστως καὶ πάσης βλάβης χωρὶς ἡ ἔκτεξις γένηται, εἰ δὲ ⟨εἰς⟩ τὴν τρίτην δεκάδα ἤτοι εἰς τὸν ἐπιμερισμὸν τῆς Ἀφροδίτης, φυλαχθήσεται τὸ βρέφος καὶ τεχθήσεται ἀκινδύνως ἐν τῷ καιρῷ αὐτοῦ · εἶθ' οὕτως πάλιν ὁ Ἑρμῆς κυριεύει τοῦ ς΄ μηνός, ἡ Ἀφροδίτη τοῦ ζ΄, διὸ τὰ ἑπταμηνιαῖα ἐξ ἅπαντος ζωογονοῦνται καὶ τελεσφοροῦσιν, διότι καὶ ἀρτιοῦται τὸ βρέφος διὰ τοῦ ζ΄ μηνὸς κατὰ τὴν τῶν ζ̄ ὡς δεδήλωται πλανωμένων περίοδον· μετὰ τοῦτο πάλιν κατὰ τὴν ἑπτάζωνον τάξιν ὁ Κρόνος κυριεύει τοῦ η´ μηνός, καὶ ὁ Ζεὺς τοῦ θ´· τότε γὰρ καὶ ὁ μέσος τῆς ἐκτέξεως δηλοῦται καὶ ἀγαθὴ ἡ γένεσις γίνεται· ὅτε δὲ συμβῇ καὶ τοῦ δεκάτου μηνὸς τὴν ἐγκυμονοῦσαν ἐφάψασθαι, ἀπλανῶς διὰ τὴν ἀδηλίαν – λέγω τῆς κινήσεως τοῦ ε´ μηνός – καὶ τὸν μέγιστον φθάσει χρόνον· ἄλλως οὐ πολυχρόνιον γίνεται τὸ τικτόμενον εἰ μὴ κατὰ τὸν καιρὸν τῆς ἐκτέξεως ἀγαθοποιοὶ ὁρῶσι τὸν Ἄρην καὶ τὸν Ἥλιον· ἰδοὺ τὸ ζητούμενον ἐλύθη καὶ ἐσαφηνίσθη παρ᾽ ἡμῶν τέως ἐκ τῆς τῶν πλανωμένων σημασίας, καθ᾽ ἣν δέδωκε γνῶσιν ὁ τῶν ἁπάντων δημιουργὸς καὶ Θεός, καὶ ἡ αἰτία καθαρῶς ἐδηλώθη δι᾽ ἣν οἱ ἄρτιοι μῆνες τῶν ἐγκυμονουσῶν γυναικῶν εἰσι κλιμακτηρικοὶ καὶ ἐπικίνδυνοι.

Ὑποδειχθήσεται δὲ τὸ τοιοῦτον κεφάλαιον καὶ ἐξ ἑτέρου τρόπου εὐλήπτως ὡς ἂν ἐκ δύο ἐντέχνων χρήσεων εὔδηλον καὶ καταφανὲς τὸ[ν] ζητούμενον γένηται.

Δεῖ γινώσκειν ὅτι ὁ πρῶτος τόπος τῆς οὐρανίας διαθέσεως, ὃς καὶ ὡροσκόπος καλεῖται, ζωῆς καὶ πνευματικῆς γονῆς ἐστι δηλωτικός. οὗτός ἐστιν ὁ σημαίνων τὰ τοῦ πρώτου μηνὸς διότι, καθάπερ ἐκ τούτου τοῦ τόπου ἡ τῶν κεκρυμμένων φανέρωσις γίνεται, οὕτω καὶ τὸ τοῦ ἀνδρὸς σπέρμα ἐκ τοῦ ἀφανοῦς πάντῃ καὶ ἀδήλου εἰς τὸ φανερὸν καὶ δῆλον ἔρχεται καὶ σωματοῦσθαι καὶ ἀναζωοῦσθαι ἀπάρχεται. καθεξῆς οὖν ὁ β´ τόπος τοῦ β´ μηνός· καὶ ἐπεὶ οὗτος ἀργὸς τόπος ἐστὶ καὶ ἀπόστροφος τῷ ὡροσκόπῳ τῷ τῆς ζωῆς δηλωτικῷ καὶ Ἅιδου πύλη καλεῖται, εἰκότως ἐστὶν ὁ τοιοῦτος μὴν ἐπικίνδυνος· ὁ γ´ τόπος τοῦ γ´ μηνός· καὶ διότι ἑξάγωνός ἐστι πρὸς τὸν ὡροσκόπον, ἀγαθός ἐστιν, ὅτι δὲ καὶ τόπος ὑπάρχει τῆς Σελήνης (χαίρει γὰρ ἐν αὐτῷ), διὰ τοῦτο καὶ κατ᾽ αὐτὸν τὸν μῆνα κίνησις ἀνεπαίσθητος γίνεται· εἰ μὲν γάρ, ὡς προείρηται, κατὰ τὴν πρώτην δεκάδα ἤτοι εἰς τὸν ἐπιμερισμὸν τῆς Σελήνης ἡ κίνησις γένηται, ἑπταμηνιαῖον τὸ βρέφος τεχθήσεται καὶ ζωογονηθήσεται, εἰ δὲ εἰς τὸν ἐπιμερισμὸν τοῦ Κρόνου ἤτοι εἰς τὴν β´ δεκάδα, φθάνει εἰς τὸν η´ μῆνα καὶ κινδυνεύσει, εἰ δὲ εἰς τὸν ἐπιμερισμὸν τοῦ Διὸς εἴτε εἰς τὴν γ´ δεκάδα, φυλαχθήσεται καὶ εἰς τὸν θ´ μῆνα τεχθήσεται· ὁ δ´ τόπος πάλιν τοῦ δ´ μηνός· καὶ διότι ἐστὶν ὑπόγειον κέντρον καὶ ἀφανὴς τόπος καὶ τετράγωνος πρὸς τὸν ὡροσκόπον, ἐπικίνδυνός ἐστιν· ὁ ε´ τόπος τοῦ ε´· καὶ διότι τρίγωνός ἐστι τοῦ ὡροσκόπου καὶ τόπος τῆς Ἀφροδίτης, ἀγαθός ἐστιν ὁ τοιοῦτος μήν· ἐν αὐτῷ γὰρ καὶ τὸ βρέφος τὴν φανερὰν ποιεῖται κίνησιν· καὶ εἰ μὲν εἰς τὴν κυβέρνησιν τῆς Ἀφροδίτης ἤτοι εἰς τὴν πρώτην δεκάδα ἡ κίνησις γένηται, ζωογονηθήσεται τὸ βρέφος καὶ καλῶς τεχθήσεται ἐν τῷ καιρῷ αὐτοῦ, εἰ δὲ εἰς τὴν β´ δεκάδα ἤτοι εἰς τὴν κυβέρνησιν τοῦ Ἑρμοῦ καὶ τύχῃ τότε κεκακωμένος ὁ Ἑρμῆς, οὐ πάσης βλάβης χωρὶς ἡ γέννησις γενήσεται, εἰ δὲ εἰς τὴν κυβέρνησιν τῆς Σελήνης ἤτοι εἰς τὴν τρίτην δεκάδα, ἀνεμποδίστως τὸ βρέφος τεχθήσεται ἐν τῷ καιρῷ αὐτοῦ· ὁ ϛ´ πάλιν τόπος τοῦ ϛ´ μηνός· καὶ ἐπεὶ ὁ τοιοῦτος ἀσύνδετός ἐστι πρὸς τὸν ὡροσκόπον καὶ τὸ ζῴδιον ἀπόστροφον πρὸς τὸ ὡροσκοποῦν ζῴδιον, ἔστι δὲ καὶ τόπος τοῦ Ἄρεως, καλεῖται δὲ καὶ τῆς νόσου τόπος, ἐναντίος ἐξ ἅπαντος καὶ κλιμακτηρικὸς ὁ τοιοῦτος μήν ἐστι καὶ παρὰ τῶν δοκιμωτέρων μάλιστα ἰατρῶν σφάκτης ὀνομάζεται· ὁ ζ´ τόπος τοῦ ζ´ μηνός· καὶ διότι ἀπὸ τοῦ ὑπογείου καὶ ἀφανοῦς εἰς τὸ ὑπέργειον καὶ φανερὸν ἔρχεται, καλεῖται δὲ ὁ τοιοῦτος τόπος ἀνθωροσκόπος καὶ ὁρᾷ πρὸς τὸν ὡροσκόπον καὶ ὁρᾶται (τριγωνίζεται παρὰ τοῦ τρίτου τόπου, καὶ παρὰ τοῦ ε′ καὶ θ′ ἐξαγωνίζεται τῶν σημαντικῶν τῶν ἀκινδύνων μηνῶν), κατὰ τοῦτο καὶ ὁ ζ′ μὴν ἀγαθός ἐστιν, καὶ τὰ κατ᾽ αὐτὸν τικτόμενα ἀκινδύνως καὶ τίκτονται καὶ ζωογονοῦνται · ὁ η′ τόπος τοῦ η′ μηνός · καὶ ἐπεὶ ὁ τόπος οὗτος ἀσύνδετός ἐστι πρὸς τὸν ὡροσκόπον καὶ θανατικὸς τόπος ἐστὶ καὶ ὀνομάζεται, διὰ τοῦτο καὶ ὁ τοιοῦτος μὴν κλιμακτηρικὸς καὶ ἐπικίνδυνος · καὶ ὁ θ′ τόπος τοῦ θ′ μηνός · καὶ ἐπεὶ ὁ τοιοῦτος τόπος τρίγωνος τοῦ ὡροσκόπου ἐστίν, ὑπάρχει δὲ καὶ τοῦ Ἡλίου τόπος, διὰ τοῦτο ἐν αὐτῷ τῷ μηνὶ καθολῶς ἡ ἔκτεξις γίνεται, καὶ ἐκ τοῦ ἀφανοῦς τὸ βρέφος εἰς τὸ φανερὸν ἔρχεται.

19 προσεκόντως B ‖ 37 ἐξ$\overset{\pi}{\alpha}$τ′ B ‖ 50 χρύσεων B ‖ 57 εἰκότος B ‖ 72 γεν$\overset{\alpha}{\nu}$ B ‖ 77 ἐξάπαῦ B .

Box 1900
Brown University
Providence, Rhode Island 02912, U.S.A.

A NEW LOOK AT *MELENCOLIA I*

Upon recently reading William Heckscher's fascinating study of Camerarius's description of Dürer's *Melencolia I*,[1] I was led to consider some interpretations of this enigmatic print (Pl. 35c) which I believe have not been noticed hitherto. I present them in this note for others more expert than I both in iconography and in Dürer's art to appreciate or to deprecate. Certainly I do not pretend to have any evidence that these meanings were intended by the artist.

The basis of the new interpretations is a consideration of the composition as a whole, whereas previously scholars have tended to concentrate on individual elements.[2] Three linked states of being are represented in the print (Pl. 35c): the celestial occupies the upper third, including both the celestial body to the viewer's left and the upper parts of the ladder and tower to his right; the terrestrial is depicted in the centre left; and an intermediate state appears in the centre right and lower third. The two main figures in the print—the block of stone and Melancholy herself—are located in the intermediate state.

I begin to justify this analysis by the observation that the celestial body in the upper left quadrant of the print must be a star or a planet rather than the comet that it is usually interpreted to be. This is indicated by the rays extending from it in all directions to the limits of the visual field. These are not, however, rays of light; for the presence of the rainbow, eccentric to the celestial body, shows that the Sun has not yet set but is in the west—in the direction from which the viewer beholds the scene—while the celestial body itself is rising in the East.[3] The rays are rather those by which, according to astral magic, the planets effect their influence in the sub-celestial world.[4] In a depiction of Melancholy, the celestial body emitting magical rays can only be Saturn.

In the same plane with Saturn are two other celestial symbols.[5] To the right is the magic square of Jupiter, whose influence must be combined with Saturn's to produce the melancholic philosopher;[6] and in the centre hang the scales of Libra, the exaltation of Saturn. The upper third of the picture, then, represents the celestial configuration under which Saturn is most effective in producing philosophers: Saturn is rising in its exaltation in association with Jupiter.

Above the frame of the print extends the supercelestial world, which is, quite literally, imperceptible to us. Into this lead the ladder and the tower. The latter is, as Heckscher shows, a House of Wisdom; more specifically, it symbolizes the intellectual mode of ascent to the supercelestial, as the ladder refers to ascent by faith.[7] The intellectual character of the tower is indicated by the four objects hung on its exterior wall, which represent the external or practical aspects of the quadrivium. The scales represent the Arithmetic of weighing and measuring; the hourglass the Astronomy of time-keeping; the bell the Music of rhythmic sound; and the magic square the Geometry of the lines, squares and

---

[1] W. S. Heckscher, 'Melancholia (1541). An Essay in the Rhetoric of Description by Joachim Camerarius', *Joachim Camerarius (1500–1574)*, ed. F. Baron, Munich 1978, pp. 31–120.

[2] *E.g.*, R. Klibansky, E. Panofsky, F. Saxl, *Saturn and Melancholy*, London 1964.

[3] The Sun is 180° from the centre of the circle of which the rainbow is an arc; that centre is below the horizon.

[4] A doctrine first enunciated by al-Kindī (see M.-Th. d'Alverny and F. Hudry, 'Al-Kindi De radiis', *Archives d'histoire doctrinale*, xli, 1974, pp. 139–260); and developed by the author of the *Ghāyat al-ḥakīm* (*e.g.*, i 3, 1; ii 7, 1–8; and iii 5, 5 in my forthcoming edition of the Latin *Picatrix*). It has often been argued that Dürer received any knowledge that he might have had of Picatrician magic through a manuscript of the early version of Agrippa's *De occulta philosophia* as found in the Würzburg manuscript (a facsimile is published in K. A. Nowotny, *Henricus Cornelius Agrippa ab Nettesheym De occulta philosophia*, Graz 1967, pp. 519–86). However, it is possible that Dürer may have seen a copy of the *Picatrix* in Italy, perhaps on his second journey. An indication of this may be the fact that he includes two Saturnine animals in *Melencolia I*, a bat and a dog; only the former is mentioned in the Würzburg manuscript (fols. 17ᵛ–18ʳ on p. 527 Nowotny), but the bat is Saturnine in *Picatrix* iii 8, 2, the dog in iv 1, 3.

For another representation of the rays of planetary influence, see ill. 37 in Klibansky, Panofsky, Saxl (n. 2 above).

[5] It may be objected that, because we do not normally find Jupiter represented by his magic square or Libra by its scales in engravings, therefore we ought not to find them so represented in *Melencolia I*. I would rephrase this objection thus: because we do not normally find Jupiter and Libra so represented, therefore we *did* not recognize them until some learned scholars pointed out to us their presence. If Dürer had expressed himself in conventional terms, the *Melencolia I* would not have puzzled its beholders. The fact that the engraving is an enigma should alert us to the fact that Dürer meant us to use the intellectual capacity symbolized by Melancholia herself in attempting to understand the meaning of her image. Our awareness of this intention should not convince us of the correctness of any particular interpretation, but should deter us from relying to any great extent on the conventions followed by artists contemporary with Dürer to help us in understanding his unique creation.

[6] See, *e.g.*, Ficino cited in Klibansky, Panofsky, Saxl (n. 2 above), pp. 271–3.

[7] Heckscher, pp. 54–59.

triangles that constitute that particular object.[8]

The seascape with its fringe of inhabited land clearly represents our sublunar world; the four elements are also indicated, fire and air above earth and water. The scene in the foreground, then, bounded by the block of stone on the left and by Melancholy herself on the right, lies between this world and the heavens. It represents those hypostases placed by some Neoplatonists and astral magicians between the spiritual and the material worlds, wherein, under celestial influence, the elements[9] and the human spiritus[10] are formed. In fact, Dürer shows under Saturn what appears to be an inchoate octahedron, the second of Plato's five perfect solids,[11] though others have interpreted it, less persuasively, as a truncated cube.[12] In one passage in the *Timaeus* (53c) the second of the elements is earth, which possesses the two qualities of Saturn, dryness and cold; in another passage (55d) the cube is assigned to earth. The artist shows under Jupiter Melancholy herself, the intellectual temperament produced by Saturn in conjunction with Jupiter. And under Libra sits a winged putto, the melancholic and philosophic native to be. Scattered on the ground are the tools of the demiurge.

One other observation deserves to be made. The ladder of faith leads up into the supercelestial from behind inchoate matter, while the tower of intellect rises up behind Melancholy and the unborn philosopher.

Excursus on Rays in Dürer's Woodcuts and Engravings

In his woodcuts and engravings Dürer rarely uses rays to indicate natural light as he seldom portrays the luminous bodies occurring in nature. The Sun may have short, fiery rays as in the *Crucifixion* in the early *Great Passion*; and a star may have short rays as in the *Adoration of the Shepherds* from the *Small Passion* (1509), or it may have no rays at all as in the *Adoration of the Magi* (1511). But usually the rays in Dürer's early works emanate from divine beings and represent their visible effulgence; *cf.*, among engravings, the *Holy Family with a Butterfly* (1495/6), the *Virgin on the Crescent* (1498?), the *Virgin and Child with St. Anne* (*c.* 1500), *Justice* (*c.* 1501), the *Virgin with a Starry Crown* (1508), and *Christ on the Mount of Olives* (1508).

This use of limited rays continued to the end of Dürer's career. But, beginning with the *Small Passion* in 1509, Dürer also sometimes showed the divine with unlimited rays. In the *Small Passion* itself one notes this in *Christ Appears to His Mother*, *Christ at Emmaus*, the *Incredulity of St. Thomas*, *Christ Appears to Mary Magdalene*, the *Descent of the Holy Spirit*, and the *Last Judgment*. In one woodcut of this series, the *Annunciation*, rays link God the Father to the Holy Spirit. Other woodcuts and engravings in which the divine have unlimited rays include the new title-page to the *Apocalypse* (1511), the title-page to the *Life of the Virgin* (1511), the *Holy Trinity* (1511), the *Resurrection of Christ* (1512), the *Virgin and the Carthusian Monks* (1515), the *Virgin Suckling the Child* (1519), the *Virgin with the Child in Swaddling Clothes* (1520), and the *Virgin Crowned by an Angel* (1520). This distinction between the short, limited rays of visible effulgence and the unlimited rays of the divine in certain contexts suggests that the latter represent not light rays, but rays of divine energy.

In many of the depictions of the unlimited divine rays those rays are 'brighter' in three directions 90° apart; this is reminiscent of the depictions of the limited visible effulgence. In *Melencolia I* the rays emanating from the celestial body are unlimited; and they are clearly 'brighter' in two directions 90° apart, just barely 'brighter' in the third. Within the context of Dürer's symbolic use of rays, then, the celestial body in *Melencolia I* does not cast light rays, but rays of divine energy. The interpretation I have given above would indicate that they are, in fact, creative rays emanating from the Divine through the planet Saturn.

David Pingree
Brown University, Providence, R.I.

[8] To construct this square, one begins with the following figure of four lines and four columns, each of which contains four segments.

(a)	16	15	14	13←
(b)	→5	6	7	8
(c)	9	10	11	12—
(d)	—4	3	2	1

The central pairs of segments in lines (a) and (d) and in lines (b) and (c) are interchanged to produce the magic square, in which the numbers in each line, each column and each diagonal add up to 34 (2 × 17). Furthermore, each of the four squares at the corners of the figure contains numbers adding up to 34. The square of four segments in the centre also contains numbers adding up to 34: the outer square of numbers that is left after removing this central square can be divided into four triangles, of which each contains three segments. The sum of the numbers in each pair of opposite triangles is 51 (3 × 17). Concerning the role of the quadrivium in the education of the perfect philosopher, see, *e.g.*, *Picatrix* ii, 1, 3.

[9] *E.g.*, *Picatrix* i 7.

[10] *E.g.*, *Picatrix* iv 1, 1.

[11] *Timaeus* 55a.

[12] See Klibansky, Panofsky, Saxl, pp. 400–02.

AN ILLUSTRATED GREEK ASTRONOMICAL MANUSCRIPT

COMMENTARY OF THEON OF ALEXANDRIA ON THE HANDY TABLES AND SCHOLIA AND OTHER WRITINGS OF PTOLEMY CONCERNING THEM

I

AMBROSIANUS H. 57. sup. (= 437)[1] is one of the few Greek astronomical manuscripts to contain miniatures. The manuscript consists of 180 parchment folios, each measuring 24 cm × 16.4 cm. The miniatures, all executed by the same artist, are found on the folios now numbered 1, 97, 106, 113ᵛ, 116ᵛ, 121, 121ᵛ, and 125ᵛ. However, the folios of the manuscript were not originally bound in their present order. This fact is indicated by the quire numbers, which are entered in the upper right margin of the recto of the first leaf in each quire and in the centre of the lower margin of the last leaf; each quire (except 19) consists of eight folios. The quires with the present foliation are: 1–14 = fols 1–112; 15 = fols 172–179; and 16–21 = fols 113–162 (quire 19, fols 137–146, consists of ten folios). Fol. 163 seems to be an extra leaf — probably the original guard leaf; and the final quire, fols 164–171, though written by the original scribe of the rest of the manuscript, is not numbered and seems to be a later addition. Fol. 180 is a guard leaf taken from another manuscript.

At the bottom of fol. 180 are two chronological notes in a hand different both from that of the original scribe of fol. 180 and from those of the scribes of fols 1–179; these notes are 'κατὰ τὸ ͵ςϡμ ἔτος παρὰ Ἕλλησιν ἀπ' Ἀλεξάνδρου ἀρχῆς ͵αψξγ εὑρέθη'; and 'κατὰ δὲ τὸ ͵ςϡμς ἔτος παρ' Ἕλλησιν ἀπ' Ἀλεξάνδρου ἀρχῆς ͵αψξθ ἐστίν'. The dates of these notes, A.D. 1431/1432 and 1437/1438, probably indicate that the manuscript was bound in its present order with the guard leaf fol. 180 before the 1430s.

The date at which the original scribe wrote is indicated by several notes in the manuscript. Three marginalia written on fol. 21ᵛ by that original scribe mention the date A.D. 1357/1358 as though it were current. They are: 'κατ' ἀρχὴν τοῦ ͵ςωξς ἔτους εἰσὶν ἔτη ἀπὸ ἀρχῆς Φιλίππου Ῥωμαϊκὰ ͵αχπα'; 'κατ' ἀρχὴν τοῦ ͵ςωξς ἔτους ἔτη ἀπὸ Διοκλητιανοῦ ͵αοδ'; and 'κατὰ τὸ ͵ςωξς ἔτος ἔτη ἀπὸ Αὐγούστου πέμπτου ἔτους ͵ατπβ'.

ABBREVIATIONS

CCAG: Catalogus codicum astrologorum graecorum, 12 vols., Brussels 1898-1954.

Halma: l'abbé Halma, *Commentaire de Théon d'Alexandrie* (vol. I) and *Tables manuelles astronomiques* (vols II–III), Paris 1822–25.

Tihon: A. Tihon, *Le 'Petit commentaire' de Théon d'Alexandrie aux Tables Faciles de Ptolémée* (*Studi e testi*, CCLXXXII), Vatican City 1978.

Tihon *List*: 'Liste des textes anonymes cités,' Tihon, pp. 359–369.

Tihon *Scholia*: A. Tihon, 'Les scolies des Tables Faciles de Ptolémée,' *Bulletin de l'Institut Historique Belge de Rome*, XLIII, 1973, pp. 49–110.

[1] Described most recently in Tihon, pp. 88–90, in whose edition of Theon it is denoted B. See also A. Martini and D. Bassi, *CCAG* III, 17 (describing fols 147ᵛ-175) and *Catalogus codicum graecorum Bibliothecae Ambrosianae*, Milan 1906, pp. 527–30; cf. P. Revelli, *I codici ambrosiani di contenuto geografico*, Milan 1929, p. 87, no. 208; M. L. Gengaro, F. Leoni, G. Villa, *Codici decorati e miniati dell'Ambrosiana: Ebraici e graeci*, Milan 1958, p. 234; and R. Cipriani, *Codici miniati dell'Ambrosiana*, Milan 1968, p. 60.

Journal of the Warburg and Courtauld Institutes, Volume 45, 1982

Furthermore, the κανὼν βασιλέων on fols 66–67 originally ended with the two entries:[2]

Ἀνδρονίκου τοῦ Παλαιολόγου	ϛ	͵αχκβ
Ἰωάννου τοῦ υἱοῦ αὐτοῦ	μβ	͵αχξδ

These rulers are Andronicus III (1328–41; crowned co-Emperor in 1325) and John V (1341–91). The dates in the last column, given in Era Philippi, correspond to 1299 and 1341; the first is obviously an error. The regnal years in the second column belong to the preceding ruler. John VI, who ruled from 1347 to 1354, is not mentioned, as John V was the sole occupant of the throne in 1358.

A second scribe has added to the κανὼν βασιλέων the entries:

Μανουὴλ ὁ υἱὸς αὐτοῦ	λε	͵αχϟθ
Ἰωάννου τοῦ υἱοῦ αὐτοῦ	κβ	͵αψκα

These rulers are Manuel II (1391–1425) and John VIII (1425–48). The second column contains the correct regnal years of these two Emperors. The years in column 3 are computed by adding these regnal years to the previous entries in column 3; since the last preceding entry, 1664 Era Philippi, is the first year of the reign of John V, it was obvious nonsense to add to it the regnal years of Manuel II. In any case, the second scribe wrote after 1448.

A third hand has added below these the entries:

Κωνσταντίνου τοῦ ἀδελφοῦ αὐτοῦ καὶ ὁ Ἀμηρᾶς ὁ Μουράτ	γ	͵αψκδ

These rulers are Constantine XI (1449–53) and the Ottoman Murād II (1421–44 and 1446–51). It seems likely that these entries were made in 1451, shortly after Murād II's death in February; Constantine XI's third regnal year had begun in January.

A fourth hand has added in the appropriate columns after Murād II:'ιγ ͵αψλζ,' and in the margin: 'μῆνες γ ἕως Αὐγούστου'. The explanation of this cryptic note is found in the first of three notes that the same fourth scribe has written on fol. 163, which was probably the guard leaf of the manuscript in its original state. This note mentions the date 29 May 1466, the fourteenth indiction, and states that there are three months until the 'whole' of August — i.e. until 29 August = 1 Thoth. The author also identifies the Amīr Murād with Muḥammad II the Conqueror (1451–81); 1466 was thirteen years after the fall of Constantinople.

The note is (I have retained the scribe's orthography but added capitalization and punctuation):

Τῶν ἀπὸ Ἀδὰμ ἐτῶν καθὼς ἐν τοῖς βίβλοις τῶν Ἰουδαίων γράφονται ἕως πάντων τῶν Κριτῶν καὶ ἕως Ἰλὴ τοῦ ἱερέως ἔναι χρόνοι ͵δτνε˙ ὁμοίως καὶ ἐν τοῖς ἀστρονομικοῖς βιβλίοις εὑρίσκονται οὕτως ἐν ταῖς κε ετηρίδαις. οὐ γὰρ τοῖς αὐτοῖς ὀνόμασι χρόνται οἱ Ἕλληνες ὥσπερ οἱ Ἰουδαῖοι, ἀλλὰ ἀπὸ Ναβονασάρου ἕως τέλους τῶν βασιλέων, ἕως Αὐγούστου καὶ Ἄννα καὶ Καϊαφᾶ καὶ Πιλάτου, ἀπὸ δὲ τῆς σταυρώσεως τοῦ Χριστοῦ ἕως ὅδε ἰνδικτιῶνος ιδ, Αὐγούστου μηνός. ἀπὸ δὲ Ἰλὴ τοῦ ἱερέως ἕως τοῦ Χριστοῦ ἔναι χρόνοι ͵αρνγ, ἀπὸ δὲ τῆς σταυρώσεως ἕως Μουράτου τοῦ Ἀμηρᾶ τοῦ τὴν πόλιν ἑλῶντος ἔναι χρόνοι. ἀπὶ Μαίου κθ ἡμέρας ͵αυξς μῆνες γ̄ ἕως ὅλου Αὐγούστου, ιδ ἰνδικτιῶνος εἰσὶν ἔτη ͵ϛϡοδ. λοιποὶ ἕως ͵ζ κς.

The final sentence, which mentions the year A.D. 1491/1492, remains inexplicable as the phrase 'ἕως ὅδε ἰνδικτιῶνος ιδ, Αὐγούστου μηνός' seems clearly to indicate that the scribe wrote in August 1466.

We may conclude, then, that the original manuscript was copied in 1358, and that a series of owners for the next century added to it, rearranged it, and annotated it. It is likely that the manuscript was copied in Byzantium, and remained there while these changes and additions were being made to it. There is no absolute proof for this supposition in the manuscript itself, but it is known that the texts in Ambrosianus H. 57. sup. were copied in part from Laurentianus 28, 7;[3] and that manuscript can be shown to be Constantinopolitan. I shall first present the evidence for the dating and localization of Laurentianus 28, 7, and then demonstrate the relation between the two manuscripts.

Laurentianus 28, 7[4] is a manuscript of 179 unnumbered paper folios, each measuring 29.1 cm × 19.9 cm. Its date is indicated by a note on fol. 32^{v} which refers to the current year A.D. 1343/1344:

[2] Tihon, p. 89 n. 4, who may well be right, states that the regnal years from Leo VI (who is the first entry on fol. 67) to John V were written by the second scribe, who also added the names of Manuel II and John VIII, and that a fourth scribe wrote the name of Murād II; this is not apparent from the prints available to me.

[3] For the Ambrosianus's dependence on the corrected state of the Laurentianus in Theon, and on another source, see Tihon, pp. 96–100; for its partial dependence on the Laurentianus in Ptolemy's *Introduction to the Handy Tables* and *Hypotheses* see J. L. Heiberg, *Claudii Ptolemaei Opera astronomica minora*, Leipzig 1907, pp. x and CLXXVII, and VIII and CLXIX–CLXX.

[4] Described most recently in Tihon, pp. 90–91. See also A. M. Bandini, *Catalogus codicum mss. Bibliothecae Mediceae Laurentianae*, Florence 1764–70, vol. II, p. 16, and A. Olivieri, *CCAG*, I, 3 (description of fols 146–175).

"Οπως εὑρίσκειν δεῖ ἑκάστην τοῦ μηνὸς ἡμέραν, ποία ἡμέρα της ἑβδομάδος ἐστίν.

Λαμβάνομεν τὰ ἀπὸ τοῦ τρίτου ἔτους τῆς α ἰνδικτιῶνος τῆς Ἡρακλείου βασιλείας μέχρι τοῦ προκειμένου, οἷον ,ϛωνβ ἔτους ἀπὸ κτίσεως κόσμου μετὰ καὶ αὐτοῦ τοῦ ἐνισταμένου, καὶ τούτων λαμβάνομεν τὸ δ, καὶ προστίθεμεν αὐτῷ, καὶ ἀφαιροῦμεν ἕν. εἶτα λαμβάνομεν τὴν τοῦ μηνὸς ἡμέραν κατὰ Ἀλεξανδρεῖς, καὶ προστίθεμεν αὐτοῖς καὶ τὸ τῶν μηνῶν ἀπὸ Θὼθ πλῆθος — καὶ διπλασιάζομεν καὶ προστίθεμεν καὶ αὐτό. καὶ ἀπὸ τῶν οὕτω συναχθέντων ἀφαιροῦμεν ἑπτάδας, καὶ τὰς ἐναπολειφθείσας ἐλάττονας τῶν ἑπτὰ κατέχομεν· οἷον, εἰ μὲν μία ἐστί, πρώτη· εἰ δὲ β, δευτέρα· εἰ δὲ τρεῖς ἢ δ, τρίτη ἢ τετάρτη· καὶ πέμπτη εἰ πέντε· καὶ ἑξῆς. <τ>ὰ ἀπὸ κτίσεως κόσμου ἔτη μέχρι τοῦ τρίτου ἔτους τῆς Ἡρακλείου βασιλεας ,ϛρμς.

This is based on chapter 28 of Stephanus of Alexandria's (or the Emperor Heraclius's) commentary on the *Handy Tables*.[5]

The place in which Laurentianus 28, 7 was probably copied is indicated by the long text on fols 178^{v}–179 on determining for a given locality the altitude of the Sun on each day of the year. The example used is ἡ βασιλὶς τῶν πόλεων, whose latitude is given as 43;12°, which is the latitude of Byzantium in Ptolemy's *Geography*.[6] That the manuscript probably remained in Byzantium is shown by a note, apparently in the corrector's hand, in the margin of fol. 23: 'οἷον ὡς ἐν Βυζαντίῳ πολλαπλασίαζε τὰς καταχθείσας ἐπὶ τὸν ιε. εἶτα ἐκ τῶν γινομένων ἀφαίρει τὰ [τὰ] τῆς ὀρθῆς σφαίρας ἑξηκοστά, ἔτι δὲ καὶ τὴν διάστασιν τῶν μοιρῶν. καὶ τὰ ἐναπολειφθέντα μέριζε αὖθις παρὰ τὸν ιε καὶ γίνονται ὧραι ἰσημεριναί'. Thus it is most probable that the Ambrosianus was executed in Byzantium.

The scribe who copied Ambrosianus H. 57. sup. arranged his material by quires:

I quires 1–5 = fols 1–40. Θέωνος Ἀλεξανδρέως εἰς τοὺς Προχείρους κανόνας τῆς <ἀστρονομίας> ἐξήγησις. Since this work ended on fol. 40, the rest of that leaf and its verso contain scholia.

II quires 6–8 = fols 41–64. Κλαυδίου Πτολεμαίου Σαφήνεια καὶ διάταξις τῶν Προχείρων κανόνων τῆς ἀστρονομίας, καὶ ὅπως χρηστέον αὐτοῖς μέθοδος ἐναργής on fols 45–54^{v}. Fols 41–44^{v} contain a continuation of the scholia on fols 40–40^{v}, and fols 54^{v}–58 contain another set of scholia on the *Handy Tables*. Then comes Κλαυδίου Πτολεμαίου Ὑποθέσεων τῶν πλανωμένων on fols 58^{v}-65^{v}, thus continuing on to the first leaf of quire 9.

III quire 9 = fols 65–72. The κανὼν βασιλέων on fols 66–67 and the κανὼν ἐπισήμων πόλεων on fols 68–72. Fol. 67^{v} is blank, and fol. 72^{v} contains only the skeleton of a table.

IV quires 10–19 = fols 73–112, 172–179, and 113–146. The *Handy Tables*.

V quires 20–21 = fols 147–162. Ἰωάννου γραμματικοῦ Ἀλεξανδρέως περὶ τῆς ἀστρολάβου χρήσεως on fols 147–159^{v}. Μέθοδος ἑτέρα τοῦ ἀστρολάβου on fol. 159^{v}–162. Fol. 162^{v}, which was originally blank, was later partially covered with α Χρησμὸς εἰς τὸν τῆς Πελοποννήσου ἰσθμὸν εὑρεθὺς ἐν Ῥόδῳ

VI quire <22> = fols 164–171. The κανὼν βασιλέων on fols 164–164^{v}, the κανὼν ὑπατῶν on fols 165–168, and the πόλεις ἐπίσημοι on fols 168^{v}–171^{v}.

The contents of at least groups I, II, and IV were largely copied from Laurentianus 28, 7, though with some additions and rearrangements. I list below the contents of Laurentianus 28, 7 with the corresponding folios of Ambrosianus H. 57. sup (= A) and references to Halma, to Tihon *List*, and to Tihon *Scholia*.[7]

fols 1–29^{v}. Θέωνος Ἀλεξανδρέως εἰς τοὺς Προχείρους κανόνας τῆς ἀστρονομίας ἐξήγησις ἢ παράδοσις. A fols 1–40.

fols 29^{v}–30. Τρόπος ἐξαναλόγου πρὸς τὰς τῶν ἀστέρων ἐπισκέψεις. Tihon *List* 96. A variant version is found on A fol 40^{v}, entitled Ἑρμηνεία τοῦ ἐξαναλόγου. Tihon *List* 22. This is close to Halma III 59–60, which is ascribed to Isaac Argyrus.

fol. 30. Ἰστέον ὅτι τὴν μίαν μοῖραν . . . A fol. 41.

fol. 30^{v}. Περὶ τοῦ ἐξαναλόγου τῆς τῶ<ν> ὁριζόντων κύκλων καταγραφῆς. A fols 41–41^{v}. Tihon *List* 76.

fols 30^{v}–31. Οἷον ὑποδείγματος χάριν . . . A fols 41^{v}–42.

fol. 31. Εἰ θέλεις εὑρίσκειν πόσας ἔχει ἰσημερινὰς . . . A fol. 42. Tihon *List* 21.

fol. 31. "Οτι καθ' Ἕλληνας ἀπὸ κτίσεως κόσμου . . . A fol. 44^{v}. Tihon *List* 75.

fol 31^{v}. Εἰ θέλεις εὑρίσκειν πόσαι ὧραί εἰσιν ἰσημεριναὶ . . . A fols 42–42^{v}. Tihon *List* 20.

fol. 31^{v}. Εἰ βούλει εὑρίσκειν ἀπὸ τῶν Αἰγυπτιακῶν μηνῶν τούς τε Ἑλληνικοὺς καὶ Ῥωμαϊκοὺς μῆνας . . . A fol. 42^{v}. Tihon *List* 16 (which includes the next item).

fol. 31^{v}. "Οταν θέλης εὑρεῖν τοῦ κατ' Αἰγυπτίους ἐνισταμένου ἔτους τὴν ἀρχὴν . . . A fol. 42^{v}.

[5] Halma III, 101; H. Usener, 'De Stephano Alexandrino', id., *Kleine Schriften*, III, Leipzig-Berlin 1914, pp. 247–322, esp. 311–12.

[6] III 11, 5 in C. F. A. Nobbe, *Claudii Ptolemaei Geographia*, I, Leipzig 1898, p. 188.

[7] Partial inventories of the *Handy Tables* preserved in these two manuscripts are given by Heiberg, op. cit. n. 3 above pp. CXCVII–CXCVIII (Laurentianus, fols 50–106) and CC (Ambrosianus, fols 66–73^{v} and 164–172).

fol. 31^{v}. Δεῖ εἰδέναι ὅτι ἐὰν μὲν αἱ ὁριζόμεναι ἀπὸ μεσημβρίας ὧραι ἐν ἑτέρᾳ πόλει . . . A fol. 44^{v}. Tihon *List* 9.
fols 31^{v}–32. Ὅπως διὰ τοῦ κανόνος τοῦ ἐξάρματος ἔστιν εὑρίσκειν ἑκάστου τόπου τὸ μέγεθος τῆς μεγίστης ἡμέρας. A fols 42^{v}–43.
fols 32–32^{v}. Οἷον ἐπὶ ὑποδείγματος . . . A fol. 43^{v}.
fol. 32^{v}. Ἰστέον ὅτι ὅταν τὸ πλάτος τῆς πόλεως ἐκείνης . . . A fol. 43^{v}.
fol. 32^{v}. Ὅπως εὑρίσκειν δεῖ ἑκάστην τοῦ μηνὸς ἡμέραν, ποία ἡμέρα τῆς ἑβδομάδος ἐστίν.
fols 33–40. Κλαυδίου Πτολεμαίου Σαφήνεια καὶ διάταξις τῶν Προχείρων κανόνων τῆς ἀστρονομίας, καὶ ὅπως χρηστέον αὐτοῖς μέθοδος ἐναργής. A fols 45–54^{v}.
fol. 40^{v}<Περὶ π>ροσθέσεως ἢ ἀφαιρέσεως <τῆς> ☾ οὕτως. A fol. 54^{v}. Tihon *Scholia* II.
fol. 40^{v}. <Τ>οῦ ζῳδιακοῦ ἐπὶ τοῦ ☉. A fol. 54^{v}. Tihon *Scholia* III.
fols 40^{v}–41. <Ἐ>πὶ δὲ τῆς ἀρχῆς τῆς ☾. A fol. 55. Tihon *Scholia* IV.
fol. 41. <Ἐ>πὶ τῆς Καρδίας τοῦ ♌. A fol 55^{v}. Tihon *Scholia* V.
fol. 41^{v}<Ἐ>πὶ δὲ τῶν λοιπῶν δ̄. A fol. 55^{v}. Tihon *Scholia* VI.
fol. 41^{v}–48. Κλαυδίου Πτολεμαίου Ὑπόθεσις τῶν πλανωμένων. A fols 58–65^{v}.
fol. 48^{v}. Blank. After this probably came three quires containing the tables found on A fols 66–91; there is now no trace of them.
fol. 49. Περὶ τοῦ Ἡλίου. A fol. 91^{v}. Tihon *List* 68.
fol. 49. Περὶ τῆς Σελήνης. A fols 91^{v}–92.
fols 49–49^{v}. Περὶ τῆς θέσεως τοῦ κέντρου τοῦ ἐπικύκλου ☾. A fol. 92.
fol. 49^{v}. Περὶ τῆς θέσεως τοῦ κέντρου ☾. A fols 92–92^{v}.
fol. 49^{v} Περὶ τῆς θέσεως τοῦ βορείου πέρατος ☾. A fol. 92^{v}.
fols 50–50^{v}. Κανὼν εἰκοσιπενταετηρίδων Ἡλίου καὶ Σελήνης[8] (to 1726 Era Philippi = A.D. 1403). A fols 93–93^{v}. Halma II 66–68.
fol. 51. Κανὼν ἁπλῶν ἐτῶν Ἡλίου καὶ Σελήνης. A fol. 94. Halma II 70.
fol. 51^{v}. Κανὼν μηνῶν Αἰγυπτίων Ἡλίου καὶ Σελήνης. A fol. 94^{v}. Halma II 72.
fol. 52. Κανὼν ἡμερῶν Αἰγυπτίων Ἡλίου καὶ Σελήνης. A fol. 95. Halma II 74.
fol. 52^{v}. Κανὼν ὡρῶν ἀπὸ μεσημβρίας Ἡλίου καὶ Σελήνης. A fol. 95^{v}. Halma II 76.
fol. 53. Κανὼν τῶν τῆς ὥρας μορίων Ἡλίου καὶ Σελήνης. A fol. 96.
fol. 53^{v}. Κανὼν τῶν τῆς ἰσημερινῆς ὥρας μερῶν. A fol. 96^{v}.
fol. 54. Κανὼν ἀνωμαλίας λέγεται διότι . . . A fol. 96^{v}.
fol. 54. Οἱ κοινοὶ ἀριθμοὶ τῶν ὁμαλῶν . . . A fol. 96^{v}
fol. 54. Τὸ τρίτον σελίδιον περιέχει . . . A fol. 96^{v}.
fols 54–56^{v}. Κανὼν ἀνωμαλίας ☉ καὶ ☾. A fols 97–99v. Halma II 78–88.
fol. 57. Κανὼν Ἡλίου λοξώσεως καὶ Σελήνης πλάτους. A fol. 100. Halma I 144.
fols. 57^{v}–58. Κανὼν Ἡλίου λοξώσεως κατὰ μονομοιρίαν. A fols 100^{v}–101.
fols 58^{v}–59. Κανὼν Σελήνης πλάτσυς κατὰ μονομοιρίαν. A fols 101^{v}–102.
fol. 59^{v}. Ἐξήγησις τῶν δύο κανόνων τοῦ τε Ἡλίου ἀπὸ ἰσημερίας καὶ τοῦ ἐξάρματος τοῦ πόλου καὶ τῆς ὑπεροχῆς τῶν ὡρῶν. A fol. 102^{v}. Halma III 144–145.
fol. 59^{v}. Κανὼν Ἡλίου ἀπὸ ἰσημερίας. A fol. 103. Halma II 92.
fol. 60. Κανὼν ἐξάρματος πόλου καὶ ὡρῶν ὑπεροχῆς. A fol. 103^{v}. Halma I 132.
fol. 60^{v}–61. Skeletons of tables.
fol. 61^{v}. Κανὼν σεληνιακῶν ἐκλείψεων. A fol. 104. Halma II 90.
fol. 62. Κανὼν ἡλιακῶν ἐκλείψεων. A fol. 104^{v}. Halma II 90.
fol. 62. Κανὼν μεγέθους ☉ καὶ ☾. A fol. 104^{v}. Halma II 94.
fol. 62^{v}. Προκανόνιον. A fol. 105. Halma I 146.
fol. 63. Ὁριζόντων καταγραφή. A fol. 105^{v}. Halma III 43.
fol. 63^{v}–64. Παράλλαξις κλίματος πρώτου τοῦ διὰ Μερόης. A fol. 106–106^{v}. Halma II 98.
fol. 64^{v}–65. Παράλλαξις τοῦ διὰ Συήνης δευτέρου κλίματος. A fols 107–107^{v}. Halma II 100.
fols 65^{v}–66. Παράλλαξις τοῦ διὰ τῆς Κάτω χώρας γ′ κλίματος. A fols 108–108^{v}. Halma II 102.
fols 66^{v}–67. Παράλλαξις τοῦ διὰ Ῥόδου τετάρτου κλίματος. A fols 109–109^{v}. Halma II 104.
fols 67^{v}–68. Παράλλαξις τοῦ δι᾽ Ἑλλησπόντου ε′ κλίματος. A fols 110–110^{v}. Halma II 106.
fols 68^{v}-69. Παράλλαξις τοῦ διὰ μέσου Πόντου ς′ κλίματος. A fols 111–111^{v}. Halma II 108.
fols 69^{v}–70. Παράλλαξις τοῦ διὰ Βορυσθένους ἑβδόμου κλίματος. A fols 112–112^{v}. Halma II 110.
fols 70^{v}–71. Παράλλαξις τοῦ διὰ Βυζαντίου παραλλήλου. A fols 172–172^{v}. Halma II 194.
fol. 71^{v}. Περὶ τῆς ἐποχῆς τῆς Καρδίας τοῦ Λέοντος. A fol. 173. Tihon *List* 58.
fol. 71^{v}. Περὶ τῆς τοῦ Κρόνου ἐποχῆς. A fol. 173.
fols 71^{v}–72. Περὶ τοῦ Διός. A fols 173–173^{v}.
fol. 72. Περὶ τοῦ Ἄρεος. A fols 173^{v}–174.
fols 72–72^{v}. Περὶ τῆς Ἀφροδίτης. A fol. 174.
fol. 72^{v}. Περὶ τοῦ Ἑρμοῦ. A fols 174–174^{v}.
fol. 72^{v}. Ἰστέον ὅτι τὸ μέσον ἀπόστημα . . . A fol. 174^{v}.
fols 73–74^{v}. Κανὼν εἰκοσιπενταετηρίδων τῶν ε̄ ἀστέρων (to 1726 Era Philippi = A.D. 1403). A fols 175–176^{v}. Halma II 112–118.
fols 75–75^{v}. Κανὼν ἐτῶν ἁπλῶν τῶν πέντε ἀστέρων. A fols 177–177^{v}. Halma II 120–122.
fol. 76. Κανὼν Αἰγυπτίων μηνῶν τῶν πέντε ἀστέρων. A fol. 178. Halma II 124.
fols 76^{v}–77. Κανὼν ἡμερῶν Αἰγυπτίων τῶν πέντε ἀστέρων. A fols 178^{v}–179. Halma II 126–128.
fols 77^{v}–78. Κανὼν ὡρῶν ἀπὸ μεσημβρίας τῶν πέντε ἀστέρων. A fols 179^{v} and 113. Halma II 130–132.
fols 78^{v}–81. Κανὼν ἀνωμαλίας Κρόνου. A fols 113^{v}–116. Halma II 134–144.

[8] In the following I give the titles of the tables in the form in which they occur in A; there are only very minor variants in the Laurentianus. For the structure and use of the tables see O. Neugebauer, *A History of Ancient Mathematical Astronomy*, Berlin-Heidelberg-New York 1975, pp. 969–1028.

fols 81^{v}–84. Κανὼν ἀνωμαλίας Διός. A fols 116^{v}–119. Halma II 146–156.
fols 84^{v}–87. Κανὼν ἀνωμαλίας Ἄρεος. A fols 119^{v}–122. Halma II 158–168.
fols 87^{v}–90. Κανὼν ἀνωμαλίας Ἀφροδίτης. A fols 122^{v}–125. Halma II 170–180.
fols 90^{v}–93. Κανὼν ἀνωμαλίας Ἑρμοῦ. A fols 125^{v}–128. Halma II 182–192.
fol. 93^{v}. Κανὼν Κρόνου βορείου πλάτους. A fol 128^{v}. Halma III 1.
fol. 94. Κανὼν Κρόνου νοτίου πλάτους. A fol. 129. Halma III 2.
fol. 94^{v}. Κανὼν Διὸς βορείου πλάτους. A fol. 129^{v}. Halma III 3.
fol. 95. Κανὼν Διὸς νοτίου πλάτους. A fol. 130. Halma III 4.
fol. 95^{v}. Κανὼν Ἄρεος βορείου πλάτους. A fol. 130^{v}. Halma III 5.
fol. 96. Κανὼν Ἄρεος νοτίου πλάτους. A fol. 131. Halma III 6.
fol. 96^{v}. Κανὼν Ἀφροδίτης βορείου πλάτους. A fol. 131^{v}. Halma III 7.
fol. 97. Κανὼν Ἀφροδίτης νοτίου πλάτους. A fol. 132. Halma III 8.
fol. 97^{v}. Κανὼν Ἑρμοῦ βορείου πλάτους. A fol. 132^{v}. Halma III 9.
fol. 98. Κανὼν Ἑρμοῦ νοτίου πλάτους. A fol. 133. Halma III 10.
fol. 98^{v}. Κρόνου κανὼν στηριγμῶν. A fol. 133^{v}. Halma III 11.
fol. 99. Διὸς κανὼν στηριγμῶν. A fol. 134. Halma III 12.
fol. 99^{v}. Ἄρεος κανὼν στηριγμῶν. A fol. 134^{v}. Halma III 13.
fol. 100. Ἀφροδίτης κανὼν στηριγμῶν. A fol. 135. Halma III 14.
fol. 100^{v}. Κανὼν Ἑρμοῦ στηριγμῶν. A fol. 135^{v}. Halma III 15.
fol. 101. Κρόνου κανὼν φάσεων. A fol. 137. Halma III 16–17.
fol. 101^{v}. Διὸς κανὼν φάσεων. A fol. 137^{v}. Halma III 18–19.
fol. 102. Ἄρεος κανὼν φάσεων. A fol. 138. Halma III 20–21.
fols 102^{v}–103. Ἀφροδίτης κανὼν φάσεων. A fols 138^{v}–139. Halma III 22–25.
fols 103^{v}–104. Ἑρμοῦ κανὼν φάσεων. A fols 139^{v}–140. Halma III 26–29.
fol. 104^{v}. Κανὼν φάσεων τῶν πέντε ἀστέρων ἐπὶ τοῦ διὰ Βυζαντίου παραλλήλου. A fol. 140^{v}. Halma III 32–33.
fol. 105. Φάσεων ἀποστάσεις πρὸς τὸν ἀκριβῆ Ἥλιον τῶν πέντε ἀστέρων. A fol. 141. Halma III 30.
fol. 105^{v}. ♀ καὶ ☿ μέγισται ἀποστάσεις. A fol. 141^{v}. Halma III 32.
fols 106–108^{v}. Ἐποχαὶ ἀπλανῶν ἀστέρων μέχρι δεκαμοιριαίου πλάτους καὶ μεγέθου δ'. A fols 142–145^{v}. Halma III 44–58.

The rest of Laurentianus 28, 7 was not used by the scribe of Ambrosianus H. 57. sup, but I continue to catalogue its contents for the sake of completeness.

fol. 109. A world map. Published by Neugebauer.[9]
fols 109^{v}–111^{v}. Blank.
fols 112–144^{v}. Πρόκλου Διαδόχου Πλατωνικοῦ Ὑποτύπωσις τῶν ἀστρονομικῶν ὑποθέσεων. This is manuscript L^{5} in the edition by Manitius,[10] which he places in the second family of his second class.[11]
fols 144^{v}-145^{v}. Ἐκ τῶν Γεμίνου περὶ ἐξελιγμοῦ = Geminus, Εἰσαγωγή 18. This manuscript is L in the edition by Manitius,[12] H in that by Aujac.[13]
fols 146–147^{v}. Originally blank, but a later scribe has written on fols 146^{v}–147 a poem to the Virgin which begins:

<Δ>ὸς τῷ στρατηγῷ τῶν βροτῶν ὑπερμάχῳ.

fols 148–172^{v}. Παύλου Ἀλεξανδρέως Εἰσαγωγὴ εἰς τὴν ἀποτελεσματικήν. This is ℒ in the edition by Boer,[14] who places it in her class γ.[15]
fols 173–177^{v}. Ῥητορίου ἔκθεσις καὶ ἐπίλυσις περί τε τῶν προειρημένων δώδεκα ζῳδίων καὶ περὶ ἑτέρων διαφόρων ἐκ τῶν Ἀντιόχου Θησαυρῶν = *Epitome* II b of Rhetorius.[16] This is manuscript a in the edition by Boll.[17]
fol. 178. Tables of the weekdays with which Roman months begin in the 28 years of a Solar cycle, and of the weekdays with which Arabic months begin in a 30-year intercalation cycle.
fols 178^{v}–179. <E>ἰ βούλει γνῶναι ἐν οἱῳδηποτοῦν τόπῳ καθ' ἑκάστην ἡμέραν πόσας μοίρας ὑψοῦται ὁ ☉. . . Tihon *List* 15.
fol. 179^{v}. Blank.

The sections of Ambrosianus H. 57. sup. that were derived from another source than Laurentianus 28, 7 or from parts of that manuscript that are now lost are:

fols 40–40^{v}. Ψῆφος τῆς κατὰ μῆκος Σελήνης. Tihon *List* 61. Edited by Zuretti.[18]
fols 43^{v}–44^{v}. Περὶ μεταβολῆς τῶν καιρικῶν ὡρῶν πρὸς τὰς ἐν Ἀλεξανδρείᾳ μεσημβρινὰς καὶ πρὸς ἰσημερινὰς ὥρας καὶ πρὸς ὁμαλὰ νυχθήμερα. Halma III 60–62.
fol. 44^{v} (after Tihon *List* 9). Μετὰ δὲ τὸ διακριθῆναι τὰς ὥρας . . .
fol. 56. Ἐπὶ τῶν βουλομένων. Tihon *Scholia* VII.

---

[9] O. Neugebauer, 'A Greek World Map', *Hommages à Claire Préaux*, Brussels 1975, pp. 312–17.
[10] K. Manitius, *Procli Diadochi Hypotyposis astronomicarum positionum*, Leipzig 1909, p. VIII.
[11] Ibid., p. XXIX.
[12] K. Manitius, *Gemini Elementa astronomiae*, Leipzig 1898, p. XIII.
[13] G. Aujac, *Géminos. Introduction aux phénomènes*, Paris 1975, pp. XCVIII–XCIX.
[14] E. Boer, *Pauli Alexandrini Elementa apotelesmatica*, Leipzig 1958, p. VII.
[15] Ibid., p. XII.
[16] D. Pingree, 'Antiochus and Rhetorius', *Classical Philology*, LXXII, 1977, pp. 203–23, esp. 209–10.
[17] F. Boll, 'Rhetorii quaestiones astrologicae ex Antiochi thesauris excerptae', *CCAG* I; 140–6.
[18] C. O. Zuretti, *CCAG* XI, 2; 113–14.

fol. 56^{v}. Ἐὰν βούλῃ ποιῆσαι. Tihon *Scholia* VIII.
fol. 56^{v}. Περὶ τῆς τοῦ Κυνὸς ἐπιτολῆς ὑπόδειγμα. Tihon *Scholia* IX.
fol. 56^{v}. Ἀπὸ συνόδου ἐπὶ σύνοδον. Tihon *Scholia* X.
fol. 57. Περὶ ἀνέμων ἀπὸ χειρός. Tihon *Scholia* XI.
fol. 57. Εὑρεῖν πότερον ἡ ☾ ἐπὶ ☾☊ ἔρχεται ἢ ἐπὶ π ☾ ἢ ἀπὸ ☾☊ ἐπὶ π ☾. Tihon *Scholia* XII.
fol. 57. Τὰ τοῦ ☊ σημεῖα ἐν ταῖς παραλλάξεσι λαμβάνομεν τὸν τρόπον τοῦτον. Tihon *Scholia* XIII.
fol. 57^{v}. Ἐπὶ τῶν φάσεων. Tihon *Scholia* XIV.
fols 66–67. Κανὼν βασιλέων. Ended originally with John V (1341–1391). Halma I 139–142.
fol. 67^{v}. Blank.
fols 68–72. Κανὼν ἐπισήμων πόλεων. Similar to the catalogues published by Honigmann from Vaticanus graecus 1291[19] and Leidensis graecus 78.[20]
fol. 72^{v}. Skeleton for geographical table.
fol. 73. Τὰ ἑξηκοστὰ τῶν ὡρῶν ἐν τῇ ὀρθῇ σφαίρᾳ κείμενα . . . Apparently added by the third scribe.
fol. 73. Εἰδέναι χρὴ ὅτι ἡ διαφορὰ τῆς τῶν ὡρῶν παραυξήσεως. Apparently added by the third scribe.
fol. 73. Ὅτι δὲ μετὰ τὸ πρῶτον ἔτος τοῦ Φιλίππου. Apparently added by the third scribe.
fols 73^{v}–75. Κανὼν ἀναφορῶν ὀρθῆς σφαίρας. Halma I 148–154.
fols 75^{v}–77. Κανὼν ἀναφορῶν τοῦ διὰ Μερόης πρώτου κλίματος. Halma II 2–8.
fols 77^{v}–79. Κανὼν ἀναφορῶν τοῦ διὰ Συήνης δευτέρου κλίματος. Halma II 10–16.
fols 79^{v}–81. Κανὼν ἀναφορῶν τοῦ διὰ τῆς Κάτω χώρας τρίτου κλίματος. Halma II 18–24.
fols 81^{v}–83. Κανὼν ἀναφορῶν τοῦ διὰ Ῥόδου τετάρτου κλίματος. Halma II 26–32.
fols 83^{v}–85. Κανὼν ἀναφορῶν τοῦ δι' Ἑλλησπόντου πέμπτου κλίματος. Halma II 34–40.
fols 85^{v}–87. Κανὼν ἀναφορῶν τοῦ διὰ μέσου Πόντου ἕκτου κλίματος. Halma II 42–48.
fols 87^{v}–89. Κανὼν ἀναφορῶν τοῦ διὰ Βορυσθένους ἑβδόμου κλίματος. Halma II 50–56.
fols 89^{v}–91. Κανὼν ἀναφορῶν τοῦ διὰ Βυζαντίου παραλλήλου. Halma II 58–64.
fols 136–136^{v}. Skeletons of tables.
fol. 146. Skeleton for star catalogue.
fol. 146^{v}. Blank.
fols 147–147^{v}. Ἰωάννου γραμματικοῦ Ἀλεξανδρέως τοῦ Φιλοπόνου Περὶ χρήσεως τοῦ ἀστρολάβου. πίναξ.
fols 147^{v}–159^{v}. Ἰωάννου γραμματικοῦ Ἀλεξανδρέως Περὶ τῶν τῆς ἀστρολάβου χρήσεως γεγραμμένων ἕκαστον τί σημαίνει. Edited by Hase.[21]
fols 159^{v}–162. Μέθοδος ἑτέρα τοῦ ἀστρολάβου. Edited by Hase.[22]
fol. 162^{v}. Χρησμὸς εἰς τὸν τῆς Πελοποννήσου ἰσθμὸν εὑρεθὺς ἐν Ῥόδῳ. Added by the third scribe? Edited by Bodnar.[23]
fol. 163. Τῶν ἀπὸ Ἀδὰμ ἐτῶν . . . Added by the fourth scribe.
fol. 163. Τὰ ἀπὸ Ἀδὰμ ἕως τῶν ὅλων Κριτῶν . . . Added by the fourth scribe.
fol. 163. Ὁμοίως λέγει καὶ ὁ Ἰωσηπὸς ἐν τῷ τῆς ἁλώσεως λόγῳ . . . Added by the fourth scribe.
fol. 163^{v}. Blank.
fols 164–164^{v}. <Κανὼν βασιλέων>. Begins with Φίλιππος ὁ μετὰ Ἀλέξανδρον τὸν κτίστην, and ends with Leo VI (886–912).[24] Halma I 139–142.
fols 165–168. Table of Consuls. Halma III 120–134. Also edited, from Leidensis graecus 78, by Usener.[25]
fols 168^{v}–171^{v}. Geographical table. Similar to that on fols 68–72.

Thus, aside from rearranging the scholia following the text of Theon in the Laurentianus, the scribe of the Ambrosianus has added three more pieces; he has also included more of the scholia that follow Ptolemy's commentary. He has, of the *Handy Tables* themselves, a more complete copy, including the royal canon, the geographical table, and the tables of right and oblique ascensions; but these may have formerly been included in the Laurentianus. An additional quire written by the first scribe of the Ambrosianus contains a second, incomplete royal canon; the table of consuls; and a second geographical table. Finally, quires 20–21 contain two works on the astrolabe.

But the designer of the Ambrosianus, who evidently had a wealthy patron, planned to add another feature — illustrations. The first folio (Pl. 30), with the incipit of Theon, was obviously intended by the scribe to be illuminated. The scribe left no space for other illustrations in the manuscript, but the illustrator has added representations of the seven planets in the lower margins of selected pages (Pls 31–33);

[19] E. Honigmann, *Die Sieben Klimata und die πόλεις ἐπίσημοι*, Heidelberg 1929, pp. 194–208.
[20] Ibid., pp. 211–24.
[21] *Joannis Alexandrini cognomine Philoponi, de Usu astrolabii eiusque constructione libellus*, ed. H. Hase, *Rheinisches Museum*, VI, 1839, pp. 127–71 (Spezial-ausgabe, Bonn 1839).
[22] Ibid.
[23] E. W. Bodnar, 'The Isthmian Fortifications in Oracular Prophecy', *American Journal of Archaeology*, LXIV, 1960, pp. 165–71. This is Bodnar's long version, probably composed between *c.* 1435 and 1445.
[24] As, e.g., in Vaticanus graecus 1291 fols 16^{v}–17, as continued; see F. Boll, 'Beiträge zur Ueberlieferungsgeschichte der griechischen Astrologie und Astronomie, *Sitzungsberichte der Bayerischen Akademie der Wissenschaften*, Phil.-hist. Kl., Munich 1899, pp. 77–139, esp. 113 and 115. The royal canon in this manuscript has been re-examined by I. Spatharakis, 'Some Observations on the Ptolemy MS. Vat. Gr. 1291: its Date and the Two Initial Miniatures', *Byzantinische Zeitschrift*, LXXI, 1978, pp. 41–49, who shows that the manuscript was copied during the reign of Theophilus (828–42).
[25] H. Usener, 'Fasti Theonis Alexandrini', in *Chronica minora*, ed. T. Mommsen, vol. III, (*Monumenta Germaniae Historica, Auctores antiquissimi*, XIII), Berlin 1898, pp. 359–81.

since the mean motion tables in the *Handy Tables* are arranged for combinations of planets (the Sun and the Moon together in one set of tables on fols 93–95^{v}, and the five star-planets in another set of tables on fols 175–179^{v} and fol. 113), the illustrator chose those tables where at least the five star-planets appear separately — that is, the equation tables. Therefore, the Sun and the Moon are depicted on fol. 97 (Pl. 31), the first page of their equation table; Saturn on fol. 113^{v} (Pl. 32a), the first page of its equation table; Jupiter on fol. 116^{v} (Pl. 32b), the first page of its equation table; Mars on fol. 121 (Pl. 32c), the *fourth* page of its equation table; Venus on fol. 122^{v} (Pl. 32d), the first page of its equation table; and Mercury on fol. 125^{v} (Pl. 33b), the first page of its equation table. The planets are represented as naked (except for Mars) humans (the Moon and Venus female, the rest male) sitting on or carrying the symbols of the zodiacal signs that are their astrological houses, exaltations, and, in some cases, dejections. The illustrator has also shown an astronomer using a parallactic instrument[26] on the lower margin of fol. 106 (Pl. 33a, c), the first page of the parallax tables.

DESCRIPTIONS OF THE ILLUSTRATIONS[27]

fol. 1 (Pl. 30). Scene at top: Bearded astronomer, tonsured, wearing long, dark brown robe with cowl, standing, facing right, in front of long, narrow building with snake on roof; holds astrolabe(?) with bent alidade at height of chest in left hand; ray extends from celestial body (Sun?) to end of alidade. Field of stars in upper right, with Sun(?) in upper right corner; disc, perhaps of Moon, to its left. Tonsured amanuensis wearing red robe, seated, facing left, to right of astronomer.

To right and left of text and below it: some northern constellations. To right, Draco (with three stars on either side); at bottom, Ursa (four central stars of constellation above and below it); and to left, Ophiuchus with snake's tail.

fol. 97 bottom (Pl. 31). To left: Sun as a naked man riding on a lion (Leo: its house) and a ram (Aries: its exaltation).

To right: Moon as a dark, naked woman sitting on a crab (Cancer: its house) and holding in her lap a bull (Taurus: its exaltation). The tail of Scorpio (its dejection) may also appear. Four stars around the picture.

fol. 106 bottom (Pl. 33a, c). Tonsured astronomer, kneeling on right knee, facing right; observes ray-projecting Moon through diopter on upper arm of parallactic instrument while holding disconnected lower arm with left hand.

fol. 113^{v} bottom (Pl. 32a). Saturn as a naked man riding on a goat (Capricorn: its house) and a stooping man pouring water from a pot (Aquarius: its house); the two wear a yoke (Libra: its exaltation) on their necks. Saturn's right hand rests on the two heads of a standing, naked two-headed man who elsewhere (fol. 125^{v}) represents Gemini. One expects instead to find a symbol of Aries, Saturn's dejection; perhaps Gemini was substituted because in it lies Saturn's perigee.

fol. 116^{v} (Pl. 32b). Jupiter as a naked man sitting on two fish facing in opposite directions (Pisces: its house) and on a centaur holding a bow in his right hand (Sagittarius: its house). Below his feet is a crab (Cancer: its exaltation). In his lap he holds a ram, which normally signifies Aries; here it is presumably a mistake for Capricorn, Jupiter's dejection.

fol. 121 bottom (Pl. 32c). Mars as a warrior in armour with a sword in his right hand, sitting on a ram (Aries: its house) and holding in his lap a scorpion (Scorpio: its house) and a crab (Cancer: its dejection). His feet rest on a goat (Capricorn: its exaltation).

fol. 122^{v} bottom (Pl. 32d). Venus as a naked woman sitting on a bull (Taurus: its house) which has a yoke on its back and neck (Libra: its house). Her left foot rests on two fish facing in opposite directions (Pisces: its exaltation); she herself may represent Virgo, Venus's dejection.

fol. 125^{v} bottom (Pl. 33b). Mercury as a naked man carried on the shoulders of a naked woman (Virgo: its house and exaltation) and a naked, two-headed man (Gemini: its house).[28]

David Pingree

Brown University, Providence, R.I.

II

EDITORIAL NOTE

One of the Editors' aims in publishing Professor Pingree's description of MS H. 57. sup. in the Biblioteca Ambrosiana is to encourage discussion of the interesting problems raised by its illustrations. Having shown photographs of

[26] *Almagest* 5, 12 in J. L. Heiberg, *Claudii Ptolemaei Syntaxis mathematica*, 2 vols, Leipzig 1898–1903.

[27] I should like especially to thank J.-M. Massing for having scrutinized the photographs of these illustrations and saved me from many blunders.

[28] Greek, Arabic and Slav occasionally use the word for twin in the singular to mean a pair. For other two-headed men as Gemini, see, e.g., F. Saxl and H. Meier, *Verzeichnis astrologischer und mythologischer illustrierter HSS.*, III, *HSS. in englischen Bibliotheken*, ed. H. Bober, London 1953, p. 268, pl. LXXXV, ill. 218; and H. Bober, this *Journal*, XI, 1948, pl. 4a.

these to a number of scholars in various fields without being able to persuade any of them to commit themselves to print, they have decided on an unconventional expression of gratitude. Some of the experts' comments have been incorporated into the text of Part I; others, with Professor Pingree's response to them, have been combined into the ensuing layman's amalgam. Any inaccuracies and faults of emphasis are inadvertent and the responsibility of the Editors.

There is no doubt, on the evidence of the script, that the codex could have been written in Constantinople in 1357–58. It is in the style associated with the Hodegoi scriptorium over a period of about half a century.[29] Western connexions of this scriptorium are not apparent. If they existed they would be more likely at this time to have been with Venice than with Southern Italy.

The consensus of opinion is that the style of the miniatures is basically Western, though with an admixture of Byzantine elements. Islamic tradition lies behind the curious iconography, in which the planets are shown with their day- and night-houses, exaltations and dejections: this is seldom, if ever, found in the West. No immediate model has been located in an admittedly cursory search. An artist active in Southern Italy or Sicily may be indicated by the mixture of Italian and French styles. It is not impossible that an artist of, say, the Neapolitan school was working in Constantinople in the mid-fourteenth century — a time when others (Barlaam of Seminara, for example) travelled freely between that city and Italy. No artist answering to this description can, however, be securely documented in Constanincple at this time. Nor can it be assumed that the manuscript was decorated in the same place as it was written: the illustrations may be later additions. Indeed, they look very much as if they are. Neapolitan connexions are suggested by parallels in the illustrations of the *Bible moralisée* in the Bibliothèque nationale in Paris (MS fr. 9561), ascribed by both Bologna and Meiss to Naples in the mid-fourteenth century.[30] A Byzantine background is also indicated by similarities with the illustrations in later manuscripts of Leo the Wise.[31] None of these parallels can be regarded as close. The ribbon decoration on the initial H (fol. 1^r) is of a simplified type resulting from a fusion of Greek elements with Western forms of a kind commonly found in Italo-Greek initials in Southern Italy during the fourteenth century. The treatment of ascenders and horizontals derives ultimately from Carolingian and post-Carolingian initials of the type described by Garrison as 'hollow-shaft', and the four-lobed, four-leaved medallions at the intersection of ascenders and horizontal are Byzantine in character. The small palmettes of simplified form at the angles of the text on fol. 106^r (Pl. 33a) have relations with the 'stile carminato' of Basilian manuscripts of the fourteenth century, but are here perhaps more reminiscent of Bolognese or even of Bolognese-derived work.

[29] L. Politis, 'Eine Schreiberschule in Kloster τῶν Ὁδηγῶν', *Byzantinische Zeitschrift*, LI, 1958, pp. 17–36, 261–87.

[30] F. Bologna, *I Pittori alla corte angioina di Napoli, 1226–1414*, Rome 1969, pls VII - 56, 67, 73; M. Meiss, *French Painting in the Time of Jean de Berry: The Late Fourteenth Century and the Patronage of the Duke*, London 1967, pp. 27–29.

[31] Irmgard Hutter, *Corpus der Byzantinischen Miniaturenhandschriften*, II, Stuttgart 1978, figs 573, 579.

MS H 57 Sup., Milan, Ambrosiana, fol. 1[r] (*pp.* 190f.)

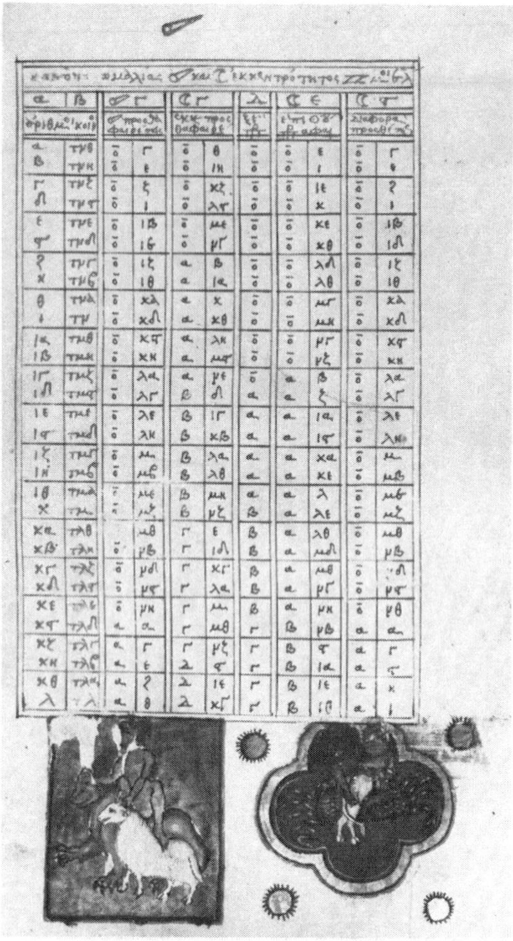

a

a—Fol. 97r (*pp.* 190f.)

b—Detail of Pl. 31a (*p.* 191)

c—Detail of Pl. 31a (*p.* 191)

b

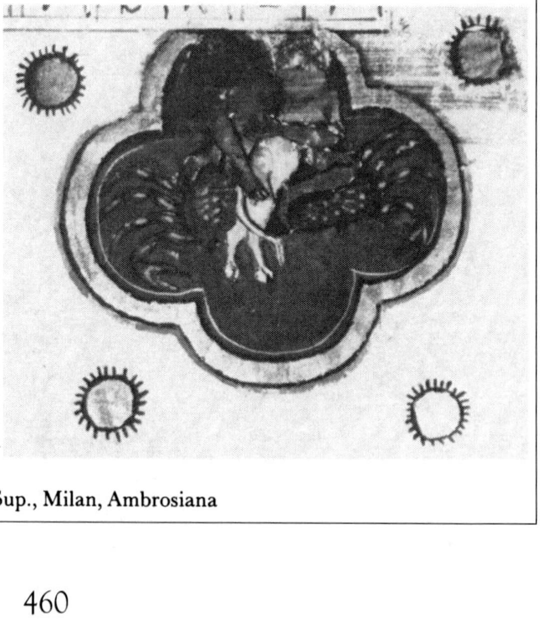

c

MS H 57 Sup., Milan, Ambrosiana

b

a

d

c

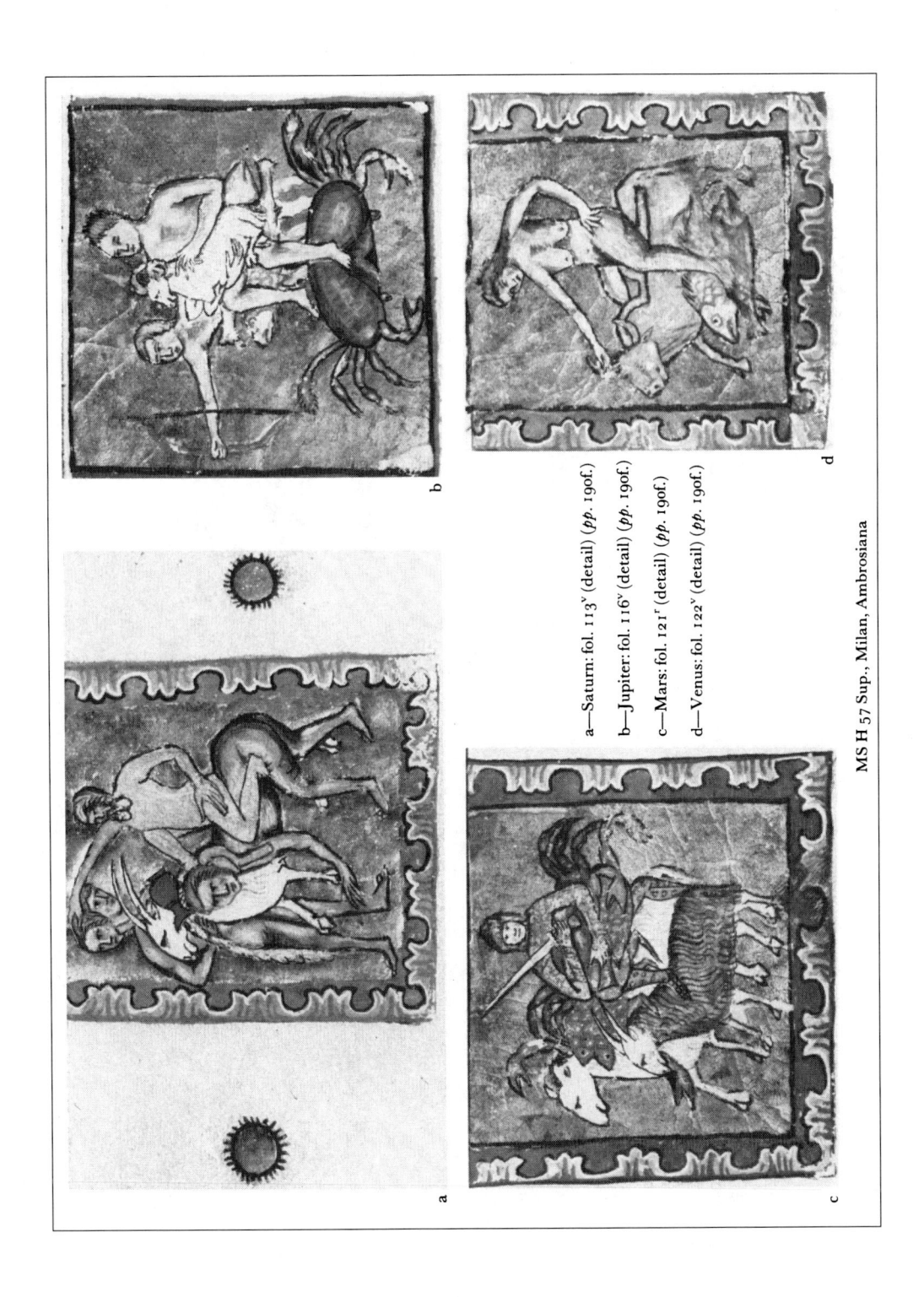

a—Saturn: fol. 113^v (detail) (*pp.* 190f.)

b—Jupiter: fol. 116^v (detail) (*pp.* 190f.)

c—Mars: fol. 121^r (detail) (*pp.* 190f.)

d—Venus: fol. 122^v (detail) (*pp.* 190f.)

MS H 57 Sup., Milan, Ambrosiana

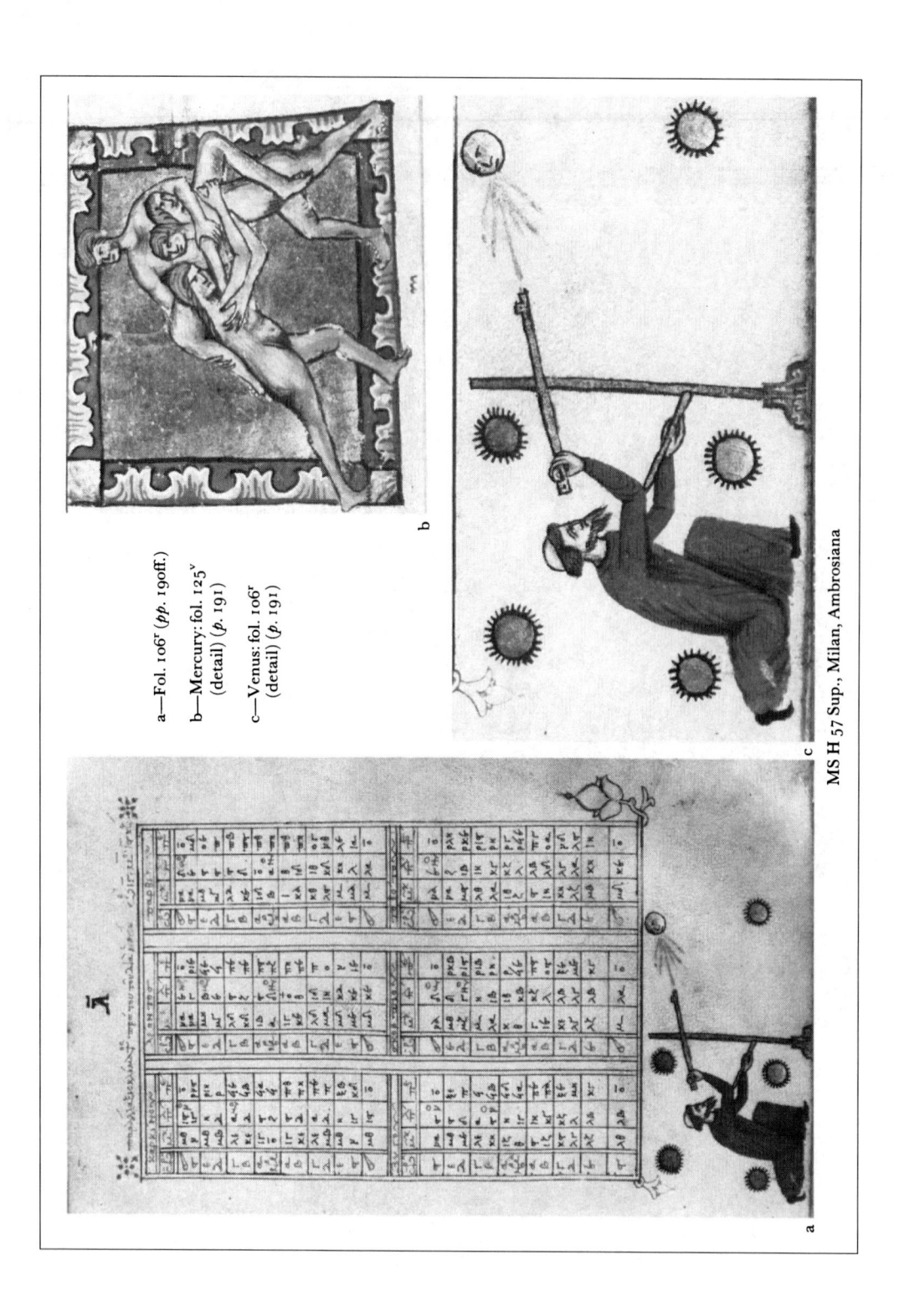

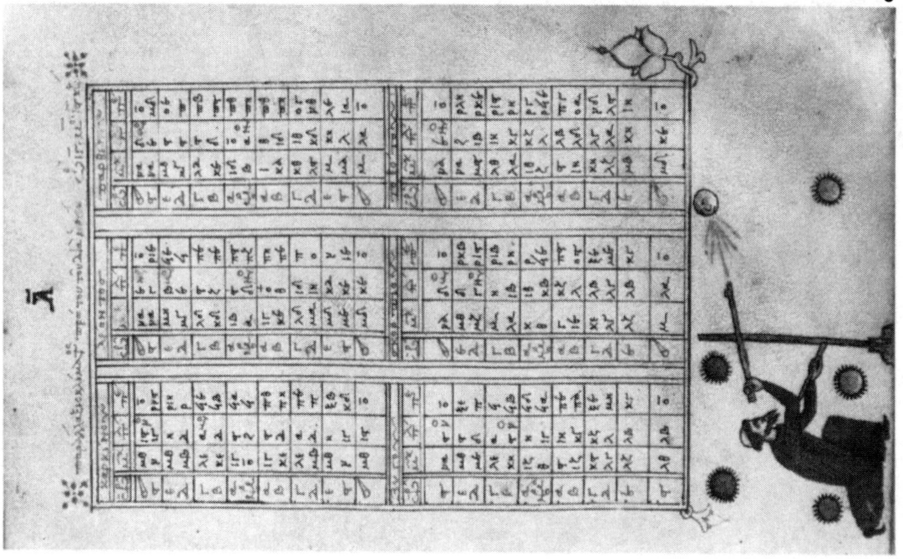
a

b

c

a—Fol. 106^{r} (*pp.* 190ff.)

b—Mercury: fol. 125^{v} (detail) (*p.* 191)

c—Venus: fol. 106^{r} (detail) (*p.* 191)

MS H 57 Sup., Milan, Ambrosiana

David E. Pingree

Plato's Hermetic *Book of the Cow*

Near the beginning of his *Disputationes adversus astrologiam divinatricem* Pico della Mirandola mentions (1) «the books of Plato concerning the Cow (which) the magi cause to circulate [...] filled with execrable dreams and figments». This paper presents an attempt to define the origins — in part Neoplatonic — of the execrable dreams contained in the *Book of the Cow*, to explain how its magic works in tandem with its *gemellus*, the celestial magic of the *Picatrix* generally so much more familiar to historians of Renaissance Neoplatonism, and finally to examine what little evidence there is for the popularity of this text during the Renaissance.

The Syrian city of Ḥarrān has long been associated with occult events (2) — if not from the time that Abraham was brought there from Ur of the Chaldaeans as recorded in *Genesis* (3), at least from about 550 B.C. when the last of the Chaldaean kings, Nabonidus, dreamed that Marduk ordered him to rebuild Ehulhul, the temple of Sin, the god of the Moon, at Ḥarrān. In order to make this restoration possible, Marduk employed Cyrus the Persian — the eventual obliterator of Nabonidus' feeble power and of the ancient kingdom of Babylon — as his instrument to destroy the army of Medes who were besieging the city (4).

A millennium and a half after the jubilant re-installation of the Moon in his temple Ḥarrān was still in ferment with the occult sciences, with some elements derived from Babylon, some from Sasanian Iran, and some from Western and Southern India, while others came from the Greeks, and especially the Neoplatonists (5). In the ninth, tenth, and eleventh centuries the planets and the noetic hierarchy of Neoplatonism were still being worshipped in Ḥarrān with rituals of mixed Mesopotamian, Zoroastrian, Indian, and Hermetic origins. The Ṣābians'

(1) Pico della Mirandola, *Disputationes adversus astrologiam divinatricem*, ed. by E. Garin, 2 vols., Firenze 1946-1952, vol. I, p. 64.

(2) The most useful collection of the Arabic traditions concerning Ḥarrān remains D. Chwolsohn, *Die Ssabier und der Ssabismus*, 2 vols., St. Petersburg 1856.

(3) *Genesis*, 11, 31-12, 4.

(4) Translated by A. Leo Oppenheim, *The Interpretation of Dreams in the Ancient Near East*, Philadelphia 1956, p. 250; see also pp. 202-203. For another auto-biographical account of Nabonidus' dream see A. L. Oppenheim in J. Pritchard, *The Ancient Near East*, Princeton 1969, pp. 562-563.

(5) D. Pingree, *Some of the Sources of the* Ghāyat al-ḥakīm, in «Journal of the Warburg and Courtauld Institutes», XLIII (1980), pp. 1-15; *Indian Planetary Images and the Tradition of Astral Magic, ibidem*, 52, 1989, pp. 1-13; and C. S. F. Burnett, *Arabic, Greek, and Latin Works on Astrological Magic Attributed to Aristotle*, in *Pseudo-Aristotle in the Middle Ages*, London 1986, pp. 84-96.

study of the κόσμος, of the νοῦς, of the ψυχή, and of σώματα, however, was based not only on Plato's *Republic*, *Timaeus*, and *Laws*, but as well on an Aristotelian corpus familiar to students of the Neoplatonic schools at Alexandria and at Athens; Ibn al-Nadīm in his *Fihrist* (6) quotes al-Sarakhsī (7), a pupil of al-Kindī, as citing their use of the *Physics*, the *De Caelo*, the *De generatione et corruptione*, the *Meteorology*, the *De anima*, the *De sensu et sensato*, and the *Metaphysics*. We shall see that other Platonic and Aristotelian works as well as some of those already mentioned provided the intellectual background to the *Book of the Cow*. The Ḥarrānians investigated the relations between the heavens and the sub-lunar world through the medium of astrology, for the correct practise of which they, like Proclus, plunged into the intricacies of Ptolemy's Σύνταξις and 'Αποτελεσματικά among many other texts, including Hermetic, Sasanian, and Indian.

Out of this rich mixture of intellectual traditions, the Ḥarrānians in the latter half of the ninth century — by then calling themselves Ṣābians and proclaiming Hermes to be their prophet in order to gain the approval of the Islamic authorities (8) — created the astral magic that is summarized in such confusion in the *Ghāyat al-ḥakīm* (9) and in its Latin translation, the *Picatrix* (10). This astral magic is based on Neoplatonic cosmology, astrology, and Ḥarrānian worship of the planets. It seeks, through rituals in part suggestive of the Hermetic art of the vivification of statues (11), performed at astrologically propitious times, to persuade the souls of the planets whom God has placed in charge of his material creation to send their subordinate spirits to occupy talismans. These latter are thereby empowered to effect changes in this sublunar world. This is a non-demonic form of magic that relies for its effectiveness on the powers granted by God to the planets and on those inherent by nature in corporeal substances and in the magician's rational soul.

But at the same time the Ḥarrānians invented another form of magic based on quite different principles. The procedures of this magic were described in a work called in Arabic the *Kitāb al-nawāmīs* (12), the *Book of the Laws*, where *nawāmīs* is the plural of the Arabic transliteration, *nāmūs*, of the Greek νόμος. Like the other Περὶ νόμων it was claimed to have been written by Plato. It is known now primarily through a Latin translation made in Spain in the twelfth century under the title *Liber aneguemis* (13), where *aneguemis* (sometimes corrupted into *neumich*) is based on a

(6) *Kitāb al-fihrist*, ed. by G. Flügel, 2 vols., Leipzig 1871-1872, vol. I, pp. 319-320.

(7) F. Rosenthal, *Aḥmed b. aṭ-Ṭayyib as-Saraḫsī*, New Haven 1943, pp. 49-50.

(8) See, *e.g.*, D. Pingree, *The Thousands of Abū Ma 'shar*, London 1968, p. 10. n. 2.

(9) *Pseudo-Mağrīṭī. Das Ziel des Weisen*, ed. by H. Ritter, Leipzig 1933; translated into German as *«Picatrix» Das Ziel des Weisen von Pseudo-Mağrīṭī*, ed. by H. Ritter and M. Plessner, London 1962.

(10) *Picatrix. The Latin Version of the Ghāyat al-Ḥakīm*, ed. by D. Pingree, London 1986.

(11) See, *e.g.*, D. Pingree, *Some of the Sources...*, cit., pp. 13-14.

(12) See F. Sezgin, *Geschichte des arabischen Schrifttums*, vol. IV, Leiden 1971, pp. 98-99; and M. Ullmann, *Die Natur- und Geheimwissenschaften im Islam*, Leiden 1972, p. 364, which is very inadequate and incorrect. Most useful of anything published up till now is P. Kraus, *Jābir ibn Ḥayyān*, 2 vols., Le Caire 1943-1942, vol. II, pp. 104-105; see also his notice on Jābir's *Kitāb al-nawāmīs wa al-rudd 'alā Iflāṭun*, in vol. I, pp. 152-153.

(13) The description by L. Thorndike, *History of Magic and Experimental Science*, vol. II, New York 1923, pp. 777-782 and 809-810, is both inaccurate (*e.g.*, he thinks that Ibn al-Jazzār's *De proprietatibus* that precedes the *Liber aneguemis* in several manuscripts is a part of pseudo-Plato's work), and, while titillating, trivializing.

Latin transliteration of the Arabic *al-nawāmīs*. The Latin translation is also known as the *Liber institucionum activarum* and the *Liber vacce* or *Book of the Cow*, the title under which Pico della Mirandola referred to it; it is presumably called the *Book of the Cow* in commemoration of its first *experimentum*, which we shall soon describe.

The *Book of the Cow* is first cited in a Latin work by William of Auvergne, in the 1220's, and a copy was in the library assembled at Amiens by Richard of Fournival in the 1240's ([14]). Though alongside some quotations given by Jābir and others I am certain of only one small fragment of the Arabic original containing three suffumigations described near the beginning of the first of the two books of the *Kitāb al-nawāmīs* ([15]), there still exist twelve manuscripts of the Latin version, some complete and some containing just excerpts, upon which my edition will be based. These manuscripts were copied between the early thirteenth century and the late fifteenth century. The Latin was also rendered into Hebrew in the fourteenth century, and of this Hebrew translation one copy exists, in a Munich manuscript ([16]) that also contains a Hebrew version of the *Picatrix*.

The *Book of the Cow* itself says nothing of the philosophical and scientific theories that justify its bizarre prescriptions. I shall attempt to remedy that lack in this paper, at least in part; and to identify some of the Platonic, Aristotelian, and Ḥarrānian elements that contributed to the rather wild φαντασία of its unknown author. I begin by quoting the equally unidentified author of the *Ghāyat al ḥakīm*, who was laboring in Spain in the twelfth century with materials largely derived from Ḥarrān. He describes the two books of the *Kitāb al-nawāmīs* thus ([17]):

> As for (our) teacher (and) leader, Plato, I have seen two books by him. The name of one of them is the *Greater Book of the Laws*, and the name of the second is the *Smaller Book of the Laws*. In the greater book he speaks of effecting by means of images abominable things like walking on water; causing the appearance in whatever shape you may wish of the forms of composite animals which are not found in the world; causing rain to fall at a time unseasonable for its falling, or causing it to be obstructed at a time (proper) for its falling; making meteors, shooting stars, and lights appear in the sky, lightning-bolts to descend at unseasonable times, and the ships of one's enemies or whatever one wishes at a great distance to burn; walking in the air; causing the stars to rise at unseasonable times, and seeing them when they have fallen from their heavenly places to the center of the earth; conversing with the dead;

([14]) D. Pingree, *The Diffusion of Arabic Magical Texts in Western Europe*, in *La diffusione delle scienze islamiche nel medio evo europeo*, Roma 1987, p. 80.

([15]) This is found on *ff.* 104-105 of Paris BN arabe 2577, wherein it follows a copy of the Hermetic *Kitāb al-ustūwaṭās*. Both the Bodleian manuscript (Ouseley 95, *ff.* 150*v.* sqq.) to which Sezgin refers and the British Library manuscript (Or. 12070, *ff. 17-32v.*) to which Ullmann refers contain a different text, entitled *Jawāmī ʿ al-maqālāt al-thalāth li-Iflāṭun fī al-nawāmīs*. This is clearly identical with the *Kitāb al-nawāmīs li-Iflāṭun* found in Leiden MCCCCXXX (Golius 169). In the Bodleian manuscript it is said to have been translated by Ḥunayn ibn Isḥāq; this may, then, be the 'translation' of Plato's *Laws* by Ḥunayn to which the Arabic bibliographers refer. P. Kraus (*op. cit.*, vol. I, p. 182) refers to some excerpts on pp. 387 sqq. of a manuscript in the possession of Muḥammed Amīn al-Khāngī's bookstore in Cairo in 1937; I am not aware of the present location of this manuscript. A sentence from it is cited by P. Kraus, *op. cit.*, vol. II, p. 105 n.

([16]) Munich Hebrew 214, which contains the *Picatrix* on *ff.* 46-101 and Ibn al-Jazzār's *De proprietatibus* followed by pseudo-Plato's *Liber vacce* on *ff.* 102-104; see D. Pingree, *The Diffusion*..., cit., pp. 71-72.

([17]) *Ghāya*, in *Pseudo Mağrīṭī*, cit., p. 147, ed. by H. Ritter; see also *Picatrix*, cit., II xii 59, ed. by D. Pingree.

dividing the Sun and the Moon into separate parts; causing sticks and ropes to appear as serpents and snakes which devour whatever they come upon; and traversing great distances over the earth in as short a time as the twinkling of an eye. All of this (is accomplished) by means of the effects of images, the employment of spiritual powers, and the implanting of their powers in the motionless forms which consist of elemental substances so that they become moving spiritual (forms) (*instituciones active*) producing marvellous effects and actions with which you are not familiar.

According to the *Ghāya*, then, the method of Platonic magic is to implant spiritual powers — that is, souls — in material bodies so that they may move; for, according to the genuine *Laws* of Plato [18], soul «is the cause of change and of all motion for all things». By means of these newly ensouled forms or images the magician can effect what he will; the actual effect will depend upon the type of soul that is implanted, and upon the type of material into which it is implanted.

The *Ghāya's* list of the effects described in the *Kitab al-nawāmīs* — and they are all found therein — is quite representative of that text; they are either cosmic, transforming, at least in appearance, the heavenly bodies and the earth's atmosphere, or they are such miracles as occur in the *Old Testament* (Moses' turning a staff into a snake), in the *New Testament* (Jesus' walking on water), or in Classical Greek literature (Socrates' walk in the air in Aristophanes' *Clouds* and Archimedes' burning mirrors). But there are many other effects as well, such as that of creating artificially rational and irrational animals, an achievement that the author of the *Ghāya* mentions later on in the paragraph from which I have been quoting. In connection with these last operations for creating living beings he also cites the *Kitāb al-tajmī'* or *Book of Putting Together* ascribed to Jābir ibn Ḥayyān, and composed in Syria in the early tenth century [19].

This text, which has been brilliantly analyzed by Kraus [20], does indeed deal with the artificial generation of rational and irrational animals. In it man's τέχνη imitates the Demiurge's φύσις in bringing about a union of soul and body. The *Kitāb al-tajmī'* teaches how to apply this principle of imitation to the artificial generation of plants, animals, and men, each of which, of course, is characterized by one of Aristotle's three types of soul: plants by the nutritive (τὸ θρεπτικόν), animals by the sensitive (τὸ αἰσθητικόν), and man by the rational (τὸ διανοητικόν) [21]. Jābir names as his chief source for the doctrine and methods of artificial generation the *Kitāb al-tawlīd* (Περὶ γεννήσεως) of Porphyrius the Syrian; though no such work is known to have been written by Porphyry. Kraus [22] is able to demonstrate a strong Neoplatonic, including a Porphyrian, influence on Jābir's magic, even though Porphyry may never actually have claimed that it is possible for man to create rational animals by artificial means.

Jābir, on the other hand, justifies the possibility of creating irrational animals by

(18) Plato, *Laws*, 896, a5-b1.

(19) See P. Kraus, *op. cit.*, vol. I, pp. 95-97. As part of the collection of *Kutub al-mawāzīn* he (vol. I, p. LXV) would date it to the beginning of the tenth century. Excerpts from the *Kitāb al-tajmī'* were edited by P. Kraus on pp. 341-391 of his *Jābir ibn Ḥayyān. Textes choisis*, Paris-Le Caire 1935.

(20) P. Kraus, *op. cit.*, vol. II, pp. 97-134.

(21) Aristotle, *De anima*, 414a29-414b19.

(22) P. Kraus, *op. cit.*, vol. II, pp. 122-134.

citing the ancient theory of the spontaneous generation of insects, of small reptiles, and of rodents from the putrefaction of corpses. While presenting several other examples he states that (23):

> serpents, especially black serpents, are born of hairs placed in a glass vessel, the glass vessel being considered the mother and the hairs the father. [...] As for scorpions, they are born in basilic confined in a glass vessel.

The second *experimentum* is described in greater detail in the *Ghāya* (24):

> For making deadly green scorpions. One should fast for one day, and at the end of the day chew the leaves of wild basilic, bind what was chewed in a glass tube, seal its mouth, and hang it in a dark, damp room which sunlight never enters. After forty days green scorpions will be born in it which, if they sting a man, will kill him.

In these two examples the imagery is clear; the material inserted into the glass vessel or tube represents the semen of the male animal while the vessel itself is the womb in which the embryo gestates. Note that the semen is here as artificial as is the womb; this is one major difference between the normal Jābirian technique of artificial generation and that of the *Book of the Cow*, which employs real semen. Another difference is that, in the experiments described in the *Kitāb al-tajmī'*, the female's nutritive blood generally plays no role. This is not always the case in Jābir's book; shortly after the experiments that have just been quoted he speaks thus of spontaneous generation:

> Hornets are born from highly decomposed flesh, that is, of a corpse, and maggots from the flesh of a slaughtered [animal]. The reason for this is that the one [the slaughtered animal] loses its blood, and the other [the corpse] retains it.

If for Jābir hornets are spontaneously generated from corpses that have not been drained of their blood, the *Book of the Cow* prescribes an elaborate ritual for manufacturing artificial bees. This involves a slaughtered heifer whose blood is placed along with its bones, nose, eyes, ears, and genitals in an enclosed chamber; thus we have the nutritive matter in its womb. After it has rotted for seven days, a dog's penis is thrown on top. At the end of another seven days worms are born in this mess; they are to be fed with a fistful of dead bees on each of the succeeding fourteen days, after which the worms grow wings and turn into bees. The reason, I believe, that the author of the *Book of the Cow* limits the spontaneous generation to the worms, which then must be transformed into bees, is found in the *De anima* (25), where Aristotle claims that ants and bees possess the faculty of φαντασία, while worms do not.

The worms, therefore, will be more likely to be produced by spontaneous generation than will be the bees who possess a more complex soul. Even so, the *Book*

(23) *Kitāb al-tajmī'*, cit., p. 359.
(24) *Ghāya*, in *Pseudo-Mağrīṭī*, cit., p. 411, ed. by H. Ritter; see also *Picatrix*, cit., IV ix 19, ed. by D. Pingree.
(25) Aristotle, *De anima*, 428a10-11.

of the Cow takes no chances, and induces the spontaneous generation by uniting symbolically semen and blood in a womb. But what is truly remarkable in the *Book of the Cow* is the affirmation that the process can be reversed so as to generate heifers out of bees!

Jābir, on the other hand, when he speaks of creating artificial animals of higher orders than those produced by spontaneous generation ([26]), relies on the insertion of a clay image of the creature to be created, whether animal, human, or chimera, into a sphere or a nesting of spheres, preferably made of glass. In these operations the sphere or spheres on one level represent the heavens, the macrocosm whose astral influences, under the guidance of the Demiurge, endow with life, ensoul, the clay image, which becomes an irrational or rational animal; but, on another level, the innermost sphere is the womb in which the enlivened embryo is nourished. This is the pure version of the Jābirian method, which he claims to have found in Porphyry's *Kitāb al-tawlīd* and in Zosimus' *Kitāb al-mīzān*. But at the end of this passage Jābir quotes the opinions of several other schools who infuse the hollow interior of the clay image with either the semen or the blood of the animal one wishes to produce, or with a mixture of bloods to create composite monstrosities. The idea of employing semen and blood in such operations is shared by the author of the *Book of the Cow*.

Toward the end of the *Kitāb al-tajmī'* Jābir cites «the book in which we refuted Plato in his book whose title is *al-nawāmīs*». The context in which Jābir refers to his refutation of the *Kitāb al-nawāmīs* is a discussion, unfortunately too corrupt in the unique manuscript to be understood completely, concerning the large figure of a snake whose operation, he says, is invalid ([27]). The artificial generation of a snake is referred to in the *Kitāb al-nawāmīs* at the end of a chapter on creating a rational animal. This rational animal is minced, sprinkled with the ash of an oak, and buried in the ground for eighteen days. By the end of that period it will have been transformed into a serpent. Here the artificial rational animal, whose principle ingredients after human semen are the blood and the gall-bladder of a black cat, seems to be a humanoid body entered into by a demonic spirit. It is altogether symbolically fitting, then, that such a one be transformed into a poisonous snake. The generation of demonic creatures, indeed, seems to be one of the principle criticisms leveled by Jābir against the *Kitāb al-nawāmīs*. In his *Kitāb al-khawāṣṣ* ([28]) Jābir states that he associates the *Kitāb al-nawāmīs* with the feigned knowledge of the people of the religion of blasphemy, the *kafirs*, who worship devils.

Clearly, since the *Kitāb al-tajmī'* was composed in the early tenth century, Jābir must have written his refutation of Plato's *Kitāb al-nawāmīs* in about 900 A.D. This, then, is the *terminus ante quem* for the date of the *Kitāb al-nawāmīs* itself. A rough *terminus post quem* is provided by the fiction of its preface that the *Kitāb al-nawāmīs* was composed by Galen as an explanation of Plato's Περὶ νόμων, and that this work

([26]) *Kitāb al-tajmī'*, cit., pp. 344-350; see P. Kraus, *op. cit.*, vol. II, pp. 110-116.
([27]) *Kitāb al-tajmī'*, cit., pp. 366-367.
([28]) P. Kraus, *op. cit.*, vol. I, p. 152 n. Jābir also refers to his own *Kitāb al-nawāmīs wa al-rudd 'alā Iflāṭun*, in the *Kitāb al-sumum* (A. Siggel, *Das Buch der Gifte des Ğābir ibn Ḥayyān*, Wiesbaden 1958, p. 142), wherein he reports that a beneficial plant, when mixed with certain ingredients, is transformed into a deadly poison.

of Galen was translated from Greek into Arabic by Ḥunayn ibn Isḥāq. From Ḥunayn's pen indeed we have a summary in three *maqālas* of Plato's *Laws* ([29]). Since Ḥunayn died in Baghdād in 873, and since it is less likely that his name was used by the forger of the magical *Kitāb al-nawāmīs* before his death than after it, it is probable that the forger did his work after 875, perhaps in the 880's. However, a more definitive *terminus post quem* is established by Ḥunayn's *Risāla* to 'Alī ibn Yaḥyā on the books of Galen that he had translated ([30]). Ḥunayn states at the end of the *Risāla* that he composed it in the year 1167 of Alexander, or 856-857 A.D., when he was 48 years old, but that he added to the bibliography in the year 1175 of Alexander, or 864-865. In the *Risāla* Ḥunayn describes as number 46 — i.e., in the section on translations made before 856 — his translation into Syriac of Galen's work *Fī ārā' Buqrāṭ wa Filāṭun* ([31]), which is the Περὶ τῶν Ἱπποκράτους καὶ Πλάτωνος δογμάτων ([32]). The forger of the *Kitāb al-nawāmīs* makes his fictitious Ḥunayn declare: «By my life, it is fitting for all men to praise Galen because he wore out his soul and his body, made his heart to be weary and his eye to remain awake in expounding the books of Plato and of Hippocrates». This clear reference to a book of Galen that the real Ḥunayn had translated before 856 — probably in the 840's — provides convincing proof that the *Kitāb al-nawāmīs* was forged after, say, 850. That the translation was made into Syriac confirms that the forger was a Syrian.

What, however, do either Plato's *Laws* or Galen's Περὶ δογμάτων have to do with the magic of the *Kitāb al-nawāmīs* and its use of spiritual powers? The clue is again found in the preface, where the forger has his fictitious Galen proclaim: «*exposui librum philosophi Platonis decem tractatuum usque ad finem eorum*». Since the *Laws* has twelve books rather than ten, this is clearly a mistranslation of the Arabic. I believe that the Arabic original said: «I explained the book of the philosopher Plato, its tenth treatise, at its end». For towards the end of book X of the *Laws* Plato is refuting those who say that all things past, present, and future occur by nature (φύσις), by art (τέχνη), and by chance (τύχη) ([33]). In turning first to the theory that φύσις, when identified with the four elements, is prior to and, indeed, the origin of soul (ψυχή) ([34]), the Athenian stranger is led to prove that soul is prior to body; and he does this by showing that soul, whose essence is motion (κίνησις), is the unique cause of generation (γένεσις), of motion (μετάβασις), and of change (μεταχόσμησις) in all things ([35]). He concludes, then, not only that soul is prior and superior to body ([36]), but also that it is the cause of «the evil, the good, and the shameful, the just and the unjust, and of all opposites». ([37]) Finally, he argues that the

([29]) See n. 15 above.

([30]) G. Bergsträsser, *Ḥunain ibn Isḥāq. Über die syrischen und arabischen Galen-Übersetzungen*, Leipzig 1925; and *Idem, Neue Materialien zu Ḥunain ibn Isḥāq's Galen-Bibliographie*, Leipzig 1932.

([31]) G. Bergsträsser, *Ḥunain*..., cit., pp. 26-27 of the Arabic, pp. 21-22 of the German translation. Ḥunayn records that an Arabic translation of this work was made by Ḥubaysh for Muḥammad ibn Mūsā.

([32]) C. G. Kuhn (ed.), *Claudii Galeni Opera omnia*, vol. V, Leipzig 1825, pp. 181-805.

([33]) Plato, *Laws*, 888e 4-6.

([34]) Plato, *Laws*, 891c 1-5.

([35]) Plato, *Laws*, 892a2-b1; cf. n. 18 above.

([36]) Plato, *Laws*, 896b10-c3.

([37]) Plato, *Laws*, 896d5-8.

Sun, the Moon, the stars, and heaven are moved by souls; and that, because these souls are good, these celestial beings are gods (38).

This analysis and exaltation of the ψυχή at the end of the tenth book of the *Laws* is indeed the theoretical foundation for much of the magic in the *Book of the Cow*; though it is never directly stated thus in that text, the ingredients in its recipes normally include one or several parts of bodies that were believed to be the *loci* of parts or functions of the animal's soul. The inclusion of one of these organs induces that part or function of a soul to effect an artificial animal's γένεσις or to affect the soul of an already existing creature by causing its μεταχόσμησις; in this latter case one can alter the soul of a plant, animal, man, or celestial body, or else one can cause the souls of men to perceive or fantasize falsely. The claimed power to move the souls of the celestial bodies places this psychic magic on an equal level with astral magic, which persuades the planetary and stellar souls to send their spirits to reside in talismans.

It is Galen's Περὶ τῶν Ἱπποκράτους καὶ Πλάτωνος δογμάτων, and especially chapter 1 of book 3, that tells one in which part of an animal's body which part or function of its soul resides. Of course Aristotle had already stated (39) that the nutritive and the sensitive souls are inherent in semen, while the νοῦς enters it from the outside; it is presumably this feature of semen that justifies its frequent use in creating artificial animals. But Galen informs us that Chrysippus located τὸ ἡγεμονικόν in the heart (40), while others place it in the chest or in the head; and that Plato in the *Timaeus* (41) locates τὸ λογιστικόν in the head, τὸ θυμοειδές in the chest, and τὸ θυμητικόν in the navel (42). In chapter 4 of book 3 Galen discusses the relation of theories of the soul to the three common adjectives «brainless», «heartless», and «gutless». In terms of Plato's theory of the tripartite soul he interprets them to signify respectively «thoughtless», «cowardly» and «lacking sensation» (43). These indications suffice to complete the *Book of the Cow's* symbolic vocabulary. The brain represents the rational soul (and, presumably, the φαντασία of irrational animals), the heart the sensitive soul (which also controls appetite and locomotion), and the guts — especially the gall-bladder — the nutritive soul. Moreover, the *Book of the Cow* assumes that one animal's organ of sense or of motion will transfer its δύναμις to another animal when properly applied.

Now that we understand its language we can begin to decipher at least the Platonic — or Neoplatonic — ingredients in the psychic magic of the *Kitāb al-nawāmīs*; many other elements, of course, are present as well. I should like to begin by summarizing with interspersed annotations the first *experimentum* in the *Book of the Cow*, the one to which it owes its name. I remind you that my summary must be based on the Latin translation in the absence of an Arabic manuscript.

«Whoever wishes to make a rational animal should take his own water (*aquam*

(38) Plato, *Laws*, 899b2-10.
(39) Aristotle, *De generatione animalium*, 733b30-734a1, and 736b8-29.
(40) *Claudii Galeni Opera omnia*, cit., vol. V, p. 287.
(41) Plato, *Timaeus*, 69a6-72d3.
(42) *Claudii Galeni Opera omnia*, cit., vol. V, p. 288.
(43) *Ibidem*, p. 316.

suam) while it is (still) warm». It seems to me clear that *aqua* here translates the Arabic *mā'*, «juice», which also signifies semen. This semen must be used while it is still warm since, according to Aristotle (44), it retains its πνεῦμα only while it is in that state. «One should mix it (that is, the warm semen) with an equal amount of the stone which is called the stone of the Sun». Note that in astrology the Sun represents the father.

> Then one should look out for a ewe or a cow, whichever of these one chooses, and insert [the mixture] in her womb and plug up her vagina with the Sunstone. The genitals of the ewe or of the cow are then smeared with the blood of the other.

This blood is to provide nourishment to the embryo (45), and represents, in the Aristotelian view (46), the body contributed by the female as the semen represents the soul contributed by the male. «She is locked in a house which sunlight does not enter». This serves as a second womb in addition to her own.

> While awaiting the birth of the animal, [the magician] mixes a powder made of crushed Sunstone, sulfur, magnet, and green zinc stirred in with the sap [*aqua* again] of a white willow. The undefined *forma* to which the ewe or cow gives birth is placed in this powder, whereupon it is instantly clothed in human flesh.

The powder presumably somehow symbolizes the clay out of which God fashioned Adam.

> The new-born homonoid is kept in a large glass or lead vessel for three days, so that it becomes exceedingly hungry; it is then fed for seven days on its mother's blood, and develops into a complete animal, with which other marvels can be performed. For instance, it can be placed on a white cloth with a mirror in its hands and suffumigated with the previously mentioned powder mixed whit human blood; then the Moon will appear to be full on the last day of the month. Or it can be decapitated, and its blood collected; if the blood is given to a man to drink, he will assume the form of a bovine or a sheep; but if he is anointed with it, he will have the form of an ape. Finally, if the animal is fed for forty days on a diet of blood, milk, and semen, and then its guts are extracted from its belly and rubbed on someone's hands and feet, he may walk on water or traverse the diameters of the earth in the winking of an eye.

The rational animal that the magician has thus artificially generated is, I believe, a demon. It is for this reason that one can so readily kill it; it is for this reason, it seems, that Jābir condemned the *Kitāb al-nawāmīs*; and it is for this reason that the animal, if kept alive for a year, can tell one of all things beyond one's perceptions.

The next chapter in the *Book of the Cow* describes the artificial generation of another rational animal by essentially the same means, except that this time the womb is provided by a female ape. Her offspring, after undergoing appropriate manipulations, appears in the form of a one-legged man. It seems also to be a demon;

(44) Aristotle, *De generatione animalium*, 735b32-37.
(45) Aristotle, *De generatione animalium*, 740a21.
(46) Aristotle, *De generatione animalium*, 738b25-26.

for if one plucks out its eyes and rubs one's own with them, thus transmitting their δυνάμεις directly, one will see all the spirits and demons. If one goes further and mixes the creature's brain in with its eyes before applying them to one's own, one will see spiritualities. And if one minces its tongue, soaks the mincemeat in urine, and imbibes the mess, one will hear the spiritualities and converse with the demons. Since the powers of the organs of sense of this artificial animal enable the magician to see, to hear, and to talk to demons, the original possessor of those organs must himself have been a demon.

Dozens of other *experimenta* are described in the *Kitāb al-nawāmīs* that illustrate the correctness of our method of interpreting at least one element of psychic magic. I can now mention only a few. To cause a person to lose his reason and fall down in an epileptic fit, one makes him enter a house in which there is an image of Satan wearing a crown made out of the gall-bladder of a fish; the ἐπιθυμητικόν part of the fish's soul located in its gall-bladder overpowers, through the mediation of Satan, the λογικόν part of the man's soul. To endow a tree with the imagination to bow down before him the magician suffumigates it with the brain of a vulture mixed with the bones of a black snake and of a man; the black snake I take to be the demonic element, while the brain of the vulture bears its φαντασία. To comprehend the chirping of birds the magician drinks a concoction of the gall-bladder of a black crow combined with its brain, guts, and palate. Or to see the demons or the Devil himself the *magus* bathes in an alcohol distilled from the smoke of a suffumigation consisting of the gall-bladder of a crane, the eye of a hoopoe, the eyes of three swallows, the eye of a crow, the heart of a black chicken, and the eye and gall-bladder of an untamed black cat.

I believe, then, that we have uncovered the hidden meaning of at least a part of the strange operations described in the *Book of the Cow*, and demonstrated their dependence on theories concerning the soul and the generation of animals found in Plato and Aristotle. We have also dated the forging of this *Kitāb al-nawāmīs* in the latter half of the ninth century A.D., most probably in the 880's, and tentatively located the forger in Ḥarrān. I would like now to strengthen this last conclusion by offering you a rendition, again from the Latin, of a remarkable passage in the *Book of the Cow* wherein its author describes without naming them the Ṣābian *magi* of Ḥarrān and their magical deeds, in terms that recall the magician of the *Picatrix* and, by extension, his imitators in the Renaissance.

> Ḥunayn says: From the knowledge of the operators of the *Laws* (47) and the benevolence of their opinions as well as the goodness of the understanding of those who have considered things, have paid attention to the operations, and have become familiar with the roots from which these operations originate and in what knowledge these preparations are prepared, by which is known their superiority over the others who lived in their times, I have rarely seen that they could accomplish this unless they worshipped the stars, constructed altars [*i.e.*, temples], and placed all of their doors to the East so that they might face the great light which is the light of the Sun,

(47) The *opifices aneguemis* are perhaps the *asḥāb al nawāmīs* who are the highest form of rational beings that can be generated artificially according to Jābir in *Kitāb al-tajmī'*, cit., p. 369. An alternative explanation of this phrase is found in P. Kraus, *op. cit.*, vol. II, pp. 133-134.

from which comes the light of the day and its splendor, and [unless] they suffumigated in these [temples] with suffumigations of great exaltation, brought nigh in them the sacrificial victims, lit candles and lamps in them, built oratories in them, and illuminated in them the stars which are their deity, whom they address with the address of one worshipped, praying for them to fulfill their needs. And they question them about one who is absent, about things already done and things to be done before they are done; and when they happen, they make them tell about each thing separately; and they ask about life, death, health, truthfulness, food, and drink, and about all things over the knowledge of which stands none but their Creator. Therefore [the stars] caused [men] to know by means of that which they possess of the power by which they have come to address the stars, until anyone of them has come to the point that he may be in Rome and perambulate Africa in the same day, or that he may be in Scythia in the morning and walk about the vast occident in the evening. For the earth is rolled up for them, and they are carried about in the air and walk on water. They know before its birth whether the fetus is male or female and whether it will live. But with all this they disagree in their intentions, their laws, and their sciences, so that some of them worship the Sun, others the Moon, and some of them the stars one by one, and they worship both fire and darkness. Those who worship fire said that it is the eye of the Sun, but it is not as they thought it is. Because there is no day except from the Sun, and we do not discover the splendor of creation without it, they said: «This is our deity». They lit candles and lamps to it, and this sufficed that they became endowed with the genius to build buildings that do not decay, and to make candles which do not burn out over the extension of time and in the passage of time, but they [the fires] remain after them even though according to their own science corruption and change are necessary for them. Because they do receive corruption and change [eventually], when [these things] are corrupted they do not change their deity because of that. Therefore they were inspired to make [their gods] those seven [luminaries] which do not burn out, which are the candles which do not cease but remain for all time.

I remember the torches which are not extinguished in water, and the suffumigations with which, when there is a suffumigation and a query about some entity, they look in [the smoke] and they hear the answer. For when they are properly placed in the temple designated for the stars, then the nature of benefit descends into them. And they do nothing by which there is not some benefit. For whenever someone tries this and asks about an entity, small or large, he hears an answer as if the god were standing on his altar holding the suffumigation in his hands. That [*magus*] addresses his star with the address of the stars and their prayers. There is a single book on these prayers of which my recollection is not gone. But, since this is something which needs an abbreviation in order to find its end — its language is prolix, and so is its explanation — I refrain from reporting about it here.

The monobiblos on the prayers to the planets may be that composed by Abū al-ʿAbbās (Abū Ḥafṣ?) al-Ṭabarī which is, after conflation with other texts, incorporated into the *Ghāya*. In its Latin form in the *Picatrix* this single book of al-Ṭabarī inspired some of the most astonishing passages in the third book of Ficino's *De vita*. It is not necessary to follow Tardieu's theory ([48]) that Simplicius retired to Ḥarrān upon his return from Persia to discern the strong influence of Neoplatonism in that fascinating city ([49]). The *Book of the Cow* increases that evidence and enlarges

([48]) M. Tardieu, *Ṣābiens coraniques et 'Ṣābiens' de Ḥarrān*, in «Journal Asiatique», CCLXXIV (1986), pp. 1-44, esp. pp. 22-29.
([49]) To cite just one recent study, I refer to C. Genequand, *Platonism and Hermeticism in al-Kindī's Fī al-nafs*, in «Zeitschrift für Geschichte der Arabisch-Islamischen Wissenschaften», 1987-1988, pp. 1-18.

magically our conceptions of what Neoplatonism may generate — or into what it may degenerate.

But the fact remains that, whatever its origin, the *Book of the Cow*, in its Latin form, seems bereft of intellectual content and, as Pico says, filled with execrable dreams. Neither Neoplatonic nor other philosophers of the Renaissance, then, would be likely to quote from it, though Pico informs us that it was circulating among magicians. Lodovico Lazarelli, indeed, had a dream of creating demons, but he intended to accomplish this by words alone rather than by the elaborate corporeal and essentially disgusting rituals of the *Book of the Cow* ([50]).

In order to assess the influence of pseudo-Plato in the West, then, among intellectuals we must examine the few stray references to it, but primarily the date and distribution of its manuscript copies.

The twelve manuscripts were all copied between about 1200 and 1500, and almost all of them in France or in Italy. Two are datable to the thirteenth century ([51]); and the *Liber vacce* was read by William of Auvergne in Paris in the 1220's while a copy was in Richard of Fournival's library at Amiens in the 1240's. Much of the second book was incorporated into the *De mirabilibus mundi* falsely ascribed to Albertus Magnus. Three manuscripts were copied within a decade or two of the year 1300 ([52]); one at Bologna and a second at Montpellier. In the early fourteenth century two copies made their way from France to St. Augustine's Abbey in Canterbury ([53]). Toward the middle of the fourteenth century another copy was made in Italy ([54]), while toward that century's end, in the 1370's, the *Liber vacce* was discussed and censured by Henry of Hesse in his *De habitudine causarum* ([55]) and by Nicolas Oresme in his *De configurationibus qualitatum et motuum* ([56]). Finally, in the fifteenth century six copies were made, mostly in Italy ([57]); this was clearly the height of the work's popularity, which almost entirely vanished after 1500. No longer was it to be copied, though there were two manuscript of it in John Dee's library toward the middle and end of the sixteenth century, and one in Thomas Allen's hands at the beginning of the seventeenth. The *Book of the Cow*, then, was most widely circulated in the early Renaissance, but soon disappeared from view, except in England.

In contrast, its companion piece, the *Picatrix*, whose sources, including the *Book of the Cow*, also lay among the Neoplatonists of Ḥarrān, but which expressed the Neoplatonic and pseudo-Hermetic alleged foundations of its magic clearly and repeatedly, first began to be popular in the 1450's, in Italy, and from there was

([50]) See D. P. Walker, *Spiritual and Demonic Magic from Ficino to Campanella*, London 1958, pp. 64-72.

([51]) Munich, CLM 22292; and Prag X. H. 20.

([52]) Munich, CLM 615; Codex Fritz Paneth (now in the Yale Medical Library); and Oxford CCC 125.

([53]) M. R. James, *The Ancient Libraries of Canterbury and Dover*, Cambridge 1903, p. 348 (n. 1275), and pp. 348-349 (n. 1277, which is now CCC 125).

([54]) BL Arundel 342.

([55]) L. Thorndike, *op. cit.*, vol. III, New York 1934, pp. 489-490.

([56]) M. Clagett, *Nicole Oresme and the Medieval Geometry of Qualities and Motions*, Madison 1968, p. 358.

([57]) Montpellier 277; Firenze BN, II iii 214 and Pal. 945; Oxford CCC 132 and Digby 71; and Vatican Pal. lat. 1892.

rapidly disseminated to Poland, Bohemia, and France, and finally, in the late sixteenth century, to England. Of the twenty complete manuscript copies of the *Picatrix* none was made before 1450, four are dated in the latter half of the fifteenth century, eight in the sixteenth century, seven in the seventeenth, and one in the eighteenth. And, of course, it was extensively used by Neoplatonizing philosophers and magicians from Ficino on, and was translated into Italian, French, and English. Yet the magic of the *Picatrix* is oftentimes equally nauseous as that of the *Liber vacce*; but in the Neoplatonic climate of the later Renaissance the *Picatrix's* claim to a philosophical basis, and one that fit in neatly with the concept of a hierarchical universe emanating from an omnipotent deity, served it much better than the *Liber vacce's* pretence of sharing Platonic and Aristotelian theories of the soul and of generation, while in fact relying on demons and even Satan himself. Even the most magically inclined of Renaissance Neoplatonists, Henry Agrippa, while he at one point nonchalantly cites Plato — by whom he means the author of the *Book of the Cow* — as an authority on the manufacture of magical lamps ([58]), when he comes to deal with the problem of the artificial union of soul to body utilizes, as does Ficino, the celestial magic of the *Picatrix*, but remarks that the celestial power lies «quiescent within material things», and

> thus certain marvelous works are produced [*procreantur*] such as are read of in the *Liber Nemith* [*i.e.*, in the *Liber Neumich*], which is also called the book of the *Laws* of Pluto, because generations of this sort are monstrous and not according to the laws of nature ([59]).

But Agrippa's punning substitution of Pluto for Plato cannot conceal the fact indicated by his use of the indicative form *procreantur*, that *he* at least thought that the magic of the *Book of the Cow* would actually work, *Satana volente*.

([58]) *Henrici Cornelii Agrippae ab Nettesheym... De occulta philosophia*, [NP] 1533, p. LVIII (I, 49).
([59]) *Ibidem*, p. XLII, (I, 36).

David Pingree

LEARNED MAGIC IN THE TIME OF FREDERICK II

When one thinks of magic in the period of Frederick II, one's thoughts naturally turn to his *astrologus*, Michael Scot[1]. For Dante, some seventy-five or eighty years after Michael's death, had Virgil point out in the fourth *bolgia* of the eighth circle of Hell[2]:

> Quell' altro che ne' fianchi è così poco,
> Michele Scotto fu, che veramente
> Delle magiche frode seppe il gioco.

But neither Henry of Avranches[3], Michael's contemporary, nor Salimbene[4], the chronicler of Frederick's reign, mention this proclivity for magic. Yet a reading of the manuscripts of Michael's *Liber introductorius*[5] and *Liber particularis*[6] quickly reveal that, while often

1. Concerning Michael see L. Thorndike, *Michael Scot*, London 1965, though the older account by C. H. Haskins, *Studies in the History of Mediaeval Science*, 2nd ed., New York 1927, 272-98, is still most valuable. A more recent survey is by L. Minio-Paluello in *Dictionary of Scientific Biography*, IX, New York 1974, 361-65.

2. Inferno XX 115-17. Concerning his slender thighs see, for example, Polemon, *De physiognomonia*, 9 (ed. G. Hoffmann in R. Foerster, *Scriptores physiognomonici*, I, Lipsiae 1893, 206): Tenuitas et paucitas carnis lumborum fraudem dolum et improbitatem indicat.

3. Haskins, *Studies*, 276.

4. *Cronica*, ed. G. Scalia, 2 vols., Bari 1966, I, 515 and 525-26, and II, 749 and 774.

5. I have used Oxford, Bodleian Library, Bodley 266 (henceforth denoted B) and München, Staatsbibliothek Clm 10268 (henceforth denoted M). M is described most recently by U. Bauer, *Der Liber Introductorius des Michael Scotus in der Abschrift Clm 10268 der Bayerischen Staatsbibliothek München*, München 1983. For the other manuscripts of the *Liber introductorius* see Thorndike, *Michael Scot*, 5-6; see also the unpublished dissertation of G. M. Edwards, *The Liber introductorius of Michael Scot*, University of Southern California 1978.

6. I have used Oxford, Bodleian Library, Canonici Misc. 555 (henceforth denoted C). This and other manuscripts of the *Liber particularis* are briefly described in Thorndike, *Michael Scot*, 6 and 122-23. In reading through the *Phisionomia* (published on pp. 219-358 of *Albertus Magnus De secretis mulierum*, Amstelodami 1648).

stating that the use of magic is against the teachings of the Roman church[1], Michael firmly believed in the efficacy of magic and was familiar, in embarrassingly intimate detail, with the modes of operating magically. In this paper I intend to relate his magic to what was known by his contemporary, William of Auvergne[2], and by the somewhat later Albertus Magnus[3], both of whom had access, in the region of Paris, to the new magic of Arabic origin that was coming into France and England from Spain[4] – traditionally from Toledo[5]; and then to describe in some detail a fifteenth-century manuscript now in the Biblioteca Nazionale in Firenze – II. iii. 214[6] – that turns out to be one of our most complete extant copies of these new magical texts, texts which are the foundation also for most of what Michael knew about learned magic.

Let us begin with William of Auvergne, who, in his *De universo*, divides magic into three types[7]: sleight of hand, optical illusions, and the employment of demons. The first two are obviously not unnatural, though they may be fraudulent. Nor are the marvelous powers inherent in certain stones, plants, or animals unnatural; they are simply virtues that we are incapable of understanding. Indeed, one of the innate powers possessed by some natural objects is that of hindering magicians or putting evil spirits to flight. Those evil spirits or demons are real for William and for his contemporaries, and they are permitted by God to deceive man and to entice him into error and sin. The *magus*

I note only two references to magic; on p. 251 Michael refers to magical uses of menstrual blood, sperm, hair from the head, blood, and tracks of feet in dust or mud (see the magical uses of parts of the human body, different from those in Jābir's *Flos naturarum*, that Michael lists in the *Liber particularis*, C ff. 51vb-52a), and on p. 270 he notes the disastrous effect of a «sign of Salomon» on infants and fetuses.

1. So, e.g., in the *Liber introductorius* condemnatory passages occur on B ff. 22b-22va, and in the *Liber particularis* on C f. Iva. Many more passages could be cited.

2. A general description of William's references to magic is available in L. Thorndike, *A History of Magic and Experimental Science*, II, New York 1923, 338-71.

3. I refer to chapter XI of his *Speculum astronomiae*, ed. S. Caroti, M. Pereira, and S. Zamponi, Pisa 1977, 27-33.

4. See D. Pingree, «The Diffusion of Arabic Magical Texts in Western Europe», in *La diffusione delle scienze islamiche nel Medio Evo europeo*, Roma 1987, 58-102.

5. See, for instance, V. Rose, «Ptolemaeus und die Schule von Toledo», *Hermes*, 8 (1874), 327-49, esp. 343-44.

6. This manuscript is very inadequately described in Mazzatinti, *Inventari*, X, Forlì 1900, 7-8.

7. *Guilielmi Alverni Opera omnia*, 2 vols., Paris 1674, repr. Frankfurt am Main 1963, I, 1059-61.

who falsely believes that he can obtain control of these demons by means of his invocations, suffumigations, prayers, and sacrifices is led not only into the error of heresy but into self-delusion as well. The demons do evil out of their own malignancy, the necromancer who attempts to direct them becomes their instrument.

This may appear to be the attitude towards magic that one ought to expect from a Bishop of Paris. What William brings in to the discussion are references, both direct and oblique, to a number of Latin translations of Arabic works on black magic. Thus he refers to the *Idea* and *Almandal* of Salomon[1]; to the seven books of the seven planets written by Mercurius – that is, by Hermes[2]; to the *De stationibus ad cultum Veneris* of Toz the Greek[3] a name variously distorted in the printed edition of William's works; to Artefius on *characteres* or magical writing[4] and to Plato's *Liber Neumich*[5] or *Book of the Cow*. We shall have more to say of most of these works later; for now it must suffice to report that William is the first Westerner to record the names of these works, save for that of Toz. For already, in 1143, Hermann of Carinthia, writing in Béziers on the border of France and Spain, cited the *thelesmatici* – makers or operators of talismans – Iorma the Babylonian and Tuz the Ionian as employing the planetary spirits in their magical arts[6] and Daniel of Morley, reporting in the last quarter of the twelfth century to John, Bishop of Norwich, concerning his studies in Toledo, speaks approvingly of «the science of images, which the great and universal *Book of Venus*, edited by Thoz the Greek, has handed down»[7]. Indeed, I suspect that all of the translations accessible to William, and others that I shall discuss later, were made in Spain toward the middle and end of the twelfth century.

A more complete description of these translations was included by Albertus Magnus in chapter eleven of his *Speculum astronomiae*, for which he probably gathered material while he was in Paris in the

1. See Thorndike, *History*, II, 280 and 351.
2. *Opera,* I, 881 and 953. For these and the following references see Pingree, «The Diffusion», 79-80.
3. *Opera,* I, 44 and 671.
4. *Opera,* I, 91.
5. *Opera,* I, 43 and 70.
6. *Hermann of Carinthia. De essentiis,* ed. C. Burnett, Leiden 1982, 182.
7. K. Sudhoff, «Daniels von Morley Liber de naturis inferiorum et superiorum», *Archiv für die Geschichte der Naturwissenschaften und der Technik,* 8 (1917), 1-40, esp. 34.

1240's[1]. Here, he distinguishes three varieties of magic, though they are quite different from William's[2]. The first is the abominable, which employs suffumigations and invocations; the demons, however, are not compelled by these ritual bindings, but God permits them to deceive the magicians. The examples Albert cites are Hermetic. The second is less reprehensible, but still detestable; it employs characters, names, and exorcisms; it is to be avoided because beneath the cover of names in foreign languages may lurk what is contrary to the Catholic faith. For this type of magic Albert refers to Salomonic texts or other works associated with figures from the Old Testament. Finally, the third type, which does not allow suffumigations and invocations, nor admit exorcisms and characters, relies on power drawn from the celestial spheres. Albert's exemplary texts – Thābit ibn Qurra's *Liber ymaginum* and Ptolemy's *Opus ymaginum* – both claim to depend on ancient Greek authorities – Thābit's on Aristotle.

Albert's classification of the Latin translations of Arabic magical works is sensible and useful; I shall follow it here. The abominable or Hermetic treatises are derivatives from the astral magic invented by the Sabaeans of Harran in the ninth and tenth centuries[3]. They are based on the construction of talismans by *magi* fitted to the task by their horoscopes, construction at astrologically suitable moments and out of the materials appropriate to the planet whose spirit is to be persuaded to empower it. This empowerment is accomplished by an elaborate ritual, in part Indian in origin, involving the use of special garments, flowers or branches, suffumigations, prayers to the planets, and animal sacrifices, carried out at times determined astrologically to be appropriate for the planetary rays to penetrate the talisman, which bears a shape, images, and characters consonant with the use to which it is to be put. Once so empowered, the talisman remains potent to accomplish its intended objective for an indefinite period of time, though that potency is released only by its being employed in

1. See Pingree, «The Diffusion», 81-87.

2. For a more rational classification of magical practices, see D. Pingree, «Some of the Sources of the *Ghāyat al-ḥakīm*», *Journal of the Warburg and Courtauld Instituts*, 43 (1980), 1-15.

3. See Pingree, «Some of the Sources»; C. S. F. Burnett, «Arabic, Greek, and Latin Works on Astrological Magic Attributed to Aristotle», in *Pseudo-Aristotle in the Middle Ages*, London 1986, 84-96; and D. Pingree, «Indian Planetary Images and the Tradition of Astral Magic», *Journal of the Warburg and Courtauld Instituts*, 52 (1989), 1-13.

the proper way. Detestable or Salomonic magic, on the other hand, makes use of secret or angelic names, rings with engraved stones, exorcisms with magical circles and squares performed by elaborately robed and equipped priests, burning candles, and the like, all designed to compel the angels or the demons to appear and to obey. Though there are many similarities and coincidences between the abominable and the detestable, especially in their shared appeal to planetary powers and use of astrology, they are sharply distinguished in the following way. For the Hermeticists the planetary and astral spirits are the intermediate powers between men and God; in fact, they are God's creatures set by him in the heavenly spheres to supervise sub-lunar events. They are not compelled, but persuaded, to empower the talismans with the power they have received from God. The Salomonists, on the other hand, taking their clue from such Christianized Jewish traditions as the *Testament of Solomon*, seek to bind the demons, the malefic spirits, or the angels to do their will, to compel them by ritual acts and threats of violence to carry out the necromancer's wishes.

The followers of the *Books of Images* also rely on stellar powers transferred to talismans bearing images. But the empowerment of these talismans depends only on their being made of the appropriate planetary metal or other substance at an astrologically propitious time and on the saying of a brief statement of purpose over them. Again, they are related to the Hermetic talismans of the Sabaeans, but the objectionable rituals have been almost entirely eliminated. Still, it is difficult to understand how Albert could have looked upon them with such tolerance as he exhibits.

Let us begin now our investigation of the manuscript, Florence BN 11. iii. 214, and the relationships of the treatises it preserves to those known to William of Auvergne, to Albertus Magnus, and to Michael Scot, and to our three types of magic. It will soon be clear, I believe, that the hypothetical thirteenth-century ancestor of the manuscript contained almost all of the magical texts known to our three authors, and much more besides.

It begins with Albert's two books on images[1], that by Thābit in the

1. F ff. 1-4. The two versions of this text, I and J, are incompetently edited by F. J. Carmody, *The Astronomical Works of Thabit b. Qurra*, Berkeley-Los Angeles 1960, 165-97. See Pingree, «The Diffusion», 74-75.

translation by John of Seville and that ascribed to Ptolemy[1], in the same order in which Albert cites them[2]. I have shown elsewhere that Albert probably saw them in this order in the manuscript from Richard of Fournival's library – Paris, BN lat. 16,204 – from which he drew most of his knowledge of Latin translations of Arabic astrological works[3]. Of the two versions of John's translation of Thabit's work the Florence manuscript contains J, while the Paris manuscript has the shorter version I. Michael Scot, in his *Liber introductorius*, quotes the first sentence from version J[4]; he also refers to Thabit in his *Liber particularis*[5]. In Thabit's *Liber ymaginum* are described seven talismans, each designed to accomplish a particular task: expelling scorpions, destroying an enemy's city, obtaining possession of something, increasing business, being put in charge of a city or region, gaining the friendship of the king, and creating friendship or enmity between two persons. This is a rather limited range of objectives that is served by Thabit's images, and most of them are or could be regarded as benign.

The *Opus ymaginum* ascribed to Ptolemy, on the other hand, describes forty-five talismans constructed for a wide variety of generally trivial purposes, such as preventing cows from abandoning their calves, binding a crocodile so that it will not harm you, or muting musical instruments – the last a most desirable talisman in this age of hard rock. The version of the *Opus* in the Florence manuscript is, this time, the same as that in the Paris copy[6]. Each talisman is made when one of the thirty-six decans is in the ascendent. The *Opus ymaginum* does not seem to have been referred to by Michael, though he is familiar with the demons of the decans who boast of their names and accomplishments in the *Testament of Solomon*[7]; for he refers in the *Liber introductorius*[8] to the thirty-six images among the twelve zodiacal signs in which are found the wisest demons who will answer questions and bring many things to completion if they are conjured «*racionabiliter*».

1. F ff. 4v-8. See Pingree, «The Diffusion», 75-76.
2. *Speculum* XI 129-39.
3. Pingree, «The Diffusion», 83 and 87.
4. B f. 20va.
5. C f. 1b.
6. Paris, Bibl. Nat. lat. 16204, 538b-543a.
7. C. C. McCown, *The Testament of Solomon*, Leipzig 1922, section XVIII, 51*-59*. For the Arabic tradition see Ibn al-Nadīm, *Kitāb al-fihrist*, ed. G. Flügel, 2 vols., Leipzig 1871-72, I, 309-10.
8. B f. 21va.

The next item in the Florence manuscript describes an *experimentum* for the relief of men rendered impotent by magical mean[1], a rather popular little work that apparently originated with Constantine the African in the late eleventh century[2] and was adopted from him by Peter of Spain and Arnald of Villanova in the thirteenth[3]. This is a piece of folk magic employing much obvious sexual symbolism both in the initial impeding *maleficium* and in the proposed counter-measures. The ultimate cure, however, when it is seen that a diabolical power is involved, is the performance of an exorcism by a bishop or a priest; at this point it is elevated to the level of Salomonic magic.

There are many other treatises of detestable or Salomonic magic in the Florence manuscript, following a group of Hermetic texts. The first of the detestable treatises is the *Super eutuntam et super ydeam Salomonis* composed, it is said, by Salomon's four disciples: Fortunatus, Eleazar, Macarius, and Toz the Greek[4]. William of Auvergne speaks of the horrible image called the *Idea Salomonis et entocta*[5] and it is referred to by Albert[6] as the *De quatuor annulis*, which is entitled with the names of his four disciples, (and) which begins: «De arte eutonica et ydaica, etc.». The text in the Florentine manuscript breaks off in the middle of a sentence and in the middle of a page; the next two pages are left blank to receive the rest of the book, if it should ever be recovered. What remains describes the preparation of a gold ring of Mercury, of the symbol or *ydea* of Salomon, of the elaborate «pontifical» mitre to be worn by the magicianpriest, his gloves, the candles, and the book in which the demons are to be compelled to write the answers to the questions posed to them. All of these preparations are accompanied by fasting, abstinence, suffumigations, exorcisms, and prayers, some of which are prescribed to be in Arabic. Finally a carefully oriented house is constructed and properly adorned and prepared for the final ceremony of summoning the demons. Here is where the text breaks off.

1. F ff. 8-8v. For the great antiquity of magic connected with problems of impotence see R. D. Biggs, *SA. ZI. GA. Ancient Mesopotamian Potency Incantations*, Locust Valley 1967.
2. L. Thorndike and P. Kibre, *A Catalogue of Incipits of Mediaeval Scientific Writings in Latin*, London 1963, 1542.
3. Thorndike, *History*, II, 850-51.
4. F ff. 26v-29v.
5. Thorndike, *History*, II, 280.
6. *Speculum* XI 76-78; see XI 23.

Though we cannot read the rest of this fascinating treatise, it appears that Michael Scot, like his colleagues in Paris, could. For he states in the *Liber introductorius*[1]: «It must be known that spirits sometimes enter into the bodies of the dead and are accustomed through them to give answers to the wise man who summons them, as is proved in the *Ars* (i. e., the *Ars notaria*) of Alphiareus, in Florieth[2], and in the *Ydea* of Salomon»; and he later mentions Salomon as one of the masters of the art along with the Sibilla, Tebit Benchoraf, Mesahalla, Dorotheus, Hermes, Boetius, Averroys, Iohannes Yspanus, Ysidorus, Zael, and Alchabicius[3], a most curious mixture of authorities.

The next detestable book in the Florence manuscript is one on the seven secret names of God[4]. It begins with a chain of authorities or *isnād*: Muhamet filius Alhascen et filius Amoemen (Muḥammad ibn al-Ḥassan ibn al-Ma' mūn) who cites Abubelver Hamet filius Habrae (Abū Bakr Aḥmad ibn Ibrāhīm), who cites Abubocher filius Mugehic (Abū Bakr ibn Mujāhid), who cites Abulacha Abdinimen filius Abduc (this name is utterly corrupt), who finally quotes Abubelver Mauhmet filius Bosir (Abū Bakr Maḥmūd ibn Bashīr). This book is evidently the «*Liber septem nominum* from the books of Muhameth» referred to by Albert[5]. The seven names, which were given to Moses on Mt. Sinai, are each associated with a week-day and a ceremony of inscription accompanied by suffumigations; there follow exorcisms for each inscribed name and the uses to which each may be put. Finally there are listed the angels of the four seasons, and the angels, winds, and names of the seven planets in the four triplicities. Some of these categories of angels, winds, and planet names are included by Albert in his description of the *Institutio* of Raziel[6]; perhaps that was our author's inspiration.

Between his notice of the seven names from the book of Muḥammad and the *Institutio* of Raziel Albert notes a *Liber quindecim nominum* also by Muḥammad[7]. This indeed follows the text that we have just described in the Florence manuscript[8]. This curious tract begins:

1. B f. 22va.
2. Compare the demon Floriget bound in the conjuration described at the end of this paper.
3. B f. 25b.
4. F ff. 38-41.
5. *Speculum* XI 83-84; see XI 26.
6. *Speculum* XI 27-31.
7. *Speculum* XI 84-85; see XI 26.
8. F ff. 41-42.

«These are the fifteen secret names which the philosophers found in the four principle books – namely, in the law of Moses, in the Gospels, in the Psalms of David, and in the book of Muḥammad». It then lists the recipients of the names, which were given by God: first to Enoch who is called Hermes, then in order to Noah, Moses, Lot, Isaiah, Abraham, and Joseph; then Moses receives the next three, followed by Joshua, Salomon (he received the name inscribed on the ring with which he controlled the demons), and Jesus the son of Mary (he used it to raise the dead and to cure lepers); and the last two God gave to Muḥammad. In this text, then, instead of being the seal of the prophets Muḥammad becomes the seal of the magicians. The text goes on to describe the object on which, the time at which, and the astrological configuration concurrent with which each of the fifteen names is to be inscribed, but it never gets around to revealing the names themselves. We are, however, told of the every-day objectives they will fulfill: one will repel mosquitos, another causes fish to swarm, and a third makes girls fall in love with the bearer. These objectives scarcely seem commensurate with the miracles wrought by Moses or Muḥammad, allegedly using the same names.

There shortly follows in the Florence manuscript a book *On the secrets of angels*[1] which, it claims, God sent by his angel Rachael to Adam after his expulsion from paradise. This recounts first several traditions concerning the names of the angels who guard the planets, and then numerous names of angels associated with each month of the Jewish calendar; it breaks off after Tabeth or December. Several of the planetary angels named in the *De secretis angelorum* are named also as angels guarding the planets by Michael Scot in the *Liber introductorius*[2]. These shared names are: Cadaciel in Michael for Cadet in the *Secrets*, Samiel in Michael for Zamael, Raphael in both, Abyael in Michael for Anael, Michael in both, and Gabriel in both. Clearly Michael Scot and the author of the *De secretis angelorum* drew upon similar Jewish traditions. Michael's angels, it should be noted, are identifiable with those inscribed on a mirror according to Jabīr's *Flos naturarum*[3] as quoted by

1. F ff. 43v-44v.
2. B f. 144vb, M f. 95a. See P. Morpurgo, «Note in margine a un poemetto astrologico presente nei codici del *Liber particularis* de Michele Scoto», *Pluteus*, 2 (1983), 5-14, esp. 8.
3. *Flos naturarum*, par. 79 in my forthcoming edition.

the Latin *Picatrix*[1]; Jabīr claims to have derived the mirror from Ptolemy of Babylon, but states that it was invented by three Indians from Egypt.

Finally, the three treatises at the end of the Florence manuscript are all representatives of Salomonic magic. The first two concern *almandal* of Salomon. This *mandal* is in shape and inscriptions completely conformable to an Indian *mandala*; it is remarkable to see the Sanskrit word transmitted so purely through Arabic, in which it is still used to refer to a magical object, to Latin. The figure is a square «wall» with a circle in the center and spokes pointing to the four cardinal directions (indicated by «gates») and to the four intermediate directions. On each of the four side walls are inscribed the names of angels. The whole *almandal* is engraved on a talismanic plate, which is suffumigated and exorcized. It then can be used for various magic acts including the expulsion of demons from the possessed and four exorcisms of the *jinn* and the *shaytān*. Subsidiary sigils molded from Toledan wax reveal the place where our versions of this text were concocted.

The first of these two versions is entitled *Liber in figura almandal et eius opere*[2]; it is perhaps the *Almandal Salomonis* of Albert[3], and the *almandal* referred to by William[4]. The second book is the *Liber de almandal* which is called the table or altar of Salomon; this has the same incipit as Albert's *De figura almandal*[5]. Here the four exorcisms are attributed to Içmile (Ishmael) «Arhginemem». I have so far observed no mention of *almandal* by Michael Scot.

Even weirder than Salomon's *almandal* are the tabernacle and the sacrificial knife described, along with Salomonic exorcisms, animal sacrifices, and invocations of winds and demons in the most curious *Liber Bileth* which fills the last pages of the Florence manuscript[6]. Of this extraordinary text I shall only report now that Bileth or Biled traces his ancestry back through ten generations to Asmoday the son of Sanron, both of whom lived in the days of Salomon and assisted him in performing his greater and less works among the malignant spirits; in fact, according to the *Testament of Solomon*[7], Asmodaeus was

1. *Picatrix*, ed. D. Pingree, London 1986, IV vii 23.
2. F ff. 74v-77.
3. *Speculum* XI 25, though this is more likely to be the next, derivative treatise.
4. Thorndike, *History*, II, 351.
5. *Speculum* XI 80-81.
6. F ff. 74v-84.
7. Section V, 21*-25*.

one of the first wicked demons to be hauled in, bound, before the throne of Solomon. Despite this ancestry, our author concludes with the boast: «Non est potentior neque maior Biled». It is interesting in this context to note that the Arabic word *balīd* means «stupid».

The pages of the Florence manuscript[1] are also richly inscribed with abominable or Hermetic texts, many of which were known to Albertus Magnus[2], a few to William of Auvergne, but virtually none to Michael Scot. I begin with a short book ascribed to Hermes in which he describes the images and characters of the seven planets, culminating in an eighth treatise, *In magisterio imaginum*[3]. Mercury's seven books of the seven planets were also familiar to William of Auvergne[4]. According to Albert each of these books, when complete, should have included sections on the images, on the characters, on the rings, and on the sigils of its planet; but he found of the *Liber Solis* only that section, on characters, that is preserved in the Florence manuscript; he suggests that the remaining sections may simply not have been translated. However, neither this *Liber Solis* nor the following *Liber ymaginum Lune* comes, in fact, from Hermes' series of eight.

The Florence manuscript's *Liber ymaginum Lune*[5] consists, after a brief preface, of the work of Belenius Apollo – that is, Apollonius – on the applications to magic of the lunar mansion occupied by the Moon, the names of the angels subservient to the images of the Moon, the names of and operations suitable for the hours of the day and the night, and on the images engraved in each of the twelve hours (with four additional images) and the magical uses of each. Clearly, much of Belenius' work is strongly influenced by Salomonic magic. This composite *Liber ymaginum Lune* of Hermes corresponds to the *Liber Lune* to which is joined a *De horarum opere* and a *De quatuor imaginibus ab aliis separatis* ascribed to Balenuz according to Albert[6]. Michael Scot offers the name Ballemich – a common variant of Belenius – in a list of astrological and magical authorities given at the beginning of the *Liber*

1. F ff. 8v-9v. For a reconstruction of the seven books of the seven planets based on what was known prior to the investigation of Florence, Bibl. Nat., II. iii. 214 see D. Pingree, «The Diffusion», 77-78.

2. *Speculum* XI 59-62.

3. Albert's discussion of what he could find of the seven Hermetic books is to be found in *Speculum* XI 47-68.

4. *Opera*, I, 881 and 953.

5. F ff. 9v-15.

6. *Speculum* XI 47-51.

particularis[1], but in the *Liber introductorius* copies much of it as if from a *Liber ymaginum Lune* written by Thābit and revised by Ptolemy[2]. In this case Michael may be deliberately trying to disguise the nature of his source by attributing it to acceptable authors.

The entire series of the seven planetary books of Hermes once followed the *Liber ymaginum Lune* in the Florence manuscript[3], though it is now imcomplete because of the loss of perhaps two gatherings between f. 21v and f. 22. What remains is a lengthy preface concerning the origin of the text and general rules for the practice of Hermetic magic; the *Liber Lune*, the beginning of the *Liber Solis*, and most of the final *Liber Saturni*. Much of the lost middle section can be recovered from other manuscripts.

The story of the origin of the seven books given in the preface is attributed to Hermes Triplex, and is parallel to Abū Ma'shar's history of science[4], in which the first of three Hermes inscribed scientific texts on the pyramids of Egypt before the Flood that inaugurated the Kaliyuga. According to Hermes Triplex the whole of science was revealed to Adam, from whom it was learned by his son Abel. Abel, the Just, anticipating the coming Flood of Noah, composed the seven books of the seven planets and inscribed them on pieces of marble. Hermes claims to have discovered these marble slabs in Hebron, and made them the basis of his books. There clearly must have been numerous tablets since the *Liber Lune* alone occupied twenty of them. Abel-Hermes instructs the magician to employ suffumigations, prayers to the planets, and images of gold, silver, tin, and lead. This work is indeed Sabaean in character. The magician's objectives, however, remain banal.

More impressive for its imaginative exuberance is the incomplete *Liber Veneris*, the fourth book in the series, which is found in a fifteenth-century manuscript originally copied in Germany, but now in the Vatican[5]; this is largely based on the two authorities we have met previously – Toz the Greek and Germa the Babylonian. It informs its reader of such things as the names of the hundred angels of Venus that are to be inscribed on a cast head of that planet; one is to do the same

1. C f. lb.
2. Mixed with other material on M ff. 108b-114b.
3. F ff. 15-23v.
4. D. Pingree, *The Thousands of Abū Ma'shar*, London 1968, 15.
5. Vat. lat. 10803, (henceforth V), ff. 55-60; for this manuscript's contribution to the reconstruction of the seven books see Pingree, «The Diffusion», 77.

on heads of the other six planets, but the names of their six hundred angels are missing. Other chapters deal with Venus' rings and mirrors; we have no time to consider them now. But we are led to wonder what connection this *Liber Veneris* might have to the *De stationibus ad cultum Veneris* by Toz that both William of Auvergne[1] and Albert[2] have mentioned. The identification is made likely by one passage in the *Speculum*[3] that speaks of, «the images of Toz the Greek and Germath the Babylonian, which include the stations to the ritual of Venus». The *Liber Martis*[4] and the incomplete *Liber Iovis*[5] as preserved in the Vatican manuscript contain little that is startling.

There follows Hermes' seven books of the seven planets in the Florence manuscript[6] a brief, anonymous treatise that is pure Ḥarrānian magic as I imagine it to be, though not as thorough as it might have been. It is entitled *Liber de ieiuniis et sacrificiis et suffumigationibus septem stellarum*, and in a short paragraph for each planet states the duration of the fast, and the bird or animal that must be sacrificed at its end. It then instructs the magician on the conjuration he must utter and the angel to whom he should address it while he munches the heart of the sacrifice and draws a series of characters. This little *opus* does not seem to have attracted anyone's attention; or, if it did, he preferred not to acknowledge the fact.

Two more rare texts are attributed to Hermes in the Florence manuscript. The first is a *Liber Mercurii*[7]. Its incipit, «This is the Book of Mercury Hermes who is swift in running in the sphere of heaven», indicates that its author was considered to be the inferior planet rather than the terrestrial Trismegistus. It describes magic rings bearing gems engraved with images and characters and the rituals of suffumigation, prayer, and sacrifice that render them potent. As usual, astrology guides the operator in his choice of the times to perfect the rings and to perform these rituals. Of the incomplete text there survive descriptions of two rings of Mercury, two rings of Saturn (for the first of which the *Liber Saturni* is quoted), and one ring of Venus.

1. *Opera*, I, 44 and 671.
2. *Speculum* XI 71-73.
3. *Speculum* XI 5-6.
4. V ff. 60-61v.
5. V ff. 61v-62v.
6. F ff. 23v-24v.
7. F ff. 24v-26v.

The second Hermetic work[1] is, unfortunately, very badly preserved; clearly the fifteenth century scribe found his archetype either to be missing pages or to be impossible to decipher in many places. Fortunately the beginning survives: «In the name of God. The conjuration is Betronos, which is also Saturnus, translated from Arabic into Latin by Theodosius, the Archbishop of Sardinia. This is the *Book of the Planets* found among the books of Hermes. Apollonius the philosopher of Egypt translated it».

I presume that Apollonius is supposed to have translated it from Egyptian into Chaldaean, since we are told later on that a sigil of Jupiter was translated from Egyptian into Chaldaean, and from Chaldaean into Latin[2]. Theodosius, the Archbishop of Sardinia – *archiepiscopus Sardiensis* – remains an utter mystery to me. He is not the Theodosius, a monk of Syracuse, who was captured with that city by the Arabs in 880 and thrown into a prison in Palermo, for that Theodosius wrote in Greek[3]. In this magical text, of course, Theodosius may be a pseudonym. But, given the fact that all of the other datable translations in the manuscript were extant in Latin by the second quarter of the thirteenth century, I would argue that this translation also antedates 1225. Its title, *Betronos*, should be emended to *Becronos*; that represents a misreading of the Arabic *yā qurūnus*, the words with which the Sabaeans conjured Saturn[4]. The *Becronos* describes planetary rings bearing engraved gems and ceremonies similar to those in the *Liber Mercurii* – ten rings of Saturn, five of Jupiter (with Jovian sigil and table as well) and three of Mars; at this point it breaks off.

Clearly both of these texts will eventually provide valuable information concerning the magical practices of the Ḥarrānian Sabaeans. Even more exciting, however, is another translation in the Florence manuscript[5], the *Liber de locutione cum spiritibus planetarum*. This turns out to be a Latin version of the original form of al-Ṭabarī's treatise on the planetary prayers of the Sabaeans, which has hitherto been known only from its citation in the *Ghāyat al-ḥakim*[6]. The author of the

1. F ff. 33v-38.
2. F f. 35v.
3. K. Krumbacher, *Geschichte der Byzantischen Literatur*, München 1891, 59.
4. *Pseudo-Mağrīṭī. Das Ziel des Weisen*, ed. H. Ritter, 204. The name appears corruptly in Ibn al-Nadīm, I, 321 as Qirqis; he is quoting from a Christian writer, Abū Sa'īd Wahb ibn Ibrāhīm.
5. F ff. 31-33.
6. *Ghāya* III 7, 195-228.

Ghāya has conflated Ṭabarī's text with material from other Sabaean sources; the Latin version presents a much more uniform and homogeneous description of the ceremonies. In the *Ghāya* the author of this treatise is called simply al-Ṭabarī, for which the Latin version produced in Alfonso el Sabio's court has Athabary; the newly found translation calls him Abuelabeç Altanarani, which is a slightly mistaken transliteration of Abū al-ᶜAbbās al-Ṭabarānī – a name not otherwise familiar to Arabic bibliographers.

This survey does not exhaust the rich resources of the Florentine manuscript. It also contains, for instance, the pseudo-Platonic *Book of the Cow*[1] composed in Syria in the late ninth century under the influence of Sabaean Hermeticism and known in Latin already to William of Auvergne[2], and several works on suspensions and ligatures applied particularly to medicine – Ibn al-Jazzār's *De proprietatibus*[3], Qusṭā ibn Lūqā's *De physicis ligaturis*[4], in this manuscript attributed to Ḥunayn and said to have been translated from Arabic into Latin by Gerard of Cremona in Toledo, and the chapter on ligatures and suspensions of stones[5] from the *Mineralia* completed by Albertus Magnus[6], probably in Italy, in 1256/7 or 1261/2.

This last piece, by Albert, appears to be the only one in the manuscript that is not a translation from Arabic. It occurs in a group of texts that are neither abominable nor detestable in Albert's terms and that are not referred to in the *Speculum*. This group, occupying ff. 49v-74v, consists of the astrological *De iudiciis partium* falsely ascribed to Ptolemy[7]; Ibn al-Jazzār's *De proprietatibus*; pseudo-Plato's *Book of the Cow*, the chapter on ligatures and suspensions from Albert's *Mineralia:* and Qusṭā ibn Lūqā's *De physicis ligaturis*. The scribe of the Florentine manuscript may well have copied this section from a manuscript other than the source or sources of his other texts. If this conjecture be considered plausible, it is also possible to surmise that the next and last group of texts – those on Salomon's *Almandal* and the *Book of Bileth* –

1. F ff. 57v-72. See Pingree, «The Diffusion», 71-72 and 80.
2. *Opera*, I, 43 and 70.
3. F ff. 54-57v. See Pingree, «The Diffusion», 70-71.
4. F ff. 73-74v. See Pingree, «The Diffusion», 68-69.
5. F ff. 72-73.
6. *Mineralia* II iii 6, 146-51 of *Albertus Magnus. Book of Minerals*, translated by D. Wyckoff, Oxford 1967.
7. F ff. 49v-54.

may have been copied from a third source. It would still remain true that none of the translations was made after ca. 1225, and that the only text in the manuscript later than that date is Albert's, of about 1260. In any case, it seems plausible to me that the collection of treatises on Hermetic, Salomonic, and celestial magic, preserved in the first section of the Florence manuscript, ff. 1-49, all of which had been translated from Arabic in the twelfth century, many of which were known to William of Auvergne, and almost all of which are mentioned by Albert in chapter 11 of his *Speculum astronomiae*, were collected into a *corpus* in the early thirteenth century, perhaps in Paris.

A number of the Salomonic treatises in this *corpus*, we have seen, were known to Michael Scot, who may have acquired this knowledge in Paris if he was in fact lecturing there on Sacrobosco's *Sphere* in about 1230[1]. He indeed reveals a rather wide knowledge of this form of learned magic in the pages of the *Liber introductorius*, as he also displays an extensive familiarity with natural magic in the *Liber particularis*, where he gives examples of the marvels of nature – of heaven, of the elements, of paradise, of lakes, of rivers, of metals, of stones, of herbs, of trees, of concoctions, of man, of animals and their parts, of birds, of fish, and of small beasts and insects – a marvellous catalogue of curiosities[2].

But his main interest as far as magic was concerned centered on demons, whom he firmly believed in. At one point[3] he even describes, along with a *Book of the Angelic Art* by Solomon and a *Book of Adam*, a *Book of the Perdition of the Soul and of the Heart* which deals exhaustively with demons. This is presumably different from the *Book of the Death of the Soul* which Albert, who presents it as a letter from Aristotle to Alexander, calls the worst of all[4]. In another place Michael gives a fantastic history of astronomical magic, which I shall attempt to summarize, adhering to his illogical order[5]. Zoroaster was the first discoverer of magic. He was born of the race of Sem[6] and taught by the demons in Persia. Nemroth the giant came to Persia after the dispersion of the

1. This is the conjecture of Thorndike, Michael Scot, 35-36.
2. C ff. 48va-54va.
3. M f. 114b.
4. *Speculum* XI 95-97; see C. B. Schmitt and D. Knox, *Pseudo-Aristoteles Latinus*, London 1985, 44.
5. B ff. 24b-25a.
6. Son of Noah: *Genesis* 10.1.

72 languages, and learned from the demons how to worship fire. Then Cham[1] tested the teachings of these spirits; and his son, Chanaam[2], who had become even more learned in the art, composed thirty volumes on divination[3]. Chanaam's son was Nemroth[4], who wrote a book on all of astronomy. But Chanaam had another son, named Abraham, who also became an expert in that art. This Abraham, coming from Africa to Jerusalem, taught Demetrius and Alexander, who in turn taught Ptolemy the king of Egypt. This is the Ptolemy who, Michael says, «either by his own wit or perhaps by the induction of some spirit» discovered astronomical tables and the astrolabe (an instrument that, in the *Liber particularis*[5], Michael claims the astronomers use to summon the evil spirits). This art together with the astrolabe was brought from Egypt to Spain by Athalax. Two clerics or scholars of France came to Athalax and, admiring the astrolabe but being ignorant of how to use it, bought it for a steep price. Bearing this precious instrument, they hastened back to France to show it to their friends. At that time master Gilbert (that is, Gerbert of Aurillac) was the best necromancer in France, one whom the demons of the air obeyed in everything because of his sacrifices, prayers, fastings, books, rings, and candles (these are all involved in the magic of the *Super eutuntam et super ydeam Salomonis* attributed to Salomon's four pupils). Gilbert took the astrolabe into his chamber, summoned the demons with the help of his consecrated books, and ordered them to reveal to him the secrets of the astrolabe and of astronomy. They, taking on corporeal bodies, went to Egypt and Jerusalem and copied out the appropriate books for him.

So was Arabic science, including magic, transmitted to Europe, according to Michael. But he goes on to relate how Gerbert's soul was saved. After, with the help of diabolic wickedness, he became Bishop of Rheims, he was illuminated by God, abandoned his magical books, and developed a taste for holy scripture. Because of his fame for *scientia* and for chastity he became first Archbishop, and then Pope.

I have given this lengthy summary of Michael's odd history not only because it demonstrates his firm belief in the efficacy of demonic

1. Sem's younger brother: *Genesis* 10.1.
2. *Genesis* 9.6.
3. See the first antediluvian Hermes' thirty pages in the story told by Abū Ma'shar; Pingree, *Thousands*, 15.
4. From here on Michael's genealogy diverges from that in *Genesis*.
5. C f. lva.

magic, however contrary to the rules of the Roman church it may be, but also because the one magical text of his that we have, the *Necromantie experimentum*[1], which he sent to the *coronatus* of the king of Cecilia, one Philip who lay ill in the city of Corduba, gives instructions remarkably similar to the procedures recommended in the *Super eutuntam et super ydeam Salomonis*, and these instructions are for summoning the demons to teach one a science just as Gerbert had done in Michael's story. This is not the only description of a ceremony of Salomonic magic in Michael's works, in the *Liber introductorius*[2] there is a detailed description of one in which a virgin of five or seven years binds a demon named Floriget to appear in human form and to reveal the person of a thief and his hiding place. As Dante wrote, Michael was well informed about magic. Though we cannot be certain that he himself actually practiced it, we can be sure that many of his contemporaries did.

1. Edited in J. W. Brown, *An Enquiry into the Life and Legend of Michael Scot*, Edinburgh 1897, 231-34, from Florence, Laur. 89 sup., 38, ff. 409v-41 (formerly ff. 256v-260).
2. B ff. 23a-23b.

Bibliography of David Pingree

Books

The Thousands of Abū Ma'shar (London, 1968).

Albumasaris De revolutionibus nativitatum (Leipzig, 1968).

Sanskrit Astronomical Tables in the United States (Philadelphia, 1968).

The Vidvajjanavallabha of Bhojarāja (Baroda, 1970).

Census of the Exact Sciences in Sanskrit, Series A, volumes 1–5 (Philadelphia, 1970–1994).

The Pañcasiddhāntikā of Varāhamihira, 2 volumes (Copenhagen, 1970–1971) [with O. Neugebauer].

The Astrological History of Māshā'allāh (Cambridge MA, 1971) [with E. S. Kennedy].

The Bṛhadyātrā of Varāhamihira (Madras, 1972).

Hephaestionis Thebani Apotelesmaticorum libri tres, 2 volumes (Leipzig, 1973–1974).

Sanskrit Astronomical Tables in England (Madras, 1973).

Babylonian Planetary Omens, Part 1: The Venus Tablet of Ammiṣaduqa (Malibu, 1975) [with E. Reiner].

The Laghukhecarasiddhi of Śrīdhara (Baroda, 1975).

Dorothei Sidonii Carmen astrologicum (Leipzig, 1976).

The Vṛddhayavanajātaka of Mīnarāja, 2 volumes (Baroda, 1976).

The Yavanajātaka of Sphujidhvaja, 2 volumes (Cambridge MA, 1978).

The Book of Reasons Behind Astronomical Tables (Delmar NY, 1981) [with F. I. Haddad and E. S. Kennedy].

Jyotiḥśāstra (Wiesbaden, 1981).

Babylonian Planetary Omens, Part 2: Enūma Anu Enlil, Tablets 50–51 (Malibu, 1981) [with E. Reiner].

A Catalogue of the Chandra Shum Shere Collection in the Bodleian Library, Part 1: Jyotiḥśāstra (Oxford, 1984).

The Astronomical Works of Gregory Chioniades, Part 1: The Zīj al-'Alā'ī, 2 volumes (Amsterdam, 1985–1986).

Vettii Valentis Anthologiarum libri novem (Leipzig, 1986).

Picatrix: The Latin Version of the Ghāyat al-ḥakīm (London, 1986).

The Rājamṛgāṅka of Bhojarāja (Aligarh, 1987).

MUL.APIN: An Astronomical Compendium in Cuneiform (Horn, 1989) [with H. Hunger].

The Grahajñāna of Āśādhara together with the Gaṇitacūḍāmaṇi of Harihara (Aligarh, 1989).

The Astronomical Works of Daśabala (Aligarh, 1990).

Levi ben Gerson's Prognostication for the Conjunction of 1345 (Philadelphia, 1990) [with B. R. Goldstein).

The Liber Aristotilis of Hugo Santalla (London, 1997) [with C. Burnett].

From Astral Omens to Astrology, From Babylon to Bīkāner (Rome, 1997).

Preceptum Canonis Ptolomei (Louvain-la-Neuve, 1997).

Babylonian Planetary Omens, Part Three (Groningen, 1998) [with E. Reiner].

Astral Sciences in Mesopotamia (Leiden, 1999) [with H. Hunger].

Arabic Astronomy in Sanskrit: Al-Birjandī on Tadhkira II, Chapter 11 and its Sanskrit Translation (Leiden, 2002) [with T. Kusuba].

A Descriptive Catalogue of the Sanskrit Astronomical Manuscripts Preserved at the Maharaja Man Singh II Museum in Jaipur, India (Philadelphia, 2003).

Catalogue of Jyotiṣa Manuscripts in the Wellcome Library: Sanskrit Astral and Mathematical Literature (Leiden, 2004).

A Catalogue of the Sanskrit Manuscripts at Columbia University (New York, 2007).

Historic Scientific Instruments of the Adler Planetarium and Astronomy Museum, Volume II: Eastern Astrolabes (Chicago, 2009).

The Reconciliation of Puranic and Siddhantic Cosmologies: The Treatises of Kevalarama and Nandarama (forthcoming) [with C. Minkowski].

Rhetorius of Egypt (forthcoming) [with S. Heilan].

Chapters in Books

"The *Apologia* of Alphonus Lyncurius," in J. Tedeschi (ed.), *Italian Reformation Studies in Honor of Laelius Socinus* (Florence, 1965), 199–214.

"Astrology," in P. P. Wiener (ed.), *Dictionary of the History of Ideas*, 4 volumes (New York, 1973–1974), I, 118–126.

"The *Karaṇavaiṣṇava* of Śaṅkara," in *Charudeva Shastri Felicitation Volume* (Delhi, 1974), 588–600.

"Māsha'allāh: Some Sasanian and Syriac Sources," in G. F. Hourani (ed.), *Essays in Islamic Philosophy and Science* (Albany NY, 1975), 5–14.

"Al-Bīrūnī's Knowledge of Sanskrit Astronomical Texts," in P. J. Chelkowski (ed.), *The Scholar and the Saint: Studies in Commemoration of Abu 'l-Rayhan al-Biruni and Jalal al-Din al-Rumi* (New York, 1975), 67–81.

"Thessalus Astrologus," in F. E. Cranz (ed.), *Catalogus Translationum et Commentariorum* (Washington DC, 1976), volume 3, 83–86.

"The Library of George, Count of Corinth," in K. Treu (eds.), *Studia Codicologica* (Berlin, 1977), 351–362.

"The Horoscope of Constantinople," in Y. Maeyama and W. Saltzer (eds.), *Πρίσματα: Festchrift für Willy Hartner* (Weisbaden, 1977), 305–315.

"History of Mathematical Astronomy in India," in C. C. Gillispie (ed.), *Dictionary of Scientific Biography* (New York, 1978), volume 15, 533–633.

"The *Gaṇitapañcaviṃśī* of Śrīdhara," in J. P. Sinha (ed.), *Ludwik Sternbach Felicitation Volume* (Lucknow, 1979), 887–909.

"The *Kheṭamuktāvalī* of Nṛsiṃha," in M. Nagatomi et al. (eds.), *Sanskrit and Indian Studies: Essays in Honor of Daniel H. H. Ingalls* (Dordrecht, 1980), 143–157.

"Mesopotamian Astronomy and Astral Omens in Other Civilizations," in H. J. Nissen and J. Renger (eds.), *Mesopotamien und Seine Nachbarn* (Berlin, 1982), 613–631.

"The Zodiac Ceilings of Petosiris and Petubastis," in *Denkmaler der Oase Dachla* (Mainz, 1982), 96–101 [with O. Neugebauer and R. A. Parker].

"The Diffusion of Arabic Magical Texts in Western Europe," in B.-M. Scarcia Amoretti (ed.), *La diffusione delle scienze islamiche nel medio evo europeo* (Rome, 1987), 57–102.

"Venus Omens in India and Babylon," in F. Rochberg-Halton (ed.), *Language, Literature, and History: Philological and Historical Studies Presented to Erica Reiner* (New Haven CT, 1987), 293–315.

"Sumatiharṣa Gaṇi and Some Other Jaina Jyotiṣīs," in *Astha aura cintana* (Delhi, 1987), 99–104.

"Indian and Islamic Astronomy at Jayasiṃha's Court," in D. A. King and G. Saliba (eds.), *From Deferent to Equant: A Volume of Studies in the History of Science in the Ancient and Medieval Near East in Honor of E. S. Kennedy* (New York, 1987), 313–328.

"Babylonian Planetary Theory in Sanskrit Omen Texts," in J. L. Berggren and B. R. Goldstein (eds.), *From Ancient Omens to Statistical Mechanics: Essays on the Exact Sciences Presented to Asger Aaboe* (Copenhagen, 1987), 91–99.

"A Babylonian Star Catalogue: BM 78161," in E. Leichty et al. (eds.), *A Scientific Humanist: Studies in Memory of Abraham Sachs* (Philadelphia, 1988), 313–321 [with C. Walker].

"MUL.APIN and Vedic Astronomy," in H. Behrens et al. (eds.), *DUBU-E$_2$-DUB-BA-A: Studies in Honor of Ake W. Sjöberg* (Philadelphia, 1989), 439–445.

"Astrology," in M. J. L. Young, J. D. Latham and R. B. Serjeant (eds.), *The Cambridge History of Arabic Literature: Religion, Learning and Science in the 'Abbāsid Period* (Cambridge, 1990), 290–300.

"The Preceptum canonis Ptolomei," in M. Fattori and J. Hamesse (eds.), *Rencontres de culture dans la philosophie médiéval* (Louvain-la-Neuve-Cassino, 1990), 355–375.

"Mesopotamian Omens in Sanskrit" in D. Charpin and F. Johannes (eds.), *La circulation des biens, des personnes et des idées dans la Proche-Orient ancient* (Paris, 1992), 375–379.

"Innovation and Stagnation in Medieval Indian Astronomy," in *17º Congreso International de ciencias historicas* (Madrid, 1992), 519–526.

"Thessalus Astrologus Addenda," P. O. Kristeller (ed.), *Catalogus Translationum et Commentariorum* (Washington DC, 1992), volume 7, 330–332.

"Venus Phenomena in *Enūma Anu Enlil*," in H. D. Galter (eds.), *Die Rolle der Astronomie in den Kulturen Mesopotamiens* (Graz, 1993), 259–273.

"Plato's Hermetic Book of the Cow," in P. Prini (ed.), *Il Neoplatonismo nel Rinascimento* (Rome, 1993), 133–145.

"La magia dotta," in P. Toubert and A. Paravicini Bagliani (eds.), *Federico II e la scienza* (Palermo, 1994), volume 2, 354–370.

"Astronomy in India," in C. Walker (ed.), *Astronomy Before the Telescope* (London, 1996), 123–142.

"Indian Astronomy in Medieval Spain," in J. Casulleras and J. Samsó (eds.), *From Baghdad to Barcelona: Studies in the Islamic Exact Sciences in Honour of Prof. Juan Vernet* (Barcelona, 1996), 39–48.

"Indian Reception of Muslim Versions of Ptolemaic Astronomy," in F. J. Ragep and S. P. Ragep (eds.), *Tradition, Transmission, Transformation: Proceedings of Two Conferences on Premodern Science Held at the University of Oklahoma* (Leiden, 1996), 471–485.

"Māshā'allāh: Greek, Pahlavī, Arabic, and Latin Astrology," in A. Hasnawi et al. (eds.), *Perspectives arabes et médiévales sur la tradition scientifique et philosophique grecque* (Leuven, 1997), 123–136.

"Legacies in Astronomy and Celestial Omens," in S. Dalley (ed.), *The Legacy of Mesopotamia* (Oxford, 1998), 125–137.

"Mathematics and Mathematical Astronomy," in J. W. Elder et al. (eds.), *India's Worlds and U. S. Scholars: 1947–1997* (New Delhi, 1998), 355–361.

"Preliminary Assessment of the Problems of Editing the Zīj al-Sanjarī of al-Khāzini," in Y. Ibish (ed.), *Editing Islamic Manuscripts on Science: Proceedings of the Fourth Conference of Al-Furqān Islamic Heritage Foundation, 29th-30th November 1997* (London, 1999), 105–113.

"Avranches 235 dans la tradition manuscrite du *Preceptum Canonis Ptolomei*," in L. Callebat and O. Desbordes (eds.), *Science antique, science medieval: Actes du colloque international (Mont-Saint-Michel, 4–7 septembre 1998)* (Hildesheim, 2000), 163–169.

"A Greek List of Astrolabe Stars" in M. Folkerts and R. Lorch (ed.), *Sic itur ad astra: Studien zur Geschichte der Mathematik und Naturwissenschaften. Festschrift für der Arabisten Paul Kunitzsch zum 70. Geburtstag* (Weisbaden, 2000), 474–477.

"The Coining of New Words to Express New Concepts in Sanskrit Astronomy," in *Harānandalaharī* (Reinbek, 2000), 217–220.

"Astrology," in P. Murdin, *Encylopedia of Astronomy & Astrophysics* (London: IOP Publishing, 2001).

"Ravikās in Indian Astronomy and the Kālacakra," in R. Torella (ed.), *La parole e I marmi: studi in onore di Raniero Gnoli nel suo 70º compleanno* (Rome, 2001), 655–664.

"I professionisti della scienza e la loro formazione," in S. Petruccioli (ed.), *Storia della scienza* (Rome, 2001), volume 2.

"Cosmologia vedica, cosmologia puranica," in S. Petruccioli (ed.), *Storia della scienza* (Rome, 2001), volume 2.

"Stelle e constellazioni," in S. Petruccioli (ed.), *Storia della scienza* (Rome, 2001), volume 2.

"Il Sole e la Luna," in S. Petruccioli (ed.), *Storia della scienza* (Rome, 2001), volume 2.

"I calendar," in S. Petruccioli (ed.), *Storia della scienza* (Rome, 2001), volume 2.

"La calendaristica vedica (Jyotisa)," in S. Petruccioli (ed.), *Storia della scienza* (Rome, 2001), volume 2.

"Astronomia," in S. Petruccioli (ed.), *Storia della scienza* (Rome, 2001), volume 2.

"Divinazione e astrologia," in S. Petruccioli (ed.), *Storia della scienza* (Rome, 2001), volume 2.

"Philippe de La Hire at the Court of Jayasiṃha," in S. M. Razaullah Ansari (ed.), *History of Oriental Astronomy: Proceedings of the Joint Discussion 17 at the 23rd General Assembly of the International Astronomical Union, organised by the Commission 41 (History of Astronomy) held in Kyoto, August 25–26, 1997* (Dordrecht, 2002), 123–131.

"Sasanian Astrology in Byzantium," in A. Carile (ed.), *La Persia e Bisanzio: Convegno internazionale Rome 14–18 Ottobre 2002* (Rome, 2004), 539–553.

"The Sarvasiddhāntarāja of Nityānanda," in J. P. Hogendijk and A. I. Sabra (eds.), *The Enterprise of Science in Islam: New Perspectives* (Cambridge MA, 2003), 269–284.

"Zero and the Symbol for Zero in Early Sexagesimal and Decimal Place-Value Systems," in A. K. Bag and S. R. Sarma (eds.), *The Concept of śūnya* (New Delhi, 2003), 137–141.

"Kevalarāma's *Pañcāṅgasāraṇī*," in V. Miśra (ed.), *Padmabhūṣaṇa: Professor Baladeva Upādhyāya Birth Centenary Volume* (Varanasi, 2004), 725–729.

"Māshā'allāh's Zoroastrian Historical Astrology," in G. Oestmann, H. D. Rutkin and K. von Stuckran (eds.), *Horoscopes and Public Spheres, Essays on the History of Astrology, Religion and Society* (Berlin and New York, 2005), 95–100.

"The Byzantine Translations of Māshā'allāh on Interrogational Astrology," in P. Magdalino and M. Mavroudi (eds.), *The Occult Sciences in Byzantium* (Geneva, 2006), 230–243.

"From Hermes to Jābir and the Book of the Cow," in C. Burnett and W. F. Ryan (eds.), *Magic and the Classical Tradition* (London, 2006), 19–28.

"Mesopotamian and Greek Astronomy in India," in K. Preisendanz (ed.), *Expanding and Merging Horizons: Contributions to South Asian and Cross-Cultural Studies in Commemoration of Wilhelm Halbfass* (Vienna, 2007), 41.

Journal Articles

"The Empires of Rudradāman and Yaśodharman: Evidence from Two Astrological Geographies," *Journal of the American Oriental Society* 79 (1959), 267–270.

"A Greek Linear Planetary Text in India," *Journal of the American Oriental Society* 79 (1959), 282–284.

"The Yavanajātaka of Sphujidhvaja," *Journal of Oriental Research, Madras* 31 (1961–1962), 16–31.

"Δάρας τὸ νῦν καλούμενον Ταυρές," *Bulletin de la Classe des Sciences de l'Academie Royale de Belgique*, CI, lettres 5, 48 (1962), 323–326.

"Historical Horoscopes," *Journal of the American Oriental Society* 82 (1962), 487–502.

"Astronomy and Astrology in India and Iran," *Isis* 54 (1963), 229–246.

"The Indian Iconography of the Decans and Horās," *Journal of the Warburg and Courtauld Institutes* 26 (1963), 223–254.

"Gregory Chioniades and Palaeologan Astronomy," *Dumbarton Oaks Papers* 18 (1964), 133–160.

"Rejoinder to Wayman's 'The Buddha's Birthdate,'" *Journal of the American Oriental Society* 84 (1964), 174–175.

"Indian Influence on Early Sasanian and Arabic Astronomy," *Journal of Oriental Research, Madras* 33 (1964), 1–8.

"Indian Influence on Sasanian and Early Islamic Astronomy and Astrology," *Journal of Oriental Research, Madras* 34–35 (1964–1965), 118–126.

"Representation of the Planets in Indian Astrology," *Indo-Iranian Journal* 8 (1965), 249–267.

"The Persian 'Observations' of the Solar Apogee in *ca.* A.D. 450," *Journal of Near Eastern Studies* 24 (1965), 334–336.

"The Astronomical Tables of Mahādeva," *Proceedings of the American Philosophical Society* 111 (1967), 69–92 [with O. Neugebauer].

"The *Paitāmahasiddhānts* of the *Viṣṇudharmottarapurāṇa*," *Brahmavidyā* 31–32 (1967–1968), 472–510.

"The Fragments of the Works of Ya'qūb ibn Ṭāriq," *Journal of Near Eastern Studies* 27 (1968), 97–125.

"The Later Pauliśasiddhānta," *Centaurus* 14 (1969), 172–241.

"On the Classification of Indian Planetary Tables," *Journal for the History of Astronomy* 1 (1970), 95–108.

"The Fragments of the Works of al-Fazārī," *Journal of Near Eastern Studies* 29 (1970), 103–123.

"The Astrological School of John Abramius," *Dumbarton Oaks Papers* 25 (1971), 191–215.

"On the Greek Origin of the Indian Planetary Model Employing a Double Epicycle," *Journal for the History of Astronomy* 2 (1971), 80–85.

"Precession and Trepidation in Indian Astronomy Before A.D. 1200," *Journal for the History of Astronomy* 3 (1972), 27–35.

"The Horoscope of Constantine VII Porphyrogenitus," *Dumbarton Oaks Papers* 27 (1973), 219–231.

"The Greek Influence on Early Islamic Mathematical Astronomy," *Journal of the American Oriental Society* 93 (1973), 32–43.

"The Beginning of Utpala's Commentary on the *Khaṇḍakhādyaka*," *Journal of the American Oriental Society* 93 (1973), 469–481.

"The Mesopotamian Origin of Early Indian Mathematical Astronomy," *Journal for the History of Astronomy* 4 (1973), 1–12.

"Concentric with Equant," *Archives Internationales d'Histoire des Sciences* 24 (1974), 26–29.

"A Neo-Babylonian Report on Seasonal Hours," *Archiv für Orientforschung* 25 (1974–1977), 50–55 [with E. Reiner].

"Vasiṣṭa's Theory of Venus: The Misinterpretation of an Emendation," *Centaurus* 19 (1975), 36–39.

"Observational Texts Concerning the Planet Mercury," *Revue d'Assyriologie* 69 (1975), 175–180 [with E. Reiner].

"The Recovery of Early Greek Astronomy from India," *Journal for the History of Astronomy* 7 (1976), 109–123.

"The Indian and Pseudo-Indian Passages in Greek and Latin Astronomical and Astrological Texts," *Viator* 7 (1976), 141–195.

"Antiochus and Rhetorius," *Classical Philology* 72 (1977), 203–223.

"The Byzantine Version of the Toledan Tables: The Work of George Lapithes?," *Dumbarton Oaks Papers* 30 (1977), 85–132.

"Political Horoscopes from the Reign of Zeno," *Dumbarton Oaks Papers* 30 (1977), 133–150.

"The *Liber universus* of 'Umar al-Farrukhān al-Ṭabarī," *Journal for the History of Arabic Science* 1 (1977), 8–12.

"Horoscopes from the Cairo Geniza," *Journal of Near Eastern Studies* 36 (1977), 113–144 [with B. R. Goldstein].

"Political Horoscopes Relating to Late Ninth Century 'Alids" *Journal of Near Eastern Studies* 36 (1977), 247–275 [with W. Madelung].

"The Astronomical Tables of al-Khwārizmī in a Nineteenth Century Egyptian Text," *Journal of the American Oriental Society* 98 (1978), 96–99 [with B. R. Goldstein].

"Islamic Astronomy in Sanskrit," *Journal for the History of Arabic Science* 2 (1978), 315–330.

"Indian Astronomy," *Proceedings of the American Philosophical Society* 122 (1978), 361–364.

"Astrological Almanacs from the Cairo Geniza," *Journal of Near Eastern Studies* 38 (1979), 153–175 and 231–256 [with B. R. Goldstein].

"Reply to B. L. van der Waerden," *Journal for the History of Astronomy* 11 (1980), 58–62.

"Some of the Sources of the *Ghāyat al-ḥakīm*," *Journal of the Warburg and Courtauld Institutes* 43 (1980), 1–15.

"A New Look at *Melancolia I*," *Journal of the Warburg and Courtauld Institutes* 43 (1980), 257–258.

Frammento astrologico (PL II(27))," *Bulletin of the American Society of Papyrologists* 18 (1981), 83–87 [with R. Pintaudi].

"Sanskrit Evidence for the Presence of Arabs, Jews, and Persians in Western India *ca.* 700–1300," *Journal of the Oriental Institute, Baroda* 31 (1981), 172–182.

"Between the *Ghāya* and *Picatrix* I: The Spanish Version," *Journal of the Warburg and Courtauld Institutes* 44 (1981), 27–56.

"More Horoscopes from the Cairo Geniza," *Proceedings of the American Philosophical Society* 125 (1981), 155–189 [with B. R. Goldstein].

"Astronomical Computations for 1299 from the Cairo Geniza," *Centaurus* 25 (1982), 303–318 [with B. R. Goldstein].

"A Note on the Calendars Used in Early Indian Inscriptions," *Journal of the American Oriental Society* 102 (1982), 355–359.

"An Illustrated Greek Astronomical Manuscript," *Journal of the Warburg and Courtauld Institutes* 45 (1982), 185–192.

"Al-Khwārizmī in Samaria," *Archives Internationales d'Histoire des Sciences* 33 (1983), 15–21.

"The Byzantine Tradition of Vettius Valens' *Anthologies*," *Harvard Ukrainian Studies* 7 (1983), 532–541.

"Brahmagupta, Balabhadra, Pṛthūdaka, and al-Bīrūnī," *Journal of the American Oriental Society* 103 (1983), 353–360.

"Additional Astrological Almanacs from the Cairo Geniza," *Journal of the American Oriental Society* 103 (1983), 673–690 [with B. R. Goldstein].

"Al-Bīrūnī's Treatise on Astrological Lots," *Zeitschrift für Geschichte der Arabische-Islamischen Wissenschaften* 1 (1984), 9–54 [with F. I. Haddad and E. S. Kennedy].

"In Defence of Gregory Chioniades," *Archives Internationales d'Histoire des Sciences* 35 (1985), 114–115.

"Some Little Known Commentators on Bhāskara's *Karaṇakutūhala*," *Aligarh Journal of Oriental Studies* 2 (1985), 158–168.

"Ibn al-Ḥātim on the Talismans of the Lunar Mansions," *Journal of the Warburg and Courtauld Institutes* 50 (1987), 57–81 [with K. Lippincott].

"The Śīghrasiddhi of Lakṣmīdhara," *Journal of the Oriental Institute, Baroda* 37 (1987–1988), 65–81.

"Classical and Byzantine Astrology in Sassanian Persia," *Dumbarton Oaks Papers* 43 (1989), 227–288.

"On the Identification of the *Yogatārās* of the Indian *Nakṣatras,*" *Journal for the History of Astronomy* 20 (1989), 99–119 [with P. Morrissey].

"Indian Planetary Images and the Tradition of Astral Magic," *Journal of the Warburg and Courtauld Institutes* 52 (1989), 1–13.

"The Purāṇas and Jyotiḥśāstra: Astronomy," *Journal of the American Oriental Society* 110 (1990), 274–280.

"A Hitherto Unknown Sanskrit Work Concerning Mādhava's Derivation of the Power Series for Sine and Cosine," *Historia Scientiarum* 42 (1991), 49–65 [with D. Gold].

"Al-Ṭabarī on the Prayers to the Planets," *Bulletin d'Études Orientales* 44 (1992), 105–117.

"On the Date of the *Mahāsiddhānta* of the Second Āryabhaṭa," *Gaṇita Bhāratī* 14 (1992), 55–56.

"Hellenophilia versus the History of Science," *Isis* 83 (1992), 554–563.

"Āryabhaṭa, the Paitāmahasiddhānta, and Greek Astronomy," *Studies in the History of Medicine and Science*, New Series 12 (1993), 69–79.

"The Teaching of the *Almagest* in Late Antiquity," *Apeiron* 27 (1994), 75–98.

"Learned Magic in the Time of Frederick II," *Micrologus* 2 (1994), 39–56.

"Sanskrit Geographical Tables," *Indian Journal of History of Science* 31 (1996), 173–220.

"Bīja Corrections in Indian Astronomy," *Journal for the History of Astronomy* 27 (1996), 161–172.

"Two *Karmavipāka* Texts on Curing Diseases and Other Misfortunes," *Journal of the European Ayurvedic Society* 5 (1997), 46–52.

"Some Fourteenth-Century Byzantine Astronomical Texts," *Journal for the History of Astronomy* 29 (1998), 103–108.

"An Astronomer's Progress," *Proceedings of the American Philosophical Society* 143 (1999), 73–85.

"Māshā'allāh's(?) Arabic Translation of Dorotheus," *Res Orientales* 12 (1999), 191–209.

"Amṛtalaharī of Nityānanda," *SCIAMVS* 1 (2000), 209–217.

"Sanskrit Translations of Arabic and Persian Astronomical Texts at the Court of Jayasiṃha of Jayapura," *Suhayl* 1 (2000), 101–106.

"From Alexandria to Baghdād to Byzantium: The Transmission of Astrology," *International Journal of the Classical Tradition* 8 (2001), 3–21.

"Nīlakaṇṭha's Planetary Models," *Journal of Indian Philosophy* 29 (2001), 187–195.

"Artificial Demons and Miracles," *Res Orientales* 13 (2001), 109–122.

"The Sabians of Harrān and the Classical Tradition," *International Journal of the Classical Tradition* 9 (2002), 8–35.

"The Logic of Non-Western Science: Mathematical Discoveries in Medieval India," *Daedalus* 132 (2003), 45–53.

"Siddhāntakaustubha of Jagannātha Paṇḍita (original versions), edited with introduction in English," *Indian Journal of History of Science* 39 (2004), S1–S74.

"Between the *Ghāya* and *Picatrix* II: The Flos Naturarum Ascribed to Jābir," *Journal of the Warburg and Courtauld Institutes* 72 (2009), 41–80 [with C. Burnett].